Basic Biotechnology

Biotechnology impinges on everyone's lives. It is one of the major technologies of the twenty-first century. Its huge, wide-ranging, multi-disciplinary activities include recombinant DNA techniques, cloning and genetics, and the application of microbiology to the production of goods as prosaic as bread, beer, cheese and antibiotics. It continues to revolutionise treatments of many diseases, and is used to provide clean technologies and to deal with environmental problems.

Basic Biotechnology is a textbook that gives a full account of the current state of biotechnology, providing the reader with insight, inspiration and instruction. The fundamental aspects that underpin biotechnology are explained through examples from the pharmaceutical, food and environmental industries. Chapters on the public perception of biotechnology and the business and economics of the subject are also included.

The book is essential reading for all students and teachers of biotechnology and applied microbiology, and for researchers in the many biotechnology industries.

Basic Biotechnology

SECOND EDITION

edited by
Colin Ratledge and Bjørn Kristiansen
University of Hull, UK *European Biotechnologists, Norway*

PUBLISHED BY THE PRESS SYNDICATE OF THE UNIVERSITY OF CAMBRIDGE
The Pitt Building, Trumpington Street, Cambridge, United Kingdom

CAMBRIDGE UNIVERSITY PRESS
The Edinburgh Building, Cambridge CB2 2RU, UK
40 West 20th Street, New York, NY 10011-4211, USA
10 Stamford Road, Oakleigh, VIC 3166, Australia
Ruiz de Alarcón 13, 28014 Madrid, Spain
Dock House, The Waterfront, Cape Town 8001, South Africa

http://www.cambridge. org

First published 2001

Printed in the United Kingdom at the University Press, Cambridge

Typeface Swift 9.5/12.25pt. *System* QuarkXPress™ [SE]

A catalogue record for this book is available from the British Library

ISBN 0 521 77074 2 hardback
ISBN 0 521 77917 0 paperback

Contents

Contributors

ALASTAIR J. ANDERSON
Department of Biological Sciences
University of Hull
Hull
HU6 7RX, UK

THORLEIF ANTHONSEN
Kjemisk Institutt
NTNU
7034 Trondheim
Norway

DAVID B. ARCHER
School of Life and Environmental Sciences
University of Nottingham
University Park
Nottingham
NG7 2RD, UK

WILLIAM BAINS
Merlin Biosciences Ltd
12 St James's Square
London
SW1Y 4RB, UK

JOAQUIM M. S. CABRAL
Laboratorio de Eugenharia Bioquimica
Instituto Superior Technico
1000 Lisboa
Portugal

YUSUF CHISTI
Institute of Technology and Engineering
Massey University
Private Bag 11 222
Palmerston North
New Zealand

MIKE CLARK
Immunology Division
Department of Pathology
University of Cambridge
Cambridge
CB2 1QP, UK

L. EGGELING
Institut für Biotechnologie
Forschungszentrum Jülich GmbH
PO Box 1913
D-52425 Jülich
Germany

SVEN-OLOF ENFORS
Department of Biochemistry and Biotechnology
Royal Institute of Technology
S-100 44 Stockholm
Sweden

CHRIS EVANS
Merlin Biosciences Ltd
12 St James's Square
London
SW1Y 4RB, UK

L. HÄGGSTRÖM
Department of Biochemistry and Biotechnology
Royal Institute of Technology
S-100 44 Stockholm
Sweden

COLIN R. HARWOOD
Department of Microbiology
The Medical School
University of Newcastle
Framlington Place
Newcastle upon Tyne
NE2 4HH, UK

RAJNI HATTI-KAUL
Department of Biotechnology
Centre for Chemistry and Chemical Engineering
Lund University
PO Box 124
S-221 00 Lund
Sweden

J. J. HEIJNEN
Delft University of Technology
Department of Biochemical Engineering
Julianalaan 67
NL-2628 BC Delft
The Netherlands

DAVID J. JEENES
Institute of Food Research
Norwich Research Park
Colney, Norwich
NR4 7UA, UK

GEORG-B. KRESSE
R & D Biotechnology
Boehringer-Mannheim GmbH
Werk Penzberg
Nonnenwald 2
D-82372 Penzberg
Germany

BJØRN KRISTIANSEN
European Biotechnologists
Gluppeveien 15
N-1614 Fredrikstad
Norway

CHRISTIAN P. KUBICEK
Department of Microbial
Biochemistry
Institute of Biochemical Technology
Getreidemarkt 9/172
Vienna a-1060
Austria

DAVID A. LOWE
Bristol-Myers Co.
Industrial Division
PO Box 4755
Syracuse
NY 13221, USA

A. LÜBBERT
Institut für Bioverfahrenstechnik und
Reaktionstechnik
Martin Luther Universitat
Halle Wittenberg
D-06099 Halle/Saale
Germany

DONALD A. MACKENZIE
Institute of Food Research
Norwich Research Park
Colney, Norwich
NR4 7UA, UK

BO MATTIASSON
Department of Biotechnology
Centre for Chemistry and Chemical
Engineering
Lund University
PO Box 124
S-221 00 Lund
Sweden

MURRAY MOO-YOUNG
Department of Chemical Engineering
University of Waterloo
Waterloo, Ontario N2L 3G1
Canada

JENS NIELSEN
Institute for Biotechnology
Danish University of Technology
DK 2800 Lyngby
Denmark

HENK J. NOORMAN
Gist-brocades NV
PO Box 1
2600 MA Delft
The Netherlands

W. PFEFFERLE
Biotechnologie
Degussa-Hüls AG
PF 1112
D-33788 Halle
Germany

COLIN RATLEDGE
Department of Biological Sciences
University of Hull
Hull
HU6 7RX, UK

H. SAHM
Institut für Biotechnologie
Forschungszentrum Jülich GmbH
PO Box 1913
D-52425 Jülich
Germany

R. SIMUTIS
Control Department
Kaunas Technical University
Kaunas
Lithuania

JOHN E. SMITH
Applied Microbiology Division
University of Strathclyde
George Street
Glasgow G1 1XW, UK

PHILIPPE VANDEVIVERE
OWS
Dok Noord 4
B 9000 Gent
Belgium

J. P. VAN DIJKEN
Department of Microbiology and Enzymology
Delft University of Technology
Julianalaan 67A, NL 2628 BC Delft
The Netherlands

WILLY VERSTRAETE
Laboratory of Microbial Ecology
University of Gent
Gent
Belgium

N. VRIEZEN
Centocor B. V.
PO Box 251
NL 2300 AG Leiden
The Netherlands

ANIL WIPAT
Department of Microbiology
The Medical School
University of Newcastle
Framlington Place
Newcastle upon Tyne
NE2 4HH, UK

JAMES P. WYNN
Department of Biological Sciences
University of Hull
Hull
HU6 7RX, UK

Preface

It is some 14 years since the first edition of this book appeared. Much has happened to biotechnology in these intervening years. Recombinant DNA technology which was just beginning in the mid-1980s is now one of the major cornerstones of modern biotechnology. Developments in this area have radically altered our concepts of health-care with the arrival of numerous products that were unthinkable 20 years ago. Such is the pace of biotechnology that it can be anticipated in the next 14 years that even greater developments will occur thanks to such programmes as the Human Genome Project which will open up opportunities for treatment of diseases at the individual level. All such advances though rely on the application of basic knowledge and the appreciation of how to translate that knowledge into products that can be produced safely and as cheaply as possible. The fundamentals of biotechnology remain, as always, production of goods and services that are needed and can be provided with safety and reasonable cost.

Biotechnology is not just about recombinant DNA, of cloning and genetics; it is equally about producing more prosaic materials, like citric acid, beer, wine, bread, fermented foods such as cheese and yoghurts, antibiotics and the like. It is also about providing clean technology for a new millennium; of providing means of waste disposal, of dealing with environmental problems. It is, in short, one of the two major technologies of the twenty-first century that will sustain growth and development in countries throughout the world for several decades to come. It will continue to improve the standard of all our lives, from improved medical treatments, through its effects on foods and food supply and into the environment. No aspect of our lives will be unaffected by biotechnology.

This book has been written to provide an overview of many of the fundamental aspects that underpin all biotechnology and to provide examples of how these principles are put into operation: from the starting substrate or feedstock through to the final product. Because biotechnology is now such a huge, multi-everything activity we have not been able to include every single topic, every single product or process: for that an encyclopedia would have been needed. Instead we have attempted to provide a mainstream account of the current state of biotechnology that, we hope, will provide the reader with insight, inspiration and instruction in the skills and arts of the subject.

Since the first edition of this book, we sadly have to record the death of our colleague and friend, John Bu'Lock, whose perspicacity had led to the first edition of this book being written. John, at the time of his death in 1996, was already beginning to plan this second edition and it has been a privilege for us to have been able to continue in his footsteps to see it through into print. John was an inspiring figure in biotechnology for many of us and it is to the memory of a fine scientist, dedicated biotechnologist and a remarkable man that we dedicate this book to JDB.

Part I

Fundamentals and principles

Chapter 1

Public perception of biotechnology

John E. Smith

1.1 Introduction

Biotechnology can be viewed as a group of useful, enabling technologies with wide and diverse applications in industry, commerce and the environment. Historically, biotechnology evolved as an artisanal skill rather than a science, exemplified in the manufacture of beers, wines, cheeses etc. where the techniques of manufacture were well understood but the molecular mechanisms went unknown. In more recent times, with the advances in the understanding of microbiology and biochemistry, all of these empirically derived processes have become better understood and as a result improved. The traditional biotechnology products have now been added to with antibiotics, vaccines, monoclonal antibodies and many others, the production of which has been optimised by improved fermentation procedures and novel downstream processing.

It is clear that biotechnology has its roots in the distant past and has large, highly profitable, modern industrial outlets of great value to society, e.g. the fermentation, biopharmaceutical and food industries. Why then has there been such public awareness and concern of this subject in recent years? The main reasons must be associated with the rapid advances in molecular biology, in particular, recombinant DNA (rDNA) technology, which is now giving bioscientists a remarkable understanding and control over biological processes. By these techniques it is increasingly possible to directly manipulate the heritable material of cells between different types of organisms, creating new combinations of characters and functions not previously achievable by traditional breeding methods.

It is most probable that this rDNA technology or genetic engineering will be the most revolutionary technology in the first part of the twenty-first century. Genetic engineering will be increasingly viewed as a branch of modern science which will have profound impacts on medicine, contributing to the diagnosis and cure of hereditary defects and serious diseases, the development of new biopharmaceutical drugs and vaccines for human and animal use, the modification of micro-organisms, plants and farmed animals for improved and tailored food production and to increased opportunities for environmental remediation and protection.

In plant and animal breeding, the new technologies are much faster and have lower costs than the traditional methods of selective breeding. Furthermore, the desired modifications can be achieved in fewer generations. Improvements to plant yields for nutritional content could result in significant increase in our ability to feed an ever-increasing world population (8 billion before the year 2025) at a reduced environmental cost. Furthermore, introduction of pest, drought and salinity resistances could permit many crops to be grown in regions previously considered as hostile to plant growth. Field trial numbers of genetically modified (GM) crops are growing exponentially. At least 10 000 have been carried out worldwide and at least 80 cash crops have been subjected to GM experiments. In 1997 approximately 12.8 million hectares of transgenic or genetically modified crops were being grown around the world. The focus of agriculture must be to adequately feed the world's population, particularly in the USA where the emphasis is on how best to apply genetic engineering to harness the undoubted crop efficiency improvement which can be achieved.

1.2 Public awareness of genetic engineering

However, genetic engineering is surprisingly being subjected to a massive level of criticism from a deeply suspicious public. While the American public seem to have a greater acceptance of the potentials of genetic engineering, in Europe the technology appears to arouse deep unease among many consumers. Consumers demonstrate concern about 'unknown' health risks, possible deleterious effects on the environment and the 'unnaturalness' of transferring genes between unrelated species. Also for many people there is an increasing concern about the ever-growing influence of technology in their lives and, in some instances, an unjustified mistrust of scientists.

While genetic engineering is an immensely complicated subject, not easily explained in lay terms, that does not mean that it must remain, in decision-making terms, only in the control of the scientist, industrialist or politician. A Royal Society report in 1985 on 'Public Understanding of Science', finished with the following statement: 'Our most direct and useful message must be to the scientists themselves –

learn to communicate with the public, be willing to do so and consider it your duty to do so!' There is no doubt that many of the public or consumers are interested in the science of genetic engineering but are unable to understand the complexity of this subject. Furthermore, genetic engineering and its myriad of implications must not be beyond debate. Public attitudes to genetic engineering will influence its evolution and marketplace applications. It is important for public confidence for everyone to recognise (including scientists) that all science is fallible – especially complex biological sciences. All too often press and TV reports on genetic engineering present the discoveries as absolute certainties when this is rarely the case.

What then must be done to advance public understanding of genetic engineering in the context of biotechnology? What does the public need to know and how can this be achieved to ensure that the many undoubted benefits that this technology can bring to mankind do not suffer the same fate as the food irradiation debacle in the UK in the early 1990s? While gamma-irradiation of foods was demonstrated to be a safe and efficient method to kill pathogenic bacteria, it was not accepted by the lay public following the Chernobyl disaster, since most were unable to differentiate between the process of irradiation and radioactivity. Effective communication about the benefits and risks of genetic engineering will depend on understanding the underlying concerns of the public together with any foreseeable technical risks.

Over recent years there have been many efforts made to gauge the public awareness of biotechnology by questionnaires, Eurobarometers and Consensus Conferences. Early studies highlighted public attitudes to the application of genetic manipulation to a wide range of scenarios (Table 1.1). While medical applications were more generally acceptable others such as the manipulation of animal and human genomes were highly unacceptable.

Eurobarometer surveys revealed a broad spectrum of opinions that were influenced by nationality, religion, knowledge of the subject and how the technology will be applied (Box 1.1). A major contributory factor is the plurality of beliefs and viewpoints that are held explicitly or implicitly about the moral and religious status of Nature and what our relationship with it should be. Do we view Nature, in the context of man's dependency on plants and animals, as perfect and complete derived by natural means of reproduction and therefore not to be tampered with by 'unnatural' methods, or do we see it as a source of raw material for the benefit of mankind? For centuries now man has been indirectly manipulating the genomes of plants and animals by guided matings primarily to enhance desired characteristics. In this way, food plants and animals bear little resemblance to their predecessors. In essence, such changes have been driven by the needs and demands of the public or consumer, and have been readily accepted by them. In the traditional methods that have been used, the changes are made at the level of the whole organism, selection is made for a desired phenotype and the genetic changes are often poorly characterised and occur together with other possibly undesired genetic changes. The new

Table 1.1 Public attitudes (in % responses) to applications of genetic manipulation in Europe

	Comfortable	Neutral	Uncomfortable
Microbial production of bio-plastics	91	6	3
Cell fusion to improve crops	81	10	10
Curing diseases such as cancer	71	17	9.5
Extension of shelf life of tomatoes	71	11	19
Cleaning up oil slicks	65	20	13
Detoxifying industrial waste	65	20	13
Anti-blood-clotting enzymes produced by rats	65	14	22
Medical research	59	23	15
Making medicines	57	26	13
Making crops to grow in the Third World	54	25	19
Mastitis-resistant cows by genetic modification of cows	52	16	31
Producing disease-resistant crops	46	29	23
Chymosin production by micro-organisms	43	30	27
Improving crop yields	39	31	29
Using viruses to attack crop pests	23	26	49
Improving milk yields	22	30	47
Cloning prize cattle	7.2	18	72
Changing human physical appearance	4.5	9.5	84
Producing hybrid animals	4.5	12	82
Biological warfare	1.9	2.7	95

Box 1.1 Eurobarometer (1997) on Public Perception of Biotechnology

- The majority of Europeans consider the various applications of modern biotechnology useful for society. The development of detection methods and the production of medicines are seen to be most useful and considered the least dangerous.
- The use of modern biotechnology in the production of foodstuffs and the insertion of human genes into animals to obtain organs for humans were judged least useful and potentially dangerous.
- Europeans believe that it is unlikely that biotechnology will lead to a significant reduction of hunger in the developing world.
- The vast majority of Europeans feel genetically modified products should be clearly labelled.
- The majority of Europeans tend to believe that we should continue with traditional breeding methods rather than changing the hereditary characteristics of plants and animals through modern biotechnology.
- Less than one in four Europeans think that current regulations are sufficient to protect people from any risk linked to modern biotechnology.
- Only two out of ten Europeans think that regulations of modern biotechnology should be primarily left to industry.
- A third of Europeans think that international organisations such as the United Nations and the World Health Organization are better placed to regulate modern biotechnology, followed by scientific organisations.

methods, in contrast, enable genetic material to be modified at the cellular and molecular level, are more precise and accurate, and consequently produce better characteristics and more predictable results while still retaining the aims of the classical breeder. A great many such changes can and will be done within species giving better and faster results than by traditional breeding methods.

1.3 Regulatory requirements – safety of genetically engineered foods

Much debate is now taking place on the safety and ethical aspects of genetically modified organisms (GMOs) and their products destined for public consumption. Can such products with 'unnatural' gene changes lead to unforeseen problems for present and future generations?

The safety of the human food supply is of critical importance to most nations and all foods should be fit for consumption i.e. not injurious to health or contaminated. When foods or food ingredients are derived from GMOs they must be seen to be as safe as, or safer than, their traditional counterparts. The concept of **substantial equivalence** is widely applied in the determination of safety by comparison with analogous conventional food products together with intended use and exposure. When substantial equivalence can be shown then normally no further safety considerations are necessary. When substantial equivalence is not clearly established the points of difference must be subjected to further scrutiny.

When such novel products are moving into the marketplace the consumer must be assured of their quality and safety. Thus there must be toxicological and nutritional guidance in the evolution of novel foods and ingredients to highlight any potential risks which can then be dealt with appropriately. Safety assessment of novel foods and food ingredients must satisfy the producer, the manufacturer, the legislator *and* the consumer. The approach should be in line with accepted scientific considerations, the results of the safety assessment must be reproducible and acceptable to the responsible health authorities and the outcome must satisfy *and* convince the consumer!

A comprehensive regulatory framework is now in place within the EU with the aim to protect human health and the environment from adverse activities involving GMOs. There are two Directives providing horizontal controls i.e. (1) contained use and (2) deliberate release of GMOs.

The contained use of GMOs is regulated under the Health and Safety at Work Act through the Genetically Modified Organisms (Contained Use) Regulations which are administered by the Health and Safety Executive (HSE) in the UK. The HSE receives advice from the Advisory Committee on Genetic Modification. These Regulations, which implement Directive 90/219/EEC, cover the use of all GMOs in containment and will incorporate GMOs used to produce food additives or processing aids. All programmes must carry out detailed risk assessments with special emphasis on the organism that is being modified and the effect of the modification.

Any deliberate release of GMOs into the environment is regulated in the UK by the Genetically Modified Organisms (Deliberate Release) Regulations, which are made under the Environmental Protection Act and implement EC Directive 90/220/EC. Such regulations will cover the release into the environment of GMOs for experimental purposes (i.e. field trials) and the marketing of GMOs. Current examples could include the growing of GM food crop plants or the marketing of GM soya beans for food processing.

All experimental release trials must have government approval and the applicant must provide detailed assessment of the risk of harm to human health and/or the environment. All applications and the risk assessments are scrutinised by the Advisory Committee on Releases into the Environment which is largely made up of independent experts who then advise the Ministers.

The EC Novel Foods Regulation (258/97) came into effect in May 1997 and represents a mandatory EC-wide pre-market approval process for all novel foods. The regulation defines a novel food as one that has not previously been consumed to a significant degree within the EU. A part of their regulations will include food containing or consisting of GMOs as defined in Directive 90/220 and food produced by GMOs but not containing GMOs in the final product.

In the UK the safety of all novel foods including genetically modified foods is assessed by the independent Advisory Committee on Novel Foods and Processes (ACNFP) which has largely followed the approach developed by the WHO and OEDC in assessing the safety of novel foods. The ACNFP has encouraged openness in all of its dealings, publishing agendas, reports of assessments and annual reports, a Newsletter and soon a Committee Website. By such means it hopes to dispel any misgivings that may be harboured by members of the public. The ultimate decisions are not influenced by industrial pressure and are based entirely on safety factors.

There is undoubtedly going to be a steady increase in the range of GM foods coming to the market in the US and in Europe (Table 1.2). A comprehensive EU regulatory framework covering GMOs is now firmly established and the specific legislation now in force will ensure the safety of GM foods.

In all of the foregoing, the risk assessments of GMO products etc. have been made by experts and judged on the basis of safety to the consumer. However, it must be recognised that subject experts define risk in a narrow technical way, whereas the public or consumer without sufficient knowledge generally displays a wider, more complex view of risk that incorporates value-laden considerations such as unfamiliarity, catastrophic potential and controllability. Furthermore, the public, in general, will almost always overestimate risks associated with technological hazards such as genetic engineering and underestimate risks associated with 'lifestyle' hazards such as driving cars, smoking, drinking, fatty foods etc. It is puzzling to note that food-related technologies

Table 1.2 Genetically modified crops completing the regulatory process in the USA as of December 1997

Crop	Trait	Company
Tomato	Modified ripening	Zeneca Plant Science
Soya beans	Glyphosate tolerance	Monsanto Co.
Tomato	Modified ripening	Monsanto Co.
Potato	Insect resistance	Monsanto Co.
Tomato	Modified ripening	DNA Plant Technology
Cotton	Bromoxynil tolerance	Calgene
Tomato	Delayed ripening	Calgene
Squash	Virus resistance	Asgrow Seed Co.
Cotton	Insect resistance	Monsanto Co.
Oilseed rape	Glyphosate tolerance	Monsanto Co.
Cotton	Glyphosphate tolerance	Monsanto Co.
Maize	Insect resistance	Ciba Geigy Corp.
Oilseed rape	High laurate – oil	Calgene
Maize	Glufosinate tolerance	AgrEvo Inc.
Oilseed rape	Glufosinate tolerance	AgrEvo Inc.
Maize	Male sterile	Plant Genetic Systems
Oilseed rape	Male sterile/fertility restorer	Plant Genetic Systems
Maize	Insect resistance	Northrup King
Potato	Insect resistance	Monsanto Co.
Maize	Insect resistance	Monsanto Co.
Maize	Insect resistance, glyphosate tolerance	Monsanto Co.
Cotton	Sulphonylurea tolerance	Du Pont
Maize	Glufosinate tolerance	Dekalb Genetics Corp.
Tomato	Modified ripening	Agritope Inc.
Soya	High oleic acid content of oil	Du Pont
Maize	Herbicide tolerance	Monsanto Co.
Maize	Herbicide tolerance and insect resistance	Monsanto Co.
Cotton	Herbicide tolerance and insect resistance	Calgene
Chicory	Male sterile	Bejo Zaden

tend to be perceived as high in risk relative to benefit when compared with other technologies. Perception of the risks inherent in genetic engineering may be moderated by recognition of the tangible benefits of specific products of genetic engineering that could be shown to have health or environmental benefits.

How do you achieve effective communication with the public about benefits and risks of genetic engineering? Trust in the reliability of the information source is of major importance in any communication about risk and this is associated with perceptions of accuracy, responsibility and concern with public welfare. In contrast, distrust may be generated when it is assumed that the facts are distorted or the information misused or biased.

1.4 Labelling – how far should it go?

Perhaps the most contentious issue related to foods derived from genetic engineering is to what extent should they be labelled. The purpose of labelling a food product is to provide sufficient information and advice, accurately and clearly, to allow consumers to select products according to their needs, to store and prepare them correctly and to consume them with safety. With respect to the principle of labelling, information should be accurate, truthful, sufficiently detailed, not misleading and above all understandable.

Labelling of a product will only be relevant if the consumer is able to understand the information printed on the labels. The Food and Drug Administration of the USA considers that labelling should not be based on the way a particular product is obtained. This is, or should be, a part of normal approval for agricultural practice or industrial processes, and if approved, then labelling should be unnecessary, which is the common practice for most food products. It can be argued that certain consumers are in principle against a certain technique (such as genetic engineering) and such consumers should have the right to know if this technology has been used to produce the particular product.

At least within the EU there is considerable evidence that there is strong support for the clear labelling of genetically engineered foods. Labelling is all about consumer choice and has nothing to do with health and safety and where there is concern about the specific safety of products labelling will not solve the problem (note tobacco products are already labelled as injurious to health). The labelling of GM foods and ingredients has become a major issue and is generating a great deal of confusion and difficulty in full application. Industry associations, especially in the USA, consider that consumers are entitled to information if the characteristics and composition of foodstuffs are only different.

The novel food regulations in the EU introduced specific labelling requirements over and above those already required for food. Labelling would be required where there are special ethical concerns such as copies of animal or human genes or if the food product contained live GMOs. However, the British Government is pressing for all foods which contain genetically modified material to be labelled clearly to enable consumers to have a real choice.

If consumers insist that they want to choose whether or not to buy GM or non-GM products then there must be an adequate supply of non-GM foodstuffs. In the future can this be achieved? Plant breeders are increasingly turning to the new genetic technologies to improve their plants and animals in order to produce the *cheap* food now demanded by the consumer.

The whole aspect of labelling of GMO-derived foods is undergoing a fierce and often bitter and ill-informed debate which can differ in approach throughout the world. It can only be expected that in the near future all major food organisms especially plants will have had some level of genetic engineering in their breeding programmes. This will be necessitated by the need to produce food for an ever increasing world population. Let us not delude ourselves. Without the addition of this technology to the armoury of the plant breeder there *will* be serious and indeed calamitous food shortages especially in the developing world. If all aspects of genetic modification must be recognised and recorded it can only lead to unacceptably complex labelling criteria.

Consumer rights, now recognised by all member states in the EU, involve a right to information and its corollary, a duty to inform. As a consequence, labelling should be meaningful but appropriate.

1.5 Policy making

Policy making on genetic engineering throughout the industrial world is strongly influenced by the varied interests of governments, industry, academics and environmental groups. After almost two decades of discussions, the dominant issue still is whether government regulations should depend on the characteristics of the products modified by rDNA technique or on the technique of rDNA. For those who support the product-based regulations, the new techniques could be considered as an extension and refinement of more conventional or traditional breeding approaches. As a result they would expect the new products to be treated in a way similar to products created by conventional technology. In contrast, there are those who profess that the regulation of the genetic modification process could be viewed as a *new* biotechnology and as an undeveloped science about which there is little knowledge and working experience. In this way the process is sufficiently different and distinct from conventional techniques and should demand unique regulatory rules.

As with other techniques, the genetic engineering debate could also prove to be a critical testing ground for efforts to insert into governmental policies socio-economic and sociocultural measures – the so-called 'fourth criterion'. The advocates of this approach consider that measures of efficacy, quality and safety are, alone, insufficient to judge the potential risk associated with such new techniques and their products and to these they would add social and moral considerations.

These new approaches are having a significant impact on the pace of agricultural and environmental applications of genetic engineering. In contrast, biomedical applications have progressed relatively rapidly. Millions of people throughout the world have accepted and benefited from diagnostics. Drugs provided by the new biotechnology companies include the GM products erythropoietin, for kidney dialysis patients, and insulin for diabetics, while diagnostics developed by genetic engineering in particular keep dangerous pathogens, such as HIV and hepatitis viruses, out of the blood supplies. Those activists who oppose genetically engineered products have studiously kept clear of these successful areas of application. In agriculture there has been concerted opposition by the activists against GM bovine somatotropin (BST) but almost total silence on the engineered chymosin enzyme used to clot milk in cheese production, which now claims about 40–45% of the US market. Millions of calves are now no longer required for this process. Indeed, the success of GM chymosin has been applauded by animal rights activists.

How far the fourth criterion will influence science policy and, in particular, genetic engineering in the making of agricultural and environmental policies are now at important cross-roads. While there is great need to increase the science literacy of the public in general, a well-informed public must still rely heavily on independent experts for the evaluation of sophisticated technical issues.

1.6 Areas of significant public concern

1.6.1 Antibiotic resistance marker genes

Antibiotic 'marker' genes are used to identify and select cells which have been successfully modified following a genetic modification process. By the use of such genes, cells which have been successfully modified can grow in the presence of the particular antibiotic. The most commonly used antibiotic resistance marker genes in GM plants confer resistance to kanamycin or hygromycin while for GM bacteria the ampicillin resistance marker gene is more often used. Could such antibiotic resistance genes be transferred from a consumed GM plant or micro-organism into the human gut microflora and so increase antibiotic resistance in the human population? Bacterial resistance to commonly used antibiotics can now be found throughout the world and it is most probable that this incidence is the result of the transfer of resistance genes between bacteria followed by the selective pressures imposed by the use of antibiotics. To date it has not been possible to demonstrate the transfer of antibiotic resistance from a marker gene in a consumed GM plant to micro-organisms normally present in the gut of humans and other animals. However, a potential does exist and should not be ignored. The Advisory Committee on Novel Foods and Processes (ACNFP) in the UK has recommended that antibiotic resistance markers should be eliminated from all GM foods and micro-organisms that will be consumed live. Researchers developing GMOs for food should develop and use alternatives to antibiotic resistance markers and/or employ methods to jettison those used in developing the early transgenic cells. By following this advice, public concern over this facet of genetic engineering should be eliminated. The recent and heated controversy over the Ciba-Geigy transgenic corn revolved around the presence of a non-functional β-lactamase (an enzyme that can inactivate penicillin) sequence. While the risk of transfer of β-lactamase DNA to gut microflora would indeed be vanishingly small, it has been a massive publicity winner for those opposed to GMOs. How easily it could have been avoided with a little foresight by the company.

1.6.2 Transfer of allergies

Food allergies arise when the immune system responds to specific allergens which are usually glycoproteins in the food. This is now a major concern especially with respect to the peanut and other nuts and severe anaphylactic reactions are not uncommon. Consequently, labelling of food products with respect to the presence of peanut, is now widely practised. Thus, it becomes essential with GM foods to ensure that transfer of allergens does not occur from donor species to recipient species. Clearly, this is a complex process and one which all producers of GMOs are now fully alerted to and due consideration is being given. There have been no recorded examples of new allergies by the process of recombinant DNA technology. Many databases that can identify proteins that could be problematic if inserted into food materials are now increasingly available.

1.6.3 Pollen transfer from GM plants

The possibility of gene transfer from transgenic crop plants to compatible wild relatives has been given serious examination. Could pest or herbicide resistance incorporated into transgenic plants be transferred into other closely related plants and increase their 'weediness'? Under normal conditions, gene transfer by way of pollen even between close relatives is exceptionally rare and there is little hard evidence that this will be different with transgenic plants. However, released transgenic plants will be routinely monitored to continue the already extensive validation that has occurred. The possible presence of toxic recombinant proteins in honey (which may contain about 2% pollen) has been shown to have no relevance.

1.6.4 Social, moral and ethical issues associated with GMOs

Initial concerns about GMOs were perceived basically on safety issues but more recently social, moral and ethical issues have become part of the decision-making process.

The control of GM crop plants and their seeds by multinational agrochemical companies and their need to recover the high investment costs could imply that only high technology farmers will be able to meet the full costs. GM seeds are normally made sterile to prevent further propagation by farmers. However, this is normal practice by seed companies where the seeds used commercially are F_1 hybrids which do not breed truly so a constant supply of seed is needed for farmers to grow the right crops. This ensures a fair return from the investment to produce them originally. The development of herbicide-resistant GM crops could well result in dependence by farmers on the specific herbicides and their producing companies. It is doubtful if the poor Third World farmers will benefit in the near future from these developments.

The concept of substitutability would also have dramatic social impact on some developing countries. For example, several novel sweeteners have been developed, many times sweeter than sucrose, which could ultimately lead to a reduction in the traditional sugar market for sugar cane and sugar beet. In this way, these economies predominantly in developing countries could experience severe financial and employment disruption with alternatives difficult to find. Similarly, increased milk production from fewer cows by regular injection of genetically engineered bovine growth hormone (BST) may well cause severe difficulty for small farmers in USA and the EU. While this development has gone through quite successfully in the USA there is a moratorium currently in operation in the EU.

It must be expected that many aspects of this new genetic technology, particularly when applied to agriculture and food production, could well lead to decreases in employment with an ensuing increase in poverty in developing economies. Different value judgements come into operation to reconcile the advantages to society against the disadvantages. The developed nations must endeavour to assist the developing nations both technically and financially to be part of this agriculture revolution. To what extent this is now happening is variable and questionable. It is sad that public awareness of such issues is seldom voiced by the western activists against genetic engineering.

It is in the area of animal transgenics that public awareness and concern is most regularly expressed. Transgenic animals are those incorporating foreign or modified genetic information resulting from genetic engineering. Genetic modification of animals with ensuing transgenesis may be considered by some on a religious basis as a fundamental breach in natural breeding barriers which nature developed through evolution to prevent genetic interplay between unlike species. Such viewpoints consider the species as 'sacred'. In contrast, in the reductional philosophy of most molecular biologists the gene has become the ultimate unit of life and is merely a unique aggregation of organic molecules (largely common to all types of living cells) available for manipulation. In this way, most molecular biologists are unable to accept that there is any ethical problem created in transferring genes between species and genera.

What are the foreseen reasons or benefits for pursuing animal transgenesis?

(1) There have been extensive animal studies aimed at understanding developmental pathways with a view to understanding gene action. The development of the Oncomouse for cancer studies could have significant value in developing anticancer drug treatment for humans. However, the fact that the mouse will suffer the development of cancer and die prematurely does raise the issue of animal morality.

(2) Improved growth rates in animals and fish have been achieved by selected gene transfer.

(3) Biopharmaceuticals (or bio-pharming) in which human genes have been inserted into lactating animals, e.g. sheep, have had dramatic and generally acclaimed results. There appears to be little effect on the animal and valuable human health proteins can be extracted from the milk of the transgenic animal. Public acceptance of such products is not seen as a contentious issue.

(4) Xenotransplantation especially from pigs is seen as an excellent source of organs for human transplants. Human complement-inactivating factors to prevent acute hyper-immune rejection of the transplant have been successfully transferred into the pig. This programme has obvious attractions and could overcome the acute short-

age of human organs for essential transplants. The main concerns have been possible transfer of pig viruses to humans in the light of the BSE scare. Also the question of breeding pigs for this purpose as opposed to the accepted current practice for eating purposes does generate ethical concerns for some!

(5) Biomarkers for detecting environmental pollution using transgenic nematodes would appear to be a worthwhile activity.

(6) Transgenic animals can be used as models for human genetic diseases with the ultimate aim to develop new drugs or gene-therapy treatments.

While many people see the aims of these studies to be of significant value to mankind others express genuine concerns for animal welfare, and whether we have the right to indulge in modifying an animal genome to human advantage.

Central to public concern is a strong feeling of 'unnaturalness' in transferring human genes into animals with the new transgenic animal containing copies of the original human gene. It is difficult for the average layperson to comprehend that while the transgene has human origin and structure it is not its immediate source. In the process of genetic manipulation, genes cannot be directly transported from the donor to the recipient but must progress through a complicated series of *in vitro* clonings. In this manner a series of amplification steps are carried out in which the original gene is copied many times during the whole process such that the original genetic material is diluted 10^{55}. Thus the original DNA is not directly used, but rather, similar DNA is synthesised artificially. Because the transgenic organ does not contain the actual human gene but only an artificial created copy of the gene, molecular biologists widely accept the status of the transgene to be that now of the new organism. It is their view that genes fulfil their biological role only by their activity within the environment of the cell and an organism.

The issue of animal rights is highly contentious – do they have intrinsic rights or not? Some would assign equal moral worth to sentient animals as to humans, but where do you draw the line?

A sub-committee of the Advisory Committee on Novel Foods and Processes (ACNFP) considered some of the main ethical concerns arising from the food use of certain transgenic organs and prepared the following exclusions:

(1) Transfer of genes from animals whose flesh is forbidden for use as food by certain religious groups to animals which they normally consume (e.g. pig genes into sheep) would offend most Jews and Muslims.

(2) While transfer of human genes to food animals (e.g. transfer of the human gene for Factor IX, a protein involved in blood clotting, into sheep) is acceptable for pharmaceutical and medical purposes, the animals should not, upon slaughter, enter the food chain.

(3) Transfer of animal genes into food plants (e.g. for vaccine production) is acceptable for pharmaceutical and medical purposes but plant remains should not then enter the animal and human food

chain as they would be anathema to vegetarians and especially vegans.

(4) Use of organisms containing human genes as animal feed (e.g. micro-organisms modified to produce pharmaceutical human proteins such as insulin).

Close consultation with a wide range of religious faiths strongly suggested that there were no overwhelming objections to necessitate an absolute ban on the use of food products containing copy genes of human origin. (This was particularly noticeable when the concept of the copy gene was understood.) The report, however, strongly advocated that the insertion of ethically sensitive genes into food organisms be discouraged especially when alternative approaches were available. If transgenic organisms containing copy genes that are unacceptable to specific groups of the population subject to religious dietary restrictions were to be present in some foods, it should be compulsory that such foods were clearly labelled.

1.7 Conclusions

The safety and impact of genetically modified organisms continues to be addressed by scientific research. Basic research into the nature of genes, how they work and how they can be transferred between organisms has served to underpin the development of the technology of genetic modification. In this way, basic information about the behaviour of genes and of GMOs will be built up and used to address the concerns about the overall safety of GMOs and their impact on the environment.

1.8 Further reading

Frewer, L. J. & Shepherd, R. (1995). Ethical concerns and risk perceptions associated with different applications of genetic engineering: interrelationship with the perceived need for regulation of the technology. *Agric. Hum. Values* **12**, 48–57.

OECD (1993). *Safety Evaluation of Food Derived by Modern Biotechnology – Concepts and Principles.* Organisation for Economic Cooperation and Development, Paris (93 93 04 1).

Rehm, H.-J. & Reed, G. (Eds.) (1995). *Biotechnology.* 2nd Edition. Vol. 12. Legal, economic and ethical dimensions. VCH, Weinheim.

Sheppard, J. (1996). *Spilling the Genes – What we Should Know about Genetically Engineered Foods.* The Genetics Forum, London.

Smith, J. E. (1996). Safety, moral, social and ethical issues related to genetically modified foods. *Eur. J. Genet. Soc.* **2**, 15–24.

The Royal Society Statement (1998). *Genetically Modified Plants for Food Use.* Pp. 1–16.

Chapter 2

Biochemistry and physiology of growth and metabolism

Colin Ratledge

2.1 Introduction

The First Law of Biology (if there was one) could be: *The purpose of a micro-organism is to make another micro-organism.*

In some cases biotechnologists, who seek to exploit the micro-organism, may wish this to happen as frequently and as quickly as possible; in other words they wish to have as many micro-organisms available at the end of the process as possible. In other cases, where the product is not the organism itself, the biotechnologist must manipulate it in such a way that the primary goal of the microbe is diverted. As the micro-organism then strives to overcome these restraints on its reproductive capacity, it produces the product which the biotechnologist desires. The growth of the organism and its various products are therefore intimately linked by virtue of its metabolism.

In writing this chapter, I have not attempted to explain the structure of the main microbial cells: the bacteria, the yeasts, the fungi and the microalgae. These are available in most biology textbooks and these should be consulted if there are uncertainties about cell structures. However, biology textbooks rarely explain the chemistry that goes on in the living cell (i.e. their biochemistry) in simple terms but, as the biochemistry of the cell is fundamental to the exploitation of the organism, it is important to be acquainted with the basic systems that microbial cells use to achieve their multiplication.

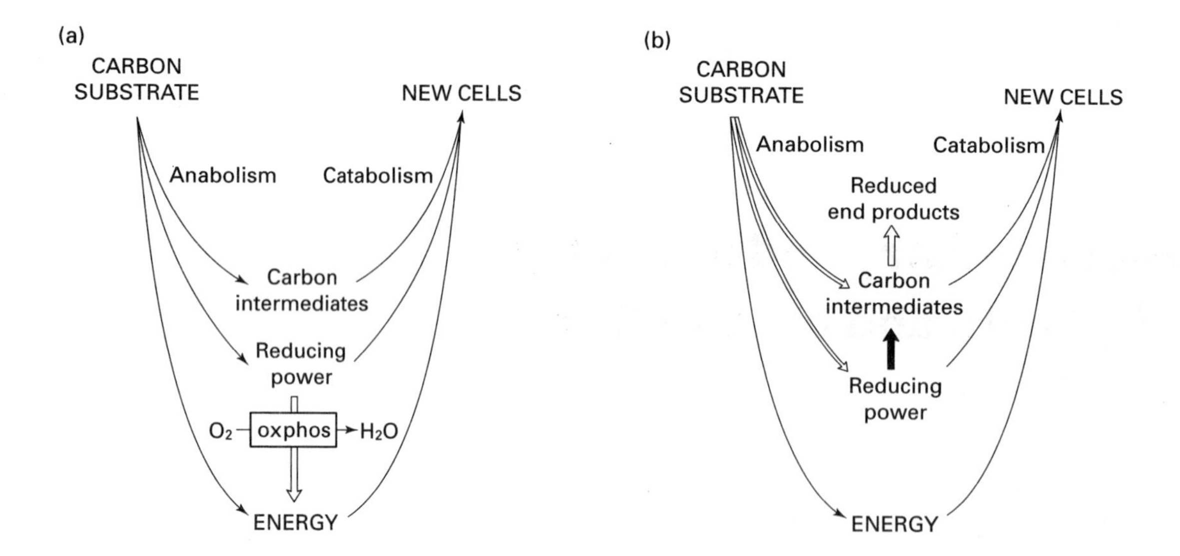

Fig. 2.1 Processes of catabolism (degradation) and anabolism (biosynthesis) linked to energy production and provision of reducing power. (a) Anaerobic metabolism; oxphos = oxidative phosphorylation (see Section 2.5); (b) Anaerobic metabolism.

The biochemistry of the cell is therefore described as an account of the chemical changes which occur within a cell as it grows and multiplies to become two cells. The physiology of the cell, however, goes beyond the biochemistry of the cell as this term extends the simple account of the flow of carbon, and the changes which occur to other elements, by describing how these processes relate to the whole growth process itself. The biochemical changes therefore are to be seen occurring in the three-dimensional array which is the cell with a fourth dimension of time being added. Not all reactions which are capable of happening will occur; some may occur during the period of its fastest growth whilst others may occur only as the growth rate of the organism is slowing down and entering a period of stasis. Physiology therefore is a complete understanding of the chemical changes within a cell related to its development, growth and life cycle.

2.2 Metabolism

2.2.1 Some definitions

Metabolism is a matrix of two closely interlinked but divergent activities (see Fig. 2.1).

Anabolic processes are concerned with the building up of cell materials, not only the major cell constituents (protein, nucleic acids, lipids, carbohydrates, etc.) but also the intermediate precursors of these materials – amino acids, purine and pyrimidines, fatty acids, various sugars and sugar phosphates. Anabolism concerns processes which are endothermic overall (they 'require energy'). They also invariably require a source of reducing power which must come by the degradation of the substrate (or feedstock).

The compensating exothermicity is provided by various **catabolic** ('energy-yielding') processes. The degradation of carbohydrates, such as sucrose or glucose, ultimately to give CO_2 and water, is the principal exothermic process whereby 'energy generation' is accomplished. During this process reducing power needed for the subsequent anabolic processes is also generated. The same considerations, however, apply to all substances that are used by micro-organisms: their degradation must yield not only carbon for new cells but also the necessary energy and reducing power to convert the metabolites into the macromolecules of the cell.

We can also distinguish between organisms which carry out their metabolism **aerobically**, using O_2 from the air, and those that are able to do this **anaerobically**, that is, without O_2. The overall reaction of reduced carbon compounds with O_2, to give water and CO_2, is a highly exothermic process; an aerobic organism can therefore balance a relatively smaller use of its substrates for catabolism to sustain a given level of anabolism, that is, of growth (see Fig. 2.1a). Substrate transformations for anaerobic organisms are essentially **disproportionations**, with a relatively low 'energy yield', so that a larger proportion of the substrate has to be used catabolically to sustain a given level of anabolism (Fig. 2.1b).

The difference can be illustrated with an organism such as yeast, *Saccharomyces cerevisiae*, which is a **facultative anaerobe** – that is, it can exist either aerobically or anaerobically. Transforming glucose at the same rate, aerobic yeast gives CO_2, water and a relatively high yield of new yeast, whereas the yeast grown anaerobically has lower yield of energy and reducing power. Consequently, fewer cells can be made than under aerobic conditions. Also it is not possible for the cells, in the absence of O_2, to oxidise all the reducing power that is generated during catabolism. Consequently, surplus carbon intermediates (in the case of yeast it is pyruvic acid) are reduced in order to recycle the reductants (see Fig. 2.2) and, in the case of yeast, ethanol is the product.
Overall, this process can be described by the simple reaction:

$$X + NADH \rightarrow XH_2 + NAD^+$$

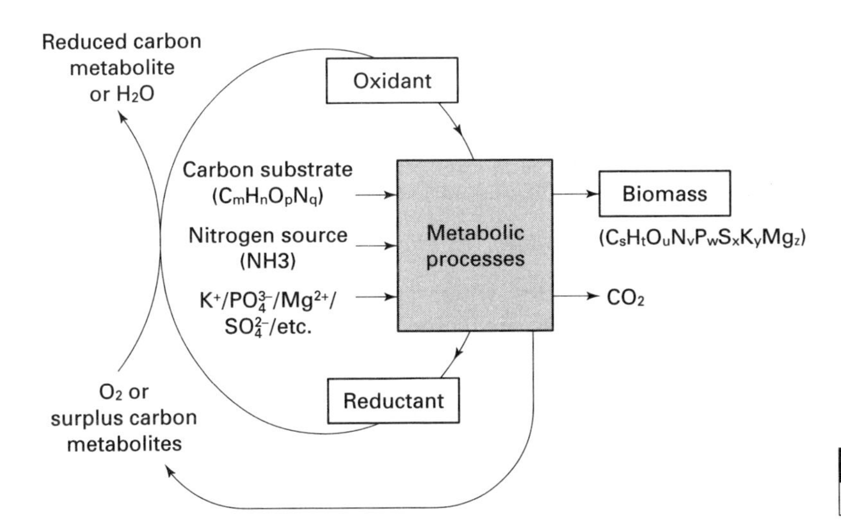

Fig. 2.2 Thermodynamic balance for a metabolising cell.

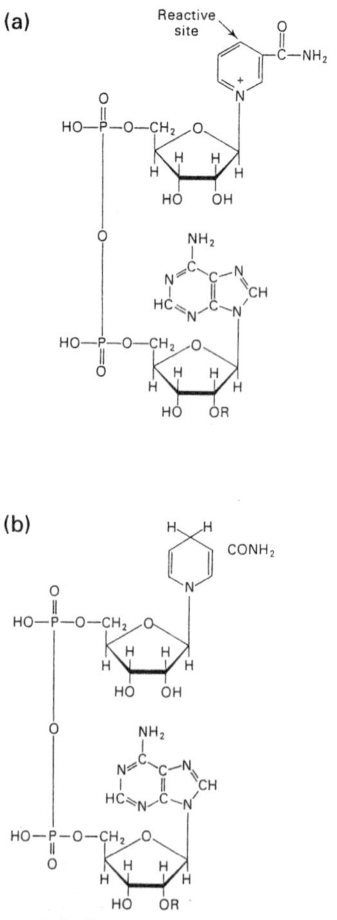

(a)

(b)

Fig. 2.3 (a) NAD^+ / $NADP^+$ (oxidised); (b) NADH / NADPH (reduced). In NAD^+ and NADH, R = H; in $NADP^+$ and NADPH, $R = PO_3^{2-}$.

where X is a metabolite and NADH is the reductant, and NAD^+ is its oxidised form (see Fig. 2.3a, b). NAD stands for nicotinamide adenine dinucleotide; NADH is therefore reduced NAD. There is also the phosphorylated form of NAD^+, NAD phosphate designated as $NADP^+$. This can also be reduced to NADPH and it can also function as a reductant but usually in anabolic reactions of the cell, whereas NADH is usually involved in the degradative reactions. All four forms (NAD^+, $NADP^+$, NADH and NADPH) occur in both aerobic as well as anaerobic cells: in the former cells re-oxidation of NADH can occur with O_2, but this cannot take place in the anaerobe, hence the need for the alternative re-oxidation strategy (see Section 2.6).

A cell that grows obviously uses carbon but many other elements are needed to make up the final composition of the cell. These will include nitrogen, oxygen – which may come from the air if the organism is growing aerobically (otherwise the O_2 must come from a rearrangement of the molecules in which the organism is growing, or even water itself) – together with other elements such as K^+, Mg^{2+}, S (as SO_4^{2-}), P (as PO_4^{3-}) and an array of minor ions such as Fe^{2+}, Zn^{2+}, Mn^{2+}, etc. The dynamics of the system are set out in Fig. 2.2.

2.2.2 Catabolism and energy

The necessary linkage between catabolism and anabolism depends upon making the catabolic processes 'drive' the synthesis of reactive reagents, few in number, which in turn are used to 'drive' the full range of anabolic reactions. These key intermediates, of which the most important is adenosine triphosphate, ATP (Fig. 2.4), have what biologists term a 'high-energy bond'; in ATP it is the anhydride linkage in the pyrophosphate residue. Directly or indirectly the potential energy released by splitting this bond is used for the bond-forming steps in anabolic syntheses. Molecules such as ATP then provide the 'energy currency' of the cell. When ATP is used in a biosynthetic reaction it generates ADP (adenosine diphosphate) or occasionally AMP (adenosine monophosphate) as the hydrolysis product:

$$A + B + ATP \rightarrow AB + ADP + P_i \quad \text{or} \quad A + B + ATP \rightarrow AB + AMP + PP_i$$

(where A and B are both carbon metabolites of the cell and P_i = inorganic phosphate, and PP_i = inorganic pyrophosphate).
ADP, which still possesses a 'high-energy bond', can also be used to produce ATP by the adenylate kinase reaction:

$$ADP + ADP \rightarrow ATP + AMP$$

Phosphorylation reactions, which are very common in living cells, usually occur through the mediation of ATP:

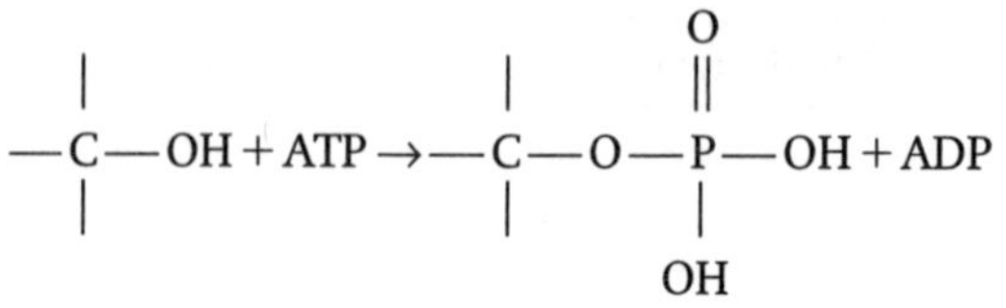

$$-\overset{|}{\underset{|}{C}}-OH + ATP \rightarrow -\overset{|}{\underset{|}{C}}-O-\overset{O}{\overset{\|}{\underset{\underset{OH}{|}}{P}}}-OH + ADP$$

The phosphorylated product is usually more reactive (in one of several ways) than the original compound.

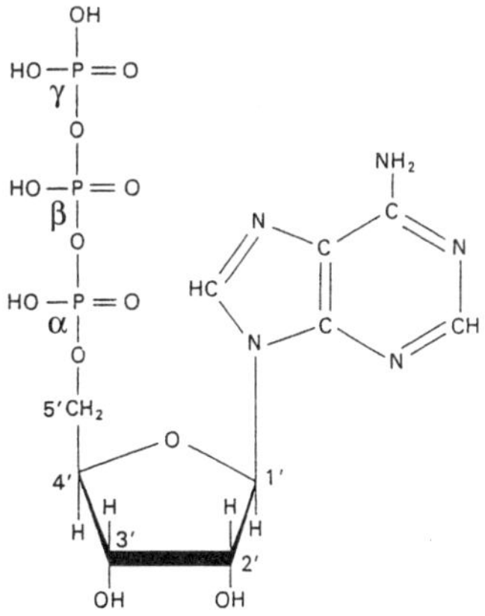

2.3 Catabolic pathways

2.3.1 General considerations of glucose degradation

The purpose of breaking down a substrate is to provide the micro-organisms with:

- building units for the synthesis of new cells;
- energy, principally in the form of ATP, by which to synthesise new bonds and new compounds;
- reducing power, which is mainly as reduced NAD (i.e. NADH) or reduced NADP (NADPH).

Both the ATP and the NAD(P)H (which can be used to denote both NADH and NADPH) act in conjunction with various enzymes in the conversion of one compound to another.

Although the microbial cell may be considered to be a vast collection of compounds, it can be simplified as being composed of:

- **proteins**, which can be either functional (such as enzymes) or structural as in some proteins associated with the cell wall or intracellular structures;
- **nucleic acids**, DNA and RNA;
- **lipids**, which are often based on fatty acids and are used in the formation of membranous structures surrounding either the whole cell or an organelle (a 'micro-organ') within the cell;
- **polysaccharides**, which are used mainly in the construction of cell walls, and cell capsules.

These in turn are made from simple precursors:

- proteins from amino acids;
- nucleic acids from nucleotide bases (plus ribose and phosphate);
- lipids from fatty acids which are, in turn, produced from acetate (C_2) units;
- polysaccharides from sugars.

It is therefore possible to identify a mere **nine** precursors from which the cell can make all its molecules and, indeed, replicate itself (see Fig. 2.5). Thus, as long as the cell can produce these basic nine molecules from any substrate, or combination of substrates, it will be able to resynthesise itself (provided of course that in the formation of these precursors, ATP and NAD(P)H are also produced).

If we consider glucose as the usual growth substrate of a microbial cell, we can show how it is degraded into these nine key precursors (Fig. 2.5). Their formations are linked through the process of glycolysis, which is sometimes called the **Embden–Meyerhof–Parnas (EMP) pathway**, which is given in more detail in Fig. 2.6, and then by the further oxidation of pyruvate, as the end-product of glycolysis, through the reactions of the tricarboxylic acid cycle (see Fig. 2.9).

In addition to the glycolytic sequence, there is also an important adjunct to this which is responsible for forming pentose (C_5) phosphates

Fig. 2.4 Adenosine triphosphate (ATP). When donating energy, the γ-bond is hydrolysed and the available energy is used to make a new bond in a molecule. Adenosine diphosphate (ADP) is without the last phospho group and adenosine monophosphate is without the last two phospho groups.

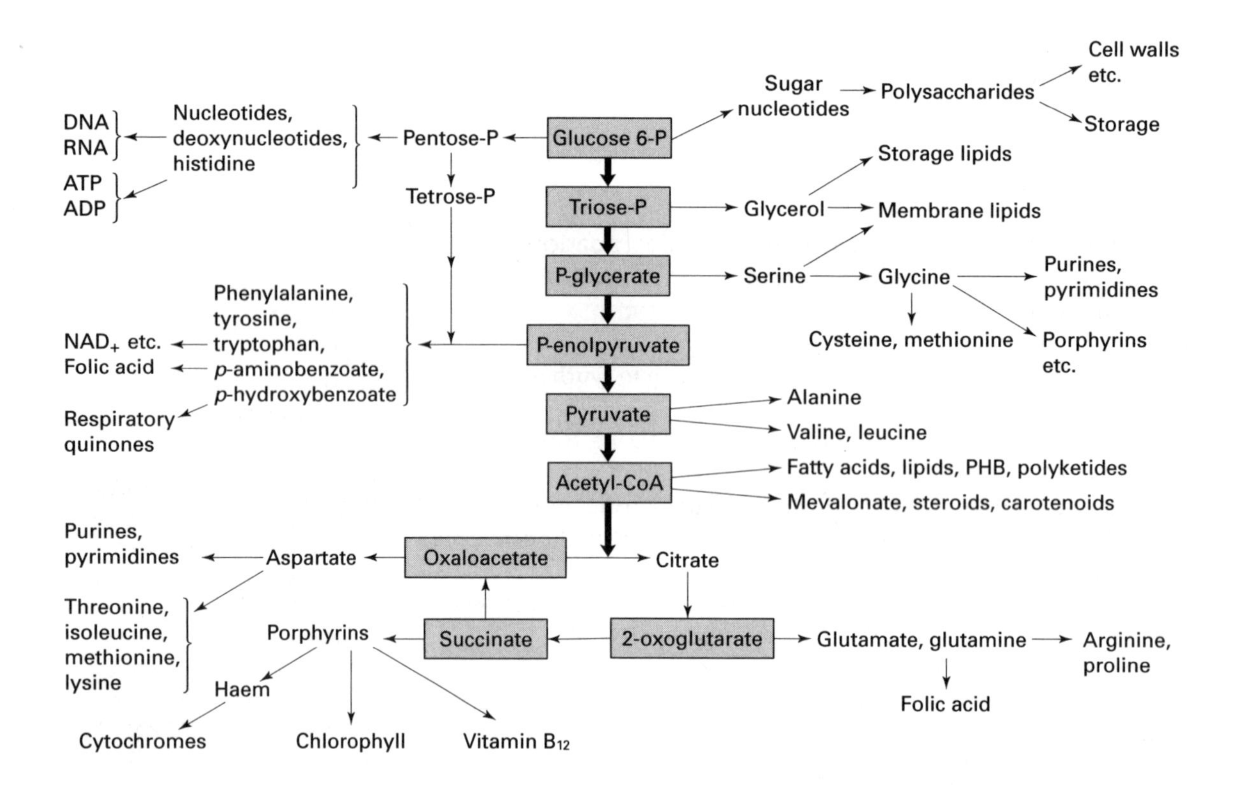

Fig. 2.5 Anabolic pathways (synthesis) and the central catabolic pathways. Only the main biosynthesis routes, and their main connections with catabolic pathways are shown, all in highly simplified versions. Connections through 'energy' (ATP) and 'redox' (NAD^+, $NADP^+$) metabolism and through the metabolism of nitrogen, etc., are all omitted. (PHB = poly-β-hydroxybutyrate; P = phospho group). The nine principal precursors are in the shaded boxes.

and tetrose (C_4) phosphates. This is the **pentose phosphate pathway**, sometimes referred to as a 'shunt' or as the **hexose monophosphate pathway** (see Fig. 2.7). The purpose of this pathway is two-fold: to provide C_5 and C_4 units for biosynthesis (see Fig. 2.5) and also to provide NADPH for biosynthesis.

Although the EMP pathway and the pentose phosphate (PP) pathway both use glucose 6-phosphate, the extent to which each route operates depends largely on what the cell is doing. During the most active stage of cell growth, both pathways operate in the approximate ratio 2:1 for the EMP pathway over the PP pathway. However, as growth slows down, the biosynthetic capacity of the cell also slows down and less NADPH and C_5 and C_4 sugar phosphates are needed so that the ratio between the pathways now moves to 10:1 or even to 20:1.

It is therefore apparent that metabolic pathways are controllable systems capable of considerable refinement to meet the changing needs of the cell. This is discussed later (see Section 2.8).

Although the EMP and PP pathways are found in most micro-organisms, a few bacteria have an alternative pathway to the former pathway. This is the **Entner–Doudoroff pathway** (see Fig. 2.8) which occurs in pseudomonads and related bacteria. The pentose phosphate pathway though still operates in these bacteria as the Entner–Doudoroff pathway does not generate C_5 and C_4 phosphates.

2.3.2 The tricarboxylic acid cycle

The degradation of glucose, by whatever route or routes, invariably leads to the formation of pyruvic acid: $CH_3.CO.COOH$. The fate of pyruvate is different in aerobic organisms and anaerobic ones. In aerobic systems, pyruvate is decarboxylated (i.e. loses CO_2) and is simultaneously activated in the chemical sense, to acetyl coenzyme A (abbreviated as acetyl-CoA) in a complex reaction also involving NAD^+:

$$\text{pyruvate} + \text{CoA} + \text{NAD}^+ \rightarrow \text{acetyl-CoA} + CO_2 + \text{NADH}$$

This reaction is catalysed by *pyruvate dehydrogenase*. (The fate of pyruvate in anaerobic cells is described later.)

Acetyl-CoA, by virtue of it being a thioester, is highly reactive. It is capable of generating a large number of intermediates but its principal though not sole fate is to be progressively oxidised through a cyclic series of reactions known as the citric acid cycle. This is also known as the **tricarboxylic acid cycle** or the Krebs cycle after its discoverer.

The reactions of the citric acid cycle are shown in Fig. 2.9. This cycle fulfils two essential functions:

- it provides key intermediates for biosynthetic reactions (see Fig. 2.5), principal of which are 2-oxoglutarate (to make glutamate and thence glutamine, arginine and proline), succinate (to make porphyrins) and oxaloacetate (to make aspartate and the aspartate family of amino acids – see Chapter 13);
- to produce energy from the complete oxidation of acetyl-CoA to CO_2 and H_2O. (This process is described in detail in Section 2.5.)

However the citric acid cycle cannot fulfil either function exclusively: if intermediates are removed for biosynthesis, then some energy production must be sacrificed; if all the acetyl-CoA is oxidised to CO_2 and H_2O there will be no intermediates left for biosynthesis. Consequently, the cycle runs as a balance between the two objectives. Pyruvate, coming from glucose, provides the input and the cycle provides the output in the way of energy *and* biosynthetic precursors (see Fig. 2.10). In meeting its twin objectives, the cycle cannot entirely replenish the initial oxaloacetate that is needed as a priming reactant to make citrate as some of the intermediates must inevitably be used for biosynthetic purposes. (If they were not so used, there would be no point in the cycle just producing energy as this could not then be used in any

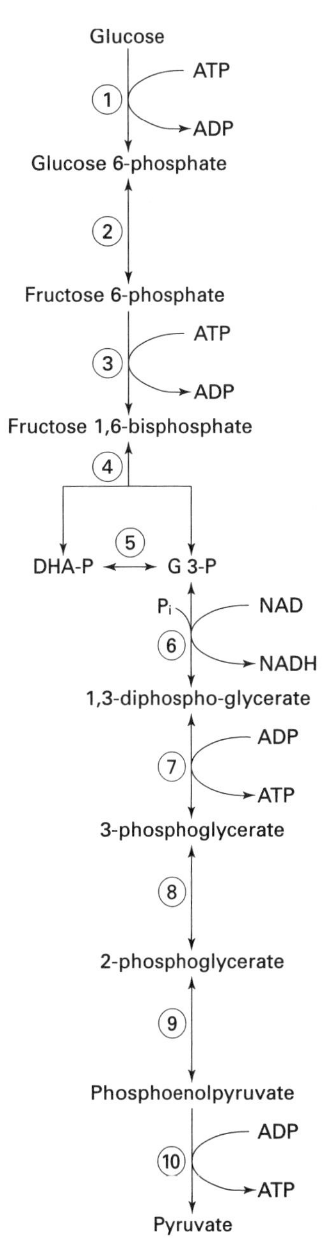

Fig. 2.6 The Embden–Meyerhof–Parnas pathway of glycolysis. Overall:

Glucose + 2 NAD^+ + 2 ADP + 2P_i → 2 pyruvate + 2 NADH + 2 ATP

The reactions are catalysed by:
(1) hexokinase, (2) glucose-6-phosphate isomerase, (3) phosphofructokinase, (4) aldolase, (5) triose phosphate isomerase, (6) glyceraldehyde-3-phosphate dehydrogenase, (7) 3-phosphoglycerate kinase, (8) phosphoglyceromutase, (9) phosphoenolpyruvate dehydratase, (10) pyruvate kinase. DHA-P = dihydroxyacetone phosphate and G3-P = glyceraldehyde 3-phosphate. P_i represents an inorganic phosphate group.

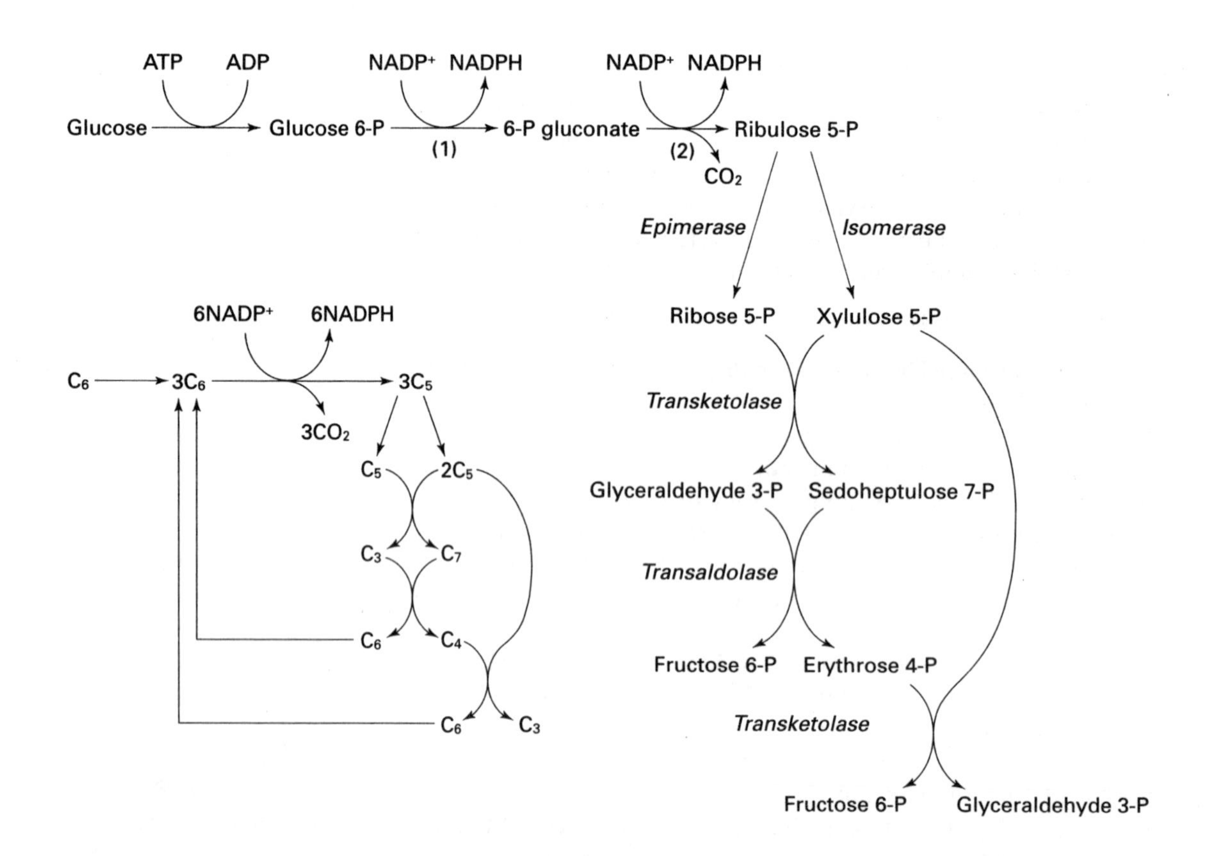

Fig. 2.7 The pentose phosphate cycle (hexose monophosphate shunt). The numbered enzymes are: (1) glucose-6-phosphate dehydrogenase, (2) phosphogluconate dehydrogenase. Inset: summary showing stoichiometry when fructose 6-phosphate is recycled to glucose 6-phosphate by an isomerase; glyceraldehyde 3-phosphate can also be recycled by reverse glycolysis (Fig. 2.6). With full recycling the pathway functions as a generator of NADPH, but the transaldolase and transketolase reactions also permit sugar interconversions which are used in other ways.
Net reaction:

glucose + ATP + 6 $NADP^+$ → glyceraldehyde 3-P + ADP + 6 NADPH

(Any removal of C_5 and C_4 sugars for biosynthesis will diminish the recycling process and thus the yield of NADPH will decrease.)

sensible manner as there can be no biosynthesis without precursors.) It is therefore essential for there to be a second pathway by which oxaloacetate can be formed and this arises principally by the carboxylation of pyruvate:

$$\text{pyruvate} + CO_2 + \text{ATP} \rightarrow \text{oxaloacetate} + \text{ADP} + P_i$$

This reaction is carried out by *pyruvate carboxylase*. However, insofar as oxaloacetate is also produced from the activity of the cycle, the carboxylation of pyruvate must be regulated so that acetyl-CoA and oxaloacetate are always produced in equal amounts. This is usually achieved by the pyruvate carboxylase being dependent upon acetyl-CoA as a **positive effector** (see Section 2.8), i.e. one which increases its activity. The more acetyl-CoA that is present, the faster becomes the production of oxaloacetate. As oxaloacetate and acetyl-CoA are removed equally (to form citrate), the concentration of acetyl-CoA will fall; pyruvate carboxylase will then slow down but, as pyruvate dehydrogenase still operates as before, more acetyl-CoA will be produced. In this way not only will citric acid synthesis always continue, but the two reactions leading to the precursors of citrate will always be balanced (see Fig. 2.11). This type of reaction catalysed by pyruvate carboxylase is referred to as an **anaplerotic reaction**, meaning 'replenishing'.

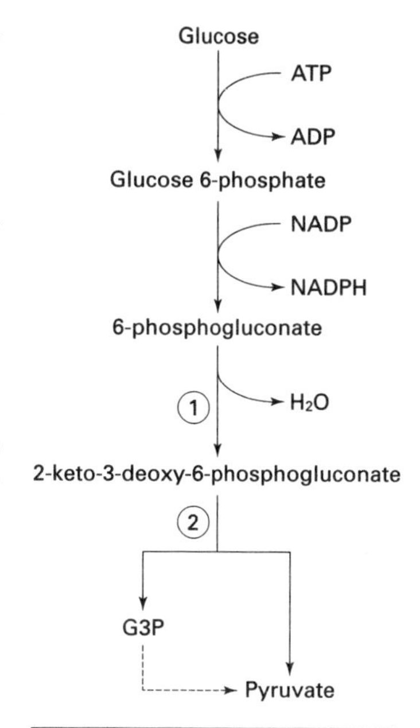

Fig. 2.8 Entner–Doudoroff pathway. This sometimes replaces the Embden–Meyerhof–Parnas pathway (see Fig. 2.6) in some pseudomonads and related bacteria. Numbered enzymes are: (1) phosphogluconate dehydratase, (2) a specific aldolase. Glyceraldehyde 3-phosphate (G3P) is converted to pyruvate by the relevant enzymes given in Fig. 2.6.

2.3.3 The glyoxylate by-pass for growth on C_2 compounds

If an organism grows on a C_2 compound, or on a fatty acid, hydrocarbon or any substrate that is degraded primarily into C_2 units (see Section 2.3.4), the tricarboxylic acid cycle is insufficient to account for its metabolism. Acetyl-CoA can be generated directly from acetate, if this is being used as carbon source, or from a C_2 compound more reduced than acetate, i.e. acetaldehyde or ethanol:

$$\underset{\text{ethanol}}{C_2H_5OH} \xrightarrow{NAD^+ \to NADH} \underset{\substack{\text{acetaldehyde}\\ \text{(ethanal)}}}{CH_3CHO} \xrightarrow{NAD^+ \to NADH} \underset{\substack{\text{acetic}\\ \text{acid}}}{CH_3COO^-} \xrightarrow[CoA]{ATP \to ADP + P_i} \text{acetyl-CoA}$$

The manner in which acetate units are converted to C_4 compounds is known as the **glyoxylate by-pass** (see Fig. 2.12) for which two enzymes additional to those of the tricarboxylic acid cycle are needed: *isocitrate lyase* and *malate synthase*. The former enzyme cleaves isocitrate into succinate and glyoxylate. The latter enzyme then uses a second acetyl-CoA to add to the glyoxylate to give malate. Both these enzymes are 'induced' (that is they are synthesised only when the specific signal is given – see Section 2.8.4) when micro-organisms are grown on C_2 compounds. The activity of both enzymes increases by some 20 to 50 times under such growth conditions. The glyoxylate by-pass does not supplant the operation of the tricarboxylic acid cycle; for example 2-oxoglutarate will still have to be produced (from isocitrate) in order to supply glutamate for protein synthesis etc. Succinate, the other product from isocitrate lyase, will be metabolised as before to yield malate, and thence oxaloacetate. Thus through the reactions of the glyoxalate cycle, the C_4 compounds can now be produced from C_2 units and are then available for synthesis of all cell metabolites (see Fig. 2.5). Their conversion into sugars is detailed in Section 2.4.

2.3.4 Carbon sources other than glucose

Any compound that is used by a micro-organism and can feed into any of the intermediates of glycolysis, or even the citric acid cycle, can be handled by the organism with its existing complement of enzymes. However a great many other substrates can be handled by micro-organisms. In other words, all natural compounds are capable of degradation and the majority of this degradative capacity is found in microbial systems. The application of micro-organisms as 'waste disposal units' is therefore paramount and this activity forms an intrinsic

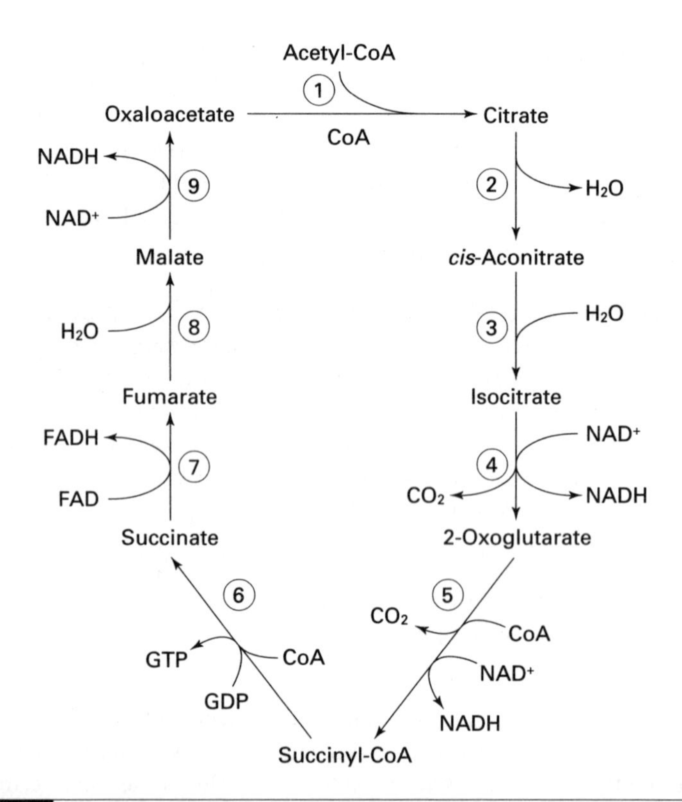

Fig. 2.9 The tricarboxylic acid cycle. (ATP / ADP may replace GTP / GDP in reaction 7.) The overall reaction is:

acetyl-CoA + 3 NAD^+ + FAD + GDP(ADP) $\rightarrow$
2 CO_2 + coenzyme A + 3 NADH + $FADH^+$ + GTP(ATP)

The numbered steps are catalysed by: (1) citrate synthase, (2, 3) aconitase, (4) isocitrate dehydrogenase, (5) 2-oxogluconate dehydrogenase, (6) succinate thiokinase, (7) succinate dehydrogenase, (8) fumarase, (9) malate dehydrogenase.

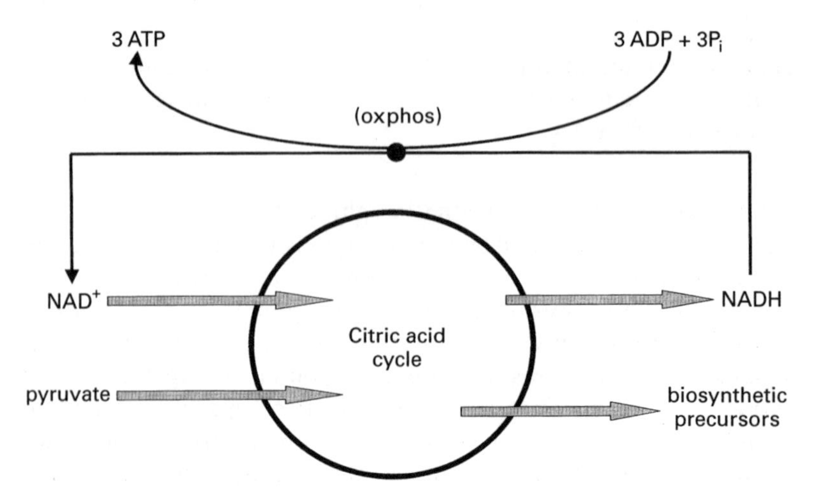

Fig. 2.10 Diagrammatic presentation of the dual roles of the tricarboxylic acid cycle: to produce intermediates and energy (ATP).

part of environmental biotechnology which is explained in detail in Chapter 24.

To illustrate this diversity, the example of microbial degradation of fatty acids will be considered. The ability of micro-organisms to grow on oils and fats is widespread. The difference between an oil and a fat is whether one is liquid or solid at ambient temperatures: they are both chemically the same, that is they are fatty acyl triesters of glycerol:

CH_2OH	$CH_2O.OC\text{-}(CH_2)_n\text{-}CH_3$
|	|
CHOH	$CHO.OC\text{-}(CH_2)_m\text{-}CH_3$
|	|
CH_2OH	$CH_2O.OC\text{-}(CH_2)_p\text{-}CH_3$
glycerol	*triacylglycerol*

where n, m and p are typically 14 or 16; the long alkyl chain may be saturated as indicated or may have one or more double bonds giving unsaturated, or polyunsaturated, fatty acyl groups.

The oils, when added to microbial cultures, are initially hydrolysed by a **lipase** enzyme into its constituent fatty acids and glycerol. The latter is then metabolised by conversion to glyceraldehyde 3-phosphate (see Fig. 2.6). The fatty acids are taken into the cell and immediately converted into their coenzyme A thioesters. The fatty acyl-CoA esters are degraded in a cyclic sequence of reactions (see Fig. 2.13) in which the

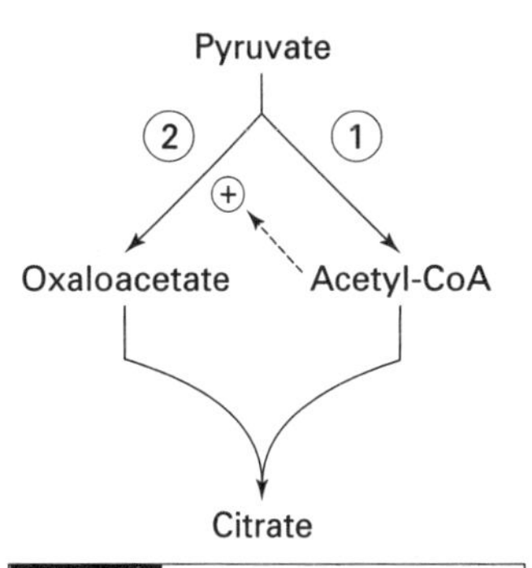

Fig. 2.11 How the cell ensures equal supplies of oxaloacetate (OAA) and acetyl-CoA (AcCoA) for citric acid biosynthesis. The activity of pyruvate carboxylase (2) is stimulated by acetyl-CoA formed by pyruvate dehydrogenase (1)

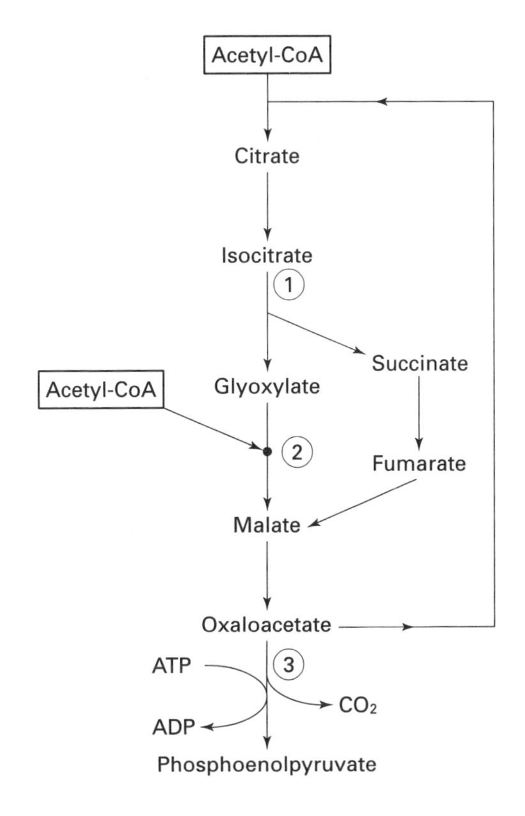

Fig. 2.12 The glyoxylate by-pass. The additional reactions, beyond those of the tricarboxylic acid cycle (see Fig. 2.10), are (1) isocitrate lyase and (2) malate synthase. The scheme also shows how the bypass functions to permit sugar formation from acetyl-CoA, with the added reaction (3) phosphoenolpyruvate carboxykinase, followed by reversed glycolysis (cf. Fig. 2.14).

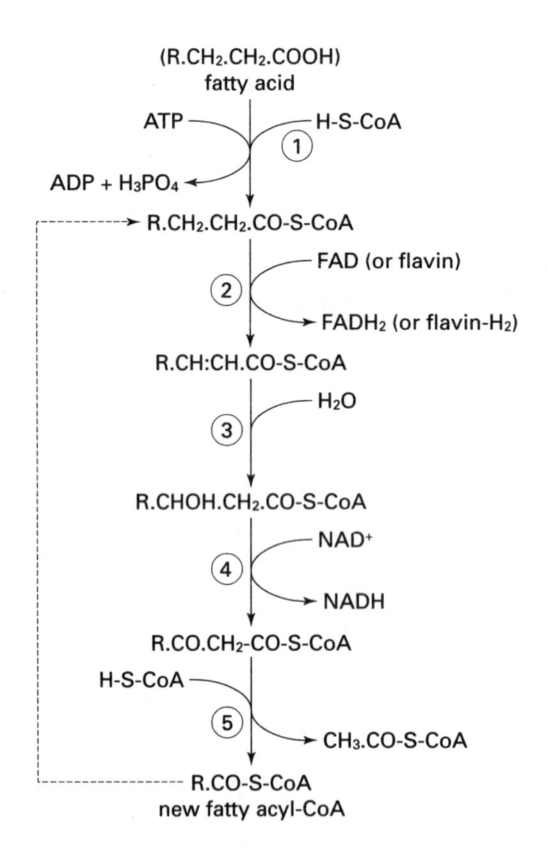

Fig. 2.13 β-Oxidation cycle of fatty acyl-CoA esters. Enzymes are: (1) fatty acyl-CoA synthetase; (2) fatty acyl-CoA oxidase (in yeast and fungi) linked to flavin, or fatty acyl-CoA dehydrogenase in bacteria linked to FAD; (3) 2,3-enoyl-CoA hydratase (also known as crotonase); (4) 3-hydroxyacyl-CoA dehydrogenase; (5) 3-oxoacyl-CoA thiolase. The new fatty acyl-CoA then recommences the cycle at reaction (2).

fatty acyl chain is progressively shortened by loss of two C atoms. These atoms are lost as acetyl-CoA units. This is known as the **β-oxidation cycle** as the initial attack on the fatty acyl chain is at the β- (or 3-) position. Each turn of the cycle produces a shorter fatty acyl-CoA ester which then repeats the sequence of the four reactions until, finally, a C_4 fatty acyl group C_4 (butyryl-CoA) is produced which then gives rise to two acetyl-CoA units.

For unsaturated fatty acids some adjustment in the position of the double bond may be needed to ensure that it is in the right position and configuration for it to be attacked by the second enzyme of the cycle, the hydratase (see Fig. 2.13).

The degradation of fatty acids liberates the energy as heat rather than metabolically usable energy in the form of ATP. This is by coupling the reoxidation of $FADH_2$ (see Fig. 2.13) to O_2 which produces H_2O_2. This is then cleaved by an enzyme, catalase, to give H_2O and $½O_2$ with the liberation of considerable heat. Thus micro-organisms growing on fatty acids and related materials, such as long chain alkanes, invariably generate considerable amounts of heat.

2.4 Gluconeogenesis

When an organism grows on a C_2 or C_3 compound, or a material whose metabolism will produce such compounds, at or below the metabolic level of pyruvate (for example, acetate, ethanol, lactate or fatty acids), it is necessary for the organism to synthesise various sugars to fulfil its metabolic needs. This is termed **gluconeogenesis**. Though most of the reactions in the glycolytic pathways (see Figs. 2.5 and 2.6) are reversible, those catalysed by *pyruvate kinase* and *phosphofructokinase* are not and it is necessary for the cell to circumvent them.

As phosphoenolpyruvate cannot be formed from pyruvate (there are, though, a few exceptions), oxaloacetate is used as the precursor:

$$\text{oxaloacetate} + \text{ATP} \rightarrow \text{phosphoenolpyruvate} + CO_2 + \text{ATP}$$

This reaction is catalysed by *phosphoenolpyruvate carboxykinase* which is the key enzyme of gluconeogenesis. (The formation of oxaloacetate has already been discussed in relation to acetate metabolism, in Section 2.3.3.) For growth on lactate, or pyruvate itself, the lactate would be oxidised to pyruvate:

$$CH_3CH(OH).COOH + NAD^+ \rightarrow CH_3.CO.COOH + NADH$$

The pyruvate will be carboxylated to oxaloacetate using pyruvate carboxylase:

$$CH_3.CO.COOH + CO_2 + ATP \rightarrow COOH.CH_2.CO.COOH + ADP + P_i$$

The oxaloacetate is then converted to phosphoenolpyruvate (see above).

The net reaction for growth in lactate is therefore:

$$\text{lactate} + NAD^+ + 2ATP \rightarrow \text{phosphoenolpyruvate} + NADH + 2ADP + 2P_i$$

The irreversibility of the second glycolytic enzyme, phosphofructokinase (producing fructose 1,6-bisphosphate), is circumvented by the action of *fructose bisphosphatase*:

$$\text{fructose 1,6-bisphosphate} + H_2O \rightarrow \text{fructose 6-phosphate} + P_i$$

From this point, hexose sugars can be formed by the reversal of glycolysis and C_5 and C_4 sugars can now be formed via the pentose phosphate pathway (Fig. 2.7). Glucose itself is not an end-product of 'gluconeogenesis' but glucose 6-phosphate is used for the synthesis of cell wall constituents and a large variety of extracellular and storage polysaccharides (see Fig. 2.14).

2.5 Energy production in aerobic micro-organisms

It has already been explained how, in the metabolism of glucose (Figs 2.6 and 2.7) and in the tricarboxylic acid cycle (Fig. 2.9), oxidation of the various metabolic intermediates is linked to the reduction of a limited

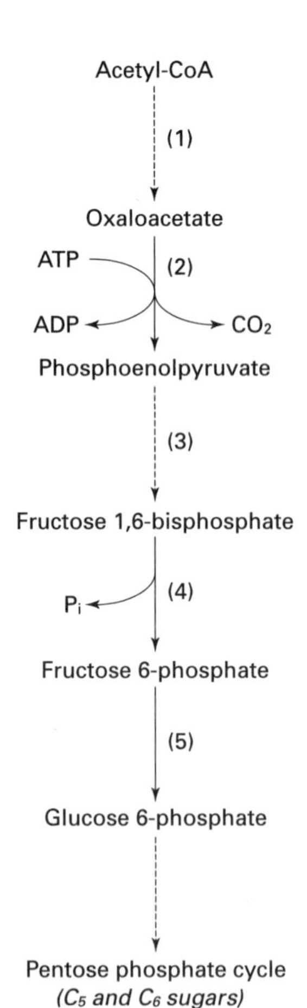

Fig. 2.14 Gluconeogenesis sequence. From a substrate such as acetyl-CoA this is converted (1) by the reactions of the glyoxylate by-pass (see Fig. 2.12) to oxaloacetate and thence to phosphoenolpyruvate by phosphoenolpyruvate carboxykinase (2). This is converted by the reversed glycolytic sequence of enzymes (3) into fructose 1,6-bisphosphate (see also Fig. 2.6) which is then hydrolysed with the release of inorganic phosphate (P_i) by fructose-1,6-bisphosphatase (4). The product is then isomerised to glucose 6-phosphate (5) which can then be fed into the reactions of the pentose phosphate pathway (Fig. 2.7) or used for the biosynthesis of cell envelope polysaccharides.

number of co-factors (NAD^+, $NADP^+$, FAD) producing the corresponding reduced forms (NADH, NADPH and $FADH_2$). The reducing power of these products is released by a complex reaction sequence which, in aerobic systems, is linked eventually to reduction of atmospheric O_2. This process is known as **oxidative phosphorylation** and the sequence of carriers that are used to convey the hydrogen ions and electrons, eventually to be coupled to O_2 to form H_2O, is referred to as the **electron transport chain**. The function of the electron-transport-coupled phosphorylation (ETP) is the synthesis of ATP. The electron transport chain, together with *ATP synthetase*, forms a multi-component system which is integrated into a membrane that is either the cytosolic membrane of a bacterial cell or, in eukaryotic cells, is the mitochondrial membrane.

During the electron transport sequence, ATP is generated from ADP and inorganic phosphate (P_i) at two, or more usually three, specific points, depending on the nature of the original reductant. This is shown in Fig. 2.15. Although there are many variations in the respiratory chains [the principal systems for mitochondria (in eukaryotic organisms) and for bacteria as typified by *Escherichia coli* are shown in Fig. 2.15], the mechanism by which ATP is produced is similar in all cases. ATP synthetase is a complex protein that sits across the membrane on one side of which are the reductants and on the other are protons (H^+) (see Fig. 2.16). As the reductants become linked into the electron transport chain, further protons move back across the membrane and they literally drive the ATP synthetase to physically revolve in its socket which has the effect of coupling ADP with inorganic phosphate to give ATP. The system is referred to as having a **proton motive force (PMF)**. Without a membrane creating two, otherwise unlinked, sides there could be no PMF, no ETP and therefore no energy production.

In the case of the three reductants, the overall reactions may be written as:

$$NADPH + 3ADP + 3P_i + \tfrac{1}{2}O_2 \rightarrow NADP^+ + 3ATP + H_2O$$
$$NADH + 3ADP + 3P_i + \tfrac{1}{2}O_2 \rightarrow NAD^+ + 3ATP + H_2O$$
$$FADH + 2ADP + 2P_i + \tfrac{1}{2}O_2 \rightarrow FAD + 2ATP + H_2O$$

One can therefore describe the ATP yield, in each case, as being 3, 3 and 2, respectively. These are sometimes referred to as the **P/O ratios** which is the amount of ATP gained from the reduction of $\tfrac{1}{2}O_2$ to H_2O, and involves the transport of two electrons.

The yields of ATP per mole of glucose metabolised by the Embden–Meyerhof pathway (Fig. 2.6) and from the resulting pyruvate, metabolised by the reactions of the tricarboxylic acid cycle (Fig. 2.9) are summarised in Table 2.1. As can be seen, the vast majority of ATP is generated with the reactions of the electron transport chain coupled to the reactions of the citric acid cycle.

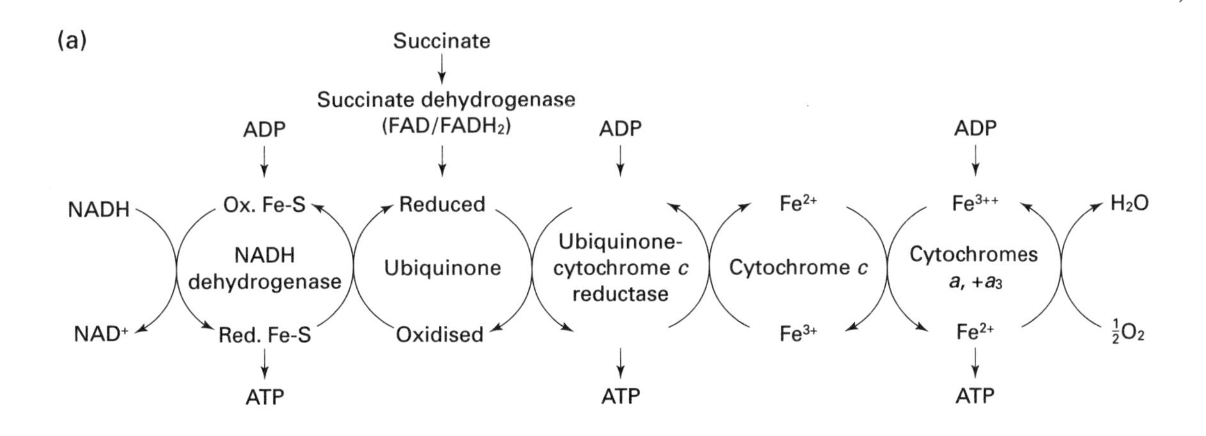

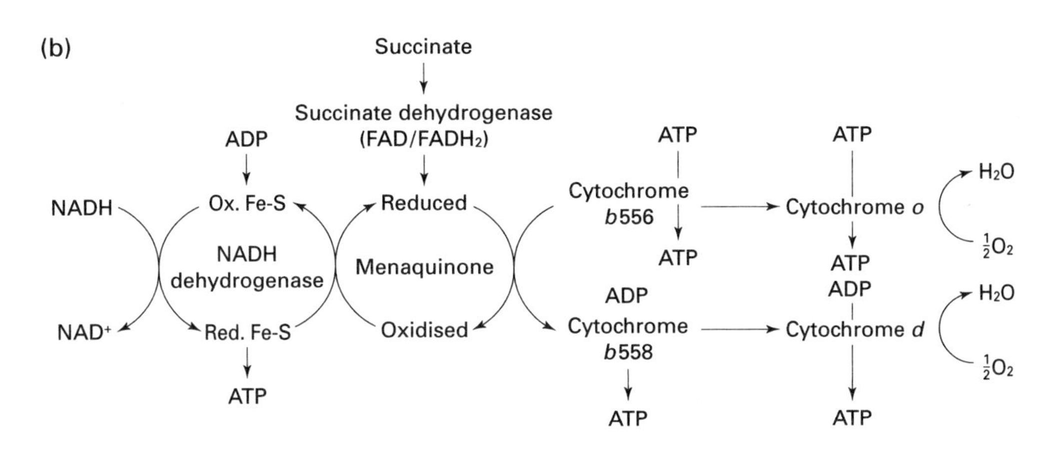

Fig. 2.15 Electron-transport-coupled phosphorylation (ETP) system of: (a) mitochondria and (b) *E. coli*. Not all the electron carriers are shown, there being at least 16 proteins involved in each case. There are also considerable variations in the electron transport carriers. In the *E. coli* system, the electron transport chain divides and electrons and protons can flow through both the systems as described. The sites of phosphorylation of ADP to ATP are only indications as the actual formation of ATP is carried out by ATP synthases that are driven by the physical movement of proteins through the membrane (see also Fig. 2.16).

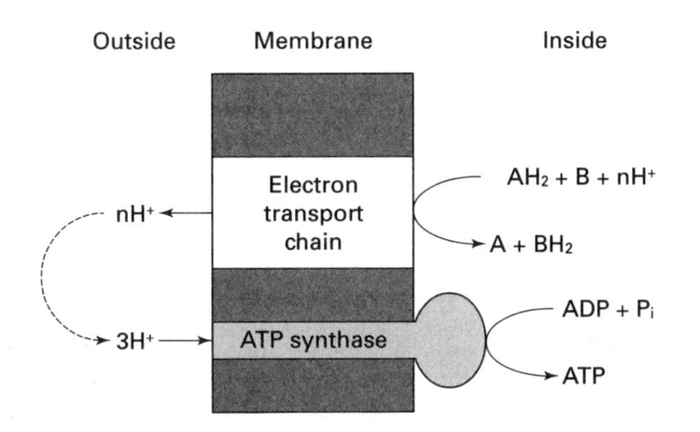

Fig. 2.16 The coupling mechanism of ETP. The electron transport carriers (see Fig. 2.15) are located within a membrane. As the reductant (AH_2) is oxidised, this sets up a movement of protons through the membrane. These protons are then pumped back across the membrane driving the ATP synthase into coupling ADP with P_i to give ATP.

Table 2.1 ATP yields for glucose metabolism

	Moles ATP produced per mole hexose
Glycolysis (glucose to pyruvate):	
Net yield of ATP = 2 mol	2[a]
NADH = 2 mol × 3	6
Pyruvate to acetyl-CoA:	
NADH = 1 mol × 3 (× 2 for 2 pyruvate)	6
Tricarboxylic acid cycle:	
NADH = 3 mol × 3 (× 2 for 2 acetyl-CoA)	18
FADH2 = 1 mol × 2 (× 2 for 2 acetyl-CoA)	4
ATP = 1 mol (× 2 for acetyl-CoA)	2
Total	38

Note:
[a] Under anaerobic conditions these 2 moles of ATP represent the maximum attainable yield ('substrate-level phosphorylation', see Section 2.6.1).

2.6 Anaerobic metabolism

2.6.1 General concepts

Under anaerobic conditions, the process of oxidative phosphorylation cannot occur, and the cell is deprived of its principal way of generating energy. Under such circumstances energy must be provided from the very process of degrading the original substrate. This process, known as **substrate level phosphorylation**, yields only about 8% of the energy that can be produced under aerobic conditions. This means that to produce a given weight of cells, much more substrate must be degraded than under aerobic conditions. Under anaerobic conditions, the cell is trying its utmost to maximise the formation of energy; the reducing power which is produced during the degradation of the substrate (see Figs. 2.1 and 2.2) has to be re-circulated for the process to continue as its supply is not limitless. The oxidation of the reducing equivalents therefore must be linked to the reduction of some of the carbon intermediates which are accumulating under these conditions:

$$X + NADH \rightarrow XH_2 + NAD^+$$

Only a little of the reducing power per se is needed for synthesis of new cells. Consequently, the reducing equivalents react with the accumulated carbon intermediates to reduce them in their turn (Fig. 2.1b). The only energy that is then available to the cell is that which comes from ATP formed during the anaerobic degradation of the substrate. Examples of substrate-level phosphorylation are given in Table 2.2.

This process of anaerobic catabolism occurs not only in microorganisms, but may occur in higher animals too: an example would be

Table 2.2 Substrate-level phosphorylation reaction in anaerobes

Enzyme	Reaction catalysed	Occurrence
1. Phosphoglycerol kinase	1,3-bisphosphoglycerate + ADP → 3-phosphoglycerate + ATP	Widespread, see Fig. 2.6
2. Pyruvate kinase	phosphoenolpyruvate + ADP → pyruvate + ATP	Widespread, see Fig. 2.6
3. Acetate kinase	acetyl phosphate + ADP → acetate + ATP	Widespread
4. Butyrate kinase	butyryl phosphate + ADP → butyrate + ATP	E.g. enterobacteria on allantoin
5. Carbamate kinase	carbamoyl phosphate + ADP → carbamate + ATP	E.g. clostridia on arginine
6. Formyl-tetrahydrofolate synthetase	N^{10}-formyl-H_4 folate + ADP + P_i → formate + H_4 folate + ATP	E.g. clostridia on xanthine

the accumulation of lactic acid in muscle tissue during hyper-activity of an athlete. In micro-organisms we can see a whole range of reduced carbon compounds being accumulated by organisms growing anaerobically. Examples may also include lactic acid itself (produced by lactic bacteria), but would include short chain fatty acids such as butyric or propionic acids, and alcohols such as butanol, propanol and ethanol (see Fig. 2.17). Some organisms, such as the methanogenic bacteria, may go even further and produce, as the completely reduced end-product, methane (not shown in Fig. 2.17).

It is important to point out that pyruvate, produced by the glycolytic pathway (Fig. 2.6), will still enter the tricarboxylic acid cycle which must continue, at least in part, to provide essential precursors for biosynthesis, principally 2-oxoglutarate, and oxaloacetate, but not to produce energy. The NADH produced in the cycle cannot be converted into ATP as the cells have no supply of O_2 to drive oxidative phosphorylation; however in certain bacteria there are terminal electron acceptors other than O_2 which are capable of being coupled into the ETP system (Fig. 2.15) and which will allow the formation of ATP to take place. This includes microbes that can use nitrate (which is reduced to nitrite), nitrite (reduced to NH_4 or in some cases N_2 in a process known as **denitrification**), CO_2 (reduced to methane by methanogens) or sulphate (reduced to H_2S by sulphate reducing bacteria) as alternatives to O_2. In all cases, although the yield of ATP is less than occurs in aerobic systems, it is much greater than obtained by substrate-level phosphorylation alone.

2.6.2 Products of anaerobic metabolism

Figure 2.17 summarises some of the main reactions leading to the formation of reduced end-products in anaerobic micro-organisms.
The major products are:

- **glycerol**, produced by yeasts when the conversion of pyruvate to ethanol is blocked;
- **lactic acid**, formed by lactic acid bacteria;
- **formic acid**, formed by enterobacteria via pyruvate-formate lyase and the formate can be converted to CO_2 and H_2 by formate dehydrogenase;
- **ethanol**, formed by yeasts (e.g. *Saccharomyces cerevisiae*), bacteria (e.g. *Zymomonas*) and by many fungi;
- **2.3-butanediol**, produced by various bacteria including *Serratia marcescens* and various *Bacillus* spp;
- **butanol** with **acetone** and some **propanol** or **isopropanol**, produced by *Clostridium* spp., some of which also produce butyric acid;
- **propionic acid**, produced by *Propionibacterium*.

Other products may arise from the anaerobic metabolism of compounds other than glucose, for example organic acids, such as citric acid, or amino acids and sometimes purines.

Methane (not shown in Fig. 2.17) is perhaps the ultimate reduced carbon compound and is produced by highly specialised *Archaebacteria* by cleavage of acetate to CO_2 and CH_4 or in some cases by reduction of CO_2,

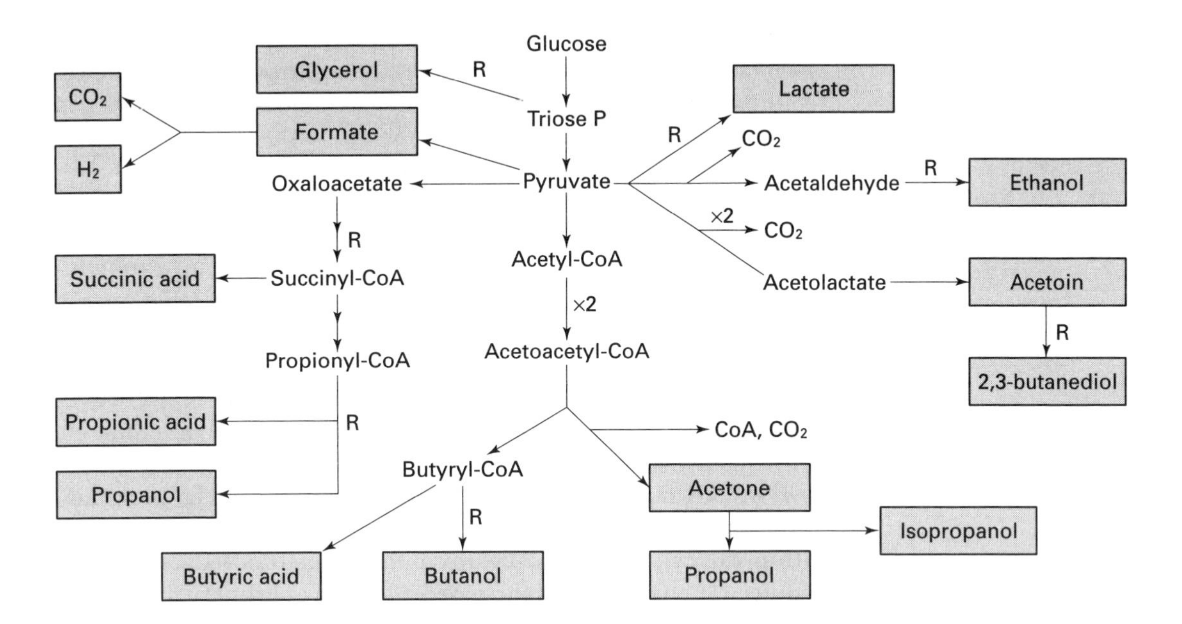

Fig. 2.17 Products of anaerobic metabolism in various micro-organisms. Reactions leading to the recycling of NADH are indicated by R. The end-products, which are given in the shaded boxes, may occur singly or in groups according to the organism.

methanol (CH_3OH), ethanol (CH_3CH_2OH) or formic acid (HCOOH) all in the presence of H_2 gas.

2.7 Biosynthesis

The provision of energy (ATP), reducing power (NADH and NADPH) and a variety of monomeric precursors (see Fig. 2.5) from the degradation of a substrate provides the cell with the necessary means of regenerating itself. The cell undertakes the biosynthesis of the macromolecules of the cell: nucleic acids (DNA and RNA), proteins (for enzymes and other functions), lipids for membranes and polysaccharides as components of the cell envelope, from these simple building blocks. As many of the biosynthetic pathways are covered elsewhere in this book (e.g. Chapters 13, 14 and 15) these need not be detailed here. However, a distinction needs to be drawn between **primary** metabolism and **secondary** metabolism as these have considerable importance for formation of biotechnological products.

2.7.1 Primary metabolism

Primary metabolism occurs during **balanced growth**, sometimes known as the **tropophase**, of the organism in which all nutrients needed by the cell are provided in excess in the medium (see Fig. 2.18). Under such conditions the cells grow at an exponential rate in keeping with their mode of reproduction. The cells will have optimum contents of all the various macromolecules of the cell – DNA, RNA, proteins, lipids etc. – but their proportions will change as growth progresses and then slows down.

Eventually, though, the cell must run out of some nutrient, even if this is only O_2, and consequently the growth rate slows and eventually ceases. However, metabolism does not cease. The only time that metabolism completely ceases is when the cell dies; thus as long as the cell retains viability it is able to carry out some metabolic processes and, conversely and most importantly, if the cell wishes to remain alive it must carry out a modicum of metabolism.

2.7.2 Secondary metabolism

The need for the cell to keep a flux (or flow) of carbon going through it when active multiplication has ceased requires that the cell diverts its core metabolites into products other than the primary ones which are not needed in the same abundance. Some maintenance of vital components, though, must be carried out: key proteins must be replaced as all proteins undergo turnover; DNA must be repaired, RNA maintained etc. Consequently, some primary metabolism must continue but the cell now switches into a secondary mode of metabolism (see Fig. 2.18). These secondary products then begin to arise sometimes as storage products within the cell (e.g. poly-β-hydroxybutyrate or triacylglycerols – see Chapter 15), sometimes as increased amounts of primary metabolites (such as organic acids, see Chapter 14), but sometimes new products are synthesised which are not normally present in any great abundance during the balanced phase of growth.

These secondary metabolites can be of considerable biotechnological importance as many of them are biologically active not only within the producing cell but also in other cells and therefore some may act as antibiotics.

As the range of secondary metabolites varies almost with the species of organism being studied, this phase of unbalanced metabolism has been referred to as the **idiophase** in contrast to the tropophase phase of growth. The secondary metabolites are usually synthesised from unwanted monomers that are still produced from glucose or fatty acid catabolism. Not surprisingly, acetyl-CoA is often used as a key starting point (Fig. 2.19).

The function of the secondary metabolites is uncertain. As they are not produced during balanced growth, they are evidentally not essential for growth and multiplication. As their type and occurrence varies enormously their roles may be equally diverse. Two schools of thought have been advanced:

- The secondary metabolites fulfil a functional role that probably has some benefit to the cell for its survival in a natural environment or produces a response in the cell that is difficult to mimic in laboratory cultivation.
- Secondary metabolites per se have no value; it is the process of their formation which is important and not what the final product may be.

Whatever is the reason for the profusion of secondary metabolites, their properties make them some of the most exploited biotechnological products of all time.

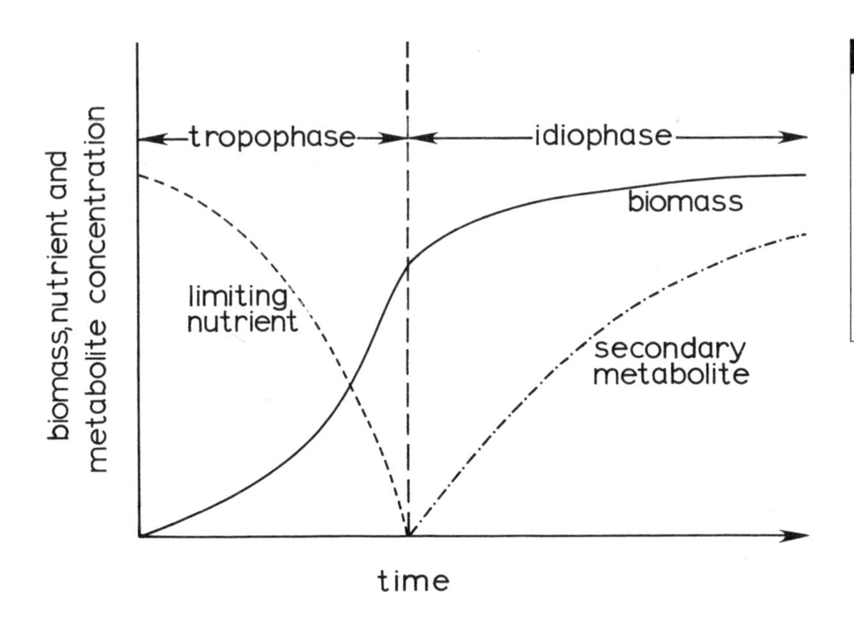

Fig. 2.18 Microbial growth: primary and secondary metabolic phases. In the initial phase (balanced growth = tropophase) all nutrients are in excess. When one nutrient (not carbon) is consumed (----------) cell growth (———) slows down and secondary metabolite(s) (– - – - – -) are formed in the idiophase.

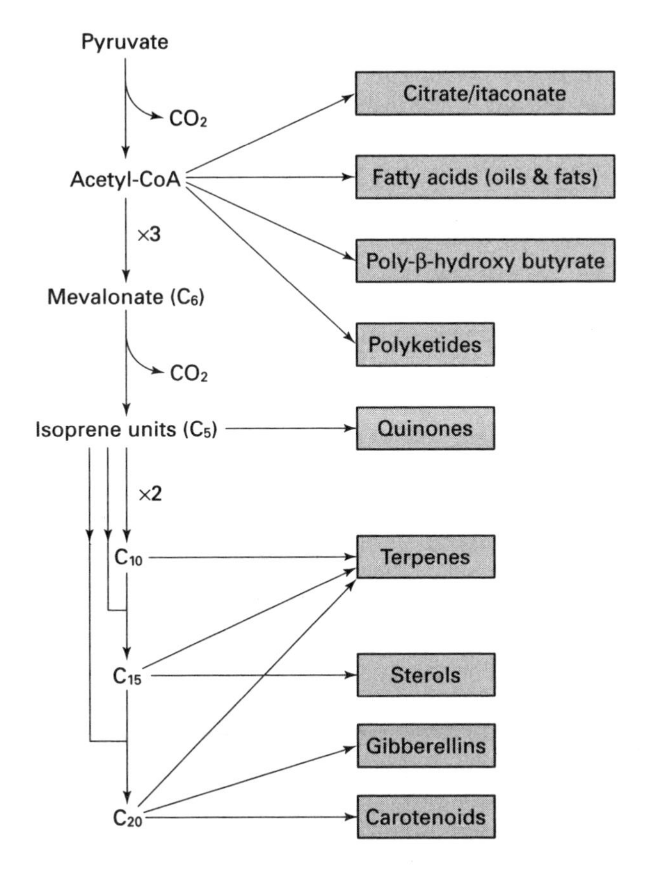

Fig. 2.19 Formation of secondary metabolites from acetyl-CoA. The secondary metabolites are shown in the shaded texts. There can be considerable variation in structure of some of these metabolites.

2.8 Control of metabolic processes

2.8.1 Metabolic flux

The concept of **metabolic flux** (or flow) has been developed that attempts to describe in mathematical terms the rate at which metabolic intermediates are moved along the various pathways. As this movement is invariably achieved by the action of the associated enzymes, then measurement of the individual enzyme reaction rates may, if one is extremely lucky, identify a single rate-controlling step. If this reaction can be de-regulated or its rate increased (or if a geneticist can change or amplify the appropriate gene coding for the enzyme to increase its activity – see Fig. 2.20 and also Chapters 4 and 5) then the rate-limiting step will be removed and increased product formation should result. Unfortunately, this rarely gives the required result as usually all the enzymes involved in a pathway are operating at similar rates and removal of one rate-limiting step merely identifies the next one. Entire pathways need to be 'engineered' (which can only be done at the genetic level) if a de-regulated pathway is needed for a particular product. Examples of removing some metabolic 'bottle-necks' are given in various chapters later in this book: over-production of organic acids (Chapter 14), over-production of amino acids (Chapter 13) and over-production of antibiotics (Chapter 16).

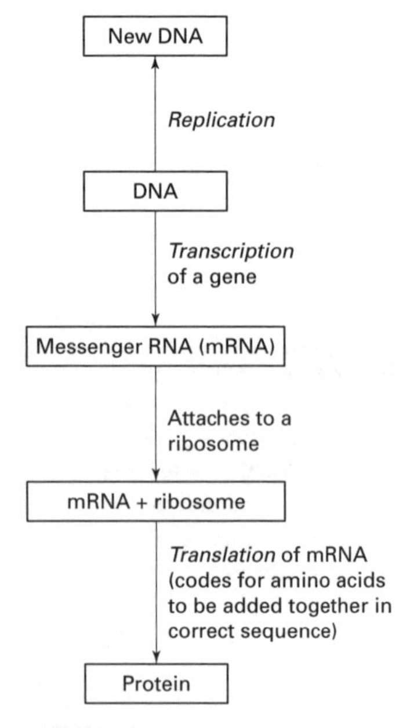

Fig. 2.20 Idealised diagram showing how DNA can either be replicated to give new DNA (for new cell synthesis) or be transcribed into messenger RNA (mRNA) that is decoded (or translated) by it becoming attached to a ribosome which then makes a protein molecule by sequential addition of amino acids. Thus the original sequence of bases along the DNA (the gene) is first converted to a corresponding sequence of bases (mRNA) that gives rise to a new protein; see also Chapter 4, Fig. 4.1.

Identification of the key enzymes that control the flux of carbon to various products is important for increasing the productivity of any process. The flux of carbon can be controlled in various ways and these are briefly described below. It should however be stated that many products have been increased by mutating micro-organisms in a random manner and then picking out the one or two improved cells from the myriad of others that have an increased capacity to produce the desired product. However, with our considerably increased knowledge of biochemistry, of genetics and genetic manipulations, the current trend is to alter specific enzymes, or groups of enzymes, in order to effect improvements in a precise manner. This rational approach means that the key enzyme must be identified with great care otherwise considerable effort will be expended for little or no return. This understanding of how enzymes may be regulated, therefore, is of considerable importance to the biotechnologist.

2.8.2 Nutrient uptake

Control of cell metabolism begins by the cell regulating its uptake of nutrients. Most nutrients, apart from oxygen and a very few carbon compounds, are taken up by specific transport mechanisms so that they may be concentrated within the cell from dilute solutions outside. Such 'active' transport systems require an input of energy. The processes are controllable so that once the amount of nutrient taken into the cell has reached a given concentration, further unnecessary (or even detrimental) uptake can be stopped. (This is also discussed in Section 2.8.5 on catabolite repression.) In some cases the rate at which a carbon source, such as glucose, is taken up into the cell may be the limiting process for growth of the whole cell and therefore should receive particular attention when evaluating potential bottle-necks to increased productivity of a bioprocess.

2.8.3 Compartmentalism

A simple form of metabolic control is the use of compartments, or organelles, within the cell wherein separate pools of metabolites can be maintained. An obvious example is the mitochondrion of the eukaryotic cell which separates (amongst others) the tricarboxylic acid cycle reactions from reactions in the cytoplasm. Another would be the biosynthesis of fatty acids which occurs in the cytoplasm of eukaryotic cells whereas the degradation of fatty acids (see Fig. 2.13) occurs in the peroxisome organelle. Separating the two sets of enzymes prevents any common intermediate being recycled in a futile manner. Other organelles (vacuoles, the nucleus, peroxisomes, etc.) are similarly used to control other reactions of the cell. Bacteria, however, do not have such compartments within their cells and therefore must rely on other means of metabolic control.

2.8.4 Control of enzyme synthesis

Many enzymes within a cell are present constitutively; that is, they are there under all growth conditions. Other enzymes only 'appear' when needed; e.g. isocitrate lyase of the glyoxylate by-pass (see Fig. 2.12) when the cell grows on a C_2 substrate. This is termed **induction** of enzyme synthesis. Conversely enzymes can 'disappear' when they are no longer required; for example, enzymes for histidine biosynthesis stop being produced if there is sufficient external histidine available to satisfy the needs of the cell. This is termed **repression**; when the gratuitous supply of the compound has gone, the enzymes for synthesis of the material 'reappear'; their synthesis is **de-repressed**. The key to both induction and repression is that the genes coding for the synthesis of the proteins by the processes of transcription (see Fig. 2.20) are either switched on (induction) or off (repression) according to the metabolites present (or absent) in the cell. These processes are shown diagrammatically in Fig. 2.21.

2.8.5 Catabolic repression

This type of metabolic control is an extension of the ideas already set out with respect to enzyme induction and repression, being brought about by external nutrients added to the microbial culture. The term **catabolite repression** refers to several general phenomena seen, for example, when a micro-organism is able to select, from two or more different carbon sources simultaneously presented to it, that substrate which it prefers to utilise. For example, a micro-organism presented with both glucose and lactose may ignore the lactose until it has consumed all the glucose. This sequential utilisation of two substrates is referred to as **diauxic growth**. Similar selection may occur for the choice of a nitrogen source if more than one is available. The advantage to the cell is that it can use the compound which provides it with

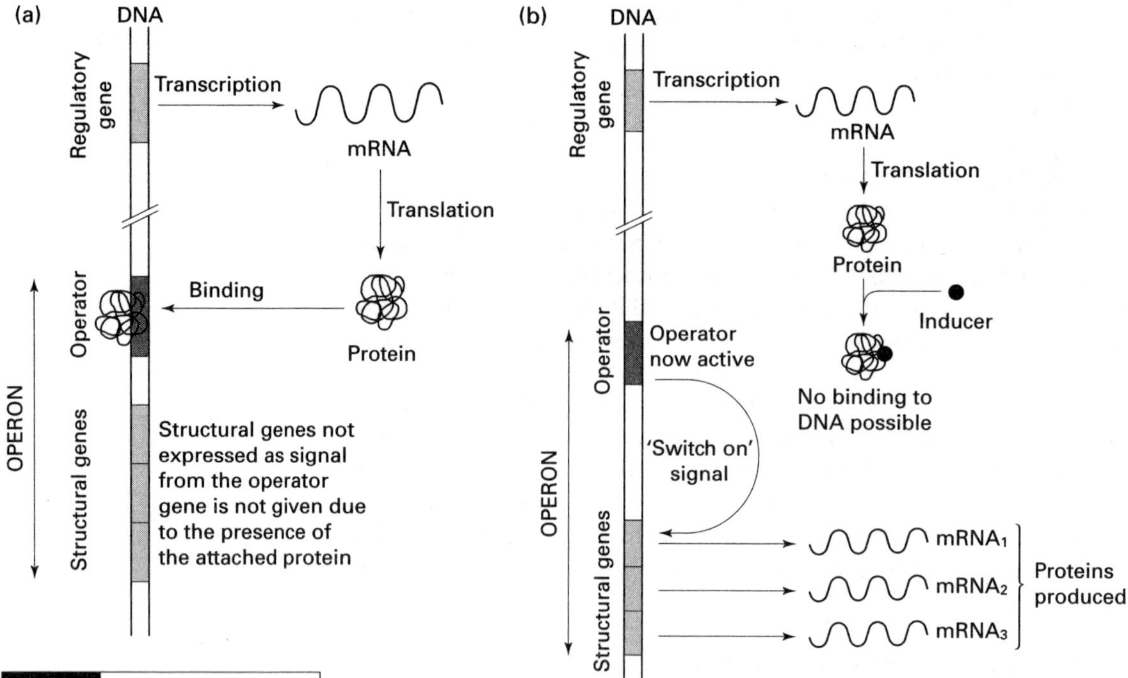

Fig. 2.21 Control of enzyme synthesis through regulation of DNA expression. (a) Repression: in the absence of any inducing molecule the messenger RNA (mRNA) from the regulatory gene produces a protein that binds to an 'operator' gene further down the DNA molecule. As a result of this binding, the operator gene is inactivated and no signal is given to allow the structural genes (that would make active enzymes) to be expressed. (b) Induction: in the presence of an inducing molecule, the protein arising from the regulatory gene is now no longer able to bind to the operator gene. Consequently, the operator 'switches' on the structural genes and active proteins (enzymes) are now made.

the most useful substrate for production of energy and provision of metabolites.

The mechanisms by which catabolite repression is achieved varies from organism to organism. A simple case is with *E. coli* where control is exerted via an effector molecule, cyclic AMP (cAMP). (In cAMP the single phospho group of AMP – see Fig. 2.4 – bridges across from the 3′-hydroxy group of ribose to the 5′-hydroxy group, thereby forming a cyclic diphosphoester.) cAMP interacts with a specific protein, **catabolite activator protein** (*CAP*) (also known as the *CRP* = **catabolite receptor protein**), and the cAMP–CAP complex binds to DNA causing the genes that follow after (or **downstream** of) the binding site to be transcribed (see Fig. 2.22). These genes may then be used to synthesise new proteins for uptake and metabolism of the next substrate (e.g. lactose if the cells are growing on a glucose/lactose mixture). This positive system of genetic control is the reverse of the negative control system described in Fig. 2.21.

The key molecule is therefore cAMP. As long as glucose or its catabolites are present, cAMP is not formed as its synthesising enzyme (adenylate cyclase) is inhibited by these catabolites and thus lactose uptake and metabolism cannot occur. The catabolites therefore repress the synthesis of new enzymes. The repression is removed when the catabolites disappear – i.e. all the glucose has been consumed.

2.8.6 Modification of enzyme activity

Once an enzyme has been synthesised, its activity can be modulated by a variety of means.

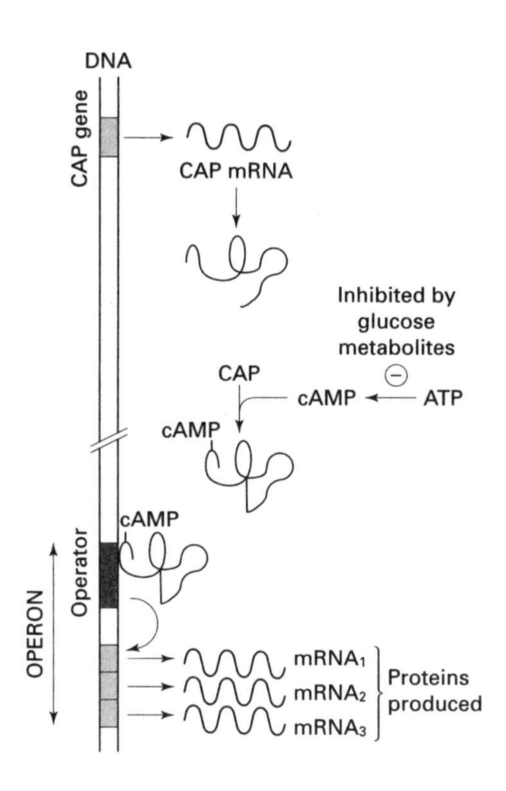

Fig. 2.22 Catabolite repression. The mechanism shown is mediated by cyclic AMP (cAMP). An operon is controlled by the operator gene being activated by a complex found between a protein (the **catabolite activator protein, CAP**) and cyclic AMP (**cAMP**). cAMP is only formed when glucose is absent. The structural genes are therefore 'switched off' (i.e., repressed) as long as glucose or its catabolites are present. Several operons may respond to the cAMP–CAP signal.

Post-transcriptional modifications

This process is so-called because it occurs after the enzyme has been synthesised, i.e. after its formation by transcription (see Fig. 2.20).

Enzymes may be modified from one form to another, one form being active and the other inactive or less active:

$$E_{(active)} \leftrightarrow E_{(inactive)}$$

This process of activating or inactivating an enzyme is carried out by an entirely separate enzyme which has nothing to do with catalysing the reaction that the original enzyme will be involved with.

A common way of achieving this conversion is by phosphorylation of the enzyme using a new enzyme – a *protein kinase.* These protein kinases, of which there can be many, usually react with only one enzyme and thus are highly specific. They add a phospho group (from ATP) to a specific hydroxyl group (normally a serine residue) on the enzyme. The activated enzyme may be either the phosphorylated form or its de-phosphorylated form. The de-phosphorylation will be carried out by a specific phosphatase enzyme. The activities of the protein kinase and the phosphatase will be obviously controlled by other factors within the cell and will work according to the metabolic status of the cell.

There are other mechanisms of altering the activities of specific proteins by the attachment (or removal) of a simple molecule to a particular amino acid residue in an enzyme, but the addition of a phospho group is by far the most common.

Action of effectors

The second way in which an enzyme's activity can be controlled is by its response to various effectors. (Effectors can act positively, i.e. are **promoters**, or negatively, i.e. are **inhibitors**.) An example is the process known as **feedback inhibition**. Here, in a sequence of biosynthesis

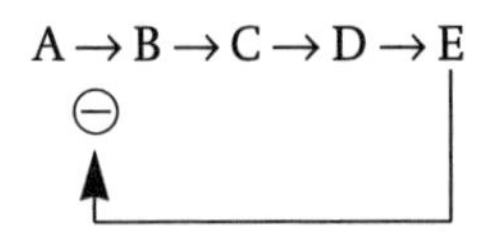

the end product, E, may be able to inhibit the first enzyme of the sequence (converting A to B). This will only occur when sufficient E has been produced by the cell for its immediate requirements and therefore no further carbon need be channelled down this pathway. As the cell continues to grow it will consume the accumulated E and thus diminish the amount of it in circulation. Thus, as E is withdrawn for the cell's own needs, the inhibitory effect will be withdrawn and the conversion of A to B will recommence with the further synthesis of E then occurring to match the cell's requirements. This process will also occur if the end-product, E, is added to the growth medium of the organism. Here, as the product is now supplied gratuitously, the cell has no need to 'waste' its resources synthesising E so the pathway is now inhibited. In addition to the feedback inhibition, a high concentration of the end-product can also lead to the **repression** of the enzymes for the entire pathway; thus there is a 'quick' response mode to a high concentration of the end-product arising: the initial enzyme of the pathway is inhibited and there is no flux of carbon along the pathway; and then there is a longer-term response whereby all the enzymes needed for the pathway stop being synthesised by repression (see above) at the DNA level as they are surplus to requirement and their continued synthesis would be a waste of valuable amino acid precursors.

This process can, of course, be quite complicated should the pathway not be linear as depicted above but be a branching pathway with multiple-end products. This is of particular importance in the biosynthesis of amino acids several of which (such as phenylalanine, tyrosine and tryptophan) share a common initial pathway. This is discussed in greater detail in Chapter 13.

2.8.7 Degradation of enzymes

Enzymes are not particularly stable molecules and may be quickly and irreversibly destroyed. Their half-lives are very variable; they may be as short as a few minutes or as long as several days. Although the syntheses of enzymes can be regulated at the genetic level (see Section 2.8.4), once an enzyme has been synthesised it can remain functional for some time. If the environmental conditions change abruptly, it may not suffice for the synthesis of the enzyme to be 'switched off', i.e. repressed; the cell may need to inactivate the enzyme so as to avoid needless, or even perhaps deleterious, metabolic activity. This may be by feedback inhibition (Section 2.8.6). Additionally, under nitrogen-limited conditions, when a cell becomes depleted of nitrogen and then ceases to

Table 2.3 Growth yields of micro-organisms growing on different substrates

Substrate	Organism	Molar growth yield (g organism dry wt per g-mol substrate)	Carbon conversion coefficient (g organism dry wt per g substrate carbon
Methane	*Methylomonas* sp.	17.5	1.46
Methanol	*Methylomonas* sp.	16.6	1.38
Ethanol	*Candida utilis*	31.2	1.30
Glycerol	*Klebsiella pneumoniae*	50.4	1.40
Glucose	*Escherichia coli*:		
	aerobic	95.0	1.32
	anaerobic	25.8	0.36
	Saccharomyces cerevisiae:		
	aerobic	90	1.26
	anaerobic	21	0.29
	Penicillium chrysogenum	81	1.13
Sucrose	*Klebsiella pneumoniae*	173	1.20
Xylose	*Klebsiella pneumoniae*	52.2	0.87
Acetic acid	*Pseudomonas* sp.	23.5	0.98
	Candida utilis	21.6	0.90
Hexadecane	*Yarrowia (Candida) lipolytica*	203	1.06

grow, **proteases**, that are **proteolytic enzymes**, may be activated to degrade surplus copies of enzymes so that the amino acids therein can be scavenged and used for the biosynthesis of new enzymes that may be still essential. Thus, enzymes may be 'turned over' more rapidly than may occur by simple denaturation.

2.9 Efficiency of microbial growth

The overall efficiency of microbial growth is discussed in strict thermodynamic terms in Chapter 3. It is usually expressed in terms of the yield of cells formed per unit weight of carbon substrate consumed. The **molar growth yield, Y_S**, is the cell yield (dry weight) per mole of substrate, while the **carbon conversion coefficient**, which allows more meaningful comparisons between substrates of different molecular sizes, is the cell yield per gram of substrate carbon.

A particular feature in Table 2.3 is the lower growth yields attained when facultative organisms are transferred from aerobic to anaerobic conditions, a phenomenon which is obviously connected with decreased energy production under these conditions.

Empirically, the growth yield of a micro-organism will depend on many factors:

(1) The nature of the carbon source.

(2) The pathways of substrate catabolism.

(3) Any supplementary provision of complex substrates (obviating the need for some anabolic pathways to operate).

(4) Energy requirements for assimilating other nutrients especially nitrogen.

(5) Varying efficiencies of ATP-generating reactions.

(6) Presence of inhibitory substrates, adverse ionic balance, or other medium components imposing extra demands on transport systems.

(7) The physiological state of the organism: nearly all microorganisms modify their development according to the external environment, and the different processes (e.g. primary and secondary metabolism) will entail different mass and energy balances.

In continuous culture systems, in which the growth rate and nutritional status of the cells are controlled (see Chapters 3 and 6), further factors can be identified:

(8) The nature of the limiting substrate; carbon-limited growth is often more 'efficient' than, for example, nitrogen-limited growth, in which catabolism of excess carbon substrate may follow routes which are energetically 'wasteful' (however useful they may be to the biotechnologist!).

(9) The permitted growth rate: whereas the growth rate is decreased, the proportion of the substrate going towards maintaining the cells increases, thereby diminishing the amount of substrate that can go to other products.

As a final factor governing all aspects of microbial performances, one might usefully add:

(10) **'The competence of the microbiologist'.**

2.10 Further reading

Hames, B. D. and Hooper, N. M. (2000). *Instant Notes: Biochemistry.* 2nd edn. Bios Scientific Publishers, Oxford.

Lengeler, J. W., Drews, G. and Schlegel, H., eds (1999). *Biology of the Prokaryotes.* Blackwell Science, Oxford.

Nicklin, J., Graeme-Cook, K., Paget, T. and Killington, R. (1999). *Instant Notes: Microbiology.* Bios Scientific Publishers, Oxford.

Madigan, M. T., Martinko, J. M. and Parker, J. (2000). *Biology of Micro-organisms, 9th Edition.* Prentice-Hall International (UK) Ltd, London.

Chapter 3

Stoichiometry and kinetics of microbial growth from a thermodynamic perspective

J. J. Heijnen

Nomenclature

Y_{sx}^{max}	maximal growth yield of biomass (X) on substrate (S) or electron donor (D)	C-mol X per C-mol S
m_i	maintenance coefficient of compound i	C-mol i per C-mol Xh
μ_{max}	maximum specific growth rate	h^{-1}
K_s	affinity constant	mol l^{-1}
Y_{ix}	yield of biomass (X) on compound i	C-mol X per mol i
r_i	reactor specific conversion rate of compound i	mol i m^{-3} per h
C_i	concentration	mol i m^{-3}
V	liquid volume of reactor	m^3
μ	specific growth rate	h^{-1}
ΔH_f°	standard enthalpy of formation	kJ mol^{-1}
ΔG_f°	standard Gibbs energy of formation	kJ mol^{-1}
$-\Delta G_{CAT}$	Gibbs energy of catabolism per C-mol organic electron donor or per mol of inorganic electron donor	kJ per (C) mol
q_i	biomass specific conversion rate of compound i	mol i per C-mol Xh
γ	degree of reduction	
X	biomass	C-mol

3.1 Introduction

Quantitative information on microbial growth is needed in many fermentation and biological waste treatment processes. Typically, growth

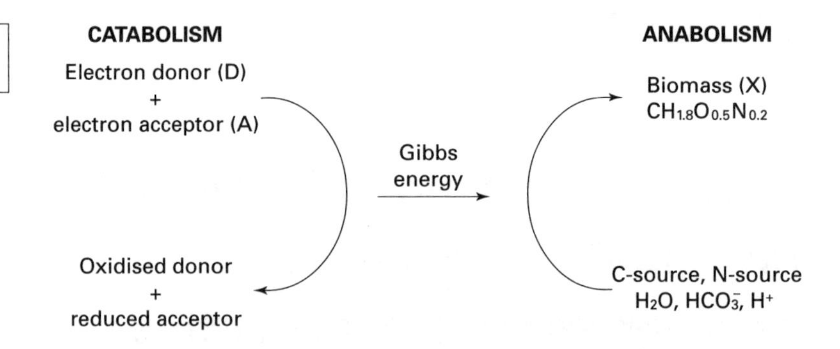

Fig. 3.1 Growth represented as coupled anabolism/catabolism.

is quantified using well known parameters such as maximum biomass yield on substrate S (also called electron donor D) (Y_{SX}^{max} or Y_{DX}^{max}), maintenance requirements for substrate S or electron donor D (m_S or m_D), μ_{max}, K_s and a threshold concentration of the electron donor. A practical problem is that the values for these parameters vary by one to two orders of magnitude, depending on the growth systems being considered. Such growth systems are generally characterised (Fig. 3.1) by their electron donor and electron acceptor, their C-source and N-source. In addition in each growth system HCO_3^-, H_2O and H^+ are involved. A practical point is that many micro-organisms have similar elemental compositions, as illustrated in Fig. 3.1. This allows the use of a standard biomass composition, in case this information is not available. However it is always preferable to determine this, using elemental analysis. For practical purposes it is important to know the complete stoichiometry of growth. In fermentation processes not only is the biomass yield from the substrate (Y_{SX} or Y_{DX}) important, but so are the oxygen requirement, CO_2 production and heat production in order to design an optimal process. Therefore, methods of stoichiometry calculation are of fundamental value. In addition, the estimation of growth stoichiometry for arbitrary growth systems is relevant, for example in biological waste treatment. In the past decades, many methods have been proposed to estimate growth parameters for arbitrary growth systems.

3.2 Stoichiometry calculations

3.2.1 Definition of the growth system

A microbial growth system is conveniently represented as an overall reaction equation (Fig. 3.2) where 1 C-mol of biomass is formed and which takes into account the role of N-source, H_2O, HCO_3^-, H^+, electron donor and electron acceptor couple. In addition it is indicated that heat and Gibbs energy are also involved. One C-mol of biomass is the amount which contains 12 g of carbon which usually amounts to about 25 g of dry matter, knowing that the biomass carbon content is about 45%. For microbial growth energy must be generated to enable the construction of the complex biomass molecules from simple carbon compounds. This energy is generated in a redox reaction between the electron donor and electron acceptor. The proper measure of energy spent in growth

$$-\frac{1}{Y_{DX}}\text{ electron donor} - \frac{1}{Y_{AX}}\text{ electron acceptor} + 1\text{C} - \text{mol biomass} + \frac{1}{Y_{HX}}\text{ kJ heat}$$

$$+\frac{1}{Y_{GX}}\text{ kJ Gibbs energy} + (\ldots)\text{ N-source} + (\ldots)\ H_2O + (\ldots)\ HCO_3^- + (\ldots)\ H^+$$

Fig. 3.2 General stoichiometric representation of the formation of 1 C-mol of biomass. Y_{iX} is the yield of 1 C-mol X on 1 C-mol of compound i; (....) are unspecified unknown stoichiometric compounds.

processes is not the released heat but Gibbs energy, because Gibbs energy (ΔG) combines the heat related enthalpic (ΔH) and entropic (ΔS) contributions ($\Delta G = \Delta H - T\Delta S$). The stoichiometric coefficients in the reaction equation (Fig. 3.2) are related to well known yield coefficients. Y_{DX}, Y_{AX}, Y_{HX} and Y_{GX} are the yields of biomass (in C-mol X) on electron donor (per C-mol for organic and per mol for inorganic compounds), on electron acceptor (per mol acceptor), on heat (per kJ) respectively Gibbs energy (per kJ). For biomass, the 1 C-mol composition is used. The minus sign shows consumption.

3.2.2 Measuring yields

It is stressed that stoichiometric yield coefficients are ratios of conversion rates (r_X is given as C-mol X m^{-3} reactor per h, r_i in mol i m^{-3} per h).

$$Y_{iX} = \frac{r_X}{r_i} \tag{3.1}$$

These rates are calculated from measurements in experiments which may be either batch, continuous or fed batch cultures, using correct mass balances.

The most frequently measured growth stoichiometric coefficient is the biomass yield on substrate (or electron donor) Y_{SX} (or Y_{DX}).

In a constant volume batch culture (0 indicating time 0), Y_{SX} will be:

$$Y_{SX} = (C_X - C_{XO})/(C_{SO} - C_S) \tag{3.2a}$$

In a chemostat, where input and outflow rates are equal, a similar equation holds, where $C_{XO} = 0$ and C_{SO} is replaced by the concentration C_{Si} of the incoming substrate.

If volume variations occur, more complex relations can be derived from the mass balances. For batch reactors with variable volume V, we get:

$$Y_{SX} = (VC_X - V_O C_{XO})/(V_O C_{SO} - VC_S) \tag{3.2b}$$

3.2.3 Maintenance effect

Initially Y_{SX} was introduced by Monod as a constant. However after the introduction of the chemostat, cultivation of micro-organisms under different growth rates showed that Y_{SX} was dependent on the specific growth rate μ. The proposed explanation is based on the maintenance concept (Herbert–Pirt). In this concept it is assumed that maintenance of cellular functions requires the expenditure of Gibbs energy (for restoring leaky gradients, protein degradation, etc.). This Gibbs energy is produced by the catabolism of a certain amount of electron donor (= substrate). If this maintenance rate is m_S or (m_D) C-mol substrate per C-mol X·h, the following equation holds

$$\frac{1}{Y_{SX}} = \frac{1}{Y_{SX}^{max}} + \frac{m_S}{\mu} \tag{3.3}$$

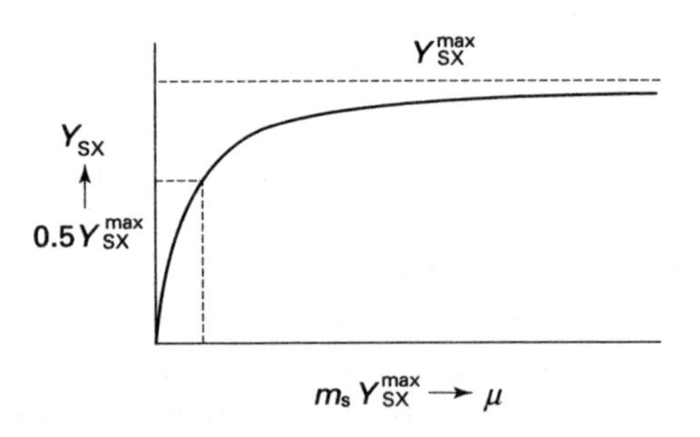

Fig. 3.3 Dependence of biomass yield Y_{SX} on specific growth rate (maintenance effect). m_s is the substrate maintenance coefficient, Y_{SX}^{max} is the maximal biomass yield on substrate.

Experimentally Y_{SX} is measured in a chemostat under different specific growth rates μ. From the obtained Y_{SX} and μ values, one calculates, using Eqn (3.3), the model parameters Y_{SX}^{max} and m_s.

Figure 3.3 shows how Y_{SX} depends on μ. For high values of μ, Y_{SX} approaches the value of the model parameter Y_{SX}^{max}. For low μ values Y_{SX} drops significantly, becoming ½ Y_{SX}^{max} for $\mu = m_s Y_{SX}^{max}$. For most conventional processes, it can be shown that at normal growth temperatures the effect of maintenance on yield can be neglected for $\mu > 0.05\,h^{-1}$. This means that in batch cultures during exponential growth (where $\mu \approx \mu_{max}$) $Y_{SX} \approx Y_{SX}^{max}$. However, in the fed-batch processes (which are the norm in most industrial processes), where $\mu < 0.05\,h^{-1}$, maintenance aspects dominate the biomass yield.

3.2.4 Conservation principles to calculate the full stoichiometry of growth

Figure 3.2 shows that besides Y_{SX} (or Y_{DX}), there are many more stoichiometric coefficients. Fortunately these need not be determined experimentally. The application of conservation principles often allows all other coefficients to be calculated if a single coefficient (Y_{SX}) is measured. This calculation is most easily explained in an example.

Example: Use of conservation principles in calculation of all stoichiometric coefficients

An aerobic micro-organism grows on oxalate using NH_4^+ as N-source. The following overall reaction equation (according to Fig. 3.2) can be written based on 1 C-mol biomass being produced with a biomass yield on oxalate of $Y_{DX} = 1/5.815$ C-mol biomass per mol oxalate

$$-5.815\,C_2O_4^{2-} + aNH_4^+ + bH^+ + cH_2O + dO_2 + eHCO_3^- + 1CH_{1.8}O_{0.5}N_{0.2}$$

There are five unknown stoichiometric coefficients for which five conservation constraints can be formulated.

C-conservation	$-11.63 + e + 1 = 0$
H-conservation	$4a + b + 2c + 3e + 1.8 = 0$
O-conservation	$-23.26 + c + 2d + 3e + 0.5 = 0$
N-conservation	$a + 0.2 = 0$
Charge-conservation	$+11.63 + a + b - e = 0$

Solving gives the full stoichiometry.

$$-5.815\,C_2O_4^{2-} + 0.2NH_4^+ + 0.8H^+ - 1.857O_2 - 5.42H_2O + 1CH_{1.8}O_{0.5}N_{0.2} + 10.63HCO_3^-$$

Thus we see that $Y_{AX} = 1/1.857$ C-mol X / mol O_2. Also $Y_{CX} = 1/10.63$ C-mol X / mol HCO_3^-.

For this overall equation one can, using ΔH_f^o and ΔG_f^o values from thermodynamic tables (Table 3.1), also calculate the $(-\Delta H_R)$ and $(-\Delta G_R)$, which then provide $1/Y_{HX}$ and $1/Y_{GX}$.

3.2.5 Balance of degree of reduction

The application of the conservation constraints is straightforward. A useful short cut of such calculations is to apply a degree of reduction balance (Roels, 1983). The degree of reduction (symbol γ) is defined for each compound and is a stoichiometric quantity, defined in such a way that its value is zero for the reference compounds H_2O, H^+, HCO_3^-, SO_4^{2-}, NO_3^-, Fe^{3+}, N-source.

The γ-value for each compound is found by calculating the redox half reaction which converts the compound into the previous defined reference chemicals and a number of electrons. γ follows as the number of produced electrons per C-mol for organic and per mol for inorganic compounds.

For example, the degree of reduction of O_2 follows from the redox half reaction $O_2 + 4H^+ \rightarrow 2H_2O - 4e^-$, and $\gamma = -4$. For glucose, the redox half reaction is $C_6H_{12}O_6 + 12H_2O \rightarrow 6HCO_3^- + 30H^+ + 24e^-$ and $\gamma = 24/6 = 4$.

Using the redox half reactions the γ-values for atoms and charge can also be calculated (Table 3.2). For example for the carbon atom one obtains $C + 3H_2O \rightarrow HCO_3^- + 5H^+ + 4e^-$ and a value of $\gamma = 4$ is found for carbon. Similarly values for H, O, S, N, + and − charge are found (Table 3.2).

It should be noted that the nitrogen atom in the biomass and in the N-source for growth has a degree of reduction, which depends on the N-source used for growth. This is needed to ensure that the molecular degree of reduction of N-source becomes 0. For example using Table 3.2 the degree of reduction for NH_4^+ follows as $-3 + 4 - 1 = 0$. The degree of reduction of a molecule (Table 3.1) represents the amount of electrons becoming available from that molecule upon oxidation to the reference compounds. For organic molecules, its value is usually normalised per C-mole, for inorganic molecules its value is per mole. Table 3.1 shows that for organic molecules the value range from 0 to 8. For biomass (standard composition) it follows that $\gamma_x = 4.2$ $(1 \times 4 + 1.8 \times 1 - 0.5 \times 2 - 0.2 \times 3)$ for NH_4^+ as a N-source and $\gamma_x = 5.8$ for NO_3^- as a N-source.

Because electrons are conserved, it is possible to calculate the balance of degree of reduction as shown in the following example.

Example: Application of the balance of degree of reduction

Consider the previous example of aerobic growth on oxalate.

Calculation of the degree of reduction (using Table 3.2) of 1 molecule of oxalate ($C_2O_4^{2-}$) gives

$$\gamma = 2 \times 4 + 4 \times (-2) + (+2) = 2.$$

Table 3.1 Degree of reduction, standard Gibbs energy and enthalpy (298 K, pH = 7, 1 bar, 1 mol^{-1}) for relevant compounds in growth systems

Compound name	Composition	ΔG_f^{01} kJ mol^{-1}	ΔH_f^0 kJ mol^{-1}	Degree of reduction
Biomass	$CH_{1.8}O_{0.5}N_{0.2}$	−67	−91	4.2 (N-source NH_4^+)
Water	H_2O	−237.18	−286	0
Bicarbonate	HCO_3^-	−586.85	−692	0
CO_2 (g)	CO_2	−394.359	−394.1	0
Proton	H^+	−39.87	0	0
O_2 (g)	O_2	0	0	−4
Oxalate $^{2-}$	$C_2O_4^{2-}$	−674.04	−824	+1
Carbon monoxide	CO	−137.15	−111	+2
Formate$^-$	CHO_2^-	−335	−410	+2
Glyoxylate$^-$	$C_2O_3H^-$	−468.6	—	+2
Tartrate^{2-}	$C_4H_4O_6^{2-}$	−1010	—	+2.5
Malonate $^{2-}$	$C_3H_2O_4^{2-}$	−700	—	+2.67
Fumarate $^{2-}$	$C_4H_2O_4^{2-}$	−604.21	−777	+3
Malate^{2-}	$C_4H_4O_5^{2-}$	−845.08	−843	+3
Citrate^{3-}	$C_6H_5O_7^{3-}$	−1168.34	−1515.	+3
Pyruvate$^-$	$C_3H_3O_3^-$	−474.63	−596	+3.33
Succinate^{2-}	$C_4H_4O_4^{2-}$	−690.23	−909	+3.50
Gluconate$^-$	$C_6H_{11}O_7^-$	−1154	—	+3.67
Formaldehyde	CH_2O	−130.54	—	+4
Acetate$^-$	$C_2H_3O_2^-$	−369.41	−486	+4
Dihydroxy acetone	$C_3H_6O_3$	−445.18	—	+5.33
Lactate$^-$	$C_3H_5O_3^-$	−517.18	−687	+4
Glucose	$C_6H_{12}O_6$	−917.22	−1264	+4
Mannitol	$C_6H_{14}O_6$	−942.61	—	+4.33
Glycerol	$C_3H_8O_3$	−488.52	−676	+4.67
Propionate$^-$	$C_3H_5O_{2-}$	−361.08	—	+4.67
Ethylene glycol	$C_2H_60_2$	−330.50	—	+5
Acetoine	$C_4H_8O_2$	−280	—	+5
Butyrate	$C_4H_7O_2^-$	−352.63	−535	+5
Propanediol	$C_3H_8O_2$	−327	—	+5.33
Butanediol	$C_4H_{10}O_2$	−322	—	+5.50
Methanol	CH_4O	−175.39	−246	+6
Ethanol	C_2H_6O	−181.75	−288	+6
Propanol	C_3H_8O	−175.81	−331	+6
n-Alkane (l)	$C_{15}H_{32}$	+60	−439	+6.13
Propane (g)	C_3H_8	−24	−104	+6.66
Ethane (g)	C_2H_6	−32.89	−85	+7
Methane (g)	CH_4	−50.75	−75	+8
H_2 (g)	H_2	0	0	+2
Ammonium	NH_4^+	−79.37	−133	+8
N_2 (g)	N_2	0	0	+10
Nitrite ion	NO_2^-	−37.2	−107	+2
Nitrate ion	NO_3^-	−111.34	−173	0
Iron II	Fe^{2+}	−78.87	−87	+1
Iron III	Fe^{3-}	−4.6	−4	0

Table 3.1 (*cont.*)

Compound name	Composition	ΔG_f^{01} kJ mol^{-1}	ΔH_f^0 kJ mol^{-1}	Degree of reduction
Sulphur	S^0	0	0	+6
Hydrogen sulphide (g)	H_2S	−33.56	−20	+8
Sulphide ion	HS^-	+12.05	−17	+8
Sulphate ion	SO_4^{2-}	−744.63	−909	0
Thiosulphate ion	$S_2O_3^{2-}$	−513.2	−608	+8
Ammonium	NH_4^+	−79.37	−133	+8

Table 3.2 Degree of reduction (γ) for atoms/charge

H	1
O	−2
C	4
S	6
N	5
+1 charge	−1
−1 charge	+1
N in biomass and	−3[a]
in N-source	0[b]
	5[c]

Notes:
[a, b, c] relate to different N-sources, being (a) NH_4^+; (b) N_2; (c) NO_3^-.

The γ value for biomass (NH_4^+ as a N-source) follows as $1 \times 4 + 1.8 \times 1 + 0.5(-2) + 0.2(-3) = 4.2$.

Similarly, the degree of reduction of O_2 will be −4. For the other compounds $\gamma = 0$, e.g. for HCO_3^- $\gamma = 1 \times 1 + 1 \times 4 + 3 \times (-2) + 1 = 0$ and for NH_4^+ the value of $\gamma = -3 + 4 - 1 = 0$.
This gives for the degree of reduction balance:

$$-5.815 \times 2 - 4d + 4.2 = 0.$$

It is seen that in this balance only the stoichiometric coefficients of substrate (or electron donor), the electron acceptor and biomass occur.

This gives $d = -1.857$, being identical to the full solution of conservation constraints obtained before. The other coefficients follow from application of the regular conservation constraints, i.e. the N-source coefficient from the N-balance, the HCO_3^- from the C-balance etc.

From the example several points must be stressed:

- the balance of degree of reduction specifies always a linear relation between the stoichiometric coefficients of electron donor, electron acceptor and biomass, making this relation extremely useful in practice;
- the balance of degree of reduction is not a new constraint, it is just a suitable combination of the C, H, N charge conservation constraints.

Other useful applications of the conservation constraints are outlined in the references and include:

- selection of the yield measurements which provide the least errors in the calculated other yields (due to error propagation in the measurements);
- improvement of the errors in all yields by measuring more than the minimal required yields (measurement redundancy allowing data reconciliation);
- use of redundant measurements to investigate the occurrence of systematic measurement errors or errors in the system definition (e.g. a product has been forgotten).

3.3 Stoichiometry predictions based on Gibbs energy dissipation

A number of methods have previously been proposed to estimate biomass yields (Y_{DX}) from correlations. A particularly simple but useful and recent method has been the thermodynamically based approach using Gibbs energy dissipation per unit biomass ($1/Y_{GX}$) in kJ per C-mol X. This is a stoichiometric quantity which can [similar to the biomass yield Y_{DX} on electron donor as in Eqn (3.3)] be written as

$$\frac{1}{Y_{GX}} = \frac{1}{Y_{GX}^{max}} + \frac{m_G}{\mu} \tag{3.4}$$

m_G is the biomass specific rate of Gibbs energy dissipation for maintenance purposes in kJ per C-mol X h and Y_{GX}^{max} is the maximal biomass yield on Gibbs energy in C-mol X kJ^{-1}.

Eqn (3.4) shows that the Gibbs energy dissipation contains a growth and a maintenance related term.

Simple correlations have been proposed for $1/Y_{GX}^{max}$ and for m_G (see Further reading, Section 3.5). These correlations cover a wide range of microbial growth systems and temperatures (heterotrophic, autotrophic, aerobic, anaerobic, denitrifying growth systems on a wide range of C-sources, growth systems with and without reversed electron transport – RET).

3.3.1 Correlation for maintenance Gibbs energy

The following correlation has been found to be valid for maintenance Gibbs energy

$$m_G = 4.5 \exp\left[-\frac{69000}{8.314}\left(\frac{1}{T} - \frac{1}{298}\right)\right] \tag{3.5}$$

This correlation holds for a temperature range of 5 to 75 °C, for aerobic and anaerobic conditions. It does not depend on the C-source or electron donor or acceptor being applied and only shows a significant temperature effect. This seems logical because maintenance only involves Gibbs energy, irrespective of the electron donor/acceptor combination which provides this Gibbs energy.

3.3.2 Correlation for Gibbs energy needed for growth

For the growth-related Gibbs energy requirement $1/Y_{GX}^{max}$, the following correlations can be used:

For heterotrophic or autotrophic growth without RET:

$$\frac{1}{Y_{GX}^{max}} = 200 + 18(6 - c)^{1.8} + \exp[((3.8 - \gamma_s)^2)^{0.16}(3.6 + 0.4c)] \tag{3.6a}$$

For autotrophic growth requiring reversed electron transport:

$$\frac{1}{Y_{GX}^{max}} = 3500 \tag{3.6b}$$

Eqn (3.6a) shows that $1/Y_{GX}^{max}$ for heterotrophic growth is mainly determined by the carbon-source used. This C-source is characterised by its degree of reduction, γ_s, and the number of C-atoms (parameter c) per mole of C-source.

Eqn (3.6a) shows that $1/Y_{GX}$ ranges between about 200 and 1000 kJ of Gibbs energy requirement per C-mol biomass dependent on the C-source used. Furthermore, it can be seen that $1/Y_{GX}$ increases for C-sources which have less carbon atoms and for which the degree of reduction is higher or lower than about 3.8.

This result is easily understood because a C-source with a low number of C-atoms requires numerous biochemical reactions to produce the required C_4 to C_6 compounds needed for biomass synthesis. In addition, with the degree of reduction of biomass being about 4, it is clear that C-sources which are more reduced or more oxidised than biomass require additional biochemical reactions for oxidation or reduction, respectively. Hence, increased values of $1/Y_{GX}^{max}$ reflect the increased requirement for additional biochemical reactions which leads to greater dissipation of Gibbs energy. For example, making biomass from CO_2 ($\gamma_s = 0$, $c = 1$) requires a Gibbs energy dissipation of 986 kJ per C-mol X, whereas use of glucose ($\gamma_s = 4$, $c = 6$) only requires 236 kJ per C-mol X, which reflects the much more extensive biochemical reactions needed for growth using CO_2 as a C-source.

To estimate the value of $1/Y_{GX}^{max}$ for autotrophic growth we need to know whether RET is required. This follows after establishing the biomass formation reaction from CO_2, using the available electron donor as electron source. If $\Delta G_R \gg 0$ for this reaction, it is clear that the energy level of the electron donor electrons is insufficient to reduce CO_2 to biomass. The micro-organism must then convert part of the donor electrons to a higher energy level, using reversed electron transport (RET). Examples of such energy deficient electron donors are Fe^{2+}/Fe^{3+}, NO_2^-/NO_3^-.

Autotrophic electron donors such as H_2/H^+ or CO/CO_2 do not need RET. For these electron donors $1/Y_{GX}^{max} \approx 1000$ kJ per C-mol X, as found from Eqn (3.6a), where $\gamma_S = 0$ and $c = 1$ (CO_2 is C-source).

For RET requiring electron donors $1/Y_{GX}^{max}$ values are 3500 kJ per C-mol X (Eqn 3.6b). This shows that RET requires many additional biochemical reactions leading to a much higher dissipation of Gibbs energy.

Example: Occurrence of RET in autotrophic growth

Consider the autotrophic aerobic microbial growth using Fe^{2+}/Fe^{3+} as electron donor. HCO_3^- is the C-source. This allows the following biomass formation reaction to be drawn up where HCO_3^- is reduced using the donor electrons:

$$HCO_3^- + 4.2\,Fe^{2+} + 0.2\,NH_4^+ + 5H^+ \rightarrow 1\,CH_{1.8}O_{0.5}N_{0.2} + 4.2\,Fe^{3+} + 2.5\,H_2O$$

Using Table 3.1 one can calculate that $\Delta G_R^{01} = +454$ kJ.
Clearly, for the electron donor Fe^{2+}/Fe^{3+} RET is needed and Eqn (3.6b) applies.

3.3.3 Stoichiometry prediction using the Gibbs energy correlations

The correlations found for m_G and $1/Y_{GX}^{max}$ in Eqns (3.5, 3.6a, 3.6b) can easily be used to estimate, for each microbial growth system, the complete stoichiometry of the growth equation as a function of:

- the C-source applied;
- the electron donor/acceptor combination;
- growth rate and temperature.

It has been shown that for a wide range of microbial growth systems the estimation of Y_{DX} is possible in a range of 0.01 to 1 C-mol X per C-mol donor with a relative accuracy of about 10 to 15% (see Further reading, Section 3.5). The calculation of the complete stoichiometry is best shown using an example.

Example: Estimation of growth stoichiometry using the Gibbs energy correlations

Consider the aerobic autotrophic growth of a micro-organism using Fe^{2+} to Fe^{3+} as electron donor at 50 °C, a growth rate of $0.01\,h^{-1}$, using NH_4^+ as N-source and growing at pH = 1.5. It is required to calculate the complete growth stoichiometry.

The following stoichiometric equation can be specified:

$$+aHCO_3^- + bNH_4^+ + cH_2O + dO_2 + eFe^{2+} + 1\,CH_{1.8}O_{0.5}N_{0.2} + fFe^{3+} + gH^+ + 1/Y_{GX}\text{ Gibbs energy.}$$

We can specify six conservation constraints and one Gibbs energy balance to calculate the seven (a to g) unknown stoichiometric coefficients. Using Eqn (3.4) $1/Y_{GX}$ follows from the correlations (knowing that RET is involved, that $\mu = 0.01\,h^{-1}$ and that $T = 323$ K) as:

$$1/Y_{GX} = 3500 + \frac{38.84}{0.01} = 7384\text{ kJ/C-molX}$$

The six conservation constraints and the Gibbs energy balance (using ΔG_f^{01}-values from Table 3.1) are as follows:

C-conservation	$a + 1 = 0$
O-conservation	$3a + c + 2d + 0.5 = 0$
Degree of reduction	$-4d + e + 4.2 = 0$

Iron-conservation $e + f = 0$
N-conservation $b + 0.2 = 0$
Charge-conservation $-a + b + 2e + 3f + g = 0$

Gibbs energy balance

$$(-586.85)a + (-79.37)b + (-237.18)c + (-78.87)e + (-67) + (-4.6)f + (-8.54)g + 7384 = 0$$

It is noted that for H^+ ΔG_f is recalculated from pH = 7 (in Table 3.1) to pH = 1.5 (change from −38.87 to −8.54 kJ mol H^+). Also the balance of degree of reduction has been used as a constraint (replacing the H-constraint). After solving the six equations one obtains the complete stoichiometry.

3.3.4 Algebraic relations to calculate stoichiometry

Because all the stoichiometric coefficients are, through the conservation constraints, related to $1/Y_{GX}$, one can derive also algebraic relations, between $1/Y_{iX}$ and $1/Y_{GX}$. For the biomass yield on electron donor Y_{DX} the following relation is obtained as an example (see for additional relations Further reading, Section 3.5):

$$Y_{DX} = \frac{(-\Delta G_{CAT})}{1/Y_{GX} + \gamma_x/\gamma_D(-\Delta G_{CAT})} \tag{3.7}$$

ΔG_{CAT} is the Gibbs energy of the catabolic reaction of 1 C-mol organic electron donor or of 1 mol inorganic electron donor in kJ per (C)-mol donor. γ_x and γ_D are the degree of reduction for biomass and electron donor (per mol or C-mol). For the previous example the catabolic reaction of 1 mol electron donor is $Fe^{2+} + \frac{1}{4} O_2 + H^+ \rightarrow Fe^{3+} + \frac{1}{2} H_2O$. Using the values of ΔG_f^{01} in Table 3.1, and using for H^+ the ΔG_f-value at pH = 1.5 of −8.54 kJ/mol leads to $\Delta G_{CAT} = -35.78$ kJ per mol Fe^{2+}. In addition $\gamma_x = 4.2$, $\gamma_D = 1$ and $1/Y_{GX} = 7384$ kJ per C-mol X giving $Y_{DX} = 0.0047$ C-mol X per mol Fe^{2+}, showing that the stoichiometric coefficient e in the previous example equals 215 mol Fe^{2+} per C-mol X.

Eqn (3.7) shows that

- Y_{DX} increases hyperbolically with increasing Gibbs energy production in catabolism $(-\Delta G_{CAT})$.

This explains why anaerobic growth systems (with low $(-\Delta G_{CAT})$-values) have lower Y_{DX} values as aerobic systems.

- Y_{DX} is higher for systems with lower $1/Y_{GX}$-values (as found for high specific growth rate μ, low temperature, favourable C-source and the absence of RET);
- Y_{DX} depends hyperbolically on μ [substitute $1/Y_{GX}$ using Eqn (3.4)] due to maintenance effects in agreement with Eqn (3.3);
- Y_{DX} has a theoretical maximal limit from the 2nd law of thermodynamics (because according to the 2nd law $1/Y_{GX}$ has a theoretical minimal value of 0 kJ per C-mol X) of γ_D/γ_X.

3.3.5 Heat aspects

Finally the aspect of heat merits some thoughts.
$1/Y_{HX}$ (= kJ heat per C-mol X) follows by calculating $(-\Delta H_R^0)$ of the

complete stoichiometric growth equation where 1 C-mol biomass is produced using ΔH_f^o values from Table 3.1.

Some relevant remarks:

- For aerobic heterotrophic growth heat production is closely related to O_2-uptake, where 1 mol O_2 = 450 kJ heat;
- There are microbial growth systems where heat uptake (not production) can be calculated to occur;
- An example is the growth of methanogenic bacteria which split acetate to CH_4 and CO_2.

3.3.6 Limitation of the yield prediction using the thermodynamic approach

The presented method provides Y_{SX}-estimates where the biochemical details of metabolism, characteristic for each micro-organism, are neglected. This is the attractive aspect of this method, because this knowledge is often not available. However one should always realise that differences in biochemistry are relevant. For example, ethanol fermentation from glucose is performed by *Saccharomyces cerevisiae* with Y_{SX} measured to be around 0.15 C-mol X per C-mol glucose. A similar Y_{SX} value is also obtained using the above method. However *Zymomonas mobilis* also performs the ethanol fermentation but with a $Y_{SX} = 0.07$. The difference is caused by a different biochemical pathway (glycolysis versus Entner–Doudoroff route, see Chapter 2). This example shows that, if the value of the estimated Y_{SX} deviates strongly from a measured Y_{SX}, one might expect that an unusual, perhaps new, biochemical pathway is being used in catabolism (or anabolism).

3.4 Growth kinetics from a thermodynamic point of view

Growth kinetics are generally characterised by the two parameters μ_{max} and K_s. It is known that K_s values can be very different, dependent on the occurrence of passive diffusion, facilitated transport or active transport for the transfer of electron donor (substrate) into the micro-organism. A general thermodynamic correlation for K_s is therefore not possible.

Also for μ_{max} a very wide range (0.001 to 1 h^{-1}) of values is found, depending on the microorganism and cultivation conditions. However, it would seem reasonable to expect that a low maximal specific rate of Gibbs energy production from catabolism leads to a lower maximal specific growth rate. Using this concept of energy limitation one can derive the following expressions for the maximal specific rate of Gibbs energy production q_G^{max} (kJ per C-mol X h).

$$q_G^{max} = 3[-\Delta G_{CAT})/\gamma_D]\exp\left[\frac{-69000}{R}\left(\frac{1}{T} - \frac{1}{298}\right)\right] \tag{3.8}$$

This relation is based on:

- A maximal electron transport rate of 3 mol electrons per C-mol X h at 298 K leading to the coefficient 3 in Eqn (3.8);

- A temperature effect on this rate according to an Arrhenius relation with an energy of activation of 69 000 J per mol (equivalent to the rate doubling for every 10 °C increase in temperature). R is the gas-constant (equal to 8.314 J per mol K);
- The maximum rate of catabolic Gibbs energy production q_G^{max} is then the rate of electron transport multiplied by $(-\Delta G_{CAT}/\gamma_D)$, which is the catabolic energy release per electron in the electron donor/acceptor reaction.

Equating the maximal rate of catabolic Gibbs energy production (Eqn 3.8) to the Gibbs energy needed for growth under maximal growth rate condition (being equal to the sum of μ_{max}/Y_{GX}^{max} and maintenance being equal to 4.5 × temperature correction, see Eqn (3.5)) gives then the μ_{max}-value (in h^{-1}) according to Eqn (3.9):

$$\mu_{max} = \frac{[3(-\Delta G_{CAT})/\gamma_D - 4.5]}{1/Y_{GX}^{max}} \exp\left[\frac{-69000}{R}\left(\frac{1}{T} - \frac{1}{298}\right)\right] \quad (3.9)$$

Eqn (3.9) can be shown to provide reasonable estimates of μ_{max}-values for a wide variety of micro-organisms (e.g. nitrifiers, methanogens, heterotrophic aerobes).

A final aspect to be discussed is the occurrence of so-called threshold concentrations. These are the concentrations of substrates below which no metabolism occurs anymore. According to Eqn (3.8), Gibbs energy production is seen to stop if $\Delta G_{CAT} = 0$ (this would represent equilibrium of the catabolic reaction).

However one should realise that the Gibbs energy released in catabolism is used to produce ATP, which involves the transmembrane transport of protons (proton motive force concept, see Chapter 2). This transport requires about 10–20 kJ per mol H^+, and therefore there must exist a minimal value of $(-\Delta G_{CAT})$ of about 10 to 20 kJ per catabolic reaction. This minimal value of ΔG_{CAT} leads to the existence of concentration levels of electron donors, below which there cannot be generation of ATP (hence metabolism). These so-called threshold concentrations of electron donor have indeed been observed for Fe^{2+} (in aerobic Fe^{2+} oxidising bacteria) and for H_2 (in anaerobic H_2-consuming reactions converting CO_2 to CH_4).

3.5 Further reading

Battley, E. H. (1987). *Energetics of Microbial Growth*. John Wiley and Sons, Chichester.

Heijnen, J. J. and van Dijken, J. P. (1992). In search of a thermodynamic description of biomass yields for the chemotrophic growth of microorganisms. *Biotechnol. Bioeng.* **39**, 833–858.

Heijnen, J. J., van Loosdrecht, M. C. M. and Tijhuis, L. (1992). A black box mathematical model to calculate auto- and heterotrophic biomass yields based on Gibbs energy dissipation. *Biotechnol. Bioeng.* **40**, 1139–1154.

Roels, J. A. (1983). *Energetics and Kinetics in Biotechnology*. Elsevier, New York.

Tijhuis, L., van Loosdrecht, M. C. M. and Heijnen, J. J. (1993). A thermodynamically based correlation for maintenance Gibbs energy requirements in aerobic and anaerobic chemotrophic growth. *Biotechnol. Bioeng.* **42**, 509–519.

van der Heijden, R. T. J. M., Heijnen, J. J., Hellinga, C., Romein, B. and Luyben, K. Ch. A. M. (1994). Linear constraint relations in biochemical reaction systems: I. Classification of the calculability and the balanceability of conversion rate. *Biotechnol. Bioeng.* **43**, 3–10.

van der Heijden, R. T. J. M., Romein, B., Heijnen, J. J., Hellinga, C. and Luyben, K. Ch. A. M. (1994). Linear constraint relations in biochemical reaction systems: II. Diagnosis and estimation of gross errors. *Biotechnol. Bioeng.* **43**, 11–20.

van der Heijden, R. T. J. M., Romein, B., Heijnen, J. J., Hellinga, C. and Luyben, K. Ch. A. M. (1994). Linear constraint relations in biochemical reaction systems: III. Sequential application of data reconciliation for sensitive detection of systematic errors. *Biotechnol. Bioeng.* **44**, 781–791.

von Stockar, U. and Marison, I. W. (1989). The use of calorimetry in biotechnology. *Adv. Biochem. Eng. Biotechnol.* **40**, 93–136.

Westerhoff, H. V. and van Dam, K. (1987). *Mosaic Non-equilibrium Thermodynamics and the Control of Biological Free Energy Transduction.* Elsevier, Amsterdam.

Chapter 4

Genome management and analysis: prokaryotes

Colin R. Harwood and Anil Wipat

4.1 Introduction

Gene manipulation is nowadays a core technology used for a wide variety of research and industrial applications. In addition to representing an extremely powerful analytical tool, it can be used to: (i) increase the yield (and quality) of existing products (proteins, metabolites or even whole cells); (ii) improve the characteristics of existing products (e.g. protein engineering); (iii) produce existing products by new routes (e.g. pathway engineering); and (iv) develop novel products not previously found in Nature. In this chapter we assume a knowledge in the reader of the basic structure and properties of nucleic acids, the organisation of the genetic information into genes and operons and the mechanisms by which bacteria transcribe and translate this encoded information for the synthesis of proteins (see also Chapter 2).

4.2 Bacterial chromosomes and natural gene transfer

4.2.1 Bacterial chromosomes

Chromosomes are the principal repositories of the cell's genetic information, the site of gene expression and the vehicle of inheritance. The

Table 4.1 The usual size range of the various classes of genetic elements found in bacterial cells

Genetic element	Size range
Transposons	800 bp to 30 kbp
Plasmids	1 kbp to 150 kbp
Prophages	3 kbp to 300 kbp
Bacteriophages	4 knt to 170 kbp
Bacterial chromosomes	600 kbp to 9.45 Mbp

Note:
bp, nucleotide base pair(s); k, thousand; M, million; nt, nucleotide(s).

term chromosome, meaning dark-straining body, was originally applied to the structures visualised in eukaryotic organisms by light microscopy. This term has now been extended to describe the physical structures that encode the genetic (hereditary) information in all organisms. The term **genome** is used in the more abstract sense to refer to the sum total of the genetic information of an organism. The term **nucleoid** is applied to a physical entity that can be isolated from a bacterial cell and that contains the chromosome in association with other components including RNA and protein. In addition to the main chromosome, other discrete types of replicating genetic material have been identified in bacterial cells, including transposable genetic elements (transposons), plasmids and proviruses. The usual size ranges of the various genetic elements found inside bacterial cells are shown in Table 4.1.

The genetic material of bacteria consists of double-stranded (ds) DNA. The nucleotide bases are usually unmodified excepting for the addition of methyl residues that function to: (i) identify the 'old' (conserved) DNA strand following replication; (ii) protect the DNA from the action of specific nucleases; and (iii) synchronise certain cell cycle events. Many viruses also use dsDNA as the genetic information (e.g. T-phages and lambda), while others have single-stranded (ss) DNA (e.g. ϕX174 and M13), ssRNA (e.g. MS2) or dsRNA (e.g. rotoviruses). Microbial chromosomes range in size over several orders of magnitude and vary in number, composition and topology (Table 4.2). Genome sizes tend to reflect the organism's structural complexity and life style. Obligate bacterial parasites, such as *Mycoplasma genitalium* (580 kbp), tend to have small genomes, while bacteria with complex life cycles, such as *Myxococcus xanthus* (9.45 Mbp), tend to have large genomes. The genomes of many eubacterial, archaeal and simple eukaryotic micro-organisms have been completely sequenced.

The chromosome of *Escherichia coli* is typical of many eubacteria. It weighs 5 femtigrams (5×10^{-15}g), is 1100 μm in length and its 4.6 Mbp of DNA codes for about 4400 proteins. The chromosome has a single set of genes (excepting for those encoding ribosomal RNA). At least 90% of the DNA encodes proteins/polypeptides while the remaining 10% is used either for controlling gene expression or has a purely structural

Table 4.2 Comparative properties of viral, bacterial and fungal chromosomes with respect to size, composition and topography

Organism	Type	Number	Size	Nucleic acid	Topology
MS2	bacteriophage	1	3.6 knt	ssRNA	circular
φX174	bacteriophage	1	5.4 knt	ssDNA	linear
lambda	bacteriophage	1	48.5 kbp	dsDNA	linear
T4	bacteriophage	1	174 kbp	dsDNA	linear
Mycoplasma genitalium	eubacterium	1	580 kbp	dsDNA	circular
Borrelia burgdorferi	eubacterium	1	1.4 Mbp	dsDNA	linear
Campylobacter jejuni	eubacterium	1	1.7 Mbp	dsDNA	circular
Rhodobacter sphaeroides	eubacterium	2	3.0 Mbp + 0.9 Mbp	dsDNA	2 × circular
Bacillus subtilis	eubacterium	1	4.2 Mbp	dsDNA	circular
Escherichia coli	eubacterium	1	4.6 Mbp	dsDNA	circular
Myxococcus xanthus	eubacterium	1	9.45 Mbp	dsDNA	ND
Methanococcus jannaschii	archaea	1	1.66 Mbp	dsDNA	circular
Archaeoglobus fulgidus	archaea	1	2.8 Mbp	dsDNA	circular
Schizosaccharomyces pombe	eukaryote	3	3.5, 4.6 and 5.7 Mbp, total 18.8 Mbp	dsDNA	linear
Saccharomyces cerevisiae	eukaryote	16	0.2 to 2.2 Mbp, total 12.43 Mbp	dsDNA	linear

Notes:
ND, not determined; dsDNA, double-stranded DNA; ssDNA, single-stranded DNA; bp, nucleotide base pair(s); M, million; nt, nucleotide(s); k, thousand.

function. Genes of related function are often, but not always, clustered together on the chromosome. Protein coding sequences can be on either strand of the DNA, although there is a preference for an orientation in the direction in which the DNA is replicated. Gene expression involves two distinct, highly co-ordinated processes. The DNA is firstly **transcribed** by the enzyme RNA polymerase into messenger (m)RNA, an unstable molecular species with half-lives (i.e. the time taken for half of the RNA to be degraded) that are measured in minutes. Even as they are being transcribed, ribosomes (large nucleoprotein complexes) attach to specific sites on the mRNA, the ribosome binding sites, and **translate** the encoded information into a linear polypeptide. To enable bacterial cells to regulate gene expression, the DNA is organised into transcriptional units or **operons** with distinct control sequences and transcriptional and translational start and stop points (Fig. 4.1).

The *E. coli* chromosome **replicates** by a bi-directional mode from the origin of replication (*oriC*) to the terminus (*terC*), primarily using the enzyme DNA polymerase III (Fig. 4.2). The rate of replication (at 37 °C) is about 800 bases per second, and consequently it takes approximately 40 minutes to replicate the entire chromosome. Since *E. coli* can divide by binary fission into two similarly sized cells every 20 minutes when growing on highly nutritious culture media, the chromosome of a single cell may have multiple sites of replication.

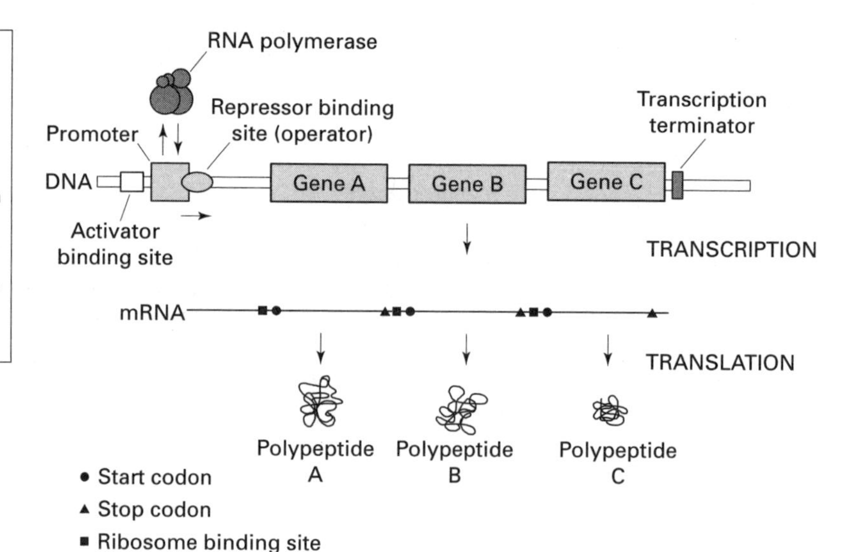

Fig. 4.1 Schematic diagram illustrating the key components of a bacterial operon. The activator and repressor binding sites are control sites at which the frequency of transcription initiation are controlled. Although start and stop codons, and ribosome binding sites can be recognised in the DNA sequence, they are only functional in the mRNA.

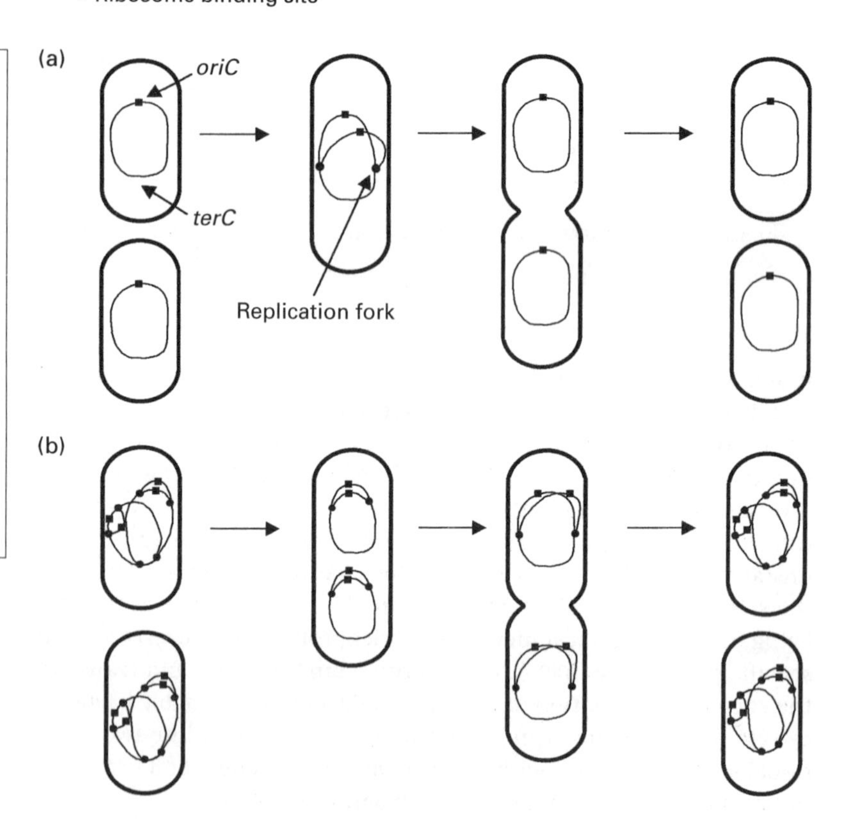

Fig. 4.2 The bi-directional replication of the bacterial chromosome. Replication is initiated at the origin of replication (*oriC*) and is completed at the terminus (*terC*). (a) In slow growing cells (doubling time >60 min), each side of the chromosome has just one replication fork or site of DNA synthesis. (b) In rapidly growing cells (doubling time ~20 min), a new round of replication is initiated before the previous one has reached the terminus. Consequently, each side of the chromosome has more than one replication fork.

4.2.2 Mechanisms of gene transfer

The ability to engineer changes to the characteristics of a bacterium dates back to 1928 and the experiments of Griffith on the mode of infection of *Streptococcus pneumoniae*. Griffith observed that of the two colonial morphologies exhibited by these bacteria on solid media, rough and smooth, only the latter was able to cause infection in mice. The rough and smooth characteristics (**phenotypes**) were due to the absence or presence, respectively, of a polysaccharide capsule that enables the bacterium to avoid the host's immune response. Griffith was able to show that if a rough (non-encapsulated) strain was injected into mice together with a heat-killed smooth strain, the rough strain could be transformed into a smooth strain to cause a fatal infection. It was some 16 years (1944) before the chemical agent responsible for this transformation was identified as DNA by Avery, MacLeod and McCarty, and another 9 years (1953) before Watson and Crick determined its structure. The mechanism for transferring isolated DNA into a bacterial cell is still referred to as **transformation**, reflecting the rough to smooth transition of Griffith's pneumonococci. Cells that have received transforming DNA are referred to as **transformants**. Genetic transformation is a natural characteristic of a wide variety of bacterial genera including *Azotobacter, Bacillus, Campylobacter, Clostridium, Haemophilus, Mycobacterium, Neisseria, Streptococcus* and *Streptomyces*. In addition, many strains that are not naturally transformable can be induced to take up isolated DNA by chemical treatment or by the application of an electric field (see Section 4.4.5).

Two other mechanisms for transferring DNA between bacterial strains have been identified since the work of Griffith, namely **transduction** and **conjugation**. Transduction is the transfer of DNA from a **donor** cell to a **recipient** cell mediated by a bacterial virus (**bacteriophage**, usually just referred to as **phage**). It was first demonstrated in *Salmonella* by Zinder and Lederberg in 1952 using phage P22. During the replication of the phage in the donor, a small proportion of the phage particles (virions) encapsulate bacterial rather than phage DNA. These so-called **transducing particles** are still infective, but instead of injecting phage DNA infect the host cell with chromosomal DNA from the donor strain. The recipients of transducing DNA are referred to as a **transductants**.

The third mechanism of gene transfer, **conjugation**, was discovered in 1946 by Lederberg and Tatum and involves cell-to-cell contact between the donor and recipient cells. Conjugation is usually mediated by a class of 'extrachromosomal, hereditary determinants' called **plasmids**. Plasmids are usually composed of covalently closed circular (*ccc*) molecules of double-stranded DNA that are able to replicate independently of the host chromosome, although occasionally they may integrate into the host chromosome. Plasmids are a common feature of bacterial strains where they confer a wide range of usually non-essential phenotypes, such as antibiotic resistance, toxin production, plant tumour formation, degradation of hydrocarbons and aromatic compounds (e.g. camphor, naphthalene, salicylate) and fertility. Plasmids tend to fall into the size range 1kbp to 150 kbp, although mega-plasmids (>150 kb) have been found in representatives of a number of bacterial genera, including *Agrobacterium*, *Pseudomonas* and *Streptomyces*. Plasmids may account for between >0.1 to about 4% of their host's genotype, although in rare cases this may be as high as 20%. Plasmids such as the F plasmid of *E. coli* that confer fertility on their host cells are referred to as conjugative plasmids.

Bacterial conjugation involves the transfer of DNA from a donor to a

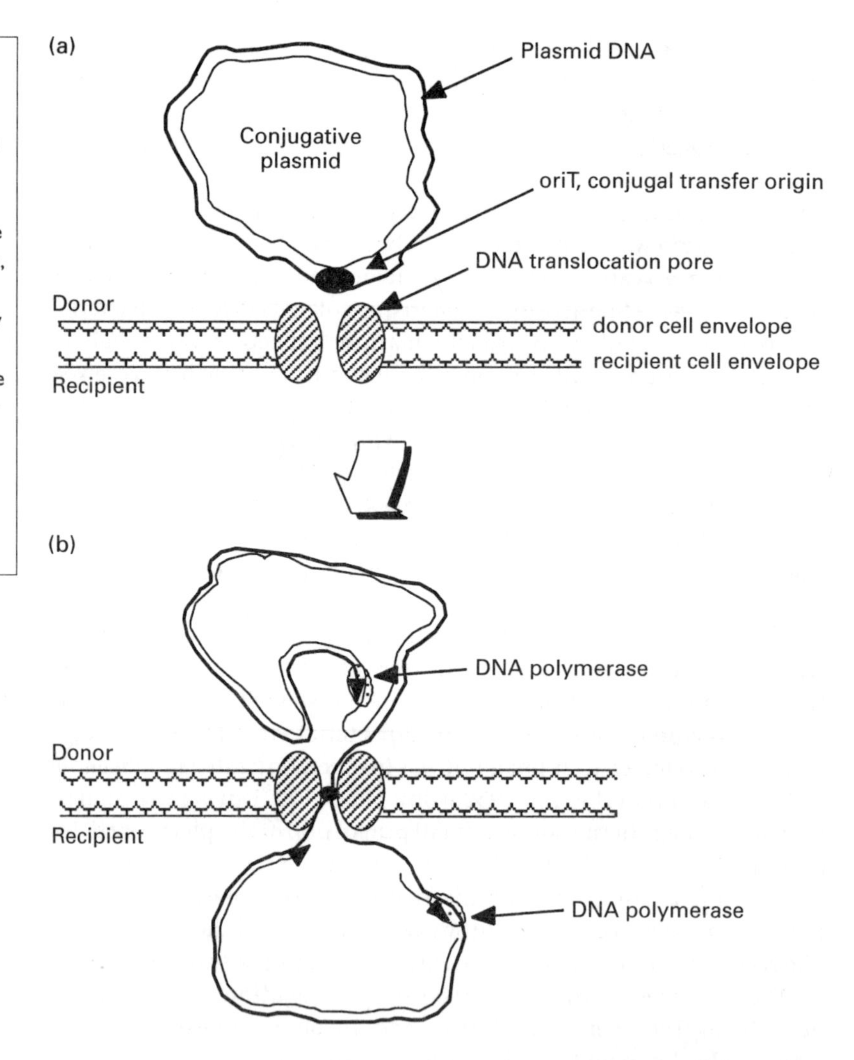

Fig. 4.3 Schematic diagram illustrating the replication and transfer of plasmid DNA during conjugation. (a) The cell envelopes of donor and recipient cells make contact and their cytoplasms are joined by a translocation pore. The conjugation-specific transfer origin, *oriT*, interacts with this pore. (b) The plasmid replicates from *oriT* by a mode that directs one of the original strands of its DNA into the recipient, where a complementary strand is synthesised. The other original strand of the plasmid remains in the donor where it too has a complementary strand synthesised.

recipient cell (Fig. 4.3). Usually this is plasmid DNA, although occasionally it is a part of the donor's chromosomal DNA that is transferred to the recipient. The machinery involved is almost exclusively encoded by the conjugative plasmid, the main exception being the enzymes involved in DNA transfer replication. The transferred DNA is always in a single-stranded form and the complementary strand is synthesised in the recipient. The transfer of host chromosomal genes usually occurs at frequencies that are very much lower than that of plasmids, although some plasmids are exceptional in being able to mobilise host chromosomal genes at a frequency approaching 1, e.g. Hfr (high frequency of recombination) strains of *E. coli*.

Many smaller plasmids are capable of conjugal transfer even though they do not possess fertility functions of their own. These plasmids, which are referred to as **mobilisable plasmids**, exploit the fertility properties of co-existing conjugative plasmids. They have an active origin of transfer (*oriT*) and mobilisation (*mob*) genes encoding proteins required for their replicative transfer. When such plasmids are used as the basis of cloning vectors, the *mob* genes are usually omitted as a requirement of containment regulations designed to avoid the dissemination of the cloned genes to wild-type populations.

Although transfer between bacterial cells is the most common type of conjugation, conjugative transfer between bacteria and fungi and between bacteria and plants has also been demonstrated. In the latter case, strains of *Agrobacterium tumefaciens* with large (> 200 kbp) tumour-inducing (Ti) plasmids can transfer part of their plasmid DNA – the so-called T-DNA (20–30 kbp) – into plant cells, where it interacts with the nuclear genome of the plant. This transfer is mediated by virulence (*vir*) genes which show similarities to the components of bacterial conjugation systems. Agrobacterial Ti plasmids have been adapted to introduce new characteristics into plant species (transgenic plants), such as resistance to specific insect pests.

Natural gene transfer methods have been used to generate genetic maps of many bacterial species that show the order and relative distances between the various genes. These classical genetic mapping techniques, highly developed in only a relatively small number of bacterial species, allowed detailed analysis of gene structure and the control of gene expression. These methods also allow strains with new characteristics to be constructed and have been adapted for use with more recently developed genetic engineering techniques.

4.3 What is genetic engineering and what is it used for?

The ability to manipulate and analyse DNA using genetic engineering techniques (**recombinant DNA technology**) was foreseen in the mid 1960s and came to fruition in the early 1970s. The technology, which is still developing rapidly, evolved from a series of basic studies in the interrelated disciplines of biochemistry and microbial genetics. Key amongst these was the elucidation of the molecular basis of bacterial **restriction and modification systems** by Werner Arber that subsequently provided enzymes for cutting DNA at precise locations (target sites). These **restriction endonucleases** (restriction enzymes) were quickly exploited for the analysis and manipulation of DNA molecules from a variety of sources. From these relatively modest beginnings, techniques for manipulating and analysing both types of nucleic acid (DNA and RNA) have become remarkably powerful and sensitive, aided by the development of key technologies such as DNA sequencing, oligonucleotide synthesis and the polymerase chain reaction (PCR). At the same time, the provision of chemicals, reagents and equipment to facilitate this technology has become a multi-million dollar industry.

The advent of recombinant DNA technologies led to the realisation that DNA could be analysed to a resolution that was unimaginable only a few years before and consequently the genomes of almost any organism, prokaryote, archaea or eukaryote, could be manipulated to direct the synthesis of biological products that were normally only produced

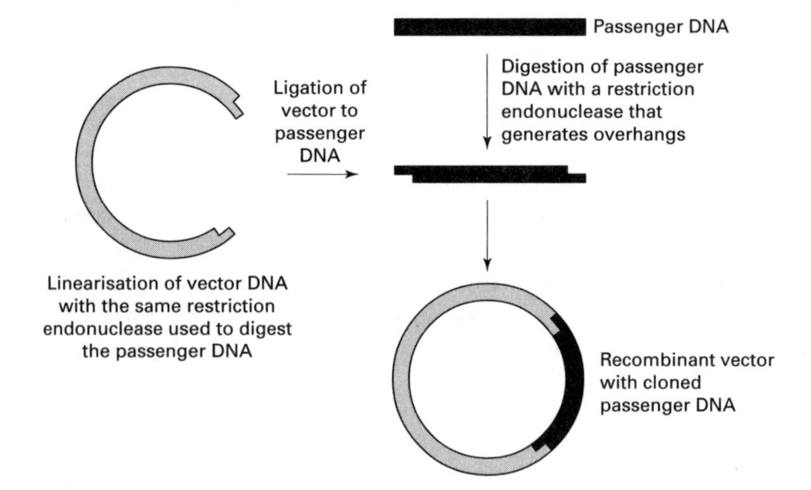

Fig. 4.4 Schematic diagram illustrating the basic concept of genetic engineering, using passenger DNA.

by other organisms. This technology, illustrated in Fig. 4.4, has not only facilitated the production of certain proteins at a quantity and quality that was not previously achievable, but also opened up the possibility of developing entirely new or highly modified bioactive products. The technology has been applied to a wide range of industries, but particularly the pharmaceutical industry where the main aims have been to produce compounds with proven or suspected therapeutic value and to produce totally new products not found in Nature.

4.4 The basic tools of genetic engineering

The techniques for isolating, cutting and joining molecules of DNA, developed in the early 1970s, have provided the foundations of our current technology for engineering and analysing nucleic acids. These allow fragments of DNA from virtually any organism to be cloned in a bacterium by inserting them into a vector (carrying) molecule that is stably maintained in the bacterial host.

4.4.1 Isolation and purification of nucleic acids

Biochemical techniques for preparing large quantities of relatively pure nucleic acids from microbial cells are an essential prerequisite for *in vitro* gene technology. The first step in the isolation of nucleic acids is the mechanical or enzymatic disruption of the cell to release the intracellular components that include the nucleic acids. Once released from the cell, the nucleic acids must be purified from other cellular components such as proteins and polysaccharides to provide a substrate of appropriate purity for nucleic acid modifying enzymes. The released nucleic acids are recovered using a combination of techniques including centrifugation, electrophoresis, adsorption to inert insoluble substrate or by precipitation with non-aqueous solvents.

4.4.2 Cutting DNA molecules

The ability to cut molecules of DNA, either randomly or at specific target sites, is a requirement for many recombinant DNA techniques. DNA may be cleaved using mechanical or enzymatic methods. Mechanical shearing results in the generation of random DNA fragments which are often used for the generation of genomic libraries (Section 4.5.5). When DNA molecules are mechanically sheared it is not possible to isolate a specific fragment containing, for example, a particular gene or operon. In contrast, when the DNA is cut using restriction endonucleases which recognise and cleave specific target base sequences in double-stranded (ds)DNA, specific fragments can be isolated. Restriction endonucleases cut the phosphodiester backbone of both strands of the DNA to generate 3′-OH and 5′-PO_4 termini. Several hundred restriction endonucleases have been isolated from a wide variety of microbial species, and their numbers continue to grow. Different classes of restriction endonucleases with distinct biochemical properties have been identified, with the type II enzymes being the main class used for genetic engineering purposes.

Restriction endonucleases are named according to the species from which they were originally isolated; enzymes isolated from *Haemophilus influenzae* are designated *Hin*, those from *Bacillus amyloliquefaciens*, *Bam etc*. When more than one type of enzyme is isolated from a particular strain or species, the strain and isolation number (in roman numerals) are added to the name. Thus the three restriction endonucleases isolated from *H. influenzae* strain Rd are designated *Hind*I, *Hind*II and *Hind*III.

The target recognition sequences (restriction sites) of type II restriction enzymes are usually short, typically four to six bases in length. The length of the restriction sites and their nucleotide composition (proportion of GCs to ATs) in relation to the rest of the target DNA, is important in determining the frequency at which the DNA is cut. For example, in DNA in which all four nucleotide bases occur randomly and with equal frequency, any given four base pair sequence will occur on average every 256 base pairs (4^4), while a six base pair recognition sequence will occur only once in every 4096 base pairs (4^6).

In most cases restriction sites are palindromic, i.e. reading the same on both strands with symmetry about a central point (Fig. 4.5). Cleavage usually occurs within the recognition sequence to generate either blunt ends or staggered ends with single-stranded overlaps. A list of commonly used restriction enzymes, their recognition sequences and cutting action is shown in Table 4.3.

4.4.3 Joining DNA fragments

DNA molecules with either blunt ends or with compatible (cohesive) overlapping ends may be joined together *in vitro* using specific 'joining enzymes' called **DNA ligases**. These enzymes catalyse the formation of phosphodiester bonds between 3′OH groups at the terminus of one strand, with the 5′PO_4 terminus of another strand. The DNA ligase encoded by bacteriophage T4 is universally used for joining DNA

Fig. 4.5 Restriction endonuclease cleavage of molecules of DNA at specific target sites to generate 3′ or 5′ overhangs (overlaps), or blunt ends.

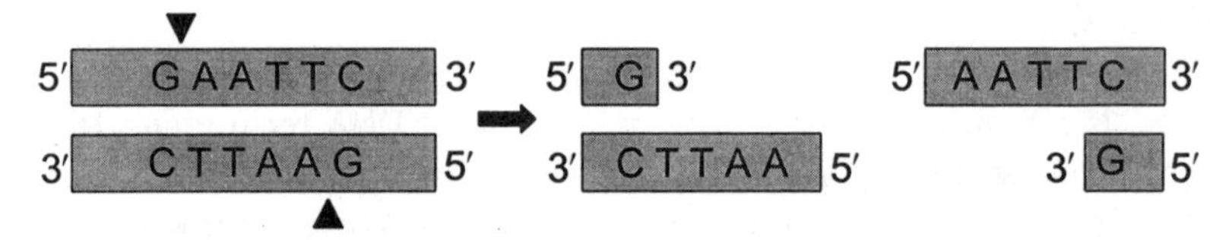

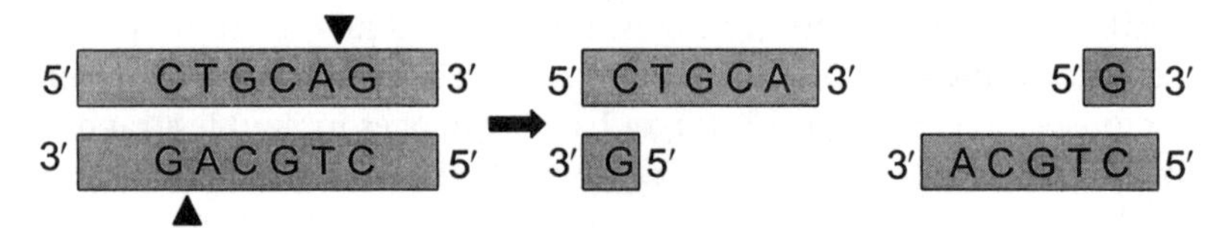

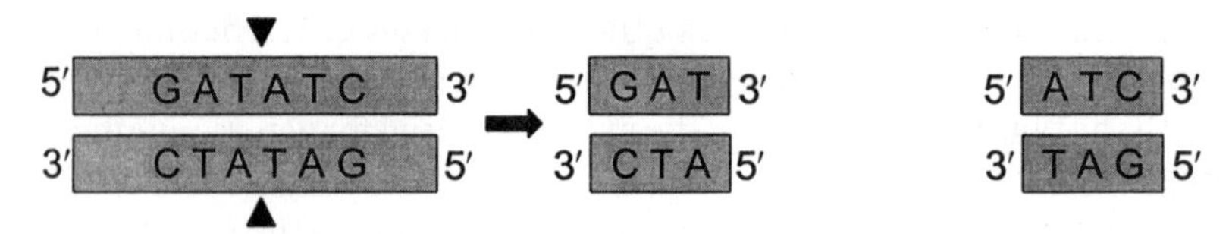

Table 4.3 Some common restriction endonucleases, their sources and recognition sites

Enzyme	Source	Recognition site
*Bam*HI	*Bacillus amyloliquefaciens* H	G↓GATCC
*Eco*RI	*Escherichia coli* RY13	G↓AATTC
*Eco*RII	*Escherichia coli* R245	↓CC[T/A]GG
*Hae*III	*Haemophilus aegyptius*	GG↓CC
*Hind*III	*Haemophilus influenzae* Rd	A↓AGCTT
*Kpn*I	*Klebsiella pneumoniae*	GGTAC↓C
*Not*I	*Nocardia otitidis-caviarum*	GC↓GGCCGC
*Pst*I	*Providencia stuartii*	CTGCA↓G
*Sau*3A	*Staphylococcus aureus* 3A	↓GATC
*Sma*I	*Serratia marcescens*	CCC↓GGG

Notes:
Bases written in parentheses indicate alternative permissible bases in the recognition sequence. Sequences are written from 5′ to 3′ on one strand only with the point of cleavage indicated by an arrow.

molecules with both blunt-ended or cohesive ends. T4 DNA ligase activity requires ATP as a cofactor to form an enzyme-AMP intermediary complex. It then binds to the exposed 3′OH and 5′PO_4 ends of the interacting DNA molecules to create the covalent phosphodiester bond (Fig. 4.6).

Ligation reactions usually involve joining a fragment of **passenger DNA** (that is the piece of new DNA to be 'carried') to a vector molecule

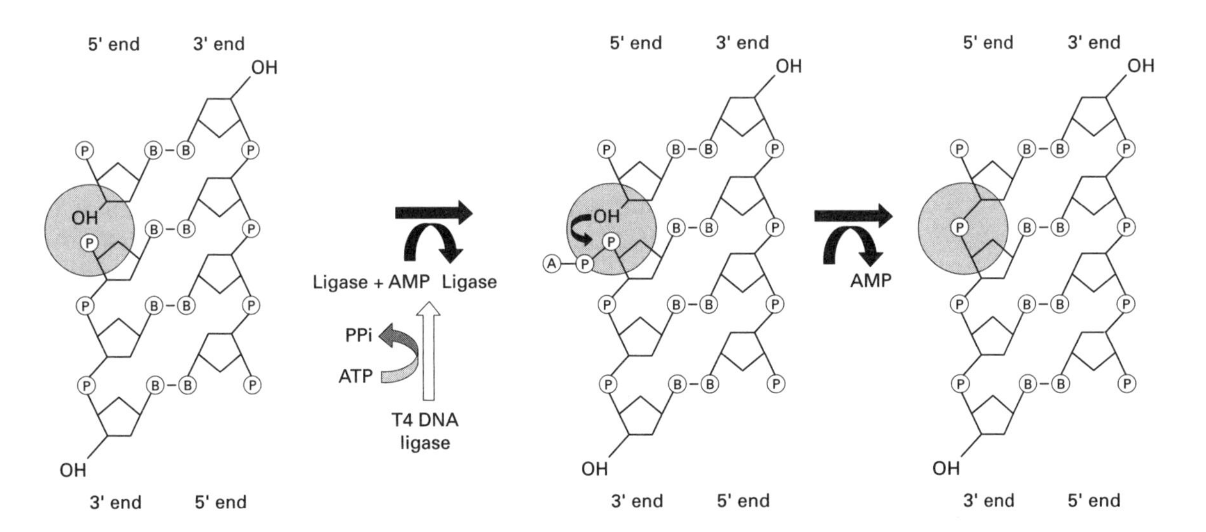

Fig. 4.6 Catalytic activity of the DNA ligase from bacteriophage T4. An enzyme-AMP complex forms that binds to breaks in the phosphodiester backbone of the DNA and makes a covalent bond between the exposed 3′OH and 5′PO_4 groups on each side of the break.

(Fig. 4.4). To increase the probability of the vector attaching to passenger DNA rather than to itself or another vector molecule, the molar ratio (i.e. number rather than mass of DNA) of passenger to vector DNA is usually about 10. An alternative strategy is to use a phosphatase (e.g. calf intestinal phosphatase or CIP) to remove phosphate groups from the 5′PO_4 ends of the linearised vector DNA. Since this phosphate group is essential for joining the two ends of the vector together, recircularisation is impossible. However, when passenger DNA is present, it can provide the 5′PO_4 ends for ligation to the 3′OH ends of the vector. This generates a circular molecule with single gaps in each of its nucleotide strands that are separated by the length of the passenger DNA. This structure is stable enough to be transformed into a cloning host where repair systems will seal the gaps.

4.4.4 The polymerase chain reaction (PCR) and its uses

Since its introduction in the mid 1980s, the polymerase chain reaction (PCR) has had a major impact on recombinant DNA technology. PCR facilitates the amplification of virtually any fragment of DNA from about 0.2 to 40 kbp in size. Because the amplification reaction is cyclical and the concentration of DNA doubles at each cycle, the total amount of DNA in the reaction increases exponentially; the theoretical yield from each original template molecule is about 10^6 molecules after 20 cycles, and about 10^9 molecules after 30 cycles. PCR requires a thermostable DNA polymerase, template DNA, a pair of oligonucleotide primers and a complete set of deoxynucleotide triphosphates (i.e. dATP, dCTP, dGTP and dTTP) substrates.

Oligonucleotide primers for PCR are synthesised chemically to be complementary to sequences which flank the region to be amplified and are usually about 20 nucleotides in length. The primers are designed to bind (anneal) specifically to the opposite strands of the template molecule, in such a way that their 3′ ends face the region to be amplified. It is the specificity of the primer annealing reaction which ensures that the PCR amplifies the appropriate region of the template

DNA. A key feature of the PCR is that the entire DNA amplification reaction is carried out in a single tube containing enzyme, template, primers and substrates. Each cycle of amplification therefore involves annealing, extension and denaturation reactions, each brought about at different temperatures (Fig. 4.7). Since the dissociation reaction may occur at temperatures as high as 95 °C, and there may be as many as 35 cycles in a single PCR, a highly thermostable DNA polymerase is a basic requirement for PCR. *Taq* polymerase, isolated from the hot-spring archaea *Thermus aquaticus*, was the first thermostable DNA polymerase to be employed in PCR.

The first step in the polymerase chain reaction (Fig. 4.7) is the denaturation of the double-stranded DNA template by heating to about 95 °C. The reaction mixture is then cooled to allow the oligonucleotide primers to anneal to the resulting single-stranded templates. The temperature at which annealing occurs is dependent on the length and G + C content of the primer sequences, but is usually designed to be in the range 50–65 °C. After the annealing step the temperature is raised to about 70 °C, the optimum temperature for the synthesis of the complementary strand by the thermostable DNA polymerase. The cycle of denaturation, annealing and synthesis is repeated 20–35 times in a typical PCR.

The maximum size of fragment that can be amplified with *Taq* polymerase using standard reaction conditions is about 4 kbp. However, recent advances in PCR technology mean that it is now possible to amplify DNA fragments of up to 40 kbp by optimising the concentrations of components in the reaction and using a mixture of thermostable polymerases.

PCR is not only useful for the *in vitro* amplification of DNA but has been developed for a whole host of other applications, including DNA sequencing, the introduction of specific nucleotide changes (site-directed mutagenesis), DNA labelling and in the fusion of DNA fragments to generate chimeric genes. Additionally, by incorporating target sites for restriction endonucleases into the 5′-ends of the oligonucleotide primers, the amplified PCR products can be digested and subsequently ligated into a specific site on the cloning vector (Section 4.5).

4.4.5 Transformation and other gene transfer methods

The ability to introduce foreign DNA into a bacterial cell host lies at the very heart of recombinant DNA technology. Transformation, in which the exogenous DNA is taken up by the host cell, is the most widely applied gene transfer technique for cloning purposes. While some bacteria possess natural transformation systems others, such as *E. coli*, require chemical pre-treatment to make them competent for the uptake of DNA.

Although transformation is usually efficient enough for most cloning purposes, there are some procedures, such as the generation of genomic libraries, for which transformation is not efficient enough. In these cases it is possible to circumvent the transformation procedure by packaging the recombinant DNA into virus particles *in vitro* (Section

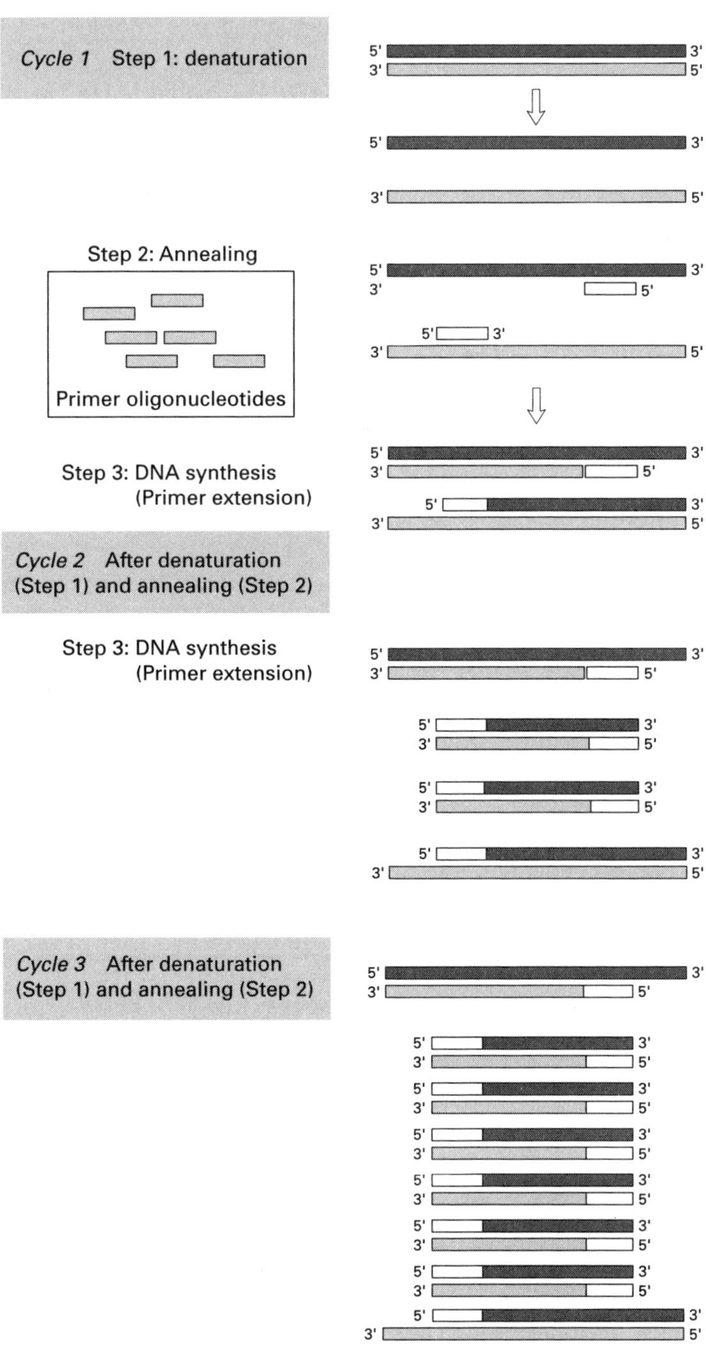

Fig. 4.7 The polymerase chain reaction, showing the cyclical nature of the annealing, synthesis and denaturation reactions which are carried out automatically in a dedicated thermocycling (PCR) machine.

4.5.2). More recently it has been discovered that bacteria are able to take up DNA when given a high voltage pulse. In this process, called **electroporation**, mixtures of cells and exogenous DNA are subjected to a brief (typically of millisecond duration) electric pulse of up to 2500 volts. The high field strength induces pores to form in the cell membrane, permitting the entry of the negatively charged DNA that is itself mobilised by the electrical gradient. In many cases electroporation is more efficient than transformation and some types of bacteria may only be transformed by this procedure.

4.4.6 Selection and screening of recombinants

After most cloning procedures it is necessary to screen the resulting clones to isolate those carrying the required gene or fragment of DNA. At the simplest level this may be done by selecting bacterial transformants that contain a copy of the vector. This is achieved by incorporating an antibiotic-resistance marker gene into the vector so that only transformed bacteria which have received a copy of the vector are able to grow on media containing the appropriate antibiotic. More advanced systems have been developed to allow the discrimination of transformants containing a vector with or without a cloned insert. These systems include the use of gene disruption methods which result in the loss of a particular trait upon insertion of foreign DNA (Section 4.5.1).

Clones containing a specific gene or fragment can be identified directly by selection techniques or indirectly by restriction endonuclease mapping, PCR or hybridisation techniques. If the target gene is expressed, its presence may be selected by complementation of a defect in the cloning host (e.g. restoration of the ability to utilise a particular substrate or to grow in the absence of an otherwise essential nutrient). In the case of restriction mapping, plasmid DNA extracted from a number of representative clones is digested with specific restriction endonucleases. Only clones containing the required gene or DNA fragment will generate the correct pattern of bands after agarose gel electrophoresis. Restriction mapping is only feasible if the target clones are likely to occur at a high frequency amongst the population to be screened. Diagnostic PCR, using oligonucleotide primers specific to the target DNA sequence, may also be used to identify clones containing the required gene or DNA fragment. Since PCR may be used directly on unprocessed samples of colonies, it is feasible to test many more clones.

If the target DNA is likely to occur at a low frequency in a population of clones, as would be the case with a genomic library (Section 4.5), a large number of clones need to be screened. In this case the method of choice is hybridisation of the bacteria colony that grows from a single cell (or in the case of phage vectors, the viral plaque). Colony or plaque hybridisation makes use of labelled nucleic acid probes (DNA or RNA) that are able to detect the presence of specific DNA sequences within individual colonies or plaques. Biomass from individual transformant colonies or plaques is transferred to a membrane onto which denatured DNA, released by breaking the cells open, will bind. The membrane is then exposed to a labelled probe (Section 4.4.7) which binds specifically to the immobilised target DNA, revealing the identity of colonies or plaques containing the appropriate cloned DNA.

4.4.7 Nucleic acid probes and hybridisation

Nucleic acid probes are used to detect specific target DNA molecules. The soluble probe binds (i.e. hybridises) to the target DNA that is immobilised onto a nylon or nitrocellulose membrane. Hybridisation is used for a variety of biotechnological applications including the detection of cloned DNA (Section 4.4.6), analysis of genetic organisation and the diagnosis of genetic diseases. Although nucleic acid hybridisation techniques are used in a wide variety of contexts, the same basic principles apply. Nucleic acid hybridisation exploits the ability of single-stranded probe nucleic acid (DNA or RNA) to anneal to complementary single-stranded target sequences (DNA or RNA) within a population of non-complementary nucleic acid molecules.

The original technique, referred to as **Southern blotting** after its inventor, Ed Southern, involved the size separation of restriction endonuclease digested fragments of DNA by gel electrophoresis, and their transfer by blotting onto nitrocellulose membranes. The probe nucleic acid is applied as an aqueous solution and, under appropriate hybridisation conditions, binds to immobilised target DNA. The location of bound probe nucleic acid on the membrane is indicated by the presence of a readily and sensitively detected label. When the technique was extended to detect target RNA it was referred to as **Northern blotting** (Section 4.7.1).

Nucleic acid probes used in hybridisation reactions must be in the form of single-stranded (ss) RNA or DNA molecules; when double-stranded DNA is used it must be denatured prior to hybridisation. Since the function of the probe is to detect the specific target sequences it means that the probe itself must be easily detected. Traditionally this is achieved by the incorporation of a radionuclide such as ^{32}P or ^{35}S and detection by exposure to photographic (e.g. X-ray) film.

In recent years, concerns over safety and pollution have led to the development of methods for labelling nucleic acids that avoid the need to use radionuclides, and nucleotide analogues containing biotin or digoxigenin are incorporated in their place. Ligand molecules with a high affinity for the incorporated nucleotide analogue (e.g. streptavidin for biotin), and which are cross-linked to enzymes such as peroxidases or alkaline phosphatases, are used to detect the probe after hybridisation to its target. Detection is based on the enzymatic cleavage of either a colourless chromogenic substrate, with release of a coloured product, or a chemiluminescent substrate, with the production of light. The latter is detected using photographic film in a similar manner to radionuclides.

RNA probes are preferred for hybridisation reactions in which the target molecule is RNA (e.g. Northern blotting) and these are synthesised *in vitro* using a phage RNA polymerase. A DNA fragment encoding all or part of the target sequence is cloned behind the appropriate phage promoter. The phage RNA polymerase is used to synthesise an RNA transcript that is complementary to the target nucleic acid. As with the labelling of DNA, radiolabelled nucleotides or nucleotide analogues are incorporated during the synthesis reaction.

4.4.8 DNA sequencing

DNA sequencing is one of the most important of the techniques available for the identification, analysis and directed manipulation of DNA. Knowledge of the DNA sequences of target DNA molecules and cloning vectors is fundamental to the construction of advanced bacterial protein production systems. It also facilitates the design of specific probes or primers and the production of computer generated transcription and restriction maps.

The Maxam and Gilbert method for DNA sequencing uses chemical reagents to bring about the base-specific cleavage of the DNA. Although still used for a limited number of applications, this chemically based technique has generally been replaced by the elegant chain-terminator method developed by Sanger and colleagues in 1977. The Sanger procedure exploits the ability of a variant of *E. coli* DNA polymerase I (the so-called **Klenow fragment**) to synthesise a complementary strand of DNA from a single-stranded DNA template, incorporating both the natural deoxynucleotides and 2′,3′-dideoxynucleotide analogues. Dideoxynucleotides lack a hydroxyl group at the 3′ position and are therefore not able to act as a substrate for further chain elongation. Their incorporation therefore terminates the synthesis of the DNA strand in question. A specific oligonucleotide primer, used to initiate the chain elongation process, determines the start point for all of the newly synthesised DNA molecules. DNA polymerase synthesises DNA from a single-stranded template in the presence of all four deoxynucleotide triphosphate substrates (dATP, dCTP, dGTP, dTTP), one of which is radiolabelled (e.g. [α-^{35}S]-dATP). Identical reactions are carried out in four tubes excepting that each tube also includes, at a lower concentration, one of the four possible dideoxynucleotide triphosphate analogues (ddATP, ddCTP, ddGTP, ddTTP). When the appropriate relative concentrations of normal and dideoxy nucleotides are used, newly synthesised DNA strands will be terminated at every possible base position and a set of fragments of all possible lengths will be generated. The products of the four reaction mixtures are separated by denaturing polyacrylamide gel electrophoresis and subjected to autoradiography. The resulting banding pattern allows the DNA sequence to be read directly (Fig. 4.8).

Recent developments in DNA sequencing technology have led to the automation of the process. This was achieved by changing the format of the chain-termination technique to allow real-time detection of the DNA bands within a polyacrylamide gel. The replacement of radiolabelled nucleotide substrates with fluorescent dye-labelled substrates allows for the use of a laser-induced fluorescence detection system. Up to 900 bases may be read from a single reaction using this technology and the data are captured directly in electronic format.

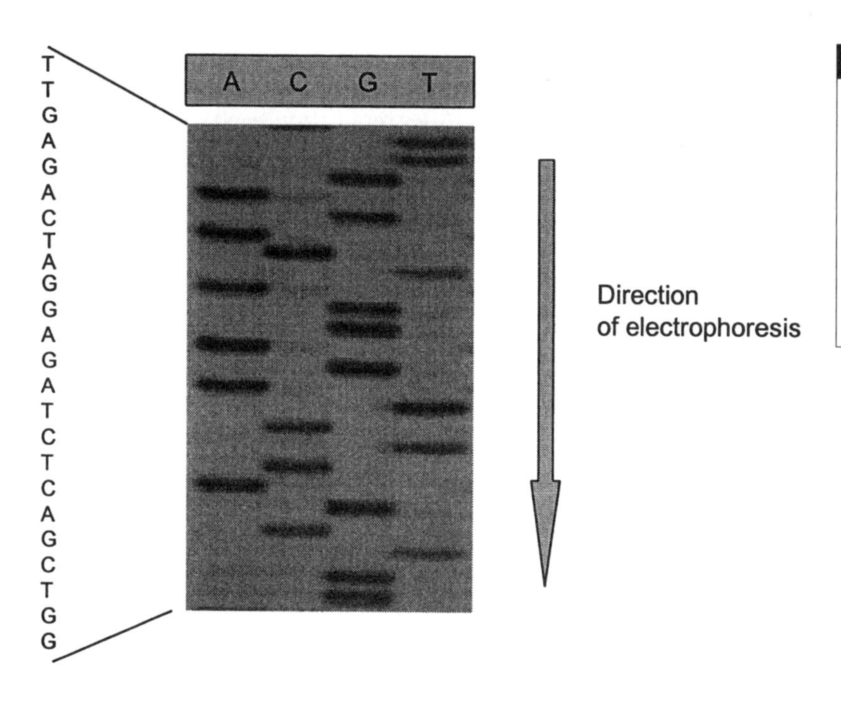

Fig. 4.8 An autoradiograph of part of a DNA sequencing gel generated by the Sanger chain termination method. The lanes are labelled according to the nucleotides at which they are terminated, namely: A, adenine; C, cytosine; G, guanine; T, thymine. The sequence specified by the gel is shown to the left of the gel.

4.4.9 Site-directed mutagenesis

Site-directed mutagenesis, the specific replacement of nucleotides in a sequence of DNA, is used to analyse or modify the activity of genes or gene products. For example, specific amino acid replacements have been use to improve the characteristics of many industrial enzymes. The mutational changes are introduced in the DNA using *in vitro* techniques which is then introduced back into the bacterium where the resulting phenotypes are observed. Oligonucleotide site-directed mutagenesis is an important tool for achieving this aim, since it permits the introduction of highly specific changes to the DNA sequence. Although many methodologies have been developed for site-directed mutagenesis, the principles are generally similar to the example shown in Fig. 4.9. The target DNA is cloned into a plasmid vector which is able to be replicated to form single-stranded DNA (Section 4.5.4). In addition to the target DNA, the vector also contains two antibiotic resistance markers. One of the resistance genes, here the ampicillin resistance gene, has been inactivated by the inclusion of a single base substitution. The single-stranded form of the plasmid is annealed with two oligonucleotide primers; one complementary to the target gene excepting for the inclusion of the desired base substitution(s), the other complementary to the mutated region of the ampicillin resistance gene but incorporating a base substitution which reverses the ampicillin sensitive phenotype by restoring the wild-type gene sequence. The addition of DNA polymerase and DNA ligase leads to the synthesis of a complementary strand of DNA. The resulting double-stranded plasmid contains two deliberate mismatches, one which corrects the mutation in the ampicillin resistance gene and the other generating the specific change(s) to the target gene. It is transformed into *E. coli* where repair of the mismatches by the host's mismatch repair systems is avoided by use of a mutant (e.g. *mutS*) that is defective in this function. After replication, each strand will form a double-stranded molecule without mismatches and these will segregate into separate daughter cells. One molecule will contain the mutations introduced by the oligonucleotides, whilst the other will be identical to the original plasmid. Cells harbouring plasmids with the desired mutation are selected by virtue of their newly acquired ampicillin resistance phenotype.

4.5 Cloning vectors and libraries

A cloning vector is a molecule of DNA into which passenger DNA can be cloned to allow it to be replicated inside a bacterial host cell. The vector and passenger DNA are covalently joined by ligation (Section 4.4.3). Cloning vectors have four basic characteristics: (i) they must be easily introduced into the host bacterium by transformation or, after *in vitro* packaging, by phage infection; (ii) they must be able to replicate in the host bacterium, preferably so that the number of copies of the vector (copy number) exceeds that of the host chromosome by between 50 and 200; (iii) they should contain unique sites for the action of a variety of restriction endonucleases; and (iv) they should encode a means for selecting or screening host cells that contain a copy of the vector. Cloning vectors are derived from naturally occurring DNA molecules, such as plasmids and phages, which are capable of replicating independently of the host chromosome. A wide variety of cloning vectors has been developed for specific applications and these are briefly described below.

4.5.1 General purpose plasmid vectors

General purpose plasmid vectors are designed for cloning relatively small (<10 kbp) fragments of DNA, usually into *E. coli*. The vectors are usually introduced into their bacterial host by transformation, and transformants that have received a copy of the vector are selected using a vector-based antibiotic resistance gene. Modern general purpose vectors have purpose-designed fragments (**multiple cloning site** or MCS) incorporating a variety of unique restriction sites.

General purpose vectors usually include a system for detecting the presence of a cloned DNA fragment, based on the loss of an easily scored phenotype. The most widely used system involves the *lacZ′* gene encoding the α-peptide from the N-terminus of *E. coli* β-galactosidase. The synthesis of this peptide, from a gene located on the vector, complements an otherwise inactive version of this enzyme, encoded by the host's chromosome. The result is an active β-galactosidase enzyme that can be detected by its ability to liberate a blue chromophore from the colourless chromogenic substrate 5-bromo-4-chloro-3-indolyl-β-D-galactoside, referred to as X-gal. Cloning a fragment of DNA within the vector-based gene encoding the α-peptide prevents the formation of an active

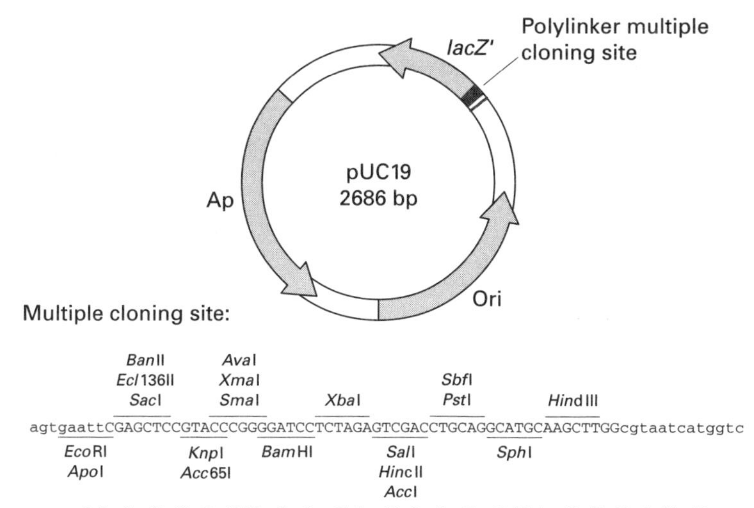

Fig. 4.9 The general purpose plasmid cloning vector pUC19 is a member of the pUC series of plasmid vectors. The pUC vectors have been generated in pairs that differ only with respect to their multiple cloning site which are located in opposite orientation, allowing the passenger DNA to be located in either the same direction or opposed to the transcription of the *lacZ'* gene. The arrows on the *lacZ'*, ampicillin resistance (Ap) and replication protein (Ori) genes indicate the direction of transcription. The sequence of the multiple cloning site (capitals) and adjacent (lower case) sequences are shown together with restriction endonuclease target sites and translated amino acids.

β-galactosidase. If X-gal is included in the selective agar plates, transformant colonies are blue in the case of a vector with no inserted DNA and white in the case of a vector containing a fragment of cloned DNA. The same detection system is used in a variety of modern vectors, including phage vectors that consequently generate blue or white plaques. The most widely used general purpose plasmid vectors are those of the pUC series (Fig. 4.9).

4.5.2 Bacteriophage and cosmid vectors

Bacteriophage lambda (λ), which infects *E. coli*, has provided the basis for the most commonly used phage vectors. Lambda has been of greatest value for cloning relatively large fragments (> 10 kbp) that are not easily cloned by general-purpose plasmid vectors. Lambda has a linear dsDNA genome of approximately 48.5 kbp in size with short 12-base pair single-stranded 5′ projections at each end to facilitate its circularisation in *E. coli*. The site generated by the circularisation reaction is known as the *cos* site (*co*hesive ends). In the development of lambda cloning vectors, non-essential genes have been removed to provide space for the insertion of DNA fragments of up to 23 kbp in size. Additionally, because the phage genome is only circularising upon infection of the host, lambda vectors can be supplied as separate 'left' and 'right' arms. Each arm has an appropriate restriction endonuclease site at one of its ends and the recombinant DNA is cloned between the arms (Fig. 4.10).

The transformation of lambda molecules into *E. coli* is relatively inefficient and this has led to the development of *in vitro* phage packaging systems for the efficient delivery of recombinant lambda genomes into their bacterial hosts.

Cosmid vectors combine the advantages of cloning in a plasmid vector (e.g. ease of cloning and propagation) with the high efficiency of delivery and cloning capacity of a phage vector. Cosmids are plasmids which have a copy of the *cos* site normally found on the lambda genome.

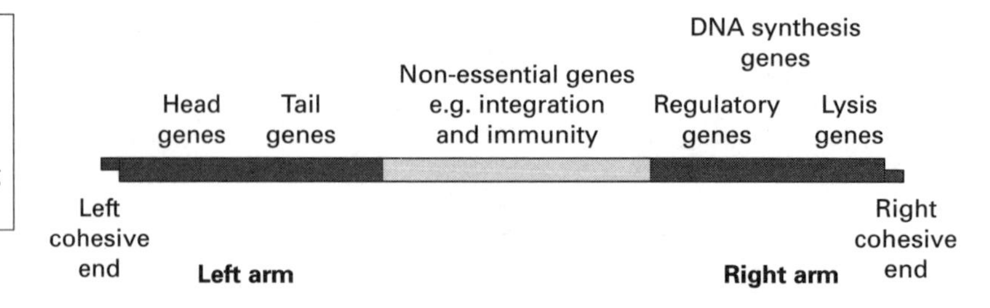

Fig. 4.10 Simplified map of bacteriophage lambda showing the left and right arms and the non-essential central region that is omitted from lambda-based cloning vectors.

The presence of the *cos* site enables these plasmids to be used in conjunction with a lambda *in vitro* packaging system. This means that molecules of cosmid DNA with large inserts of passenger DNA can be introduced efficiently into an appropriate *E. coli* host. Since the cosmid vector itself is typically only about 5 kbp in size, inserts of about 32–47 kbp can be accommodated in the phage. The packaged cosmid DNA is injected into the bacterial cell as if it was lambda DNA where it circularises and starts to replicate using its plasmid replication functions.

4.5.3 Bacterial artificial chromosomes

Bacterial artificial chromosomes (pBACs where 'p' refers to plasmid) have been developed for cloning very large (> 50 kbp) sequences of DNA. BAC vectors are usually based on the F plasmid of *E. coli* and are able to accept DNA inserts as large as 300 kbp. pBACs are maintained as single copy plasmids in *E. coli*, excluding the replication of more than one pBAC in the same host cell. Ordered pBAC libraries of bacterial genomes may be constructed in which the entire genome sequence is represented by a series of clones with overlapping inserts.

4.5.4 Special purpose vectors

In addition to the vectors described above, a range of special purpose vectors have been developed and are described briefly below.

Expression vectors

Expression vectors are designed to achieve high level, controlled expression of a target gene with resulting production of a protein product at concentrations as high as 40% total cellular protein. Expression vectors often incorporate a system that adds an affinity tag to the protein to facilitate its purification by affinity chromatography. Expression vectors are mostly plasmid-based and often use the tightly controlled and highly efficient phage T7 RNA polymerase gene expression system. The target gene is cloned downstream of the transcriptional (promoter) and translational (ribosome binding site) control signals derived from T7. The vector is transformed into an *E. coli* host with a chromosomally located gene encoding the T7 RNA polymerase. Switching the polymerase gene on leads to high-level synthesis of the target gene.

If the target gene is not fused to an affinity tag, the protein must be purified using expensive methodologies. However, a number of affinity tag systems have been developed in recent years that provide for highly specific purification protocols. A variety of tags have been used, including a tag of six histidine residues (6 × His) that binds to nickel. The

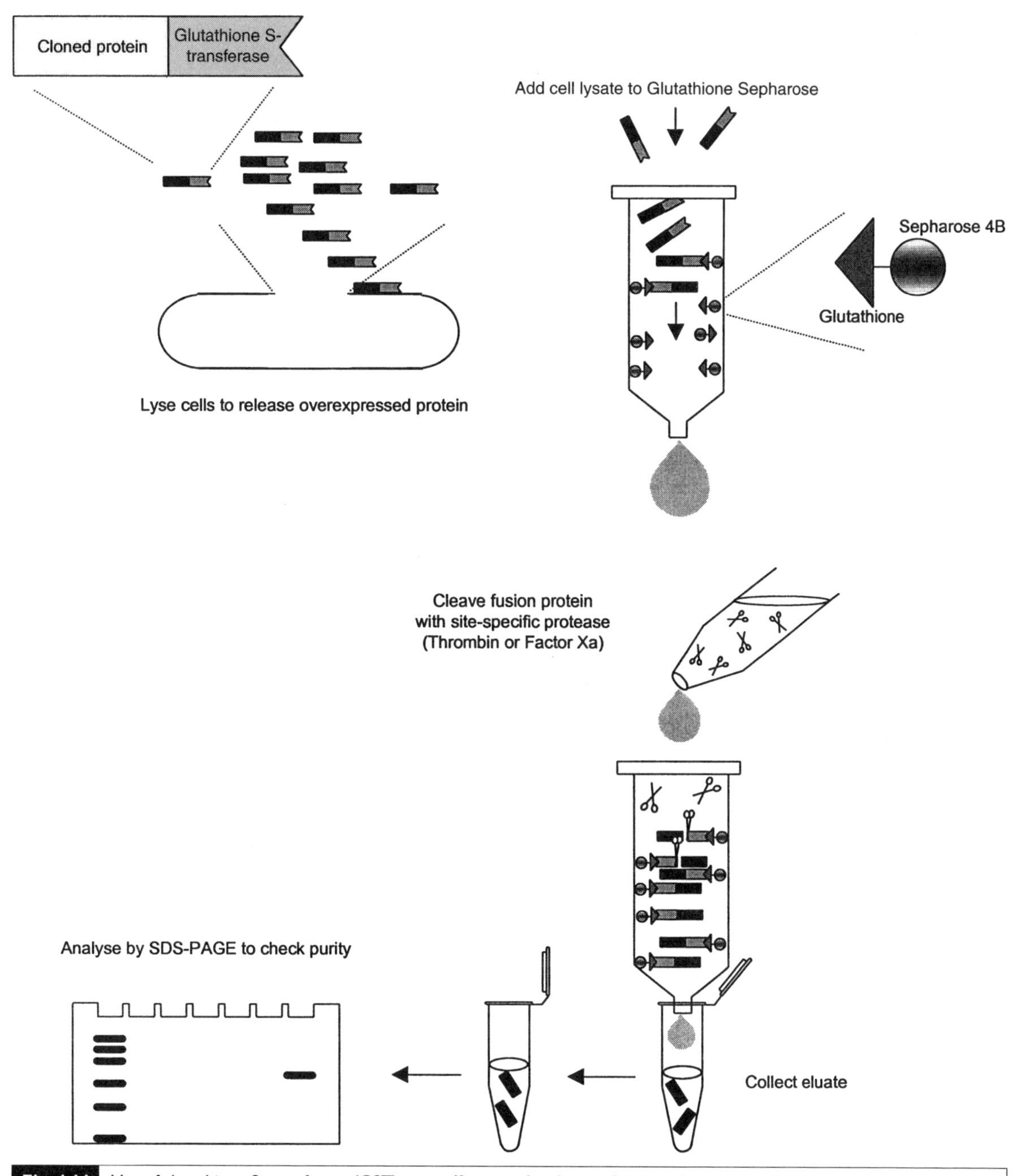

Fig. 4.11 Use of glutathione S-transferase (GST) as an affinity tag for the purification of proteins. The target protein is synthesised as an N-terminal fusion to GST. The producer cells are lysed and the target protein/GST fusion trapped on a Glutathione Sepharose column. After extensive washing, the target protein is released from the column by adding a protease that cleaves the target protein from the GST moiety.

ligand is attached to an insoluble resin and the tagged fusion protein recovered by passing through a chromatographic column containing the resin-bound ligand (Fig. 4.11). The bound fusion protein is eluted as a virtually pure protein and the affinity tag removed by treatment with a chemical or by digestion with a protease.

Fig. 4.12 Properties of a secretion vector. (a) Structure of a typical *E. coli* signal (or leader) peptide required to target a protein across the cytoplasmic membrane. (b) Organisation of a secretion vector with cloning site immediately downstream of the signal sequence. (c) Secretion vector with the DNA sequence of the target protein fused in-frame downstream of the signal sequence.

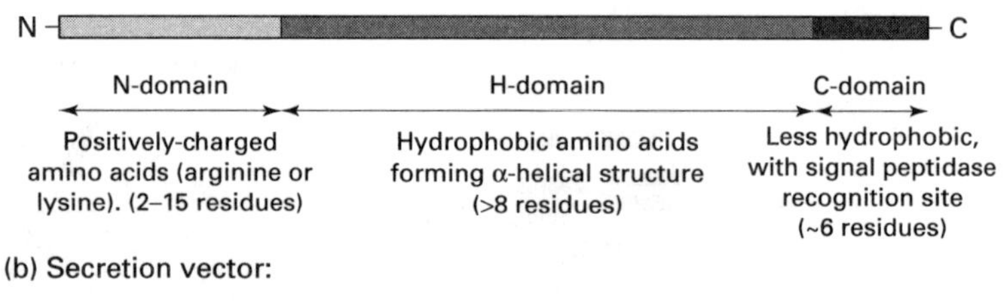

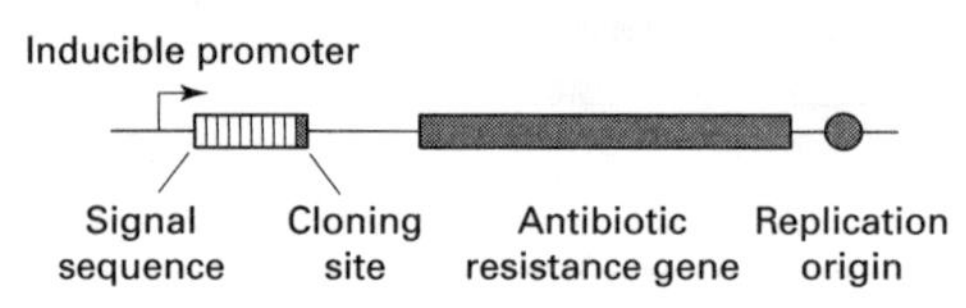

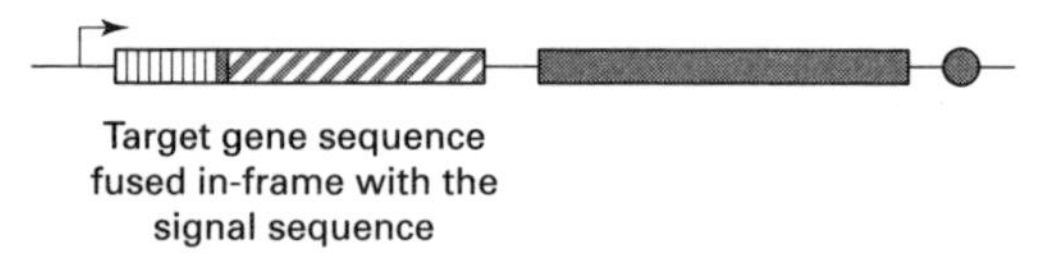

Secretion vectors

Currently most systems for the production of recombinant proteins lead to the intracellular accumulation of the product. However, intracellular accumulation can lead to lower production levels, protein aggregation, proteolysis and permanent loss of biological activity (Section 4.9.4). This can sometimes be overcome by secreting the target protein directly into the culture medium since secreted proteins can potentially be accumulated to higher concentrations and are usually correctly folded. For a protein to be secreted into the culture medium it needs to be directed to the secretion apparatus located in the cytoplasmic membrane. This requires the use of a secretion vector (Fig. 4.12) in which the target protein is synthesised as a fusion protein with an *N*-terminal signal peptide. The signal peptide directs the protein to a secretory translocase located in the membrane. Once translocated, the signal peptide is removed by cleavage with a signal peptidase located on the outer surface of the cell membrane.

Shuttle or bifunctional vectors

Rapid advances in microbial genetics and the development of cloning vectors mean that is now possible to manipulate the genomes of a broad range of micro-organisms. However, in many cases transformation efficiency remains stubbornly low and it is often expedient to use *E. coli* as an intermediate cloning host. This can be achieved by using a shuttle or bifunctional vector that has replication origins that are functional in *E. coli* and the target micro-organism. More recently it has been found that the replication functions of certain plasmids naturally have a broad host range, and these have been used for a new generation of shuttle plasmids.

Single-stranded phage and phagemid vectors

It is sometimes necessary to generate single-stranded DNA, particularly for DNA sequencing and oligonucleotide-directed mutagenesis. Messing developed a series of vectors based on bacteriophage M13, a filamentous ssDNA phage of *E. coli*. The *mp* series of M13 vectors incorporates a multiple cloning site within a gene encoding the α-peptide fragment of β-galactosidase. They are therefore amenable to blue/white plaque selection (Section 4.5.1) for the detection of cloned inserts. M13-based vectors infect F^+ strains of *E. coli* via the F pilus. The double-stranded replicative form of the phage genome is used for cloning purposes.

Phagemids are plasmids that contain the origin of replication for a ssDNA phage (Fig. 4.13), usually that of bacteriophage f1 which is closely related to M13. *E. coli* is able to maintain a phagemid as dsDNA by virtue of a plasmid origin of replication. However, if the cell is infected with a so-called helper f1 phage, the f1 replication origin is activated and the vector switches to a mode of replication that generates ssDNA that is packaged into phage particles as they are extruded from the host cell.

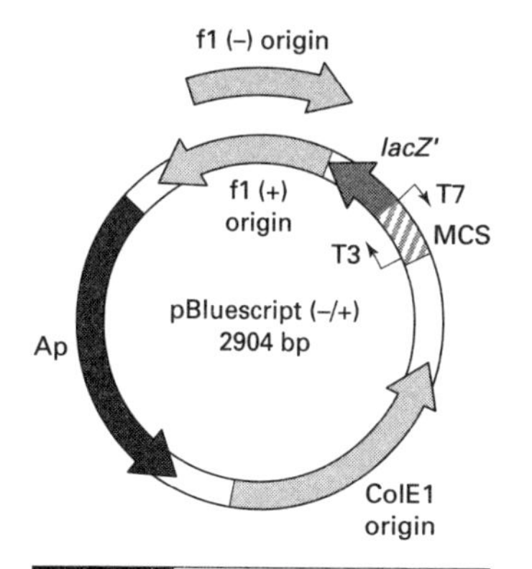

Fig. 4.13 The phagemid pBluescript is a plasmid vector used for the generation of single-stranded DNA molecules. Cells containing the phagemid are infected with an fl 'helper' phage that stimulates the fl replication origin to generate ssDNA that is assembled into phage particles and released from the bacterium. Two versions of pBluescript, the (+) or (−) derivatives, allow for replication of either the positive or negative strand as required. T3 and T7 promoters either side of the multiple cloning site allow the single-stranded DNA to be used as a substrate from *in vitro* RNA synthesis using the cognate RNA polymerase and NTP substrates. An ampicillin resistance gene (Ap) is used to select for and to maintain the phagemid in *E. coli*.

Integration vectors

Integration vectors are designed to integrate into the chromosome of a host bacterium. They are used in a variety of specific contexts including cloning genes at low copy numbers, the generation of insertion mutations, gene replacements and the generation of gene fusions.

4.5.5 Genomic and gene libraries

A genomic library is a collection of recombinant clones containing, at a theoretical level, representatives of all of the genes encoded by the genome of a particular organism. In practice, the best that can be achieved is a library with a high probability (usually $>95\%$) that a particular gene will be represented. Libraries are produced by 'shotgun cloning' randomly generated DNA fragments into a suitable cloning vector. These fragments may be generated by mechanical shearing or by enzymatic digestion. Libraries may also be generated using copy DNA which has been synthesised from the mRNA of a particular tissue or organism using the enzyme reverse transcriptase.

4.6 Analysis of genomes/proteomes

The genetic information of a bacterial cell is physically located on its chromosome(s). The physical organisation of bacterial chromosomes has recently been shown to be more variable than was previously supposed, and both linear and circular chromosomes have been recognised as a result of the application of physical mapping techniques such as pulsed-field gel electrophoresis (PFGE).

4.6.1 DNA fingerprinting/physical mapping/pulsed-field gel electrophoresis

Prior to whole genome sequencing, various physical mapping methods were developed to determine the physical structure of bacterial genomes and determine relatedness between individual bacterial strains. The latter is useful for strain identification and epidemiological studies.

Physical maps of the chromosome can be constructed by use of PFGE, a method developed specifically to resolve very large (30 to 2000 kbp) fragments of DNA. To avoid mechanical shearing, the DNA is extracted directly from bacterial cells in agarose blocks and, when required, digested *in situ* with restriction endonucleases that cleave the genome sequences infrequently (e.g. 10 to 30 times). The agarose block is then incorporated into a slab of agarose and subjected to an oscillating electric field. Separation is based on the time taken for the individual molecules of DNA to re-orientate in the modulating electric field; larger molecules taking longer than smaller molecules. The various fragments are aligned using a variety of strategies that include digesting the DNA with two rare-cutting enzymes, hybridisation between fragments isolated from separate digests and transformation of purified fragments into mutants with specific lesions.

Genome fingerprinting methods have been developed for examining the relationships between strains for epidemiological studies (e.g. monitoring outbreaks of disease) or heterogeneity in natural populations. The banding patterns generated by restriction endonucleases (e.g. *r*estriction *f*ragment *l*ength *p*olymorphism or RFLP) or PCR-based techniques using specific or random oligonucleotide primers (e.g. *r*andom *a*mplified *p*olymorphic *D*NA or RAPD) can either be used diagnostically or for revealing relationships between strains.

4.6.2 Analysis of the proteome

The term **proteome** has been devised to describe the *prote*ins specified by the gen*ome* of an organism. The characterisation of the proteome provides a link between an organism's genetics and physiology. It helps to validate the genome sequence, aids the identification of **regulons** (genes or operons controlled by the same regulatory proteins) and **stimulons** (genes or operons controlled by the same inducing signal) and allows the response of the organism to its environment to be evaluated. *E. coli*, with a genome size of 4.6 Mbp, encodes about 4400 proteins. A combination of experimental evidence and homology with proteins in other organisms has led to an identification of the functions of about 60% of these proteins. The biological role of the remaining proteins is unknown. A similar proportion of proteins of unknown function are observed even for bacteria such as *Mycoplasma genitalium* which have very much smaller genomes (0.58 Mbp).

The expression of individual proteins for which antisera are available can be detected by Western blotting. Released cellular proteins are separated by SDS-polyacrylamide gel electrophoresis, and are then transferred and bound to a nitrocellulose or nylon membrane. Specific proteins are revealed by reacting (probing) with an antiserum which is then detected using a secondary antibody or probe.

The current system of choice for analysing the proteome involves a combination of two-dimensional polyacrylamide gel electrophoresis (2-DPGE) and polypeptide micro-sequencing or mass spectrometry techniques. 2-DPGE facilitates the separation of hundreds of polypeptides extracted from whole cells. Polypeptides are initially separated in the first dimension on immobilised pH gradient gels on the basis of their pI. These gels are then separated in the second direction by SDS-polyacrylamide gel electrophoresis in which the rate of migration is primarily based on their size. The rates of migration of polypeptides in the two dimensions are extremely reproducible. Individual polypeptides can be detected by staining techniques or by radiography, using radiolabelled amino acids. Data from separate gels can be compared and, for example, the polypeptides that are induced in response to a particular growth phase, stress or change of substrate can be catalogued. Moreover, individual protein spots can be excised from the gel and identified either by micro-sequencing or by high resolution mass determination [e.g. matrix-assisted laser desorption/ionisation – time of flight (MALDI-TOF) or electro-spray mass spectrometry]. In each case, enough information can be obtained to identify the protein in question.

4.7 Analysis of gene expression

Promoters influence the frequency of transcription initiation rather than the rate of transcription, and strong promoters have a high affinity for RNA polymerase binding. A comparison of a number of *E. coli* promoters has led to the recognition of a consensus promoter sequence: 5′–TATAAT–3′, centred around 10 nucleotides upstream of (prior to) the transcription initiation site (−10 region) and 5′–TTGACA–3′, located about 35 nucleotides upstream (−35 region). The strongest promoters are those which show the closest identities to this sequence. Additionally, the spacing between the −10 and −35 regions is important, the optimal being 17 nucleotides. The ability to analyse gene expression is an important prerequisite for optimising the biotechnological potential of bacteria and many highly sensitive and precise techniques have been developed for this purpose.

4.7.1 Analysis of messenger (m)RNA transcripts

Three methods are used for the analysis of mRNA transcripts: Northern blotting, S1-nuclease mapping and primer extension analysis. The latter two techniques have the potential to identify the transcription initiation sites.

Northern blotting involves the separation of mRNA species by electrophoresis through agarose or polyacrylamide. Formamide, urea or other denaturants are included to avoid the single-stranded molecules forming secondary structures (e.g. duplexes, loops) that might affect their mobility. The separated mRNA species are transferred to activated

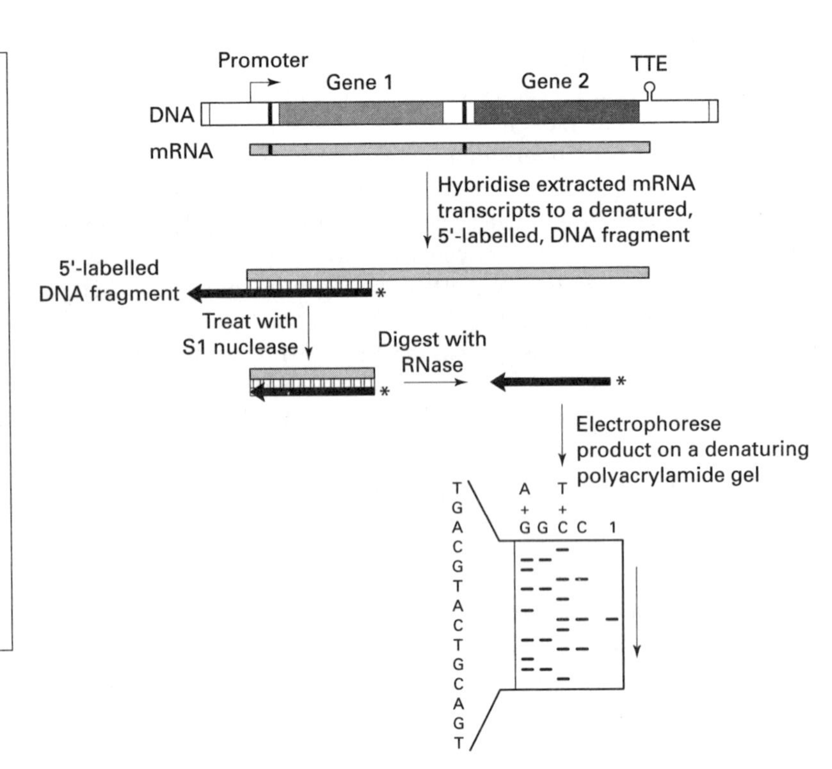

Fig. 4.14 Mapping of transcription initiation points using S1 nuclease. mRNA is hybridised to a denatured DNA fragment that has been labelled at its 5′-end. The DNA fragment is chosen so that its 5′-end is internal to the target mRNA, while the 3′-end extends beyond the putative mRNA start point. The RNA/DNA hybrid molecule has single-stranded extensions that are degraded by the single-strand-specific activity of S1 nuclease. The 3′-end of the DNA fragment is determined by running it on a denaturing gel against a DNA sequencing of the original fragment generated by the Maxam and Gilbert chemical cleavage method.

nylon membranes by blotting and then covalently cross-linked. Specific mRNA species are detected by hybridisation (Section 4.4.7), using labelled oligonucleotide, DNA or RNA probes. The use of markers with different molecular sizes allows the sizes of specific transcripts to be estimated which provides clues as to the organisation of the transcriptional unit from which the transcript was synthesised.

S1-nuclease and primer extension analyses facilitate the identification of the 5′-prime ends of mRNA transcripts or the processed products of primary transcripts. In the case of S1-mapping (Fig. 4.14), mRNA is hybridised to a specific species of ssDNA that overlaps the start of the target transcript. The resulting RNA/DNA hybrid molecule has an overlap of DNA at the 3′-end that is digested by the single-strand specific S1-nuclease. The size of the processed ssDNA molecule, which is labelled at its unmodified 5′-end, is determined by denaturing polyacrylamide gel electrophoresis using a DNA sequence ladder as molecular size marker.

In the case of primer extension analysis (Fig. 4.15), a 5′-end labelled oligonucleotide, hybridising about 60–100 nucleotides downstream of the predicted transcription initiation site, is used to prime the synthesis of a DNA copy of the mRNA transcript, using the enzyme reverse transcriptase. Synthesis of the complementary DNA strand terminates at the 5′-end of the transcript, to generate a product of defined length. Again this can be sized using a DNA sequence ladder, generated using the same primer oligonucleotide, as molecular size marker.

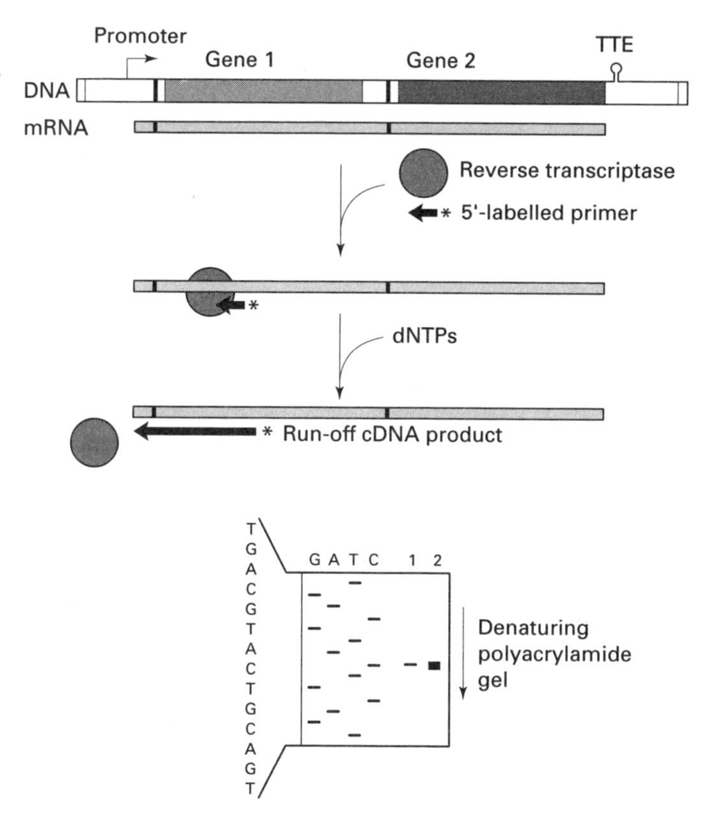

Fig. 4.15 Primer extension analysis of a specific mRNA transcript. An oligonucleotide primer, radiolabelled at its 5′-end, is annealed to extracted mRNA about 60–100 nucleotides downstream of the putative transcript start point. Reverse transcriptase (an RNA-dependent DNA polymerase) and deoxyribonucleotide triphosphate (dNTP) substrates are added, and copy (c)DNA synthesis initiated. cDNA synthesis terminates at the 5′ end of the mRNA transcript and the size of the run-off cDNA product is determined on a sequencing gel (lanes 1 and 2 on the gel) using a DNA sequence ladder generated with the same primer as a molecular size marker. This allows the precise nucleotide at which the transcript was initiated to be identified. Primer extension analysis is semi-quantitative, and the strength of the signal from each primer extension reaction reflects the amount of the specific mRNA. The reactions can also be used to compare the strength of adjacent promoters.

4.7.2 Gene fusion technology

One of the most powerful and widely used techniques for analysing gene expression is **reporter gene technology**. Reporter genes are generally used when the gene or operon under investigation does not encode an easily assayed product. Fusion to a reporter gene therefore allows the factors controlling gene expression to be identified. In recent years, gene fusion technology has been developed to facilitate the study of target gene expression at the single-cell level, allowing visualisation of population heterogeneity, sites of synthesis and the location of specific proteins within the cell.

The use of reporter gene technology involves three main variables: (i) the type of fusion constructed (i.e. transcriptional or translational); (ii) the type of reporter gene used; and (iii) the methods of detection (e.g. enzyme, immuno- or cytochemical assay).

Types of gene fusion

Reporter genes can be fused as either transcriptional or translational fusions (Fig. 4.16). In the case of a transcriptional fusion, the reporter gene is cloned downstream of the promoter controlling the transcription of the target gene or operon and must include a ribosome binding site for translation initiation that is functional in the host bacterium. Transcriptional reporters can only be used for monitoring the activities of the target promoter and associated control sequences; post-transcriptional control cannot be detected with such a fusion.

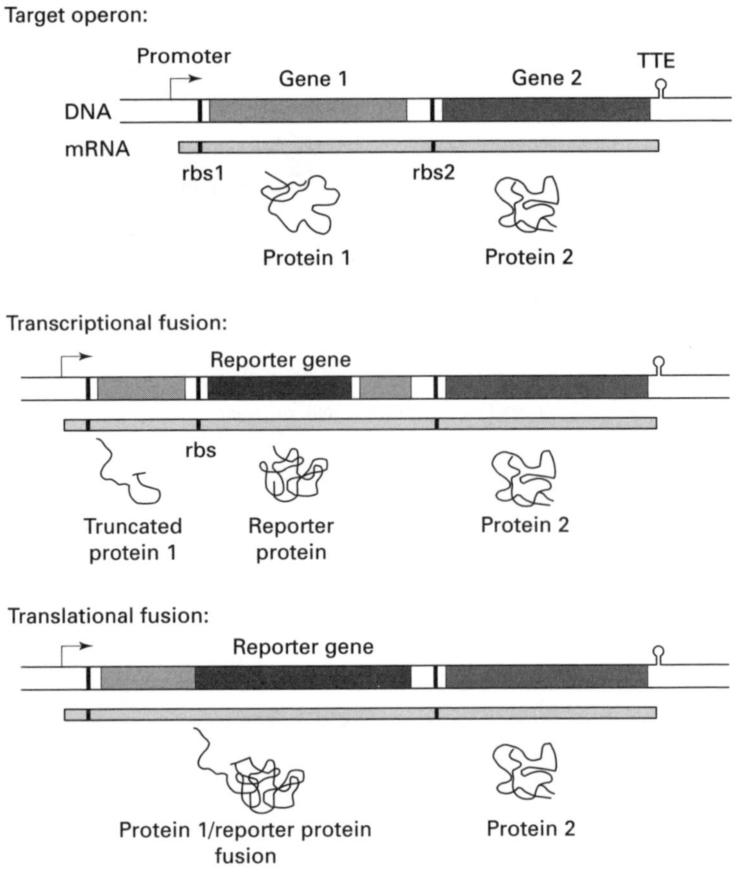

Fig. 4.16 Construction of reporter gene fusions. In a transcriptional fusion the reporter gene is transcribed from the target gene promoter, but is translated from its own ribosome binding site. In a translational fusion, the reporter gene lacks its own ribosome binding site and is fused in frame with the target gene (gene 1). The product of this fusion event is a truncated protein 1 fused to a complete copy of the fusion gene protein. TTE, transcription termination element.

Target operon:
Promoter
Gene 1
Gene 2
TTE
DNA
mRNA
rbs1
rbs2
Protein 1
Protein 2

Transcriptional fusion:
Reporter gene
rbs
Truncated protein 1
Reporter protein
Protein 2

Translational fusion:
Reporter gene
Protein 1/reporter protein fusion
Protein 2

In the case of a translational fusion (Fig. 4.16), the reporter gene is fused in the same codon reading frame as the target gene so that when the target gene is transcribed and translated, the product is a hybrid protein consisting, for example, of a portion of the target protein at the *N*-terminus and the entire reporter at the *C*-terminus. The length of the target protein depends on the type of analysis. If the purpose is simply to understand mechanisms controlling synthesis, the portion of target may be as little as 10 amino acid residues. However, if the purpose is to identify the cellular location of the target protein, virtually all of the target protein may be required to be fused to the reporter protein.

Reporter genes

A variety of reporter genes have been developed for particular applications and most can be adapted for use in a wide range of bacterial species. These include chromogenic reporters (e.g. *E. coli lacZ*, encoding β-galactosidase) that are detected on the basis of colour change, enzyme assay or immuno-detection techniques; antibiotic resistance gene reporters (e.g. *cat*, encoding Chloramphenicol Acetyl Transferase that confers resistance to chloramphenicol) that are detected using selective techniques or enzyme assay; or fluorescent and luminescent reporters (e.g. *luxAB*, encoding luciferase which catalyses a light-emitting reaction or *gfp*, encoding a green fluorescent protein) that are detected by spectrometry, luminometry or video/photo-microscopy.

Hybridisation array technology

The genes being transcribed by a bacterium at any particular time are referred to as the **transcriptome**. The availability of complete microbial genome sequences, together with advances in microfabrication technology, has led to the development of a powerful technology which allows the transcription of every gene in a bacterium to be monitored in a single experiment. Oligonucleotide hybridisation probes, or PCR products for all the genes coding for proteins on the genome, are bound in ordered arrays to glass or nylon supports. Miniaturisation of the process, using techniques developed for the construction of electronic microchips, means that oligonucleotides can be synthesised *in situ* and arrays in excess of 100 000 probes can be attached to a single **DNA chip**. Messenger RNA can be extracted from cells grown under a variety of conditions, fluorescently labelled and then applied to the DNA chip. The pattern of fluorescence obtained on the chip can be used to monitor the transcription of every gene in a genome.

4.8 Engineering genes and optimising products

Many commercially important biotechnological processes have been developed using micro-organisms found in the environment. These include traditional processes such as the production of milk products (e.g. yoghurt production by *Lactococcus casei*), the synthesis of organic solvents (e.g. acetone production by *Clostridium acetobutylicum*), and the production of enzymes for domestic and industrial catalysis (e.g. α-amylases, proteases and penicillin acylases from *Bacillus* species). These processes are usually distinct from those found in Nature and may place particular requirements on the catalyst, be it a whole organism or an industrial enzyme, that are not encountered in natural environments. The organism or enzyme may therefore have to be engineered as part of the process optimisation.

4.8.1 Protein and pathway engineering

A well studied example of enzyme optimisation is that of subtilisin, an alkaline protease isolated from *B. subtilis* and close relatives (e.g. *B. licheniformis*, *B. stearothermophilus*) and used as a stain remover in the detergent industry. Subtilisin is used in 95% of washing detergent formulations and allows protein stains to be removed more effectively and at lower temperatures than are usually needed for laundry processing. The ideal requirements for such an enzyme are stability up to 70 °C and within the pH range 8–11, resistance to non-ionic detergents and oxidising reagents such as hydrogen peroxide, and the absence of metal ion requirements. Based on an extensive knowledge of its catalytic activity and three-dimensional structure, subtilisin was engineered by site-directed mutagenesis (Section 4.4.9) to produce variant enzymes with combinations of these improved characteristics.

An alternative approach has been used to improve the enzymatic characteristics of *Bacillus* α-amylases. In this case natural recombination was used to generate functional hybrid amylases from genes encoding closely related enzymes from *B. amyloliquefaciens, B. licheniformis* and *B. stearothermophilus*. The genes for these enzymes were cloned in pair-wise combinations and then allowed to undergo rounds of reciprocal recombination. This generated a population of cells with a large number of hybrid α-amylases, which were then screened for cells producing amylases with improved catalytic or structural characteristics. More recently, knowledge of the three-dimensional structure of these amylases, together with information on the relationship between structure and functional characteristics such as thermostability, has enabled more directed approaches to be used. These have included the use of PCR gene splicing techniques for the construction of specific hybrid α-amylases.

Comparative DNA and protein sequence studies have demonstrated the importance of recombination of blocks of sequence rather than point mutation alone in the evolution of protein structure. These studies have led to the development of molecular techniques such as **DNA shuffling** or **sexual PCR** to facilitate the rapid evolution of proteins. The principle involves mixing randomly fragmented DNA encoding closely related genes and then using PCR (Section 4.4.4) to reassemble them into full length fragments, with the individual fragments acting as primers. The PCR products are used to generate a library (Section 4.5) of chimaeric genes for the selection of proteins with modified or improved characteristics. This system has been used to generate a variant of an *E. coli* β-galactosidase with a 60-fold increase in specific activity for sugar that is normally a poor substrate for this enzyme and a variant of the green fluorescent protein (Section 4.7.2) with a 45-fold increase in fluorescent signal.

Bacteria produce a number of compounds that, if synthesised at suitable concentrations, represent commercially viable products. Traditionally, bacteria that make a significant amount of a potentially commercial product can be directed to synthesise larger amounts. This may be achieved by randomly mutagenising a population of the target organism, and screening for mutants producing higher concentrations of the product, or by cloning the synthetic genes together and placing them under the control of efficient transcription and translation signals. While there are examples that testify to the success of such approaches (e.g. antibiotic production, synthesis of amino acids), recently more rational approaches have been made by engineering metabolic pathways, either to increase productivity or to direct synthesis towards specific products.

4.9 Production of heterologous products

Traditionally, genetic techniques have been applied by industry to increase the production of natural products such as enzymes, antibiotics and vitamins. Only a limited number of protein products were produced commercially, and these were produced using existing technologies from their natural hosts (e.g. proteases from *Bacillus* species). Specific genes can now be isolated from virtually any biological material and cloned into a bacterium or other host system. However, cloning a gene does not, per se, ensure its expression, nor does expression ensure the biological activity of its product. Many factors need to be considered to ensure commercially viable levels of production and biological activity. For example, the choice of host/vector systems determines both the strategy used for the cloning and expression, and these in turn can affect the quantity and fidelity of the product. In some cases it is not possible to use a bacterial system to produce a biologically active product or one that is acceptable for pharmaceutical purposes. Instead host/vector systems based on higher organisms, for example mammalian or insect cell culture systems, may be used.

4.9.1 Host systems and their relative advantages

Prior to the 1970s, the only methods for obtaining proteins or polypeptides for analysis or for therapeutic purposes were to isolate them from natural sources or, in a limited number of cases (e.g. bioactive peptides), to synthesise them chemically. Recombinant DNA technology opened up the possibility to clone the gene responsible for a particular product and to produce it in unlimited amounts in a bacterium such as *E. coli*. Initially, the only eukaryotic genes that could be cloned were those encoding products that were already available in relatively large amounts and that had very sensitive assays (e.g. insulin, human growth hormone, interferon). This was because it was necessary to have extensive information about their amino acid sequence, and protein sequencing techniques available at that time required relatively large amounts of the purified protein. Current technical improvements permit almost any characterised protein to be cloned, either directly via copy DNA synthesis from mRNA extracted from biological material, or indirectly by gene synthesis.

Although recombinant technology was developed in bacterial systems, it has been increasingly expanded into a range of eukaryotic organisms, using a wide variety of interesting and novel technologies (see Chapter 5). However, from an economic point of view, bacteria are still the organisms of choice because of the ease with which they can be genetically manipulated, their rapid growth rate and relatively simple nutritional requirements.

Proteins derived from recombinant technology are expected to meet the same exacting standards as conventionally produced drugs; particularly with respect to product potency, purity and identity. Sensitive analytical techniques are used to characterise the products, with particular attention being paid to undesirable biological activities such as adverse immunogenic and allergenic reactions. It is therefore the exacting requirements of regulatory authorities, such as the US Food and Drug Administration (FDA), for increasingly authentic pharmacological products that has led producers to switch from bacterial to eukaryotic systems for the production of certain products. These requirements have been driven by the increasing sensitivity of analytical techniques such as high performance liquid chromatography (HPLC), mass spectrometry, nuclear magnetic resonance (NMR) and circular dichroism, that are able to reveal subtle differences in the structures of natural and recombinant proteins.

Often, a great deal of effort can be put into devising a cloning strategy only to discover that the level of production of a particular protein is much lower than is anticipated. In many cases, the reasons for these failures are not easily determined. Some of the causes of low recombinant protein productivity are given below.

4.9.2 Transcription

The cloning of a DNA fragment encoding a protein of interest is not in itself sufficient to ensure its expression. Instead, the cloning strategy has to include the inclusion of sequences designed to express the DNA at an appropriate time and level in the host bacterium. Transcription is facilitated by cloning the target gene downstream of a promoter sequence and associated translation signals. Promoter strength is a key factor affecting the level of expression of a cloned gene and a wide range of promoters are currently available, including those that are tightly regulated and others whose expression is constitutive. In *E. coli*, expression from a strong promoter can lead to the production of a recombinant protein at levels that are equivalent to about 50% of the total cell protein.

A number of naturally occurring promoters have been used to direct high level expression of recombinant genes. However, with an increased knowledge of the factors required to generate strong promoters, hybrid or totally synthetic promoters have been constructed. Promoters that are induced by expensive and potentially toxic chemicals are not suitable for large-scale fermentations, and consequently promoters have been designed in which induction is mediated by increasing the temperature of the culture medium. Other promoters have been developed that are induced at low oxygen tension or with cheap, readily available sugars.

Expression from strong promoters imposes a high metabolic load on the host cell and consequently their transcriptional activity must be tightly controlled. One aspect of this control is the incorporation of efficient **transcription termination elements** (TTE). The failure to include such TTEs can lead to diminished levels of expression of the target gene, since energy is unnecessarily diverted into the synthesis of non-productive mRNA species. This is especially important when the target gene is expressed from a multi-copy plasmid vector since highly active transcriptional activity over regions necessary for plasmid replication and/or segregation can lead to greatly increased vector instability.

4.9.3 Translation

The efficient translation of mRNA transcripts requires the incorporation of an efficient ribosome binding site (RBS), located about 5 bp upstream of (prior to) the translational start codon. The structure of the RBS tends to vary from bacterium to bacterium according to the sequences at the 3′OH end of the 16S ribosomal (r)RNA that interact with the mRNA.

The genetic code is degenerate, that is many amino acids are specified by more than one codon (triplet of nucleotide bases). Codons that specify the same amino acid are said to be **synonymous** but are not necessarily used with similar frequencies. In fact, most bacterial species exhibit preferences in their use of codons, particularly for highly expressed genes. Variations in codon usage are, in part, a reflection of the %GC content of the organism's DNA, with favoured codons corresponding to the organism's most abundant transfer (t)RNA species. Since most codons are recognised by specific aminoacyl tRNA molecules, the use of non-favoured codons results in a reduction in the rate of translation and an increase in the mis-incorporation of amino acids above that of the normal error rate of about 1 in 3000 amino acids. **Codon bias** can be determined by calculating the Relative Synonymous Codon Usage (RSCU) of individual codons:

$$\text{RSCU} = \frac{\text{observed number of times a particular codon is used}}{\text{expected number if all codons are used with equal frequency}}$$

therefore:

if RSCU equals 1, the codon is used without bias
if RSCU is *less* than 1, the codon is 'non-favoured'
if RSCU is *more* than 1 the codon is 'favoured'.

Gene synthesis technology allows genes to be constructed so as to optimise the codon usage of the host producer strain. The significance of codon usage was demonstrated in studies on the production of interleukin-2 (IL-2) by *E. coli*. When the native IL-2 gene (399 bp) was analysed for its codon usage only 43% of the codons were 'favoured' by *E. coli*. When an alternative copy of the IL-2 gene was generated by gene synthesis, it was possible to adjust the codon usage such that 85% of the codons corresponded to those favoured by *E. coli*. When the two versions were cloned and expressed in *E. coli* on identical vectors, despite their producing identical amounts of mRNA, eight times more biologically active IL-2 was produced from the synthetic gene as compared with the native gene.

4.9.4 Formation of inclusion bodies

Many recombinant proteins, particularly when produced at high concentrations, are unable to fold properly within the producing cell and instead associate with each other to form large protein aggregates referred to as **inclusion bodies**. Inclusion bodies are particularly common in bacteria expressing mammalian proteins. The proteins in inclusion bodies can vary from a native-like state that are easily dissociated, to completely misfolded molecules that are dissociated only under highly denaturing conditions. The size, state and aggregation density of inclusions are affected by the characteristics of the recombinant protein itself and factors that affect cell physiology (e.g. growth rate, temperature, culture medium etc.). In some cases aggregation can be prevented or reduced by modulating the fermentation conditions.

Inclusion body formation can lead to: (i) the production of biologically inactive proteins, (ii) sub-optimal yields, (iii) extraction and purification problems. However, in some cases inclusion body formation can be advantageous since they are comprised of relatively pure protein that is insoluble under mild extraction conditions. It is therefore possible to devise protocols that recover the target protein in this insoluble state and refold it under conditions which favour the biologically active conformation. In some cases problems associated with inclusion body formation can be overcome by use of a secretion vectors system (Section 4.5.4) in which the target protein is directed into the periplasm or culture medium.

4.10 *In silico* analysis of bacterial genomes

The availability of entire genome sequences for a significant number of micro-organisms opens up new approaches for the analysis of bacteria, and the rapidly expanding field of bioinformatics has the potential to reveal relevant and novel insights on bacterial evolution and gene function. It has the potential to provide answers to long-standing questions relating to evolutionary mechanisms and to the relationships between gene order and function.

One of the most significant advances in methods for studying and analysing micro-organisms has come about through the availability of powerful personal computers which, together with the development of the internet, provides researchers with access to powerful bioinformatical tools. Bioinformatics, often referred to as *in silico* analysis, nicely complements *in vivo* and *in vitro* methodologies. An ultimate goal is to model the behaviour of whole organisms, including aspects of their evolution. Although not currently a substitute for *in vivo* and *in vitro* experimentation, bioinformatics has already demonstrated its potential to direct the focus of more traditional approaches.

Several types of computer program are available for analysing bacterial genome sequences. These include programs that attempt to identify protein-encoding genes, sequence signals such as ribosome binding sites, promoters and protein binding sites, and relationships to previously sequenced DNA of whatever source. Programs that attempt to identify protein-encoding genes translate the consensus DNA sequence in all six reading frames and then analyse the resulting data for the presence of long stretches of amino acid-encoding codons, uninterrupted by termination codons. These so-called **open reading frames** (ORF) are usually at least 60 amino acids long, but may be several thousand amino acids in length. The more advanced programs for predicting protein coding genes are able to search for the presence of ribosome binding sites located immediately upstream of a putative start codon and even to identify potential DNA sequencing errors that generate frame-shift mutations.

Once putative proteins have been identified, other bioinformatical tools can be used to determine their relationships to previously identified proteins or putative proteins. A prerequisite for this type of analysis is the availability of data libraries which act as repositories of currently available DNA and protein sequences. The databases can be routinely accessed via the internet, using programs such as FASTA and BLAST that provide a list of DNA sequences and proteins, respectively, showing homology to all or part of the query sequence.

The internet is also a source of molecular biological tools that facilitate a wide range of analyses including the identification of putative transmembrane domains, secondary structures and the signal peptides of secretory proteins.

4.11 Further reading

Davies, J. E. and Demain, A. L. (1999). *Manual of Industrial Microbiology and Biotechnology. 2nd Edition.* American Society for Microbiology, Washington DC.

Glazer, A. N. and Nikaido, H. (1995). *Microbial Biotechnology: Fundamentals of Applied Microbiology.* W. H. Freeman and Company, New York.

Lewin, B. (2000). *Genes VII.* Oxford University Press, Oxford.

Old, R. W. and Primrose, S.B. (1994). *Principles of Gene Manipulation: An Introduction to Genetic Engineering. 5th Edition.* Blackwell Scientific Publications, Oxford.

Snyder, L. and Champness, W. (1997). *Molecular Genetics of Bacteria.* American Society for Microbiology, Washington DC.

Chapter 5

Genetic engineering: yeasts and filamentous fungi

David B. Archer, Donald A. MacKenzie and David J. Jeenes

Glossary

Auxotrophic mutation A mutation in a gene that confers the requirement for a growth factor to be supplied rather than synthesised by the organism. A gene that **complements** this auxotrophic mutation is one that can return the organism to its normal phenotype (i.e. not requiring the growth factor).

cDNA Single-stranded DNA with complementary sequence to messenger RNA (mRNA), synthesised *in vitro*. Double-stranded cDNA can then be made. cDNA libraries contain double-stranded cDNA molecules, each of which forms part of a **vector**. The collection of cDNA molecules, each in a separate vector, forms the cDNA library.

Chaperone A protein which assists the folding of another protein.

Chromatin A highly organised complex of protein and DNA.

Chromosome A discrete unit of DNA (containing many genes) and protein. Different species have different numbers of chromosomes (see Table 5.1).

Complementation The ability of a gene to convert a mutant phenotype to **wild-type**.

Cosmid Plasmid containing sequences (phage lambda *cos* sites) which permit packaging of the plasmid into the proteinaceous phage coat.

Cross-over See homologous recombination.

Dimorphism Ability to exist as two structurally distinct forms.

Endoplasmic reticulum (ER) Intracellular membrane structure in eukaryotes forming the early part of the protein secretory pathway.

Exons The coding regions of a gene (excluding the **introns**).

Expressed sequence tag (EST) A portion of a cDNA, usually of sufficient length to provide useful sequence information to enable identification or cloning of a full-length gene.

Expression A term used to describe the process of protein synthesis from a gene.

Expression cassette An arrangement of DNA which permits the transcription of a gene for protein production, i.e. includes a **promoter**, the **open reading frame** and a **transcriptional terminator**.

Gene A region of DNA which is transcribed into RNA.

Genome The entire DNA in an organism.

Genotype/genotypic The information contained in the DNA of an organism.

Heterologous Pertaining to a gene or protein which is foreign in relation to the system being studied.

Homologous recombination The breakage and re-joining of two double strands of DNA that have closely related sequences. The process of homologous recombination, also called crossing-over, permits the integration of genes into specific genomic loci (see Fig. 5.4) but the term is used more widely, for example in exchange of genetic material between chromosomes during meiosis. In **transformations** using circular **plasmids**, a single cross-over leads to incorporation of the entire plasmid into a chromosome. A double cross-over employs a linearised DNA molecule and leads to the replacement of a chromosomal gene by a gene with sequence similarity to the chromosomal gene at the cross-over regions.

Intron A segment of RNA which is excised before the mRNA is translated into protein. Introns are common in eukaryotic genes but very rare in prokaryotes.

Library A collection of cloned genomic DNA fragments or cDNA molecules.

Linkage The tendency of genes which are physically close to be inherited together.

Linkage group A group of genes that is inherited or transferred as a single unit and represents an entire chromosome.

Mycelium Typical vegetative growth form of filamentous fungi, consisting of branched filaments.

Nucleotide Building block for DNA and RNA, a base-(deoxy)ribose-phosphate molecule. The nucleotides are referred to by base. Purines: A (adenine), G (guanine). Pyrimidines: C (cytosine), T (thymine), U (uracil). DNA has AGCT, RNA has AGCU.

Open reading frame (ORF) Stretch of codons uninterrupted by stop codons.

Origin (*ori*) The site of initiation of DNA replication.

Phenotype Properties (e.g. biochemical or physiological) of an organism which are determined by the **genotype**.

Plasmid DNA molecule that replicates independently of the chromosomes.

Ploidy The number of complete sets of chromosomes in a cell, e.g. haploid (1), diploid (2), polyploid (>2).

Polymerase chain reaction (PCR) A procedure for exponential amplification of DNA fragments (see Fig. 4.7).

Promoter The region of DNA upstream (i.e. 5′) of a gene which contains signals for initiating and regulating transcription of the gene.

Protoplasts Cells from which the cell wall has been removed by the action of carbohydrase enzymes. These cells are bounded by the plasma membrane and are osmotically fragile.

Recombination The exchange of DNA between two DNA molecules or the incorporation of one DNA molecule into another, e.g. between the chromosome and introduced DNA.

Restriction enzyme Enzyme which cleaves at or near a specific, short DNA sequence (restriction site).

Shuttle vector A vector that can replicate independently in more than one type of organism, e.g. can 'shuttle' between a bacterium and a yeast.

Southern blotting DNA fragments are separated according to size by electrophoresis, transferred to a membrane and probed with a labelled DNA segment to detect complementary sequences (named after E. Southern).

Splicing The excision of **introns** from RNA molecules which is followed by re-joining of the **exons**. This normally occurs in the nucleus.

Synteny Gene order in a chromosome being physically the same in different organisms.

Transcription The synthesis of RNA coded by the DNA sequence.

Transcription factor A protein which binds to DNA in the promoter region, or binds to other protein factors, and regulates the transcription of one or more genes.

Transcription terminator A DNA sequence which causes the RNA polymerase to stop synthesis of RNA.

Transcriptional start point (*tsp*) The site at which RNA synthesis is initiated.

Transformation (genetic) The uptake and stable incorporation of exogenous DNA into a cell.

Vector A DNA molecule (usually a plasmid) used for transferring DNA into an organism.

Wild-type Strain of an organism not deliberately mutated or modified by genetic manipulation. The term can also be used to describe the **phenotype** of such a strain.

5.1 Introduction

Fungi are eukaryotes classified as either yeasts or filamentous fungi primarily by their predominant form of growth in culture. Those species which are normally unicellular are the yeasts and thus superficially distinguished from their filamentous relatives. This distinction is not always adhered to by the organisms themselves as several are able to grow in both forms and are called **dimorphic** (Fig. 5.1). More discriminatory **genotypic** methods are now enabling a more soundly based classification of the fungi but the substantial similarities between yeasts and filamentous fungi enable us to treat them together in one chapter. Despite the similarities, their differences have given rise to a wide diversity of biotechnological applications which is increasing as the number of different species examined rises and as their exploitation is extended by genetic engineering.

In this chapter, we focus on the molecular biology involved in genetic engineering of fungi and describe applications of the fungi as hosts for the production of proteins encoded by introduced foreign **genes**. Many of the experimental approaches adopted in this work have been described in the preceding chapter (Chapter 4) with prokaryotes and are essentially the same with fungi. Their descriptions are not repeated here except where differences between bacteria and fungi require alterations to the protocols. Differences arise primarily because fungi are larger in both size and **genome** than bacteria and the organising principles for many of their cellular functions are typical of complex higher organisms rather than simple bacterial cells. Thus, fungi have defined membrane-bound nuclei containing several **chromosomes** and other membrane-bound organelles including a membranous intra-cellular system called the **endoplasmic reticulum**

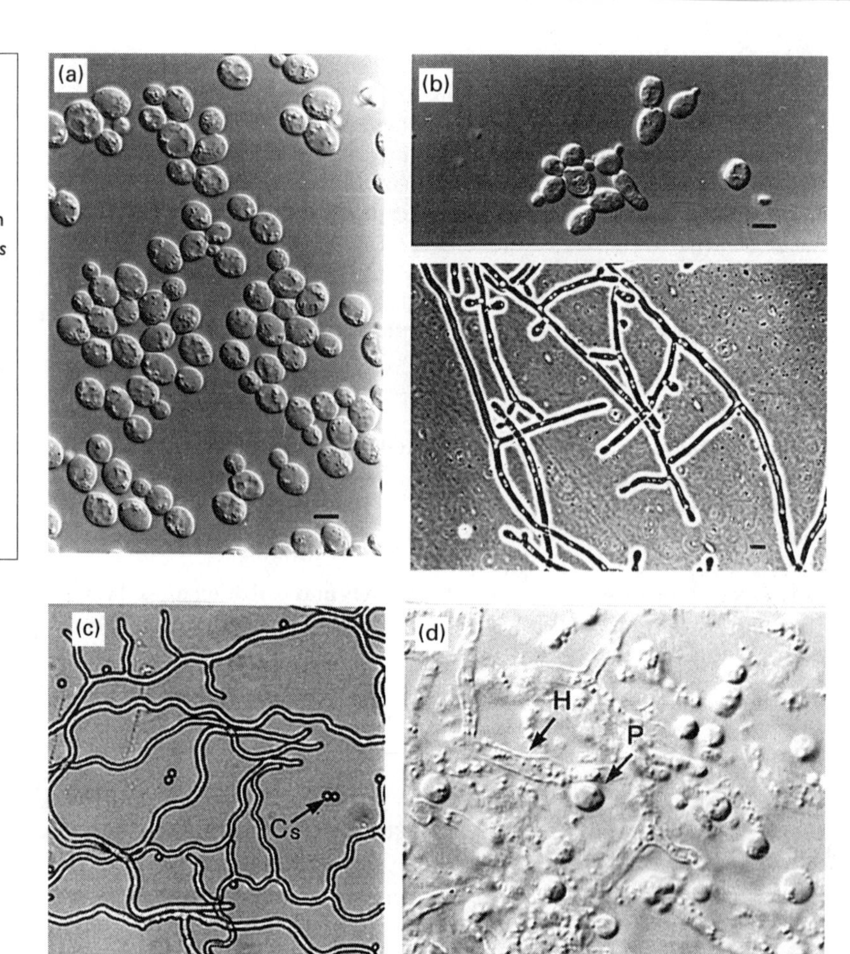

Fig. 5.1 Phase contrast micrograph of (a) the yeast *Saccharomyces cerevisiae*, (b) the yeast (above right) and hyphal (below right) forms of *Yarrowia lipolytica*, (c) bright field micrograph of the filamentous fungus *Aspergillus niger* grown on a cellophane sheet, (d) differential interference contrast micrograph of protoplast formation from *Aspergillus nidulans*. Note the branching of hyphae. Bar markers = 10 μm. Cs, conidiospores (asexual); H, hypha; P, protoplast. Micrographs a to c are courtesy of Linda and James Barnett.

(ER) into which proteins are transported for targeting either to the cell exterior or to other sub-cellular organelles. Fungal cell walls do not contain peptidoglycan which is found only in bacteria. Rather, their walls are composed primarily of various polysaccharides (e.g. glucans, chitin, chitosan) and glycoproteins depending on the species. Fungi can have complex life cycles, exhibiting normal (vegetative) cell growth as a **mycelium** followed by morphogenesis (change in form) with the formation of either sexual (resulting from the mating of two strains) or asexual spores. For most biotechnological purposes, fungi are grown vegetatively with associated mitotic nuclear division. Meiosis will not concern us here. For the interested reader reference to information on the fungal cell cycle can be found in general texts cited at the end of the chapter.

Fungal genes are organised within chromosomes, which can be separated by electrophoresis. The number of chromosomes varies with species and equates to the number of **linkage groups** where the **linkage** of genetic markers has been studied. Some examples are given in Table 5.1 together with examples of fungal genome sizes. The amount of DNA

Table 5.1 | Mapping and sequencing of fungal genomes

Organism	Number of chromosomes	Genome size (Mb)[a]	Web site[b]
Saccharomyces cerevisiae	16	12.1	http://genome-www.stanford.edu/saccharomyces
Schizosaccharomyces pombe	3	14	http://www.sanger.ac.uk/Projects/S_pombe/
Candida albicans	8	16	http://alces.med.umn.edu/Candida.html http://sequence-www.stanford.edu/group/candida/index.html
Neurospora crassa	7	43	http://www.unm.edu/~ngp/
Aspergillus nidulans	8	31	http://gene.genetics.uga.edu

Notes:
[a] Haploid genome complement excluding mitochondrial genome
[b] World Wide Web sites for physical linkage, genome sequencing or other (e.g. cDNA) information databases.

in fungi is much larger than that in prokaryotes (c.f. 4.7 Mb – that is 4.7 million nucleotide bases in length – for *Escherichia coli*). This is due to fungi having more genes and more DNA which does not code for proteins, either within genes (as **introns**) or as 'spacer' DNA between genes. Bacteria economise on non-coding DNA in comparison with fungi. For example, some bacterial genes can be physically adjacent to each other with **transcription** directed and regulated by a single **promoter**, to form an **operon** (see Fig. 4.3). Operons are not found in fungi and even where genes encoding functionally related enzymes are clustered (i.e. grouped together), the transcription of each gene is regulated by an independent promoter. Knowledge of the regulatory mechanisms of transcription is necessary to design rational approaches for genetic modification of an organism in order to achieve the desired **phenotype**. We therefore discuss in this chapter examples of wide-domain (multi-pathway) and pathway-specific transcriptional regulatory mechanisms.

Unlike the DNA found in bacteria, chromosomes from higher organisms are composed of **chromatin** which is an approximately equal weight of DNA and protein (mainly histones). Histones bind to DNA and help to package it into complex structures. For example, DNA associates with particular histones to form nucleosomes which can be visualised by electron microscopy as 'beads on a string'. Nucleosome arrangements can control access of proteins (**transcription factors**) that are required to initiate transcription at a promoter, a level of regulation not seen in prokaryotes. Discussion of the regulation of transcription at the level of chromatin is beyond the scope of this chapter but its mention is a reminder of the level of complexity in eukaryotes. Fortunately, many important fungi are amenable to investigation of their biology at the molecular level and we are able to genetically modify these species in order to alter either the types or amounts of product formed.

5.2 Introducing DNA into fungi (fungal transformation)

5.2.1 Background

The first fungal **transformation** experiments in the early 1970s were performed by transferring chromosomal DNA fragments from **wild-type** strains into mutant hosts which had a specific growth requirement for an amino acid or **nucleotide** base (examples of **auxotrophic mutants**). Successful transformation relied on converting the fungal cells into **protoplasts** (Fig. 5.1d) by digesting away the cell wall (a major barrier to DNA uptake) with carbohydrase enzyme mixtures before introducing the transforming DNA. Transformants could then be selected by their ability to grow without supplementation for the auxotrophic requirement due to the activity of the wild-type gene, a process known as genetic **complementation.** The fate of the transforming DNA in the cell could not be studied in great detail until the advent of more sophisticated molecular techniques in the late 1970s and early 1980s when it was subsequently shown that the DNA in these early experiments had integrated by genetic **recombination** into the fungal chromosomes.

Subsequently, transformation methods were developed for the yeast *Saccharomyces cerevisiae* using **shuttle vectors** which could be propagated as intact **plasmids**, without chromosomal integration, in both *E. coli* and yeast. Manipulation of these plasmids and 'bulking up' the amount of vector DNA were more effectively performed in *E. coli* prior to their introduction into yeast. Similar vectors have been designed for the transformation of filamentous fungi but in most cases the introduced DNA integrates into the fungal chromosomes rather than replicating as a plasmid. Transformation was accomplished first in genetically well-characterised organisms such as *S. cerevisiae*, *Neurospora crassa* and *Aspergillus nidulans* because these fungi had been studied extensively in the laboratory and a number of suitable auxotrophic mutants and their corresponding wild-type genes were available. Methods are now continually being modified or adapted for more biotechnologically important fungi in which genetic systems may be less well defined. In cases where auxotrophic complementation of industrial strains is not feasible, other selection markers such as resistance to an antibiotic can be used in the transformation.

5.2.2 Transformation protocols

For the transformation of many filamentous fungi, the method of choice still relies on converting the fungal mycelium into protoplasts which are prepared in a buffer containing an osmotic stabiliser to prevent cells from bursting (Fig. 5.2). Vector DNA is then added to the protoplast suspension in the presence of calcium ions and DNA uptake is induced by the addition of polyethylene glycol (PEG). Not all protoplasts can regenerate into viable fungal cells by producing new cell walls and only a proportion of these will contain the transforming DNA.

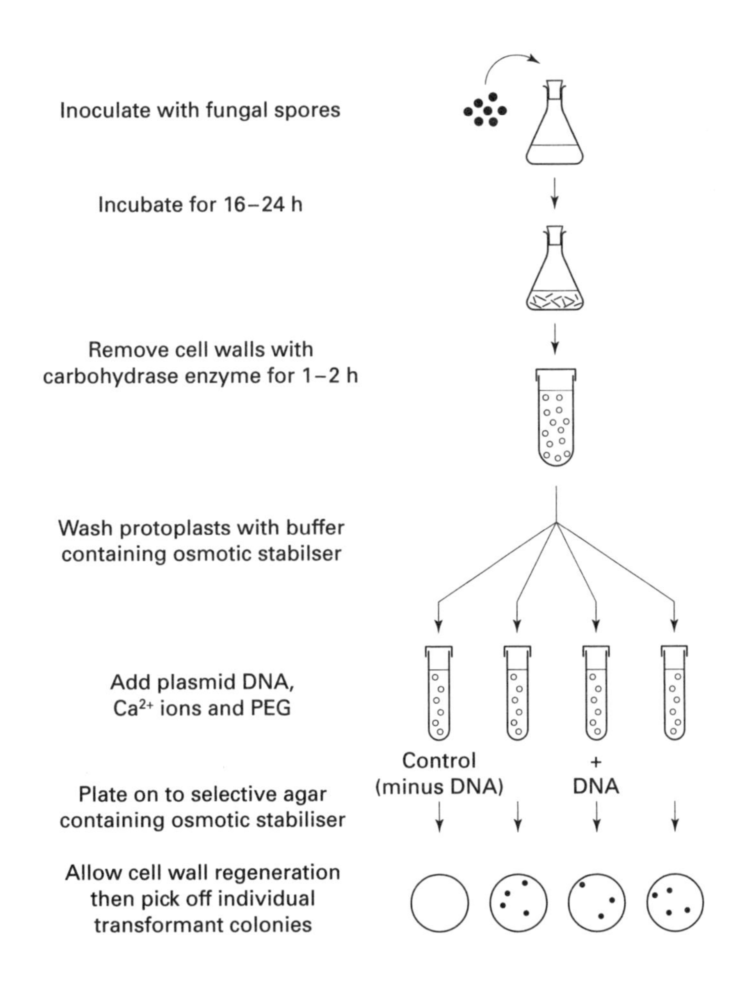
Inoculate with fungal spores
Incubate for 16–24 h
Remove cell walls with
carbohydrase enzyme for 1–2 h
Wash protoplasts with buffer
containing osmotic stabilser
Add plasmid DNA,
Ca2+ ions and PEG
Control
(minus DNA)
+
DNA
Plate on to selective agar
containing osmotic stabiliser
Allow cell wall regeneration
then pick off individual
transformant colonies

Fig. 5.2 Typical transformation protocol using protoplasts derived from a filamentous fungus. Following harvesting, the filtered mycelium is resuspended in a buffer containing an osmotic stabiliser such as sorbitol or KCl to prevent protoplasts from bursting. The selective agar can either be a minimal medium (lacking the required growth supplement which is now supplied by the activity of the introduced gene) or it can contain an antibiotic. For antibiotic selection, the transformants are normally allowed to regenerate their cell walls and express the antibiotic-resistance protein before being challenged with the antibiotic. PEG, polyethylene glycol.

Thus, there is a need to have effective selection methods to obtain those cells containing the introduced DNA. Transformation frequencies, in terms of the number of transformants obtained per microgram of vector DNA added, can vary significantly depending on the organism and transformation protocol used. Since the first reports of fungal transformation, various modifications to this basic method have been described which improve transformation frequencies by orders of magnitude. Alternative methods which obviate the need for protoplast formation, such as the lithium acetate/yeast whole cell method, electroporation of germinating spores or biolistic transformation of fungal mycelia, have had varied success but these suffer from the limitations of suitable host range and the need for specialised equipment in some cases. A list of transformation methods is given in Table 5.2.

5.2.3 Transformation vectors

Transformation vectors can be designed to introduce DNA which either integrates into the recipient organism's genome (integrative transformation) or can be maintained as a plasmid. Integrative transformation is used to insert DNA either at site(s) in the chromosome which show significant sequence similarity to a region on the plasmid (integration by

Table 5.2 Fungal transformation methods

Method or treatment		Examples of fungi transformed	Transformation frequency[a]	Remarks
Protoplasts	PEG[b]/$CaCl_2$	*Saccharomyces cerevisiae*, *Pichia pastoris*	10^2–10^5	Most widely used method but frequencies generally lower with filamentous fungi
		Aspergillus nidulans, *A. niger*, *Trichoderma reesei*, *Mucor circinelloides*	10^0–10^3	
Protoplasts	Electroporation[c]	*S. cerevisiae*, *A. niger*, *T. reesei*	10^0–10^3	Can be as efficient as the PEG/$CaCl_2$ method
Whole cells	Electroporation	*S. cerevisiae*, *A. niger*, *A. oryzae*, *Neurospora crassa*	10^0–10^3	For filamentous fungi, best results are obtained using germinating conidiospores with weakened cell walls
Whole cells	LiAc[d]/PEG	*S. cerevisiae*, *Yarrowia lipolytica*	10^3–10^7	Only applicable to a few yeast species and not effective with filamentous fungi
Whole cells	Biolistics[e]	*S. cerevisiae*, *A. nidulans*, *N. crassa*, *Trichoderma harzianum*	10^0–10^3	Most effective with intact yeast cells or conidiospores but mycelium can also be used
Protoplasts or whole cells	*Agrobacterium tumefaciens*[f]	*S. cerevisiae*, *Aspergillus awamori*, *T. reesei*, *N. crassa*	10^2–10^4[g]	Equally effective with protoplasts or conidiospores but transformation frequency is species-dependent

Notes:

[a] Expressed as the number of transformants obtained per μg of vector DNA added. Values depend on the organism, type of vector and method used.

[b] Polyethylene glycol.

[c] Introduction of DNA into cells by applying a short-pulse, high voltage electric field.

[d] Lithium acetate.

[e] Shooting DNA-coated metal particles (normally gold or tungsten) at high velocity into cells under vacuum.

[f] DNA is transferred to the fungus from the bacterium *A. tumefaciens* in a manner similar to plant transformations.

[g] Transformation frequency in this case is expressed as number of transformants obtained per 10^7 recipient cells.

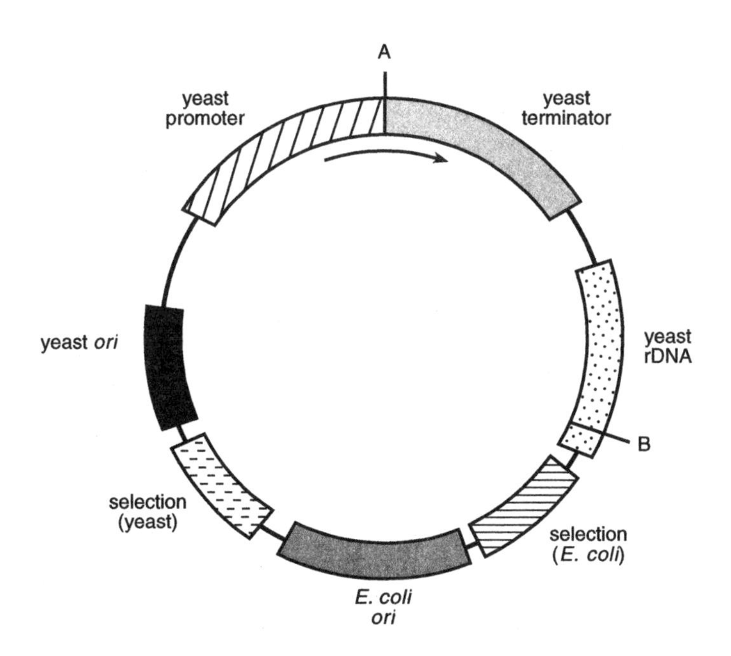

Fig. 5.3 Stylised yeast-*E. coli* shuttle expression vector showing origins of DNA replication (*ori*) which function either in yeast or *E. coli*, selection markers and cloning site A for inserting the gene to be expressed. For targeted integration of the plasmid into a chromosome, a DNA sequence from the ribosomal RNA-encoding region (rDNA) could be included in the vector. Site B in this rDNA sequence is a **restriction** site required to convert the vector from a circular molecule to a linear one which increases the efficiency of DNA integration by homologous recombination into the rDNA region of the chromosome.

homologous recombination) or randomly at one or more locations (ectopic integration). Yeast shuttle vectors contain genes which allow for their selection in both bacterial and yeast cells. They also contain bacterial and yeast **origins** of replication (***ori***), sequences which are essential for the initiation of plasmid DNA replication in both organisms. A stylised yeast shuttle vector which is capable of replication without integration into the yeast chromosomes (extrachromosomal vector) or which can be specifically targeted into a region of a chromosome (integrative vector) is illustrated in Fig. 5.3.

In order to find those cells which are 'transformed', i.e. contain the introduced DNA, the transformation vector is designed to contain a gene which confers a selectable characteristic on the transformed cells. These 'selection markers' fall into three main groups. First, a number of genes from wild-type fungi have been cloned which complement auxotrophic growth requirements, as already explained. In many cases, genes from one fungus are able to complement the appropriate mutations in a different fungus (**heterologous** transformation). However, for some fungi, only their own genes are able to complement mutations (homologous transformation). If the required auxotrophic mutant is not available, a second class of selection marker, based on resistance to antibiotics such as hygromycin B, phleomycin, kanamycin or benomyl, can be used. One drawback with this type of marker is that the fungus must be reasonably sensitive to the antibiotic in question. For example, a bacterial gene that encodes resistance to kanamycin can be expressed weakly from its own promoter in yeast. Higher levels of drug resistance can be achieved, however, if a yeast promoter is used instead for expression. For filamentous fungi too, optimisation of antibiotic resistance systems normally requires the use of a fungal promoter. Finally, a third group of vectors exploit genes which confer the ability to grow on carbon or nitrogen sources which the host strain would not normally be able to use. A good example is the acetamidase gene (*amdS*) of *A. nidulans* which allows growth of the recipient strain on acetamide or acrylamide as sole nitrogen source. This marker has been introduced into a number of *Aspergillus* and *Trichoderma* spp. and is particularly useful in generating transformants which contain multiple copies of the integrated vector.

In yeast, plasmid vectors are maintained within the cell provided the transformants are grown under selective conditions, e.g. in the presence of an antibiotic. When the selective pressure is removed, plasmids can be lost relatively rapidly from the cells because there is no advantage to the cells in maintaining a plasmid. All plasmids must contain an *ori* which can be derived either from naturally occurring plasmids or from chromosomal DNA sequences. Vectors with an *ori* from one yeast can normally function in a variety of different yeast hosts, albeit not always with the same degree of efficiency. Plasmid vectors can be present at up to 200 copies per cell and thus provide a simple system for increasing the number of copies of the introduced genes. This often leads to higher yields of the protein encoded by the introduced gene. The disadvantage is that selective pressure is normally required to avoid significant plasmid loss, i.e. maintaining the desired characteristic can be a problem, particularly in *S. cerevisiae*. Conversely, in other yeasts, such as *Kluyveromyces lactis*, some extrachromosomal vectors are comparatively stable without the need for continuous selection.

Integration of the plasmid into the yeast genome brings enhanced stability but lower numbers of the introduced gene. One way of achieving this in *S. cerevisiae* has been to target the plasmid to ribosomal DNA sequences (rDNA) (which can be present at about 150 tandemly repeated copies per genome). Incorporation of rDNA sequences into a vector (Fig. 5.3) enhances integration of that plasmid into the chromosomal rDNA region by recombination of homologous sequences, especially when the vector has been linearised in the rDNA region. The number of gene copies can be increased by placing the gene used for selection under the transcriptional control of weak, or deliberately weakened, promoters. This approach encourages the selection of multiple gene copies through selection pressure for a critical level of gene product. This approach is used together with rDNA targeting to obtain high numbers of integrated gene copies.

Vectors that integrate into the chromosomes have several important uses in the molecular manipulation of fungi. In addition to increasing the number of copies of a particular gene in a chromosome, they can also be used to disrupt or replace a desired gene. In *S. cerevisiae*, the use of 'replacement cassette' vectors has permitted the deletion of each of the 6000 or so genes identified by the Yeast Genome Sequencing Project to test the function of each gene in the cell. Each cassette, which consists of a kanamycin (G418) selection gene flanked at each end by short regions of gene-specific sequence, is released from the vector by cutting with a restriction enzyme. Homologous recombination between the ends of the cassette and the target gene in the chromosome leads to the

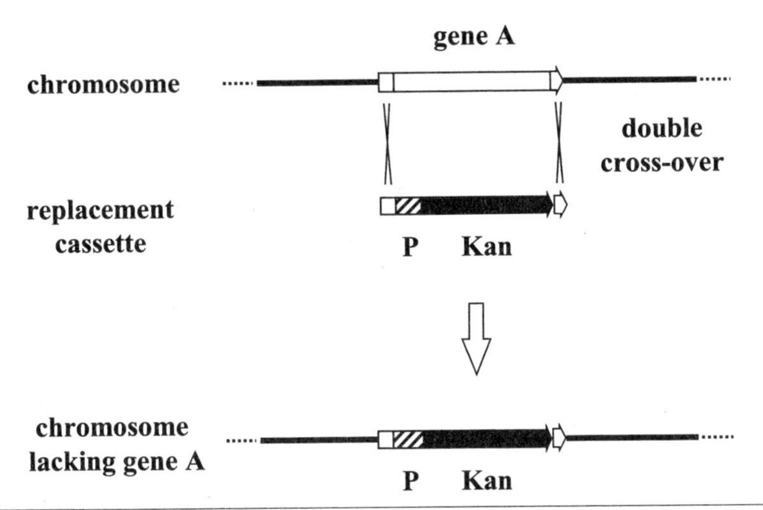

Fig. 5.4 Gene deletion in *S. cerevisiae* using the kanamycin (G418) 'replacement cassette'. The cassette consists of the kanamycin (G418)-resistance gene (Kan) under the control of a fungal promoter (P) and short flanking sequences of only about 40 bp (□) which correspond to the ends of the gene to be deleted. A double **cross-over** by homologous recombination between these ends and the chromosome leads to the deletion of the gene. At the cross-over sites, chromosomal DNA molecules are broken enzymically and DNA strands exchanged with that of the incoming cassette DNA via a DNA repair mechanism which re-joins the DNA molecules. Because many genes are essential for survival, gene deletion is first carried out in diploid cells where only one of the two copies of the gene is deleted. Yeast diploid transformants containing the deleted gene are selected on medium containing the antibiotic G418, which is more effective against yeast cells than kanamycin, and these are then forced to undergo sporulation (involving meiosis) to produce haploid spores, 50% of which contain the deleted gene. Growth tests can then be carried out to determine if the gene is essential and to discover the gene's function in the cell.

specific deletion of that gene (Fig. 5.4). In these studies, individual genes, or groups of genes, were deleted from the yeast and changes in phenotype were then looked for. In this way, a particular gene might be associated with a particular function although, in practice, the deletion of a gene does not always give a detectable phenotype.

The majority of vectors used for transforming filamentous fungi rely on random, integration events which occur at a relatively low frequency in most cases. One approach to increase transformation frequencies has been the use of Restriction Enzyme-Mediated Integration (REMI). In this method, the plasmid DNA is added to the cells along with a **restriction enzyme** (see Chapter 4) which cuts once in the vector. Under these conditions, the DNA is targeted to the corresponding restriction sites in the genome. With standard integrative transformation in filamentous fungi, the vector is found predominantly as multiple copies at one or more sites in the genome but with REMI the proportion of single copies at several different sites is increased (Fig. 5.5).

The extent to which an introduced gene is able to lead to production of the protein it encodes is affected by its site of integration in the host's genome. Therefore, methods have been developed to target genes to

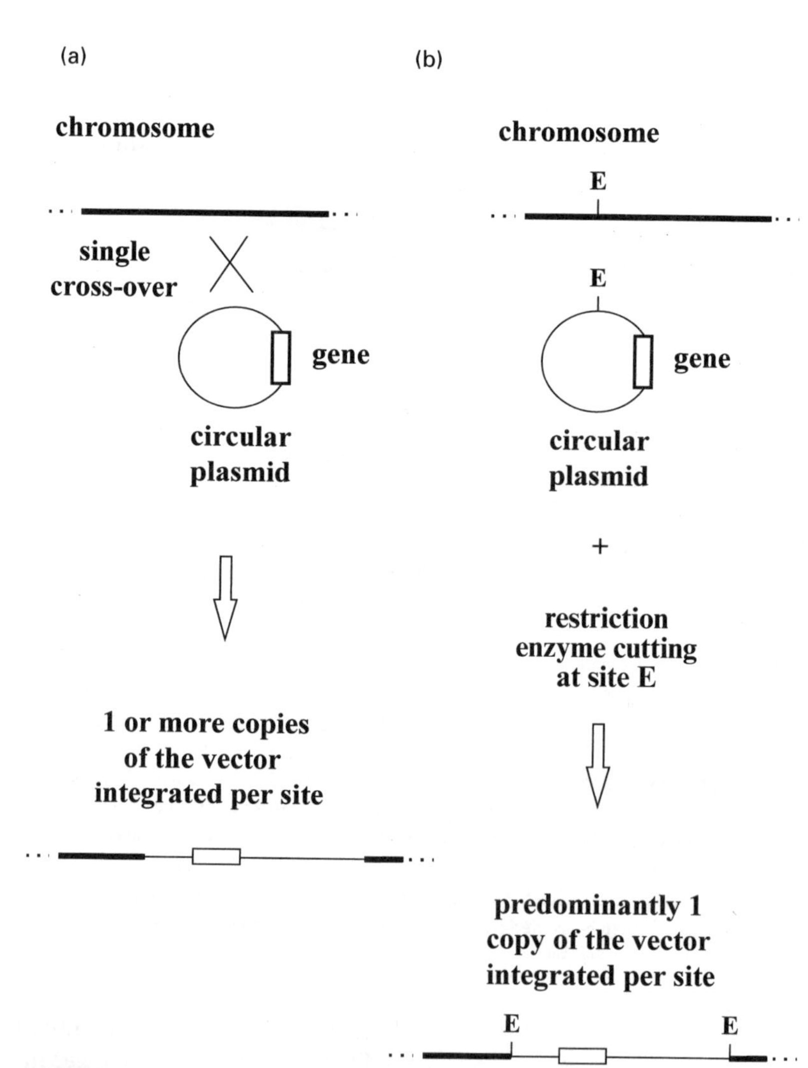

Fig. 5.5 Random plasmid integration into the chromosomes of filamentous fungi. (a) Random (ectopic) integration probably arises by crossing-over between DNA sequences which share a low level of identity and leads to integration at a few chromosomal positions. Due to the nature of the recombination event, integration often results in the duplication or further amplification of the entire vector at the site of integration. (b) Restriction Enzyme-Modified Integration (REMI) involves the addition of a vast excess of a particular restriction enzyme which cuts the plasmid only at one site (E), along with the plasmid DNA during the transformation procedure. It is thought that the enzyme enters the cell's nucleus and cuts genomic DNA randomly at several sites in the chromosomes. Conversion of the circular vector molecule to a linear one by cutting at the same restriction site allows integration of the entire plasmid at several chromosomal sites. This process results in a higher proportion of vector molecules integrated as single copies at many more sites in the fungal genome but, due to its random nature, not all chromosomal sites contain the vector.

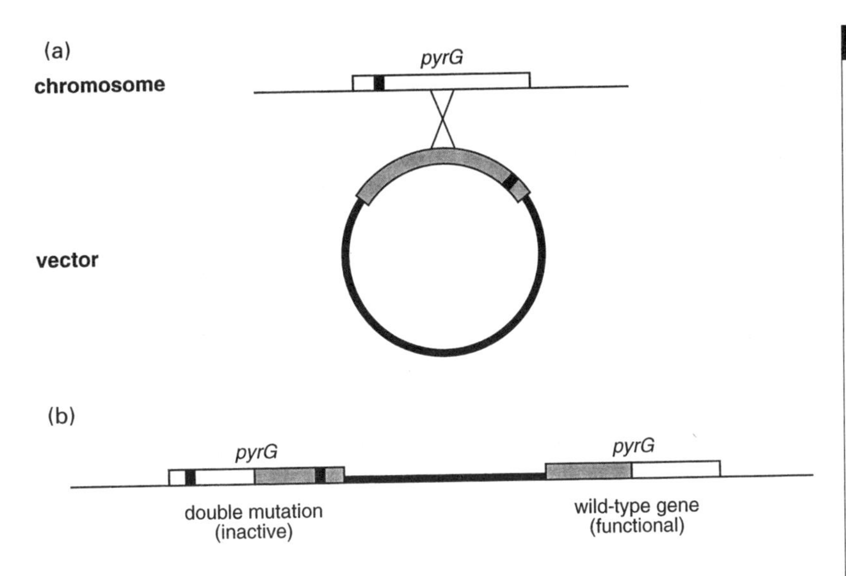

Fig. 5.6 Targeted integration at the *pyrG* locus (encoding orotidine-5′-phosphate decarboxylase) in *Aspergillus awamori*. The host strain of *A. awamori* contains a mutant (non-functional) *pyrG* gene conferring a growth requirement for uridine. The transformation vector contains a non-functional *pyrG* with a mutation in a different part of the gene. By a single homologous recombination event (a single cross-over) (a), the entire vector is integrated at the *pyrG* locus and a functional *pyrG* gene is created (b). The cross-over event involves breakage and re-joining of DNA molecules both in the vector and chromosome by a DNA repair mechanism similar to that outlined in Fig. 5.4. Fungal transformants are selected by their ability to grow without the need for added uridine in the growth medium.

specific chromosomal locations that ensure good **expression**. One method for gene targeting, developed in *Aspergillus* spp., relies on transforming a strain which has one mutation in a particular gene with an integrative plasmid which contains the same gene with a different mutation. A functional gene will only be formed if vector integration occurs at the site of the gene in question by a single cross-over event (Fig. 5.6). Up to 40% of transformants can contain the vector inserted as a single copy at the correct gene locus but transformation frequencies are normally quite low. Specific gene disruption in fungi, using circular plasmids with the appropriate marker, is quite a rare event. For gene deletion or replacement in filamentous fungi, regions adjacent to the gene in question, usually larger than 1 kilobase (kb) in size, have to be incorporated into the vector which is linearised with a restriction enzyme prior to transformation. Gene replacement by a double cross-over event occurs at a frequency of anything from 1% to 50% but is species- and gene-dependent (Fig. 5.7).

5.3 Gene cloning

The success with which fungal genes are isolated and characterised is still heavily dependent on the species being studied. Several different approaches have been adopted to overcome problems presented by genetically poorly characterised, but biotechnologically useful, fungi. The use of mutant strains has been the cornerstone of fungal gene cloning. In this approach, mutant strains are transformed to the wild-type phenotype by introduction of DNA and, when that DNA contains only one gene, it identifies the function of the gene. Other options are now available which exploit the rapidly expanding amount of gene sequence information available in gene databases although a functional assessment of a cloned gene must always be made. For some

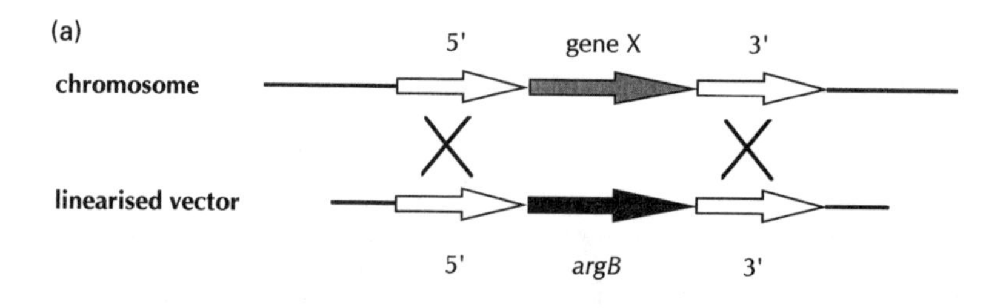

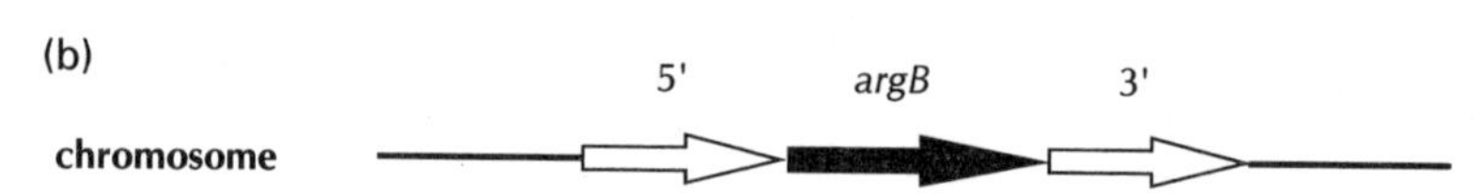

Fig. 5.7 Specific gene deletion in filamentous fungi. Upstream (5′) and downstream (3′) regions of at least 1 kb in size, immediately adjacent to the chromosomal gene to be deleted (gene X), are incorporated into a vector containing a suitable selection marker, e.g. the *argB* gene of *A. nidulans* which encodes ornithine carbamoyl transferase, an enzyme involved in arginine biosynthesis. The plasmid which is normally a circular molecule of DNA is converted to a linear one by cutting with a restriction enzyme that cuts the plasmid once. This linearised vector is then transformed into an *argB* mutant strain. A double homologous recombination event (double cross-over, as outlined in Fig. 5.4) in the 5′ and 3′ regions (a) replaces gene X in the chromosome with the *argB* selection marker (b). Fungal transformants are selected by their ability to grow without the need for added arginine in the growth medium.

genes, 'expression cloning' is a convenient and efficient approach that is based on the introduction into yeast of a new metabolic activity detectable by a simple plate assay. We describe the basic principles involved in these and other strategies below.

5.3.1 Mutant isolation

The complementation of defined mutants remains the most effective means for isolating genes of known function although the generation of such mutants remains a problem for many fungi of interest. Physico-chemical mutagens, e.g. UV or nitrosoguanidine, are still the most common means for generating fungal mutants which may, on occasion, be characterised quickly when simple growth tests are available. Because mutations in other genes in a biosynthetic pathway can confer the same growth requirement, the mutant is tested for its ability to metabolise different substrates to confirm the precise location of the mutation.

A disadvantage of physico-chemical mutagenic methods is that treatment may induce more than one mutation per cell. This may then obscure the real factors underlying an observed property of the mutant. Thus, alternative methods for obtaining mutants have been developed in model organisms such as *S. cerevisiae*, *A. nidulans* and *N. crassa* which have provided useful systems applicable to genetically less well-characterised species. Mutagenesis by means of inserting a piece of DNA into a gene so that the normal DNA sequence is altered (this is called **insertional inactivation**) or by complete gene deletion are preferred

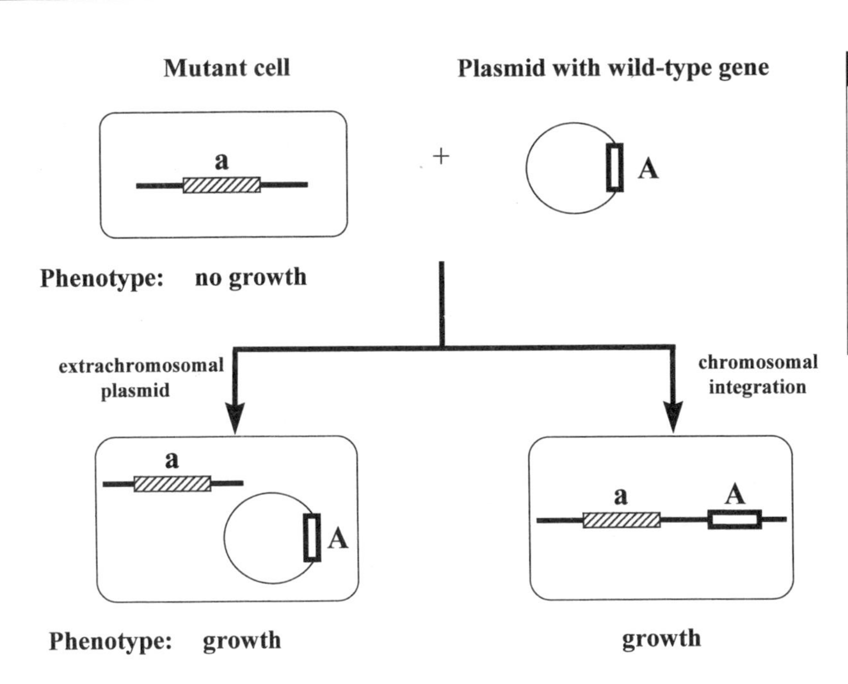

Fig. 5.8 Complementation of a mutant gene 'a' in a cell with the corresponding wild-type gene 'A' on a plasmid. [hatched box] represents the mutant gene 'a' and [open box] represents the wild-type gene 'A'. The presence of gene A within the cell, either as a plasmid or integrated into the chromosome, allows growth.

methods where the need for a defined genetic background is paramount. Fragments of DNA, termed **transposons**, with the ability to move from one site to another within the host genome occur naturally in many species and can be used for this purpose. The transposon Ty is used in this way in *S. cerevisiae* and similar elements have been identified in *Schizosaccharomyces pombe*, *Candida albicans*, *Yarrowia lipolytica* and *Pichia membranaefaciens* although these have not yet been further developed as genetic tools. Other insertional inactivation systems are available and include restriction enzyme-mediated integration (REMI), as already described in Section 5.2.3.

Of necessity, mutants must provide an easily assayed phenotype so that large numbers of colonies can be quickly tested. Screening for the production of extracellular products provides a simple and rapid means to do this. Agar plates containing substrates which, for example, allow the detection of clearing zones or colour changes are frequently used to isolate genes encoding those extracellular enzymes associated with the saprophytic lifestyle of many filamentous fungi. Thus, agar plates containing starch as the sole carbon source can be used to test for the production of amylases. Often, specific strategies can be employed to isolate particular classes of mutant. For example, the location of a protein within a cell can be exploited: if an essential protein is normally located in the cytoplasm but has been genetically engineered to contain signals which will divert it through the secretory pathway to the outside of the cell, then mutants in the secretory pathway which prevent this occurring can be selected for.

5.3.2 Mutant complementation

DNA fragments which represent the entire genome of an organism can be cloned into a vector to produce a population of DNA molecules

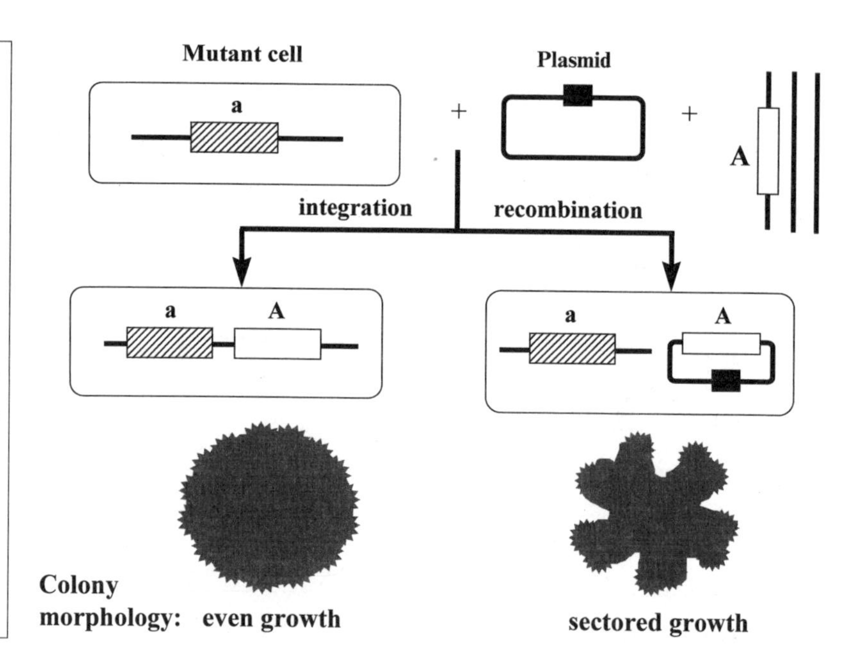

Fig. 5.9 Isolation of a gene by complementation of a defined mutant in a filamentous fungus. [hatched box] represents the mutant gene 'a' and [open box] represents the wild-type gene 'A' found in genomic DNA (———). [black box] represents the fungal origin of replication (*ori*) on the plasmid. Recombination between the genomic DNA and the plasmid results in an unstable inheritance of the wild-type gene 'A' which causes a sectored colony morphology, i.e. growth occurs where the plasmid is present but is prevented when parts of the colony lose the plasmid. The plasmid containing gene 'A' can be recovered from the colonies showing sectored growth.

termed a **library**. Complementation of defined fungal mutants by transformation with a library of fungal DNA is a well-established procedure only for those fungi, e.g. *S. cerevisiae* and *A. nidulans*, which can be transformed at high frequency (Fig. 5.8). Alternatively, some gene products are sufficiently well conserved across species boundaries that they will function in a bacterial background. Thus, a number of fungal genes have been isolated via complementation of *E. coli* mutants although transformation of fungal hosts remains the norm. The surest route to assess gene function relies on the integration of a single copy of the gene in question into the fungal host DNA. Many yeasts, such as *S. cerevisiae*, and some filamentous fungi allow the introduction of a gene on a circular DNA molecule (plasmid) which can be present at several copies per cell. In some cases, this allows closely related (but not identical) genes to complement the mutation and can therefore provide a misleading picture of the gene's function.

Plasmids which replicate extrachromosomally are much less common in filamentous fungi than in yeasts. Despite this, such vectors can still be exploited to identify fungal genes which are able to complement specific mutations (Fig. 5.9). For example, if an *A. nidulans* mutant is transformed with a mixture of a plasmid, containing only a bacterial selection marker and a fungal *ori*, and linear genomic DNA isolated from the fungus under study, two distinct classes of transformant are obtained. The first shows a stable, wild-type phenotype on plates resulting from direct integration of the complementing DNA into the *A. nidulans* chromosomes. The second class of transformant shows uneven growth within colonies on plates where certain sectors of each colony appear wild-type whilst other sectors display the mutant phenotype. This 'sectored' morphology results from recombination between the two source DNAs producing plasmids which carry the complementing gene but which can be lost during the process of cell division. Extraction of the total DNA from such unstable *A. nidulans* colonies, and transformation into *E. coli*, permits the isolation of the plasmids containing the fungal DNA fragment which complemented the original *A. nidulans* mutation. Advantages of the method are that it is rapid and it is not necessary to construct a library of individual genes from the total DNA of the fungus.

5.3.3 Gene isolation by the polymerase chain reaction

Extensive use of the **polymerase chain reaction (PCR)** (see Fig. 4.7) is now made to isolate many specific genes. This approach requires proteins with the same function to have been identified in other organisms and for the gene sequences to be available in a DNA sequence database (e.g. on the internet). Alignment of the protein sequences encoded by these genes against one another can then be used to identify highly conserved regions and allows short sequences of single-stranded DNA (PCR primers) corresponding to these regions to be designed. Because the genetic code is redundant, i.e. more than one triplet of nucleotide bases (codon) can encode the same amino acid, these primers are often mixtures of different DNA sequences which nevertheless encode the same amino acid sequence. Ideally, those conserved regions would contain amino acids which are encoded by a low number of codons to avoid the generation of many different PCR fragment species. If this is not possible, a number of strategies may be used to overcome this problem. The most obvious example is to employ 'nested' primers, i.e. a second set of primers which are internal to the first set, and are designed from additional conserved regions, in a second round of PCR (Fig. 5.10). These can be successful in amplifying fragments of a specific size from a first round reaction which gives a smeared appearance when the PCR products are electrophoresed through an agarose gel. The temperature at which a single-stranded primer binds to its complementary sequence in the template DNA is termed the annealing temperature and depends upon the DNA base composition; therefore, there may only be a narrow range of possible annealing temperatures when a mixture of primers is used. Also, in practice, several temperatures and buffer component concentrations should be examined in order to optimise yields of fragments specific only to those reactions containing both sets of primers. Potential candidate fragments should be cloned if necessary, then sequenced to verify identity to the desired gene and used as probes to isolate clones containing the complete gene sequence from a library.

The major disadvantage of any approach to isolate a gene, other than complementation of a known mutation, is that information on the functionality of the gene is missing. Identifying a gene by a comparison of its DNA sequence with sequences of genes from other organisms rarely provides a conclusive answer, especially when dealing with novel and poorly characterised organisms. Gene deletion experiments should be done to ascertain whether the gene is essential for growth. However, if no clear phenotype emerges from this type of experiment, complementation of a defined mutant should be used. An additional strategy,

Fig. 5.10 'Nested primer' PCR. Redundant primer mixes (① and ②) designed against conserved regions of the target gene are used in the initial PCR. Primer mixes (③ and ④) designed against further conserved regions, internal to the first primer set, are used in a second PCR using some of the first PCR product as template. This should enrich for PCR-amplified products specific to the target gene.

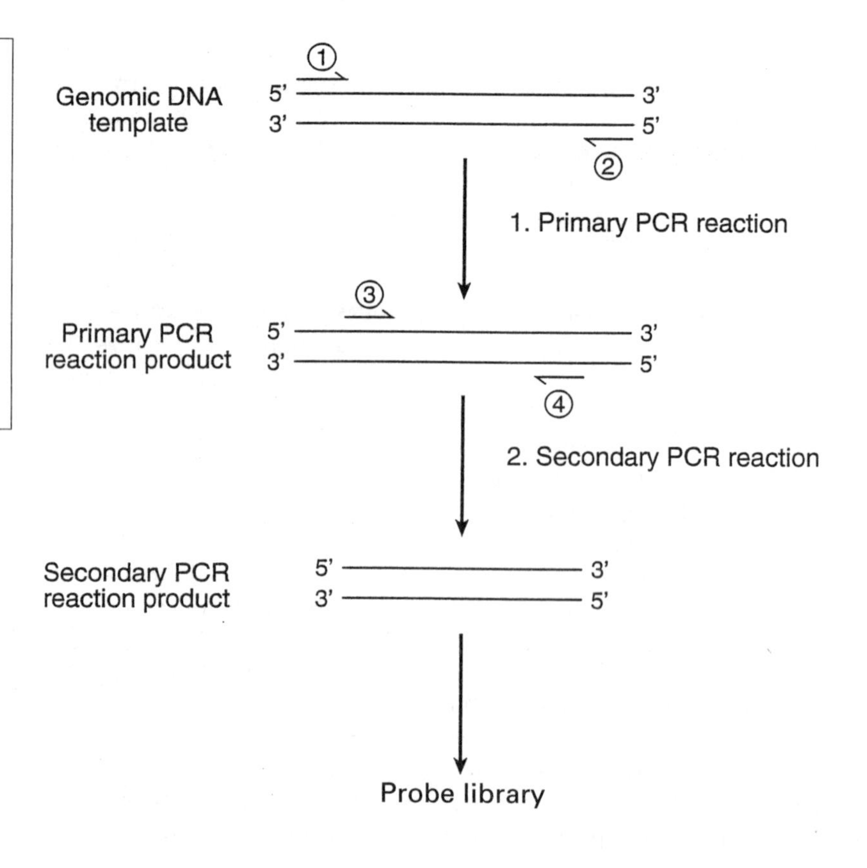

Fig. 5.11 Possible mechanism for the down-regulation of protein production by 'antisense' RNA transcripts. Open boxes represent the target gene X cloned in either of two orientations and flanked by promoter (P) and terminator (T) sequences. Transcription provides sense and antisense mRNA molecules (both with polyA tails) depending on the orientation of the gene relative to the promoter. These two complementary RNA molecules bind together to form a duplex RNA. Formation of duplex RNA in the nucleus reduces the amount of mature mRNA available for translation in the cytoplasm.

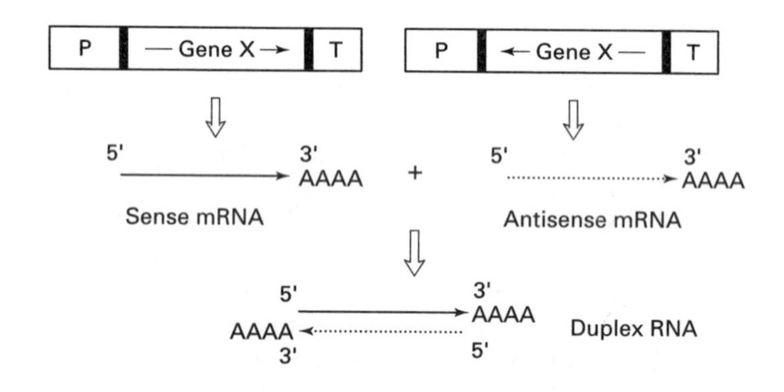

which is especially useful when the gene in question is essential for growth, is to decrease the amount of a protein produced within cells. This uses a process known as 'antisense' in which a messenger RNA (mRNA) sequence complementary to the sense mRNA interferes with the production of the protein encoded by the target gene (Fig. 5.11). This approach is becoming more commonly used for the functional assessment of cloned genes in yeasts and filamentous fungi although it has not always proved successful.

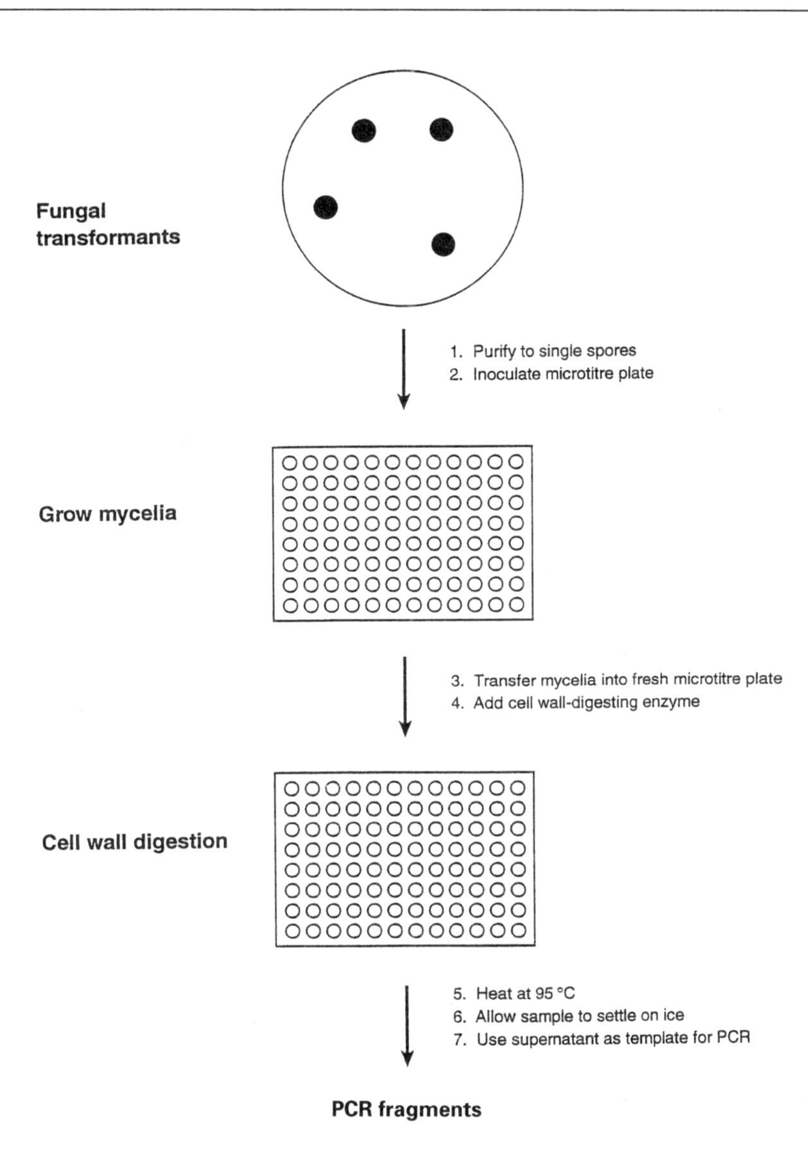
Fungal transformants
1. Purify to single spores
2. Inoculate microtitre plate
Grow mycelia
3. Transfer mycelia into fresh microtitre plate
4. Add cell wall-digesting enzyme
Cell wall digestion
5. Heat at 95 °C
6. Allow sample to settle on ice
7. Use supernatant as template for PCR
PCR fragments

Fig. 5.12 Screening fungal transformants by PCR. Spores from purified fungal transformant colonies on agar plates are inoculated into growth medium in microtitre plates. After 16–24 h growth, fungal mycelia are transferred from the original microtitre plate to a new plate containing a KCl/citrate buffer and an enzyme which digests away the fungal cell wall is added to convert some of the cells to protoplasts. After 1 h at 37 °C, 5–10 μl of supernatant is taken and the DNA which is liberated from the resulting protoplasts by heating acts as a template in the PCR.

5.3.4 PCR and fungi

In addition to its use in the isolation of genes, PCR is also commonly used in screening for the outcome of specific DNA manipulations in many systems. For some filamentous fungi, however, the cell wall has proved a substantial barrier to obtaining DNA of sufficient quantity or purity for large scale screening by PCR. A protocol has now been developed in which fungal colonies are grown in liquid culture in microtitre plates and their cell walls removed by enzymic digestion to release protoplasts. The 'hot' denaturing step of the PCR cycle lyses the protoplasts providing DNA which can be used to screen for the presence of specific DNA fragments (Fig. 5.12). Other protocols, which use mycelia, spores or cells and incorporate a short heating step to burst the cells prior to PCR amplification, work well with many fungal species.

5.3.5 Heterologous gene probes

A third strategy for the isolation of specific genes involves the use of radioactively labelled DNA fragments (probes) from other organisms which encode the same protein as that of the fungal gene required. These are termed heterologous gene probes. Initial experiments in which single-stranded (denatured) fungal genomic DNA is transferred to a nylon membrane after agarose gel electrophoresis (**Southern blotting**) can be used to determine the efficacy of this approach and to determine optimal conditions for hybridisation. Because the sequence of the heterologous probe may be substantially different from the fungal gene due to its different origin, the temperature at which the radioactive probe is annealed to the fungal DNA (hybridisation) is a critical parameter. Improved specificity can be gained by altering the salt concentration of the washes used following hybridisation. Once hybridisation conditions have been optimised, a DNA library from the organism under study can be screened.

5.3.6 Database and linkage-based methods for gene isolation

Finally, two other strategies may be used for gene isolation. The first, termed **synteny**, relies upon the fact that even where there are significant differences between the DNA sequences of genes from different species, gene linkages may be physically maintained. Thus, if in organism X, genes A and B are neighbours on the same chromosome then, if the same case pertains in the fungal system, screening for gene B may allow one to isolate gene A. The main requirement for such a method to work is a detailed genome map of the model organism (X). A number of genomes are currently being sequenced and maps established which should provide a basis for this approach (Table 5.1). The *S. cerevisiae* genome sequence is already complete, that of *S. pombe* is expected to be finished shortly, and those of *C. albicans*, *N. crassa*, *Aspergillus fumigatus* and *A. nidulans* are all under way.

An alternative approach employs partial complementary DNA (**cDNA**) libraries which can be made from fungal RNA samples using readily available cDNA synthesis kits. These cDNAs are termed **Expressed Sequence Tags (ESTs)** since they comprise the ends of transcribed sequences. When allied to automated sequencing, this approach rapidly provides information which can be used to identify those genomic sequences in the gene sequence databases which are transcribed.

5.3.7 Expression cloning

The life style of fungi means that they secrete high concentrations of protein, many of which have application in a variety of industries. A quick and effective method, termed **expression cloning**, has recently been developed to isolate genes encoding extracellular enzymes (Fig. 5.13). In this method, the fungus is grown under conditions expected to induce expression of the target gene(s), and the mRNA is extracted and used to construct a cDNA library in *E. coli*. This library is then used to

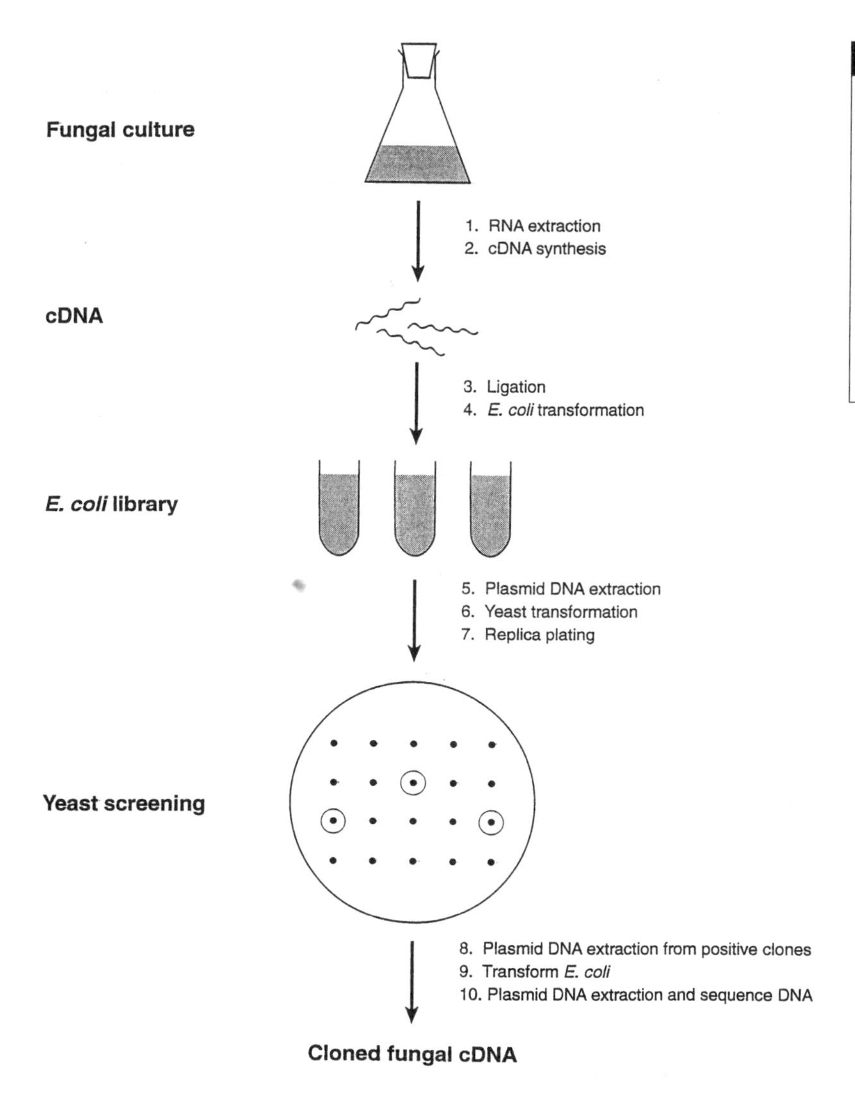

Fig. 5.13 Expression cloning. cDNAs from cultures secreting a range of proteins are cloned into a yeast expression vector capable of replicating in *E. coli*. DNA molecules from *E. coli* transformants are pooled before transforming yeast (*S. cerevisiae*) cells. The yeast transformants are then screened by simple plate assays for the relevant enzymic activities.

transform *S. cerevisiae* and the transformants are grown on media that provide a simple assay for the newly acquired enzyme activity. Other yeast hosts such as *Y. lipolytica*, *K. lactis*, *Pichia angusta* (formerly *Hansenula polymorpha*) and *S. pombe* may be preferred for the transformation step. The effectiveness of this system has been demonstrated by its use to clone genes encoding over 150 fungal enzymes including arabinanases, endoglucanases, galactanases, mannanases, pectinases, proteases and many others. This approach requires that the cDNA step synthesises full length products to ensure the presence of a protein secretion signal which is also functional in the yeast host. Expression of the enzymes in the yeast host and an easily assayed phenotype are also prerequisites.

5.4 Gene structure, organisation and expression

In general, transcriptional signals in genes from higher organisms are more complex than those found in bacterial systems. Within fungi, gene organisation shares many common features across a wide number of genera. The three main organisational units in fungal genes can be split up into those signals which (a) control the switching on or off of gene transcription (promoters), (b) control the termination of transcription (**terminators**) and (c) control the necessary excision of introns from mRNA.

Unlike their bacterial counterparts, fungal promoters can extend a substantial distance (>1 kb) upstream (i.e. 5′) of the **transcriptional start point** (***tsp***). Early cloning experiments in *S. cerevisiae* suggested that native *S. cerevisiae* promoters were required for the expression of foreign genes although subsequent experiments have shown that some promoters from other yeasts, e.g. *K. lactis*, can also function in *S. cerevisiae*. In filamentous fungi, promoters tend to function well within their own genus but are less predictable in more distantly related species.

Promoters can generally be defined as either constitutive, i.e. they are switched on permanently, or inducible, i.e. contain elements which allow them to be switched on or off in a regulated fashion. In both types of promoter, one or more *tsp* can exist. Sequences in the promoter region define both the *tsp* and the binding sites for regulatory proteins. TA-rich regions, known as TATA-boxes, are often involved in determining the *tsp* and also maintaining a basal level of transcription. An illustration of their function is provided by the *S. cerevisiae HIS4* promoter where transcription can occur either with or without a TATA-box (the *HIS4* gene encodes histidinol dehydrogenase involved in histidine biosynthesis); high transcription levels are only observed from the *HIS4* promoter containing a TATA-box. However, there are also strong yeast promoters which lack a TATA-box. Since many yeast and filamentous fungal promoters do not contain this sequence, other motifs, such as the pyrimidine-rich tracts (CT-boxes) found in filamentous fungal promoters, can perform the same function. These sequences can be used to describe what is essentially a 'core' promoter. A third sequence, CCAAT, is also often associated with core promoter function. However, most (>95%) *S. cerevisiae* genes appear not to require CCAAT-boxes for function although this motif is known to function in *A. nidulans*. It should be stressed that in the majority of fungal promoters that have been isolated, the functional significance of sequences identified in them has not been determined and therefore remains unclear.

With a constitutive promoter, the basal level of transcription is determined by the binding to the 'core promoter' of a protein complex which contains RNA polymerase and the so-called general or upstream factors. In contrast, the expression of an inducible gene can change by orders of magnitude. The regulatory elements responsible for mediating this switch are usually found upstream of the core promoter sequences and are termed upstream activation or upstream repression

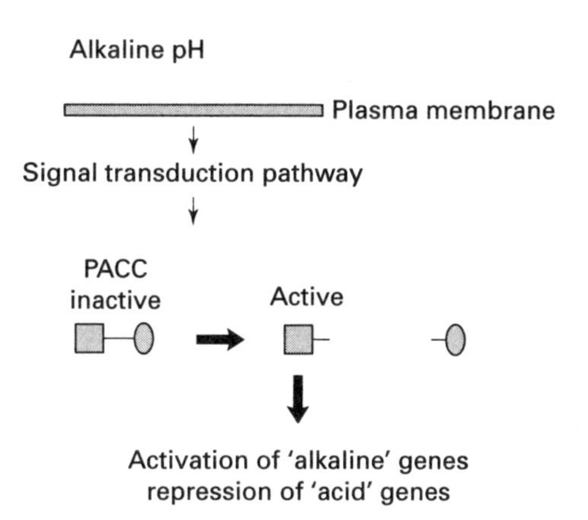

Fig. 5.14 Regulation of transcription by pH in filamentous fungi. The external pH is sensed by cells and leads to changes in gene expression. In fungi, the transcription of genes which encode proteins that are necessary for survival at alkaline pH is mediated by a transcription factor (PACC). Alkaline pH is sensed and leads, through a signal transduction pathway, to the cleavage of an inactive form of PACC to produce the active form. The activated PACC stimulates the transcription of genes leading to enzymes such as alkaline phosphatase that are active at alkaline pH ('alkaline' genes) and represses the transcription of 'acid' genes. The truncated form of PACC activates genes by binding to its target promoters at the sequence 5′-GCCARG-3′ (where R = G or A).

sequences (UAS or URS). These UAS/URS sequences bind regulatory proteins which are thought to stabilise (or destabilise), either directly or indirectly, the transcriptional complex bound to the core promoter thus elevating/decreasing the rate of transcriptional initiation.

Regulatory sequences which control expression of an inducible promoter may often be found in promoters of several genes that encode proteins of unlinked function, all of which are controlled by a single regulatory protein. Such a network of co-regulated genes may respond to physiological parameters affecting the cell such as pH, carbon or nitrogen source. For example, in the filamentous fungus *A. nidulans*, a protein (termed PACC) which binds to a specific sequence in the promoters of genes encoding pH-responsive proteins has been identified (Fig. 5.14). When the ambient pH is alkaline, PACC is activated by a protease to a form which permits expression of a wide range of alkaline-expressed genes (e.g. isopenicillin N synthase) and represses many genes normally expressed under acid conditions (e.g. acid phosphatases). Similarly, in *S. cerevisiae* and some *Kluyveromyces* spp., the DNA-binding protein, MIG1p, binds in a glucose-dependent fashion to GC-rich regions (GC-boxes) of many promoters of genes involved in carbon source utilisation. MIG1p forms a complex with other proteins that represses transcription of genes required for the catabolism of carbon sources which are less efficient in producing energy than glucose and related sugars, a process known as carbon catabolite repression. In filamentous fungi, such as *A. nidulans* and *Trichoderma reesei*, a similar function is performed by the CREA and CRE-1 proteins, respectively, which bind to GC-boxes in the promoters of genes involved in carbon catabolism. Another network of genes, whose control is mediated by the regulatory protein

AREA in *A. nidulans* (NIT2 in *N. crassa*), is that involved in the synthesis of enzymes for nitrogen catabolism. In the presence of the preferred simple nitrogen sources, ammonium and glutamine, genes responsible for breaking down complex nitrogen sources are switched off.

In addition to these regulatory systems of broad specificity covering networks of genes, there are also control mechanisms specific to a particular metabolic pathway. Thus, the positive regulatory protein, AFLR, co-ordinates the expression of at least 25 different genes responsible for the synthesis of the fungal toxin, aflatoxin. All these genes examined so far contain a specific sequence in their promoters to which the AFLR protein binds and activates transcription. Although in this instance the genes are clustered on a small region of one chromosome, there is no known *requirement* for genes which are co-ordinately regulated (either in a global or pathway-specific context) to be physically contiguous or even be on the same chromosome.

What knowledge there is regarding termination of transcription in fungi has derived mainly from studies with *S. cerevisiae* where it is tightly coupled to a process called polyadenylation, that is the addition of long stretches of adenine nucleotides to form polyA mRNA tails. A major function of polyadenylation is to increase the stability of mRNA; in mutants of *S. cerevisiae* where the rate of polyA removal is slowed down, mRNA stability is increased. Most yeast genes lack the sequence AATAAA associated with polyadenylation in higher eukaryotes although two other motifs have been associated with terminator function: a TTTTTAT motif which functions only in one orientation, and a tripartite signal based on a TAG..(T-rich)..TA(T)GT..(AT-rich)..TTT sequence which can function in either orientation. It seems likely that a number of signals are used to terminate transcription in *S. cerevisiae*.

The genes of higher organisms also differ from their bacterial counterparts through the presence of introns which must be excised from the mRNA before it is translated into protein; a process known as **splicing**. *S. cerevisiae* and filamentous fungal genes show considerable differences with regard to the presence and nature of their introns. The majority of *S. cerevisiae* genes lack introns altogether and those that do contain introns often contain only one. Genes from filamentous fungi, and those of *S. pombe*, often have several introns, usually between 50–100 bp in size. The fact that the introduction of filamentous fungal genes into *S. cerevisiae* results in incorrect splicing suggests that the splicing mechanisms differ significantly between some fungal species, although the *amdS* gene from *A. nidulans*, which has three introns, is spliced correctly in several filamentous fungal species. Intron position, though not sequence, is often conserved across filamentous species and introns need not necessarily be within the coding region but can be found in the region between the *tsp* and the point at which translation of the mRNA starts.

(a)

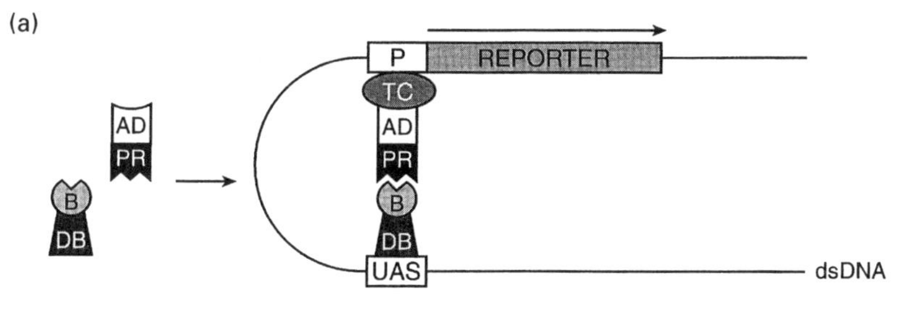

(b)

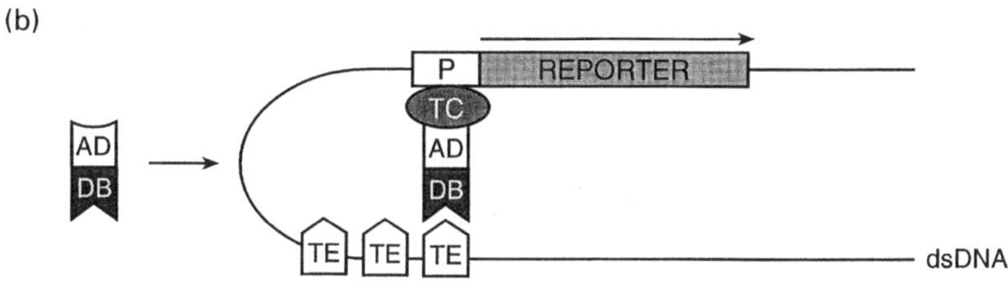

Fig. 5.15 (a) Yeast two-hybrid system based on the GAL4p transcription factor. This protein contains regions for DNA-binding (DB) and for transcription activation (AD) and is involved in switching on galactose utilisation in *S. cerevisiae*. P, GAL4p-activated yeast promoter; TC, transcription complex including the RNA polymerase; DB, the GAL4p DNA-binding region fused to B, the 'bait' protein of known function; PR, the 'prey' protein under study or the products of a cDNA library fused to AD, the GAL4p activation region; UAS, upstream activation sequence required for initiation of transcription at the promoter; dsDNA, double-stranded DNA. Expression of the reporter gene, e.g. the *lacZ* gene encoding β-galactosidase, is switched on if the AD and DB regions of the GAL4p transcription activator are brought together through interaction of the 'bait' and 'prey' proteins. (b) Yeast one-hybrid system which is also based on the GAL4p transcription factor. TE, target element (DNA sequence), normally constructed as at least three repeated copies; AD-DB, the GAL4p activation region fused to the DNA-binding protein of interest or to the products from a cDNA library; other labelling as in (a). Expression of the reporter gene (e.g. *lacZ*) is switched on if the putative DB region binds to the TEs.

5.5 Special methodologies

5.5.1 Yeast two-hybrid system

This genetic-based assay was developed in *S. cerevisiae* to test whether proteins of interest interact within the cell (Fig. 5.15a). The method can be used either to study two proteins whose genes have already been isolated or, more importantly, to identify genes from a cDNA library whose gene products will interact with a known protein of interest, termed the 'bait' protein. The two-hybrid assay relies on the fact that most eukaryotic transcription factors are single proteins which contain regions involved either in promoter DNA-binding or transcription initiation. A common system used is based on the GAL4p transcription activator protein which regulates galactose utilisation in *S. cerevisiae*. Typically, the 'reporter' gene, whose expression is measured in the yeast transformants, is the *lacZ* gene which encodes β-galactosidase, an easily measured enzyme activity. Positive interactions between the proteins under study result in an increased production of β-galactosidase compared with negative controls. This flexible system can be used to investigate interactions between proteins from any organism but once a positive result has been obtained, the interaction in question must be verified biochemically. The two-hybrid system is now commercially available in a number of different kits.

5.5.2 Yeast one-hybrid system

The one-hybrid system was developed from the two-hybrid version to identify genes which encode proteins recognising known specific DNA sequences, such as transcription factors which regulate gene expression or proteins which bind to other sequences such as DNA replication origins (Fig. 5.15b). Positive clones which show increased 'reporter' gene activity are analysed by searches through DNA databases for sequence identity to known DNA-binding proteins and their functional activity verified by protein-DNA binding assays.

5.5.3 Cosmids and artificial chromosomes

With the development of vectors which can carry large fragments of chromosomal DNA, up to 50 kb in size, it is now possible to clone entire fungal pathways, provided all the genes are clustered in one region of the genome, and to transfer these into another organism. **Cosmid** vectors have been constructed which can be propagated in *E. coli* or in filamentous fungi. Like other vectors, these DNA molecules can be designed to either integrate into the chromosome or replicate extrachromosomally when introduced into the fungus. Cosmids with overlapping inserts have been used to clone the entire pathway for aflatoxin/sterigmatocystin biosynthesis, a total of at least 25 coregulated genes, from a number of *Aspergillus* spp. In another case, a cosmid containing the three structural genes for penicillin biosynthesis from *Penicillium chrysogenum* has been integrated into the chromosomes of *Aspergillus niger* and *N. crassa*, both of which then gained the ability to synthesise penicillin.

Yeast artificial chromosomes (YACs) are now also available for cloning large DNA fragments. These are large linear molecules up to 620 kb in size and can be considered to behave as small chromosomes. Problems of YAC stability have led to the development of alternative bacterial artificial chromosome systems (BACs) but YACs still prove useful in several applications. They have been used to analyse foreign gene expression in mammalian cell lines and also in whole animals. YACs have also been used to study chromosome damage in higher cells, to map ESTs in the genome and to clone large DNA fragments for genome sequencing projects.

5.6 Biotechnological applications of fungi

Genetic engineering of yeasts and filamentous fungi is now commonly used for investigating aspects of their biological function and also for

Table 5.3 Useful products from yeasts and filamentous fungi. These products provide targets for genetic manipulation. GM strains are most advanced for the production of antibiotics and enzymes

Product	Uses	Yeasta	Filamentous fungus[a]
Biomass	Foods	*Saccharomyces cerevisiae*	*Agaricus bisporus* *Fusarium venenatum*
Ethanol	Beer, wine	*Saccharomyces cerevisiae*	
CO_2	Bread, wine	*Saccharomyces cerevisiae*	
Sulphite	Preservative (beer)	*Saccharomyces cerevisiae*	
Flavours (e.g. lactones, peptides, terpenoids)	Foods, beverages	*Saccharomyces cerevisiae* *Pichia guilliermondii* *Sporobolomyces odorus*	*Trichoderma viride* *Gibberella fujikuroi* *Mucor circinelloides* *Phycomyces blakesleeanus*
Polyunsaturated fatty acids	Foods	*Cryptococcus curvatus*	*Mortierella alpina* *Mucor circinelloides*
Organic acids (e.g. citric, gluconic, itaconic)	Preservatives, food ingredients, chemical synthesis	*Yarrowia lipolytica*	*Aspergillus niger* *Aspergillus terreus*
Antibiotics (e.g. penicillin, cephalosporin, polyketides)	Health		*Penicillium chrysogenum,* *Acremonium chrysogenum* *Penicillium griseofulvum* *Aspergillus tamorii*
Homologous enzymes (e.g. amylases, cellulases, proteases)	Food processing, paper production, detergents	*Kluyveromyces lactis*	*Aspergillus* spp. *Rhizopus* spp. *Trichoderma* spp.
Heterologous proteins	Foods, therapeutics	*Saccharomyces cerevisiae* *Kluyveromyces lactis* *Pichia pastoris* *Pichia angusta* *Yarrowia lipolytica*	*Aspergillus niger* *Aspergillus oryzae* *Aspergillus nidulans* *Trichoderma reesei*

Notes:
[a] Main species only

constructing strains for particular biotechnological applications. Although genetic engineering is well developed for only a few species, the technology can normally be developed for most species provided sufficient effort is expended. Some of the commercially important products and species are shown in Table 5.3. Each product is a target for improved production using genetically modified strains: this is most advanced in the production of enzymes and antibiotics.

The use of yeasts and filamentous fungi to produce heterologous proteins, i.e. proteins not naturally produced from that species and encoded by a gene derived from another organism, serves as a convenient topic for a comparative discussion. This is because both yeasts and filamentous fungi are used as hosts for heterologous protein production, the technology has advanced sufficiently for commercial use and research is still active in order to improve the systems. To be effective, a production system must deliver sufficient yields (whether it be for commercial viability or to provide research material) of the target protein and the protein must be authentic in its properties, i.e. the same as, or close to, those of the protein from its natural source.

5.6.1 Protein production: the importance of secretion

Most of the commercially available enzymes are secreted from their source organisms although some important enzymes are extracted from cell biomass. The main advantage of producing secreted, as opposed to intracellular, enzymes to a commercial enterprise is the ease (and therefore relative cheapness) of purifying the enzyme. Secreted enzymes are normally correctly folded and active because this is a function of the quality control system within the secretory pathway. Enhanced production of intracellular proteins can lead to the accumulation of improperly folded and inactive protein; or the extraction process itself, which is expensive, may inactivate a proportion of the protein thus reducing recoverable yields. The secretory pathway is also the site of protein glycosylation which may contribute to the function or stability of some proteins. Both *N*-linked (at asparagine) and *O*-linked (at serine or threonine) glycosylation of proteins occur, sometimes with both forms within the same protein. The composition of the glycan component of secreted glycoproteins varies according to species and the glycosylation of heterologous proteins will differ according to the production host. Whether this presents a difficulty in practice will depend upon the intended application of the enzyme. It is likely to be most acute if the heterologous protein is intended for therapeutic use in humans since the glycans of human and fungal proteins have different compositions and their antigenicity can lead to their rapid clearance from the bloodstream. Enzyme activity per se need not be affected by alterations in glycosylation.

A high secretion capacity in the producing organism is desirable although not essential. Those species that naturally secrete a range of enzymes as part of their lifestyles might therefore be expected to be systems of choice for use as hosts for the production of heterologous proteins. Many species of filamentous fungi secrete enzymes to degrade polymeric organic matter and, in laboratory and commercial scale culture, secrete prodigious amounts of enzymes. It is not surprising that the filamentous fungi are the sources of about 10 times the number of commercial (homologous) enzymes that yeasts produce. For the production of heterologous proteins, however, both yeasts and filamentous fungi have been developed as hosts and each system has its attractions and adherents.

5.6.2 Heterologous proteins from yeasts

Saccharomyces cerevisiae is widely used in the production of bread and alcohol, and is regarded as safe. Gene transfer and gene regulation/expression have been extensively studied in *S. cerevisiae* and its widespread familiarity makes it a superficially attractive host organism for heterologous protein production. A large number of different heterologous proteins have been produced following gene transfer into *S. cerevisiae* and the proteins are produced at sufficient yields for commercialisation (e.g. human serum albumin and human insulin). Despite its advantages, there are two major difficulties. The first problem is that many target proteins are 'hyperglycosylated', i.e. the *N*-linked carbohydrate chains are often extremely long and of a high-mannose type which is not characteristic of human glycans. Human serum albumin, on the other hand, is not naturally glycosylated and is not glycosylated when produced by yeast. The second problem is that secreted yields are often initially very low.

A number of alternative yeast expression systems have been developed for the utilisation of cheap, widely available substrates or for more efficient protein secretion. *K. lactis*, grown on lactose-containing whey, is used commercially for the production of lactase (β-galactosidase) and the strong, inducible promoter of the encoding gene is used to drive heterologous gene expression. Chymosin from *K. lactis* is now sold commercially and various other proteins have also been produced with yields of secreted proteins reported in the literature generally higher than those of *S. cerevisiae*. Two methanol-utilising yeasts, *Pichia angusta* and *P. pastoris*, possess the strong, methanol-inducible promoter from the methanol oxidase gene which is used to drive the expression of introduced genes. Impressive secreted yields of several heterologous proteins have been reported from both species particularly when high cell densities are obtained in bioreactors. Also, hyperglycosylation appears not to be a major problem in proteins secreted at high yields from *P. pastoris*. *Y. lipolytica* (Fig. 5.1b) is unusual in being capable of growth on some hydrocarbons although this capacity has not yet been exploited through provision of promoters to drive heterologous gene expression. Rather, the regulated promoter of the gene encoding the major secreted alkaline protease has been used to drive expression but other promoters are also available. A wide range of proteins is naturally secreted by *Y. lipolytica*, unlike some other yeasts, which indicates a well-developed secretory capacity.

The **expression cassettes** used for heterologous protein production have several interesting features (Fig. 5.4). The foreign gene (usually a cDNA gene to avoid the possibility of splicing introns incorrectly in a heterologous host organism) is flanked by a yeast promoter and transcriptional terminator. In general, a promoter derived from the host yeast is preferred, although some promoters operate effectively across species (e.g. the promoter from the *S. cerevisiae PGK* gene is used in *K. lactis*). A short (15–30 amino acids) peptide sequence, called a signal sequence, at the *N*-terminal end of the protein to be produced is a requirement for the secretion of proteins (see Fig. 4.12). Cleavage of the signal sequence is carried out in the cell by a specific intracellular endopeptidase upon entry of the protein to the endoplasmic reticulum (ER) which is the start of the secretory system in eukaryotes. The cleavage site is not sequence-specific but is governed primarily by the size and charge distribution (usually a positively charged *N*-terminal region and a hydrophobic core) of the signal sequence. Several heterologous, homologous and synthetic signal sequences have been examined with many working effectively. It has become more common to include a short 'pro-sequence' after the signal sequence and before the *N*-terminus of the target protein. Pro-sequences are naturally present in many secreted homologous proteins and can aid folding. In *S. cerevisiae*, the signal sequence and pro-sequence of the secreted α-mating factor protein is often employed. This pro-sequence ends in a dibasic pair of amino acids, lysine-arginine, and an endopeptidase that is located in the Golgi body (an organelle in the secretory pathway) cleaves after the lysine-arginine to release mature target protein with its correct *N*-terminal sequence. This endopeptidase is called KEX2 in *S. cerevisiae* and equivalent enzymes are found in other yeasts and filamentous fungi.

Many heterologous proteins are produced at yields which are too low (for commercial viability or for experimental purposes) or are structurally and functionally different from the authentic protein. Yeasts are no different in this regard to other expression systems and modifications to the standard procedures have been pursued in order to overcome the difficulties. The use of vectors which replicate at high copy number can titrate out the necessary transcription factors so that they become limiting. This occurs in *S. cerevisiae* with the galactose-inducible *GAL1* promoter. Increasing the expression of the associated regulatory protein (GAL4p) overcomes this titration effect.

Many of the limitations leading to low secreted yields are post-translational and relate to the secretory pathway or to proteolytic degradation. Thus, protease-deficient mutants have been examined and, conversely, enhancement of the specific proteolytic activities of the signal peptidase and KEX2 protease have also been examined. Each approach has shown some promise without wholly overcoming the bottle-necks to high yields. Folding of proteins during secretion occurs within the ER and is assisted by resident **chaperone** proteins. Other proteins, termed foldases, also catalyse folding by the formation of disulphide bonds within the ER. Up-regulated expression of foldases and chaperones has increased the secreted yield of some, but not all, heterologous proteins from *S. cerevisiae*. Mutagenesis of strains has been used to increase secreted protein yields and in a few cases mutations have been localised to particular genes. Mutations that affect all aspects, including transcription, proteolysis, secretion and glycosylation, have been recorded. Although mutagenesis will continue to be used as a tool for improving yields, many mutations are recessive and not easily incorporated into polyploid commercial strains which have multiple copies of each chromosome. Thus targeted gene manipulations provide a complementary and valuable approach.

5.6.3 Heterologous proteins from filamentous fungi

Typical fungal expression vectors are not dissimilar to those described already for yeast. As discussed in Section 5.2.3, the main difference is that autonomous replication is not normally an option with commercial filamentous fungi and all vectors, with the exception of some used for research purposes, are designed to integrate into the fungal genome. As with yeast integration vectors, genomic integration of the transforming DNA brings added, though not necessarily complete, stability but some uncertainty about the levels of gene expression expected and a limit to the number of gene copies that can be integrated. From a practical standpoint, a fungal transformation produces transformed strains which differ in their level of heterologous protein produced. One of the reasons for variation in gene expression upon genomic integration of the vector is that some parts of the genome are more transcriptionally active than others. Multiple copies of transforming DNA can be achieved through selection pressure and an elegant example in *A. niger* is the use of the *A. nidulans amdS* gene described earlier in Section 5.2.3. It is not, however, advantageous to increase the copies beyond a certain limit because there is no advantage in terms of protein yields. For example, beyond about 20 copies of the *A. niger glaA* promoter or three copies of the *T. reesei* cellobiohydrolase I promoter (*cbh1*) essential transcription factors become limiting. Analogous observations were made in yeast (Section 5.6.2) and, there, the limitation was overcome by up-regulating the expression of the transcription factor(s). This strategy may work also in filamentous fungi.

Aside from the gene dosage strategy just described, mutagenesis of protein-producing strains and screening the resulting mutants for improved protein production is another effective method for strain optimisation. Indeed, the combination of targeted genetic modification (GM) with conventional mutagenesis-based strain improvement provides a very powerful approach for improved production of heterologous proteins. Two further GM-based strategies are now routinely used. The most commonly exploited fungi secrete proteases which might degrade the target heterologous protein. Therefore, one strategy is to use either protease-deficient mutant strains or strains in which specific protease genes have been eliminated by gene disruption or gene deletion. The second strategy is to employ gene fusions whereby the target gene is fused downstream of a gene encoding a naturally well-secreted 'carrier' protein such as glucoamylase in *A. niger* or cellobiohydrolase in *T. reesei*. When this tactic was used for production of calf chymosin, the chymosin (which is a protease) cleaved itself from the glucoamylase-prochymosin fusion at the pro-sequence boundary to release mature chymosin. For most heterologous proteins this approach is not possible and a cleavable protease site is included at the fusion junction. The site usually employed is the dibasic lysine-arginine sequence which, in yeast, is the site for cleavage by the KEX2 protease. This works well too in filamentous fungi (Fig. 5.16). The fusion protein strategy appears to increase secreted protein yields by enhancing mRNA stability and by easing the passage of protein through the secretory pathway, although the underlying mechanisms are not known.

Recognition of bottle-necks in achieving high yields of secreted proteins which have the activities expected of the authentic proteins is the first step towards defining strategies which overcome the problems. It is

(a) Entry of nascent protein into the ER lumen

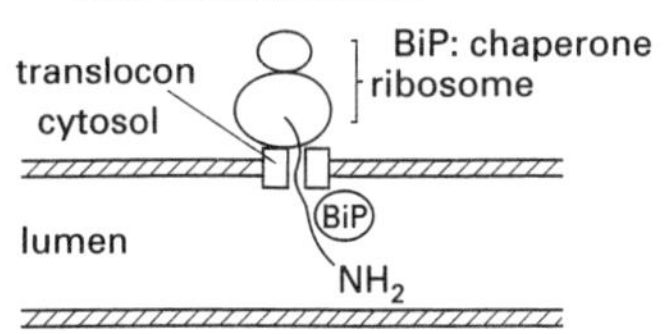

(b) Assisted protein folding in the ER lumen

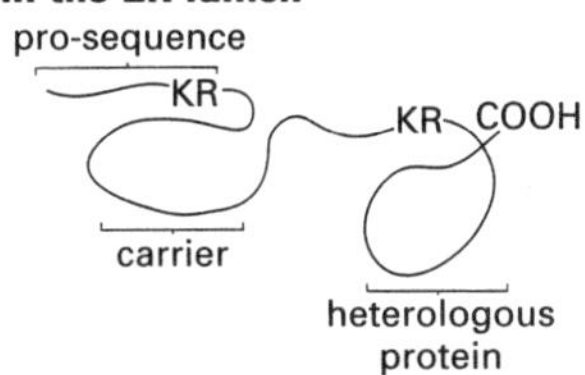

(c) Endopeptidase cleavage in the late secretory pathway (Golgi) and release of carrier protein and heterologous protein to the cell exterior

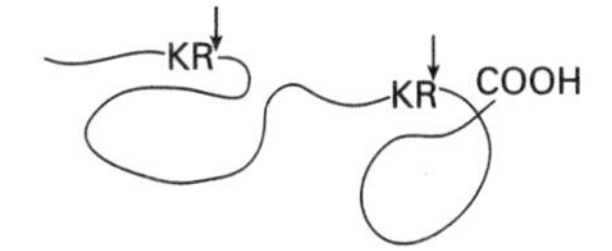

Fig. 5.16 Folding and processing of secretory fusion proteins in filamentous fungi. (a) Entry of nascent polypeptide into the lumen of the endoplasmic reticulum (ER). The signal sequence which directs entry of the polypeptide is removed by signal peptidase so that the emerging polypeptide within the lumen lacks the signal sequence. BiP is an abundant chaperone within the lumen which is associated with early protein folding events. Other chaperones and foldases (see text) are also present. (b) Folding of the full-length fusion protein within the ER. (c) The fusion protein is cleaved within the Golgi body by a specific peptidase (KEX2 in *S. cerevisiae*) to release the heterologous protein to the cell exterior following transport of the protein by membrane-bound vesicles (d).

already clear that, as with yeast expression, several factors can conspire to present a bottle-neck and that their relative importance depends on the heterologous protein. Foreign genes which use codons not common in fungi, the presence of sequences which destabilise mRNA, differences in the protein folding/secretory pathway and the abundance of proteases all contribute to the observed bottlenecks. In addition, although hyperglycosylation of heterologous proteins is not such a problem with filamentous fungi as it is with *S. cerevisiae*, it can still be a difficulty. In addition, the patterns of glycosylation differ from those seen in mammalian cells which could be important for therapeutic protein production. The glycan structures in fungal glycoproteins are being analysed and the genes that encode enzymes responsible for glycan assembly are being cloned, providing the possibility in the future of manipulating glycan synthesis.

The essential details of the secretory pathway in filamentous fungi appear to be qualitatively very similar to those in the yeast system which has been studied more extensively. Some of the genes that encode chaperones and foldases have been cloned, as have genes that encode proteins involved in vesicular transport. Although successful manipulation of the protein secretory pathway using these genes has not yet been reported, the necessary tools to do so are becoming available.

5.7 Further reading

Ausubel, F. M., Brent, R., Kingston, R. E., Moore, D. D., Seidman, J. G., Smith, J. A. and Struhl, K. (1995). *Current Protocols in Molecular Biology*. John Wiley, New York.

Broda, P., Oliver, S. G. and Sims, P. F. G. (1993). *The Eukaryotic Genome: Organisation and Regulation*. Cambridge University Press, Cambridge.

Gellissen, G. and Hollenberg, C. P. (1997). Application of yeasts in gene expression studies: a comparison of *Saccharomyces cerevisiae, Hansenula polymorpha* and *Kluyveromyces lactis* – a review. *Gene* **190**, 87–97.

Gow, N. A. R. and Gadd, G.M. (eds.) (1995). *The Growing Fungus*. Chapman and Hall, London.

Kinghorn, J.R. and Turner, G. (eds.) (1992). *Applied Molecular Genetics of Filamentous Fungi*. Blackie Academic & Professional, Glasgow.

Luban, J. and Goff, S. P. (1995). The yeast two-hybrid system for studying protein-protein interactions. *Curr. Opin. Biotechnol.* **6**, 59–64.

Oliver, R. P. and Schweizer, M. (eds.). (1999). *Molecular Fungal Biology*. Cambridge University Press, Cambridge.

Wolf, K. (ed.) (1996). *Non-conventional Yeasts in Biotechnology*. Springer-Verlag, Berlin.

Chapter 6

Microbial process kinetics

Jens Nielsen

Nomenclature

a	constant in defined by Eqn (6.62)
A	constant
B	constant
c	concentration ($g\,l^{-1}$ or moles l^{-1})
c_s	concentration of a substrate essential for growth ($g\,l^{-1}$ or moles l^{-1})
$c_{s,0}$	initial concentration of the limiting substrate ($g\,l^{-1}$ or moles l^{-1})
$c_{s,e}$	concentration of a growth enhancing substrate ($g\,l^{-1}$ or moles l^{-1})
$c_{s,i}$	concentration ($g\,l^{-1}$ or moles l^{-1})
D	dilution rate (h^{-1})
D_{crit}	critical dilution rate (h^{-1})
e^-	active form of the enzyme
e, e^{2-}	inactive forms of the enzyme
E_g	activation energy of the growth process (kJ $mole^{-1}$)
F	flow rate ($l\,h^{-1}$).
F_{out}	flow rate out of the bioreactor ($l\,h^{-1}$)
ΔG_d	change in free energy (kJ $mole^{-1}$)
K_I	inhibition constant ($g\,g^{-1}$)
K_1	dissociation constant
K_2	dissociation constant
K_s	saturation coefficient ($g\,l^{-1}$ or moles l^{-1})
m_s	maintenance coefficient ($g\,g^{-1}\cdot h$)
P_x	productivity ($g\,l^{-1}\cdot h$ or moles $l^{-1}\cdot h$)
r_i	specific rates ($g\,g^{-1}\cdot h$ or moles $g^{-1}\cdot h$)
r_p	specific production rate ($g\,g^{-1}\cdot h$ or moles $g^{-1}\cdot h$)
r_s	specific substrate uptake rate ($g\,g^{-1}\cdot h$ or moles $g^{-1}\cdot h$)
T	temperature (K)
t	time (h)
t_d	doubling time (h)
V	volume (l)

x biomass concentration ($g\,l^{-1}$).
Y_{ji} yield coefficient specifying the amount of *i* produced per unit of *j* consumed ($g_j\,g_i^{-1}$)
Y_{ij} yield coefficient specifying the amount of *j* converted per unit of *i* produced ($g_i\,g_j^{-1}$)
Y_{sx} yield coefficient specifying the amount of biomass formed per unit substrate consumed ($g\,g^{-1}$)
Y_{xp} yield coefficient specifying the amount of product formed per unit biomass formed ($g\,g^{-1}$)
Y_{xATP} yield coefficient specifying the amount of ATP consumed per unit of biomass produced ($g\,g^{-1}$)
Y_{xs} yield coefficient specifying the amount of substrate used per unit biomass formed ($g\,g^{-1}$)

Subscripts
e growth enhancing compound
i *i*-th substrate or product
j essential growth compound
o initial conditions
s substrate
x biomass
p product

Superscripts
f feed

Greek letters
α,β coefficients in equation 6.11
ΔG_d free energy change
μ *specific growth rate* of the total biomass ($g\,g^{-1}\cdot h$ or simply h^{-1})
μ_{max} maximum *specific growth rate* of the total biomass ($g\,g^{-1}\cdot h$ or simply h^{-1})

6.1 Introduction

Quantitative description of cellular processes is an indispensable tool in the design of fermentation processes. The two most important quantitative design parameters, **yield** and **productivity**, are quantitative measures that specify how the cells convert the substrates to the product. The **yield** specifies the amount of product obtained from the substrate, and it is of particular importance when the raw material costs make up a large fraction of the total costs, as exemplified in the production of solvents, antibiotics, alcohol, and other primary metabolites. The **productivity** specifies the rate of product formation, and is particularly important when the capital investments play an important role, such as in a growing market where there is an increasing demand for producing the product by a given capacity (or factory). These two design parameters can easily be derived from experimental data but, what is more difficult to predict, is how they change with the operating conditions, e.g. if the medium composition changes or the temperature changes. To do this it is necessary to set up a mathematical model.

A model is a **set of relationships** between the **variables** in the system being studied. These relationships are normally expressed in the form of mathematical equations, but they may also be specified as logic expressions (or cause/effect relationships) which are used in the operation of a process. The variables include any property that are of importance for the process, such as the agitation rate, the feed rate, pH, temperature, concentrations of substrates, metabolic products and biomass, and the state of the biomass – often represented by the concentration of a set of key intracellular compounds.

To set up a mathematical model it is necessary to specify a **control volume** wherein all the variables of interest are taken to be uniform. For fermentation processes the control volume is typically the whole bioreactor, but for large bioreactors the medium may be non-homogeneous due to mixing problems and here it is necessary to divide the bioreactor into several control volumes. When the control volume is the whole bioreactor it may either be of constant volume or it may change with time depending on the operation of the bioprocess.

When the control volume has been defined, a set of **balance equations** can be specified for the variables of interest. These balance equations specify how material is flowing in and out of the control volume and how material is converted within the control volume. Rate equations (or kinetic expressions) specify the conversion of material within the control volume. They may be anything from a simple empirical correlation that specifies the product formation rate as a function of the medium composition to a complex model that accounts for all the major cellular reactions involved in the conversion of the substrates to the product.

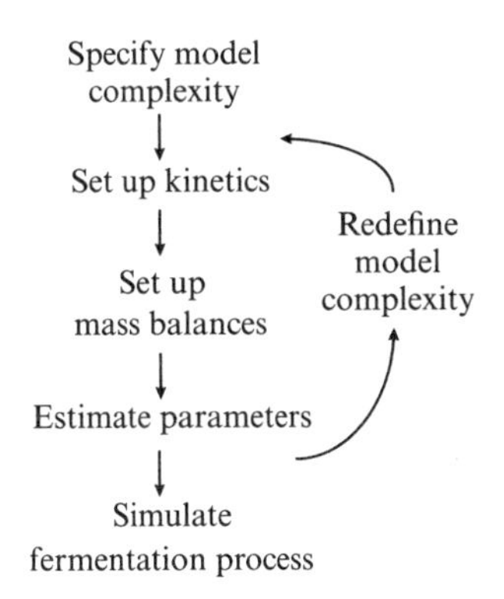

Fig. 6.1 Different steps in quantitative description of fermentation processes.

Independent of the model structure, the process of defining a quantitative description of a fermentation process involves a number of steps, as shown in Fig. 6.1.

A key aspect in setting up a model is to specify the model complexity. This depends on what the model is going to be used for (see Section 6.2.1). Specification of the model complexity involves defining the number of reactions to be considered in the model, and specification of the stoichiometry for these reactions. When the model complexity has been specified, rates of the cellular reactions considered in the model are described with mathematical expressions, i.e. the rates are specified as functions of the variables; namely the concentration of the substrates (and in some cases the metabolic products). These functions are normally referred to as **kinetic expressions**, since they specify the kinetics of the reactions considered in the model. This is an important step in the overall **modelling cycle** and, in many cases, different kinetic expressions have to be examined before a satisfactory model is obtained.

The next step in the modelling process is to combine the kinetics of the cellular reactions with a model for the reactor in which the cellular process occurs. Such a model specifies how the concentrations of substrates, biomass, and metabolic products change with time, and what flows in and out of the bioreactor. These **bioreactor models** are normally represented in terms of simple mass balances over the whole reactor, but more detailed reactor models may also be applied, if inhomogeneity of the medium is likely to play a role. The combination of the kinetic and the reactor model makes up a complete mathematical description of the fermentation process and this model can be used to simulate the profile of the different variables of the process, e.g. the substrate and product concentrations. However, before this can be done it is necessary to assign values to the parameters of the model. Some of these parameters are operating parameters, which are dependent on how the process is operated, e.g. the volumetric flow in and out of the bioreactor, whereas others are kinetic parameters which are associated with the cellular system. To assign values to these parameters it is necessary to compare model simulations with experimental data and hereby estimate a parameter set that gives the best fit of the model to the experimental data. This is referred to as **parameter estimation**. The evaluation of the fit of the model to the experimental data can be done by simple visual inspection of the fit, but generally it is preferential to use a more rational procedure, such as minimising the sum of squared errors between the model and the experimental data.

In the following we will consider the two different elements needed for setting up a bioprocess model, namely kinetic modelling and mass balances. This will lead to a description of different types of bioreactor operation, and hereby simple design problems can be illustrated. Even though parameter estimation is an important step in the overall modelling cycle, we will not consider this, since the tools available for this are extensively described elsewhere.

6.2 Kinetic modelling of cell growth

All researchers in life sciences use models when results from individual experiments are interpreted and when results from several different experiments are compared with the aim of setting up a model that may explain the different observations. During the last 10 years there has been a revolution in experimental techniques applied in life sciences, and this has made possible far more detailed modelling of cellular processes. Furthermore, the availability of powerful computers has made it possible to solve complex numerical problems with a reasonable computational time; even complex mathematical models for biological processes can be handled and experimentally verified. However, often such detailed (or mechanistic) models are of little use in the design of a bioprocess, whereas they mainly serve a purpose in fundamental research of biological phenomena. In this presentation we will focus on models which are useful for design of bioprocesses, but in order to give an overview of the different mathematical models applied to describe biological processes we start the presentation of kinetic models with a discussion of model complexity.

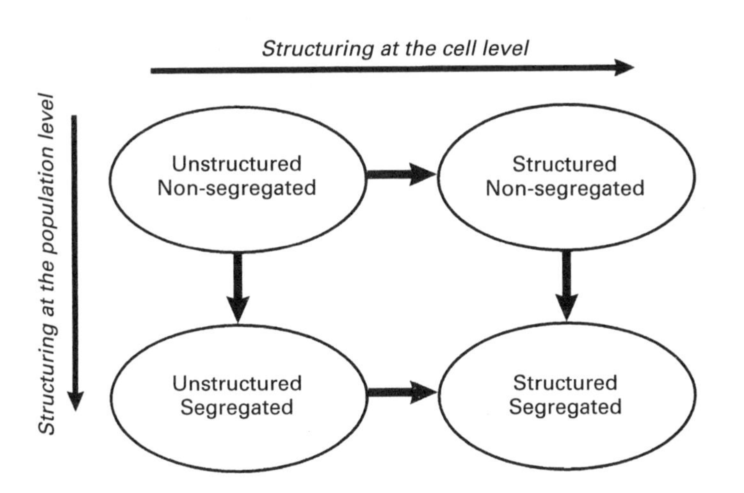

Fig. 6.2 Different types of model complexity, with increasing complexity going from the upper left corner to the lower right corner. When there is structuring at the cell level, specific intracellular events or reactions are considered in the model, and the biomass is structured into two or more variables. When there is structuring at the population level, segregation of the population is considered, i.e. it is accounted for that not all the cells in the population are identical.

6.2.1 Model structure and model complexity

Biological processes are per se extremely complex. Cell growth and metabolite formation are the result of a very large number of cellular reactions and events like gene expression, translation of mRNA into functional proteins, further processing of proteins into functional enzymes or structural proteins, and sequences of biochemical reactions leading to building blocks needed for synthesis of cellular components (see Chapter 2).

It is clear that a complete description of all these reactions and events cannot possibly be included in a mathematical model. In fermentation processes, where there is a large population of cells, non-homogeneity of the cells with respect to activity and function may add further to the complexity. In setting up fermentation models lumping of cellular reactions and events is therefore always done but the detail level considered in the model, i.e. the degree of lumping, depends on the aim of the modelling.

Fermentation models can roughly be divided into four groups depending on the detail level included in the model, see Fig. 6.2. The simplest description is the so-called **unstructured models** where the biomass is described by a single variable (often the total biomass concentration) and where no segregation in the cell population is considered. These models can be combined with a **segregated** population model, where the individual cells in the population are described by a single variable, e.g. the cell mass or cell age, but often it is relevant to add further structure to the model when segregation in the cell population is considered. In the so-called **structured models** the biomass is described with more than one variable, i.e. structure in the biomass is considered. This structure may be anything from a few compartments to a detailed structuring into individual enzymes and macromolecular pools.

It is clear that a very important element in mathematical modelling of fermentation processes is defining the model structure (or specifying the complexity of the model), and for this, a general rule can be stated:

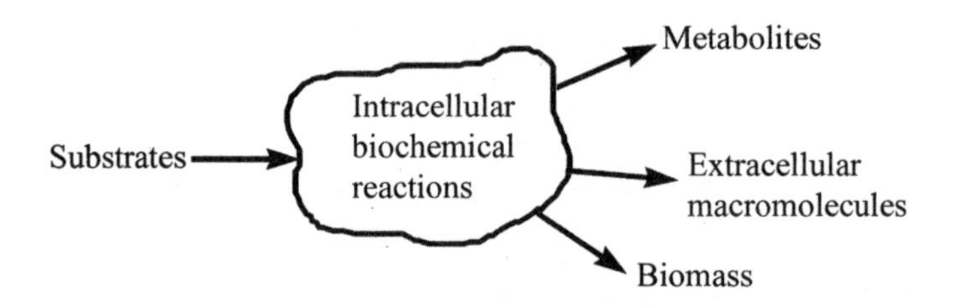

Fig. 6.3 General representation of cellular growth and product formation. Via a large number of intracellular biochemical reactions, substrates are converted into metabolic products, e.g. ethanol, acetate, lactate, or penicillin (and other secondary metabolites), extracellular macromolecules, e.g. a secreted enzyme, a heterologous protein, or a polysaccharide, and into biomass constituents, e.g. cellular protein, lipids, RNA, DNA, and carbohydrates.

As simple as possible but not simpler. This rule implies that the basic mechanisms always should be included and that the model structure depends on the aim of the modelling exercise.

6.2.2 Definitions of rates and yield coefficients

Before we turn to describing different unstructured models, a few definitions are needed. Figure 6.3 is a representation of the overall conversion of substrates into metabolic products and biomass components (or total biomass). The rates of substrate consumption can be determined during a fermentation process by measuring the concentration of these substrates in the medium. Similarly, the rates of formation of metabolic products and biomass can be determined from measurements of the corresponding concentrations. It is therefore possible to determine what flows into the total pool of cells and what flows out of this pool. The inflow of a substrate is normally referred to as the **substrate uptake rate** and the outflow of a metabolic product is normally referred to as the **product formation rate**. From the direct measurements of the concentrations, one obtains so-called **volumetric rates**. Often it is convenient to normalise the rates with respect to the amount of biomass present, since the rates hereby easily can be compared between fermentation experiments, even when the amount of biomass changes. Such normalised rates are referred to as **specific rates**, and these are often represented as r_i, where the subscript indicates whether it is a substrate (s) or a metabolic product (p). The **specific growth rate** of the total biomass is also a very important variable, and it is generally designated μ. The specific growth rate is related to the **doubling time** t_d (h) of the biomass through:

$$t_d = \frac{\ln 2}{\mu} \tag{6.1}$$

The doubling time t_d is equal to the generation time for a cell, i.e. the length of a cell cycle for unicellular organisms, which is frequently used by life scientists to quantify the rate of cell growth.

The specific rates, or the flow in and out of the cell, are very important design parameters since they are related to the productivity of the cell. Thus, the specific productivity of a given metabolite directly indicates the capacity of the cells to synthesise this metabolite. Furthermore, if the specific rate is multiplied by the biomass concentration in the bioreactor one obtains the volumetric productivity, or the capacity of the biomass population per reactor volume. In simple kinetic models the specific rates are specified as functions of the variables in the system, e.g. the substrate concentrations. In more complex models where the rates of the intracellular reactions are specified as functions of the variables in the system, the substrate uptake rates and product formation rates are given as functions of the intracellular reaction rates.

Another class of very important design parameters are the yield coefficients, which quantify the amount of substrate recovered in biomass and the metabolic products. The yield coefficients are given as ratios of the specific rates, e.g. for the yield of biomass on a substrate:

$$Y_{sx} = \frac{\mu}{r_s} \tag{6.2}$$

and similarly for the yield of a metabolic product on a substrate:

$$Y_{sp} = \frac{r_p}{r_s} \tag{6.3}$$

The yield coefficients are clearly determined by how the carbon in the substrate is distributed among the different cellular pathways towards the end products of the catabolic and anabolic routes. These parameters can be considered as an overall determination of **metabolic fluxes**, a key aspect in modern physiological studies where methods to quantify intracellular, metabolic fluxes have become an important tool in defining the activity of the different pathways within the complete metabolic network. In the production of low-value added products, e.g. ethanol, antibiotics, amino acids and baker's yeast, it is generally of utmost importance to optimise the yield of product on the substrate and the target is therefore to direct as much carbon as possible towards the product and minimise the carbon flow to by-products (including biomass in metabolite production processes). In this process the yield coefficient is the most important design parameter, both for characterising different mutants and for characterising different fermentation schemes.

For aerobic processes the yield of CO_2 from O_2 is often used to characterise the metabolism of the cells. This yield coefficient is referred to as the **respiratory quotient** (RQ). With complete respiration the RQ is close to 1 whereas if a metabolite is formed it deviates from 1 (see also Section 6.2.4).

The yield coefficients are always given with a double index that indicates the direction of the conversion, i.e. the yield for the conversion of substrate to biomass ($s \rightarrow x$) has the index *sx*. Thus, the yield coefficient Y_{xs} specifies the amount of substrate converted per unit biomass formed and, similarly, the yield coefficient, Y_{xp}, specifies the amount of product formed per unit biomass formed.

6.2.3 Black box models

The simplest mathematical presentation of cell growth is the so-called **black box** model, where all the cellular reactions are lumped into a single overall reaction. This implies that the yield of biomass on the substrate (as well as the yield of all other compounds consumed and

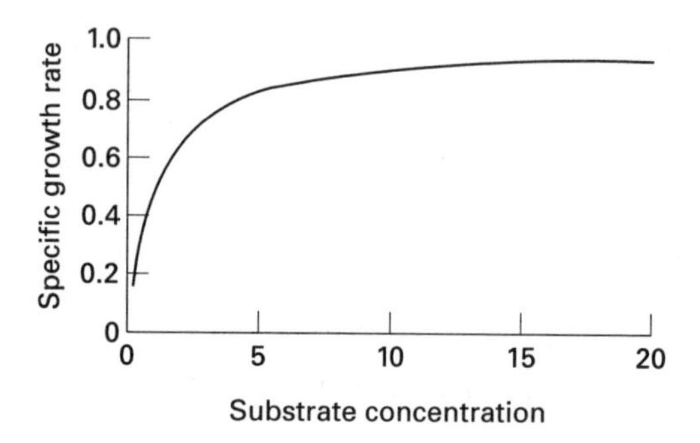

Fig. 6.4 The specific growth rate, μ, as function of the concentration of the limiting substrate, s, when the Monod model is applied.

produced by the cells) is constant. Consequently the specific substrate uptake rate can be specified as a function of the specific growth rate of the biomass, simply by rewriting Eqn (6.2):

$$r_s = Y_{xs}\mu \tag{6.4}$$

Similarly, the specific uptake rate of other substrates, such as O_2, and the formation rate of metabolic products are proportional to the specific growth rate. In the black box model, the kinetics reduces to a description of the specific growth rate as a function of the variables in the system. In the most simple model description, it is assumed that there is only one limiting substrate, typically the carbon source, and the specific growth rate is therefore specified as a function of the concentration of this substrate only. A very general observation for cell growth on a single limiting substrate is that at low substrate concentrations (c_s) the specific growth rate, μ, is proportional with c_s, but for increasing concentrations there is an upper value for the specific growth rate. This verbal presentation can be described with many different mathematical models, but the most often applied is the Monod model, which states that:

$$\mu = \mu_{max}\frac{c_s}{c_s + K_s} \tag{6.5}$$

K_s is the substrate concentration at which the specific growth rate is 0.5 μ_{max}, and is sometimes interpreted as the affinity of the cells towards the substrate s. Since the substrate uptake often is involved in the control of substrate metabolism, the value of K_s is also often in the range of the K_m values of the substrate uptake system of the cells. However, K_s is an overall parameter for all the reactions involved in the conversion of the substrate to biomass, and it is therefore completely empirical and has no physical interpretation. The influence of the substrate concentration on the specific growth rate with the Monod model is illustrated in Fig. 6.4. Table 6.1 summarises the K_s value for different microbial systems.

The Monod model is not the only kinetic expression that has been proposed to describe the specific growth rate in the black box model. Many different kinetic expressions have been presented as shown in Table 6.2. Except for the Moser model, all the kinetic expressions contain two adjustable parameters as in the Monod model. In the

Table 6.1 Compilation of K_S values for different microbial cells growing on different sugars

Species	Substrate	K_s (mg l^{-1})
Aspergillus oryzae	Glucose	5
Escherichia coli	Glucose	4
Klebsiella pneumoniae	Glucose	9
Aerobacter aerogenes	Glycerol	9
Klebsiella oxytoca	Glucose	10
	Arabinose	50
	Fructose	10
Penicillium chrysogenum	Glucose	4
Saccharomyces cerevisiae	Glucose	180

Table 6.2 Compilation of different unstructured, kinetic models

Name	Kinetic expression
Tessier	$\mu = \mu_{max}(1 - e^{-c_s/K_s})$
Moser	$\mu = \mu_{max}\frac{c_s^n}{c_s^n + K_s}$
Contois	$\mu = \mu_{max}\frac{c_s}{c_s + K_s x}$
Blackman	$\mu = \begin{cases} \mu_{max}\frac{c_s}{2K_s}; & c_s \leq 2K_s \\ \mu_{max} & c_s \geq 2K_s \end{cases}$
Logistic law	$\mu = \mu_{max}\left(1 - \frac{x}{K_x}\right)$

Contois kinetics, an influence of the biomass concentration, x, is included, i.e. at high biomass concentrations there is an inhibition of cell growth. It is unlikely that the biomass concentration as such inhibits cell growth but there may well be an indirect effect, e.g. the formation of an inhibitory compound by the biomass or high biomass concentrations may give a very viscous medium that results in mass transfer problems. Similarly the Logistic Law expresses a negative influence of the biomass concentration on the specific growth rate. These different expressions clearly demonstrate the empirical nature of these kinetic models, and it is therefore futile to discuss which model is to be preferred, since they are all simply data fitters, and one should simply choose the model that gives the best description of the system being studied.

All the kinetic expressions presented in Table 6.2 assume that there is only one limiting substrate, but often more than one substrate concentration influences the specific growth rate. In these situations, complex interactions can occur which are difficult to model with unstructured models unless many adjustable parameters are included. Several different multiparameter, unstructured models for growth on multiple substrates have been proposed where it is often difficult to distinguish between whether a second substrate is growth enhancing or limiting growth. A general kinetic expression that accounts for both types of substrates is:

$$\mu = \left(1 + \sum_i \frac{c_{\mathrm{si,e}}}{c_{\mathrm{si,e}} + K_{\mathrm{e,i}}}\right) \prod_j \frac{\mu_{\mathrm{max,j}}\, c_{\mathrm{s,j}}}{c_{\mathrm{s,j}} + K_{\mathrm{s,j}}} \tag{6.6}$$

The presence of growth-enhancing substrates increases the specific growth rate whereas the essential substrates are necessary for growth to take place.

A special case of Eqn (6.6) is the growth in the presence of two essential substrates, $c_{\mathrm{s,1}}$ and $c_{\mathrm{s,2}}$:

$$\mu = \frac{\mu_{\mathrm{max,1}}\mu_{\mathrm{max,2}}\, c_{\mathrm{s,1}} c_{\mathrm{s,2}}}{(c_{\mathrm{s,1}} + K_1)(c_{\mathrm{s,2}} + K_{\mathrm{s}})} \tag{6.7}$$

If both substrates are at concentrations where the specific growth rate for each substrate reaches 90% of its maximum value, i.e. $c_{\mathrm{s,i}} = 0.9\ K_{\mathrm{i}}$, then the total rate of growth is limited to 81% of the maximum possible value. This is hardly practical and several alternatives to Eqn (6.7) have therefore been proposed, and one of these is:

$$\frac{\mu}{\mu_{\mathrm{max}}} = \min\left(\frac{c_{\mathrm{s,1}}}{c_{\mathrm{s,1}} + K_1}, \frac{c_{\mathrm{s,2}}}{c_{\mathrm{s,2}} + K_2}\right) \tag{6.8}$$

Growth on two or more substrates that may substitute each other, e.g. glucose and lactose, cannot be described by any of the unstructured models described above. Consider for example growth of *E. coli* on glucose and lactose. Glucose is the favoured substrate and will therefore be metabolised first. (The biochemistry of this is described in Chapter 2, Section 2.3.) Only when this sugar is exhausted will the metabolism of lactose begin. The bacterium needs one of the sugars to grow but, in the presence of glucose, there is not a growth-enhancing effect of lactose. Application of Eqn (6.6) to this example of multiple substrates will clearly not be feasible. To describe this so-called **diauxic growth** it is necessary to apply a structured model and, in general, it is advisable always to consider only a single limiting substrate in black box models.

In some cases growth is inhibited either by high concentrations of the limiting substrate or by the presence of a metabolic product. In order to account for these aspects the Monod kinetics are often extended with additional terms. Thus, for inhibition by high concentrations of the limiting substrate:

$$\mu = \mu_{\mathrm{max}} \frac{c_{\mathrm{s}}}{c_{\mathrm{s}}^2/K_{\mathrm{i}} + c_{\mathrm{s}} + K_{\mathrm{s}}} \tag{6.9}$$

and for inhibition by a metabolic product:

$$\mu = \mu_{max} \frac{c_s}{c_s + K_s} \frac{1}{1 + p/K_i} \tag{6.10}$$

Equations (6.9) and (6.10) may be a useful way of including product or substrate inhibition in a simple model. Extension of the Monod model with additional terms or factors should, however, be done with some restraint since the result may be a model with a large number of parameters but of little value outside the range in which the experiments were made.

6.2.4 Linear rate equations

In the black box model all the yield coefficients are taken to be constant. This implies that all the cellular reactions are lumped into a single, overall growth reaction where substrate is converted to biomass. A requirement for this assumption is that there is a constant distribution of fluxes through all the different cellular pathways under different growth conditions. In 1959, Denis Herbert clearly demonstrated this was not the case since he found that the yield of biomass on substrate was not constant. In order to describe his observations, he introduced the concept of **endogenous metabolism** and specified substrate consumption for this process in addition to that for biomass synthesis, i.e. substrate consumption takes place in two different reactions. In the same year Luedeking and Piret found that lactic acid bacteria produce lactic acid under non-growth conditions, which was consistent with an endogenous metabolism of the cells. Their results indicated a linear correlation between the specific lactic acid production rate and the specific growth rate:

$$r_p = \alpha r_x + \beta x \tag{6.11}$$

In 1965, John Pirt introduced a similar linear correlation between the specific substrate uptake rate and the specific growth rate. He suggested use of the term **maintenance**, which now is the most commonly used term for endogenous metabolism. The linear correlation of Pirt takes the form:

$$r_s = Y_{xs}^{true} \mu + m_s \tag{6.12}$$

The maintenance coefficients quantify the rate of substrate consumption for cellular maintenance, and it is normally given as a constant. In principle, this gives rise to a conflict since this may result in substrate consumption even when the substrate concentration is zero ($c_s = 0$), and in some cases it may therefore be necessary to specify m_s as a function of c_s.

With the introduction of the linear correlations the yield coefficients can obviously not be constants. Thus for the biomass yield on the substrate:

$$Y_{sx} = \frac{\mu}{Y_{xs}^{true} \mu + m_s} \tag{6.13}$$

which shows that Y_{sx} decreases at low specific growth rates where an increasing fraction of the substrate is used to meet the maintenance

Table 6.3 True yield (g substrate needed to produce 1 g biomass) and maintenance coefficients (g substrate consumed for maintenance metabolism per g biomass per hour) for different microbial species and growth on glucose or glycerol

Organism	Substrate	Y_{xs}^{true} (g g^{-1})	m_s (g g^{-1}·h)
Aspergillus awamori	Glucose	1.92	0.016
Aspergillus nidulans		1.67	0.020
Candida utilis		2.00	0.031
Escherichia coli		2.27	0.057
Klebsiella aerogenes		2.27	0.063
Penicillium chrysogenum		2.17	0.021
Saccharomyces cerevisiae		1.85	0.015
Aerobacter aerogenes		1.79	0.089
Bacillus megatarium		1.67	—
Klebsiella aerogenes	Glycerol	2.13	0.074

requirements of the cell. For large specific growth rates the yield coefficient approaches the reciprocal of Y_{xs}^{true}, i.e. Y_{sx} becomes equal to Y_{sx}^{true}. This corresponds to the situation where the maintenance substrate consumption becomes negligible compared with the substrate consumption for biomass growth, and Eqn (6.12) can be approximated with Eqn (6.4). Despite its simple structure the linear rate equation (6.12) of Pirt is found to hold for many different species, and Table 6.3 compiles true yield coefficients and maintenance coefficients for various microbial species.

The empirically derived, linear correlations are very useful to correlate growth data, especially in steady state continuous cultures where linear correlations similar to Eqn (6.12) are found for most of the important specific rates. The remarkable robustness and general validity of the linear correlations indicates that they have a fundamental basis and this basis is likely to be the continuous supply and consumption of ATP, since these two processes are tightly coupled in all cells. Thus, the role of the energy producing substrate is to provide ATP to drive both the biosynthetic and polymerisation reactions of the cell and the different maintenance processes according to the linear relationship:

$$r_{ATP} = Y_{xATP}\mu + m_{ATP} \tag{6.14}$$

which is a formal analogue to the linear correlation of Pirt, and states that the ATP being produced is balanced by its consumption for growth and for maintenance. If the ATP yield on the energy-producing substrate is constant, i.e. r_{ATP} is proportional to r_s, it is quite obvious that Eqn (6.14) can be used to derive the linear correlation Eqn (6.12). Y_{xATP} used in Eqn (6.14) is a true yield coefficient but it is normally specified without the superscript 'true'.

With the linear rate equations the cellular reactions can be structured into several individual reactions. This concept can, in principle, be extended to consider individual reactions for different cellular pathways, as illustrated in the Sonnleitner and Käppeli model for baker's yeast (see also Chapter 17). During aerobic growth of baker's yeast (*Saccharomyces cerevisiae*) there may be a mixed metabolism with both respiration and fermentation being active. At high glucose uptake rates there is a limitation in the respiratory pathway which results in an overflow metabolism towards ethanol. The point at which the glucose uptake rate initiates fermentative metabolism is often referred to as the **critical glucose uptake rate**, and this is dependent on the oxygen concentration in the bioreactor. Thus, at low dissolved oxygen concentrations the critical glucose uptake rate is lower than at high dissolved oxygen concentrations (and clearly at anaerobic conditions there is only fermentative metabolism corresponding to the critical glucose uptake rate being zero). A model for this fermentation process can be found in Chapter 17.

6.2.5 Effect of temperature and pH

The reaction temperature and the pH of the growth medium are other process conditions with a bearing on the growth kinetics. It is normally desired to keep both of these variables constant (and at their optimal values) throughout the cultivation process, hence they are often called **culture parameters** to distinguish them from other variables such as reactant concentrations, stirring rate, oxygen supply rate etc. which can change dramatically from the start to the end of a cultivation. The influence of temperature and pH on individual cell processes can be very different, and since the growth process is the result of many enzymatic processes the influence of both variables (or culture parameters) on the overall bioreaction is quite complex.

The influence of temperature on the maximum specific growth rate of a micro-organism is similar to that observed for the activity of an enzyme: an increase with increasing temperature up to a certain point where protein denaturation starts, and a rapid decrease beyond this temperature. For temperatures below the onset of protein denaturation the maximum specific growth rate increases in much the same way as for a normal chemical rate constant:

$$\mu_{max} = A\exp\left(-\frac{E_g}{RT}\right) \tag{6.15}$$

Assuming that the proteins are temperature denatured by a reversible chemical reaction with free energy change ΔG_d and that denatured proteins are inactive one may propose (Roels, 1983) an expression for μ_{max}:

$$\mu_{max} = \frac{A\exp(-E_g/RT)}{1 + B\exp(-\Delta G_d/RT)} \tag{6.16}$$

Figure 6.5 is a typical Arrhenius plot (reciprocal of the absolute temperature on the abscissa and log μ on the ordinate) for *E. coli*. The linear portion of the curve between approximately 294 and 300.5 K is accurately represented by Eqn (6.27) while the sharp bend and rapid decrease

Fig. 6.5 The influence of temperature on the maximum specific growth rate of *Escherichia coli* B/r. (●) Growth on a glucose-rich medium; (■) growth on a glucose-minimal medium. The lines are calculated using the model in Eqn (6.28) with the parameters: $E_g = 58$ kJ mole^{-1}, $\Delta G_d = 550$ kJ mole^{-1}, $A = 10^{10}$ h^{-1}, $B = 3.0\ 10^{90}$.

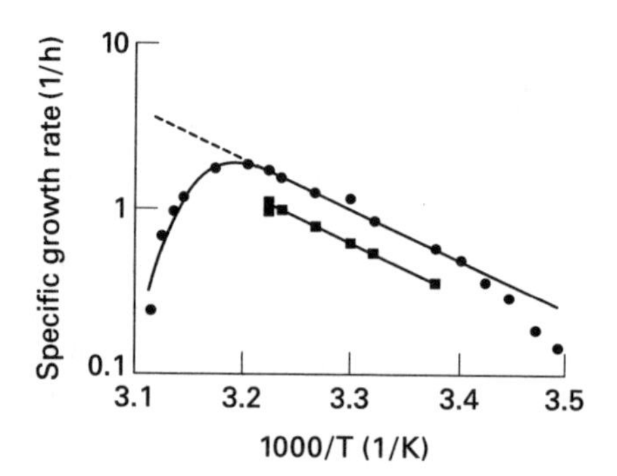

Fig. 6.6 The influence of pH on the maximum specific growth rate of the filamentous fungus *Aspergillus oryzae*. The line is simulated using Eqn (6.31) with $K_1 = 4 \cdot 10^{-3}$, $K_2 = 2 \cdot 10^{-8}$, and $ke_{tot} = 0.3$ h^{-1}.

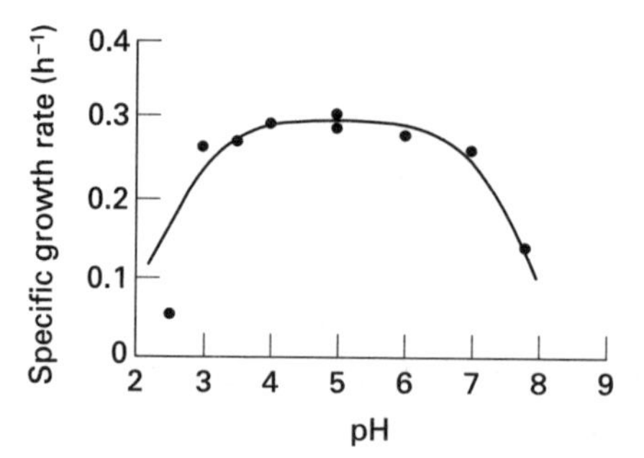

of μ for T > 312 K (= 39 °C) shows the influence of the denominator term in Eqn (6.16).

The influence of pH on the cellular activity is determined by the sensitivity of the individual enzymes to changes in the pH. Enzymes are normally only active within a certain pH range, and the total enzyme activity of the cell is therefore a complex function of the environmental pH. As an example, we shall consider the influence of pH on a single enzyme which is taken to represent the cell activity. The enzyme is assumed to exist in three forms:

$$e \leftrightarrow e^- + H^+ \leftrightarrow e^{2-} 2H^+ \tag{6.17}$$

where e^- is taken to be the active form of the enzyme while the two other forms are assumed to be completely inactive. With K_1 and K_2 being the dissociation constants for e and e^- respectively. The fraction of active enzyme e^- is calculated to be:

$$\frac{e^-}{e_{tot}} = \frac{1}{1 + [H^+]/K_1 + K_2/[H^+]} \tag{6.18}$$

and the enzyme activity is taken to be $k = k_e e^-$. If the cell activity is determined by the activity of the enzyme considered above the maximum specific growth rate will be:

$$\mu_{max} = \frac{1}{1 + [H^+]/K_1 + K_2/[H^+]} \tag{6.19}$$

Although the dependence of cell activity on pH cannot possibly be explained by this simple model it is, however, found that Eqn (6.19) gives an adequate fit for many micro-organisms, and Fig. 6.6 shows fit of the model for some data of the filamentous fungus *Aspergillus oryzae*.

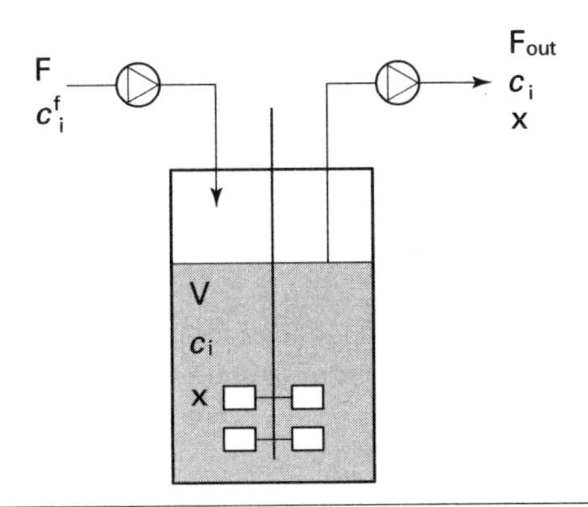

Fig. 6.7 General representation of a bioreactor with addition of fresh, sterile medium and removal of spent medium. c_i^f is the concentration of the *i*'th compound (typically a substrate) in the feed and *ci* is the concentration of the *i*'th compound in the spent medium. The bioreactor is assumed to be very well mixed (or ideal), whereby the concentration of each compound in the spent medium becomes identical to its concentration in the bioreactor. In small volume bioreactors (< 50 l) (including shake flasks) this can generally be achieved through aeration and agitation. In larger bioreactors there may, however, be significant concentration gradients throughout the bioreactor; see Chapter 8.

6.3 Mass balances for ideal bioreactors

The last step in modelling of fermentation processes is to combine the kinetic model with a model for the bioreactor. A bioreactor model is normally represented by a set of dynamic mass balances for the substrates, the metabolic products and the biomass, which describes the change in time of the concentration of these state variables. The bioreactor may be any type of device ranging from a test tube or a shake flask to a well-instrumented bioreactor. Figure 6.7 is a general representation of a bioreactor. The feed is normally assumed to be sterile, i.e. the biomass concentration in the feed is zero.

The bioreactor may be operated in three different modes:

- **batch**, where $F = F_{out} = 0$, i.e. the volume is constant;
- **continuous**, where $F = F_{out} > 0$, i.e. the volume is constant;
- **fed-batch** (or semi-batch), where $F > 0$ and $F_{out} = 0$, i.e. the volume increases.

The mass balances for the different bioreactor modes can all be derived from a set of general mass balances, and we therefore start to consider these general balances.

6.3.1 General mass balance equations

The basis for derivation of the general dynamic mass balances is the mass balance equation:

$$\text{Accumulated} = \text{Net formation rate} + \text{In} - \text{Out} \tag{6.20}$$

The term **Accumulated** specifies the rate of change of a compound in the bioreactor, such as the rate of increase in the biomass concentration during a batch fermentation. For substrates, the term **Net formation rate** is given by a substrate uptake rate (that is regarded as negative being the withdrawal of carbon from the system), whereas for metabolic products and biomass this term is given by the formation rate of these variables. The term **In** represents the flow of the compound into the bioreactor and the term **Out** the flow of the compound out from the bioreactor.

For the ith substrate, which is added to the bioreactor via the feed and is consumed by the cells present in the bioreactor, the mass balance is:

$$\frac{d(c_{s,i}V)}{dt} = -r_{s,i}xV + Fc_{s,i}^{f} - F_{out}c_{s,i} \tag{6.21}$$

The first term in Eqn (6.21) is the accumulation term, the second term accounts for substrate consumption (or net formation), the third term accounts for the inlet, and the last term accounts for the outlet. Rearrangement of Eqn (6.21) gives:

$$\frac{dc_{s,i}}{dt} = -r_{s,i}x + \frac{F}{V}c_{s,i}^{f} - \left(\frac{F_{out}}{V} + \frac{1}{V}\frac{dV}{dt}\right)c_{s,i} \tag{6.22}$$

Since for a fed-batch reactor:

$$F = \frac{dV}{dt} \tag{6.23}$$

and $F_{out} = 0$ the term within the parentheses becomes equal to the so-called **dilution rate** given by:

$$D = \frac{F}{V} \tag{6.24}$$

For both a continuous and a batch reactor, the volume is constant, i.e. $dV/dt = 0$, and $F = F_{out}$, and also for these bioreactor modes the term within the parentheses becomes equal to the dilution rate. Eqn (6.24) therefore reduces to the mass balance (6.25) for any type of operation.

$$\frac{dc_{s,i}}{dt} = -r_{s,i}x + D(c_{s,i}^{f} - c_{s,i}) \tag{6.25}$$

The first term on the right hand side of Eqn (6.25) is the volumetric rate of substrate consumption, which is given as the product of the specific rate of substrate consumption and the biomass concentration. The second term accounts for the addition and removal of substrate from the bioreactor.

Dynamic mass balances for the metabolic products are derived in analogy with those for the substrates and takes the form:

$$\frac{dc_{p,i}}{dt} = -r_{s,i}x + D(c_{p,i}^{f} - c_{p,i}) \tag{6.26}$$

where the first term on the right hand side is the volumetric formation rate of the ith metabolic product. Normally the metabolic products are not present in the sterile feed to the bioreactor and $c_{p,i}^{f}$ is therefore often zero.

With sterile feed the mass balance for the total biomass is:

$$\frac{d(xV)}{dt} = \mu xV - F_{out}x \tag{6.27}$$

which in analogy with the substrate balance can be rewritten as:

$$\frac{dx}{dt} = (\mu - D)x \tag{6.28}$$

6.3.2 The batch reactor

This is the classical operation of the bioreactor that is used extensively. The disadvantage is that the experimental data produced are difficult to interpret since there are dynamic conditions throughout the experiment, i.e. the environmental conditions experienced by the cells vary with time. In well-instrumented laboratory bioreactors many variables, e.g. pH and dissolved oxygen tension, may be kept constant and this allows study of the effect of a single substrate on the biomass growth and product formation.

The dilution rate is zero for a batch reactor and the mass balances for the biomass and the limiting substrate (in a batch fermentation the limiting substrate is defined as the substrate that is first exhausted) therefore take the form:

$$\frac{dx}{dt} = \mu x;\ x(t=0) = x_0 \tag{6.29}$$

$$\frac{dc_s}{dt} = -r_s x;\ c_s(t=0) = c_{s,0} \tag{6.30}$$

According to these mass balances the biomass concentration will increase and the substrate concentration will decrease until its concentration reaches zero and growth stops. Assuming Monod kinetics, the mass balances for biomass and the limiting substrate can be rearranged into one first-order differential equation in the biomass concentration and an algebraic equation relating the substrate concentration to the biomass concentration. The algebraic equation is given by:

$$c_s = c_{s,0} - Y_{xs}(x - x_0) \tag{6.31}$$

and the solution to the differential equation for the biomass concentration is given by:

$$\mu_{max}t = \left(1 + \frac{K_s}{c_{s,0} + Y_{xs}x_0}\right)\ln\left(\frac{x}{x_0}\right) - \frac{K_s}{c_{s,0} + Y_{xs}x_0}\ln\left(1 + \frac{x_0 - x}{Y_{sx}c_{s,0}}\right) \tag{6.32}$$

Using these equations the profiles of the biomass and the glucose concentrations during a typical batch culture are easily derived, as shown in Fig. 6.8. Since the substrate concentration is zero at the end of the cultivation the overall yield of biomass on the substrate can be found from:

$$Y_{sx}^{overall} = \frac{x_{final} - x_0}{c_{s,0}} \tag{6.33}$$

Fig. 6.8 Simulation of the biomass and glucose concentration during a batch culture. The simulation has been carried out using the Monod model with $\mu_{max} = 0.5\ h^{-1}$, $K_s = 0.05\ g\ l^{-1}$, and $Y_{sx} = 0.50$.

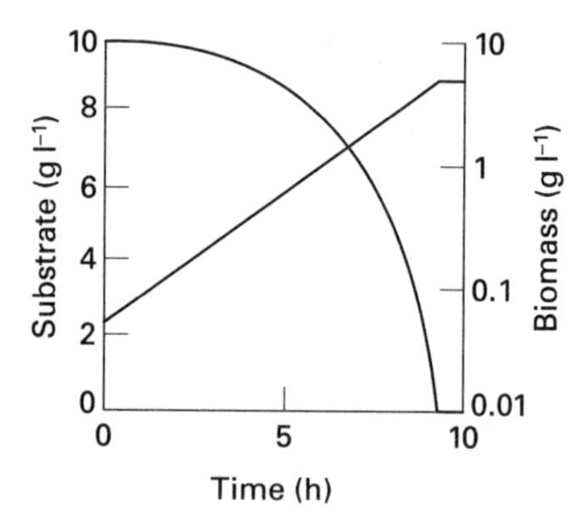

Normally $x_0 \ll x_{final}$, and the overall yield coefficient can therefore be estimated from the final biomass concentration and the initial substrate concentration alone.

Notice that the yield coefficient determined from Eqn (6.33) is the overall yield coefficient and not Y_{sx} or Y_{sx}^{true}. The yield coefficient, Y_{sx}, may well be time dependent since it is the ratio between the specific growth rate and the substrate uptake rate (see Eqn 6.2). However, if there is little variation in these rates during the batch culture (e.g. if there is a long exponential growth phase and only a very short declining growth phase) the overall yield coefficient may be similar to the yield coefficient. If there is maintenance metabolism the true yield coefficient is difficult to determine from a batch cultivation since it requires information about the maintenance coefficients, which can hardly be determined from a batch experiment. However, in a batch cultivation the specific growth rate is close to its maximum throughout most of the growth phase and the substrate consumption due to maintenance is therefore negligible and, according to Eqn (6.14), the true yield coefficient is therefore close to the observed yield coefficient determined from the final biomass concentration.

6.3.3 The chemostat

A typical operation of the continuous bioreactor is the so-called **chemostat**, where the added medium is designed such that there is a **single limiting substrate**. This allows for controlled variation in the specific growth rate of the biomass. By varying the feed flow rate to the bioreactor, the environmental conditions can be varied and thereby valuable information concerning the influence of the environmental conditions on the cellular physiology can be obtained. For industrial applications, the continuous bioreactor is attractive since the productivity may be high. However, often the titre, i.e. the product concentration, is lower than what can be obtained in the fed-batch reactor. Furthermore, it is rarely used in industrial processes since it is sensitive to contamination, e.g. via the feed stream, and to the appearance of spontaneously formed mutants that may out-compete the production strain. Other examples of continuous operation besides the chemostat are the **pH-stat**, where the feed flow is adjusted to maintain the pH constant in the bioreactor, and the **turbidostat**, where the feed flow is adjusted to maintain the biomass concentration at a constant level.

From the biomass mass balance (6.40), it is easily seen that in a steady state continuous reactor the specific growth rate equals the dilution rate:

$$\mu = D \tag{6.34}$$

Thus, by varying the dilution rate (or the feed flow rate) in a continuous culture different specific growth rates can be obtained. This allows detailed physiological studies of the cells when they are grown at a specified specific growth rate (corresponding to a certain environment experienced by the cells). At steady state the substrate mass balance (6.25) gives:

$$0 = -r_s x + D(c_s^f - c_s) \tag{6.35}$$

which upon combination with Eqn (6.34) and the definition of the yield coefficient directly gives:

$$x = Y_{sx}(c_s^f - c_s) \tag{6.36}$$

Thus, the yield coefficient can be determined from measurement of the biomass and the substrate concentrations in the bioreactor. From measurements of the substrate concentration and the biomass concentration at steady state the specific glucose uptake rate can easily be calculated using Eqn (6.35), and similarly the specific rates of product formation can be determined from measurement of the product concentration and the biomass concentration.

If the Monod model applies the mass balance for the biomass gives:

$$D = \mu_{max} \frac{c_s}{c_s + K_s} \tag{6.37}$$

or

$$c_s = \frac{DK_s}{\mu_{max} - D} \tag{6.38}$$

Thus, the concentration of the limiting substrate increases with the dilution rate. When substrate concentration becomes equal to the substrate concentration in the feed the dilution rate attains its maximum value, which is often called the critical dilution rate:

$$D_{crit} = \mu_{max} \frac{c_s^f}{c_s^f + K_s} \tag{6.39}$$

When the dilution rate becomes equal to or larger than this value the biomass is washed out of the bioreactor. Equation (6.38) clearly shows that the steady state chemostat is well suited to study the influence of the substrate concentration on the cellular function, e.g. product formation, since by changing the dilution rate it is possible to change the substrate concentration as the only variable. Furthermore, it is possible to study the influence of different limiting substrates on the cellular physiology, e.g. glucose and ammonia.

Besides quantification of the Monod parameters the chemostat is well suited to determine the maintenance coefficient. Since the dilution rate equals the specific growth rate, combination of Eqn (6.13) and (6.36) gives:

$$x = \frac{D}{Y_{xs}^{true} D + m_s}(c_s^f - c_s) \tag{6.40}$$

which shows that the biomass concentration decreases at low specific growth rates, where the substrate consumption for maintenance is significant compared with that for growth. At high specific growth rates (high dilution rates) maintenance is negligible and the yield coefficient becomes equal to the true yield coefficient, see Fig. 6.10. Since $\mu = D$ at steady state, Eqn (6.12) expresses that there is a linear relation between the specific substrate uptake rate and the dilution rate. In this linear relationship the true yield coefficient and the maintenance coefficient can easily be estimated using linear regression.

For production of biomass, e.g. baker's yeast or single cell protein, and growth-related products the chemostat is very well suited since it is possible to maintain a high productivity over very long periods of operation. The productivity of biomass is given by:

$$P_x = Dx \tag{6.41}$$

and, in Fig. 6.9, the productivity is shown as a function of the dilution rate. By inserting the expression for the biomass concentration (6.40) in Eqn (6.41), with Eqn (6.38) inserted for the substrate concentration, it is possible to calculate the dilution rates which give the maximum productivity. If there is no maintenance, i.e. $m_s = 0$, the optimal dilution rate is given by:

$$D_{opt} = \mu_{max}\left(1 - \sqrt{\frac{K_s}{c_s^f + K_s}}\right) \tag{6.42}$$

It is important to emphasise that this optimum only holds for Monod kinetics without maintenance. When maintenance is included finding the optimum dilution rate will involve solving a third-degree polynomial. This polynomial will have one solution in the possible range of dilution rates. However, instead of solving the third-degree polynomial it is generally easier to find the solution numerically.

6.3.4 The fed-batch reactor

This operation is probably the most common operation in industrial processes, since it allows for control of the environmental conditions, e.g. maintaining the glucose concentration at a certain level; it also enables formation of very high titres (up to several hundred grams per litre of some metabolites), which is of importance for subsequent downstream processing. There is striking similarity between the fed-batch reactor and the chemostat, and for the fed-batch reactor the mass balances for biomass and substrate are given by the general mass balance Eqns (6.25) and (6.28). Normally the feed concentration c_s^f is very high, i.e.

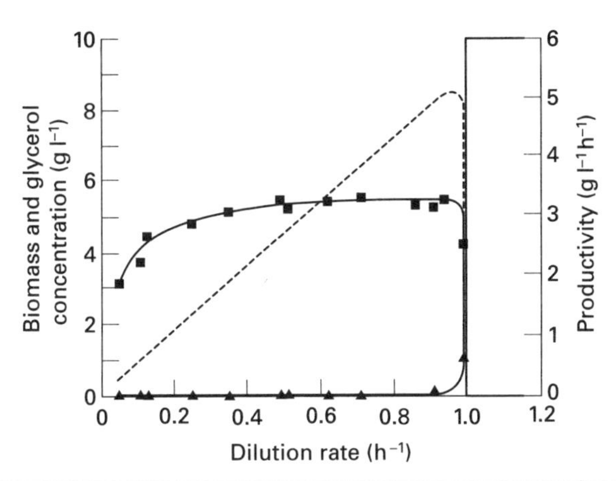

Fig. 6.9 Growth of *Klebsiella pneumoniae (Aerobacter aerogenes)* in a chemostat with glycerol as the limiting substrate. The biomass concentration (■) decreases at low dilution rates due to the maintenance metabolism, and when the dilution rate approaches the critical value the biomass concentration decreases rapidly. The glycerol concentration (▲) increases slowly at low dilution rates, but when the dilution rate approaches the critical value it increases rapidly. The lines are model simulations using the Monod model with maintenance, and with the parameter values: $c_s^f = 10$ g l^{-1}; $\mu_{max} = 1.0$ h^{-1}; $K_s = 0.01$ g l^{-1}; $m_s = 0.08$ g g$^{-1}\cdot$ h; $Y_{xs}^{true} = 1.70$ g g^{-1}. The broken line is the productivity according to Eqn (6.41).

the feed is a very concentrated solution, and the feed flow is low giving a low dilution rate. The dilution rate is given by:

$$D = \frac{1}{V}\frac{dV}{dt} \tag{6.43}$$

and if D is to be kept constant there needs to be an exponentially increasing feed flow to the bioreactor.

If the yield coefficient is constant combination of the mass balances for the biomass and the substrate gives:

$$\frac{d(x + Y_{sx}c_s)}{dt}\left\{(\mu - D)x - Y_{sx}r_s x + Y_{sx}D(c_s^f - c_s)\right\} \tag{6.44}$$

or since $\mu = Y_{sx}r_s$

$$\frac{d(x - Y_{sx}(c_s^f - c_s))}{dt} = -D(x - Y_{sx}(c_s^f - c_s)) \tag{6.45}$$

Through combination with Eqn (6.43) this differential equation can easily be solved with the solution given by:

$$\frac{Y_{sx}(c_s^f - c_{s,0}) - x_0}{Y_{sx}(c_s^f - c_s) - x} = \frac{V}{V_0} \tag{6.46}$$

where x_0, $c_{s,0}$ and V_0 define the biomass concentration, the substrate concentration and the reactor volume at the start of the fed-batch process. As mentioned above the substrate concentration in the feed c_s^f is normally very high and much higher than both the initial substrate concentration and the substrate concentration during the process (c_s). Furthermore, a very high c_s^f means that $Y_{sx}c_s^f$ is larger than the biomass concentration, both initially and during the process. Consequently the increase in volume can be kept low even when there is a very large increase in the biomass concentration.

If there is an exponential feed flow to the bioreactor there will be substantial biomass growth and, since the biomass concentration increases, this may lead to limitations in the O_2 supply. The feed flow is therefore typically increased until limitations in the O_2 supply set in and thereafter the feed flow is kept constant. This will give a decreasing specific growth rate. However, since the biomass concentration normally will increase, the volumetric uptake rate of substrates (including oxygen) may be kept approximately constant. From the above it is clear that there may be many different feeding strategies in a fed-batch process and optimisation of the operation is a complex problem that is difficult to solve empirically; and, even when a very good process model is available, calculation of the optimal feeding strategy is a complex optimisation problem. In an empirical search for the optimal feeding policy the two most obvious criteria are: (1) keep the concentration of the limiting substrate constant, and (2) keep the volumetric growth rate of the biomass (or uptake of a given substrate) constant.

A **constant volumetric growth rate** (or uptake of a given substrate) is applied if there are limitations in the supply of oxygen or in heat removal. A **constant concentration of the limiting substrate** is often applied if the substrate inhibits product formation, and the chosen concentration therefore depends on the degree of inhibition and the desire to maintain a certain growth of the cells. The required feeding profile to maintain a constant substrate concentration $c_{s,0}$ corresponding to a constant specific growth rate μ_0 is quite simple to derive. From Eqn (6.27) with $F_{out} = 0$,

$$\frac{d(xV)}{dt} = \mu_0 xV \tag{6.47}$$

or

$$xV = x_0 V_0 e^{\mu_0 t} \tag{6.48}$$

Since the substrate concentration is constant the substrate balance gives:

$$-Y_{xs}\mu_0 x + D(c_s^f - c_s) = 0 \tag{6.49}$$

or

$$F(t) = \frac{Y_{xs}\mu_0}{c_s^f - c_s} xV = \frac{Y_{xs}\mu_0}{c_s^f - c_s} x_0 V_0 e^{\mu_0 t} \tag{6.50}$$

Finally, the biomass concentration $x(t)$ is obtained from Eqn (6.46) with $c_s = c_{s,0}$:

$$\frac{x(t)}{x_0} = \frac{e^{\mu_0 t}}{1 - ax_0 + ax_0 e^{\mu_0 t}} \tag{6.51}$$

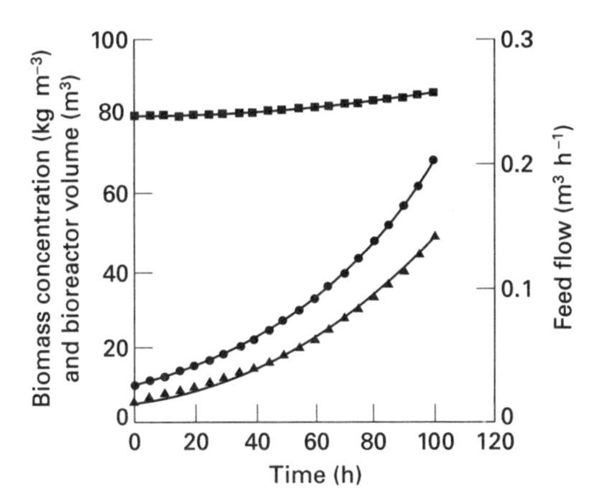

Fig. 6.10 The biomass concentration (◆), the bioreactor volume (■), and the feed flow rate (▲) for a fed-batch reactor operated with a constant substrate concentration. The yield coefficient Y_{sx} is 0.5, the constant specific growth rate μ_0 is $0.02\,h^{-1}$, and the substrate concentration in the feed c_s^f is $400\ kg\ m^{-3}$. The substrate concentration is assumed to be much less than c_s^f. The initial biomass concentration x_0 and the initial bioreactor volume are taken to be $10\ kg\ m^{-3}$ and $80\ m^3$, respectively

where

$$a = \frac{Y_{xs}}{c_s^f - c_s} \tag{6.51}$$

The bioreactor volume is given by:

$$\frac{V}{V_0} = 1 - ax_0 + ax_0 e^{\mu_0 t} \tag{6.53}$$

Figure 6.10 illustrates typical profiles for the biomass concentration, the bioreactor volume and the feed flow rate during a fed-batch process with constant substrate concentration.

Fed-batch processing is applied to processes where control of culture conditions is required, mainly to achieve high yields. Baker's yeast, secondary metabolites (where penicillins are the most prominent group of compounds), industrial enzymes and many other products are derived from fermentation processes.

6.4 Further reading

Herbert, D. (1959). Some principles of continuous culture. *Recent Prog. Microbiol.* **7**, 381–396.

Monod, J. (1942). *Recherches sur la Croissance des Cultures Bacteriennes.* Hermann et Cie, Paris.

Monod, J., Wyman, J. and Changeux, J.-P. (1965). *J. Mol. Biol.* **12**:88–118

Nielsen, J. and Villadsen, J. (1994). *Bioreaction Engineering Principles.* Plenum Press, New York.

Pirt, S. J. (1965). The maintenance energy of bacteria in growing cultures. *Proc. Royal Soc. London. Series B* **163**, 224–231.

Roels, J. A. (1983). *Energetics and Kinetics in Biotechnology.* Elsevier Biomedical Press, Amsterdam.

Sonnleitner, B. and Käppeli, O. (1986). Growth of *Saccharomyces cerevisiae* is controlled by its limited respiratory capacity. *Biotechnol. Bioeng.* **28**, 927–937.

Stephanopoulos, G., Nielsen, J. and Aristodou, A. (1998). *Metabolic Engineering. Principles and Methodologies.* Academic Press, San Diego.

Chapter 7

Bioreactor design

Yusuf Chisti and Murray Moo-Young

Nomenclature

Roman letters

A_d cross-sectional area of the downcomer (m^2)
A_H area for heat transfer (m^2)
A_r cross-sectional area of the riser (m^2)
a parameter in equation (7.8) (–)
CIP clean-in-place
C_p specific heat capacity of the broth (J kg^{-1}·°C)
c dimensionless constant (–)
d characteristic length dimension (m)
d_i diameter of the impeller (m)
d_p particle diameter (m)
d_T diameter of bubble column or tank (m)
d_w fermenter wall thickness (m)
E energy dissipation rate per unit mass of fluid (J kg^{-1})
Gr Grashof number (–)
g gravitational acceleration (m s^{-2})
h_f jacket side fouling film heat transfer coefficient (J m^{-2}·°C)
h_i film heat transfer coefficient for the cooling water film on the jacket side (J m^{-2}·°C)
h_L height of gas-free liquid (m)
h_o broth film heat transfer coefficient (J m^{-2}·°C)
k parameter in equation (7.8) (m^{-1})
k_i impeller-dependent constant (–)
k_T thermal conductivity of the culture broth (J m^{-1}·°C)
k_w thermal conductivity of the fermenter wall (J m^{-1}·°C)

l mean length of the energy dissipating fluid eddy (m)
N rotational speed of the impeller (s^{-1})
Nu Nusselt number (–)
n flow behaviour index of a fluid (–)
P power input in gas-free state (J)
P_G power input in presence of gas (J)
Po power number (–)
Pr Prandtl number (–)
Q volume flow rate of gas ($m^3 s^{-1}$)
Q_H heat transfer rate ($J s^{-1}$)
Re Reynolds number (–)
Re_i impeller Reynolds number (–)
SG sight glass
ΔT temperature difference (°C)
U_G superficial gas velocity based on the total cross-sectional area of the vessel ($m s^{-1}$)
U_{Gr} superficial velocity of gas in riser ($m s^{-1}$)
U_H overall heat transfer coefficient ($J m^{-2} \cdot °C$)
U_L superficial liquid velocity ($m s^{-1}$)
V_L volume of liquid in the reactor (m^3)

Greek letters

β coefficient of volumetric expansion ($m^3 kg^{-1} \cdot °C$)
γ average shear rate (s^{-1})
γ_{max} maximum shear rate (s^{-1})
ε_L volume fraction of liquid (–)
μ_L viscosity of liquid ($kg m^{-1} \cdot s$)
μ_w viscosity of water ($kg m^{-1} \cdot s$)
μ_{Lw} viscosity of liquid at wall temperature ($kg m^{-1} \cdot s$)
ρ_L density of liquid or slurry ($kg m^{-3}$)
τ shear stress ($N m^{-2}$)

7.1 Introduction

Bioreactors, or fermenters, are the core of any biotechnology-based production process be it for vaccines, proteins, organic acids, amino acids, antibiotics, enzymatic or microbial biotransformations, bioremediation and biodegradation, and microbial inoculants for use as biofertilisers. Biocatalysts, micro-organisms, animal or plant cells, are produced and maintained in bioreactors. A production facility typically has a train of bioreactors ranging from 20 litres to 250 m^3. Still larger vessels are encountered in certain processes. In a great majority of processes, the reactors are operated in a batch mode, under sterile or mono-septic conditions. The most common operational practice starts with culturing micro-organisms or cells in the smallest bioreactor. After a predetermined batch time the content of this reactor is transferred to a larger, presterilised, medium-filled reactor and this process is repeated until the production fermenter, the largest reactor in the train, is reached. Most commercial processing is carried out as submerged culture with the biocatalyst suspended in a nutrient medium in a suitable reactor. Irrespective of the specific reactor configuration used, features such as sterility considerations, vessel design and surface finishes, clean-in-place issues, and aspects of bioreactor performance are often common to bioreactors. The bioreactor types used extensively are the airlift, stirred tank and bubble column bioreactor.

7.2 Bioreactor configurations

7.2.1 Stirred tank reactors

Stirred tank bioreactors consist of a cylindrical vessel with a motor-driven central shaft that supports one or more agitators. The shaft may enter through the top or the bottom of the reactor vessel. A typical stirred tank reactor is shown in Fig. 7.1. Microbial culture vessels are generally provided with four baffles projecting into the vessel from the walls to prevent swirling and vortexing of the fluid. The baffle width is $\frac{1}{10}$ or $\frac{1}{12}$ of the tank diameter. The aspect ratio (i.e. height-to-diameter ratio) of the vessel is 3–5, except in animal cell culture applications where aspect ratios do not normally exceed 2. Often, the animal cell culture vessels are unbaffled (especially small-scale reactors) to reduce turbulence that may damage the cells. The number of impellers depends on the aspect ratio. The bottom impeller is located at a distance about $\frac{1}{3}$ of the tank diameter above the bottom of the tank. Additional impellers are spaced approximately 1.2 impeller diameter distance apart. The impeller diameter is about $\frac{1}{3}$ of the vessel diameter for gas dispersion impellers such as Rushton disc turbines and concave bladed impellers (Fig. 7.2). Larger hydrofoil impellers (Fig. 7.2) with diameters of 0.5 to 0.6 times the tank diameter are especially effective bulk mixers

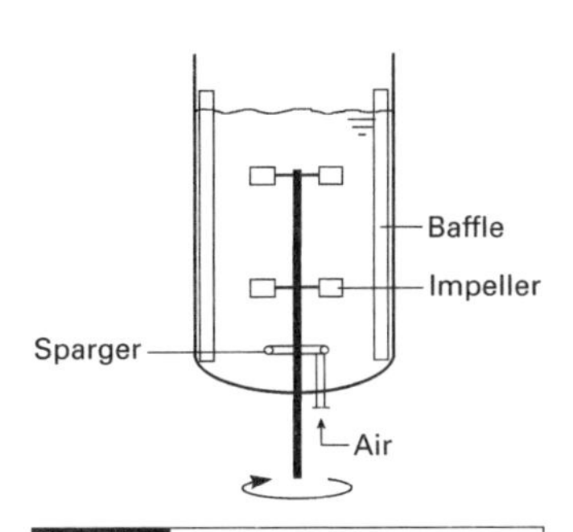

Fig. 7.1 A stirred tank bioreactor.

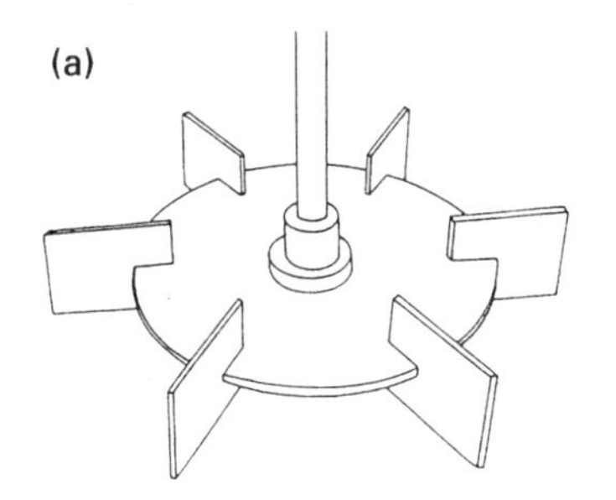

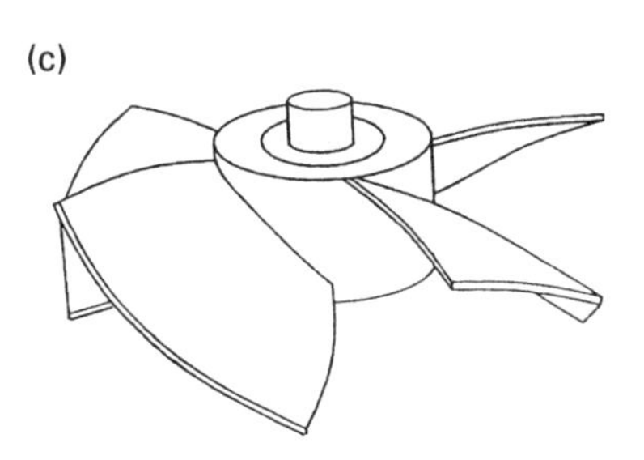

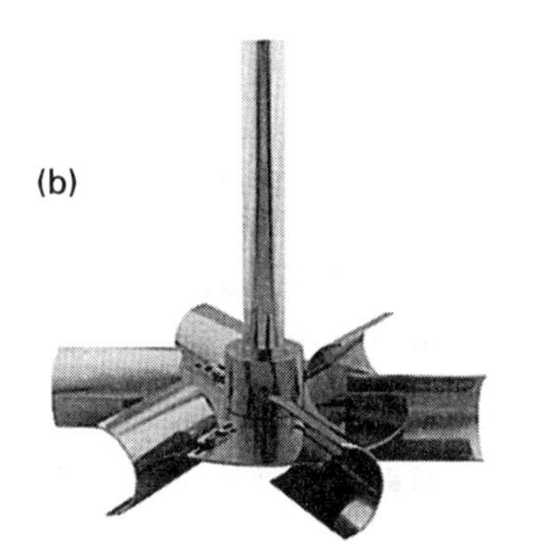

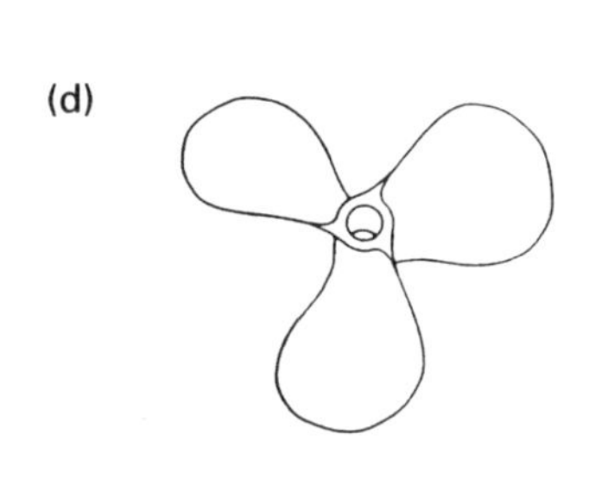

Fig. 7.2 Some commonly used impellers: (a) Rushton disc turbine; (b) a concave bladed turbine; (c) a hydrofoil impeller; and (d) a marine propeller.

and are used in fermenters for highly viscous mycelial broths. Animal cell culture vessels typically employ a single, large diameter, low-shear impeller such as a marine propeller (Fig. 7.2). Gas is sparged into the reactor liquid below the bottom impeller using a perforated pipe ring sparger with a ring diameter that is slightly smaller than that of the impeller. Alternatively, a single hole sparger may be used.

In animal or plant cell culture applications, the impeller speed generally does not exceed about 120 rpm in vessels larger than about 50 litres. Higher stirring rates are employed in microbial culture, except with mycelial and filamentous cultures where the impeller tip speed (i.e. 3.143 × impeller diameter × speed of rotation) does not in general exceed 7.6 m s^{-1}. Even lower speeds have been documented to damage certain mycelial fungi. The superficial aeration velocity (i.e. the volume flow rate of gas divided by the cross-sectional area of the vessel) in stirred vessels must remain below the value needed to flood the impeller. (An impeller is flooded when it receives more gas than it can effectively disperse.) A flooded impeller is a poor mixer. Superficial aeration velocities do not generally exceed 0.05 m s^{-1}.

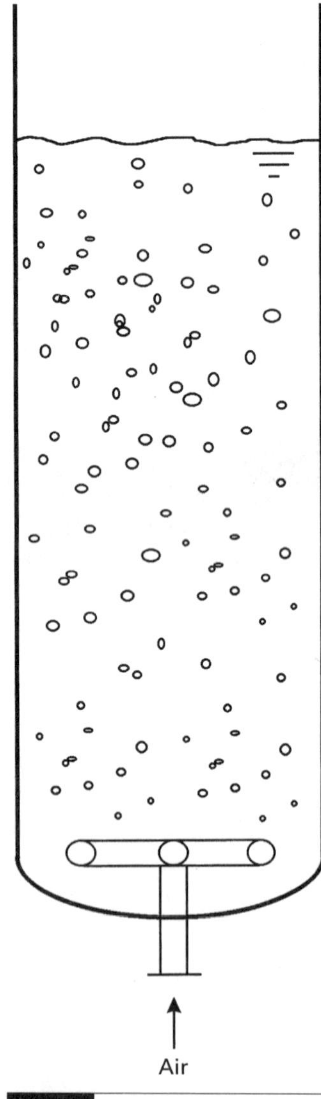

Fig. 7.3 A bubble column.

7.2.2 Bubble columns

A bubble column bioreactor is shown on Fig. 7.3. Usually, the column is cylindrical with an aspect ratio of 4–6 (height-to-diameter). Gas is sparged at the base of the column through perforated pipes, perforated plates, or sintered glass or metal micro-porous spargers. O_2 transfer, mixing and other performance factors are influenced mainly by the gas flow rate and the rheological properties of the fluid. Internal devices such as horizontal perforated plates, vertical baffles and corrugated sheet packings may be placed in the vessel to improve mass transfer and modify the basic design. The column diameter does not affect its behaviour so long as the diameter exceeds 0.1 m. One exception is the axial mixing performance. For a given gas flow rate, the mixing improves with increasing vessel diameter. Mass and heat transfer and the prevailing shear rate increase as gas flow rate is increased. In bubble columns the maximum aeration velocity does not usually exceed 0.1 m s^{-1}. The liquid flow rate does not influence the gas–liquid mass transfer coefficient so long as the superficial liquid velocity remains below 0.1 m s^{-1}.

7.2.3 Airlift bioreactors

In airlift bioreactors, the fluid volume of the vessel is divided into two interconnected zones by means of a baffle or draft tube as shown in Fig. 7.4. Only one of the two zones is sparged with air or other gas. The sparged zone is known as the riser; the zone that receives no gas is the downcomer. The bulk density of the gas–liquid dispersion in the gas-sparged riser tends to be lower than the bulk density in the downcomer, consequently the dispersion flows up in the riser zone and downflow occurs in the downcomer. Sometimes the riser and the downcomer are two separate vertical pipes that are interconnected at the top and the bottom to form an external circulation loop. For optimal gas–liquid mass transfer performance, the riser-to-downcomer cross-sectional

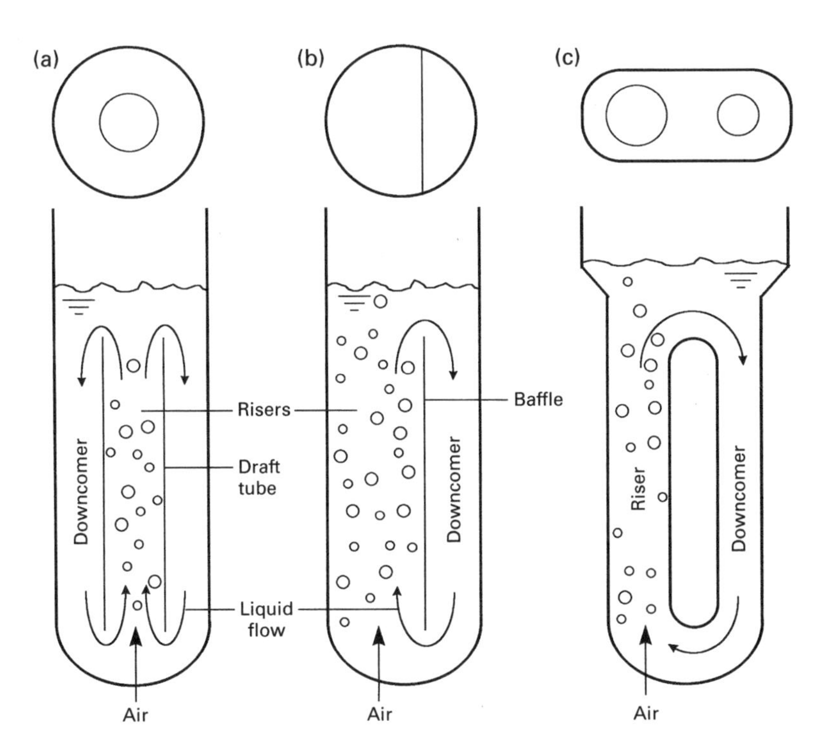

Fig. 7.4 Airlift bioreactors: (a) draft–tube internal loop configuration; (b) a split cylinder device; and (c) an external loop system.

area ratio should be between 1.8 and 4.3. External-loop airlift reactors are less common in commercial processes compared to the internal-loop designs. The internal loop configuration may be either a concentric draft-tube device or a split cylinder.

Airlift bioreactors are highly energy-efficient relative to stirred fermenters, yet the productivities of both types are comparable. Being especially suited to shear-sensitive cultures, airlift devices are often employed in large-scale manufacture of biopharmaceutical proteins obtained from fragile animal cells. Heat and mass transfer capabilities of airlift reactors are at least as good as those of other systems, and airlift reactors are more effective in suspending solids than are bubble columns.

All performance characteristics of airlift bioreactors are linked ultimately to the gas injection rate and the resulting rate of liquid circulation. In general, the rate of liquid circulation increases with the square root of the height of the airlift device. Consequently, the reactors are designed with high aspect ratios. Because the liquid circulation is driven by the gas hold-up difference between the riser and the downcomer, circulation is enhanced if there is little or no gas in the downcomer. All the gas in the downcomer comes from being entrained in with the liquid as it flows into the downcomer from the riser near the top of the reactor. Various designs of gas–liquid separators are sometimes used in the head zone to reduce or eliminate the gas carry-over to the downcomer. Relative to a reactor without a gas–liquid separator, installation of a suitably designed separator will always enhance liquid circulation, i.e. the increased driving force for circulation will more

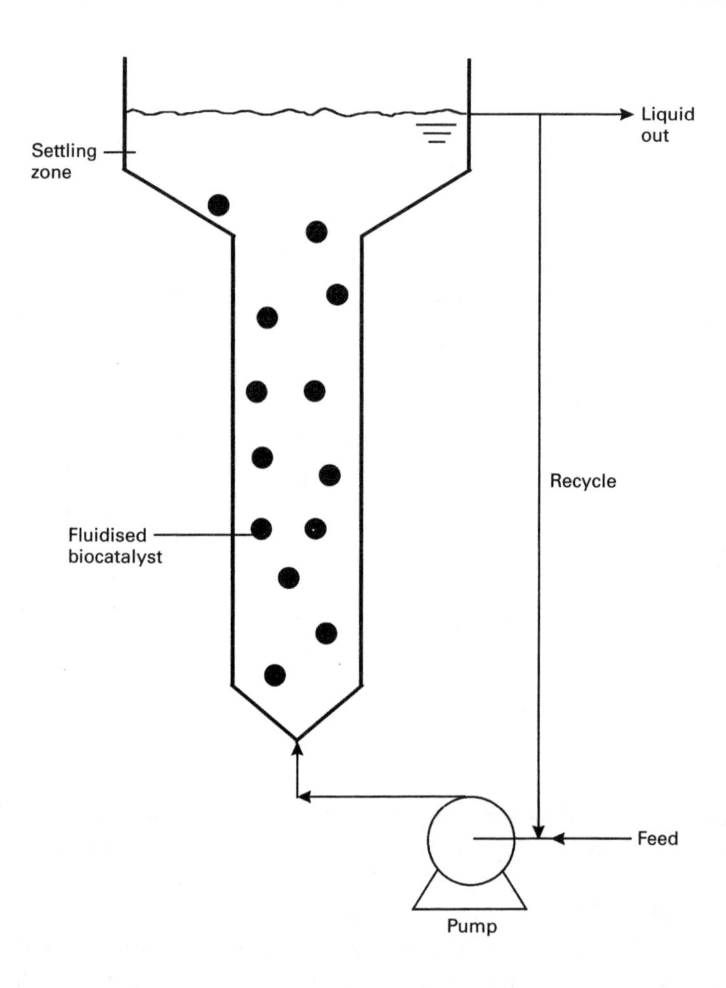

Fig. 7.5 A fluidised bed bioreactor.

than compensate for any additional resistance to flow due to the separator.

7.2.4 Fluidised beds

Fluidised bed bioreactors are suited to reactions involving a fluid-suspended particulate biocatalyst such as the immobilised enzyme and cell particles or microbial flocs. An up-flowing stream of liquid is used to suspend or 'fluidise' the solids as in Fig. 7.5. Geometrically, the reactor is similar to a bubble column except that the top section is expanded to reduce the superficial velocity of the fluidising liquid to a level below that needed to keep the solids in suspension. Consequently, the solids sediment in the expanded zone and drop back into the narrower reactor column below; hence, the solids are retained in the reactor whereas the liquid flows out.

A liquid fluidised bed may be sparged with air or some other gas to produce a gas–liquid–solid fluid bed. If the solid particles are too light, they may have to be artificially weighted, for example by embedding stainless steel balls in an otherwise light solid matrix. A high density of solids improves solid–liquid mass transfer by increasing the relative velocity between the phases. Denser solids are also easier to sediment but the density should not be too high relative to that of the liquid, or fluidisation will be difficult.

Liquid fluidised beds tend to be fairly quiescent but introduction of a gas substantially enhances turbulence and agitation. Even with relatively light particles, the superficial liquid velocity needed to suspend the solids may be so high that the liquid leaves the reactor much too quickly, i.e. the solid–liquid contact time is insufficient for the reaction. In this case, the liquid may have to be recycled to ensure a sufficiently long cumulative contact time with the biocatalyst. The minimum fluidisation velocity – i.e. the superficial liquid velocity needed to just suspend the solids from a settled state – depends on several factors, including the density difference between the phases, the diameter of the particles, and the viscosity of the liquid.

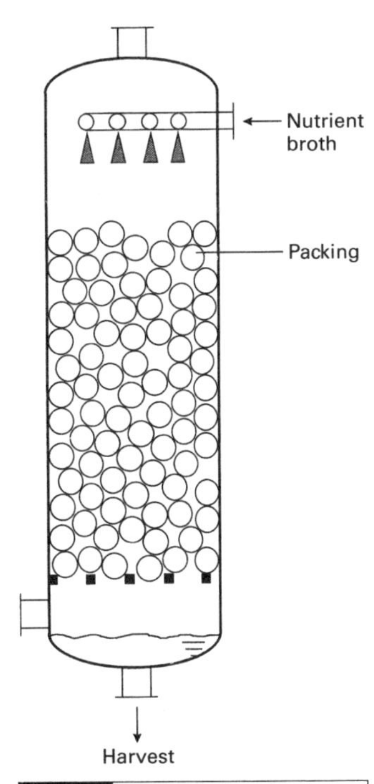

Fig. 7.6 A packed bed bioreactor.

7.2.5 Packed bed columns

A bed of solid particles, usually with confining walls, constitutes a packed bed (Fig. 7.6). The biocatalyst is supported on, or within, the matrix of solids that may be porous or a homogeneous non-porous gel. The solids may be particles of compressible polymeric or more rigid material. A fluid containing nutrients flows continuously through the bed to provide the needs of the immobilised biocatalyst. Metabolites and products are released into the fluid and removed in the outflow. The flow may be upward or downward, but downflow under gravity is the norm. If the fluid flows up the bed, the maximum flow velocity is limited because the velocity cannot exceed the minimum fluidisation velocity or the bed will fluidise.

The depth of the bed is limited by several factors, including the density and the compressibility of the solids, the need to maintain a certain minimal level of a critical nutrient, such as O_2, through the entire depth, and the flow rate that is needed for a given pressure drop. For a given void volume (i.e. solids-free volume fraction of the bed) the gravity-driven flow rate through the bed declines as the depth of the bed increases. Nutrients and substrates are depleted as the fluid moves down the bed. Conversely, concentrations of metabolites and products increase. Thus, the environment of a packed bed is non-homogeneous but concentration variations along the depth can be decreased by increasing the flow rate. Gradients of pH may occur if the reaction consumes or produces H^+ or OH^-. Because of poor mixing, pH control by addition of acid and alkali is nearly impossible. Beds with greater void volume permit greater flow velocities through them but the concentration of the biocatalyst in a given bed volume declines as the voidage (void volume) is increased. If the packing, i.e. the biocatalyst-supporting solids, is compressible, its weight may compress the bed unless the packing height is kept low. Flow is difficult through a compressed bed because of a reduced voidage. Packed beds are used extensively as immobilised enzyme reactors. Such reactors are particularly attractive for product inhibited reactions: the product concentration varies from a low value at the inlet of the bed to a high value at the exit; thus, only a part of the biocatalyst is exposed to high inhibitory levels of the product.

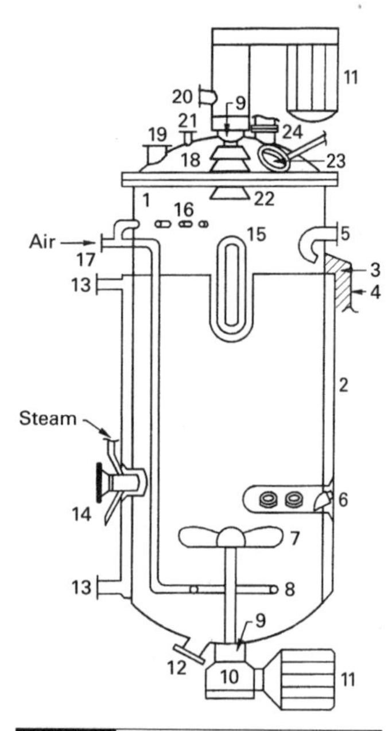

Fig. 7.7 A typical bioreactor: (1) reactor vessel; (2) jacket; (3) insulation; (4) shroud; (5) inoculum connection; (6) ports for pH, temperature and dissolved oxygen sensors; (7) agitator; (8) gas sparger; (9) mechanical seals; (10) reducing gearbox; (11) motor; (12) harvest nozzle; (13) jacket connections; (14) sample valve with steam connection; (15) sight glass; (16) connections for acid, alkali and antifoam chemicals; (17) air inlet; (18) removable top; (19) medium or feed nozzle; (20) air exhaust nozzle; (21) instrument ports (several); (22) foam breaker; (23) sight glass with light (not shown) and steam connection; (24) rupture disc nozzle.

7.3 Bioreactor design features

Irrespective of the specific bioreactor configuration used, the vessel must be provided with certain common features. Some of the principal features are illustrated in Fig. 7.7. The reactor vessel is provided with a vertical sight glass and side ports for pH, temperature, and dissolved O_2 sensors as minimum requirements. Retractable sensors that can be replaced during operation are increasingly used. Connections for acid and alkali (for pH control), antifoam agents, and inoculum are located above the liquid level in the reactor vessel. Air (or other gases, such as CO_2 or ammonia for pH control) is introduced through a sparger situated near the bottom of the vessel. The agitator shaft is provided with steam-sterilisable single or double mechanical seals. Double seals are preferred but they require lubrication with cooled, clean steam condensate. Alternatively, when torque limitations allow, magnetically coupled agitators may be used thereby eliminating the mechanical seals.

Aeration and agitation will inevitably produce foam that is controlled with a combination of chemical antifoam agents and mechanical foam breakers. Foam breakers are used exclusively when the presence of antifoam in the product is not acceptable or if the antifoam interferes with downstream processing operations such as membrane-based separations or chromatography. The shaft of the high-speed mechanical foam breaker must also be sealed using double mechanical seals.

In most instances, the bioreactor is designed for a maximum allowable working pressure of 377–412 kPa (absolute). Although the sterilisation temperature generally does not exceed 121 °C, the vessel is designed for a higher temperature, typically 150–180 °C. The vessel is designed to withstand full vacuum, or it could collapse while cooling after sterilisation. The reactor can be sterilised in place using saturated clean steam at a minimum absolute pressure of 212 kPa. Over-pressure protection is provided by a rupture disc located on top of the bioreactor. Usually this is a graphite burst disc because it does not crack or develop pinholes without failing completely. The rupture disc is piped to a contained drain. Other items located on the head plate of the vessel are nozzles for media or feed addition and for sensors (e.g. the foam electrode), and instruments (e.g. the pressure gauge).

The vessel should have as few internals as practically possible and the design should take into account the needs of clean-in-place and sterilisation-in-place procedures. The vessel should be free of crevices and stagnant areas where pockets of liquids and solids may accumulate. Attention to design of such apparently minor items as the gasket grooves is important. Easy-to-clean channels with rounded edges are preferred. As far as possible, welded joints should be used in preference to couplings. Steam connections should allow for complete displacement of all air pockets in the vessel and associated pipework. Even the exterior of a bioprocess plant should be cleanly designed with smooth contours and minimum bare threads.

The reactor vessel is invariably jacketed. In the absence of especial requirements, the jacket is designed to the same specifications as the vessel. The jacket is covered with chloride-free fibreglass insulation that is fully enclosed in a protective shroud as shown in Fig. 7.7. The jacket is provided with over-pressure protection through a relief valve located on the jacket or its associated piping. For a great majority of applications, austenitic stainless steels are the preferred material of construction for bioreactors. The bioreactor vessel is usually made in Type 316L stainless steel, while the less expensive Type 304 (or 304L) is used for the jacket, the insulation shroud and other surfaces not coming into direct contact with the fermentation broth. The L grades of stainless steel contain less than 0.03% carbon, which reduces chromium carbide formation during welding and lowers the potential for later intergranular corrosion at the welds. The welds on internal parts should be ground flush with the internal surface and polished.

7.4 Design for sterile operation

Most commercial fermentation processes are mono-cultures. To establish and maintain aseptic conditions are vital for the success of these processes. Hence, a bioreactor must be sterilised prior to inoculation and contamination during operation must be prevented. Contamination during culture is a common cause of process failure.

7.4.1 Sterilisation-in-place

A bioreactor intended for *in situ* sterilisation requires a complex arrangement of pipework, valves, and filters to enable initial sterilisation and maintenance of sterility. A typical arrangement for *in situ* sterilisation is shown in Fig. 7.8. Because almost all biopharmaceutical production processes involve aeration, the figure includes aeration and exhaust groups that must also be sterilised. The air inlet and exhaust lines have *in situ* steam-sterilisable gas filters. Typically, hydrophobic membrane cartridge filters are used. These filters are rated for removing particles down to 0.45 μm or even 0.1 μm. Often the gas streams have two filter cartridges in series; with the first serving to protect the final filter.

A good system is designed so that the different sections can be sterilised independently of any of the others, thus sterilisation of any section during fermentation can be carried out if, and when, required. Saturated clean steam (1.1–1.4 bar gauge) is used for sterilisation. The air inlet and exhaust groups are sterilised first, and then, in a second step, the bioreactor. The system is designed so that the filters, valves and the associated pipe-work reach sterilisation temperature (~121 °C) very quickly (~1 minute), and are held at the temperature for the required time (25–30 minutes).

Apart from the harvest valve, all other valves shown in Fig. 7.8 should either be diaphragm or pinch valves. The harvest valve is usually a piston valve with a metal bellows sealed stem. The valve closes flush with the internal surface of the bioreactor and there is an unobstructed flow

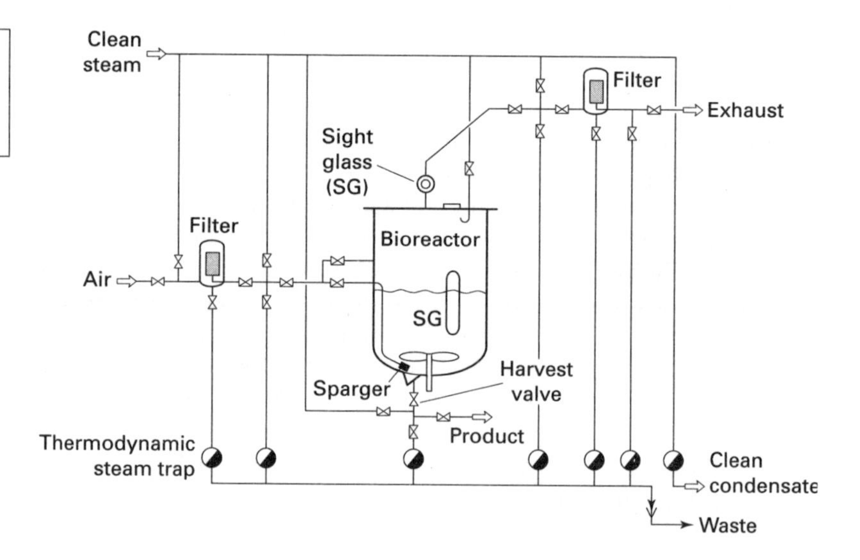

Fig. 7.8 A bioreactor with air inlet and exhaust groups arranged for in-place sterilisation with steam.

path through the valve body. The valves may be operated manually, but pneumatic operation under automatic control is more efficient and reproducible. The bioreactor is sterilised either filled with the medium or without. Empty sterilisation is the norm in cell culture applications where the media are invariably heat sensitive. In this case, filter sterilisation is used to sterilise the medium.

Proper closing and opening sequence of the various valves is important for attaining sterility and preventing recontamination from the adjacent non-sterile areas. Once the sterilising steam supply is shut off, the bioreactor is immediately pressurised with sterile air through the air inlet filter so that any leakage from the outside to the sterile vessel is prevented. In bioreactors with stirrer or foam breaker shaft penetrations, the shaft seals require suitable piping and valves for steam sterilisation and maintenance of a sterile barrier fluid between the contents of the fermenter and the outside.

7.4.2 Clean-in-place considerations

Industrial bioreactors and much of the other processing equipment are cleaned in-place using automated methods. Automation ensures consistency of cleaning and reduces down-time (i.e. unproductive time of a machine). Attaining an acceptable state of cleanliness is essential to prevent contamination and cross-contamination of biopharmaceuticals and food products. An effective and trouble-free cleaning capability requires attention to design of the bioreactors and the clean-in-place (CIP) systems. At any given time a plant may have several bioreactors at different stages of processing and some empty reactors which need to be cleaned along with any associated transfer piping. The CIP devices and procedures must be matched to the specific configuration of the bioreactor and to the fermentation process to ensure satisfactory cleaning. Generally, a bioreactor which has processed hybridoma or other animal cell culture broth is far easier to clean than one which has processed broths of *Streptomyces* or mycelial fungi such as *Penicillium*.

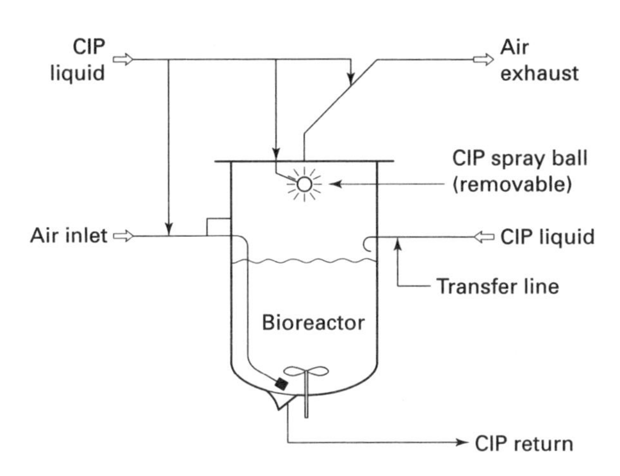

Fig. 7.9 Delivery of the clean-in-place (CIP) liquids to the bioreactor. The flow of CIP solutions is sequenced through the transfer line, the air inlet and exhaust groups and the spray ball of the bioreactor.

Design aspects

To ensure removal of solid particles and avoid sedimentation, the minimum flow velocity through piping should be 1.5 m s^{-1}, but a higher value of 2.0 m s^{-1} is preferred. In addition, the piping should be free of dead spaces as much as possible; if unavoidable, the depth of the dead zone must be less than two pipe diameters to ensure adequate cleaning using CIP techniques. Only valves with a metal-bellows-sealed stem, or diaphragm and pinch valves are recommended as all other valves carry a significant risk of contaminating reactors with accumulated debris during the final rinse cycle. For adequate cleaning, the CIP solutions are sprayed into the reactor through one or more removable, static or dynamic spray balls, or dynamic spray nozzles (see Fig. 7.9). In addition, the piping for air exhaust, which is upstream of the exhaust gas filter, and the air inlet piping, should also receive the cleaning solutions. For cleaning with jet spray, pressures of 308 to 377 kPa (absolute) are optimal. Permanently installed spray heads are not recommended for bioreactors because of potential difficulties with sterilisation. These devices must be inserted into the reactor through one of the ports on the head plate. Often, the spray heads are designed to spray the upper one-third of the tank and the falling liquid film irrigates the remaining surface.

For bioreactors for parenteral (injectable) products and other biopharmaceuticals, potable quality deionised water is recommended for all pre-rinsing and detergent formulations. Pre-rinse should be on a once-through basis without recirculation. A five minute pre-rinse is usually sufficient for bacterial, yeast and animal cell culture reactors. Following pre-rinse, 1% (w/v) NaOH at 75–80 °C should be circulated through the equipment so that all product contact surfaces are exposed to this solution for 15–20 minutes. The alkali should be discarded afterwards. Dilution, contamination with soil and microbial spores that can survive for long periods and loss of quality definition of the starting material for the next cleaning, are some of the arguments against re-use of cleaning chemicals. A deionised or reverse osmosis water rinse at 25–35 °C is used to remove all alkali from the system. Process equipment for products that are injected into the body must undergo a final wash with hot water-for-injection grade water. This ensures that all residual water complies with the requisite quality standards.

In mechanically agitated bioreactors, the spray of cleaning solutions may be unable to achieve proper cleaning of the agitators, magnetic couplings, mechanical seals and the lower portions of baffles. Therefore, filling of the vessel to at least above the level of the lowermost impeller and agitation at impeller Reynolds numbers (see Section 7.6) of 10^8–$10^{8.5}$ is recommended during pre-rinse, alkali recirculation and the final rinse. Agitation for 2–3 minutes is sufficient to dislodge adhering dirt or soil. These recommendations assume that reactors are being cleaned in-place soon after use and caking of dirt has not occurred.

7.5 Photobioreactors

Certain micro-algae and cyanobacteria provide important chemicals, such as astaxanthin and β-carotene, in addition to being used as aquaculture feed for bivalves and fish hatchlings. Cyanobacteria, such as *Spirulina*, are also grown as human health-foods. Photosynthetic cultures require sunlight or artificial illumination. Although some algae may be grown heterotrophically, i.e. without sunlight, this type of growth requires an alternative organic energy source, usually glucose. Heterotrophically grown cultures often lack photosynthetic pigments and may not yield the same products as a photosynthesising population. However, much progress is being made in developing heterotrophic strains for production of commercially valuable products.

Artificial illumination is impracticably expensive; only outdoor photobioreactors appear to be promising for large-scale production. Open ponds and 'raceways' are often used to culture micro-algae, but mono-septic culture requires fully closed photobioreactors. Because it needs light, photosynthesis can occur only at relatively shallow depths. Algal ponds are typically no deeper than 0.15 m. However, too much light causes photoinhibition; a situation in which slightly reducing the light intensity will actually improve the rate of photosynthesis. With increasing cell population, the self-shading effect of cells further limits light penetration. In addition to light, photosynthesising algal cells need a source of carbon, usually carbon dioxide. Part of the carbon may be derived from dissolved bicarbonate species. Cells convert CO_2 to carbohydrate and other cellular components. Too high a concentration of CO_2 can reduce photosynthetic productivity.

Closed photobioreactors for monoculture consist of arrays of transparent tubes that may be made of glass, or more commonly, a clear plastic. The tubes may be laid horizontally, or arranged as long rungs on an upright ladder; see Fig. 7.10. A continuous single run tubular loop configuration is also used, or the tube may be wound helically around a vertical cylindrical support. In addition to the tubes, flat or inclined thin panels may be employed in relatively small-scale operations. An array of tubes or a flat panel constitutes a 'solar receiver'. The culture is

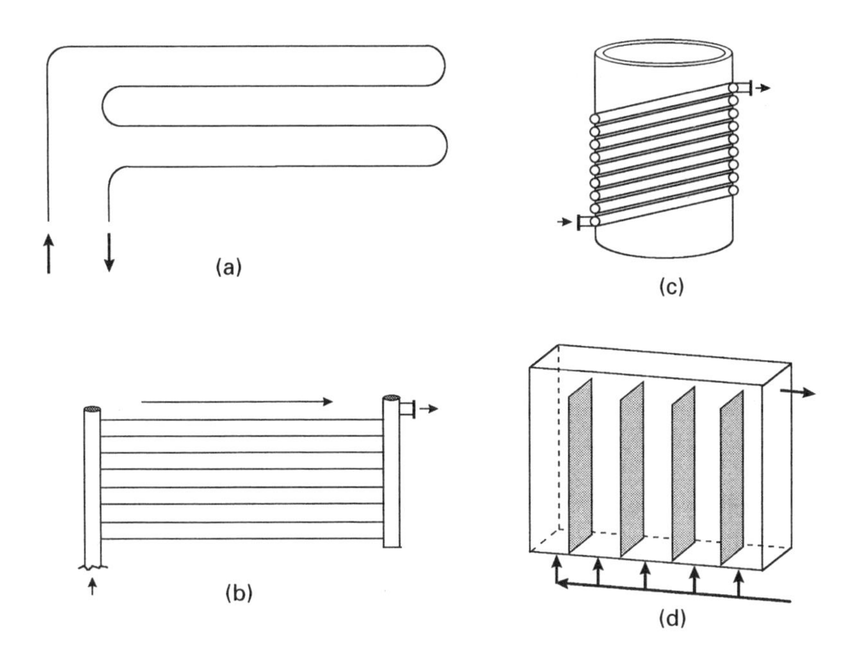

Fig. 7.10 Photobioreactors for mono-culture: (a) continuous run tubular loop; (b) a solar receiver made of multiple parallel tubes; (c) helical wound tubular loop; (d) flat panel configuration. Configuration (a) and (b) may be mounted vertically, or parallel to the ground.

circulated through the solar receiver by a variety of methods, including centrifugal pumps, positive displacement mono pumps, Archimedean screws and airlift devices. Airlift pumps perform well, have no mechanical parts, are easy to operate aseptically and are suited to shear sensitive applications.

The flow in a solar receiver tube or panel should be turbulent enough to aid periodic movement of cells from the deeper poorly lit interior to the regions nearer the walls. Generally, a minimum Reynolds number value of 10^4 is recommended. While turbulence is needed to improve radial mixing, too much turbulence can be harmful. The velocity everywhere should be sufficient to prevent sedimentation of cells. Typical linear velocities through receiver tubes tend to be 0.3–0.5 m s^{-1}.

Because of the need to maintain adequate sunlight penetration, a tubular solar receiver cannot be scaled up by simply increasing the tube diameter. The diameter should not exceed 6 cm, although this constraint may be relaxed somewhat by deploying specially designed static mixers that improve radial mixing inside the tube. Without the mixers, reducing the tube diameter from 6 cm to 5 cm and further to 2 cm noticeably improves culture performance. Light penetration depends on biomass density, cellular morphology and pigmentation, and absorption characteristics of the cell-free culture medium.

Photobioreactors are normally operated in continuous mode. Microalgae grow slowly; the doubling time of *Chlorella pyrenoidosa* is approximately 9 hours. Consequently, the dilution rate is kept low; about one culture volume per day, confined to daylight hours. Although the biomass grows only during daylight, certain products are produced predominantly during the dark hours. Biomass productivity of outdoor cultures is generally less than 2 g l^{-1}·day, often no more than 1.5 g l^{-1}·day. Culture is generally carried out at 22–37 °C. During daylight hours, the solar receiver tubes need to be cooled to prevent temperature rise to damaging levels.

A solar receiver is more amenable to scale-up by increasing the length of a continuous run tube; however, the maximum permissible length should not exceed 50–100 m because the photosynthetically generated O_2 builds up along the tube and high oxygen concentrations inhibit photosynthetic productivity. An alternative to a long continuous run tube is an array of multiple parallel tubes. For minimal costs, each tube of the multi-tube arrangement should be as long as possible but within the threshold where O_2 levels become inhibitory. The rate of photosynthesis and, consequently, the rate of O_2 generation depend on the amount of sunlight available. With *Spirulina platensis* under intense artificial illumination O_2 production rates have been estimated at 0.35–0.5 $g\,l^{-1}\cdot h$ for radiation intensity levels of 1500–2600 $\mu E\,m^{-2}\cdot s$. In a long tube, the rate of photosynthesis varies along the tube because of several factors: (i) the light intensity declines because of increasing cell concentration; (ii) the CO_2 is depleted; and (iii) O_2 concentration may increase to inhibitory levels.

7.6 Heat transfer

All fermentations generate heat; in submerged cultures, 10–50 MJ m^{-3} of the heat output typically comes from microbial activity. Heat production is especially large when the biomass is growing rapidly in high-density fermentations and when reduced carbon sources such as hydrocarbons and methanol are used as substrate. The metabolic heat generation rate in $kJ\,l^{-1}\cdot h$ is typically about 12% of the O_2 consumption rate expressed in mmol $O_2\,l^{-1}\cdot h$.

Heat removal in large vessels becomes difficult as the heat generation rate approaches 20 MJ m^{-3}, corresponding to an O_2 consumption rate of about 5 kg $m^{-3}\cdot h$. In addition to metabolic heat, mechanical agitation of the broth produces up to 50 MJ m^{-3}. In air driven fermenters, all energy input due to gassing is eventually dissipated as heat. Consequently, a fermenter must be cooled to prevent temperature rise and damage to culture. As the scale of operation increases, heat transfer and not O_2 mass transfer becomes the limiting process in bioreactors because the available surface area for cooling decreases as the fermenter volume increases. Temperature is controlled by heating or cooling through external jackets and internal coils. Less frequently, additional double walled baffles, draft tubes or heat exchangers located inside the fermentation vessel are needed to provide sufficient heat transfer surface area.

The rate of heat removal, Q_H, is related to the surface area, A_H, available for heat exchange and the mean temperature difference, ΔT, thus,

$$Q_H = U_H A_H \Delta T \tag{7.1}$$

The overall heat transfer coefficient, U_H, is the inverse of the overall resistance to heat transfer. During cooling, heat flows from the broth side to the cooling water in a jacket or cooling coil. The transferring heat

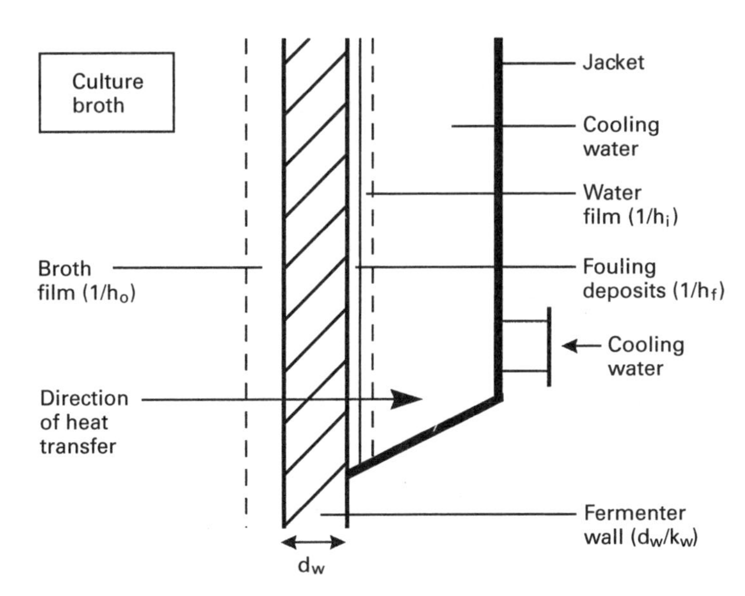

Fig. 7.11 Heat transfer resistances near a fermenter wall.

encounters several resistances in series as illustrated in Fig. 7.11: the thin stagnant film of broth on the inside wall of the fermenter; the metal wall of the fermenter or cooling coil; scale or 'fouling' deposits on the cooling water side; and a thin stagnant film of the cooling fluid on the jacket side of the fermenter wall. These individual resistances are related to the overall resistance as follows:

$$\underbrace{\frac{1}{U_H}}_{\substack{\text{overall}\\\text{resistance}}} = \underbrace{\frac{1}{h_o}}_{\substack{\text{broth film}\\\text{resistance}}} + \underbrace{\frac{d_w}{k_w}}_{\substack{\text{vessel wall}\\\text{resistance}}} + \underbrace{\frac{1}{h_f}}_{\substack{\text{fouling film}\\\text{resistance}}} + \underbrace{\frac{1}{h_i}}_{\substack{\text{water film}\\\text{resistance}}} \tag{7.2}$$

The film heat transfer coefficient is influenced by numerous factors, including the density and the viscosity of the fluid, thermal conductivity and heat capacity, the velocity of flow or some other measure of turbulence (e.g. power input, gas flow rate, etc.), and the geometry of the bioreactor. The many variables that affect heat transfer can be grouped into a few 'dimensionless numbers' to greatly simplify the study and description of those effects. The groups relevant to heat transfer and the corresponding fluid dynamics (e.g. turbulence) are as follows:

$$\text{Nu (Nusselt number)} = \frac{\text{total heat transfer}}{\text{conductive heat transfer}} = \frac{h_o d}{k_T} \tag{7.3}$$

$$\text{Pr (Prandtl number)} = \frac{\text{momentum diffusivity}}{\text{thermal diffusivity}} = \frac{C_p \mu_L}{k_T} \tag{7.4}$$

$$\text{Re (Reynolds number)} = \frac{\text{inertial force}}{\text{viscous force}} = \frac{\rho_L U_L d}{\mu_L} \tag{7.5}$$

$$\text{Gr (Grashof number)} = \frac{\text{gravitation force}}{\text{viscous force}} = \frac{d^3 \rho_L g}{\Delta T \beta \mu_L^2} \tag{7.6}$$

In these equations, d is a characteristic length (e.g. diameter of tube or impeller). The above noted dimensionless groups express the relative significance of the various factors influencing a given situation. The value of the Nusselt number tells us about the relative magnitudes of total heat transfer and that transferred by conduction alone. The Grashof number is important in situations where flow is produced by density differences that may themselves be generated by thermal gradients (hence the $\Delta T\beta$ in the Grashof number). The Reynolds number is employed in describing fluid motion in situations where forced convection is predominant. Correlations for estimating the film heat transfer coefficient, h_o, are often given in terms of these dimensionless groups.

Equations for quantifying the heat transfer resistances of the fouling films and films of heating and cooling fluids are discussed in readily available process engineering handbooks. Suitable correlations for estimating the **heat transfer coefficient**, h_o, for the film of liquid or culture broth in various configurations of bioreactors are summarised in Table 7.1. Note that the correlations given for stirred vessels utilise a Reynolds number that has been defined in terms of the tip speed of the impeller. In some cases, the correlations in Table 7.1 require the thermal conductivity, k_T, and the specific heat capacity, C_p, of the fermentation broth for estimation of the heat transfer coefficient. For most broths, the values of those parameters are close to those of water.

The film coefficient generally increases with increasing turbulence, gas flow rate, and the agitation power input. The coefficient typically declines with increasing viscosity of the culture broth. The geometry of the bioreactor affects the film heat transfer coefficient mainly by influencing the degree of turbulence or related parameters such as the induced liquid circulation rate in airlift vessels. In bubble columns the film coefficient is independent of the column diameter so long as the diameter exceeds about 0.1 m. Similarly, in bubble columns the h_o value is not affected by the height of the gas-free fluid. The value of h_o increases with increasing superficial gas velocity, or power input, but only up to a velocity of about 0.1 m s^{-1}. Furthermore, for identical specific power inputs, bubble columns and stirred vessels provide quite similar values of the heat transfer coefficient.

Literature on heat transfer in airlift reactors is sparse. Equations developed for bubble columns (Table 7.1) may be used to provide a low estimate of h_o in airlift vessels when the induced liquid circulation rates are small. Under other conditions, the coefficient in airlift reactors can be more than two-fold greater than in a bubble column. When liquid flow velocity does not exceed about 0.015 m s^{-1}, the film heat transfer coefficient is largely independent of liquid velocity; however, for higher liquid velocities h_o increases with liquid velocity as follows:

$$h_o \alpha U_L^{1/4} \; [0.015 \le U_L(\mathrm{ms}^{-1}) \le 0.139] \quad (7.7)$$

A large amount of published data is available on heat transfer in vertical two-phase flows. Some of this information may be applicable to airlift reactors provided that the fluid properties, gas hold-up and relative velocities of the two phases are identical in the airlift and the

Table 7.1 Correlations for the broth-side film heat transfer coefficient in various bioreactor geometries

Bioreactor configuration	Correlation	Ranges
1. Stirred tanks (coils)	$\frac{h_o d_T}{k_T}\left(\frac{\mu_{Lw}}{\mu_L}\right)^{0.14} = 0.87\left(\frac{d_i^2 N \rho_L}{\mu_L}\right)^{0.62}\left(\frac{C_p \mu_L}{k_T}\right)^{\frac{1}{3}}$	For cooling coils; Newtonian fluids
2. Stirred tanks (jacketed)	$\frac{h_o d_T}{k_T}\left(\frac{\mu_{Lw}}{\mu_L}\right)^{0.14} = 0.36\left(\frac{d_i^2 N \rho_L}{\mu_L}\right)^{0.67}\left(\frac{C_p \mu_L}{k_T}\right)^{\frac{1}{3}}$	For jacketed vessels; Newtonian fluids
3. Bubble columns	$h_o = 9391 U_G^{0.25}\left(\frac{\mu_w}{\mu_L}\right)^{0.35}$	Newtonian broths $10^{-1} < \mu_L$ (kg m$^{-1}\cdot$s) < 5; $U_G \leq 0.1$ m s^{-1}; $0.1 \leq d_T$ (m) ≤ 1
4. Bubble columns	$\frac{h_o}{\rho_L C_p U_G} = 0.1\left(\frac{U_G^3 \rho_L}{\mu_L g}\left(\frac{\mu_L C_p}{k_T}\right)^2\right)^{-\frac{1}{4}}$	Newtonian broths
5. Airlift vessels	$h_o = 8710 U_{Gr}^{0.22}\left(\frac{A_r}{A_d}\right)^{0.25}\left(\frac{C_p \mu_L}{k_T}\right)^{-0.5}$	Newtonian broths
(draft–tube sparged)		$\mu_L = 0.78$–5.27 mPa s; $0.008 \leq U_{Gr} \leq 0.16$ m s^{-1}; $0.25 \leq A_r/A_d \leq 1.20$. The h_o varied from 600 to 2400 W m$^{-2}\cdot$ °C
6. Airlift vessels (annulus sparged)	$h_o = 13340\left(1 + \frac{A_d}{A_r}\right)^{-0.7} U_G^{0.275}$	Air–water; $0.01 \leq U_G \leq 0.04$ m s^{-1}; $A_d/A_r = 0.242$ and 0.452
7. Fluidised beds (gas–liquid–solid)	$\frac{h_o d_p \varepsilon_L}{k_T(1-\varepsilon_L)} = 0.044\left[\frac{d_p U_L \rho_L}{\mu_L(1-\varepsilon_L)}\frac{C_p \mu_L}{k_L}\right]^{0.78} + 2.0\left(\frac{U_G^2}{g d_p}\right)^{0.17}$	All properties are for the liquid phase

vertical two-phase flow device. Fungal mycelia-like solids may enhance or reduce heat transfer depending on hydrodynamic conditions in the airlift device. Whereas in microbial fermentations the temperature control tolerances are fairly narrow, animal cell cultures demand even more closely controlled temperature regimens. Typically, cells are cultured at 37 ± 0.2 °C. The cells generate little heat and the heat produced by agitation is also small. In addition, the almost water-like consistency of cell culture broths means that heat transfer is relatively easy; however, the temperature differences between the heating/cooling surface and the broth must remain small, or the cells may be damaged.

7.7 Shear effects in culture

Shear rate is a measure of spatial variation in local velocities in a fluid. Cell damage in a moving fluid is sometimes associated with the magnitude of the prevailing shear rate. But the shear rate in the relatively turbulent environment of most bioreactors is neither easily defined nor easily measured. Moreover, the shear rate varies with location within the vessel. Attempts have been made to characterise an average shear rate or a maximum shear rate in various types of bioreactors. In bubble columns an average shear rate has been defined as a function of the superficial gas velocity as follows:

$$\gamma = kU_{\mathrm{G}}^{\mathrm{a}} \tag{7.8}$$

where the parameter a equals 1.0 in most cases, but the k value has been reported variously as 1000, 2800, 5000 m^{-1}, etc. Equation (7.8) has been applied also to airlift bioreactors using the superficial gas velocity in the riser zone as a correlating parameter; however, that usage is incorrect. A more suitable form of the equation for airlift reactors is

$$\gamma = \frac{kU_{\mathrm{Gr}}}{1 + \dfrac{A_{\mathrm{d}}}{A_{\mathrm{r}}}} \tag{7.9}$$

Depending on the value of k equations such as (7.8) and (7.9) produce wildly different values of the supposed shear rate. In addition, the equations fail to take into account the momentum transfer capability, i.e. the density and the viscosity, of the fluid. Both of these will influence the shear rate.

An average shear rate in stirred fermenters is given by the equation:

$$\gamma = k_{\mathrm{i}} \left(\frac{4n}{3n + 1} \right)^{n/(n-1)} N \tag{7.10}$$

where n, the flow index of a fluid, equals 1.0 for a Newtonian liquid such as water and thick glucose syrup. Some typical k_{i} values are: 11–13 for 6-bladed disc turbines, 10–13 for paddle impellers, ~10 for propellers, and ~30 for helical ribbon impellers. The maximum shear rate on a Rushton disc turbine in Newtonian fluids has been expressed as:

$$\gamma_{max} = 3.3N^{1.5}d_i\left(\frac{\rho_L}{\mu_L}\right)^{1/2} \tag{7.11}$$

Equation (7.11) applies when $100 \leq (Nd_i^2\rho_L/\mu_L) \leq 29\,000$. It also applies to non-Newtonian fluids if μ_L is taken as the zero shear viscosity. The shear rate can be converted to a parameter known as shear stress τ, where

$$\tau = \gamma\mu_L \tag{7.12}$$

The susceptibility of some animal cells to shear stress levels has been characterised in well-defined laminar flow environments.

Another method of deciding whether the turbulence in a fluid could potentially damage a suspended biocatalyst is based on comparing the dimensions of the cell or the biocatalyst floc with the length scale of the fluid eddies. The mean length, ℓ, of the fluid eddy depends on the energy dissipation rate per unit mass of the fluid in the bioreactor; thus,

$$\ell = \left(\frac{\mu_L}{\rho_L}\right)^{3/4} E^{-1/4} \tag{7.13}$$

In most cases, all the energy input to the fluid is dissipated in fluid eddies and E equals the rate of energy input. Methods for calculating the energy input rate in the principal kinds of bioreactors are noted in Box 7.1. Equation (7.13) applies to isotropically turbulent fluid, i.e. one in which the size of the primary eddies generated by the turbulence producing mechanism is a thousand-fold or more compared to the size of the energy dissipating micro-eddies. The size of the micro-eddies is calculated with Eqn (7.13). The length scale of the primary eddies is often approximated as the width of the impeller blade or the diameter of the impeller in a stirred tank. In bubble columns and airlift bioreactors, the length scale of primary eddies is approximated as the diameter of the column (or the riser tube) or the diameter of the bubble issuing from the gas sparger. Generally, if the dimensions of the biocatalyst particle are much smaller than the calculated length, ℓ, of the micro-eddies, the particle is simply carried around by the fluid eddy; the particle does not experience any disruptive force. On the other hand, a particle that is larger than the length scale of the eddy will experience pressure differentials on its surface and if the particle is not strong enough it could be broken by the resulting forces.

In addition to turbulence within the fluid, other damage-causing phenomena in a bioreactor include inter-particle collisions; collisions with walls, other stationary surfaces, and the impeller; shear forces associated with bubble rupture at the surface of the fluid; phenomena linked with bubble coalescence and break-up; and bubble formation at the gas sparger. Effects of interfacial shear rate around rising bubbles and those due to bubble rupture at the surface can be minimised by adding non-ionic surfactants to the culture medium. These surfactants reduce adherence of animal cells to bubbles; hence, fewer cells experience interfacial shear and rupture events at the surface of the liquid.

In micro-carrier culture of animal cells where spherical carriers as

Box 7.1 | Energy input in bioreactors

Depending on the type of the bioreactor, the energy input per unit mass of the fluid is estimated variously as detailed below:

(a) Bubble columns

$$E = gU_G$$

where g is the gravitational acceleration.

(b) Airlift bioreactors

$$E = \frac{gU_{Gr}}{1 + \frac{A_d}{A_r}}$$

where A_d and A_b are the cross-sectional areas of the downcomer and the riser, respectively.

(c) Stirred tanks

(i) *Laminar flow*

In stirred vessels the flow is laminar when the impeller Reynolds number Re_i is less than 10. The Re_i can be calculated using the equation

$$Re_i = \frac{\rho_L N d_i^2}{\mu_L}$$

In laminar flow the stirrer power number Po is related to the impeller Reynolds number as follows

$$Po = cRe_i^{-1}$$

where the constant c is ~100 (6-bladed disc turbine) or ~40 (propeller). Because the power number equals $(P/\rho_L N^3 d_i^5)$, the power input P for the unaerated condition can be calculated. In presence of aeration, the power input is lower. The gassed power input P_G is calculated using the previously determined P value in the equation

$$P_G = 0.72\left(\frac{P^2 N d_i^3}{Q^{0.56}}\right)^{0.45}$$

where Q is the volumetric aeration rate. Now E is obtained as

$$E = \frac{P_G}{\rho_L V_L}$$

(ii) *Turbulent flow*

The flow in stirred vessels is turbulent when $Re_i > 10^3$. In turbulent flow the power number is a constant which depends on the geometry of the impeller. Some constant Po values are: 0.32 (propeller), 1.70 (2-bladed paddle), 6.30 (6-bladed disc turbine), and 1.0 (5-bladed Prochem® impeller). The unaerated power input P can be calculated using the constant value of the power number. Now the P_G and the E values are calculated as explained for laminar flow.

small as 200 μm in diameter are suspended in the culture fluid to support adherent cells on the surface of the carrier, inter-particle collisions are generally infrequent under the conditions that are typically employed. However, the size of the fluid eddies in micro-carrier culture systems may be similar to or smaller than the dimensions of the carriers; hence, the adhering cells may experience turbulence-related damage. Freely suspended animal cells are generally too small to be damaged by fluid turbulence levels that are typically employed in cell culture bioreactors. In micro-carrier culture, shear stress levels as low as 0.25 N m^{-2} may interfere with the initial attachment of cells on micro-carriers.

7.8 Further reading

Chisti, Y. (1999). Shear sensitivity. In *Encyclopedia of Bioprocess Technology: Fermentation, Biocatalysis, and Bioseparation, Vol. 5* (M. C. Flickinger and S. W. Drew, eds.), pp. 2379–2406. John Wiley, New York.

Chisti, Y. (1999). Solid substrate fermentations, enzyme production, food enrichment. In *Encyclopedia of Bioprocess Technology: Fermentation, Biocatalysis, and Bioseparation, Vol. 5* (M. C. Flickinger and S. W. Drew, eds.), pp. 2446–2462. Wiley, New York.

Chisti, Y. and Moo-Young, M. (1999). Fermentation technology, bioprocessing, scale-up and manufacture. In *Biotechnology: The Science and the Business, 2nd Edition* (V. Moses, R. E. Cape and D. G. Springham, eds.), pp. 177–222. Harwood Academic Publishers, New York.

Doran, P. M. (1995). *Bioprocess Engineering Principles*. Academic Press, London.

Grima, E. M., Fernández, F. G. A., Camacho, F. G. and Chisti, Y. (1999). Photobioreactors: light regime, mass transfer, and scaleup. *J. Biotechnol.* **70**, 231–247.

Lydersen, B. K., D'Elia, N. A. and Nelson, K. L. (eds.) (1994). *Bioprocess Engineering: Systems, Equipment and Facilities*. John Wiley, New York.

Van't Riet, K. and Tramper, J. (1991). *Basic Bioreactor Design*. Marcel Dekker, New York.

Chapter 8

Mass transfer

Henk J. Noorman

Nomenclature

Roman

a	interfacial area per unit liquid volume	m^{-1}
a'	interfacial area per unit total reaction volume (gas plus liquid)	m^{-1}
C	concentration in liquid phase	$mol\ m^{-3}$
C_i	concentration at liquid side of interface	$mol\ m^{-3}$
C^*	saturation (= equilibrium) concentration in liquid phase (= p/H)	$mol\ m^{-3}$
C_x	biomass concentration	$kg\ m^{-3}$
d	liquid film thickness	m
$\mathbf{D}$	diffusion coefficient or effective diffusivity	$m^2\ s^{-1}$
D	impeller diameter	m
H	Henry coefficient	$bar\ m^3\ mol^{-1}$
H_V	liquid height	m
J	molar mass flux	$mol\ m^{-2}{\cdot}s$
J_g	molar mass flux across gas film	$mol\ m^{-2}{\cdot}s$
J_l	molar mass flux across liquid film	$mol\ m^{-2}{\cdot}s$
k	mass transfer coefficient	$m\ s^{-1}$
k_g	gas film mass transfer coefficient	$m\ s^{-1}$
k_l	liquid film mass transfer coefficient	$m\ s^{-1}$
k_la	volumetric mass transfer coefficient	$1\ s^{-1}$
K	overall mass transfer coefficient	$m\ s^{-1}$
K	consistency index	
n	power law index	
N	impeller rotational speed	s^{-1}
OTR	O_2 transfer rate (= $J\,a$ for O_2)	$mol\ m^{-3}{\cdot}s$

p	pressure	bar (N m^{-2})
p_0	reference pressure (= 1 bar)	bar
p_i	pressure at gas side of interface	bar
p_{in}	inlet gas pressure	bar
p_{out}	outlet gas pressure	bar
P	power input	W
P_s	power input by stirrer	W
q	consumption rate	mol m^{-3}·s
t	time	s
T_V	tank diameter	m
x	distance	m
v_g	superficial gas velocity	m s^{-1}
V	volume	m^3
V_g	gas volume	m^3
V_l	liquid volume	m^3

Greek

α	power law index	
$\dot{\gamma}$	average shear rate	s^{-1}
ε	hold-up or void fraction	
μ	dynamic viscosity	kg m^{-1}·s
μ_0	reference dynamic viscosity	kg m^{-1}·s
ρ_l	liquid phase density	kg m^{-3}

Dimensionless numbers

P_o	impeller power number
Re	Reynolds number

8.1 Introduction

In a bioreaction process, substrates are consumed and products are formed by action of a micro-organism, or catalytic parts of organisms, for example enzymes. Typical substrates for a living cell are carbon sources such as sugar and oil, nitrogen sources such as ammonia and amino acids, and electron acceptors such as O_2. Products can be all kinds of organic compounds, biomass and CO_2. For an optimal rate of reaction, the micro-organism, the academic researcher or the industrial process engineer should see to it that transfer of substrates to the enzyme or cell surface (or the site of reaction inside the cell) and removal of products away from the enzyme or organism is as rapid as possible, and preferably not rate-limiting. Usually this transfer involves a chain of mass transfer steps as shown in Fig. 8.1. The slowest of these steps will determine the overall mass transfer rate, and its value is to be compared with the slowest kinetic reaction step in order to find out if mass transfer will affect the overall process performance or not.

In this chapter, attention will be focused on reactions involving whole cells. In enzymatic biotransformations, cells are absent and the number of mass transfer steps is decreased, but the same concepts can be applied.

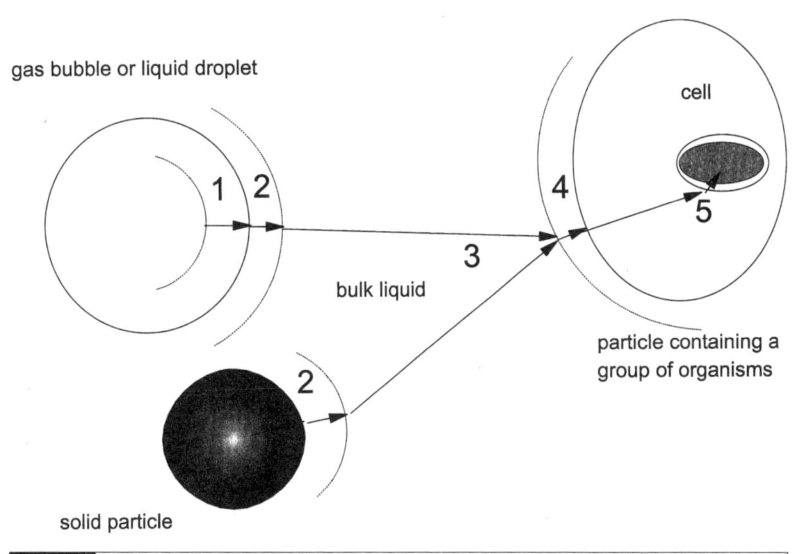

Fig. 8.1 Chain of mass transfer steps for a substrate or nutrient from a gas bubble, liquid droplet or solid particle towards the site of reaction inside a cell: 1: Transfer (mainly by diffusion) of substrates from gas, liquid or solid phase to the interface with the liquid water phase; 2: Transport (most often by a combination of diffusion and convection) across a thin, rather stagnant layer of water phase that surrounds the gas bubble, liquid droplet or solid particle; 3: Transport (usually by convection or turbulence) through the bulk liquid phase to a thin layer surrounding a single micro-organism or a particle (clump, pellet, immobilisation carrier) containing a group of organisms; 4: Transport (diffusive) across this layer to the cell surface; 5: Transport (passive by diffusion and/or active with a transport enzyme) over the cell envelope to a site inside the cell where the reaction takes place. NB: Products formed take the reverse route.

8.2 The mass transfer steps

8.2.1 Effects of transfer limitations

If one mass transfer step is slower than the key kinetic reaction step, it will limit the formation of a desired product from a selected substrate. As a result, two effects may be observed, both with freely suspended cells as well as organisms immobilised inside cell aggregates or solid particles:

- *The overall reaction rate is below the theoretical maximum, and the process output is slower than desired*. This is the case in the formation of gluconic acid from glucose by the aerobic bacterium, *Gluconobacter oxydans*. Here, the overall reaction rate is determined by the rate at which O_2 is transferred to the liquid phase. After relieving the limitation, there is no irreversible effect on this particular micro-organism. Another example is a limited supply of sugar to immobilised cells due to slow diffusion inside an immobilisation carrier. The overall rate of production is often reversibly reduced. However, there are also examples of systems where the biosynthetic capacity of a cell is irreversibly damaged after imposing an O_2

transfer limitation (e.g. in penicillin fermentation). Such processes are very sensitive to mass transfer limitations.

- *The selectivity of the reaction is altered.* For example, O_2 serves as an electron acceptor in the formation of baker's yeast from glucose. In the absence of O_2 the electrons will be directed to pyruvate resulting in the formation of ethanol and CO_2 instead of more yeast. *Bacillus subtilis* cultures produce acetoin and 2,3-butanediol when devoid of O_2. The ratio of the two products is greatly dependent on the dissolved O_2 concentration, and thus on the ratio of O_2 transfer and O_2 consumption rates. Again, the damage can be either reversible or irreversible.

8.2.2 Transfer between phases

The transfer of O_2 from an air bubble to the micro-organism in an aerobic bioprocess is a relatively slow transport step. Oxygen, and other sparingly soluble gases in aqueous solutions (such as hydrocarbons with up to four carbon atoms), may become rapidly depleted when it is consumed. If not replaced at the same high rate the situation will be detrimental for the micro-organism. Transfer of material over a liquid–liquid or liquid–solid boundary is comparable with gas–liquid mass transfer. An example is the growth on higher hydrocarbons ($>C_6$). The oil phase is present in the form of small droplets and mass transfer resistance is at the side of the surrounding water layer. Also, the exchange of material between a solid phase (substrate particles, particles that contain micro-organisms) and the liquid phase obeys similar principles.

8.2.3 Transfer inside a single phase

Inside a gas bubble or oil droplet there is usually enough motion to guarantee a quick transfer of molecules to the interface with the water phase, so the resistance is at the water side of the interface. If the distances in the bulk liquid phase to be bridged are relatively large, a transport resistance can occur in this phase. Such a situation is encountered in large bioreactors where bulk liquid mixing is usually sub-optimal. In industrial practice, it is important to realise that one has to live with this potential limitation. Therefore its effects on the microbial reaction system should be borne in mind during process development work.

Mass transfer limitations inside a solid phase can occur within biocatalyst particles that contain immobilised micro-organisms, either as a surface biofilm attached to a carrier, or dispersed throughout the carrier material. Alternatively, the micro-organism (usually filamentous) itself may be present as a clump or pellet. A substrate entering the particle or pellet may be consumed so fast that nothing enters the inner part of the particle, so that the efficiency of the catalyst is below maximum. Also, the reaction may be slowed down because a toxic or inhibiting product cannot move away quickly enough.

8.2.4 Transfer across the cell envelope

The micro-organism itself can also be considered as a separate (solid or liquid) phase. Transport across the cell envelope (mostly a combination

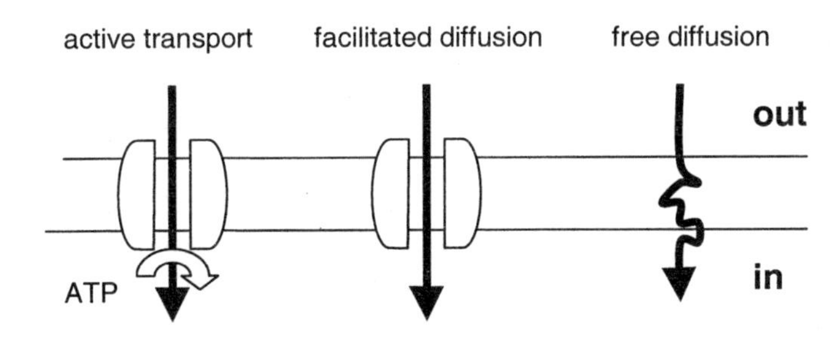

Fig. 8.2 Three mechanisms of mass transfer across the cell envelope.

of cell wall and cytoplasmic membrane) can be limited, depending on the size and physical properties (hydrophobicity, electrical charge) of the molecule and whether the organism is equipped with a specific transport mechanism or not. Generally three mechanisms can be distinguished (Fig. 8.2):

- free diffusion: passive transport down a concentration gradient;
- facilitated diffusion: as above but speeded up by a carrier protein;
- active transport: transport by a carrier protein with input of free energy.

The diameter of the microbial cell itself is very small (order of magnitude 1–5 μm) so that diffusion inside the cell is more rapid than transport across the cell envelope. Additionally, in eukaryotic cells there are intracellular organs (vacuoles, mitochondria) which can present another transport barrier. However, in quantitative terms this type of transport is much more rapid than the consumption rate inside the cell and will normally not limit the overall rate in the chain of transport steps.

8.3 Mass transfer equations

8.3.1 Fundamental principles

The Fick equation (8.1a) states that the mass transfer, J, of a component in single phase will be proportional to the concentration gradient in the direction of the transport. The phenomenological expression for **steady-state** mass flux is:

$$J = -\mathbf{D}\, dC/dx \qquad (8.1a)$$

When mass transfer in a solid phase is considered, $\mathbf{D}$ is the *effective* diffusivity, a function of the diffusion coefficient, the porosity of the solid and the shape of the channels inside the solid. For the geometry of a flat sheet boundary layer with thickness d in a stationary fluid, the relationship between mass flux and concentration difference, ΔC, becomes:

$$J = \mathbf{D}\, \Delta C \,/\, d \qquad (8.1b)$$

$\mathbf{D}/d$ is the mass transfer coefficient and the inverse, $d/\mathbf{D}$, can be interpreted as the resistance against transport. ΔC is the driving force for the transfer.

For the **unsteady-state** situation, the solution of a mass balance over a layer with diameter dx results in:

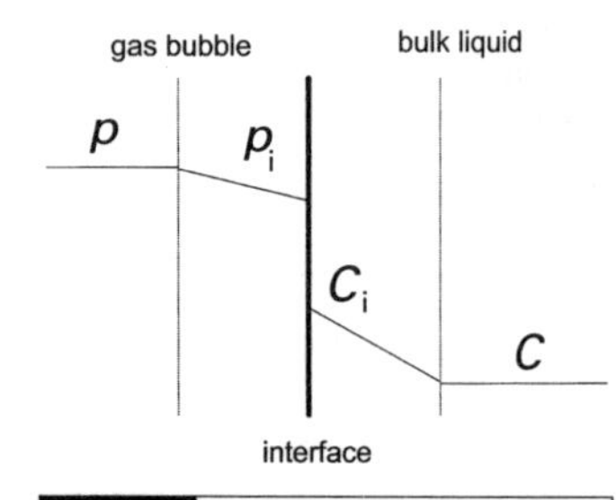

Fig. 8.3 Two-film theory: mass transfer across an interfacial boundary is composed of two steps in series.

$$D\, \delta^2 C / \delta x^2 = \delta C / \delta t \tag{8.2}$$

These fundamental, but theoretical, equations can be used to calculate mass transfer by a diffusion process. A prerequisite is that convective transport is absent but this is rarely the case. More often, a combination of diffusion and convection with phase transfer is encountered. Now we have the additional problem that the velocity pattern of the liquid flow is not known. Thus, for gas–liquid and liquid–particle mass transfer in real bioreactors a more empirical approach is preferred.

For **mass transfer between liquid and gas phases** or **liquid and solid phases**, the well-known *two-film theory* (see any standard chemical engineering textbook on mass transfer) can be adopted. Mass flux in both phases must be separately described, whereas the overall transfer is determined by two steps in series across the film, as shown in Fig. 8.3. For gas–liquid transfer the mass flux is described by:

$$\text{gas film transport:} \quad J_g = k_g (p - p_i) \tag{8.3}$$

$$\text{liquid film transport:} \quad J_l = k_l (C_i - C) \tag{8.4}$$

The concentrations at both sides of the interface p_i and C_i are not identical, but related through the Henry coefficient, H:

$$p_i = \mathrm{H}\, C_i \tag{8.5}$$

In practice it is not possible to measure the interfacial values, so it is better to eliminate these from Eqns (8.3), (8.4) and (8.5) and write the mass flow as a function of the concentrations in both bulk phases:

$$J = K (C^* - C) \tag{8.6}$$

where C^* $(= p/H)$ is the saturation value in the liquid phase. Note that Eqn (8.6) is of the same form as Eqn (8.1b). $(C^* - C)$ is the overall driving force and the overall transfer coefficient, K, results from the sum of the transfer resistances:

$$1/K = 1/(\mathrm{H}\, k_g) + 1/k_l \tag{8.7}$$

Often this general equation can be simplified as $1/(\mathrm{H}\, k_g) \ll 1/k_l$ (i.e. the gas phase film resistance is negligible compared to the resistance in the liquid film). Usually, mass transfer is expressed per unit of volume of the bioreactor, rather than per unit of interfacial area. This is because in many cases, such as when dealing with a sparged and agitated bioreactor, or a packed bed tower reactor, the interfacial area available for mass transfer is not easily determined. The volumetric mass transfer rate then follows from:

$$Ja = k_l a (C^* - C) \tag{8.8}$$

where a is the gas–liquid interfacial area per unit liquid volume, or area per unit gas/solid plus liquid volume, or area per unit gross vessel volume in the bioreactor. When dealing with the transfer of O_2 from gas to liquid, the product Ja is usually called the *OTR*, or O_2 transfer rate.

For an accurate estimate of $k_l a$, assumptions must be made on the values of C^* (or p) and C. For a laboratory scale bioreactor (<10 litres

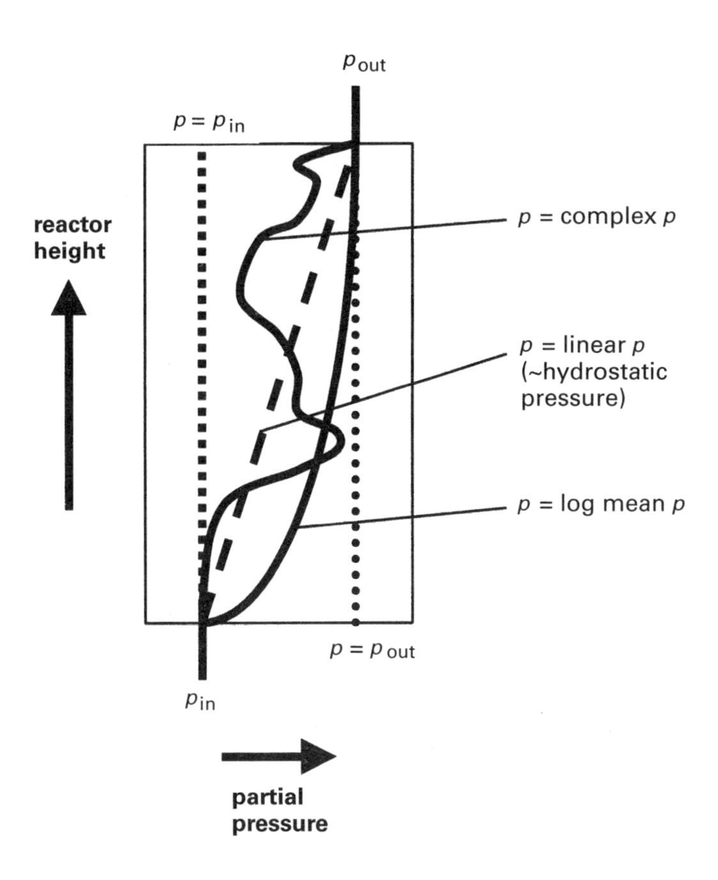

Fig. 8.4 Five possible curves of the changes in the partial pressure in the gas phase as the air bubbles travel up through a reactor vessel.

operating volume) the bulk liquid phase is assumed to be well-mixed and hence C is constant throughout the liquid. However, in pilot plants or production scale vessels (>100 litres) this will not be the case, and local concentration variations need to be taken into account (see also Section 8.5). Therefore, for the gas phase we must assume that one of the following cases will be applicable (Fig. 8.4):

(1) $p = p_{in} =$ constant; there is little or no depletion of the inlet gas, which is often true for small reactors with high gas flow and relatively little transfer.

(2) $p = p_{out} =$ constant; the gas phase is perfectly mixed, which also applies to small systems, but the transfer rate is high compared to the flow of the gas phase.

(3) p varies inside the reactor; for small reactors p is given by its logarithmic mean value, in larger reactors with adequate mixing the hydrostatic pressure determines p, and in large vessels with poor mixing more complex models must be used.

Note that the value of H, and hence C^*, is a function of liquid composition and temperature (gases are generally less soluble at higher temperature). Henry's law also implies that C^* is linearly dependent on p.

There are a number of theories with which the values of k_l and a can be separately estimated. These theories all suffer from questionable assumptions when applied to moving fluids in a real bioreactor but do give quantitative insight into fundamental aspects of mass transfer. Details on such theories can be found in most (bio)chemical textbooks.

8.3.2 Gas–liquid mass transfer in real systems

There has been a lot of focus on O_2 transfer from the gas phase into the liquid phase in bioreactor processes. Since it is experimentally very difficult to estimate the values of k_l and a separately, k_la is often treated as a lumped parameter. In bioprocess engineering literature, one can find a large number of expressions for this (volumetric mass transfer) coefficient. Here, a division should be made between the dominant types of reactors used; **bubble columns**, **air-lift reactors** and **stirred tank reactors** (see Chapter 7). In each case the physical properties of the liquid may influence the magnitude of mass transfer. Extreme values are given by:

- A liquid which greatly stimulates bubble coalescence, i.e. a coalescing liquid. Here mass transfer is poorest.
- A liquid which suppresses coalescence to a large extent, i.e. a non-coalescing liquid. This gives the highest mass transfer rates.

In a **bubble column** (see Chapter 7, Section 7.2.2) the gas enters through the sparger orifices. If the broth is coalescent and non-viscous, e.g. distilled or tap water, the bubbles will rapidly take their equilibrium average diameter of ca. 6 mm. When the air flow rate is high enough the vessel is operated in the heterogeneous flow regime and the hold-up is a function of the superficial gas velocity (= gas flow per unit cross-sectional area of the reactor), corrected for pressure differences (p_0 is a reference pressure of 1 bar):

$$\varepsilon = 0.6\,(v_g\,p_0/p)^{0.7} \tag{8.9}$$

For the mass transfer coefficient, the following correlation has been experimentally verified:

$$k_la = 0.32\,(v_g\,p_0/p)^{0.7} \tag{8.10}$$

In non-coalescing liquids, e.g. ionic solutions and some fermentation broths, the bubbles that originate from the sparger will rise and not mix with other bubbles, provided that the size is smaller than ca. 6 mm. The interfacial area, and hence k_la, will be higher than when larger bubbles are present. If the bubbles are larger they will disperse and take the same equilibrium value as in coalescing liquids. It is noted that in a large bubble column (>50 m^3), the bubbles will significantly expand as they rise through the reactor because of the decreasing hydrostatic pressure. This will influence mass transfer.

In an **air-lift reactor** (Chapter 7, Section 7.2.3), there is a riser section, in which the sparging of bubbles results in an upward liquid flow, a top section, where the bubbles escape from the liquid, and a downcomer section, in which the liquid is recirculated downward. Although the riser resembles a bubble column, the gas hold-up is lower than predicted by Eqn (8.9) due to the interaction with the liquid flow. Correspondingly, k_la will be lower, up to one third of the bubble column value. A precise quantification, however, cannot be easily made.

In a **stirred tank reactor** (Chapter 7, Section 7.2.1) the flow phenomena are determined by the balance between aeration forces and agitation forces, and large local variations in combination with a number of flow regime transitions make a precise quantification of mass transfer difficult. Sparged gas is usually rapidly collected in the gas cavities behind the rotating impeller blades. The cavities flow in a highly turbulent vortex and the gas is dispersed into small bubbles. These follow the liquid flow, but will also rise to the surface. They will coalesce in areas that are relatively calm and redisperse in places where the shear stress is high. A part of the bubbles is recirculated into the cavities and the rest escapes at the surface.

There is a correlation available for the average gas hold-up in coalescing liquids:

$$\varepsilon = 0.13\,(P_s/V_l)^{0.33}\,(v_g\,p_0/p)^{0.67} \qquad (8.11)$$

For non-coalescing liquids the hold-up value is considerably higher.

Correlations for k_la are fully empirical. Usually, the extremes of coalescing and non-coalescing liquids are given with the following coarse expressions (accuracy ca. 30%) valid for P_s/V_l between 0.5 and 10 kW m^{-3}:

$$\text{coalescing:} \quad k_la = 0.026\,(P_s/V_l)^{0.4}\,(v_g\,p_0/p)^{0.5} \qquad (8.12)$$

$$\text{non-coalescing:} \quad k_la = 0.002\,(P_s/V_l)^{0.7}\,(v_g\,p_0/p)^{0.2} \qquad (8.13)$$

In a coalescing liquid the influence of aeration is larger than agitation, while for a non-coalescing liquid the opposite is true. (Note that the correlations are independent of the agitator type.) The energy input by agitation is a vital variable in Eqns (8.11), (8.12) and (8.13). The amount of power drawn by a stirrer of diameter D with a rotational speed N is usually expressed as:

$$P = P_o\,\rho_l\,N^3\,D^5 \qquad (8.14)$$

The impeller power number is a function of the aeration rate, the impeller Reynolds number ($=\rho_l\,N\,D^2\,/\,\mu$) and the impeller type (see Fig. 8.5). When a broth is aerated, P_o generally falls due to a growing size of the cavities behind the blades. A typical value for a Rushton impeller is a factor 0.5, while for a Scaba type (6SRGT) there is hardly any drop.

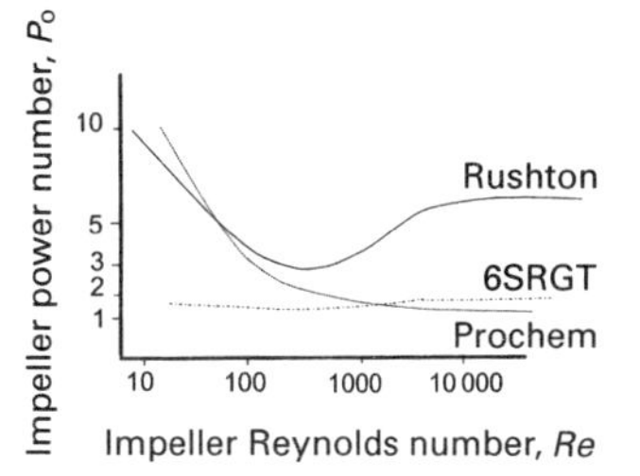

Fig. 8.5 Some unaerated impeller power numbers, P_o, as a function of the impeller Reynolds number, *Re*. In the turbulent flow regime ($Re > 4000$) in the absence of aeration the value of P_o for a standard six-bladed Rushton turbine is constant at ca. 5, for a Scaba 6SRGT impeller it is ca. 1.7 and for a Prochem agitator it is ca. 1.0. In the laminar flow regime ($Re < 1000$), there is an inverse proportionality with *Re* (e.g. for a Rushton impeller $P_o = 64/Re$).

Example

The O_2 transfer performance of a stirred tank reactor is generally better than of a bubble column with similar geometry and aeration rate. For a coalescing liquid in a vessel of 100 m^3 reaction volume, with tank diameter 3.5 m, aeration rate 1 vvm (or 1.67 Nm^{3}s^{-1}), head-space pressure 2 bar, impeller diameter 1.75 m, and power input per unit volume $P/V_l = 2$ kW m^{-3}, we get:

Bubble column; Eqn (8.10): $k_la = 0.05$ s^{-1}

Stirred tank; Eqn (8.12): $k_la = 0.14$ s^{-1}

Assuming $(C^* - C) = 0.59$ mol m^{-3} $-$ 0.10 mol m^{-3} $=$ 0.49 mol m^{-3}, the bubble column *OTR* is 0.024 mol m^{-3}·s, whereas for the stirred tank it is

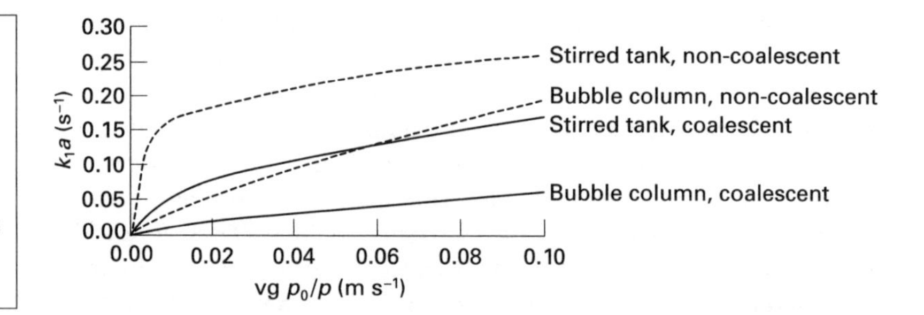

Fig. 8.6 Illustration of O_2 transfer in different reactors and media as a function of superficial gas velocity using Eqns (8.10), (8.12) and (8.13). Here it is assumed that k_la in a bubble column with non-coalescent media is three times the value of coalescent media.

0.070 mol m^{-3}·s. In spite of this, bubble columns are frequently used in bioprocesses, due to other advantages such as simple construction, even distribution of shear rate, lower energy input, etc. (see also Fig. 8.6).

When high concentrations of filamentous micro-organisms are used, the branched filaments of the mycelium interact with each other and form larger aggregates which decrease the free flow of liquid. This results in high viscosity and pseudoplastic or elastic broth behaviour. Similar observations are made when a micro-organism excretes polymeric substances, such as xanthan. The decrease of mass transfer is partly due to the stimulation of bubble coalescence, which leaves large bubbles in the broth. Also the gas hold-up is reported to be lower. As a consequence the interfacial area will be small. In large bioreactors under extreme conditions, bubbles of 1 metre diameter can be observed. Sometimes this effect is partly compensated by the simultaneous presence of a large number of very small bubbles (diameter $<$ 1 mm) which have a long residence time in the broth (15 min or more). There is also an effect of viscosity on k_l. This is due to a decrease of liquid velocity and not a direct consequence of the viscosity itself. Furthermore, in aerated broths the power input may be lowered because the size of the cavities behind the impeller blades are larger. This will diminish the shear rates and bubble break-up.

In the literature, the complex, combined effects are described by an extension of Eqns (8.10), (8.12) and (8.13) with a factor μ^{-n}, where n usually ranges from 0.5 to 0.9. For stirred tanks and non-coalescent media:

$$k_la = 0.002\,(P_s/V_l)^{0.7}\,(v_g\,p_0/p)^{0.2}\,(\mu/\mu_0)^{-0.7} \tag{8.15}$$

provided that $\mu > \mu_0$, where $\mu_0 = 0.05$ *Pa* s.

If the broth behaves like a pseudoplastic fluid (viscosity is decreased at higher shear rates) or dilatant fluid (viscosity is increased with higher shear rates), an average viscosity can be taken over the reactor:

$$\mu = K\,\dot{\gamma}^{n-1} \tag{8.16}$$

The average shear rate can be estimated from:

$$\dot{\gamma} = 10\,N \tag{8.17}$$

The parameters K and n in the rheology model depend on the biomass concentration. Typically K is proportional to C_x^{α}, with the value of α ranging from 1.5 to 4. For pseudoplastic broths $n < 1$, for dilatant liquids $n > 1$, whereas for Newtonian media $n = 1$.

Example
A pseudoplastic broth in a bioreactor has the following properties: $C_x = 30\ g\,l^{-1}$, $K=1$, $n=0.4$. The stirrer speed is 3 revolutions per s. Using Eqn (8.16), the apparent viscosity is 0.13 *Pa* s. According to Eqn (8.15), $k_l a$ is reduced to 51% of the value in a low viscosity broth. What would happen if the biomass concentration or the impeller speed doubled? (Answers: with double biomass concentration and $\alpha=2$, $k_l a$ is reduced to 19%; with double impeller speed, $k_l a$ is reduced to 69%.)

8.3.3 Liquid–solid mass transfer

The description of mass transfer through a liquid film to or from a solid surface is much simpler than for gas-liquid mass transfer. k_l is dependent on the liquid/solid properties, and the value of the interfacial area can be determined experimentally. Usually, liquid-solid transfer applies to situations where mass transfer and reaction are interacting: a substrate is transported through a liquid film to the surface of a particle where micro-organisms are present to consume the substrate. Products formed are transported back through the film to the bulk liquid. These situations are often treated in terms of apparent kinetics, i.e. the observed rate of reaction is described with standard kinetic expressions, but an *effectiveness factor* (ranging from 0 to 1) is introduced to describe the performance of the reaction compared to the same reaction with no transport limitations.

8.3.4 Mass transfer inside a solid particle

When there are micro-organisms active inside a particle or cellular aggregate, the transport (by diffusion) within the particles may present another resistance. This situation is found in biocatalytic processes when dealing with cells immobilised in alginate or porous solid particles. A proper description of this phenomenon is difficult because the kinetics of microbial reactions inside the particle may greatly differ from those with free suspended cells, and hence be unknown. This is due to altered physiological conditions for the cells. In addition to an effectiveness factor for external mass transfer, an overall effectiveness factor may be used, also including intraparticle resistance.

8.4 Determining the volumetric mass transfer coefficients

If one is interested in approximate $k_l a$ values, or when measurements with real bioreaction systems are impractical or economically not feasible, such as in large bioreactors, model fluids can be used. Information is provided from literature, either more theoretical or more empirical as described above, or a new series of experiments can be designed. The use of model fluids, such as water or salt solutions, if necessary with paper pulp or polymers added to mimic viscous broths, gives extremes of what can be expected in real systems, and qualitative trends as a function of changes. There are a few possible measurement methods:

- *Chemical reaction method.* In order to mimic a microbial reaction, the transferred component can be consumed or produced in a chemical reaction. For O_2 transfer studies, sulphite can be used, as it rapidly oxidises to sulphate in the presence of O_2 and a catalyst. $k_l a$ is found from Eqn (8.8) by measuring the rate of sulphite consumption, Ja (= $dC_{sulphite}/dt$) and C^* (= p/H for O_2). For most accurate results, conditions should be selected so that C equals zero. Alternatives for sulphite are glucose in combination with the glucose oxidase (O_2 is consumed), or a mixture of H_2O_2 and catalase (O_2 is produced). Similarly NaOH solutions can be used to study CO_2 transfer (CO_2 rapidly reacts with OH^-).
- *Physical replacement method.* Consider a gas-sparged liquid in steady state, so that C^* equals C. The experiment is started when the mole fraction of O_2 in the gas phase is rapidly changed from the steady state to another value. For example, the N_2 gas in a N_2 sparged liquid is replaced by air. Then Ja in Eqn (8.8) equals dC/dt and by continuously measuring the O_2 concentration in the liquid, $k_l a$ can be evaluated from this equation. Another possibility is a sudden change in pressure of the component to be transferred. This method is usually very fast, and a rapidly enough responding O_2 electrode is therefore an absolute requirement.

If it is required to have real values for $k_l a$, theories, model systems or published correlations should not be used. In actively growing cultures, $k_l a$ can be experimentally determined using the following methods:

- *Steady state bioreaction method.* When using a microbial consuming system in (pseudo) steady state, $k_l a$ can be calculated from Eqn (8.8), when Ja, C^* and C are known, and the difference between C^* and C is large enough. For example, for O_2, Ja (OTR) will be equal to the difference in the O_2 mole fractions in the inlet and outlet gas phases multiplied by the inlet and outlet gas flow rates (O_2 transfer rate = O_2 uptake rate, or OTR = OUR). C^* follows from the partial O_2 pressure ($C^* = p/H$), and C can be read from a dissolved O_2 electrode.
- *Dynamic bioreaction method.* When the air flow is temporarily reduced, shut off, or replaced by N_2 in a respiring culture, $k_l a$ for O_2 becomes zero and the dissolved O_2 concentration will fall rapidly. The rate of depletion, dC/dt, will become equal to the consumption rate (q). After turning on again the air flow rate, the concentration will return to its initial value. Again, Eqn (8.8) can be used to determine $k_l a$, as Ja will be equal to $dC/dt + q$ (see also Fig. 8.7). This method is only applicable in small vessels (< 100 l), because in larger vessels the gas

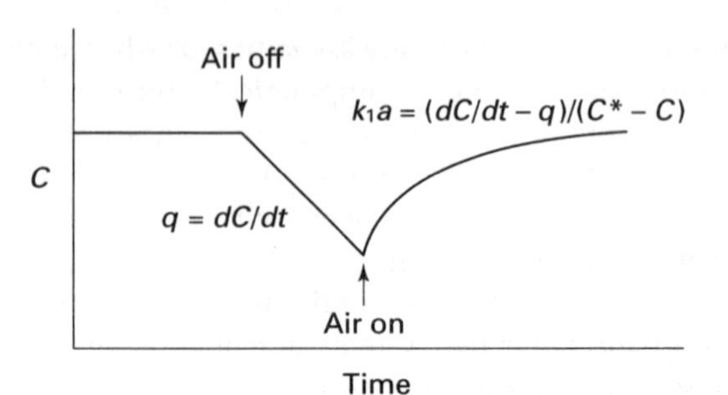

Fig. 8.7 Course of the oxygen concentration during a dynamic $k_l a$ experiment. From the linearly falling part the oxygen consumption rate, q, can be evaluated. From the upward curve then the value of $k_l a$ can be found, for example using a logarithmic plot of ($C^* - C$) vs. time.

phase composition will not be uniform after sparging with N_2 (or the gas hold-up must be built up again after shut down of the air flow). In any case, a rapid O_2 electrode is a prerequisite for accurate results.

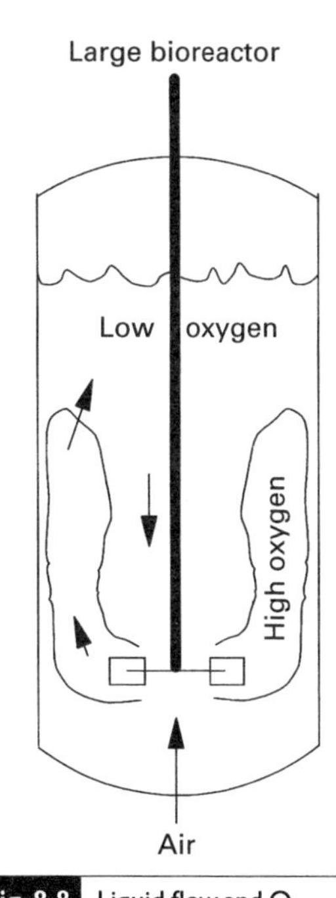

Fig. 8.8 Liquid flow and O_2 transfer in a large bioreactor. Most of the O_2 is transferred in the region near the impeller. In the circulation loop, i.e. the path the broth travels from the impeller, out into the body of the reactor and back to the impeller, more O_2 is consumed than transferred and the O_2 concentration will decrease. In a full-scale stirred tank reactor the liquid circulation loop can be as long as 10 m. With a liquid velocity of 1 m s^{-1} the mean circulation time will be 10 seconds. As a worst case estimate (no transfer at all outside the impeller region), it will take 10 s before the O_2 will become depleted in the loop. Therefore, the O_2 concentration in the bottom compartment should be so high that local depletion, which can be detrimental for the microbial state and product formation rate, is avoided. Note that Eqn (8.8) predicts that this will reduce the mass transfer rate because the driving force (C^*-C) in the bottom part is low as both C^* and C will be high.

8.5 The effect of scale on mass transfer

8.5.1 Scale-up

In large-scale bioreactors, O_2 transfer is usually better than in a laboratory scale or pilot plant reactor. This is due to a larger contribution of the gas phase (higher superficial gas velocity), and a larger driving force (high headspace pressure and hydrostatic head).

Example

Consider two geometrically similar, ideally mixed stirred tank reactors, one of 0.1 m³ reaction volume, and one of 100 m³. The H_V/T_V ratio is 3.0, and the D/T_V ratio 0.5. Scale-up is carried out according to a constant power input, P/V_l, of 2 kW m^{-3}, a constant relative air flow rate of 1 vvm, and a constant average pressure in the broth (determined by the headspace pressure plus the hydrostatic head) of 2.45 bar. Assuming that there is no depletion of the gas phase, the following comparison is made (coalescent broth, $C^* = 0.24$ mol m^{-3} at 1 bar, $C = 0.10$ mol m^{-3}):

$$0.1\ \mathrm{m}^3: v_s\, p_0/p = 0.007\ \mathrm{m\,s^{-1}} \quad k_l a = 0.046\ \mathrm{s^{-1}} \quad OTR = 0.022\ \mathrm{mol\,m^{-3}{\cdot}s}$$

$$100\ \mathrm{m}^3: v_s\, p_0/p = 0.071\ \mathrm{m\,s^{-1}} \quad k_l a = 0.145\ \mathrm{s^{-1}} \quad OTR = 0.071\ \mathrm{mol\,m^{-3}{\cdot}s}$$

What would be the OTR difference for a non-coalescing broth? (Answer: $OTR = 0.07$ or 0.12 mol m^{-3}·s, resp.)

In a large bioreactor the OTR has maximum limits due to the following restrictive conditions.

- There may be mechanical construction difficulties in very large fermenters (>300 m³). Furthermore, liquid transport and mixing will become very slow compared to mass transfer and reaction, and thus rule the overall reaction rate. Cooling limitations may become more significant.
- The average power input should not exceed 5 kW m^{-3}. If higher, the micro-organisms may be mechanically damaged in areas where the value is locally much higher; in addition the energy costs and investment costs for the motor will become excessively high.
- The pressure corrected superficial air velocity should be below 0.10 m s^{-1}. Compressor costs are restricting, and high gas hold-up will increase at the cost of broth space.
- The head-space pressure has a maximum for mechanical reasons. In addition CO_2 partial pressure will also increase, and inhibit growth and production.
- The gas phase cannot be considered ideally mixed. The O_2 partial pressure will fall as the bubbles travel up through the reactor and this reduces the driving force for mass transfer.

In reactors larger than ca. 10 m^3, the processes of liquid transport and mass transfer become comparably slow. Mass transfer and liquid circulation will interfere, and should be treated together. In a stirred tank reactor with one single impeller, most of the O_2 transfer takes place in the impeller zone. A comparison of O_2 transfer in the bubble column and the stirred tank (see example in Section 8.3.2) revealed that outside the impeller zone the OTR may be only one third of the value around the impeller. The importance of circulation loops is described in Fig. 8.8 on previous page.

8.4.2 Scale-down

As illustrated above, micro-organisms can experience a continuously changing environment when travelling around in a large bioreactor. This may give undesired scale-up effects. To avoid these problems the large scale should be taken as the point of reference, and the possible effects should be studied by simulation of the large-scale variations in a small-scale experimental set-up (think big, act small). Limiting factors on a large scale, such as mass transfer, are thus scaled down and can be studied and minimised in a practical and economic way.

In reality, scale-down cannot be precise, because the large-scale conditions are difficult to determine, and also too complex to be fully understood. Several tools are available to find adequate solutions to down-scaling:

- Two-compartment reactor set-ups can be used, mimicking the two most important reactor zones, and recirculation of broth between the zones (see Fig. 8.9). The size of the two compartments and the circulation rate are critical, but also the flow type in each compartment, i.e. ranging from well-mixed to plug flow.
- Development and (large-scale) verification of simple or more sophisticated mathematical flow models for the large vessel can be carried out, and then these used to design the scaled-down experiment.
- Well-characterised microbial test systems can be used, in which the sensitivity to selected variations is known.

In the literature these approaches have been extensively studied for the improvement of O_2 transfer and substrate (feed) mixing in large bioreactors.

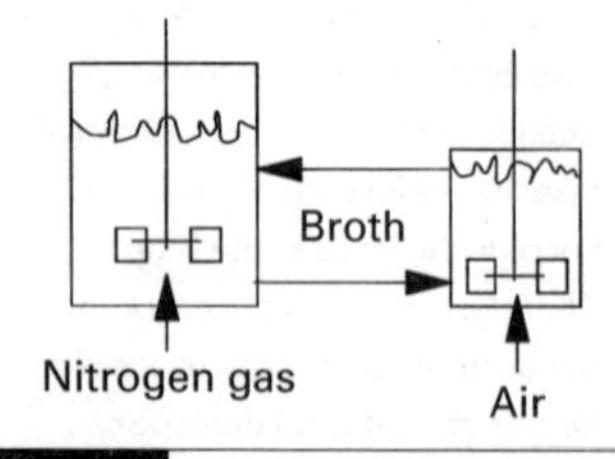

Fig. 8.9 An example of a small-scale 2-compartment reactor set-up for scaling-down the conditions in the large bioreactor. As an alternative, the N_2 sparged vessel may be in plug flow mode (with some dispersion).

8.6 Further reading

Bailey, J. E. and Ollis, D. F. (1986). *Biochemical Engineering Fundamentals, 2nd edition*. McGraw-Hill, New York.

Kossen, N. W. F. (1994). Scale-up. In *Advances in Bioprocess Engineering*. E. Galindo and O. T. Ramirez, eds. Kluwer Academic, Dordrecht.

Merchuk, J. C., Ben-Zvi, S. and Niranjan, K. (1994). Why use bubble column bioreactors? *Trends Biotechnol.* **12**, 501–511.

Nielsen, J. and Villadsen, J. (1994). *Bioreaction Engineering Principles*. Plenum Press, New York.

Nienow, A. W. (1990). Agitators for mycelial fermentations. *Trends Biotechnol.* **8**, 224–233.

'van t Riet, K. and Tramper, J. (1991). *Basic Bioreactor Design*. Marcel Dekker, New York

Chapter 9

Downstream processing in biotechnology

Rajni Hatti-Kaul and Bo Mattiasson

9.1 Introduction

The isolation and purification of a biotechnological product to a form suitable for its intended use is popularly termed '**downstream processing**' (DSP). In most cases this means recovery of a product from a dilute aqueous solution. The complexity of downstream processing is determined by the required purity of the product, in turn determined by its application. The products of biotechnology include whole cells, organic acids, amino acids, solvents, antibiotics, industrial enzymes, therapeutic proteins, vaccines, gums, etc. As the products vary greatly in size and nature, different separation principles are required for their recovery and purification. The lability or sensitivity of many of the bioproducts, particularly the proteins, to the environmental conditions, places further demands on the characteristics of the separation processes used for their production. It is also imperative to minimise the number of steps and to maintain high yields in the different stages, since the ultimate result will be poor if several low-yielding steps are combined.

9.2 Downstream processing: a multistage operation

Many of the steps in DSP are traditional unit processes used extensively in chemical industry and have been described in chemical engineering literature. This chapter is devoted to DSP processes especially adapted for biotechnology. The downstream processing scheme normally employed for isolation and purification of biomolecules can be divided into the following stages:

(1) solid–liquid separation or clarification
(2) concentration
(3) purification
(4) formulation.

Sometimes, as in the case of industrial enzymes, the concentration stage may provide purification sufficient for the final product quality, obviating the need for a separate purification stage. A wide range of unit operations is available for each stage, see Fig. 9.1. (The reader should be aware that the industrial reality of DSP starts upstream at the medium preparation stage, as many raw materials require pretreatment to remove impurities that would have increased the demands on the purification stages.)

9.3 Solid–liquid separation

Solid–liquid separation is a primary recovery operation for the separation of whole cells from the culture broth, removal of cell debris, collection of protein precipitate, collection of inclusion bodies, etc. The unit operations commonly used are **centrifugation** and **filtration**. Table 9.1 lists the characteristics of some of the culture broths used for production of biomolecules. Yeasts and bacteria are usually homogeneously suspended in the fermentation broths. Some bacteria may form slime layers depending on the strain and fermentation conditions and this will lead to separation problems. Filamentous fungi are frequently characterised by a network of intertwined filaments, producing viscous fermentation broths that may be difficult to dewater. Under certain conditions these fungi will form agglomerates called 'pellets', which are relatively easy to recover owing to their large size (100–4000 μm).

9.3.1 Filtration

A filter medium constitutes the separating agent, which retains the particles according to size while allowing the passage of the liquid through the filter. In **cake filtration**, the particles are retained as a cake on the filter medium. The flow through the filter layers is dependent on the area of the filter and flow resistance provided by the filter medium and the cake. Provided the particles do not penetrate the filter medium, the flow resistance of the latter should remain unchanged; however the cake-layer, as it grows thicker, will provide increasing resistance. In many cases of biomass separation, the filter cakes obtained are compressible and the changing effective pressure difference will influence the flow through the filter. Examples of filter media are perforated sintered metal, cloth, synthetic fibres, cellulose, glass wool, ceramics and synthetic membranes.

Many types of filtration equipment are available. Because of

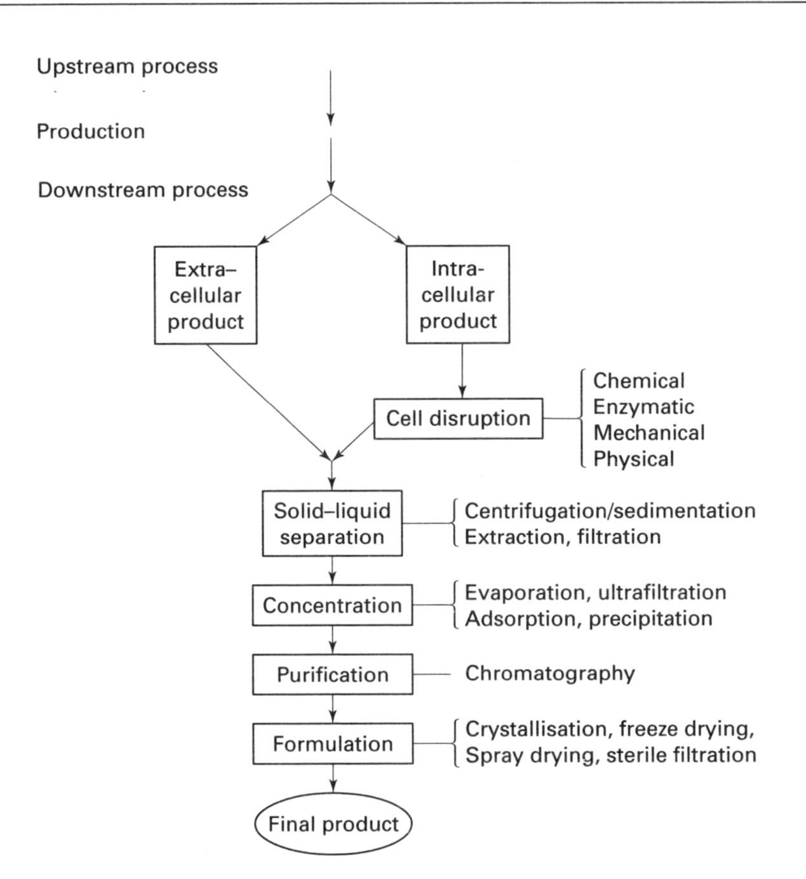

Fig. 9.1 Downstream processing. Different stages in the isolation and purification of the products of biotechnology.

Table 9.1 Size and specific gravity of representative solids in culture broths

Solids type	Size (μm)	Density difference between solids and broth (kg m^{-3})	Cost of recovery
Cell debris[a]	0.2×0.2	0–120[a]	Highest
Bacterial cells	1×2	70	<
Yeast cells	7×10	90	<
Mammalian cells	40×40	70	<
Plant cells	100×100	50	<
Fungal hyphae	1×10×(matted)	10	<
Microbial flocs	100×100	–	Lowest

Notes:
[a] Cell debris density depends on composition, e.g. lipid content.

simplicity of operation and low costs, **vacuum filters** are frequently used for clarification of fermentation broths containing 10–40% solids by volume and particles with sizes of 0.5–10 μm. The best known are **rotary drum vacuum filter** and **filter press**. The former type is commonly employed for filtration of filamentous fungi and yeast cells, and is schematically presented in Fig. 9.2. Besides simplicity and effectiveness of filtration, its attractive features are low power consumption and

Fig. 9.2 A rotary drum vacuum filter. The filtration element comprises a rotating drum, maintained under reduced internal pressure, partially immersed in a tank of broth. Rotating with a speed of 0.25–5 rpm, the drum picks up the biomass, the filtrate is drawn and the cake is deposited on the drum surface. The continuous rotation of the drum allows subsequent operations of dewatering, washing, and drying to be performed on the filter cake prior to its discharge from the drum surface. Depending on the properties, the biomass filter cake is discharged by either: knife-, string-, belt-, or roller-discharge. The drums could be single- or multi-compartment and are often coated with a filter aid that helps to prevent blocking and maintain a constant pressure. Rotary drum vacuum filters are available with filter areas from 2 to 80 m^2.

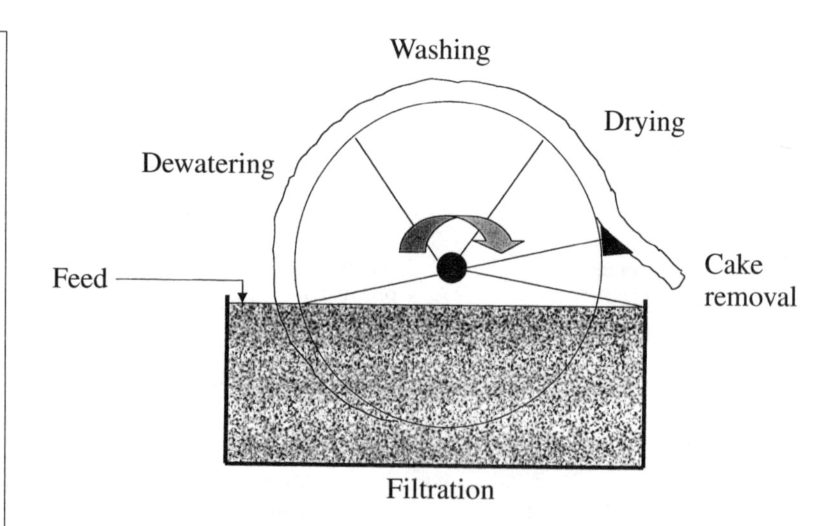

a contained operation. The filter press is built up of a sequence of perforated plates alternating with hollow frames mounted on suitable supports. The plates are covered with a filter medium (cloth) to create a series of cloth-walled chambers into which slurry can be forced under pressure. The solids are retained within the chambers, while the filtrate discharges into hollows on the plate surface, and hence to drain points. At the end of the filtration cycle, the hydraulic pressure is released and the cake manually removed from the cloth. Typical filter sizes available range from 120 cm^2 for laboratory scale to 14400 cm^2 (per plate) for a production scale equipment. In **membrane filter presses**, the cake chamber is covered with a rubber membrane that is inflated using air or water, allowing *in situ* compacting of the cake. These provide higher yield and drier cake, but involve a higher capital investment. Filtration using microporous membrane operated under pressure has developed into a viable alternative to centrifugation as a means to remove suspended particles from process fluids. The technique is described in Section 9.5.3.

Fig. 9.3 Centrifuges. (a) The tubular bowl type is the simplest sedimentation centrifuge used in pilot plants, which can be operated both in batch or continuous mode, with manual discharge of the solids. High centrifugal field can be achieved owing to its slender shape and small volume. (b) The multichamber bowl type is a modification of the tubular type, containing a number of concentric tubes connected to allow zigzag flow of the particulate feed. The centrifugal force is increasing towards the periphery of the bowl with the result that the smallest particles are collected in the outermost chamber. An important application of these centrifuges is in the fractionation of human blood plasma. (c) The disc stack centrifuge is used widely in biotechnology. It incorporates numerous discs (typically between 30 and 200) at an angle of 35–50° and kept 0.4–2 mm apart, dividing the bowl into separate settling zones. The feed enters the centrifuge through a central feed pipe leading to the feed chamber at the bottom of the stack. Solids settle on the lower surface of each disc and migrate toward the bowl periphery, while the clarified liquid moves inward and upward to reach the annular overflow channel situated at the neck of the bowl around the feed pipe. (d) The decanter or scroll centrifuge is mainly used to concentrate slurries with high dry solids concentrations. It consists of a rotating horizontal bowl tapered at one end, having a length-diameter ratio of 1:4, and fitted with a close-fitting helical screw that rotates at a slightly different speed from the bowl. The solids deposit on the wall of the bowl and are scraped off by the screw and discharged from the narrow end of the bowl. The centrifuge can be used for feeds with biomass content of 5–80% v/v.

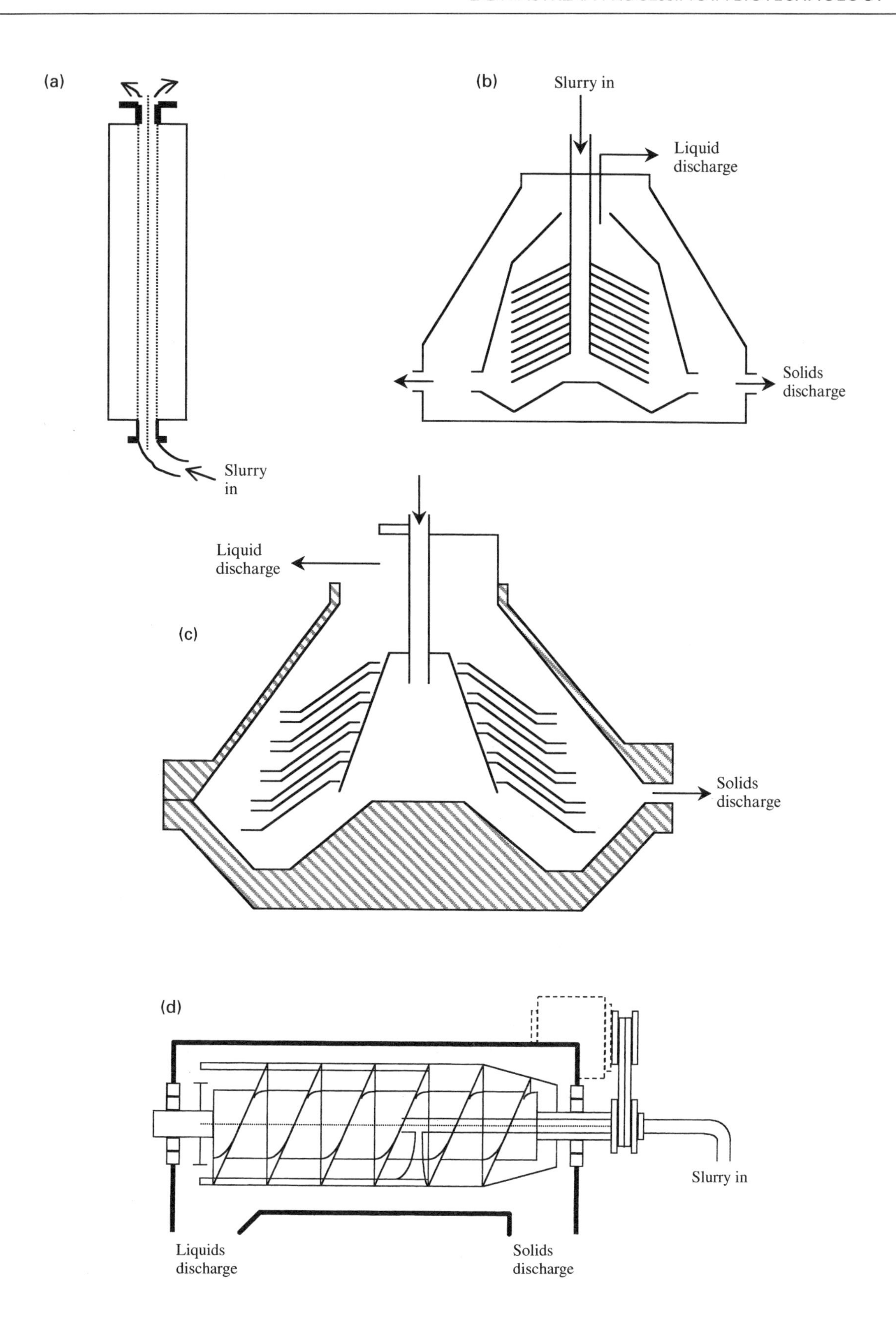
(a)
Slurry in
(b)
Slurry in
Liquid discharge
Solids discharge
Liquid discharge
(c)
Solids discharge
(d)
Slurry in
Liquids discharge
Solids discharge

9.3.2 Centrifugation

In centrifugation the removal of solids relies on the density difference between the particles to be separated and the surrounding medium. Batch centrifugation is commonly applied at a laboratory scale. Although providing high centrifugal forces, low processing capacity limits the use of such centrifuges on an industrial scale, where continuous flow centrifuges are the norm. Here, feeding of the slurry and collection of the clarified solution is continuous, whilst the deposited solids can either be continuously or intermittently removed from the centrifuge. Some of the centrifuge types used in the isolation of bioproducts are shown in Fig. 9.3. Traditionally, centrifugation is used for removal of microbial cells and other discrete large particles. The supernatant obtained by centrifugation is not free of cells (10^3–10^5 cells ml^{-1}), and costs of maintenance and power consumption are both high. Separation of particulate debris from cell homogenates is quite inefficient by centrifugation.

9.3.3 Pretreatment of broth to facilitate clarification

Pretreatment or conditioning of the broth by changing the biomass particle size, the fermentation liquor viscosity, and the interactions between the biomass particles are sometimes required to ensure efficient solid–liquid separation. This implies use of a **filter aid** (body-feed) in the broth and/or for precoating the filter medium. Filter aids are incompressible, discrete particles of high permeability with size ranging from 2–20 μm. Filter aids should be inert to the broth being treated; the most frequently used are Diatomite (skeletal remains of aquatic plants), Perlite (volcanic rock processed to give expanded structure) and inactive carbon.

Agglomeration of individual cells or cell-particles into large flocs, which can be easily separated at low centrifugal forces, is done by the addition of **flocculating agents** such as polycations, either cellulosic or based on synthetic polymers, inorganic salts or mineral hydrocolloids. Besides the flocculating agent itself, factors like the physiological state of the cells, the ionic environment, temperature, and nature of the organism influence flocculation. Interestingly, cationic filter aids also reduce the load of pyrogen, nucleic acid and acidic protein, which normally foul chromatography columns.

9.3.4 Flotation

In flotation, particles are adsorbed on gas bubbles, get trapped in a foam layer and can be collected. The gas may either be sparged into the particulate feed, or very fine bubbles can be generated from dissolved gases by releasing the overpressure or by electrolysis. Formation of stable foam is supported by the presence of 'collector substances' such as long chain fatty acids or amines.

9.4 Release of intracellular components

The primary goal for the first step in a sequence for release of intracellular compounds is to liberate maximum amount of the product in an active state. The inactivating effects of shear, temperature and proteases need to be borne in mind at this stage. The choice of the disruption method has to be made empirically, at the same time taking into consideration the subsequent processing steps.

9.4.1 Disruption of microbial cells

The different strategies that may be used for rupture of microbial cells include breaking the cells' structure by mechanical forces, damaging preferentially the cell wall e.g. by drying or enzymatic lysis, or lysing primarily the membranes e.g. by treatment with chemicals (see Table 9.2).

Mechanical disruption of cells is the most common means of releasing intracellular products both at the laboratory and industrial scale. **Ultrasonication** disrupts the cells by cavitation, and is commonly used at laboratory scale; removal of the heat generated is difficult on a larger scale. Industrial scale disruption of microbial cells is achieved by **high-pressure homogenisation** or by **vigorous agitation with abrasives.** In the former case, the cell suspension is forced at high pressure through an orifice of narrow internal diameter to emerge at atmospheric pressure. The sudden release of pressure creates a liquid shear capable of disrupting the cells. Agitation with glass in bead mills ruptures the cells by a combination of high shear and impact with the cells. The size of the beads varies from 0.2–0.5 mm for bacteria to 0.4–0.7 mm for yeasts.

Table 9.2 Methods for disruption of cells and tissues

Mechanical	Non-mechanical
Microbial cells	
Ultrasonication	Drying
High-pressure homogenisation	Heat shock
Agitation with abrasives	Osmotic shock
	Freeze-thaw
	Organic solvents
	Chaotropic agents
	Alkali
	Detergents
	Enzymes
Animal tissue	
Homogenisation	
Plant tissue	
Maceration	Enzymes
Freezing and grinding	

Recently **microfluidisation** has been introduced for cell disintegration. Here, the cell suspension is fed under high pressure through a chamber where it is split into two streams, to be directed at each other at high velocity before emerging at atmospheric pressure.

Non-mechanical disruption of cells is possible by physical, chemical or enzymatic means. **Drying** is a widely applied method of cell lysis, causing changes in the structure of the cell wall and making it possible to subsequently extract the cell contents in buffer or salt solution. **Osmotic shock** ruptures the cell membrane and is particularly effective for animal cells that lack a cell wall. The presence of a cell wall in the case of microbial cells may necessitate application of additional measures. Cell breakage by **heat shock** (thermolysis) is relatively easy and cheap but is only useful for heat-stable products. Treatment with **organic solvents** and **detergents** to render cells more permeable is often used in the laboratory scale. Toluene is used frequently, acting by dissolving membrane phospholipids and creating pores in the cell membrane. A selective liberation of enzymes from the periplasmic space is possible by treatment with water-miscible solvents such as methanol, ethanol, isopropanol or tert-butanol. The use of inflammable compounds, however, requires spark-proof equipment and special precautions with regard to fire safety.

The advantage of using **enzymes** for cell lysis, besides selectivity during the product release, is their use under mild conditions. The lytic enzyme used extensively, including large-scale operation, is lysozyme. This enzyme damages the peptidoglycan layer and, as a result, the internal osmotic pressure of the cells bursts the periplasmic membrane. Gram-positive bacteria are more susceptible to attack by lysozyme, while the lysis of Gram-negative bacteria requires passage of lysozyme through the outer membrane, which is aided by the addition of EDTA. Glucanase and mannanase, in combination with proteases, are used for the degradation of yeast cell walls. Availability, costs and the need for removal in subsequent purification steps presently limit application of lytic enzymes. A combination of enzymatic/chemical lysis with mechanical disintegration has been suggested in enhancing the efficiencies of the respective methods, with savings in time and energy and the facilitation of subsequent processing.

9.4.2 Homogenisation of animal/plant tissue

Animal cells are normally easy to break because of the absence of cell walls. A typical procedure for homogenising animal tissue is to cut it into small pieces, suspend in ice-cold homogenisation buffer, and grind in a blender. Plant cells are more resistant particularly when they have tough cell walls. The extract of a fleshy or non-fibrous plant tissue is prepared by rapid homogenisation of the material suspended in a partially frozen suitable buffer in a Waring blender pre-cooled to $-20\,°C$. The more fibrous material, which is difficult to macerate, is frozen and ground to a dry powder before adding the extraction buffer for homogenisation. Concentrated buffers with pH values around 6.5–7.2 are used in order to neutralise the acidic materials including the phenols, and also contain phenol scavengers such as polyvinylpyrrolidone (PVP) and/or Amberlite, a hydrophobic polystyrene-based adsorbent, and reducing agents such as ascorbate and thiols to prevent the accumulation of quinones and hence the inactivation of enzymes during extraction. Use of cellulases and pectinases for digestion of plant cell walls is an attractive alternative for achieving selective release of the protoplasmic material and not the vacuole contents but is limited on the large scale by the high enzyme costs.

9.5 Concentration of biological products

After separating the cells from the whole broth, the filtrate contains 85–98% of water, with the product forming only a minor constituent. Removing large amounts of water is costly. It is done in different ways: **evaporation, membrane filtration, liquid–liquid extraction, precipitation** and **adsorption**. The choice of technique is normally dictated by the nature of the product, with minimising the loss of product activity as the main criterion.

9.5.1 Evaporation

Evaporation is a simple, but in most cases, an energy-consuming way of water removal. Because of its reliability and simplicity it is often applied on a large scale, normally using steam as the heat source. In biotechnology, the evaporator must often serve as a multi-purpose equipment, and should be able to handle a broad range of product viscosity (1–10 000 mPa), heat sensitive products, and give a minimum of scale formation, fouling and foaming. The basic unit of evaporators comprises a heating section to which the steam is fed, a section where the concentrate and vapour are separated, a condenser for condensing the vapour, and the required vacuum and product pumps, control equipment, etc. Several different types of equipment have been developed, ranging from laboratory scale (0.5–1.0 l h^{-1} water evaporation capacity) to very large industrial scale (150 m^3 h^{-1}). To decrease the steam consumption, multiple-effect (stage) evaporators have been designed, where the liquid flows through a number of stages (maximum seven) using the vapour of one stage as a heating source for the next.

In **falling film evaporators**, the liquid to be concentrated flows down long tubes and distributes uniformly over the heating surface as a thin film. The vapours flowing in the same direction increase the linear velocity of the liquid, thereby improving the heat transfer. Residence times in the evaporator are in the order of minutes. These evaporators are suited for concentrating viscous products (up to 200 mPa), and are frequently used in the fermentation industry. **Plate evaporators**, where the heating surface is a plate as opposed to a tube, have a relatively large evaporation area in a small volume, but the possibility of treating viscous and solids-containing fluids is limited. For higher viscosities, **forced film evaporators** with mechanically driven liquid films are suitable, in some cases producing a dry product. The residence time ranges from a few seconds to a few minutes. **Centrifugal forced-film evaporators** permit a further reduction of the residence time so that even the heat labile substances can be concentrated under gentle conditions. Evaporation takes place on a heated conical surface or plates over which the transport of the liquid takes place through the centrifugal force produced by the rotating bowl.

9.5.2 Liquid–liquid extraction

Liquid–liquid extraction is applied on a large scale in biotechnology both for concentration and for purification. It involves the transfer of solute from one liquid phase to another. The efficiency of an extraction process is governed by the distribution of substances between the two phases, defined by the partition coefficient K (= concentration of substance in extract phase/concentration of substance in raffinate phase). A partition coefficient well removed from unity is desirable. The physico-chemical properties of the product influence the demands on a liquid–liquid extraction process, as illustrated in the following sections.

Extraction of low molecular weight products

Small lipophilic target molecules are extracted using an organic solvent, whereas for hydrophilic compounds it may be difficult to design an efficient extraction process. Extraction in organic solvent can be done in one of three ways:

Physical extraction: The compound distributes itself between the two phases according to its physical preference. This applies to non-ionising compounds and the extraction is optimised by screening for the solvents that would lead to a high K value and also show a maximal difference in K for the different components present in the crude mixture.

Dissociative extraction: Difference in the dissociation constant of the ionisable components is exploited to achieve separation, and these differences are often large enough to overcome an adverse ratio of partition coefficients. Extraction of penicillin and some other antibiotics are typical for this type of extraction principle.

Reactive extraction: A carrier, such as an aliphatic amine or a phosphorous compound, is added to the organic solvent which forms selective solvation bonds or stoichiometric complexes that are also insoluble in the aqueous phase. Thus, the compound is carried from the aqueous to the organic phase. This type of extraction is advantageous for compounds that have a high solubility in aqueous medium, e.g. organic acids.

In most cases, cells and other particulates are removed prior to extraction to avoid the formation of emulsions at the interface. After extraction, the product is recovered from the solvent either by distilling off the product in case of a high-boiling solvent or distilling off the solvent when this is low boiling. If the product is heat sensitive, it is recovered by back-extraction into a new aqueous phase under conditions different from the first extraction, e.g. penicillin is extracted into butyl acetate or amyl acetate from the fermentation medium at pH 2.5–3.0 and back-extracted into aqueous phosphate buffer at pH 5–7.5.

For high extraction yields, multi-step extraction in a counter-current mode is used, which provides also savings in both solvent and time. Different kinds of extraction equipment are available including mixer-settlers, columns and centrifugal extractors. The latter are often used in extraction of antibiotics and steroids; representative examples being the Podbielniak extractor, Delaval contactor and Westfalia extraction-decanter.

Supercritical fluid (SCF) extraction has been considered as an alternative technique to conventional extraction which has the disadvantage of toxicity and flammability of organic solvents. SCFs are materials that exist as fluids above their critical temperature and pressure, respectively. Many of the properties of SCFs are intermediate between those of gases and liquids; e.g. their diffusivity is higher than those of liquids while viscosity is lower. The attractive feature of SCFs as extractants is that their solvent properties are highly sensitive to changes in both temperature and pressure, which provides the opportunity of tailoring the solvent strength to a given application. Supercritical CO_2 is most commonly used for extractions because of its relatively low critical temperature (31.3 °C) and pressure (72.9 bars). Supercritical extraction is still rather expensive though a few large-scale applications in the area of food processing have demonstrated the possibility of economically viable operations. By adding a co-solvent, such as ethanol, in small amounts, it is possible to modify the properties of the system and thereby influence the extraction behaviour. In a typical extraction process, the compressed SCF is contacted with the feedstock to be extracted in an extraction column, and the loaded SCF is transferred via an expansion valve to a separator. On lowering the pressure, the liquid is turned into gas, releasing the product as a precipitate. The gas is re-pressurised and recycled to the extraction column.

Extraction of proteins

Aqueous two-phase systems (ATPS), prepared by mixing two different polymers, or a polymer and a salt above certain concentrations with water as the major component (80–95%), present suitable extraction systems for proteins. The two phases formed are enriched in the respective phase components, e.g. polyethylene glycol (PEG) is the main component in the top phase, while dextran or salt constitutes the bottom phase. The interfacial tension between the two phases is significantly lower than that in water-organic solvent systems. Phase separation is also slower, varying between a few minutes and 1–2 hours; but is often speeded up by centrifugation at low g values. Partitioning of a component in aqueous two-phase systems is based on its surface characteristics, nature of the phase components and the ionic composition. A common observation is that small molecules are more or less equally distributed between the two phases; the partitioning of particulate matter is invariably one-sided and that of macromolecules covers a wide range.

Use of extraction circumvents many of the shortcomings of centrifugation and filtration for large-scale isolation of proteins from crude

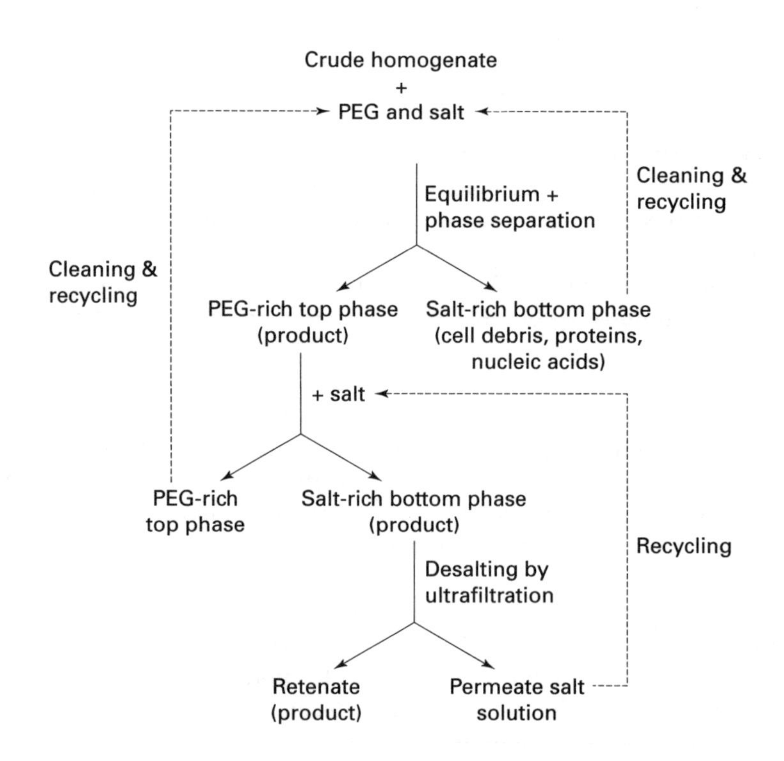

Fig. 9.4 Extraction in aqueous two-phase system. The components of the two-phase system are mixed directly with the crude homogenate. Equilibration is done by gentle mixing and is followed by phase separation. The protein product partitioned to the top phase is subsequently recovered either by a second extraction step, where it is transferred into a new salt phase, or by direct adsorption onto a chromatography matrix. Recycling of the phase components is normally possible, thus minimising the material costs.

homogenates having high viscosity and heterogeneous distribution of particle sizes. Other attractive features of the technique are high capacity (biomass to volume ratio) and straightforward scale-up. The process is easily adapted to the extraction equipment used for water-organic solvent systems. Figure 9.4 shows a schematic presentation of protein isolation and partial purification in ATPS. For industrial scale separations, PEG/salt systems are used because of their relatively low cost. Inexpensive and biodegradable polymers may also be considered as replacement for salt in the bottom phase. The selectivity of protein extraction in ATPS may be increased by incorporation of a suitable ligand molecule into the extracting phase (by being coupled to the polymer forming the phase) which selectively pulls up the target protein; the latter is subsequently released into a new bottom phase made up of either salt or fresh polymer supplemented with free ligand.

Another extraction system having potential in the downstream processing of proteins is that of **reverse micelles** which are thermodynamically stable aggregates of surfactant molecules and water in organic solvents. The polar surfactant head groups point toward the interior of the aggregate forming a polar core in which water can be solubilised to generate water pools. The aliphatic chains of the surfactant protrude into the surrounding organic phase. These systems have generated great interest among biotechnologists as they can be used to solubilise enzymes without loss of activity, which could then be used to catalyse many organic transformations. Reverse micellar systems have also shown potential for the separation and recovery of proteins by extraction from aqueous feed. The extraction seems to involve electrostatic

Table 9.3 Applications of membrane processes in biotechnology

Type of membrane	Driving force	Application
Microfiltration	Hydrostatic pressure Δp 0.5–2 bar	Concentration of bacteria and viruses Harvesting of cells Clarification of fermentation broth
Ultrafiltration	Δp 2–10 bar	Fractionation of biomolecules Desalting Production of enzymes Processing of whey
Hyperfiltration	Δp 20–100 bar	Concentration of pharmaceuticals Production of lactose Part desalination of solutions
Electrodialysis	ΔElectric field	Purification of charged small molecules e.g. organic acids
Pervaporation	Δpartial vapour pressure	Selective removal of solvents (ethanol, pressure acetone-butanol) during fermentation Purification of solvents from azeotropic mixtures
Perstraction	Partition	Extraction of small molecules from aqueous/organic solutions

interactions between the protein surface and charged surfactant, and is thus dependent on pH and ionic strength.

9.5.3 Membrane filtration

The use of membrane technology for separation of biomolecules and particles and concentration of process fluids has expanded dramatically in recent years. The membrane function has been made more versatile by integrating it with other separation principles.

Microfiltration and ultrafiltration

During separation, a semi-permeable membrane acts as a selective barrier retaining the molecules/particles bigger than the pore size while allowing the smaller molecules to permeate through the pores. Membrane filtration processes can be distinguished according to the type of force driving the transport through the membrane, and are named by the size of pores in the filter (see Table 9.3). **Microfiltration** is used for separation of particles, typically 0.02–10 μm in diameter, while **ultrafiltration** separates polymeric solutes in the 0.001–0.02 μm range. **Reverse osmosis**, or **hyperfiltration**, separates ionic solutes typically less than 0.001 μm. Microfiltration and ultrafiltration are widely used in the primary recovery stages of downstream processing. Selectivity of membranes, often expressed in terms of molecular weight cut off (MWCO), is mainly determined by the sieve action of the pores, but also hydrophilic/hydrophobic and charge interactions.

Separation with membranes was initially done by **dead-end filtration**, in which the feed flows on to the membrane. Settling of particles, deposition of colloidal species, adsorption of macromolecular solutes, precipitation of small solutes, etc. on the surface and in the pores of the filter leads to a deposition of a cake which grows in thickness with time, reducing the flow through the filter. **Cross-flow** or **tangential flow filtration** has now completely replaced the dead end filtration for harvesting cells on a large scale. Here, a flow of the feed stream is maintained parallel to the separation surface, with the aim to provide sufficient shear force close to the membrane surface, thereby preventing particulate matter from settling on, or within, the membrane structure. In practice, the membranes in cross-flow filtration are also subject to fouling, but the cake thickness is limited to a thin layer as compared to the dead end mode. Although most filtration media are relatively inert, the formation of gel layer is inevitable. The problem of permeate flux reduction can be minimised by optimising the filter selection, operating pressure, flow properties of feed, and frequent back-flushing.

Microfilters and ultrafilters are available in materials such as ceramics and steel that can be aggressively cleaned and sterilised in place. Membranes composed of polymeric materials such as polyvinyldifluoride (PVDF) and polyethersulphone (PES) are also used, but are more difficult to clean and may require chemical rather than steam sterilisation. Membrane filters are commonly plate and frame systems, employing cartridge filters within which the membrane is present in a highly folded format. This gives a large filtration surface area in a compact space with no dead spaces. Another form is the **hollow-fibre system**, which comprises a bundle of hollow capillaries packed in a tube. The liquid to be filtered is pumped through the central core of the hollow fibres. The permeate passing through the capillary walls can be drained as a permeate from one end of the cartridge, while the concentrated retentate emerges from the other end.

Membrane adsorbers

New micro/macroporous membrane matrices with ion exchange groups and affinity ligands, called membrane adsorbers, have been developed which bind proteins from the clarified feed pumped over them. Desorption of the protein is later performed using solutions as in chromatography (see Section 9.6). A stack of membranes provides a total surface area for adsorption equivalent to chromatography gels, giving similar high resolution separation as chromatograpic methods. In membranes, liquid transport is by convection as opposed to the diffusional flow in gels (see Section 9.6), which increases the speed of separation tremendously.

Pervaporation

Pervaporation is a membrane based process having potential for recovery and concentration of volatile products, see Fig. 9.5. The major limitations of the technique for large-scale work are high energy consumption, insufficient selectivity of the membrane and difficult process design due to a temperature drop across the membrane.

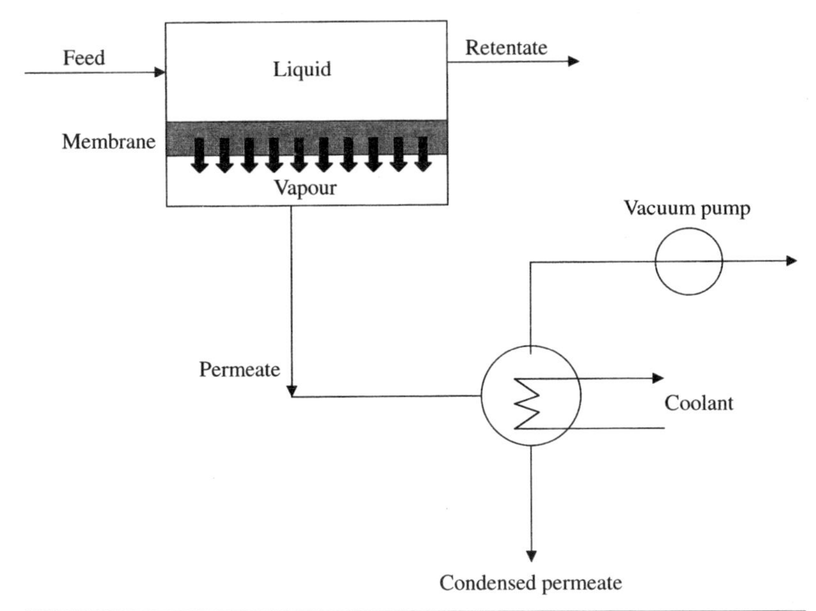

Fig. 9.5 Pervaporation. The process allows separation of volatile products by a combination of permeation through a membrane and evaporation achieved by a low pressure on the downstream side of the membrane. The trans-membrane flux of various components in a mixture and their separation is determined by differences in their vapour pressure and by the permeability of the membrane; the latter being a function of (a) diffusion of the component to be separated through the membrane and (b) membrane thickness. Pervaporation membranes can be considered as homogeneous swollen polymers, e.g. poly-(dimethylsiloxane). The solubility of the component in the membrane material and the extent to which it is swollen determine the diffusion coefficient. Selectivity of the membrane is another factor influencing the performance of the process.

Perstraction

Perstraction is another sophisticated technique for product concentration and recovery that combines membrane processes and solvent extraction. By using the membrane as a barrier between an aqueous feed and an organic solvent it is possible to transfer molecules that partition into the liquid filling the membrane pores. This technology has been used for recovery of hydrophobic substances during fermentations. The membrane protects the cells from the toxic or inhibitory effects of the extraction solvent. Depending on the direction of mass transfer, the membranes could be hydrophobic or hydrophilic. The hydrophobic membranes, which have pores filled with the organic phase, are used to extract the non-polar product from the aqueous medium. In hydrophilic membranes the pores are filled with a suitable aqueous buffer, which facilitates the removal of the product from the solvent.

9.5.4 Precipitation

Precipitation is an established separation technique in industry for concentration of proteins and polysaccharides. Often it constitutes the only

Table 9.4 Modes of protein precipitation

Mode	Example	Comments
Addition of neutral salt	$(NH_4)_2SO_4$	Increased hydrophobic interactions between neutral protein molecules; salt is removed prior to next purification step (except for hydrophobic interaction chromatography) by dialysis, ultrafiltration or gel filtration
Addition of organic solvent	Acetone, ethanol	Reduced dielectric constant enhances electrostatic interactions between protein molecules; low temperature required for operation
Addition of non-ionic polymer	Polyethylene glycol	Reduction in the effective quantity of water available for protein solvation; polymer has often a stabilising effect on proteins
Addition of charged polymer	Polyethyleneimine, polyacrylic acid	Complex formation between oppositely charged molecules leads to charge neutralisation and precipitation
Increase in temperature		Increased hydrophobic interactions; used for precipitation of heat sensitive proteins
Change in pH		Low solubility of protein at isoelectric point; extremes of pH denature and precipitate sensitive proteins

unit operation after solid–liquid separation in the recovery of bulk proteins. Precipitation can also be used to achieve the removal of undesired by-products such as nucleic acids, pigments and other residual components from a crude extract. Precipitation is usually carried out in a batch mode in stirred tanks. Aggregates settle to the bottom of the tank, the mother liquor is removed and the aggregate slurry is fed to a centrifuge or a filter.

Precipitation of proteins is usually based on a decrease in solubility induced by external factors (Table 9.4). Salts and organic solvents are the precipitating agents commonly employed in industry. These precipitation processes are non-specific in the sense that they exploit the ionic and hydrophobic interactions, which are common to all proteins. There are a few examples of precipitations that are semi-specific, e.g. protein–carbohydrate complexes have been precipitated by means of borax additions. Selectivity in precipitation has been introduced by use of affinity interactions. Creation of large complexes as a result of affinity interactions, as between antigen and antibody, is one mode of **affinity precipitation**, which is used in immunoprecipitation. This concept has been generalised to some extent for selective precipitation of multimeric proteins (having more than one binding site for a ligand) by using **homobifunctional ligands** (synthesised by coupling two ligand molecules by a spacer). The modified ligand is able to bridge different protein molecules thereby forming aggregates. The precipitation of the affinity complex occurs only at a definite ratio of the ligand and the protein. Transition metal ions are able to precipitate proteins by bonding with the surface histidine residues, the more the residues the easier is the precipitation.

With **heterobifunctional ligands**, where one functionality is responsible for the affinity binding and the other is exploited for the precipitation, it becomes possible to operate affinity precipitation in a more general mode. The precipitating component of the heterobifunctional ligand is a 'smart polymer' which responds to minor changes in an environmental parameter, e.g. pH, temperature, ionic strength, etc. by a visible change in solubility. Examples of such polymers are chitosan, alginate, galactomannan, poly(*N*-isopropylacrylamide), hydroxypropyl-methyl cellulose, etc.

9.5.5 Adsorption to chromatographic particles

Another strategy often used for concentrating a particular molecule from a crude extract is capturing on high-capacity solid adsorbent particles. Active charcoal was the first reported adsorbent material for product concentration. Ion exchange resins have since been widely used for initial capturing of low molecular weight products and proteins from crude extracts, because of their high binding capacity, their applicability to harsh cleaning-in-place (CIP) protocols and their relatively low cost. Synthetic adsorbents with hydrophobic adsorption characteristics have also been developed which can be used for extraction of organic compounds. The advantage over conventional solvent extraction is that adsorbent processes require much smaller amounts of the toxic and flammable organic solvents. Additionally, the adsorbents have a molecular sieving function based on their pore structure so that fractionation by molecular size can take place during the process. Likewise, hydrophobic adsorbents may also be used for protein concentration.

A broad range of adsorbent matrices is used for industrial processes. For capture of low molecular weight compounds like antibiotics, vitamins, and peptides, polystyrene, methacrylate and acrylate based matrices are used. Cellulose-based adsorbents are commonly employed for protein concentration. Owing to the crude nature of the feed, the adsorption is mostly carried out in a batch mode, after which the slurry is transferred to a column for elution of the product. It may be possible to achieve significant purification by employing selective elution. Adsorption from a whole broth is also conveniently done in a fluidised or expanded bed mode (see Fig. 9.6). Here, the culture broth is pumped through the resin packed column, equipped with sieve plates, in an upward direction at a velocity high enough to fluidise/expand the adsorbent particles and allowing the particulate matter to pass through. Fouling of the adsorbent by irreversible adsorption of proteins and pigments could be a problem. This necessitates intermittent cleaning of the resin under strong conditions e.g. using NaOH/propan-2-ol, or sodium hypochlorite.

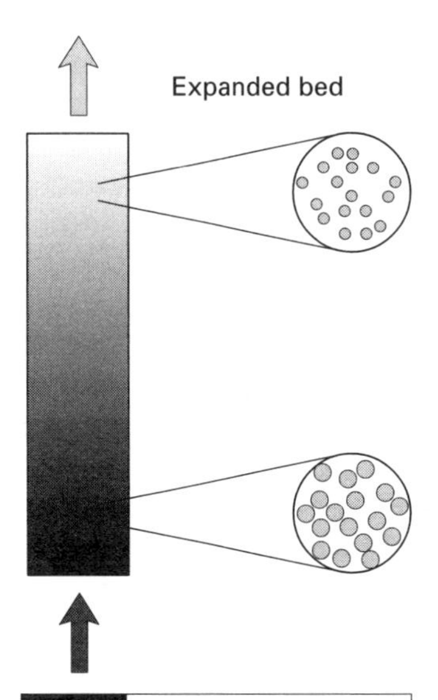

Fig. 9.6 Expanded bed adsorption chromatography. The bed of adsorbent beads in a column is expanded by an upward flow of liquid. A stable bed is easily obtained by using an adsorbent with a distribution of particle sizes. The larger, heavier particles are confined to the bottom of the bed while the lighter particles move to the top. Conventional fluidised beds, on the other hand, are well-mixed systems with back mixing of both the liquid and adsorbent phases. The space created between the beads allows unrestricted passage of particulates while the target product is captured on the adsorbent. After thorough washing, the product is desorbed from the adsorbent in a subsequent elution step.

9.6 Purification by chromatography

Chromatography is the technique used for high-resolution purification. The components to be separated are distributed between a stationary and a mobile phase. The stationary phase, usually composed of uniformly sized particles, is packed in a column and equilibrated with a suitable solvent. The mixture to be separated is loaded on the column followed by the mobile phase. Elution of the components is achieved either in an isocratic mode, i.e. the same mobile phase is maintained throughout and separation of the components depends on differences in retention time in the column, or by gradient elution where the mobile phase is continuously changed to facilitate the release of the components bound to the stationary phase by non-covalent forces. The eluate from the column can be monitored continuously, e.g. by monitoring the ultraviolet absorption at 280 nm for proteins, and is collected in fractions of definite volume using a fraction collector. Various modes of chromatography are used for protein separations, see Table 9.5. These separations may be based on size (size exclusion chromatography), characteristics of surface groups, or recognition properties of protein molecules (adsorption chromatography).

Some of the matrices available for protein chromatography are listed in Table 9.6. Ideally, the chromatographic matrix should be: *inert* – to prevent any non-specific adsorption, *rigid* – to resist compression at high flow rates, *chemically stable* – to withstand harsh cleansing procedures, *bead shaped* – for good flow properties, and *porous* – to provide a high surface area and to allow free passage of molecules. The chromatographic materials are both organic and inorganic, synthetic and natural. During the past several years, impressive developments in chromatography media with improved selectivity and physical properties allowing high flowthroughs, have taken place. Fast protein liquid chromatography (FPLC) and high performance liquid chromatography (HPLC), utilising rigid matrices for operation with relatively high pressure, are now common tools for purification of commercial proteins. All the normal chromatographic techniques are commercially available in the HPLC-mode.

Flow-through or **perfusion chromatography** uses a chromatography matrix based on poly(styrene-divinylbenzene) or agarose or methyl methacrylate containing two types of pores; the big transecting through pores allowing convective flow through the particle and smaller diffusive pores lining the through pores provide a large adsorption surface area. This combination provides a rapid mass transport of sample molecules without sacrificing resolution and capacity, in contrast to conventional chromatography where the sample transport is diffusion-limited.

An innovative design of columns is used in **radial flow chromatography** in which the sample is applied to the outer wall and allowed to move radially by sideways flow of the eluant. The capacity of the column is thus determined by both length and radius of the column; the latter influencing also the resolution. Scale-up to larger volumes does not

Table 9.5 | Various chromatographic techniques for protein separation

Chromatography	Separation principle
Size-exclusion (gel filtration)	Size and shape
Ion-exchange chromatography	Net charge
Chromatofocusing	Net charge
Hydrophobic interaction chromatography	Hydrophobicity
Affinity chromatography	Molecular recognition
Immobilised metal-ion affinity chromatography	Metal ion binding
Covalent chromatography	Content of free –SH groups

Table 9.6 | Matrices used for large-scale chromatography

Matrix	Trade name
Cross-linked dextran	Sephadex
Cellulose	Whatman TM, Cellufine, Sephacel, Cellex
Agarose	Sepharose, Sepharose CL, Fast flow Sepharose, Ultrogel, Superose
Cross-linked polyacrylamide	BioGel P
Composite of polyacrylamide and dextran	Sephacryl
Composite of agarose and dextran	Superdex
Composite of agarose and polyacrylamide	Ultrogel AcA
Composite of porous kieselguhr and agarose	Macrosorb KA
Hydroxylated acrylic polymers	Trisacryl
Hydroxyethyl methacrylate polymer	Spheron
Ethyleneglycol-methacrylate co-polymer	Fractogel TSK, Toyopearl
Polyacrylamide	Eupergit C
Porous silica	Spherosil, Accell
Rigid organic polymers	Monobeads, TSK-PW
Polystyrene/divinyl benzene	Poros

need to be done by increasing the bed height, and thus the problem of bed compression of soft gels obtained in conventional downward flow of eluant is overcome.

9.6.1 Size-exclusion chromatography

This involves partitioning of proteins between the stationary liquid held by pores of the gel particles and the mobile liquid in the void volume between the particles. The gel matrices used for size exclusion chromatography have a defined pore size range to allow small molecules into the pores while the larger molecules are excluded and pass through the column with the mobile liquid. Size-exclusion chromatography is effective only with small sample volumes (equivalent to 2–5% of the total bed volume), and is thus suitably used as a final polishing step in a purification protocol to separate the protein from aggregates, degradation products etc., and even for buffer exchange.

9.6.2 Adsorption chromatography

In adsorption chromatography, resolution of the macromolecules is a surface-mediated process i.e. there is differential adsorption of the molecules at the surface of the matrix (see Table 9.5). The matrices employed are derivatised to contain covalently attached functional groups, which adsorb and separate proteins by different mechanisms.

Ion exchange chromatography

Ion exchange chromatography is by far the most widely used technique because of its general applicability, good resolution and high capacity. Moreover, it is insensitive to sample volumes and is often used in the initial phase of downstream processing to provide both product purification and volume reduction of the process fluid. Compounds are separated according to the difference in their surface charges. Hence, pH of the medium is one of the most important parameters for binding of the target molecule, as it determines the effective charge on both the target molecule and the ion exchanger. Ionic bound molecules are eluted from the matrix, either by increasing the concentration of salt ions which compete for the same binding sites on the ion exchanger, or by changing the pH of the eluant so that the molecules lose their charges. Displacement chromatography is an alternative mode wherein a more heavily charged molecule is used to 'displace' the bound material.

Ion exchangers are grouped into anion exchangers with positively charged groups like diethylaminoethyl (DEAE) and quarternary amino (Q), and cation exchangers which have negatively charged groups like carboxymethyl (CM) and sulphonate (S).

Hydrophobic interaction chromatography

Hydrophobic interaction chromatography (HIC) is also a robust, high capacity method. It may be used early with dilute product streams, leading to concentration and purification. HIC is analogous to reverse phase chromatography (RPC) but relies on comparatively weak hydrophobic interactions between hydrophobic ligands, alkyl or aryl side chains on the gel matrix and the accessible hydrophobic amino acids on protein surface. Differences in the content of these amino acid residues can be used for separation of proteins. Binding is performed in a medium favouring hydrophobic interactions, e.g. a solution of high salt concentration. Elution of bound material is achieved by reducing the hydrophobic interactions, either by lowering the salt concentration, or the temperature, or by decreasing the polarity of the medium (by inclusion of solvents like ethylene glycol or ethanol in the buffer). Matrices with different hydrophobic groups like butyl-, octyl- and phenyl- are commercially available.

Affinity chromatography

Molecular recognition forms the basis of adsorption and separation by affinity chromatography. One of the reactants in an 'affinity pair', the ligand, is immobilised on a solid matrix and is used under suitable conditions to fish out the complementary structure (the ligate). The binding

Table 9.7 Examples of ligands used for affinity chromatography of proteins

Ligand type	Protein type
Biospecific ligands	
Mono-specific	
Receptor	Hormone
Antibody	Antigen
Hapten	Antibody
Substrate/substrate analogue, Inhibitor, cofactor	Enzyme
Group-specific	
Cofactor	Enzymes
Lectins	Glycoproteins
Sugar derivatives	Lectins
Protein A/G	Immunoglobulins
Heparin	Coagulation factors, protein kinases
Pseudobiospecific ligands	
Triazine dyes	Dehydrogenases, kinases and other proteins
Metal ions	Metal ion binding proteins
Hydrophobic groups	Various proteins

is reversible and can be broken by changing the buffer conditions. Potentially, such methods possess very high resolving power. Table 9.7 shows that a variety of ligands, with specificity for one or a group of proteins, can be used for affinity chromatography. A trend in downstream processing has been to exploit the specificity of affinity interactions earlier in the separation train so as to reduce the number of purification steps; however this puts extra demands on the ligands with respect to chemical and biological stability. This has led to increased interest in the use of pseudobiospecific ligands like dyes and metal ions.

The chemical coupling procedure for immobilisation of a ligand is chosen so as to provide satisfactory yields, strong linkage to minimise ligand leakage during chromatographic operation, and minimal non-specific interactions with biomolecules. The adsorbed protein is generally eluted under conditions that minimise its interactions with the ligand, e.g. by increasing the ionic strength or changing the pH of the buffer, or by a free ligand molecule. In the latter case, a subsequent step would be to separate the protein from the free ligand.

9.7 Product formulation

The commercial viability of a biotechnological product is dependent on the maintenance of its activity and stability during distribution and storage. Low molecular weight products such as bulk solvents, bulk organic acids, etc. are formulated as concentrated solutions after removing most of the water. When high purity is required, small

Table 9.8 | Some of the frequently used drying equipment

Type of dryer	Mode of heat transfer	Movement of the product
Belt dryer	Convection	Intensive due to gas flow
Fluidised bed dryer	Convection	Intensive due to gas flow
Spray-dryer	Convection	Intensive due to gas flow
Freeze-dryer	Contact and radiation	None or mechanical
Drum-dryer	Contact	Slight mechanical
Chamber dryer	Convection and contact	None

molecules such as antibiotics, citric acid, sodium glutamate etc., are crystallised from solution by addition of salts once they have reached the required degree of purity. Proteins are particularly sensitive to loss of biological activity during downstream processing and subsequently during storage. This could be due to factors such as oxidation, temperature, presence of proteases etc. Protein products are formulated as solutions, suspensions or dry powders. A variety of stabilising additives are included in the formulations in order to prolong the product shelf life. The choice of a suitable stabiliser is done empirically. Among the non-specific chemical additives used regularly as stabilisers in protein formulations are salts (ammonium sulphate or sodium chloride), sugars (sucrose, lactose etc.), polyhydric alcohols (sorbitol, glycerol etc.) or polymers (polyethylene glycol, bovine serum albumin etc.).

Bulk enzymes are commonly sold as concentrated liquid formulations. It is, however, often preferred to dry the product to decrease the volume as well as the denaturing reactions that are enhanced in aqueous solution. Bioproducts, often being sensitive to heat, require gentle drying methods. Table 9.8 lists some of the commonly used dryers. Depending on the mechanism of heat transfer, these may be broadly classified as **contact-**, **convection-**, and **radiation-dryers**. Batchwise drying in many contact dryers is facilitated using mechanically moved layers. The advantage is the uniform thermal stress exerted on the material being dried, high throughput and possibility for development of continuous processes. A common feature of convection dryers is that the movement of the material to be dried is promoted by a flow of gas. Drying of large streams of product is achieved in a very short time.

Spray drying essentially involves generating an aerosol of tiny droplets by passing the product containing liquid through a nozzle or a rotating atomising disc and directing it into a stream of hot gas. The water in the droplets evaporates leaving behind solid product particles. Spray-dryers are used for drying large volumes of liquids. For smaller quantities, a **chamber dryer** may be useful, where the product is placed on shelves and the transfer of heat takes place partly by contact and partly by convection. The material will be exposed to severe non-homogeneity in thermal stress, making such a method recommendable only at relatively low temperatures.

Freeze-drying or **lyophilisation** represents one of the least harsh methods of protein drying. It is used for the drying of pharmaceutical products, diagnostics, foodstuffs, viruses, bacteria etc. The drying principle is based on sublimation of the liquid from a frozen material. The liquid containing the product is frozen, ideally to a temperature below its glass transition temperature, and subjected to vacuum in a freeze-dryer. While maintaining the internal vial temperature still below the glass transition value, the shelf temperature is increased to a temperature above zero to promote efficient sublimation of the crystallised water. After the primary drying, the protein cake still retains a significant amount of water which is removed by sublimation during secondary drying by increasing the internal vial temperature. Subsequently, the vacuum is released and the vials are sealed. Under a high vacuum, the heat transfer is solely via contact, and not by convection, therefore one needs to operate at low vacuum for an efficient drying process. Various additives are normally included in the solution prior to freeze-drying. These may be compounds for maintaining the activity of the protein in the finished product, to protect the protein during the lyophilisation process (lyoprotectants), to prevent product blow out during lyophilisation (e.g. mannitol) and to enhance product solubility. All these additives influence the glass transition value of the solution and hence the freezing temperature prior to drying.

9.8 Monitoring of downstream processing

There may be several reasons for monitoring a DSP step. One such reason is a desire to keep control over the presence and concentration of the target molecule which, if monitored directly e.g. by using some kind of a sensor, would enable one to take appropriate measures quickly in the event of an unexpected performance. Furthermore, when the performance of a process is well established, the signal from the sensor may be used for control purposes, thereby increasing the level of automation. Continuous monitoring of DSP is also desirable in processes dealing with the production of pharmaceuticals, as a means to facilitate establishing an improved level of documentation, and hence the approval process. Furthermore, implementation of measuring and control in DSP will increase the reliability concerning reproducibility of repeated batches. From a practical point of view, fraction collection and other aspects of sampling and sample handling can be substantially facilitated by on-line monitoring combined with pre-set conditions for collection/rejection of the stream. Some of the signals that are used for monitoring DSP-events are listed in Table 9.9. Important features when selecting signals for process monitoring are response-time, selectivity and sensitivity.

In process monitoring it is important to carry out the analyses under sterile conditions so that infection of the bioprocess does not take place. An often-used method is off-line analyses. Flow injection analysis (FIA) has proven superior in this context. A small liquid sample from the

Table 9.9 Signals used in monitoring DSP processes

Measuring principle	Sensitivity	Selectivity	Response time	Comments
UV-absorbance	Medium	No	Fast	Commonly used
Conductivity	Low	No	Fast	
pH	Medium	No	Fast	
Molecular size	Low	No	<10 s	
Enzyme activity	Medium/high	Yes	Medium/fast	Needs addition of substrate
Protein monitoring	High	No	Fast/medium	Needs addition of reagents
Biospecific binding reactions	High	High	Slow/medium/ fast	Range of new binding reactions

medium to be analysed is introduced into a continuous flow of buffer. The sample is transported over a biosensor or a reaction site and is then analysed in a flow detector.

9.9 Process integration

Low productivity and the high cost of producing biomolecules in general have been limiting factors in the development of biotechnological processes. As the isolation and purification stage contributes substantially to this scenario, process integration has been suggested as a means to overcome these bottlenecks, wherein two or more different processing stages are integrated into one step. The outcome would be a decrease in the number of necessary steps for the complete process, leading to lower costs and higher product recovery.

Integrating a product isolation stage with the fermentation process is possible for recovery of both small molecules and proteins. *In situ* adsorption of the product is a working example. Integration of extraction in organic solvents with fermentation for *in situ* recovery of low molecular weight products has also been investigated; it is however limited by the toxicity of the solvent to the producing cells. Extraction using an interface such as a membrane (perstraction) has shown potential to reduce the solvent toxicity problem, but the technique is not yet used for large-scale production.

Integration of two or more separation stages into one unit operation can lead to tightening of a purification protocol. This has been achieved for protein purification by introduction of technologies such as extraction in aqueous two-phase systems and lately, expanded bed adsorption, which can combine clarification, concentration and significant purification in one step. Direct application of an unclarified sample to a purification method is a significant simplification, especially on a large scale, see Figs. 9.4 and 9.6. Another approach to decreasing the number of processing steps is to introduce specificity, e.g. by integrating affinity interactions, into techniques used for primary separation, e.g. as in affinity precipitation, membrane affinity filtration etc.

9.10 Further reading

Albertsson, P. Å. (1986). *Partition of Cell Particles and Macromolecules, 3rd edition.* Wiley Interscience, New York.

Asenjo, J. A., ed. (1990). *Separation Processes in Biotechnology*, Marcel Dekker, New York.

Belter, P. A., Cussler, E. L. and Hu, W-S. (1988). *Bioseparations: Downstream Processing for Biotechnology*, John Wiley, New York.

Brandt, S., Goffe, R. A., Kessler, S. B., O'Connor, J. L. and Zale, S. E. (1988). Membrane based affinity technology for commercial scale purifications. *Bio/Technol* **6**, 779–782.

Cheryan, M. (1986). *Ultrafiltration Handbook*. Technomic Publ, Lancaster, USA.

Deutscher, M. P., ed. (1990). *Methods in Enzymology, Vol. 182. Guide to Protein Purification*. Academic Press, San Diego.

Hatti-Kaul, R., ed. (2000). *Aqueous Two-Phase Systems. Methods and Protocols.* Humana Press, New Jersey.

Janson, J-C. and Rydén, L., eds. (1998). *Protein Purification. Principles, High Resolution Methods, and Applications*. John Wiley, New York.

Kaul, R. and Mattiasson, B. (1992). Secondary purification. *Bioseparation* **3**, 1–26.

Krijgsman, J. (1992). *Product Recovery in Bioprocess Technology*. BIOTOL Series (Jenkins, R. O., ed.). Butterworth Heinemann, Oxford.

Lo, T. C., Baird, M. H. I. and Hanson, C. (1983). *Handbook of Solvent Extraction*. John Wiley, New York.

Matsumura, M. (1991). Perstraction. In *Extractive Bioconversions* (Mattiasson, B. and Holst, O., eds.), pp. 91–131. Marcel Dekker, New York.

Mattiasson, B., ed. (1999). *Expanded Bed Chromatography.* Kluwer Academic Publishers, Dordrecht, NL.

McGregor, W. C., ed. (1986). *Membrane Separations in Biotechnology*, Marcel Dekker, New York.

Schmidt-Kastner, G. and Gölker, C. (1987). Product recovery in biotechnology, In *Fundamentals of Biotechnology* (Präve, P., Faust, U., Sittig, W. and Sukatsch, D. A., eds.), pp. 279–321. VCH, Weinheim.

Scopes, R. (1994). *Protein Purification. Principles and Practice, 3rd Edition*. Springer-Verlag, New York.

Stephanopoulos, G., ed. (1993). *Biotechnology Vol. 3, Bioprocessing* (Series editors: Rehm, H-J., Reed, G., Pühler, A. and Stadler, P.), VCH, Weinheim.

Strathman, H. and Gudernatsch, W. (1991). Continuous removal of ethanol from bioreactor by pervaporation. In *Extractive Bioconversions* (Mattiasson, B. and Holst, O., eds.), pp. 67–89. Marcel Dekker, New York.

Street, G., ed. (1994). *Highly Selective Separations in Biotechnology*. Blackie Academic & Professional, London.

Verrall, M. (1996). *Downstream Processing of Natural Products. A Practical Handbook*. John Wiley & Sons, Chichester.

Walter, H. and Johansson, G., eds. (1994). *Methods in Enzymology, Vol. 228. Aqueous Two-Phase Systems*. Academic Press, San Diego.

Chapter 10

Measurement and control

A. Lübbert and R. Simutis

Nomenclature

A	concentration of a metabolic by-product produced by micro-organisms.
a	delay time
A_i	individual components of a biochemical reaction system (can be represented by means of reaction equations $\Sigma_i \nu_i A_i = 0$, where the component A_i is usually represented by its sum formula, and ν_i are the usual stoichiometric coefficients)
ANN	artificial neural network
c	concentration of components
$\mathbf{c}$	concentration vector (vectors are written in bold type)
$\mathbf{c}_d(t)$	desired path for the process
$\mathbf{c}_F$	concentration vector of the solution fed to the reactor
CER	CO_2 evolution rate
$\mathbf{E}$	coefficient matrix ('Element-Composition-Matrix')
F, or $F(t)$	the rate at which the material is fed to the reactor as a function of time
$K=A/B$	amplitude ratio
K_{ai}	inhibition constant for the inhibitor A
K_c	controller gain
k_La	mass transfer coefficient
K_s	Monod constant
K_{si}	the substrate inhibition constant
m	maintenance term in Eqn (10.28)
n_i	amount of component A_i

n	vector of amounts of components
pO_2	dissolved O_2 concentration in the medium
pO_2^*	the solubility of O_2
OTR	O_2 transfer rate
OWR	O_2 uptake rate
RQ	respiration coefficient
SCR	substrate consumption rate
t	time
ε	deviation
q_m	maintenance term
P	product concentration
R_i	conversion rates
R_o	absolute O_2 consumption rate
R	vector of volumetric rates of change of the amounts of the components being created or consumed during a process
S	substrate concentration
T	response time
$\mathbf{u}_c$	manipulable variable
V, or $V(t)$	volume
X	biomass concentration
Y_{ij}	yield coefficient of component i on component j
Δn_i	amount of component i synthesised or consumed in the reaction
$\boldsymbol{\Phi}$	vector of the rates by which the amounts of the components change as a consequence of the mass transport across the boundaries of the domain
α_g	specific acetate production rate
α_c	specific acetate consumption rate
μ	specific growth rate
π	specific product production rate
τ_D	time constant for derivative controller
τ_I	time constant for integral controller
ν_i	stoichiometric coefficient
σ	specific substrate uptake rate
σ_{crit}	critical specific substrate uptake rate
ω_m	specific O_2 uptake rate used for maintenance
ω_x	specific O_2 uptake rate used for biomass growth

10.1 Introduction

In industrial production plants, process control is on everybody's agenda when the cost/benefit-ratio of a process must be improved. It is desirable to guide the process along a path, which guarantees the process to produce the product in such a way that it meets predefined quality specifications. This confirmed, the aim is to produce this product at a minimum of cost. The determination of such optimal process paths is the essential part of **open-loop control**.

The two key elements of process monitoring and control are: (i) **measurements** by which information about the current process state is being acquired, and (ii) **models** that dynamically interrelate the various process variables, which are important with respect to the task to be solved. Of particular importance are those variables by which the state of the process can be described unambiguously. These variables, however, are not necessarily the most important ones from the practical point of view. Of immediate practical importance are the variables which describe the performance of the process. In order to get access to the performance, its relationships to the variables which can be measured directly and to the variables that can be manipulated are of importance. Thus, modelling for process supervision and control needs a quantitative definition of the objectives of the process and the particular task to be solved. For supervision and control applications in industrial environments, the complexity of the models must be kept as low as possible to minimise the expenses of manpower needed to maintain them. It only makes sense to implement complex process controllers, after it was made sure that they will work significantly better than conventional simpler ones. It is the cost/benefit-ratio that is the final criterion for whether simpler or more complex controllers are used and this must include the cost of providing the relevant manpower.

It is of advantage to formulate the multi-dimensional problems of process modelling, supervision and control using a vector representation. This not only helps to keep things clear, but helps to translate them into modern software tools which are mainly matrix based. The matrix notation used in this article was adapted to software products available on the market such as MATLAB or SCILAB where the variable, x, is generally assumed to be a matrix. Vectors and the usual scalar quantities are considered matrices of special dimensions.

10.2 Structure of process models

Process identification is the procedure of developing a process model from prior knowledge and experimental data. The classical approach to process modelling is the development of a mathematical model in the form of a dynamic differential equation system derived from mechanistic considerations. The prior knowledge usually leads to the structure of a parameterised model, leaving the parameters associated with this model structure to be estimated from process data. However, suitable model structures for some parts of the process may not always be known. Then, 'blackbox' identification methods that make only minimal assumptions about the structure of these sub-processes are alternatives.

The state of a bioprocess is mainly determined by the amount, $\mathbf{n}$, (measured in mol) of its key components. The vector, $\mathbf{n}$, may be composed of the amounts of the substrate, biomass, product etc. The basis of a bioprocess model is a balance equation that can describe the changes of $\mathbf{n}$ as a function of time. Please note that bold characters or abbreviations such as $\mathbf{n}$ indicate that the corresponding quantity is a vector-valued quantity.

$$\mathbf{n} = [\text{biomass } X; \text{substrate } S; \text{product } P; \ldots] \tag{10.1}$$

As it is more convenient to formulate these balance equations in terms of concentrations, **c**, which are related to the amounts, **n**, of the components by $\mathbf{c}=\mathbf{n}/V$, our general balance **mass balance equation** reads:

$$\frac{d\mathbf{n}}{dt}=\frac{d(\mathbf{c}V)}{dt}=\mathbf{R}V+\boldsymbol{\Phi} \tag{10.2}$$

$\boldsymbol{\Phi}$ is the rate by which the various components are transported across the borders of the balance volume, V, and $\mathbf{R}V$ the rate at which these components are synthesised or consumed within V. In bioprocesses, the rates, **R**, are usually non-linear functions of the various components of **n** or **c**.

In the fed-batch operational mode, which is the usual industrial production mode, the amount of substrate fed to the reactor leads to an increase in the total volume so that the volume itself is variable. Hence, an additional differential equation must be spent for the (scalar) volume, V.

$$\frac{dV}{dt}=F(t) \tag{10.3}$$

After resolving the equation for the concentration vector, **c**, we obtain the more convenient form of the balance.

$$\frac{d\mathbf{c}}{dt}=\mathbf{R}+\frac{\boldsymbol{\Phi}}{V}+\mathbf{c}\frac{dV}{dt} \tag{10.4}$$

or, with $\boldsymbol{\Phi}=F\,\mathbf{c}_F$

$$\frac{d\mathbf{c}}{dt}=\mathbf{R}+\frac{F(t)}{V(t)}(\mathbf{c}_F-\mathbf{c}) \tag{10.5}$$

A short remark is required about the **balance domain**. In non-linear systems, the essential criterion for justifying to represent the concentration of a particular component, e.g., the substrate concentration, by a single value, S, is homogeneity across the region over which the balance is drawn. It is possible to use the entire culture volume, V, as the balance volume when the bioreactor can be considered to be an ideal stirred tank reactor. When this assumption is not justified, complex transport processes must be taken into account which lead to much more complex process models.

10.3 Kinetic rate expressions

10.3.1 Considerations of stoichiometry

The first part in Eqn (10.5), **R**, describes the rates of the biochemical conversion of the components in the system. As the products are synthesised from the substrates, the different elements of **c** do not change independently from each other. The interrelationships are determined by biochemical stoichiometry. The individual components, A_i, of a biochemical reaction system can be represented by means of reaction equations

$\Sigma_i \nu_i A_i = 0$

where the components A_i are usually represented by their sum formula, and ν_i are the usual stoichiometric coefficients. As elemental analyses showed, the components, A_i, can be considered to be mainly composed of a few different elements only. Often it suffices to consider four elements [C H O N] only. If this is assumed to be sufficient, then the individual components can be identified by their index vectors, e.g.:

		[C	H	O	N]		
1.	Glucose =	[6	12	6	0]′	*Substrate*	
2.	O_2 =	[0	0	2	0]′	*Oxygen*	
3.	Ammonia =	[0	3	0	1]′	*Nitrogen source*	(10.6)
4.	Water =	[0	2	1	0]′	*Water*	
5.	CO_2 =	[1	0	2	0]′	*Carbon dioxide*	

(Column header row: *Basic elements considered*)

where [. . .]′ means the transposed of the row vector, [. . .]. Here and in the following text, the vector representation by components is adopted from MATLAB.

Problems appear with the corresponding representations of biomass and other complex biopolymers. Here it is straightforward to represent them as molecules with the same relative composition as the originals, but with the C-index fixed to 1, e.g.

6. Yeast = [1 1.75 0.38 0.25]′ Biomass

With this assumption we can write down concrete reaction equations, e.g. for aerobic yeast (*Saccharomyces cerevisiae*) production we get:

$$\nu_s\, C_6H_{12}O_6 + \nu_n\, NH_3 + \nu_o\, O_2 \Rightarrow \nu_x\, C\,H_{1.75}\, O_{0.38}\, N_{0.25} + \nu_w\, H_2O + \nu_c\, CO_2 \tag{10.7}$$

Using matrix representation this equation can be reformulated into a homogeneous linear equation system:

$$\mathbf{E}\,\nu = 0 \tag{10.8}$$

The coefficient matrix, **E**, in literature referred to as the 'Element-Composition-Matrix', is defined by the index vectors, using the definitions in Eqn (10.6):

$$\mathbf{E} = [-\text{Glucose} - O_2 - \text{Ammonia Yeast Water } CO_2] \tag{10.9}$$

The problem of solving the linear Eqn (10.8) for the stoichiometric coefficients ν requires five equations, since one of the ν-elements, e.g. that for glucose, can be arbitrarily set to one. If this substrate is chosen as the reference component, then $\nu_S = 1$. Then we can transform the homogeneous equation system (10.8) into a non-homogeneous one

$$\mathbf{E}_C\,\nu = \text{Glucose} \tag{10.10}$$

with the coefficient matrix

$$\mathbf{E}_C = [-O_2 - \text{Ammonia Yeast Water } CO_2]$$

and, according to Eqns (10.7) the five component vector:

$$\nu = [\nu_o \; \nu_n \; \nu_x \; \nu_w \; \nu_c] \tag{10.11}$$

As Eqn (10.10) contains only four linear equations and five unknowns, we need another equation, or an estimated value for one of the unknown coefficients, to solve the linear equation system.

Additional equations can be generated, for example, by exploiting relationships between the stoichiometric coefficients, ν_i, in Eqn (10.7) and available measurement values. The ν_i in (10.11) are defined as differences, Δn_i, in the amounts, n_i, of the species, A_i, synthesised or consumed, relative the corresponding amount, Δn_s, of the substrate consumed.

$$\nu_i = \frac{\Delta n_i}{\Delta n_s} = \frac{R_i}{R_c} \tag{10.12}$$

which is equal to the ratio of the corresponding consumption or development rates, R_i.

For instance, $R_c = \text{CER}$, the CO_2 evolution rate and $R_o = \text{OUR}$, the O_2 uptake rate that are measured during many fermentations, can often be assumed to be proportional. The proportionality constant is referred to as the RQ-coefficient.

$$\nu_c = \text{RQ}\,\nu_o = \frac{\text{CER}}{\text{OUR}}\nu_o \tag{10.13}$$

This can be reformulated by a scalar product

$$[\text{RQ}\;0\;0\;0\;-1] \times \nu = \text{RQV} \times \nu = 0 \tag{10.14}$$

which is a new linear equation in ν. Hence, the row vector, RQV, defined in Eqn (10.14), can be used to extend the coefficient matrix, $\mathbf{E}_C$. In the MATLAB representation we get:

$$\mathbf{E}_{Cf} = [\mathbf{E}_C; \text{RQV}] \tag{10.15}$$

Correspondingly, we must also extend the column vector on the right hand side of Eqn (10.10):

$$\mathbf{b} = [\text{Substrate};0] \tag{10.16}$$

Now we have five linear equations for the five unknown stoichiometric coefficients, ν.

$$\mathbf{E}_{Cf}\,\nu = \mathbf{b} \tag{10.17}$$

Equation (10.17) can now be solved directly using linear algebra. This can be done with a single MATLAB statement $\nu = \mathbf{E}_{Cf}/\mathbf{b}$.

From the solution, ν, we can determine the yields, which are defined as:

$$Y_{ab} = \frac{\partial n_a}{\partial n_a} \approx \frac{\Delta n_a\,MW_a}{\Delta n_b\,MW_b}\left[\frac{kg}{kg}\right] = \frac{\nu_a\,MW_a}{\nu_b\,MW_b}\left[\frac{kg}{kg}\right] \tag{10.18}$$

This requires the molecular weights MW_a of the species involved to be known. The most important yields are the biomass per substrate yield telling how much biomass is generated per unit of substrate mass consumed:

$$Y_{XS} = \frac{\nu_X \, MW_X}{\nu_S \, MW_S} \left[\frac{kg}{kg}\right] \tag{10.19}$$

and the oxygen per biomass yield

$$Y_{OX} = \frac{\nu_O \, MW_O}{\nu_X \, MW_X} \left[\frac{kg}{kg}\right] \tag{10.20}$$

For measurement and control, stoichiometric relationships are often used with advantage, not only to estimate yields, but also to check measured data for consistency. As a rule, the carbon balance must be close to about 5% before it makes sense to compare any model of the process with that data.

10.3.2 Conversion rates

While stoichiometric considerations provide information about the relative rates by which the different components are consumed or produced, the basic conversion rate expressions, **R**, are a matter of kinetics. Since the performance of the micro-organism is central in this respect, it is straightforward to discuss conversion rates per biomass concentration. These quantities are referred to as specific rates $\mathbf{q} = \mathbf{R}/X = [\mu, \sigma, \pi, \ldots]$. Central to our discussion is the **specific substrate consumption rate component**, σ, which is the quotient of the rate, R_s, of substrate consumption and the biomass concentration, X. Once again we start with simple macroscopic models when we are dealing with measurement and control, hence, we take the simple Monod expression to specify the specific substrate consumption rate as a function of substrate concentration, S:

$$\sigma = \frac{1}{X} R_s = \frac{\sigma_{max} S}{K_s + S} \tag{10.21}$$

With increasing S, σ approaches σ_{max}. At larger S, σ remains constant. Thus, the main idea behind the equation is to describe the decrease of σ at low values of S, i.e. under substrate-limiting conditions.

The cell channels the substrate into several metabolic paths. Mathematically, we may express this by the differential expression:

$$\sigma = \frac{1}{X}\frac{dS}{dt} = \frac{1}{X}\frac{\partial S}{\partial X}\frac{\partial X}{\partial t} + \frac{1}{X}\frac{\partial S}{\partial P}\frac{\partial P}{\partial t} + q_m \tag{10.22}$$

or with $\frac{1}{Y_{xs}} = \frac{\partial S}{\partial X}$ and $\frac{1}{Y_{ps}} = \frac{\partial S}{\partial P}$

$$\sigma = \frac{1}{X}\frac{dS}{dt} = \frac{\mu}{Y_{XS}} + \frac{\pi}{Y_{ps}} + q_m \tag{10.23}$$

which is of high practical value in modelling for process supervision and control.

The Monod equation often does not describe substrate degradation accurately enough. Then, further mechanisms must be taken into account, e.g. substrate inhibition and inhibition by undesirable by-products, A. These mechanisms can be considered by attaching additional factors to the original Monod equation for σ:

$$\begin{aligned}\sigma = \quad & \sigma_{max} && \text{Maximal specific uptake rate (capacity)} \\ & \times \frac{S}{K_s + S} && \text{Conventional substrate limitation term} \\ & \times \frac{K_{si}}{K_{si} + S} && \text{Substrate inhibition term} \\ & \times \frac{K_{ai}}{K_{ai} + A} && \text{Inhibition by by-product } A \qquad (10.24)\end{aligned}$$

The parameters, σ_{max}, K_s, K_{si} and K_{ai} must be determined from process data. The formation of the by-product, A, needs an additional kinetic consideration.

Cells are assumed to be able to take up more substrate than they can fully oxidise. In terms of the specific substrate uptake rates, σ, there is a critical rate, σ_{crit}, above which all substrate taken up cannot be oxidised directly. In such situations, the cell produces a by-product, *A*, which usually, at least at higher concentrations, acts as an inhibitor. In *E. coli* this by-product is acetate, while in the yeast *Saccharomyces cerevisiae* it is ethanol or in mammalian cells it may be lactate. In the example of an *E. coli* model, we may assume that at $\sigma > \sigma_{crit}$, acetate, *A*, is generated with the **specific by-product generation rate,** α_g:

$$\alpha_g = (\sigma - \sigma_{crit}) Y_{as} \quad \text{for } \sigma > \sigma_{crit} \quad \text{and} \quad \alpha_g = 0 \text{ else} \qquad (10.25)$$

When $\sigma < \sigma_{crit}$, the acetate already formed, can be consumed by the cells. This occurs when the concentration, *S*, of the main substrate drops below a critical value. The **specific by-product consumption rate,** α_c, can also be described by a Monod-like expression:

$$\alpha_c = \alpha_{max} \frac{A}{K_a + A} \frac{K_{ai}}{K_{ai} + S} \text{ for } \sigma < \sigma_{crit} \text{ and } \alpha_c = 0 \text{ else} \qquad (10.26)$$

where the repression of the by-product consumption by the substrate, *S*, is described by the third term. The net specific by-product production rate α is obviously the difference

$$\alpha = \alpha_g - \alpha_c.$$

When we assume that no further product is being synthesised, the **specific growth rate,** μ, can be determined. In order to keep the description as transparent as possible, it is of advantage to distinguish two cases: The first is the situation where the specific substrate consumption rate is lower than the critical value, σ_{crit}. Cells then grow on the substrate, *S*, and on the by-product, *A*, when available:

$$\mu = \sigma Y_{xs} + \alpha_c Y_{xa} - m \qquad \text{for} \qquad \sigma < \sigma_{crit} \qquad (10.27)$$

In the second case, at $\sigma > \sigma_{crit}$, the cells convert part of the substrate into biomass with a high biomass/substrate yield, Y_{xs}, and use the rest of the

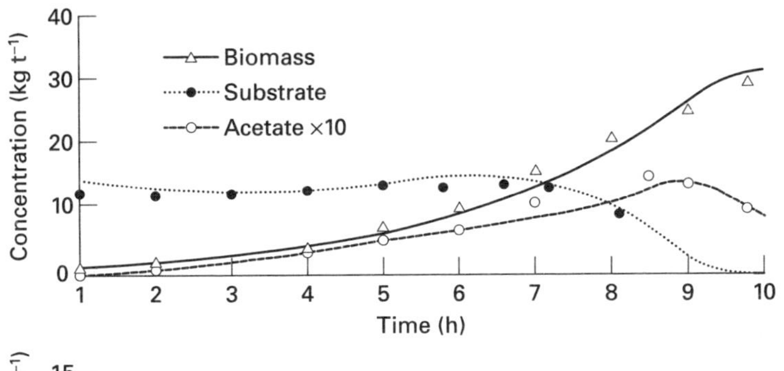

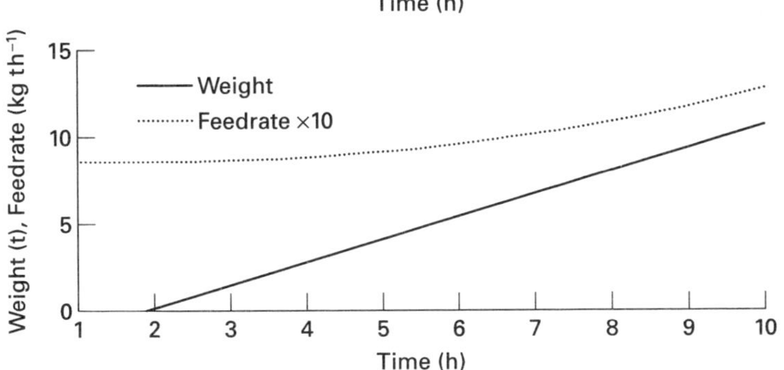

Fig. 10.1 Typical result for a fed-batch cultivation of modified *E. coli*. In the upper part the computed trajectories for biomass, substrate and acetate are compared with measured data, depicted by the symbols. In the lower part the corresponding feedrate (glucose solution) and culture weight profiles are shown.

entire specific substrate consumption rate with a significantly lower yield, Y_{xsa}. Hence we obtain

$$\mu = \sigma_{crit} \times Y_{xs} + (\sigma\text{-}\sigma_{crit}) \times Y_{xsa} - m \qquad \text{for} \qquad \sigma > \sigma_{crit} \tag{10.28}$$

Finally, the **specific product development rate,** π, must described. It most often primarily depends on the specific biomass growth rate, μ, in some more or less complex relationship. The desired product may be a recombinant protein. Then, a highly non-linear relationship, such as e.g.:

$$\pi = \frac{\pi_{max}}{1 + \left(\frac{K_\mu}{\mu}\right)^\gamma} \tag{10.29}$$

may be used in the simulation. The sigmoidal form of this rate expression reflects the experience that the cells need some growth to start protein production and that there is a final specific production rate limit, π_{max}, that is asymptotically approached at higher specific growth rates, μ. The parameters of this expression must be determined from experimental data, in particular from $\mu(t)$ and $\pi(t)$ estimations discussed later. With these specific rate expressions, the absolute volumetric rate vector, **R**, can be determined as:

$$\mathbf{R} = X\,[\mu, -\sigma, \alpha, \pi] \tag{10.30}$$

With **R**, the entire dynamic system is determined. Hence, when the required parameters, kinetic constants and yields can be supplied, we only need to solve Eqn (10.5). This, once again, is very easy when a modern software tool like MATLAB is used.

Figure 10.1 shows a typical result of a process simulation where these model components were used. The kinetic bottleneck mechanism is of general use not only for the accompanying example of *E. coli* growth, but also for most other systems of practical interest.

10.4 Advanced modelling considerations

10.4.1 Methods of lean modelling

When a model is to be used in process optimisation or for on-line process measurement or control, it must be solvable quickly, since the model evaluation must be frequently repeated. This requires that the model must be strictly restricted to those aspects that directly influence the process performance. Restriction to these most important variables is not the only possibility to reduce the computing time. An important further measure is to identify variables that are important, but change with much smaller time constants in comparison with the key variables substrate, biomass or product concentration. These variables can be assumed to be at any time in an equilibrium state with the key variables. Hence, their dynamic changes need not be considered separately. In other words, they can be statically related to the other state variables. In this way the number of differential equations can be decreased and thus the computing time for the simulation.

An important example is the dissolved O_2 concentration, pO_2, one of the most important variables in aerobic production processes. Since pO_2 is immediately adapting to the biomass growth rate, its rate of change can be neglected.

The specific O_2 consumption rate, ω, is usually a linear function of the specific biomass growth rate, μ, where the slope is the yield, Y_{ox}, and the absolute term is the specific oxygen uptake rate, ω_m, needed for maintenance of the cells:

$$\omega = \mu Y_{ox} + \omega_m \tag{10.31}$$

The absolute O_2 uptake rate, $R_o = OUR = \omega\ X$, is balanced by the O_2-transfer rate OTR, which is most often modelled by:

$$OTR = k_L a\,(pO_2^* - pO_2) \tag{10.32}$$

Thus, we can combine Eqns (10.31) and (10.32) to get:

$$pO_2 = pO_2^* - OUR/(k_L a) = pO_2^* - (\mu Y_{ox} + \omega_m)X/(k_L a) \tag{10.33}$$

The computational advantage (in terms of the computing time) of this approach is not alone due to the reduced number of differential equations. An even greater effect on the computing time is the fact that a rapidly changing component has been eliminated from the dynamical equation system. Thus, the solution of the reduced ordinary differential equation system can be performed with considerably larger time increments. Hence, much less computing time is required for its integration.

The same general approach can also be applied to fed-batch cultivation, where the substrate concentration is usually kept at a very low quasi-constant level. Then, all the substrate fed to the reactor is immediately consumed by the cells.

10.4.2 Representations of the kinetic expressions by artificial neural networks

In comparison with the knowledge behind the mass balance equations, the knowledge about details of the biochemical reaction rates is rather limited. This is the reason why simple formal relationships such as the Monod expression are used to describe the biochemical conversion rates, **R**. The parameters of such formal representations cannot be derived from first principles and must thus be determined from experimental data through parameter fitting.

In cases where simple approaches such as the Monod expression do not accurately enough describe the kinetics and no further mechanistic knowledge can be made available, other data driven representations might better represent the kinetic rates. Artificial neural networks (ANNs) proved to be very flexible tools to describe strongly non-linear complex relationships between multiple variables. As such they were often applied to represent the components of the conversion rate vector, **R**, in Eqn (10.5), which are high-dimensional and in particular strongly non-linear. In many practical cases, **R** can be determined by means of a simple feedforward artificial neural net with only a single hidden layer.

ANNs are crisp parametric relationships. Their parameters, in this connection termed weights, must be determined by numerical fitting to available process data. Such parameter fits are referred to as network training. Although there are available a number of techniques to train artificial neural networks, the simple algorithms offered by standard software suppliers, unfortunately, cannot be used directly, since they would require measurement values of the specific rates, such as μ, σ, π, etc. for the network training. The quality of such data, however, is usually rather poor, since they must be extracted numerically from differences between noisy concentration measurement data, which are furthermore not available with sufficiently high sampling rates. Thus, alternative techniques must be applied for the training of hybrid combinations of networks and mass balances. One well tested technique is the 'sensitivity equation technique'.

ANNs require extended data bases for their training. However, when sufficiently many data records are available they are very powerful in determining the rate expression, **R**. One of their particular advantages in comparison with the classical Monod expressions is that they can easily consider the influence of many other variables besides the substrate concentration on the absolute rates, **R**, such as other concentrations as well as temperature and pH.

A problem that generally appears with 'black box' representations and in particular with ANNs is that they cannot be safely used for extrapolation outside the area of data with which they were trained. In this respect they do not depict a global representation feature comparable with the Monod model and its derivates. Hence it is straightforward to combine the advantage of ANNs to describe local kinetic details with the advantage of classical Monod expressions to describe the global rates, **R**, by using both simultaneously.

Because of the flexibility of ANNs in representing complex non-linear relationships, such networks can not only be used as an alternative representation for **R**, but also as a means to correct available classical representations for missing relationships to additional variables. This approach has the advantage to keep and extend the already conventionally obtained knowledge.

10.4.3 Fuzzy expert systems

When there is already available some concrete mechanistic knowledge about the rate expressions which, however, could not yet be transformed into mathematical expressions, then another alternative to a mathematical model representation is possible. This is the representation by **fuzzy rule systems** where the dependencies between the relevant quantities are represented by heuristic rules-of-thumb in the form 'if {conditions} then {actions}'.

The variables considered in these rules assume fuzzy 'values' like small, large, normal, etc. What is understood by these linguistic terms must be defined individually in the context of the particular process quantities. This is done by means of membership functions, which are used to describe to what extent a concrete measurement value will be considered 'small' or 'large'. The fuzzy rules representing heuristic knowledge are processed using fuzzy logic operators. The main advantage of fuzzy rule systems is that the large amount of heuristic knowledge available in practice can be activated for measurement and control purposes. This technique is widely applied even in household washing machines. Software for formulating and processing of fuzzy rules is commercially available in many software packages, e.g. in the MATLAB package already mentioned.

It should be stressed, however, that both techniques, the ANNs and the fuzzy rule systems, should not replace already available mechanistic models represented in the form of mathematical equations. Instead, these new techniques should extend our capabilities for representing the processes.

A particular difficulty of biochemical production processes is that their dynamics change with time. Mechanisms that dominate at the beginning might not be significant at the end of the fermentation. For example, in batch processes, substrate inhibition is of importance only at the beginning, while product inhibition might become a problem only at the end. This fact has an immediate consequence on identifying process model parameters as model components that are of no significance in a particular process phase, and cannot obviously be identified from data measured in this phase. In such cases it is straightforward to design the process model in a modular way i.e. to divide the process into phases which are modelled individually, considering only those mechanisms which actively contribute to the process dynamics in the phase

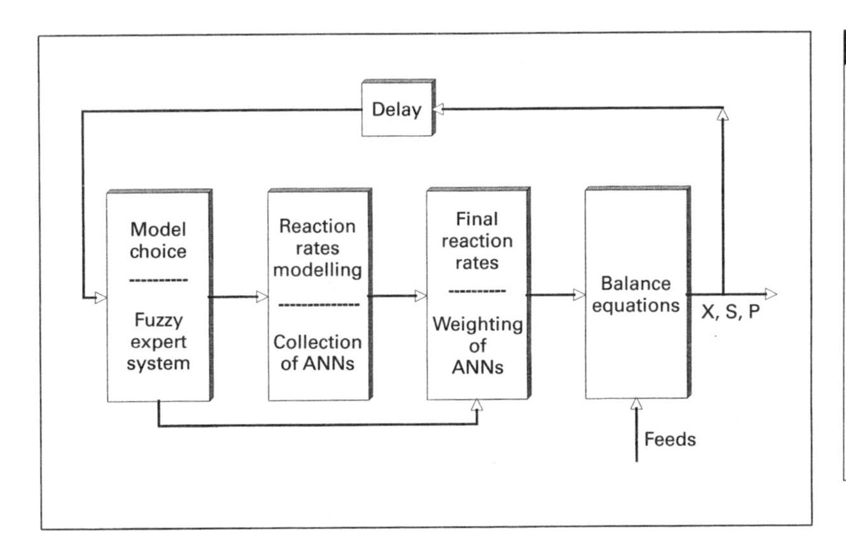

Fig. 10.2 Structure of a typical hybrid model in which different phases of the process are modelled separately and a fuzzy rule system is applied to smoothly switch between the modules of the model. The modules process information and knowledge on different levels: the fuzzy expert system processes heuristic rules, the artificial neural networks (ANN) process data records and the balance equations are ordinary differential equation systems based on mechanistic knowledge.

under consideration. Then the question appears: how to decide when situations are changing and which model components are to be activated.

For such questions, experienced process engineers usually have heuristic answers, which they often formulate by rules of thumb. Such rules can be formulated and exploited in process computers by means of fuzzy rule systems. An example for the structure of such an approach is sketched in Fig. 10.2.

This example shows the combination of set of ANN modules describing the rate expressions, **R**, for the individual process phases and a balance equation into which the rates must fit for the modelling of a production process. A **fuzzy expert system** selects the kinetic modules in such a way that there is a smooth transition from one phase to the next one. A typical result of a simulation using this kind of model is shown in Fig. 10.3.

There are two essential advantages with respect to model accuracy of this approach:

- The first is that the modules can be more closely related to the process properties in the corresponding process phases. Consequently they need not be globally applicable and can describe the particular situation more accurately.
- The second is that the models for the individual process phases can be kept smaller and more transparent than the comprehensive model and, as such, they can be identified with a much higher accuracy.

10.4.4 Measurements

So far we stressed the model structure. An essential step to finalise the model is to determine the relevant model parameters. Initial estimates for the model parameters can be taken from literature. However, without measurement information it is seldom possible to obtain a process model that is accurate enough to monitor and control a

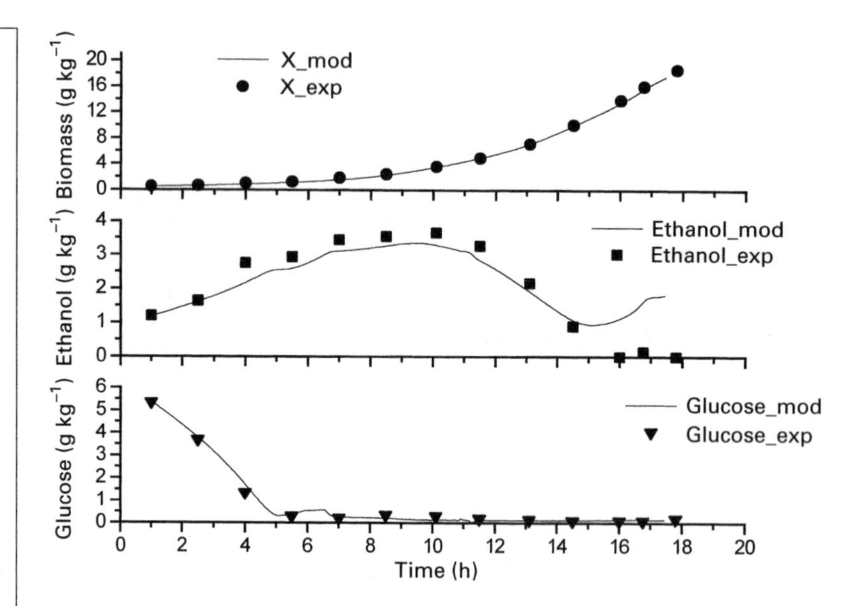

Fig. 10.3 Typical example of a result from a yeast cultivation process obtained with the hybrid model depicted in Fig. 10.2. The uppermost curve compares the measured and simulated biomass, X, profile, then the ethanol concentration and the corresponding glucose profiles are compared. At high growth rates ethanol is developing during the fermentation but it is consumed again in a later fermentation phase. The glucose curve depicts the typical substrate concentration behaviour of a fed-batch cultivation: after an initial adaptation phase, the concentration approaches a small, nearly constant value. Mod, model prediction; exp, experimental results.

biochemical production process. Consequently, a minimum of exact measurement information is always necessary, how much depends on the problem to be solved. Generally, as much measurement data as possible would be welcome, however, the number of measurement devices at a given fermenter is much restricted for several reasons.

With respect to bioprocess measurements we must distinguish on-line measurements from off-line measurements. The key variables, such as biomass, substrate or product concentrations, usually cannot directly be measured on-line since appropriate on-line sensors are not yet available. Measurement data for these quantities are usually provided by off-line analysis of samples from the cultivation broth. There are a number of off-line measurement techniques. Biomass concentration values are most often measured via dry weight measurements, substrates with enzymatic analysis techniques, and products by chromatography or electrophoresis. For all such measurements, many analysers are commercially available. Common for all these measurements is that the sampling frequency is limited, at best, to a few events per hour and that the resulting measurement values usually become available after considerable time delays. Consequently, these data cannot be used to control the process in a feedback manner.

For on-line measurements the situation is quite different. Here, the measurement quantities are mainly restricted to basic physical quantities, like temperature, culture weight, feed rates, and a few basic concentrations of small molecular entities like pH and pO_2. Also, some components of the gas in the vent line (e.g. O_2, CO_2, ethanol, etc.) can be monitored. The values of these quantities are available on-line as continuous signals with no significant time delays.

A first point to recognise is that industrial production reactors are rather sparsely equipped with measurement devices, at least when compared to bioreactors in the laboratory. There are several reasons for this. Measurement devices are possible sources for contamination as their ports make 'cleaning-in-place' more difficult. Hence, their number is kept as low as possible. Secondly, only very reliable and robust measurement devices can be justified, when decisions in production environments are based on their values. Furthermore, the maintenance of the devices and the immediate utilisation of the data are considerable cost factors. There are only a few measurements for which the cost/benefit-ratio is considered to justify the investment and operating costs, including the costs that can be associated with the risk of contamination, against the benefits of having the information. The possible accuracy of the measurements may have a considerable influence on the model structure and vice versa. This fact has practical importance when a single effect or process property can be characterised by different quantities.

An important example appears in connection with the accuracy by which volumes, and hence concentrations, can be measured. Culture volumes can be measured by means of the heights of the dispersions within the reactor. However, the accuracy of such measurements is low since the surfaces of fermentation broths are often covered by foam. Also, the level depends on the gas hold-up in the reactor, which itself is dependent on the concentration of surface active agents dissolved in the broth. Since these may vary in an unpredictable way, it is very difficult to measure volumes during fermentation processes. The practical consequence is to discard volumes from the models and replace them by weights, which can be determined much more accurately by commercially available balances. In most cases the entire reactor is placed on load cells, which allow the culture weight to be determined with a small relative percentage error. In practice, it is convenient to redefine the term, concentration, to:

$$c = \frac{\text{Mass of the species dissolved in the culture}}{\text{Mass of the entire system}} [\text{kg kg}^{-1}] \qquad (10.34)$$

Consequently, all other volume related quantities must also be redefined, e.g. the flow rate must be defined as mass transported per time unit etc. It is customary to speak of weights rather than of masses, since the corresponding measurements are most often weight measurements. We will use concentrations as weight-based quantities in the following discussions.

10.4.5 Model validation

Before any far-reaching decision can be based on a model, it must be validated. In model validation we take data from the process to test the assumption that the model describes the process correctly. In other words, we test whether the model could have produced the observed data. When such tests are performed with independent data that were not used during the model development this procedure is known as **'cross validation technique'**. As the criterion for this comparison, one usually takes the root mean square (rms) deviation between the test data and the corresponding data computed by the model. This rms-value can be taken as the model performance index.

Consider the case that data records from 10 individual production runs are available for model development. Then, some part of the available process data e.g. typically only seven records are used for model parameter estimation, while the rest (test data records) are used for independent comparisons with the modelling results. Since the credibility attributed to a model, rises with the number of successful applications to process situations not considered during the model development, the number of test data records should not be too small. However, on the other hand it is not possible to make this number very high, since a sufficient number of records is required to derive a reliable model. Hence, a reasonable compromise is required for dividing the available data into model development and test data sets. A ratio of 7 to 3 has proved to be a useful working principle.

10.5 Process supervision and control

Process supervision or monitoring includes any activity that is concerned with the ability to track the critical variables that affect the fermentation in order to detect deviations from the predetermined process path early enough to initiate measures which can compensate for them. Process control then entails the use of this monitored information to make decisions that affect the process in some desirable way. As the key variables of biochemical production processes are usually not directly measurable on-line, the basic problem of process supervision is to use the possible on-line measurements in order to estimate the values of the variables indirectly. This is the field of **state estimation procedures**.

10.5.1 On-line measurements

Some of the most important quantities with respect to process supervision and control that can be measured on-line are: temperature, pH, pO_2, culture weight, feed rates, vessel pressure, as well as the partial pressures of O_2 and CO_2 in the off-gas. Consequently, many of these quantities are measured on-line in industrial production reactors. The first two are of primary importance and are most often separately controlled. From the point of view of process supervision, the measurement information they deliver critically depends on the signal-to-noise ratio, provided the measures are accurate, i.e. the measurement errors are mainly due to random signal fluctuations around the true value, or, distortions without a significant systematic error.

Temperature is usually only measured at a single position in the reactor and this is normally done using a platinum layer resistant thermometer. pH is most often measured by single electrochemical pH electrodes. Clark electrodes are most often applied for pO_2 measurements. All these probes are standard equipment of bioreactors.

To measure the weight of the culture, the entire reactor is usually placed on load cells which allow the culture weight to be continuously monitored. Such measurements are important as the reactor weight is one of the key variables in the models. The feeding rates, which are important since they directly influence the reactor weight and thus all concentrations, are most often determined from the mass of the feed solution pumped to the reactor as a function of time. A sufficient accuracy can be obtained when the mass of the storage vessel for the individual component is continuously monitored. The feeding rates are then determined from the rates of change of the mass values. Gaseous flow rates are measured in a different way. The air flow rate through a sparger into the reactor is often measured by thermal mass flow meters. Since this quantity does not often change, a simple rotameter may also serve as a suitable device.

There are several methods to determine the concentrations of O_2 and CO_2 in the vent line of bioreactors. Mass spectrometers or quadrupole mass filters are being used, however, they are not only expensive in terms of investment cost but also in terms of manpower required to maintain them in a proper state. Thus, most often, paramagnetic O_2 and infrared-based CO_2 analysers are used in practice.

10.5.2 State estimation

The straightforward approach to estimate the current status of a fermentation is to take the on-line measurement signals and to make use of the process model to relate these signals to the key process variables of primary interest, i.e. the state (or status) variables, biomass, substrate, product concentrations etc. The problem of state estimation is the determination of a parameter while the process is running and of knowing the values of quantities which essentially describe the state of the current process from available on-line measurement data. Several different techniques, distinguished by the quality and the amount of knowledge they use about the particular process, are discussed in literature. In the following list, an increasing amount of knowledge is used from the first to the last example cited:

- **Filters** (low-, high-, and band pass-) eliminate in its most simple approach the noise in the measurement data, i.e. they filter off the noise.
- **Indirect measurements** use measurement data to determine noise-free values of quantities, which cannot be measured directly, from (static) non-linear model relationships between the different quantities involved.
- **Observers** additionally make use of a dynamic model in order to estimate the values of non-measurable quantities.
- **Extended Kalman filters** additionally consider information about the uncertainties of the model as well as of the measurement data.
- **Sequential state/parameter estimators** additionally adapt the parameters of the process model to the currently observed process data.

10.6 Open-loop control

In open-loop control, the pattern of the manipulable variable, as determined beforehand, is used directly to adjust the corresponding actuator. For instance, in a fed-batch cultivation the pattern of the feed rate profile is used directly to adjust the pump which provides the feed. In open-loop control, there is no feedback from the process to the controller. The process state is consequently assumed to be dominated by variables which are adjusted by means of the actuator. Possible distortions must not sensitively influence the process behaviour.

The essential problem in open-loop control is thus to determine the optimal profiles of the control variables. This requires a reliable process model. The following issues must be carefully defined before a control profile, e.g. a feeding profile optimisation, can be worked out:

- All objectives of the process must be described quantitatively: The **performance measure** must be defined as a function of those variables that can be manipulated. The variables which might be of influence on the process performance should include economic variables such as costs for substrates, electricity and man-power.
- A clear description of the **constraints** to the process variables is required that encompass the performance parameters of the equipment, e.g. the maximal $k_L a$-value that can be obtained with the reactor, the maximal broth weight allowed, the maximal and minimal feed rates possible with the given equipment etc., but also all other constraints, e.g. economical (costs), ecological (environmental requirements) etc.
- The properties of the **process model** must be defined and the accuracy of the process model must be estimated.
- **Starting values** for the manipulable/adjustable variables are required to guarantee that the non-linear numerical optimisation procedures quickly converge to the desired optimum.
- **Initial values** of the state variables are required in order to solve the general mass balance equations.

For optimisation, a couple of different software procedures are available. In the case of small models that contain a few differential equations only, the well known simplex optimisation method can be used with advantage. All major software libraries provide ready-to-use modules for the simplex procedure, e.g. the MATLAB software provides the famous Nelder–Mead algorithm. With such algorithms one can easily solve the optimisation problem.

In more complicated situations, in particular when the available data are rather noisy, random-search algorithms, in particular the so-called evolutional algorithms, may lead to better results, since they do not suffer so much from the general problem of non-linear fitting procedures, namely to end up in local minima, away from the true global minimum of the least square deviation. However, this advantage must be paid for by considerably larger computing times.

Apart from single closed loop control or regulation circuits for temperature, pH and vessel pressure, most other manipulable variables in industrial production processes are usually controlled in an open-loop fashion along fixed predefined patterns. As everyday practice shows, these patterns most often can be tuned up in such a way that the processes can be run at a sufficient cost/benefit-ratio. It requires, however, that the plant personnel monitors the process and interferes when significant distortions appear.

10.7 Closed-loop control

Closed-loop control is used to keep the running process on the track that has been found desirable beforehand despite significant distortions appearing in reality. Variables like pH, temperature, and sometimes also pO_2 are variables that are controlled by means of closed-loop or feed-back controllers during cultivation processes. As mentioned, with respect to many key variables in cultivation processes such as biomass, specific growth rate, product concentration, most industrial processes can be successfully run by open-loop control procedures. Closed-loop control of these variables becomes important, when the accuracy requirements become higher, i.e. when process improvements can only be obtained by reducing the fluctuations of the process around the predetermined process control path.

In closed-loop control, the deviations between the actual value of the control variable and its desired set point are used to change the process by appropriate changes in one or more manipulable variables in such a way that the deviation becomes smaller.

10.7.1 Proportional integral derivative (PID) controllers

Instead of a manual reaction on a deviation between the actual measurement value of the control variable from its desired value (set point), one often tries to use automatic controllers to reduce the deviations. For some control variables it suffices to use very simple controllers, e.g. temperature in the fermenters is usually regulated to a constant set point by low-level, three-point controllers that switch the cooling system on when the measured temperature reaches an upper temperature level, off when temperature is within the predefined temperature window, and activates heating coils when the temperature falls below a fixed lower temperature level.

In many cases, e.g. with pH control or substrate flow rates, the processes under consideration react improperly on such simple step changes in the manipulable variable, m: in unfavourable cases, the control variable may begin oscillating with considerable amplitude around its set point. Such an instability is a sign that the process dynamics must be taken into account. The dynamics of the process may be described by an appropriate process model. Often, the dynamical behaviour of the process can be approximately described in very simple way as a combination of a *pure* transport time lag element and one with a *first order* response lag characteristics. The behaviour of both can be

Table 10.1 PID-controller tuning for uncontrolled processes. Step response experiments lead to two characteristic times A and T (see Fig. 10.4). The table shows how to determine the PID-controller parameters K_c, τ_I and τ_D from them. In the control mode P only the proportional part is switched on, in the PI-mode, the intergral part is additionally switched on

Control mode	K_c	τ_I	τ_D
P	(1/K) T/a	–	–
PI	(0.9/K) T/a	3.33a	–
PID	(1.2/K) T/a	2.0a	0.5a

Table 10.2 PID-controller tuning for a controlled process. Experimentally the K_c parameter of the controller is increased starting from a low value until the controller becomes unstable at some value $K_{c,crit}$. This table shows how to determine the PID parameters from this value and the period, P, of the appearing oscillations

Control mode	K_c	τ_I	τ_d
P	0.5 $K_{c,crit}$	–	–
PI	0.45 $K_{c,crit}$	$P_c/1.2$	–
PID	0.6 $K_{c,crit}$	$P_c/2.0$	$P_c/8.0$

characterised by simple time constants, which can be identified using experimental data.

In the widely applied **Proportional Integral Derivative** (PID) controllers, use is made of a simple assumption about the process dynamics. A PID controller responds to a non-zero deviation, ε, between the real measurement value of the variable to be controlled and its corresponding set-point value with changes, m, in the manipulable variable, which are proportional to a combination of terms that are (i) directly proportional (P) to ε; (ii) proportional to an integral (I) over $\varepsilon(t)$; and (iii) proportional to the time derivative of $\varepsilon(t)$ in the following form:

$$m(t) = K_c\left(\varepsilon + \frac{1}{\tau_I}\int_0^t \varepsilon dt + \tau_D\frac{d\varepsilon}{dt}\right) \tag{10.35}$$

The vast majority of closed-loop controllers used in biochemical engineering processes are PID controllers. The **Ziegler/Nicols procedure** may be used to determine the three parameters, K_c, τ_I and τ_D, in the expression $m(t)$ from simple experiments. Two cases are distinguished, experiments with controlled and uncontrolled processes.

Procedure for an uncontrolled process: In the uncontrolled process, a simple step change B in the manipulable variable (actuator variable) is performed. This step change leads to a change A in the control variable as depicted in Fig. 10.4. Ziegler and Nicols showed how to make use of the characteristic time measures as defined in Fig. 10.4: namely the delay time, a, the response time, T, and the amplitude ratio, $K = A/B$, to determine the three parameters, K_c, τ_I and τ_D, of the PID controller defined in Eqn (10.35). The relations are summarised in Table 10.1.

Procedure for a controlled process: When the process is already under control, another tuning procedure is required. Here Ziegler and Nicols proposed experimentally to

- find a critical controller gain $K_{c,crit}$ in the following way. First the integral and the differential part of the controller are switched off. Starting with a small value, K_c is then systematically increased until the control variable depicts a stable oscillation.
- Then, this critical gain $K_{c,crit}$ and the corresponding period P_c of the sustained oscillation are used to determine the three parameters of the PID controller. Table 10.2 shows how to proceed.

In the literature, the two ways of determining the controller parameters are shown to lead to equivalent results.

Example

Assume the task is to control the specific growth rate, μ, in a fed-batch culture of *E.coli* to $\mu = \mu_{\text{set-point}}$ by appropriately changing the feed rate, F, with a PID controller. In order to determine, F, the controller uses values of μ determined from measured O_2 uptake rates, OUR. By rearrangement of Eqn (10.31) we get

$$\mu(\text{OUR}) = \frac{\text{OUR-}\omega_{\text{m}}}{Y_{\text{xo}}X} \tag{10.36}$$

State estimation using the balance Eqn (10.5) delivers the current value for X.

The PID controller responds to the deviation

$$\varepsilon = \mu_{\text{set-point}} - \mu(\text{OUR}) \tag{10.37}$$

according to Eqn (10.35). Its output delivers the actual feed rate, F. In this case, $m(t)$ in Eqn (10.35) must be replaced by $F(t)$.

The quality of control crucially depends on the controller parameters, K_c, τ_I and τ_D, defined in Eqn (10.35). Assume the PID controller is tuned once at the beginning of the cultivation. According to the Ziegler–Nicols procedure for uncontrolled processes, the response in the specific growth rate, μ, upon a step change in the feed rate, F, is observed and the corresponding time constants as well as the ratio K are extracted from the response curve (Fig. 10.4).

After some hours of fermentation time, the controller performance will no longer be sufficient. As can be seen in Fig. 10.5, the process cannot be kept close to the pre-defined set point profile for μ. Obviously, after several hours of fermentation time the PID controller is no longer able to keep the control variable close to its set point.

This example demonstrates the most important problem with applications of PID controllers in biochemical production processes namely that these processes do change their dynamical behaviour with time. Hence, the controller parameters must be updated continuously. Consequently, the question appears, whether it is possible to relate the

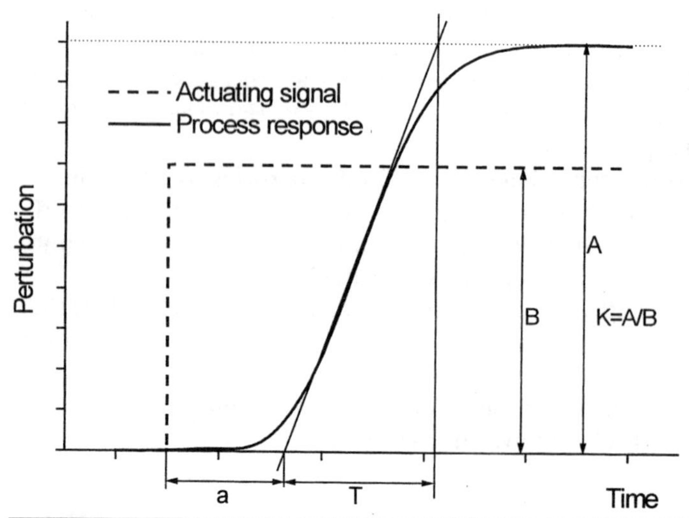

Fig. 10.4 Typical response in terms of some control variable in a simple process on a step change in manipulable variable (actuating signal). This graph shows how to estimate the time constants of the process that are needed to tune a PID controller at an uncontrolled process. The first reaction observed is that the response is delayed by some time span, a, with respect to the actuating signal. The second reaction of interest with respect to PID controllers is that the process is not able to follow the actuating signal immediately; instead it rises in a delayed way to a level, A. The corresponding time constant, T, can be estimated from the abscissa values of the intersects of the tangent through turning point of the response curve and the abscissa and its parallel at the level finally approached by the response curve.

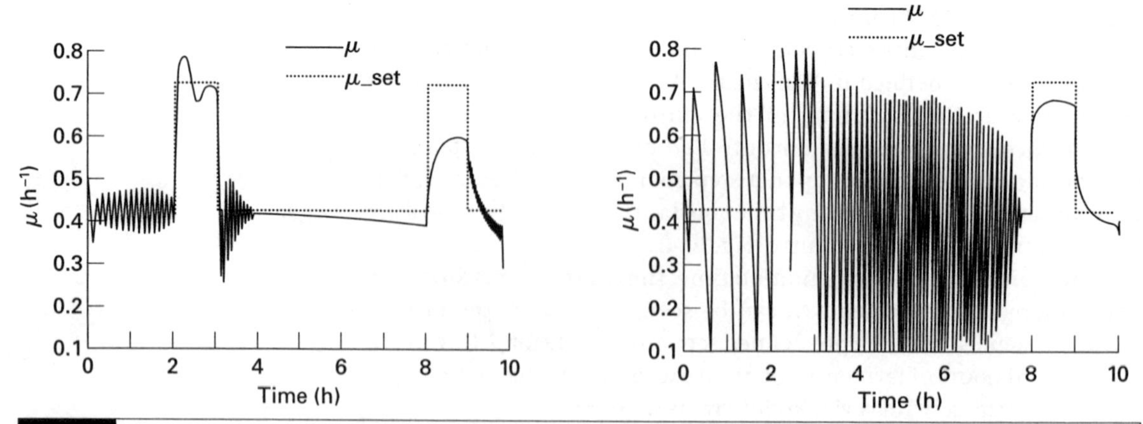

Fig. 10.5 Result obtained with a standard PID controller for the specific growth rate, μ, at an *E. coli* cultivation on glucose, where the controller parameters were determined in different phases of the process. The left part of the figure was obtained with parameters determined during the initial phase of one of the previous runs of the process, and the right with parameters determined during the end-phase of one of the previous runs of the process. As can be seen particularly in the right part where μ is strongly oscillating, the controller is not able to perform the expected task, namely to keep the specific growth rate close to the set-point profile.

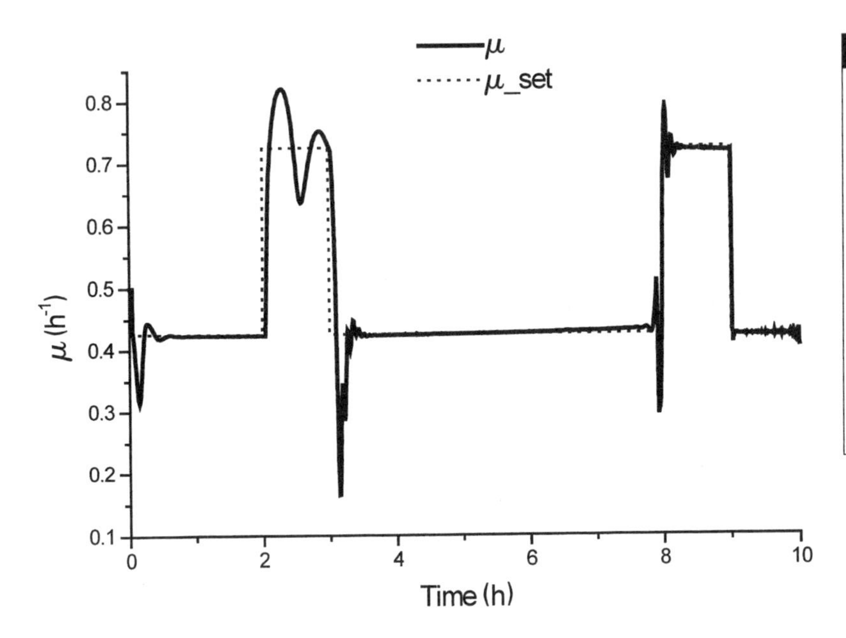

Fig. 10.6 Typical result of a PID controller for the specific growth rate, μ, with parameter adaptation. The improvement in comparison with the usual PID controller, depicted in Fig. 10.5, becomes obvious. Once again, the dashed curve is the set point profile, while the full line depicts the controlled specific growth rate. Apart from times around the steep changes in the set-point profile, the controller is able to keep the specific growth rate on the desired level.

changes in the dynamics of the process to some easy-to-measure quantities. This problem is discussed next.

10.7.2 Adaptive control

Adaptive controllers continuously adjust the parameters of simple controllers upon changes of the process dynamics during the running process. To illustrate how one could proceed to adapt the parameters of a PID controller for the specific growth rate, the example discussed in the last section is continued.

From experience, the O_2 uptake rate (OUR) is known to be an important indicator of changes in the bioprocess dynamics. Hence, one can try to directly relate the measured values of OUR to the optimal parameters of the PID controller.

A first approach could be to assume linear relationships. After some tuning, one can arrive at the following concrete set of linear equations:

$$\begin{aligned} K_c &= 0.81 + 0.108\ \text{OUR} \\ \tau_I &= 0.072 - 0.01\ \text{OUR} \\ \tau_D &= 0.019 - 0.0024\ \text{OUR} \end{aligned} \tag{10.38}$$

This leads to significant improvements of the controller as can be seen in Fig. 10.6.

When the process dynamics is not too complex it is possible to adapt the controller parameters to the changing dynamics of the processes even with some simple black-box assumptions. When such simple black-box approaches do not work, the process dynamics must be considered by means of an appropriate process model. Then we are speaking about model-supported control which is described next.

10.7.3 Model predictive control (MPC)

There are many different approaches to model-supported controllers. One particularly effective example, which is conceptually easy and easy to implement, is the concept of **model predictive control**. Model predictive control, as the name states, uses a model-based prediction of the variable, $\mathbf{c}_c$, to be controlled, to keep the process on the desired path, $\mathbf{c}_d(t)$. The prediction allows to determine beforehand the expected reaction of the process upon an envisaged change in the manipulable variable, $\mathbf{u}_c$. From the possible changes, the controller selects the best one and then changes the manipulable variable, $\mathbf{u}_c$, accordingly. It suffices to restrict the prediction of the process behaviour to some time interval, $\{t, t+t_H\}$, referred to as the **time horizon.**

As compared to simple controllers, the model predictive controller does not merely determine its action from a deviation between the actual value, $\mathbf{c}_c$, of the control variable and the corresponding desired value, $\mathbf{c}_d(t)$, at a single time instant, t, only, but from deviation of the entire path segment within the finite time horizon $\{t, t+t_H\}$ from the desired profile. This deviation can be quantified by the mean square deviation $\langle(\mathbf{c}_c - \mathbf{c}_d)^2\rangle$ between the desired trajectory $\mathbf{c}_d(t)$ of the control variable and that $(\mathbf{c}_c(t))$ to be expected when the process is running with the pre-defined profile segment $\mathbf{u}_c(t)$ of the manipulable variable within the time horizon. When the time horizon is not too long, one can proceed in a very simple way in order to determine the minimal deviation: a finite set, $\{\Delta\mathbf{u}_{c,i}(t)\}$, of possible changes of the pre-determined profile of the manipulable variables can be tested individually and the best of them can be taken to determine the actual control action.

A particularly simple change or correction of the pre-defined profile, $\mathbf{u}_c(t)$, of the manipulable variable is a proportional shift $\Delta\mathbf{u}_c(t) = \alpha\, \mathbf{u}_c(t)$. The behaviour of the process within the time horizon can easily be simulated for a set of constants, α. The best of the simulated paths (that determined with α_b) which led to the least deviation from the desired path, is then used to determine the actual control action. Since at every time step, t_k, we only need to know what to do at the next time step, t_{k+1}, the corrected value control of the predefined profile is simply

$$\mathbf{u}_{c,corr}(t_{k+1}) = \mathbf{u}_c(t_{k+1}) + \mathbf{u}_c(t_{k+1}) \times \alpha_b \qquad (10.39)$$

A typical result obtained with a simple model predictive controller is depicted in Fig.10.7.

In the current control engineering literature, several variants of model predictive control algorithms have been discussed. The different approaches may have one or the other advantage or disadvantage in particular applications, however, the essential point to note is that the model quality is the crucial factor of model predictive control in real applications.

10.8 Conclusion

Sophisticated control procedures make sense particularly in cases where it becomes necessary to run the process outsides the areas in the state space where stability can be obtained in a natural way. For example, when a higher performance can be obtained near states from which the process can easily run out of control, a sophisticated control

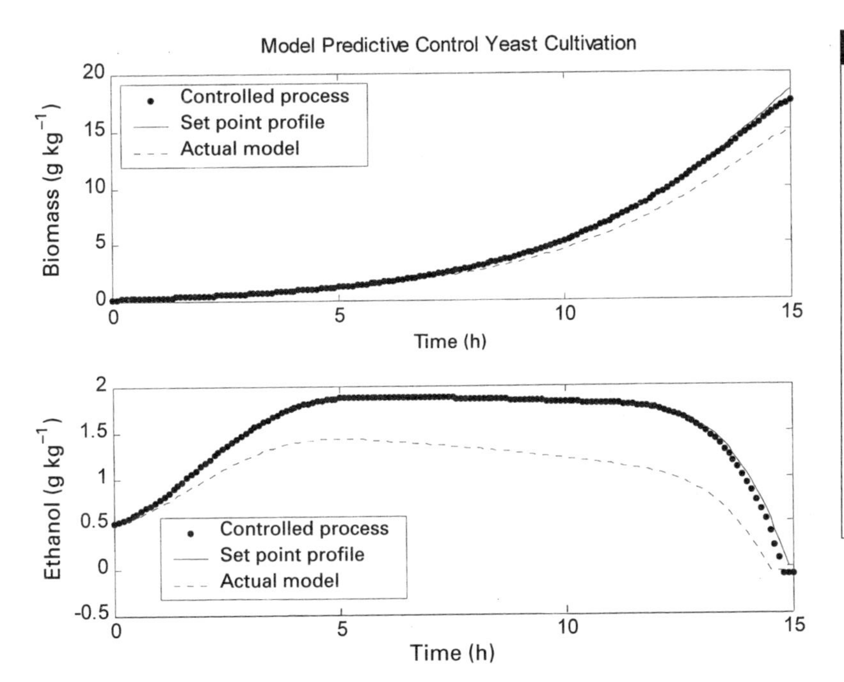

Fig. 10.7 Typical example of model predictive control applied to a yeast cultivation process. Open-loop optimisation resulted in the set point profiles depicted by the full lines. Unfortunately the glucose concentration in the feed deviates from the assumed one. With the actual feed concentration, the process would not run in an optimal way (dashed line). The model predictive controller is used to correct the feed rate in such a way that the desired profiles are approached. The symbols are the on-line estimated values of the controlled process.

procedure makes sense. In bioreactors such states are often associated with process regimes that are running close to process constraints such as maximal reactor O_2 transfer capability or maximal reactor cooling capacity.

In order to make use of modern model-supported process monitoring and control procedures, appropriate software implementations must be available. Such implementations can only be based on detailed knowledge about the process, i.e. it must be done in close co-operation with the responsible process engineers. In order to keep the cost/benefit-ratio of such implementations within reasonable limits, well-performing development tools are required. Essential for the development of models and control algorithms are powerful tool boxes. There are only a few packages available for this purpose, most popular are MATLAB/SIMULINK and SCILAB/SCICOS.

Process supervision and control is important to ensure product quality in the sense that the product meets the quality requirements necessary to sell it profitably. On the other hand, supervision and control aims to guarantee that the process can be run at minimum cost. The cost/benefit-ratio is the iron hand that rules all the related activities. Modern model-supported techniques allow activation of most of the available knowledge in order to keep the product in the pre-defined quality margins at a minimum of cost.

The advantages to be expected from process supervision and control depend on three main issues: (i) measurement data providing information about the actual state of the process; (ii) a process model that allows the relevant process quantities to be related to each other and, in particular, to the process's performance; and finally, (iii) actuators by which the desired values of the manipulable quantities can be precisely and accurately adjusted to pre-determined values. All three components are subject to errors and it is essential to take care that none of these errors become too large compared with the others. That means, if the knowledge about the process is poor, then high precision measurements cannot be exploited and, hence, do not make much sense; or when the manipulable quantities cannot be adjusted accurately, a sophisticated modelling is a waste of effort. In other words it makes sense to take care of the weakest element in a set of issues which often is the measurement data.

10.9 Further reading

Lübbert, A. and Simutis, R. (1994), Adequate use of measuring data in bioprocess modeling and control. *Trends Biotechnol.* **12**, 304–311.

Royce, P. N. (1993). A discussion of recent developments in fermentation monitoring and control from a practical perspective. *Crit. Rev. Biotechnol.* **13** (2), 117–149.

Schubert, J., Simutis, R., Dors, M., Havlik, I. and Lübbert, A. (1994). Bioprocess optimization and control: Application of hybrid modeling. *J. Biotechnol.* **35**, 51–68.

Schügerl, K. (Vol. Ed.) (1991), *Measurement and Control, Vol. 4 of Biotechnology, 2nd Edition*, VCH, Weinheim.

Chapter 11

Process economics

Bjørn Kristiansen

11.1 Introduction

Most of the chapters in this book are concerned with technological aspects – how to make things work. Our present society claims that technological processing only works if it makes money. Assuming the biotechnological advances described throughout the book are technical successes, is it possible to make money with them or, in more conventional terms, can they lead to economically viable processes? This is where this chapter comes in.

Every emerging technology needs investments for society to reap the benefits that the technological developments promise. Chapter 12, entitled 'The business of biotechnology', describes how to obtain funding to start up an enterprise in biotechnology. It stresses the need for blending scientific enthusiasm with economic awareness. Biotechnological processes are costly, both in terms of the capital required to set up the production facility and to run the processes themselves. The aim of this chapter is to show the tools used to estimate and calculate how much money will be required to set up a production unit and how much money it can be expected to make. To illustrate this, we will go through the process of designing and costing a plant.

Readers will appreciate that process economics is a major subject and a short treatise such as this can only hope to be a mere introduction. For more details, the reader should refer to the reading list at the end of this chapter. This contains further information that will demonstrate that much processing understanding will be gained by viewing a process from an economic rather than scientific point of view.

11.2 The starting point

The basic assumptions for this chapter are that:

- The project you are working on seems so interesting that you have been asked to prepare a case for building a plant to produce your product.
- The technology works as specified.
- All permits (production, effluents and product approval) will be, or have been, granted.

There can be many reasons for wanting to start producing or increase existing production of a specific product:

- The market is increasing and will absorb another X kg (or tonnes).
- You have a technology that allows you to compete with existing producers.
- Your existing process, or that of a rival company, is becoming out of date.
- There is a clear trend that a new market is opening up.

For the production engineer, there are primarily two issues that must be dealt with:

(1) what will be the price of your product, and
(2) what will be the production volume.

The price is determined by the cost of putting together the hardware to build the plant, collected in the **capital costs**, and the cost of running the plant to support the operations, collected into the **operating costs** and what you hope to earn. Both the capital and operating costs are dependent on the scale of operation.

The starting point is therefore: How much of your product are you going to produce?

Having decided this, the rest is relatively straightforward. There are procedures for costing the design, constructing and operating a production plant, including steps to ensure that the plant becomes profitable, as indicated in the further reading list. The list also contains references to some personal computer (PC) based simulation programmes that can be used to do the calculations and design the plant, once you have supplied process details.

11.3 Cost estimates

The decision to invest will be based on cost estimates for the proposed production process. Without these, no rational decision concerning the investment can be taken. There are different methods to calculate cost estimations, all depending on how near you are to taking a final decision on whether to invest (see Table 11.1).

The figures in Table 11.1 must be regarded as indicative; the costs

Table 11.1 Cost estimation for a €50 M project

Type of estimate	Cost of preparing estimate (€K)	Accuracy required (%)	Comments
Order of magnitude	10 to 30	± 25 to 50	Project at R&D stage
Study	30 to 70	± 20 to 40	Details of major equipment units known
Preliminary	70 to 150	± 15 to 20	
Project	150 to 500	± 10 to 15	
Detailed (Firm)	500 to 1 000	± 5 to 10	Will decide if the plant is to be built

Table 11.2 Typical items included in capital cost estimates

All pieces of equipment for production, recovery, supply of utilities including spares
Instrumentation
Installation – equipment, instruments, piping
Labour for installation
Land purchase, preparation and buildings
Supervision
Insurance and tax
Site preparation
Contractor's fee
Contingency

depend on factors such as the scale of the project, the strategic importance of the project, the split between internal and external expertise used etc. The message is that (a) the nearer the time for investing large sums of money the stricter the demand for accuracy in cost estimations, and (b) estimating costs is expensive. However, all decisions on whether to invest will be based on a cost estimate.

There are two main parts to the cost estimate: **capital costs** and **operating costs**. The former covers the fabrication of the complete plant, including production units, buildings, preparing the land etc. The latter concerns the cost of operating the plant, in terms of manpower, raw materials, utilities (steam, water and electricity) etc. The number of items included in capital and operating cost estimates will depend on the method used for the estimate.

11.3.1 Estimating the capital costs

Conventionally, the capital costs are separated into:

- Direct, or fixed capital, costs – the amount of money required for establishing, building and furbishing the plant.
- Indirect, or working capital costs – the working capital required for constructing the plant (overheads, transport, engineering, taxes, etc.).

Each will have a number of items as shown in Table 11.2. There are many reasons for separating the direct and indirect costs, the most important

Table 11.3 Items typically included in an operating cost estimate
Raw materials
Utilities
Waste treatment
Labour (including training)
Supervision
Maintenance
Overheads
Royalties
R&D
Cost of sales
Site maintenance
Tax and insurance

being tax and duty and control of the finance. (These aspects are beyond the scope of this chapter; here we are primarily concerned with the total amount of capital required.) Table 11.2 is a non-exclusive list of items that contribute to the capital costs.

11.3.2 Operating costs

The **operating**, or **manufacturing**, **costs** are simply a measure of how much money you spend to produce your product, including development work to improve the process as well as the cost of marketing and selling it. The operating costs are divided into:

- Fixed costs – items that are not related to the volume of production, i.e. taxes, depreciation, overheads.
- Variable costs – related to output by way of raw materials, labour and energy.

For our purposes, the division into fixed and variable costs will not be considered in detail, but for profitability analysis, this may be important, see Section 11.8.1. Items contributing to the operating costs are given in Table 11.3.

Note that all costs related to an item are included. Thus, the labour costs include salaries, overtime, fringe benefits, holidays, training, sick pay, etc. It can also include phones, office supplies, professional fees (and large expensive cars for the senior executives!). Similarly, raw materials will include all medium ingredients for all growth stages, chemicals for process monitoring and control, analysis, gas supplies, catalysts etc.

To carry out a cost estimate the production facility, be it an expansion of existing facilities or a completely new plant, must be designed.

11.4 Process design

To get an overview of the plant, the process has to be designed in terms of unit processes required, essential services and instrumentation. The total amount produced, together with concentration of product in the reactor and the productivity, will give the total reactor volume. An educated guess will give the number of reactors required and will make it possible to fix the sizes and number of the seed and inoculum vessels. The latter will only become important in the more demanding cost estimates. It is desirable to keep the number of reactors as small as possible. In addition, from overall engineering considerations, big is beautiful. The reason for this is the following general relationship between costs and reactor size:

$$\text{Reactor costs} \propto (\text{Reactor volume})^{0.7-0.8}$$

Thus, one 100 m^3 reactor will be better than two 50 m^3 reactors.

It must be pointed out that we will not always be in a position to act according to this. Whilst it may be easy to accommodate in the design of a new plant, factors such as the capacity of existing plant and utilities will be more prominent when expanding existing plants.

Knowing the product concentration and productivity will allow you to decide on feed composition and energy requirement to supply all the substrates required. In the case of aerobic fermentations, you must know the mass transfer characteristics of the reactor to calculate the optimum air flow-rate and thus the size of the compressors that must be purchased. Once the product has been produced to the desired concentration, you must decide on the level of product isolation and purification required (see Chapter 9). A schematic presentation of this procedure is given in Fig. 11.1.

A generalised layout of a biotechnology plant is given in Fig. 11.2. A major part of a cost-estimate exercise is to decide which unit processes are best suited to your particular process. To do this, you will require processing skills described in the other chapters. However, as there are so many similarities between different biotechnology plants, it is relatively easy to find information for an order of magnitude cost estimate (see Table 11.1). The real problems arise when accuracy of 15% or below is necessary.

To carry out an order of magnitude cost estimate you require:

- Some knowledge of the major processing steps.
- Some knowledge of the material balance.
- Some knowledge of the energy balance.
- Some knowledge of the kinetics.

The estimate is then obtained from:

- an analogy with an old process or comparing it to similar processes; or,
- a capital cost estimate from cost of the major items of equipment or process steps; and,
- working capital and manufacturing costs from yields, energy requirements and cost of capital.

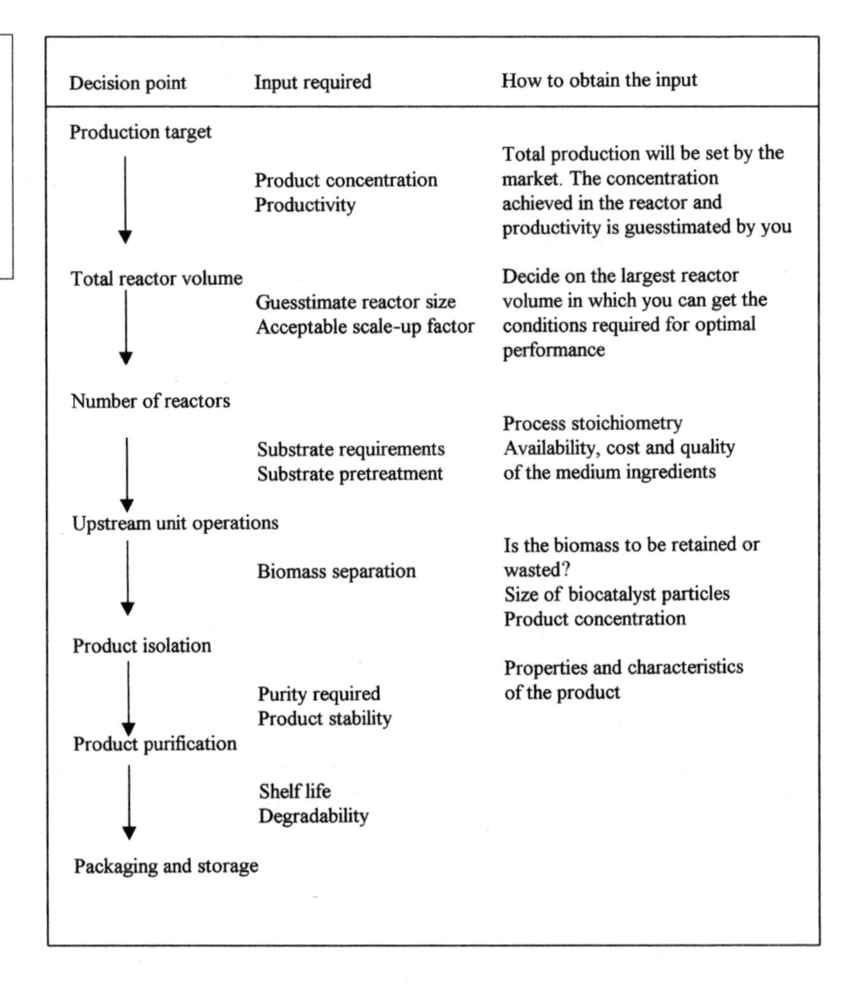

Fig. 11.1 Schematic approach to plant design for an order of magnitude cost-estimate. Even at this level of accuracy it can be seen that input from many aspects of plant operations, including scientific, engineering, sales and marketing will be required.

11.5 Design exercise

The next step is to design and cost a real process, as indicated in Fig. 11.1. The process we are going to work on is the production of fabulase, an imaginary intracellular enzyme used in the fragrance industry. The goal is to produce 10 tonnes per year.

11.5.1 Process details

The process details are given in Table 11.4. The figures are obtained in laboratory and pilot-plant tests, and it is assumed that the technology can be scaled up successfully to production scale. Please note that there are no defined volume restrictions for a pilot plant. It merely refers to a reactor volume, which is traditionally one order of magnitude less than your production vessel, whatever the volume this may be. A pilot plant is the reactor stage where you try out your laboratory results under (semi) production conditions. If you succeed here, you can assume that you can achieve the same (and sometimes better) result in the production-scale reactors.

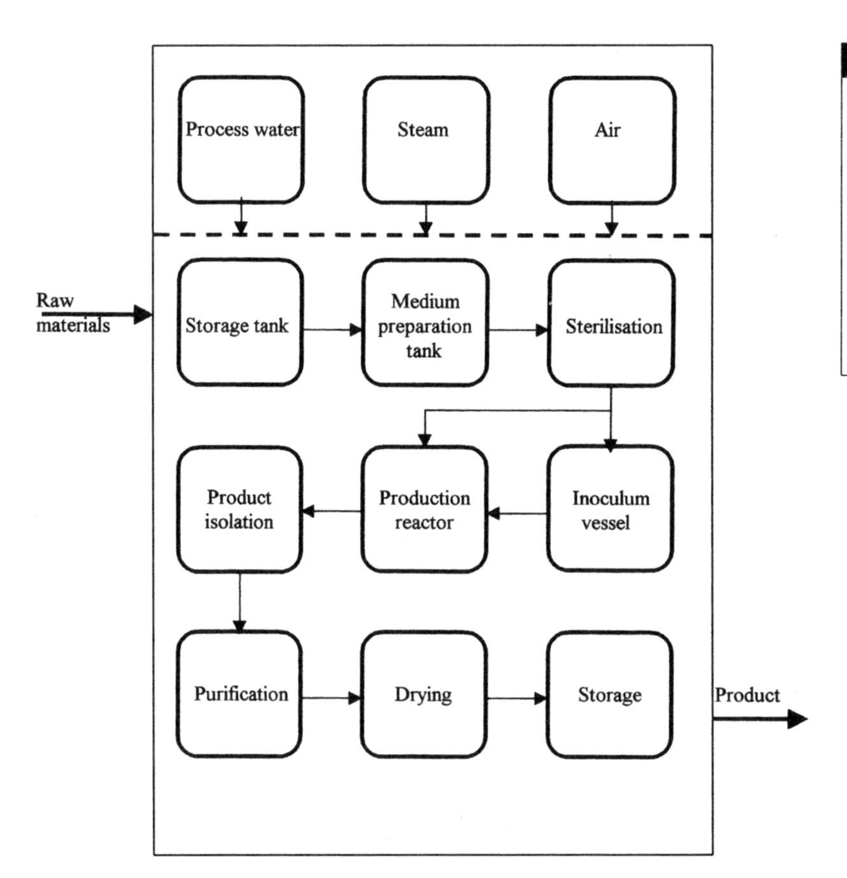

Fig. 11.2 General layout of a production plant with unit processes and utilities. Biotechnology plants are in general very similar, the main difference being the nature of the catalyst, whether it is an enzyme or micro-organism. Compared to general chemical plants, biotechnology plants are demanding in terms of utilities and cleanliness.

Table 11.4 Process yields and operating parameters for production of fabulase

Production goal	10 tonnes per year
Product	Fabulase[a]
Producer micro-organism	*Aspergillus oryzae*[b]
Main carbon source	Glucose
Other nutrients	Soy-flour, potassium and magnesium salts
Batch fermentation time	120 h
Down time	8 h
Number of batches per year	61
Final biomass concentration in the reactor	45 kg m^{-3}
Final fabulase in the reactor	2 kg m^{-3}
Yield of biomass on glucose	0.36
Operating temperature	28 °C

Notes:
[a] An intracellular enzyme.
[b] The genes for fabulase were found in a bacterium, but the company has expressed the genes in *Aspergillus oryzae* as this is used for most of their other products. Having one producer micro-organism simplifies plant operation and is a standard procedure for most enzyme producers, although the host micro-organism may vary.

11.5.2 Calculated process details

Reactors

There will be some product losses during product recovery and purification, so we will assume that the total amount of enzyme produced in the reactor is 12 tonnes (a loss of 20%), i.e. 196.7 kg per batch. The number of batches arises from the total annual operating hours divided by batch processing time (= fermentation hours plus downtime). Note that the total annual operating hours is not the same as the number of hours in a year, as the plant will be shut for overall maintenance once a year. With the product concentration given above, the total liquid volume per batch will be 98.4 m^3. Normally, the liquid volume occupies between 70 and 80% of the reactor volume, and using a reactor occupancy [= (liquid volume/total reactor volume) × 100%] of 75%, we get a reactor volume of 131.1 m^3. Thus we need one 130 m^3 reactor, which is a medium-sized industrial reactor and a stirred tank should suffice. The seed and inoculum development for most *Aspergillus* processes is relatively straightforward, so a 200 litre seed fermenter and a 10 m^3 reactor for inoculum preparation will be adequate. The 200 litre fermenter is inoculated using a 10 litre laboratory fermenter which, in turn, will be inoculated with a 200 ml shake flask culture. (However, if you want to be clever and want to save money and time, you will reduce the number of scale-up steps by, for example, omitting the 200 litre fermenter. This is a level of fine-tuning your process which we will leave for another occasion.)

Medium preparation

Biomass of 4428 kg will be produced in each batch. This will require 12 295 kg glucose (assuming a biomass yield factor of 0.36), which is kept as a syrup in a holding tank containing one month's supply, at a concentration of 500 kg m^{-3}. The other ingredients are stored as powders and must be mixed into a blending tank for every batch. The medium ingredients are sterilised in a continuous steriliser.

Utilities

The consumption of steam, water and electricity can be calculated but this is a detail beyond the scope of the chapter. Medium sterilisation, aerating and stirring the fermenter, and drying the product are normally the principal energy-demanding steps. The subject is covered well in the reading list. The cost of utilities is often related to a company's energy policy and some companies have very efficient heat recovery programmes to keep the overall energy cost down. As our process will have a relatively large energy demand, primarily because of the aeration and mixing of the content in the production reactor, we will assume that the cost for utilities will be 10% of operating costs.

Product isolation

Normally, separating the biomass from the process liquor is the first step in downstream processing. In our case, the product of interest is intracellular and the biomass concentration is so high that we will not remove the process liquor, but we will process the fermentation broth as it is. The first step is to break open the cells to release the intracellular enzyme.

Product purification

The enzyme is separated from the cell debris and organelles in a centrifuge and is then precipitated from the supernatant solution using an ammonium salt. This is a conventional method employed in the enzyme industry. It renders a product sufficiently pure for bulk application. Further purification will be required before the enzyme can be applied in the cosmetics industry. This will often require different technologies, all described in Chapter 9, and for the purposes of this example we will assume that we do not have this technology in-house, consequently we will sell the product as the ammonium salt (it is not unusual for fermentation companies to produce bulk products and leave final purification to specialist companies).

Drying

The nature of the product, tonnage and its application area render freeze drying as the best method.

The plant is shown in Fig. 11.3.

11.6 Capital costs estimates

The estimates for the capital costs are given below. The values have been obtained from manufacturers and the literature. Table 11.5 illustrates the purchase cost for the major plant equipment items.

Having bought the equipment, it must be shipped, insured, and installed and connected up in suitable buildings. This adds substantially to the costs, as indicated in Table 11.6. The main purpose of this table is to highlight that estimating the capital cost is much more than costing the processing equipment. Note that starting capital, normally estimated at 5% of capital cost must be included to get the total investment required.

11.7 Operating costs estimates

The main contributing items to the operating costs have been given in Table 11.3.

Raw materials

In most biotechnology-based production processes, the medium ingredients are major cost factors, ranging from 10 to 60% of operating costs. Normally, the carbon source, being the only bulk ingredient, contributes 60–90% of the raw materials costs. In our case, we will operate with a refined glucose price of 100 € per tonne.

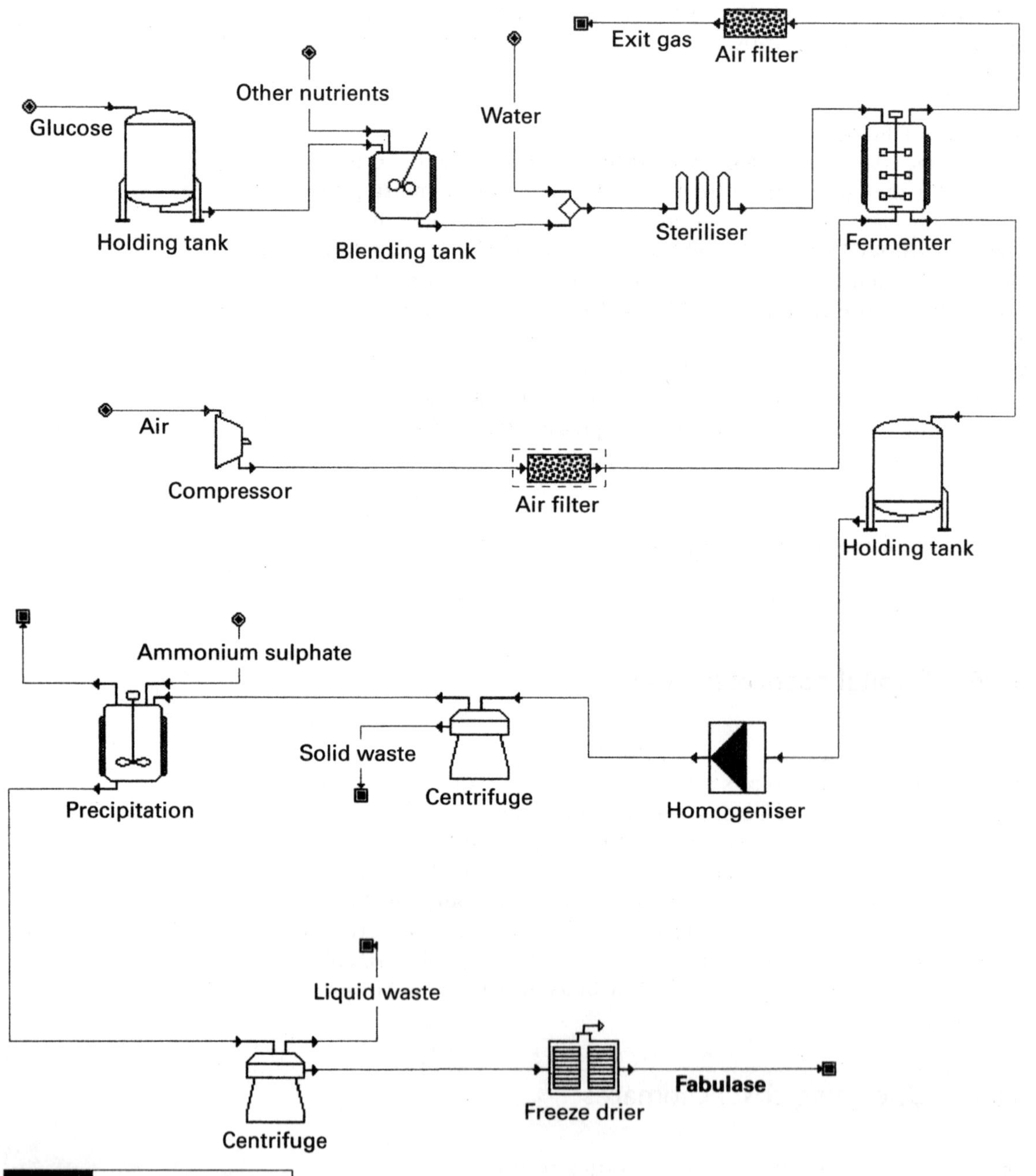

Fig. 11.3 Diagram of the plant for production of fabulase (drawn with permission from Intelligen Inc, New Jersey, USA).

Table 11.5 Equipment purchase costs

Item	Number	Cost (K€)
Holding tanks	2	70
Blending tanks	2	100
Continuous steriliser	1	125
Production reactor	1	2324
Seed and inoculum reactors	2	240
Homogeniser	1	185
Precipitation tank	1	700
Centrifuges	2	335
Compressors	2	1100
Freeze dryer	1	145
Air filters	2	39
Auxiliary processing equipment		1300
Total purchase costs		6663

Table 11.6 Capital cost estimations

Item	Cost (K€)
Equipment	6663
Installation cost (250% of equipment cost – includes instrumentation)	16658
Total direct costs	23321
Construction expenses (70% of direct costs)	16325
Total direct and indirect costs	39646
Contingency (10% of direct and indirect costs)	3964
Total capital costs	43610

Utilities

Assumed to be 10% of the production costs, see above.

Waste treatment

The assumed costs for treatment of liquid and solid wastes are 0.001 €/kg and 0.01 €/kg, respectively. Venting of gas streams will in this case not involve any costs.

Labour costs

It is estimated that the 24 operators (three shifts of six and one stand-by shift) will be required to run the plant, working a 37.5 hour week with 4 weeks holiday a year. Additional labour costs will be supervision (10% of operator costs), laboratory (15% of operator costs) and maintenance and social costs (50% of total labour costs).

Table 11.7 Estimated production costs

Item	Cost (K€)
Raw materials	762
Utilities	1173
Waste treatment	11
Labour (@ 20 € per hour)	2160
Administration and overheads (40% of labour)	864
Depreciation (10% of capital costs)	4361
Contingency (2% " " ")	872
Insurance (1% " " ")	436
Taxes (2.5% " " ")	1090
Total production costs	11729

Table 11.8 Profitability analysis for fabulase plant

Item	Cost (K€)
Capital costs	43610
Start up costs (5% of capital costs)	2181
Total investment	45791
Income from sales @1700 € kg^{-1}	17192
Production costs	11729
Gross profit	5463
Taxes (@ 40%)	2185
Net profit	3278
Expected return on investment	16.2%

Other

Costs of sales, R&D expenses, patent and royalties costs will not be included. These items are very product specific with typical figures of 10%, 5% and 5% of production costs respectively.

The production cost for the production of 10 tonnes fabulase per year is given in Table 11.7.

11.8 The costs case – to build or not to build

To obtain information on which a decision to build the plant can be based, we will carry out a profitability analysis. The results of such an analysis based on a plant life of 15 years is given in Table 11.8.

In some texts, the term 'operating profit' is used. This is used to describe the profit generated from plant operations and is the same as gross profit in Table 11.8 and does not include taxes, cost of capital, depreciation etc.

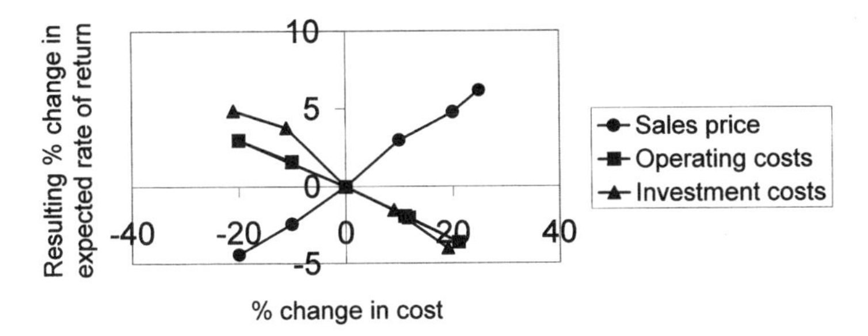

Fig. 11.4 Result of a cost sensitivity analysis, in which the effect of changes in important cost parameters are studied.

The expected rate of return on the capital invested helps to decide whether to invest in the process. For existing processes, the generated cash flow may be a better indication of the health of the company. The cash flow is obtained by adding money spent on the depreciation of the plant to the net profit. For our plant, the cash flow will be (in kiloEuros):

Net profit	K€ 3278
Depreciation	K€ 4134
Cash flow	**K€ 7412**

An expected return on investment of 16.2%, equivalent to a pay-back period of 6.2 years, is relatively low for the biotechnological industry and it is unlikely that our factory will be built. However, the return is sufficiently high not to be discarded immediately and is accordingly subjected to a **cost sensitivity analysis**. Here, the effect of changes in important cost parameters, such as sale price, investment and operating cost, are studied. The result of such an analysis is given in Fig. 11.4.

The figure shows that the process is sensitive to changes in all three parameters, although it is more sensitive to changes in the sale price and investment (or capital) costs as the slopes of these curves are approximately the same and steeper than the slope of the operating costs. It may be difficult for us to influence the sale price as this is set by a number of external factors over which we have little control and will depend on such things as the number of players in the market, the age of the market, competing products, etc. However, the figure shows that a decrease in the investment costs will also lead to a higher rate of return, and we will therefore go back to our design to see if we can cut the costs without affecting plant performance. Thus we must find out:

- Are all the processing steps required?
- Can we alter the capacity of the units?
- Can we reduce downtimes (a common fault for a first design effort is to overestimate the required downtime)?
- Can we use cheaper materials of construction?
- Can we use multi-purpose units?

Whilst doing this we must remember that the new plant must give the same performance as before. If we can cut the capital cost by 10%, we will get a return of around 20%. The process is now beginning to look rather attractive and will warrant further study. It is at this point that the reader takes over. Good luck!

11.9 Further reading

AspenPlus, Aspen Technology Inc, Massachusetts. [Simulation software.]

Peters, M. S. and Timmerhaus, K. D. (1991). *Plant Design and Economics for Chemical Engineers*. McGraw-Hill International Editions.

Reisman, H. B. (1988). *Economic Analysis of Fermentation Processes*. CRC Press, Boca Raton, Florida.

Seider, W. D., Seader, J. D. and Lewin, D. R. (1998). *Process Design Principles*. John Wiley, New York.

SuperProDesigner, Intelligen Inc, New Jersey, New York. [Simulation software.]

Turton, R., Baille, R. C. Whiting, W. B. and Shaeiwitz, J. (1998). *Analysis, Synthesis and Design of Chemical Processes*. Prentice Hall International Series in the Physical and Chemical Engineering Series.

Part II

Practical applications

Chapter 12

The business of biotechnology

William Bains and Chris Evans

12.1 Introduction

Biotechnology is the application of biological processes. 'New' biotechnology is when this is driven by systematic knowledge of biological processes. In this chapter we will discuss the 'new' biotechnology industry's most spectacular commercial manifestation – the 'biotech start-up company' – and what factors contribute to the success and failure of the entrepreneurial application of the science described elsewhere in this volume. This industry appeared in the 1970s, initially in the USA (which consequently has more and larger companies) but soon afterwards in Europe and Asia as well (Table 12.1). Our aim is to give you a feel for what you need to have, and to do, if you are to start up a biotechnology company.

12.2 What is biotechnology used for?

As a start, we will look at what those companies do now to see why those areas have been successful and others have not. This section will review briefly the main areas that 'biotechnology companies' work in.

12.2.1 The applications – medicine

The majority of biotechnology investment since the mid 1970s has been in health-care, and specifically in the discovery of new drugs. An effective new drug can be sold at good profits for as long as the patent on it

Table 12.1 Comparative numbers of biotechnology companies: numbers of biotechnology companies in different countries and their employment and R&D expenditure compared to the countries' populations

<table>
<tr><th>Country</th><th>Population (millions): mid 1990s average</th><th>Number of biotechnology companies (1998)</th><th>Total biotech employees (1998)</th><th>Total biotech R&D spend (million €)</th></tr>
<tr><td>USA</td><td>260.5</td><td>1274</td><td>140000</td><td>8268</td></tr>
<tr><td>UK</td><td>57.9</td><td>245</td><td rowspan="5">39000</td><td rowspan="5">1910</td></tr>
<tr><td>Germany</td><td>80.9</td><td>165</td></tr>
<tr><td>France</td><td>57.6</td><td>141</td></tr>
<tr><td>Sweden</td><td>8.7</td><td>85</td></tr>
<tr><td>Rest of EU and Scandinavia</td><td>175.3</td><td>400</td></tr>
</table>

Source: Ernst and Young's European Life Sciences 99, 6th Annual Report. Ernst and Young International, London, 1999.

prevents someone else from selling it at a lower price. Once a drug comes 'off patent' it can be manufactured as a '**generic**', and profit margins on it plummet. Patents do not last for ever and, if a drug takes 15 years to develop, a patent will only protect its manufacturer from competition for a further 5 years. This means that the original inventor of the drug must have invented a new one every five years (ideally more often), if they are to sell high value, high profit drugs. Thus discovery or invention of new drugs is critical to the commercial strategy of many big pharmaceutical companies. In fact, the drug 'super-companies' formed by the mergers that created such companies as Glaxo–Wellcome and Novartis must launch at least three new drugs each year to keep their competitive position.

The majority of this book is about the technologies of biotechnology, not about their application to drug discovery, so we will summarise here how the latter is done. The most commonly used approach is outlined in Fig. 12.1. A wide range of discovery techniques can identify a molecular target, although genomics-based discovery is currently considered one of the most powerful. This 'target' is a molecular entity whose activity is considered important in a disease. The rest of the discovery process then searches for a small molecule compound that will interfere with the effects of that target. The result is a candidate drug, which is developed into the active ingredient of a medicine.

This process has a high failure rate – only about 4% of programmes succeed in producing a drug that is approved by the regulators, and probably less than 25% of these are profitable. And it is very expensive: a drug typically takes around $250 million to bring it to market, and most of them fail along the way (these are similar costs and success rates to a Hollywood blockbuster film, a new make of car, or re-launching Pepsi in blue cans). Typical failure rates, times and costs are listed in Table 12.2, but the reality is even worse than this, because you have to run at least 12 target discovery programmes at $3.5 million each to have an even chance of getting one drug at the end. This means that the pharmaceutical industry spends nearly 20% of its aggregate $20.9 billion

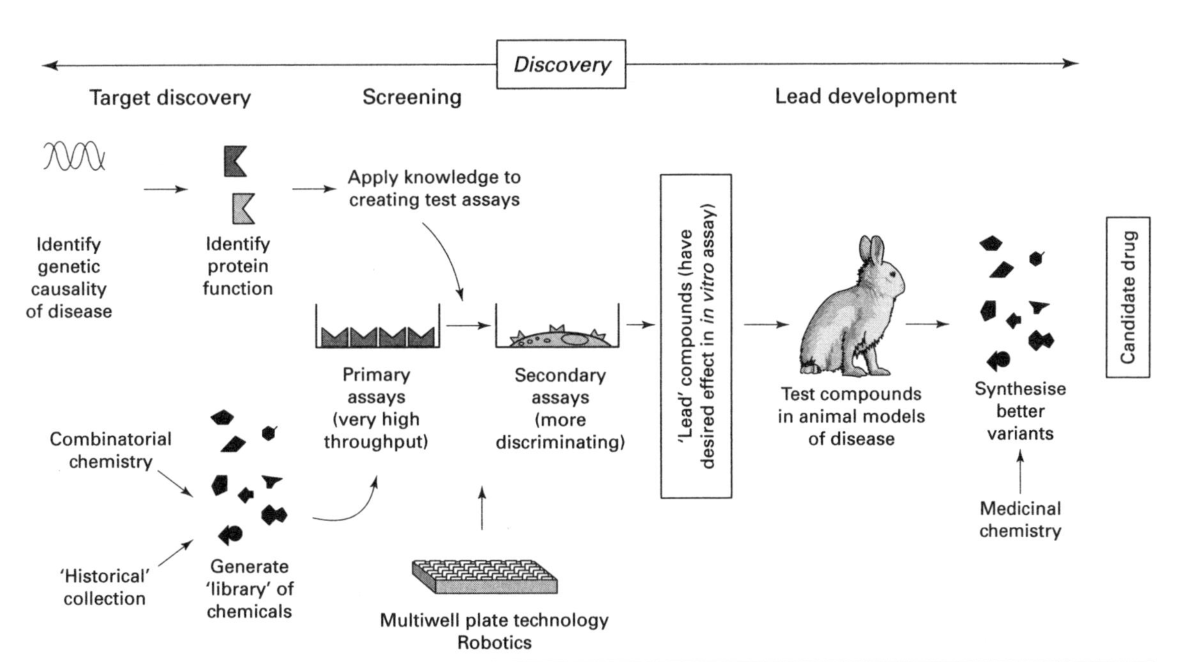

Fig. 12.1 Drug discovery path. A current model of the drug discovery process. Process flows from left to right. The process starts with genomics-driven discovery of a 'target' gene, and hence protein, and with the generation of a diverse set of chemicals from combinatorial libraries or from collections of chemicals accumulated during a company's history. The chemicals are assayed for their ability to block (or sometimes enhance) the target protein's action initially in a high-throughput, usually biochemical assay, and then in more complex 'secondary' assays, usually cellular function assays. The result is a screen 'lead'. These are tested in whole animal disease models, and tested for pharmacological properties, and if necessary modified by directed medicinal chemistry to produce a candidate drug.

Table 12.2 Typical costs and success rates for drug discovery

Stage	Cost (M$)	Time (years)	Success rate (%)
Target discovery	3.5	3	65
Screening	5	1	
Medicinal chemistry	7	1	60
Pre-clinical development	6	1	50
Phase I clinical trial	10	} 5	} 25
Phase II clinical trial	10		
Phase III clinical trial	140		
Total	180	10	4

Notes:
Column 1: stage in drug discovery and development process (see Figs 12.1 and 12.2).
Column 2: cost in US dollars.
Column 3: Time taken for this phase.
Column 4: typical success rate for that stage of the process for a project.
Source: Figures compiled by Merlin from several pharmaceutical company sources, 1997–1999.

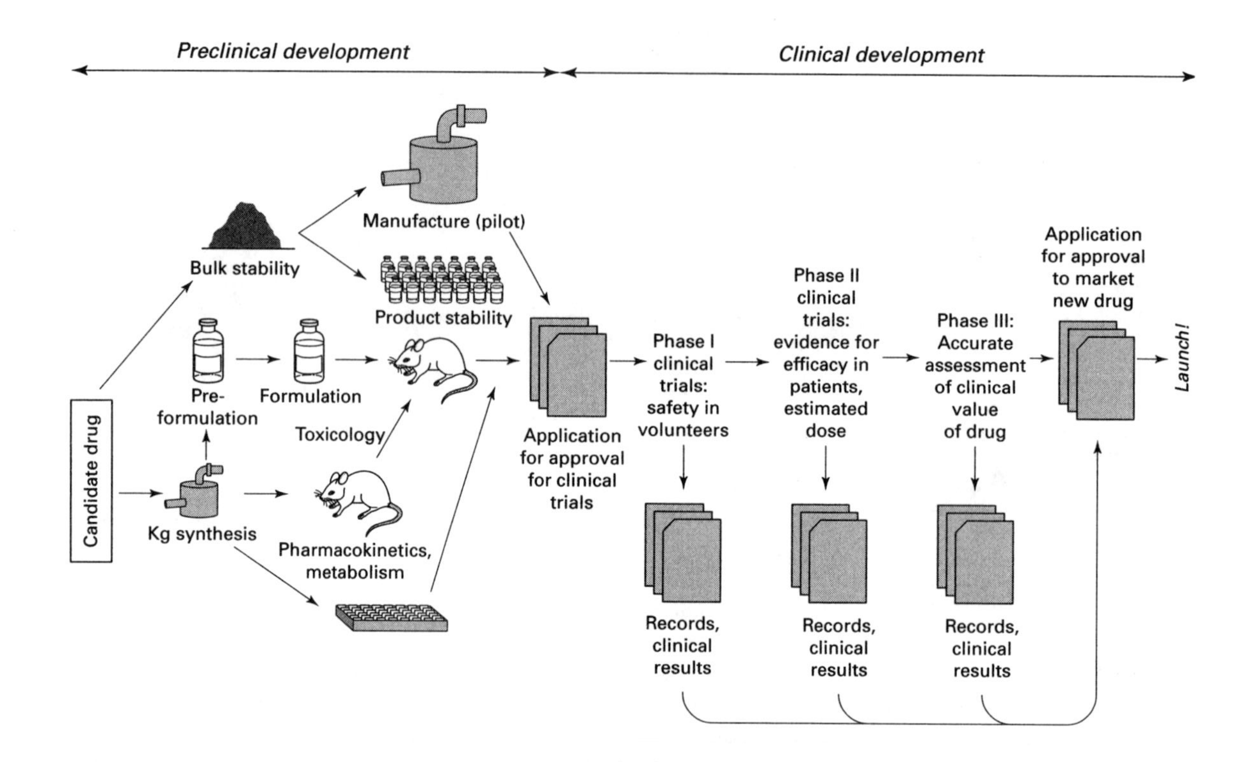

Fig. 12.2 Drug development path. A current model of drug development. Process flows from left to right. The compound is formally tested for metabolism, toxicity, bioavailability and other pharmacological properties, traditionally in animals but increasingly in *in vitro* model assays. Successful compounds are then entered into an escalating series of clinical trials, producing systematic and extensive records which are used in the submission for permission to market the product as a drug.

1998 R&D expenditure on things that do not work. They are therefore willing to pay very large sums to biotechnology companies that can provide science or technology that

- enhances the understanding of the disease (and hence lowers the inherent risk in this approach);
- increases the efficiency of the discovery process (and hence means you can do more discovery for less cost);
- has already been proven in clinical trials to be superior to existing therapy.

This is a continuum of activity from basic biomedical research to commercial drug development, and the drug discovery biotechnology industry occupies the middle of this continuum. Thus some companies are essentially applied extensions of academic groups, others are indistinguishable from small drug companies. Inbetween are companies providing specific technological skills or services, such as companies providing genomics, combinatorial chemistry, or molecular design technology, or companies specialising in screening. In addition, some companies are seeking to radically alter the order in which these steps are done, for example performing aspects of the conventional development (Fig. 12.2) as part of discovery (Fig. 12.1).

Medical diagnostics have a quite different dynamic. While it is hard for an academic researcher to discover a new drug, it is relatively easy to discover a new diagnostic 'marker' for the difference between sick and healthy people. The limitation on their commercialisation is making them reliable and simple enough to be used on a large scale, and ideally to be performed by automated machinery, thus removing the need for skilled assay technicians. As a result, the diagnostics industry is dominated by a small number of companies with powerful marketing and distribution abilities, usually allied to their 'platform' instrumentation – large automated instruments that can perform a wide range of tests. Small companies can only gain a foothold in this market by finding specialist niches, such as specialist 'over-the-counter' tests (for pregnancy, cholesterol etc.), or unusual medical specialties that do not fit into the mainstream of diagnostics.

Genomics-driven drug discovery may change this, with drugs being increasingly targeted according to diagnostic tests that have been developed for those drugs (an idea called the RxDx tandem). This will create a need for many new technologies and systems for testing.

12.2.2 The applications – food and agriculture

Food and agriculture is more important economically than health-care, even in Western countries, and is clearly of much greater concern to the rest of the world. However these areas have not attracted so many biotechnology companies. At root, this is because a new food cannot be sold at $1000 a meal in the same way that a new drug can be sold at $1000 a bottle. Food is price sensitive – the higher the price, the less you sell. Above a certain price, you sell none (price limited). So it is hard to justify expending very substantial amounts of money on developing new food materials because that money cannot be reclaimed in a premium price on the food.

The main exception is in breeding, where the cost of generating a new strain of plant can be offset both by sales of a very large amount of seed-stock and in the premium the farmer can charge for the resulting produce, or the savings in production. In principle, the cost of developing a transgenic crop plant that is resistant to pests (an exercise costing tens to hundreds of millions of dollars) can be recovered by charging extra for the seed – farmers would pay more for the seed because they would have to spend less on pesticides. In practice these economic arguments have proven hard to make in many cases.

A similar argument makes animal reproduction technologies valuable, either in the generation of 'transgenic' animals or, more recently, cloning them. The scientific and commercial value of such 'cloning' explains some of the excitement over the 1997 announcement of 'Dolly' the cloned sheep. Dolly is not a product in her own right, however, but a demonstration of a technology for making tools that themselves will end up in products.

Product-orientated biotechnology in agriculture has been most successful when it focuses on added value in the final product (rather than increased bulk). Typical of food and agricultural biotechnology programmes are the use of genetically engineered enzymes in food processing (added value can be the development of more reliable food flavour, for example), transgenic fruit and vegetables to prolong shelf life (the Flavr Savr tomato was the first such product), and bacterial silage additives and nodule stimulants for legumes to increase productivity.

Even so, the raw material cost in many consumer products is a small fraction of costs of packaging, transport, storage and selling: for example, in the 'over-the-counter' pregnancy tests, the majority of the manufacturing cost lies not in the antibody reagents, but in the plastic casing. And this is itself a small fraction of the cost of storage and transport of the packaged tests. So the biotechnological product must add exceptional value to be worth developing.

Two other areas of biotechnology have had successful application in plant sciences. Both are applications of the 'new' biotechnology to very extensive, established 'old' industries. The first area is in the use of enzymes and, to a lesser extent, micro-organisms in food preparation. The other is in horticulture, where micro-propagation technologies have now become so widely accepted for developing new decorative plant types that they are mainstream horticultural practice. Gardeners will tolerate levels of pesticide use and 'crop failure' greatly in excess of those allowed a farmer: their 'crop' only has to look pretty. For crop plants these techniques have proven only occasionally successful in large-scale production, although they are part of the panoply of technologies used in plant breeding.

12.2.3 The applications: other industries

Many other industries could, in principle, benefit from biotechnology. The fabric and textiles industries are using biotechnology quite substantially, using enzymes to treat textiles and leather, for example. The paper pulp industry is taking up biotechnology rapidly as a cleaner (and hence cheaper) alternative to chemical and mechanical processes. The plastics industry uses the polymers made by micro-organisms, although in practice materials such as the poly-hydroxyalkanoates (such as polyhydroxybutyrate mixtures – 'Biopol') have gained only marginal industrial use (see Chapter 15).

Other biomaterials such as xanthan gums (see Chapter 15) are used in some specialised industrial applications, but this is rare, and opportunistic, and usually does not exploit our systematic knowledge of biological systems, but only our accidental knowledge of their properties and products. This is because oil is very cheap, and the industry for converting it into products is flexible, efficient and sophisticated.

12.3 Biotechnology companies, their care and nurturing

The 'biotechnology company' is a company that is set up specifically to turn the science of biotechnology into a commercial product and sell the result. It is the *science* base of the company that is defining. In the next section we will discuss what it takes to take a biotechnology company from that initial scientific idea to a flourishing commercial enterprise.

12.3.1 General rules

Successful biotechnology companies must combine scientific creativity with market need.

Scientific creativity

The science in a new biotechnology company generally falls into '**discovery**' – you have discovered something wonderful – or '**platform technology**' – you can do something wonderful. In either case, first rate science is needed to found a first-rate company. Because living systems cannot be exactly modelled or predicted, genuinely new products must be created, at least in part, by experimental research. This must in turn be based on knowledge, creativity, and rigorous and systematic investigation, the hallmarks of good science in any context. We shall return to this theme several times below, because it is crucial.

Good science is not necessarily 'leading edge' science. Research has 'fashions' and, to an extent, the biotechnology industry follows the fashion because these areas of research or technologies are where senior researchers have chosen to work. But they are not the only, or even the most productive, areas where creativity can be exercised. It can be 'old science', carried out rigorously.

Nor, unfortunately, does it necessarily mean science that is captivating for the bench scientist to perform. However, it must conform to what most people would recognise as scientific 'good practice'. This is taking care that your experiments test your hypothesis rigorously, and using all the data and knowledge available to put the results in context. Several high-profile failures of biotechnology companies, notably some of the early 'products' for the treatment of sepsis using monoclonal antibodies, are now recognised as being due to companies pushing poor science in order to achieve funding goals.

Market need

Science on its own is not enough. We must sell it to someone – a 'market'. But what is a 'market need'? A general statement that, for example, 'People want a cure for AIDS' is not useful. Which people? Who will pay for it? How? How much? Will your product cure all cases of AIDS or only some? Just as scientific creativity cannot occur in a vacuum, so market research must research something specific. Biotechnology research and development is expensive, so it is important that a market for the intended product is big enough to give a return on all the investment needed.

12.3.2 The basic components

The market is the environment in which the company works – it is not a component of the company. Science is a central, critical component, but it is not the only one. For the biotechnologist, it is important to remember that the scientist does not have to provide all of the other features we will discuss below, but someone does. If the team initiating a biotechnology programme cannot provide an aspect of the successful commercialisation of a piece of science, then they should team up with someone else who can. This is the role that **seed venture companies** can provide, as can '**business angels**' – individuals who can bring their own wealth and business experience to a company as joint investors and directors.

12.3.3 People

A new company's need for excellent, motivated people who have commitment as well as skill and knowledge is paramount. Who is going to be the entrepreneur who makes this company happen? It may be the founding scientists, but they are not going to do it in any spare time left from an academic job. It is not going to be the scientific advisory board, who are there to advise and support the scientists. It is not going to be the Board of Directors. It needs someone to jump with both feet into the science and business and make sure that things happen.

In Europe the fear of failure has severely limited academics' inclination to do this. To a limited extent, the USA supports entrepreneurship even at the cost of failure – it is seen as meritorious to have 'had a go' and failed, because it proves motivation and drive, and the scientist who has tried and failed is unlikely to fail again in the same way, thus increasing the chances of success. In Europe, cultural conservatism means that failure is considered more significant than effort. People are therefore not willing to try for a major success if there is a significant risk of failure. This cultural barrier is disappearing slowly; the high media profile of successful scientific entrepreneurs is encouraging this cultural change and an increasing number of scientists are 'having a go'. But, in our experience, the large majority of researchers who want to see their science commercialised also are unwilling to jump whole-heartedly into that commercialisation themselves.

Although the central, driving entrepreneur is often a founding scientist with a 'good idea', it need not be. Packard Inc. were turned into a leader in the field of scientific analytical instrumentation by two business school graduates who, at the start, knew almost no science at all. Against the background of failures and successes in Europe and the USA in the last 10 years, experience shows that both business and scientific skills are essential for the success of a company, and that it is a rare scientist indeed who can combine both roles. No biotechnology company has been a commercial success when one person tried to combine both roles for more than the first 2–3 years.

12.3.4 Attitudes and culture

This 'jump-in-feet-first' approach from academia requires a major culture change. Academic science focuses on the subject, commercial science on the object. Academics typically address a topic or discipline, and follow it wherever it goes. It is the process of research that is important. Commercial science addresses a specific problem, and uses any tools or disciplines that are appropriate. It is the product that is king.

These apparently small differences in emphasis have major cultural effects. For example, there is little reward for an academic to be part of a multi-disciplinary team but it is essential for most commercial programmes. It is impossible for an academic scientist to be 'redundant', as by definition what they do is what they are meant to be doing. (They may be incompetent or unfundable, but that is different.) Industrial scientists can most definitely be redundant in the sense that their science, no matter how excellent, is no longer needed to achieve the company's aims. This is made more acute by the need for a company to focus on a small number of products or projects, while it is worth an academic group having at least as many projects as it has PhD students.

This is not the same as the choice between 'blue sky' and 'applied' research. Many companies carry out highly speculative research, and much academic work in biomedicine is, in essence, applied.

Some academics believe that these differences make science, in a commercial context, less attractive to the scientist. This old-fashioned view is now not very widely held because commercial science, and especially commercial science in a small company, can be an extraordinary place to do science for several reasons:

- the environment is intellectually stimulating, with hard problems to solve and many different disciplines being brought to bear to solve them;
- problems change fast;
- money is not usually a limit in developing excellent science, and state-of the art equipment and materials are in plentiful supply;
- there is a real opportunity for career development into any or all of the areas of science, technology or business the company is involved with;
- there is the chance of making substantial financial gains from your inventiveness (although not usually a large salary in the short term in a small company).

12.3.5 Strategy

Having found the people, and the great science, you must decide what you are going to do. This is your strategy, what you mean to do in the longer term, beyond the exigencies of day-to-day research. The strategy of a company is, of course, specific to that company, but we can frame the things that the strategy should address as questions. Some key strategic questions for a small company start-up are:

- What is your company's *specific* aim? 'Cure cancer' is not a reasonable strategic goal for a health-care company.
- What is your first product going to be? This is absolutely essential. Out of the cornucopia that your science could create, you must chose **one** thing to start with and focus most of your energies on that one. This 'focus' is critical for new science-driven companies. This means hard choices and it means dumping some 'pet projects'.
- How do you deal with success? Success in a research programme usually means having to start a development programme. Do you have the skills or funds to do this? If not, how are you going to get them?
- What will you do next? After your initial research programme has finished (with success or failure), do you have to fire all the scientists

or do you have another programme for them to move on to? Remember that a company science is focused on a particular problem, not on a discipline or process. Whatever those scientists have to do, it must fit in with the overall aims of the company and the company's competitive advantage (see below). Define what the key scientific advantage of the company is and hence what the scientists are going to be doing.

Do not confuse **'strategy'** with **'mission statement'**. The latter is a single phrase that encapsulates what you think you are about but it says nothing about why, how, or what you are going to do to get there. Some people think 'mission statements' are purely public relationship exercises for the company brochure.

12.3.6 Product vs. service vs. technology

A key aspect of your strategy is how your company is going to make money. In the 1980s it was every biotechnology company's stated dream to become a FIPCO – a 'Fully Integrated Pharmaceutical Company', like Pfizer or Roche, doing everything from basic discovery to trucking boxes of pills to doctors. This is very unrealistic. There are several more realistic goals

- **Product company.** You discover or invent products, take them as far through development as your funding allows, and then sell or license them to someone with experience in manufacture, distribution etc. Examples include all the larger 'first wave' biotech companies such as Amgen, Celltech and Chiroscience. In rare cases, these companies may seek to manufacture their own product. In even rarer ones they may go round doctor's surgeries selling it. But there are probably 400 000 general practitioners in Europe alone. Are you going to personally sell your pills to them all? If not, someone must work with you on that end of the business.
- **Tools company.** You develop tools or technologies that help other people develop products. Examples of such 'toolsets' include genomics and combinatorial chemistry. These are often also called 'technology platform' companies.
- **'Solution providers'.** You integrate several tools into one company. This is often achieved by the merger of two or more companies as any one start-up company may be excellent at one technique or approach but is unlikely to master all the 'tools' that a collaborator needs. Mergers and acquisitions are becoming more common in biotechnology and are an effective way of welding a lot of small brilliant companies into one (hopefully brilliant) whole. If this is your goal, you should say so from the start.

Your strategy as to which of these companies you wish to emulate will probably change. But you should at least have some idea today as the approaches the different companies take to product discovery and development are radically different, and it is that first product that will make or break your new company.

12.3.7 Success

Success is hard to define. Intellectual leadership is not the same as company success (witness the commercial success of Mutant Ninja Turtles and The Spice Girls). A strategy must be careful to define 'success' in a useful, meaningful way. What is your ultimate goal? (Remember 'cure cancer' or 'enhance shareholder value' are too vague to be useful.) What are significant steps along the way, and how will you show that you have passed them to the outside world? Defining criteria for success is very important, as it defines your commercial goals, and hence shapes your strategy in getting there.

By different criteria, the biotechnology industry as a whole either has been very successful or a dismal failure. Less than 1% of the first tier of (almost all US) biotechnology companies have become profitable on the basis of sales of products. However over 90% are still in existence as active, science-based companies. Over 60% would have given their initial investors an IRR of over 10% ('IRR' is 'Internal Rate of Return', a measure of the financial success of the investment – see below). For your start-up, these might be rather long-term goals. You might define success in terms of milestones along the way, such as flotation on a public stock market, signing a major collaboration with a pharmaceutical company, or entering your first product into Phase II clinical trials.

12.3.8 Competitive advantage

This is a trendy phrase from the management manuals of the 1980s that means that you can do something better than your competitors. What is it that you can do and no-one else (or, more realistically, very few other people) can do? What is it that, rather than merely being good at, you *excel* at? 'Excellence' is the watchword here, and it may come from one of five reasons.

You hold the patent on doing it. This is a powerful argument. Scientists should always patent an idea, process or invention that they think might be of some use to someone. The patent prohibits anyone else from 'practising' your patented invention without your agreement. It does not physically prevent anyone from copying your invention, but it makes it illegal to do so, and you can sue them if you have the time and money, and it is worth the effort. An example is the patent on polymerase chain reaction owned by Hoffman–La Roche, to whom anyone in the world using PCR for commercial purposes must pay a license fee or Roche will sue them.

You have the tools necessary to do it. This is as good as holding the patent in the short term, as it means that, while someone else could copy your process or invention in theory, in practice they cannot. Examples would be owning key cell lines, gene clones or production equipment. This, however, is only a competitive advantage until your competitors can either duplicate your tools, or find a way round using them, for example, by building their own production plant. A good tool to own is therefore one that inherently cannot be duplicated, like a unique genetic population.

You have the skills necessary to do it. Early practitioners in the science of *in vitro* fertilisation were in that position, as are those able to produce clones of mammals from adult cells today. This is a powerful competitive weapon until someone else learns how to do it. The skills base of this sort is sometimes called the company's 'intellectual capital'.

You have a lot of resources or money to do it. This is a weaker form of competitive advantage in biotechnology, because the industry is a knowledge-based one, not a resources based one. Many companies have lots of resources, especially major pharmaceutical or agricultural companies, and if your sole competitive advantage is that you have bought 20 DNA synthesisers and lots of computers and technicians to run them then you will shortly be out-competed by another company which can afford 30 DNA synthesisers.

You are the first to do it. This is the least attractive of all but it is often where biotechnology companies start. They see an opportunity and set up a company to exploit it. Their advantage is that they can move faster than anyone else. This only lasts while you keep moving.

Other forms of competitive advantage that apply to companies, such as Pepsico or Ford, such as having efficient factories or a recognised 'brand name', rarely apply to biotechnology companies.

The only way to show you have a competitive advantage is to demonstrate that what you want to do can actually work. In therapeutics discovery, this means proving that your material has some effect in people (remember that most drug discovery programmes fail). As it takes tens of millions of pounds to develop a product to the point of proving therapeutic efficacy, companies often have to accept a less rigorous proof, such as a sound scientific reason for supposing that it will work, evidence from animal tests that it works in animals, evidence that it is not actually harmful in humans (Phase I data). Each step along the path in Figs. 12.1 and 12.2 adds to the evidence for your competitive advantage.

12.3.9 Competitive intelligence

Part of proving that you have a competitive advantage is knowing how good you are compared to how good you have to be. This is competitive intelligence. Is there a medical need that you are going to satisfy and is someone else already filling that need? Is that need still going to be there in 10 years' time? This is a combination of finding out what the competition is doing, and what the market is. A surprising number of business proposals we have seen contain no evidence that their authors realise that the outside world exists, even less that it might contain competitors.

12.3.10 The business plan

Much of the above goes beyond 'strategy' and into tactics. Tactical planning should be carried out by a team of people bringing scientific, product development, business and financial skills, because all of these things are essential. The end product of this planning is a detailed plan of what your business is going to do – a **business plan**. But the business plan is a *product of planning*, not an end in itself. No matter how colourful or typographically creative it is, it is worthless if the planning behind it is not rigorous.

During the construction of a business plan, scientists must be aware that not only will bankers, accountants and the like be telling them what experiments they can and cannot do in the company, but that these people actually have a valid and useful viewpoint, and can sharpen and focus a company's plan substantially. (The decision on what is and is not good science, though, must rest with the scientists unless the bankers have post-doctoral laboratory experience, which is rare.) Typically, the stages that this process goes through are summarised below: we have considered several of them already.

- Identifying the science that will go into the company, according to the criteria summarised above.
- Defining what you are going to do with that science. This is the first part of the 'business plan', a document that should literally describe what the business plans to do. It should include consideration of
  - what can the science *really* do?
  - who is going to do it, and where?
  - who is going to manage them (ie make sure that everything happens)?
  - are there bits the company cannot do, or it is not sensible to do, and if so who is going to do them, and how will you pay them?
  - who will own the new intellectual property?
  - who will manage the development programmes?
  - what are the key milestones?
- Identifying the company's competitive advantage.
- How will the company be funded, and specifically,
  - how much money do you need to get started
  - where will you be when that runs out, and who will give you some more then?
- Who do you sell your product to, and by implication, what is your product. Is your strategy to generate intellectual property that you sell to another company, to develop drugs to Phase II clinical trials and then sell them to a drug company, to provide a service, or to get a product all the way to selling it to a high-street store. These paths take very different skills, and very different amounts of money.

and most importantly:

- **What happens if (when) it doesn't work?**

The last point can be hard for some scientists to accept but it is a fact that most science fails and you have to ask what happens to your company then? If it is a 'one product company' then when the product fails, the company fails, and everyone is out of a job. So it is wise to look round for other technologies you can bring in to the company. It is possible that, after a year, half of the brilliant science that led to forming the company has been abandoned! This should be viewed as evidence of growth and evolution, not failure, providing it has been replaced by something better.

The whole process boils down to identifying the shortest route

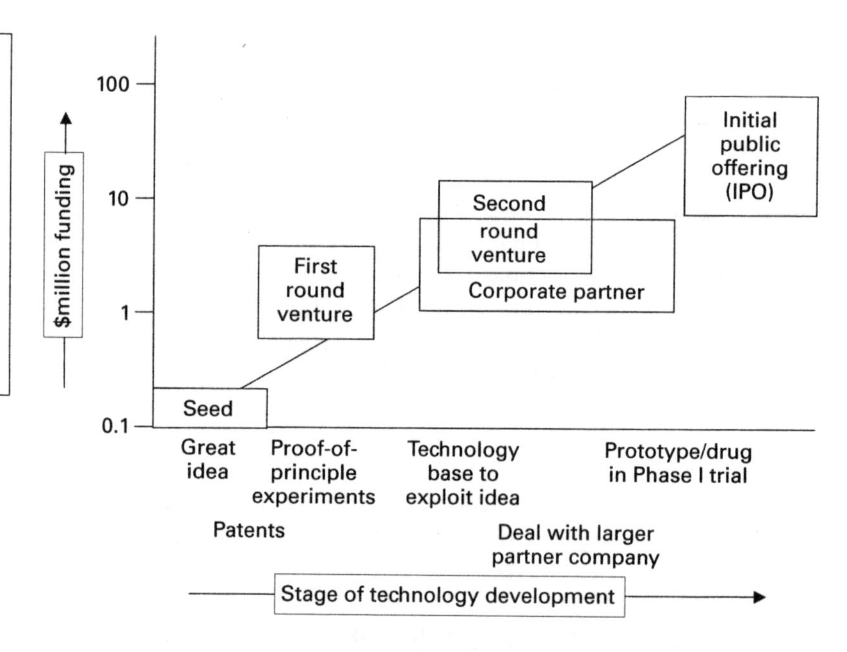

Fig. 12.3 Funding for European biotechnology start-ups. X axis: stage of development of the technology of a new, small biotechnology company. Y axis: level of funding the company can expect to receive following a venture capital funding route. The boxes illustrate the range of funds typically provided to European companies at different stages of technical development.

between where you are and where you want to go (but not taking scientifically unjustified short-cuts), with suitable cut-outs for when it does not work. This is why the strategy is important – you cannot define the shortest path to where you want to be until you know where that is. It also shows that in a company you cannot separate science from business issues.

The result of this process is a detailed plan of what the business will be doing, why it will survive and, preferably, flourish if given a certain amount of money. This forms the basis of an **investment proposal** – you take the plan to a funder, and say 'I propose you invest in this company because it will do *this* with the money'.

The business plan will almost certainly be wrong. Unforeseen events, from scientific breakthroughs or failures to stock market crashes, will derail your carefully laid plans. This should be expected, even embraced. But if you cannot think through where you might go if all goes well, the chances are that you will go nowhere.

12.4 Investment in biotechnology

Having defined what your company is to do, you need money to allow you to start. Very few biotechnology ideas can be realised in a way that requires no investment. Sometimes the investment is 'only' a few tens of thousands of pounds (or dollars or ecus) to make the first material you can sell. More usually it will take tens or hundreds of millions. Very few individuals can afford such sums, so you must convince other people to put money into your idea. These other people are the investors. There are several different types of investor depending on the stage your company is at and they will fund you in stages. Figure 12.3 illustrates the stages on a typical company funding path. Understanding this path and the motivations of the people that you will meet along its way is important if you are to get your biotechnology company funded.

An investor should not only be a source of funding. From an early stage, the investor should also help you found and run the business. They should provide help with issues such as employment contracts, location, finding funding for expensive equipment, securing the intellectual property and having discussions (or arguments) with all the other parties involved such as university technology transfer officers, patent lawyers, and the many researchers at the founding institution who have not had the courage to do this themselves but now want a piece of the action. All this can be pushed on to the business co-founder.

12.4.1 Seed investment

An increasingly common route to developing an idea into a company (which includes all the processes we alluded to above) is to seek **seed funding**.

Seed funding provides enough money to set the company up, acquire key patents, negotiate the graceful exit of the founding scientists from their current job and create a corporate entity. It also pays for planning and writing the business plan, a time consuming and skill-intensive business, often in terms of the investor's time and involves hiring lawyers, patent agents and accountants. Such seed funding is provided by private investors (see below) or specialist, professional seed funding companies, which are still rare in Europe, although more common in the USA. There is a general dearth of seed funding to take potential companies from 'I have this great idea' to 'This is a company you can believe in'. In part this is because the risks at this stage are huge and the rewards very uncertain. Merlin Ventures is the largest dedicated seed funding enterprise in Europe (as of mid 1998), and in two years we were able to seed only eight of the hundreds of companies that probably lie nascent in the British research establishment alone.

12.4.2 Private funding for biotechnology

Once you have a company it will probably need substantial amounts of money to pursue its product development goals. Start-up companies are usually funded privately, through investment by private transactions between the company and individuals or groups of individuals. Typically, such investments are through the issue of new shares, so new investors become shareholders and all the previous shareholders are 'diluted' (i.e. have their share in the company reduced).

Once the company exists, it can receive more substantial funding to carry out its plans as articulated in its business plan. This 'first round finance' (seed funding does not count as real money) comes from one of two sources.

Private investors

These are people with sufficient wealth to be able to put substantial amounts (usually at least $250000) into the company, take some active part in helping the company in financial or commercial terms and, most importantly, take the risk that they will lose their investment.

Venture capitalists (VCs)

These are people or companies who specialise in investment in risky propositions, usually early stage companies. They set up a fund into which people put their money and then the venture capital 'fund managers' invest that in high-risk ventures. This is exactly analogous to the investment trusts and funds that are common savings routes for the general public but with far higher risks and, the investors hope, far higher returns.

Both types of investor will want to check your business over for criteria which include the people involved, their 'due diligence' study on the science, the return on investment, exit route, and whether anyone else is willing to fund the idea.

People

Most VC groups will invest in people as much as in science. This is because the science on which a company is founded will almost certainly go wrong, and when it does it is the people who will either put it right or give up. The former is preferable. How someone goes about this is as much a matter of their general experience and personality as the specifics of the programme they want to get funded. 'Good' things in the scientific founders of a company are a *demonstrated* willingness to change fields, learn new expertise, collaborate with people from different disciplines, think of lateral things to do when challenged. A desire to make a lot of money is also good and a desire to be famous is usually bad, because it often ends up as being famous at the expense of the company rather than to its benefit.

Often a VC will also look for 'management', specifically a Chief Executive Officer (CEO). This person will have experience in running a science-based, commercial operation of some sort, will be accepted by the scientists as their company leader and is a credible person to put in front of bankers, accountants and other city professionals. A few VCs can perform the role of 'the suit' themselves (Merlin Ventures can, and does, for example), but most will not be willing to run the day-to-day operations of the dozen companies in their portfolio, and so will insist on at least a skeleton management being in the company from the start. This role can be played by the business angel or other seed investor.

Due diligence

After assuring themselves that the people are at least potentially suitable, the VC will carry out an external test of the science, by calling up experts, having any patents checked out by lawyers, asking around at meetings and conferences, and checking the perceived strength of the company's science, technology and people. This is known as 'due diligence' (after a legal phrase meaning in essence 'I have done whatever I can'). Due diligence can vary from a few chats in a bar to a full-scale consultancy project costing hundreds of thousands of pounds.

The due diligence process gives the VC an estimate for how reliable the current science is and what the market might be. Usually this will differ materially from the scientists' view. It is critical for the calculation of the 'value' of the company today, and hence a calculation of the **Return on Investment** (ROI, also sometimes called the Internal Rate of Return – IRR). This is the amount of money they will get out compared to the amount they put in, and is usually expressed as a percentage annual growth rate (like a bank might offer 8% to its savers). VCs usually look for ROIs of 50% per annum or more: this is not greed (or not only greed), but reflects the fact that this is the ROI they will get if everything works – usually, of course, it does not and they get an ROI of less than 0%.

It is also worthwhile for an entrepreneur to do 'due diligence' on the VC to see what they have done for people in the past, in terms of help with management, guidance in business and scientific strategy, building the company up so it can cope with its own success and contacts in the world of finance. Most say that they can do this, although it is the sad truth that few do.

Exit route

No-one putting money into a start-up biotechnology company expects to be paid back from the company's profits, at least for a minimum of 5 years, so there must be some other 'exit route' by which they get their money back. This can be

- privately selling your share in the company to someone else;
- getting bought by another company (merger or acquisition being different versions of a similar process);
- floating on the stock exchange (when the company is big and stable enough) and so in effect selling your shares to the general public.

All are possible for any company, if they are successful, so the question here is when will it happen. The 'when' is critical for calculation of ROI – a 100% gain in value in 1 year is 100% per annum ROI: a 200% gain in 4 years is a 50% ROI, even though the absolute amount of the latter is higher.

Funding stages

VCs usually invest when a company has already gained some seed funds, has developed its business plan, hired a couple of people, but has not got seriously under way. This is known as 'first round' or 'first stage' finance, and is typically between £0.5 million and £3 million. This is the riskiest end of venture capital. This money will typically take a company engaged in drug discovery and development through 1.5 to 3 years' work, and take the science from some basic research to a proof of principle. Then the company will need to raise more money, arranged in a second round finance with companies that specialise in that stage of investment. Second stage finance houses tend to lean more heavily on formal due diligence studies, look for an experienced management team in place, and look to the detailed timing of when they can float the company and so get their money back. Second round funding usually raises between £8 million and £15 million.

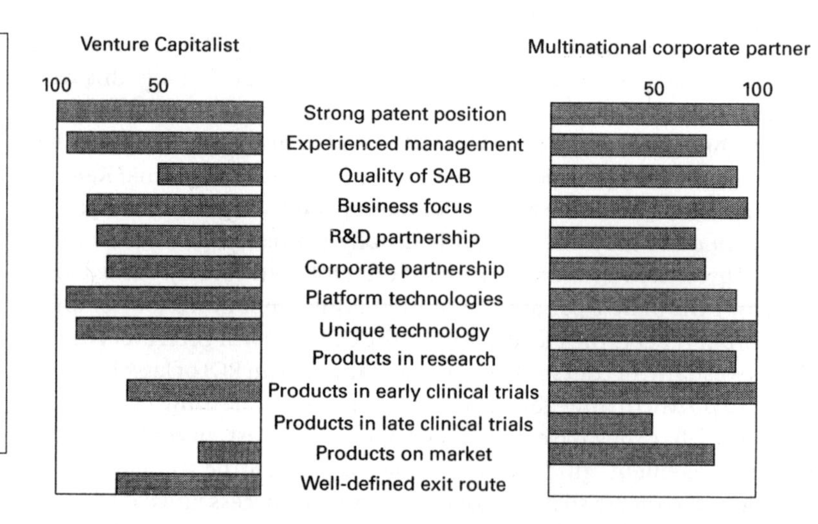

Fig. 12.4 What they look for in biotechnology companies. Summary of Ernst and Young survey of the fraction of Venture Capital investors (left) and Multinational partner company licensing negotiators (right) who have stated that specific aspects of a biotechnology company are essential to consideration of funding or collaborating with them respectively. Length of bar is proportional to percentage who considered this aspect important.

If all goes well, the company will then be floated on a stock exchange in another two or three years. This should raise £10–30 million. However it may need a 'top-up' funding to get it there – this is termed **Mezzanine financing**. Alternatively, things may go wrong with the science, requiring another round of private funding.

12.4.3 Corporate partners

The other main source of funds for your new company is other companies, and usually much larger ones. These may be clients (i.e. one who buy your products) but, in the early days, most biotech companies have no products. So larger companies may become partners with you in order to help you develop products. They benefit because you have something that can help them innovate. You benefit because they provide skills or infrastructure you do not have, such as the problem alluded to previously of distributing. Often corporate partners will also fund, organise and perform later stage clinical trials (which can be hugely expensive and complicated) as well.

In essence a corporate partner is a combination of collaborator and client. You get funds and resources, they get new programmes or products.

There are a huge variety of corporate partnership arrangements, from simple purchase of goods to outright purchase of the company. However the things that corporate partners will look for in your company are surprisingly similar to those a VC will look for, as illustrated in Fig. 12.4. Bear in mind that few companies have all of these – however if your start-up has *none* of them, you will have some problems getting it funded.

12.4.4 Grants

Occasionally agencies that provide grants to academics to perform research will also provide grants to biotechnology companies. However much more common is other types of government grant support aimed at such 'SMEs' (Small to Medium-sized Enterprises). The biotechnology industry is knowledge-based, clean, rapidly growing, and based in the West's long investment in its scientific and technological infrastructure. Also, much of it is addressing healthcare, probably the only sector of the economy to grow every decade this century. So biotechnology is seen as 'good', both socially and economically, and is encouraged with various degrees of vigour by governments.

This has led to a profusion of types of government support for new biotechnology from which start-up companies can benefit. These include:

- **Technology transfer schemes.** These are schemes to help transfer science or technology from (usually) a university setting into a commercial one. Following US models, European technology transfer offices are now becoming better equipped and better skilled to find the most appropriate route for commercialisation.
- **Small company support schemes**, such as SMART and SPUR in the UK. These are generic schemes to help small companies get off the ground with government financial assistance.
- **Regional development support.** This is government support to try to encourage industry to settle in one region rather than another. Only occasionally are these places where the best science and scientists are already based.
- **National and international co-ordination efforts.** These are attempts to get the technology policy or regional support geographically integrated. There are some commercially based schemes, such as EUREKA to encourage pan-European commercial developments.
- **Major infrastructure programmes with biotechnology relevance.** In both Europe and the USA the government supports major programmes of work which have biotechnology spin-offs, such as genome projects. These are usually 'pre-competitive'.

These can provide very substantial sums for companies especially if they are in areas of Europe targeted for economic development such as Liverpool or Sicily. However it is generally true that if the company is depending on grants to survive, then it was a bad economic idea from the start.

12.4.5 The stock market and biotechnology

More established companies can raise money from the general public by selling shares on a stock market, where suitably regulated brokers trade shares on behalf of their clients. Public funding in this way has very different constraints from private funding. It is very closely regulated to stop companies or brokers defrauding the public.

Shareholders have statutory rights that mean that they are the ultimate arbiters of the company's future, and many company brochures will talk about 'increasing shareholder value' as recognition that these people actually own the company. In principle shareholders can fire the board of directors (see below), or demand the company accounts for its actions, although in practice only major investment funds, which hold large blocks of shares, are in a position to exert any control over how the company is run.

Table 12.3 Major biotechnology public stock exchanges: major stock markets on which US and biotechnology companies' stocks are traded (i.e. are available for the general public to buy)

Stock exchange	Characteristics
NASDAQ	New York-based exchange for specialist, high-technology and higher-than-average risk companies, which lists several hundred biotechnology companies from around the world. The largest stock exchange worldwide for biotechnology companies.
London Stock Exchange (LSE)	Main London exchange ('The Stock Exchange' in London) that lists some larger companies such as Chiroscience. Usually companies must have a sales record or have at least two products in clinical trials to be allowed to list on LSE
Alternative Investment Market (AIM)	London-based attempt to have a market for smaller companies, in fact trades mostly in very small or very young companies, including some UK biotechnology companies
EASDAQ	Very new European 'clone' of NASDAQ, has yet to prove itself but promises well. Belgian biotechnology company Innogenetics was one of the first to list on EASDAQ
Frankfurt	Neuer Markt
Paris	Nouveau Marche

In order to get a biotechnology company 'listed' (i.e. have their name put on the list of shares available for trade), the company has to demonstrate that it is suitably stable. In the UK this means having a trading record for several years, or having at least two products in clinical trials, or a number of other criteria. It also means having a prospectus that has been verified by lawyers to say that every statement in it is demonstrably true, even down to the definition of chemical or medical terms. Part of this process requires an external group of experts write a report on the company, which in essence says that they, experts in the field, agree that what the company says makes sense – this is known (unsurprisingly) as **'the experts' report'**. Accounts have to be presented and audited, company directors have to sign legal forms that they are suitable people, and have to be checked out for past fraud offences and so on. This is all to protect the public from the worst excesses of entrepreneurship.

When and where you float your company is an arcane art. There are many different stock markets that can list a company (Table 12.3), and listing in market one does not imply listing in any of the others, as they all have slightly different rules and constituencies. Although their enthusiasm for biotechnology investments waxes and wanes roughly in unison, there can be substantial differences.

12.4.6 Valuing biotechnology companies

Public and private financing is by selling shares in your company. In essence, you sell a part of the company to someone in exchange for funds. But what are your shares worth? If someone is willing to give you £4 million, does this buy 5% of your company, or 95%? It depends on whether your company is worth £80 million or £4.2 million. Valuing your company appropriately is therefore important.

The details of how a value is placed on a company is beyond the scope of this article. In summary:

- There is no rational way of valuing a start-up company. You have some ideas, some patents, some people, and no premises, products, established programmes or track record. The overwhelming objective factor is the chance that your crucial first product will fail, scientifically or commercially, and this probability is a matter of opinion. Values are dominated by 'feel' and your credibility.
- When the company has been in business 3–4 years, has 40 employees and two products in late development, we can 'guesstimate' its value by working out what the company will be worth when it reaches its final goal and the chances it will make it there. Your goal may be to be bought out for $550 million or to generate a stream of new drugs that you will sell to a big pharmaceutical company. This gives you a final figure, and a guess for how long it will take to get there. You then multiply this by the probability of achieving it $\ll 1$), divide it by return that you could have earned investing the same money in a 'safe' investment over the same time (>1), and that is your value.
- If you are a public company, your value is the number of outstanding shares multiplied by whatever people will pay for them. This can lead your company to suddenly losing value because the share price drops and is the reason that reporters say that falls in the stock market have 'wiped billions off the value of industry'.

12.5 Who needs management?

'Management' is a word that has come up several times above. Why are investors so keen on management?

The scale of operations of a company is larger than in a research group. A drug discovery and development company can expect to grow to 30–50 people in 18 months, to over 100 people in 3 years, all working on essentially the same product or related groups of products. This cannot happen by chance – it must be organised. It must also be focused on a very specific goal. Company funding is based on success not on activity. If a line of research is not working someone has to make the hard decisions about what to do about it including, *in extremis*, firing the scientists involved.

This needs professional management – people who know how to organise and run a scientific programme with such defined goals. Sometimes the scientists can grow into this role. Sometimes they can accept it from an outsider recruited to the company specifically to manage it. But filling that role is an absolute condition of setting up a company, and companies without effective management almost always fail. Sometimes they take a long time and a lot of money to fail, which is why investors look for good management as part of the company team: without it, there is a very high probability that their money will be wasted.

This imposition of management is sometimes resented by scientists used to academic freedom, because they feel that they are 'giving up control' of their science. This is a fallacy for three reasons.

- They are not giving up control of anything – before the company was founded there was nothing there to control. No-one was developing the product, hiring the scientists, performing the work.
- It is not 'their' science. A successful company must be assembled from many scientific and technical strands, for reasons outlined above. They are a contributor, not a sole author.
- No one person is in control of a small company, if it is to perform with the energy, flexibility and enthusiasm that will carry it to success. It must be a team, not a dictatorship.

12.5.1 Where is management?

Finding appropriate management is difficult. You need quite different sorts of people at different stages of a company. The senior management, and particularly the Chief Executive Officer (CEO), of a new start-up with 10 employees must be able and willing to do everything, to do without formal reporting structures, and to know everything that goes on in the company. The manager of a public company of 400 employees must delegate nearly all of that and instead control a reporting and responsibility system that has several layers between him and the bench scientist. A company's management becomes more obvious, more structured and includes more people as the company grows.

As a company grows, the people who ran the company very well at one stage have to give way to ones who are competent to run it in the next. One of the skills of the entrepreneur who starts a small company is to know when their skills should be replaced by someone suited to run a more mature organisation.

Finding people who can perform these tasks, and particularly the many-sided and changing task of running a new start-up company, is hard. As in science, the only evidence that you can do it is a 'track record' of having done it before. The CEO is particularly critical, as he has overall responsibility for making the company work. CEOs for new biotechnology companies come from a variety of backgrounds, where their experience in management, in directing science and in relating to the needs and concerns of the board of a company fit them for the role. Academic research does not usually fit a scientist for such a role. Neither does being a management consultant (criticising how someone is performing is not the same as performing well yourself), nor does 'experience' of business gained solely through an MBA (Masters of Business Administration) course.

Critical tests for a CEO of a new biotechnology start-up could be caricatured as:

- **The *Nature* Test** – can they read *Nature* and understand what they are reading? This is critical, as the fundamental of the start-up is good science. (They probably will not have time to read *Nature* but that is another problem.)
- **The Lightbulb Test** – can they (and are they willing to) change the

lightbulb if it blows, i.e. do anything practical needed to keep the company running. There may be no-one else around to fix it.

- **The Cat-Herder Test** – can they convince a group of disparate scientists that what the CEO wants them to do is more worthwhile in scientific terms than what the scientists thought they wanted to do? (He or she can threaten to fire them but that will not capture the creativity and dedication of which the best scientists are capable.)
- **The Deal Test** – can the CEO go out and make deals that will bring the company money in return for a small amount of its technology of products? Such deals are critical for funding, but also to show that someone else has faith in you.
- **The Suit Test** – can the CEO put on a metaphorical (or literal) dark suit and convince investors that he or she is really on their side, so their investment is safe in his or her hands?

These general criteria apply to all the senior people in a small company. The 'head of molecular biology' in a start-up may find themselves watching the pilot plant or presenting to an investment banker who does not know what DNA stands for; there are few well-defined job descriptions in such an environment. This is half the fun of it.

As well as people who run the company as a whole, your start-up will need more specialist management functions such as financial and personnel management. Initially these will be provided by someone outside the company, such as the venture capital company backing the company, or by the CEO in his 'spare time'. As the company develops, a more specialised type of manager with less concern for science and more concern for management as a process and skill in its own right is needed. Scientists should note that these people *are needed*: they are not hired purely to make your life at the bench harder. Without them you might wake up one day and find that the company has run out of money.

This brings us into the realm of general management theory and practice, which this chapter will not discuss further. There are many books and courses on this available, some of them relevant to the unique environment of a small, science-based company.

12.5.2 Directors and others

In law, every company must have a Board of Directors. These people do exactly what their name implies – they direct the company so that its value to its shareholders is maximised. There are stringent laws about what company directors can and cannot do in general terms, and some financial scandals in major companies have involved directors abusing their position for their own gain.

The directors should add substantial value to the company, in terms of contacts, experience, advice and business acumen. Their role should emphatically not be just to rubber stamp what the CEO wants to do and for that reason it is considered bad practice for the Chairman of the Board to be the CEO. A venture capitalist looking to fund a company or a scientist looking to join it at a senior level will look at the Board to see whether they are there as window dressing or whether they will really help the company flourish.

Parallel to the Board, and often answering to it, most biotechnology companies have a Scientific Advisory Board (SAB). It should advise the CEO and directors on any technical aspect that the company needs guidance over and, specifically, provide perspective, contacts and advice on *all* areas of science that might be relevant to the company. For example, an agricultural genetics company might have an agrochemicals expert and a farmer among theirs.

12.6 Patents and biotechnology

Patents are critical to a small company based on knowledge. If you make an invention and it is not patented, anyone with suitable resources is free to come along and copy it. For a small company, many of your competitors will have far greater resources than you do and so can simply take your ideas and use them themselves. For this reason, investors and professional management are very eager to protect your intellectual property (IP) with suitable legal walls, and ideally with patents.

The process of patenting in the UK is beyond this article. In summary, the scientist, advised by someone who knows the language and law of patenting, has to submit ('file') a description of the invention to the patent office. The office's own examiners then check that the patent fulfils the three critical criteria:

- **Novelty** – no-one has done it before, or even talked about it in a realistic way.
- **Utility** – it must be useful for something. This means that the gene you discovered is not patentable in its own right but must be related to a product.
- **Enablement** – you must describe how someone else could do it (whatever it is).

If it passes, then the patent is granted. The process takes a long time, and costs a lot of money. In a novel field like biotechnology there is also a lot of debate (much of it carried out in the law courts) about just what 'an invention' is.

The critical part of the patent is the *exact* wording of the claims. The claims are a set of statements, usually at the end of the patent, which define exactly what it is you are patenting. If your claims are very 'broad' (i.e. general), you can succeed in getting a patent on a number of potential applications of your idea, not only one. The wording here is critical. Thus 'a nucleic acid' is broader than 'DNA' or 'a gene'; 'a molecule' is broader than 'an alcohol' which is broader than '2-methylbutan-1-ol' and so on. Of course, if your claim is too broad then the patent office will not allow it because it will not be novel: use of 2-methylbutan-1-ol as a cure for cancer may be novel but use of 'a molecule' certainly is not.

Your role in patenting is therefore to make sure that you have described your invention in terms of a final product – in the case of a gene sequence, maybe a diagnostic test for a genetic defect or a drug that blocks the action of that gene's protein product – and in terms that meet the criteria above. A good patent agent can be very helpful in this process and their help should be encouraged.

12.7 Conclusion: jumping the fence

This chapter has not been about 'entrepreneurship'. An **entrepreneur** is someone who can see a way to make all the things we describe above actually happen, and does it. The former needs knowledge, breadth of experience and contacts, but the latter is the most important, and really can be summed up in three words – Just Do It. Three words do not a chapter make, so we have focused on what an entrepreneur or business person should do to create a successful biotechnology business rather than the personal characteristics that make a scientist into an entrepreneur. But without the will to make it happen, none of this is relevant. This is why we have returned many times to the nature of the people involved rather than a formal process that will lead them gently and inevitably to success.

This is an extremely favourable time in history for new, fast-moving companies in high technology. Investors, regulators and governments all want to see your small company succeed. It is also a time of unprecedented technological change in the life sciences, and so the environment has never been better to build a life-science based company. For all its commercial failings and public misconception, the biotechnology industry will continue to be as dynamic and exciting a business sector as any in the next decade. It is a superb environment for a scientist to enter, for good science, the potential of substantial reward, and just plain fun.

12.8 Further reading

Glaser, V. and Hodgson, J. (1998). Before anyone knew the future nature of biotechnology. *Nature Biotechnol.* **16**, 239–242.

Lahteenmaki, R., Michael, A. and Hodgson, J. (1998). Public biotech: the numbers. *Nature Biotechnol.* **16**, 425–427.

Lee, K. B., Burrill, S. *et al.* (1997). *Biotech 97: alignment. The Eleventh Industry Annual Report*. Ernst and Young.

Mott, G. (1998). *Accounting for Non Accountants, 5th Edition*. Kogan Page Ltd., London.

Muller, A. *et al.* (1998). *European Life Sciences 98: Continental Shift. The Fifth Annual Ernst and Young Report on the European Entrepreneurial Life Sciences Industry*. Ernst and Young.

Pressman, D. and Elias, S. (1999). *Patent It Yourself, 7th Edition*. Nolo Press, New York.

Stacey, R. D. (1996). *Strategic Management and Organisational Dynamics, 2nd Edition*. Pitman, London.

White, S., Evans, P., Mihill, C. and Tysoe, M. (1993). *Hitting the Headlines: A Practical Guide to the Media*. British Psychological Society Books, Leicester, UK.

Williams, S. (1999). *Lloyds TSB Small Business Guide, 12th Edition*. Penguin Business, London.

Chapter 13

Amino acids

L. Eggeling, W. Pfefferle and H. Sahm

13.1 Introduction

The story of amino acid production started in 1908 when the chemist, Dr K. Ikeda, was working on the flavouring components of kelp. Kelp is traditionally very popular with the Japanese due to the specific taste of its preparations, kombu and katsuobushi (Fig. 13.1). After acid hydrolysis and fractionation of kelp, Dr Ikeda discovered that one specific fraction he had isolated consisted of glutamic acid which, after neutralisation with caustic soda, developed an entirely new, delicious taste. This was the birth of the use of monosodium glutamate as a flavour-enhancing compound. The production of monosodium glutamate (MSG) was soon commercialised by the Ajinomoto company based on its isolation from vegetable proteins such as soy or wheat protein. Since less than 1 kg MSG could be isolated from 10 kg of raw material the waste fraction was high. The chemical synthesis of D,L-glutamate, which had been partially successful, was also of little use since the sodium salt of the D-isomer is tasteless.

The breakthrough in the production of MSG was the isolation of a specific bacterium by Dr S. Udaka and Dr S. Kinoshita at Kyowa Hakko Kogyo in 1957. They screened for amino-acid-excreting micro-organisms and discovered that their isolate, No. 534, grown on a mineral salt

Fig. 13.1 The ideogram for kombu as it appears on kelp preparations used as a food component. The painting was kindly provided by Dr T. Ikeda (Ajinomoto), the grandson of Dr K. Ikeda.

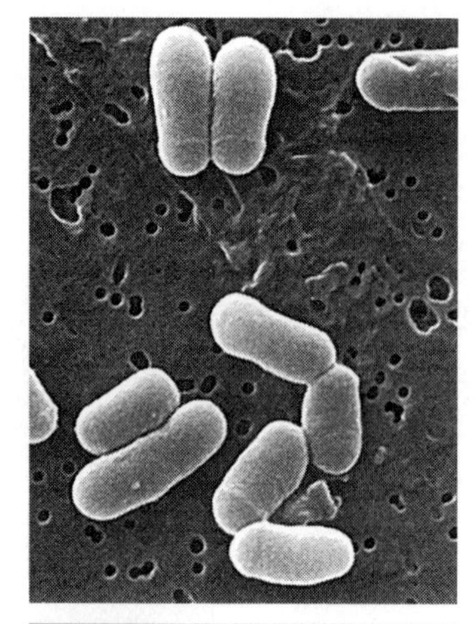

Fig. 13.2 Electron micrograph of *Corynebacterium glutamicum* showing the typical V-shape of two cells as a consequence of cell division.

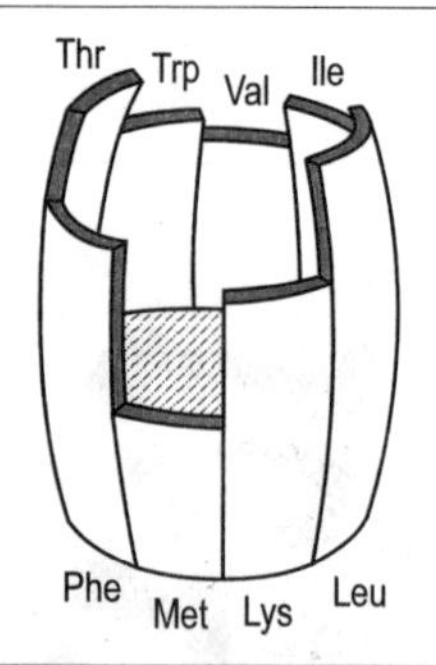

Fig. 13.3 The barrel represents the nutritive value of soybean meal, which is first limited by its methionine content.

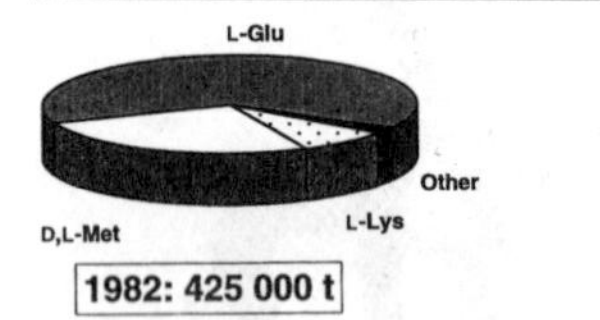

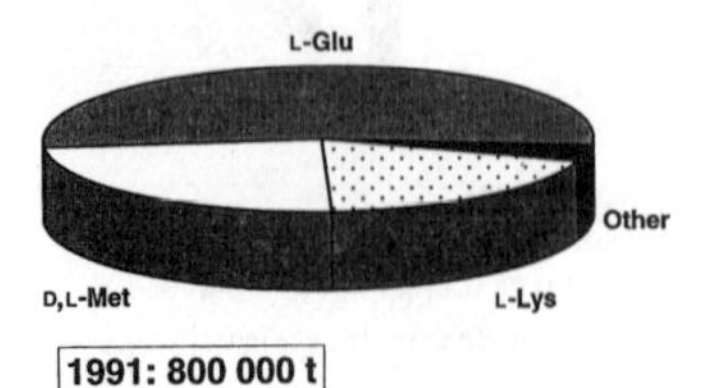

Fig. 13.4 The amino acid market doubles about every ten years. t = tonnes

medium excreted L-glutamate. It soon became apparent that the isolated organism needed biotin and that L-glutamate excretion was triggered by an insufficient supply of biotin. A number of bacteria with similar properties were also isolated, which are today all known by the species name *Corynebacterium glutamicum* (Fig. 13.2). *C. glutamicum* is a Gram-positive bacterium which can be isolated from soil. Together with genera like *Streptomyces*, *Propionibacterium* or *Arthrobacter*, it belongs to the actinomycetes subdivision of Gram-positive bacteria. The successful commercialisation of MSG production with this bacterium provided a big boost for amino acid production with *C. glutamicum* and later with other bacteria like *E. coli* as well. Nucleotide production for use as flavour enhancers also developed rapidly in the 1970s with *C. ammoniagenes*, which is closely related to *C. glutamicum*. The production mutants and the processes developed also resulted in a demand for sophisticated fermentation devices. Consequently, the development of amino acid technology was an incentive for the fermentation industry in general.

13.2 Commercial use of amino acids

Amino acids are used for a variety of purposes. The food industry requires L-glutamate as a flavour enhancer, and glycine as a sweetener in juices, for instance (Table 13.1). The chemical industry requires amino acids as building blocks for a diversity of compounds. The pharmaceutical industry requires the amino acids themselves in infusions – in particular the essential amino acids – or in special dietary food. And last but not least, a large market for amino acids is their use as animal feed additive. The reason is that typical feedstuffs, such as soybean meal for pigs, are poor in some essential amino acids, like methionine, for instance. This is illustrated in Fig. 13.3 where the nutritive value of soybean meal is given by the barrel but the use of the total barrel is limited by the stave representing methionine. Methionine is added for this reason, and considerably increases the effectiveness of the feed. The addition of as little as 10 kg methionine per tonne increases the protein quality of the feed just as effectively as adding 160 kg soybean meal or 56 kg fish meal. The first limiting amino acid in feed based on crops and oil seed is usually L-methionine, followed by L-lysine, and L-threonine. Another aspect of feed supplementation is that with a balanced amino acid content the manure contains less nitrogen thus reducing environmental pollution.

Over the years the demand for amino acids has increased dramatically. The market is growing steadily by about 5 to 10 per cent per year. Thus, within 10 years the total market has approximately doubled (Fig. 13.4). Some amino acids, such as L-lysine, which is required as a feed additive, display a particularly great increase. The world market for this amino acid has increased more than 20-fold in the past two decades. Other amino acids have appeared on the market, like L-threonine, L-aspartate or L-phenylalanine, the latter two being required for the synthesis of the newly developed sweetener aspartame. Estimates for

Table 13.1 Current amounts of amino acids produced

Production scale (tonnes y^{-1})	Amino acid	Preferred production method	Main use
800000	L-Glutamic acid	Fermentation	Flavour enhancer
350000	L-Lysine	Fermentation	Feed additive
350000	D,L-Methionine	Chemical synthesis	Feed additive
10000	L-Aspartate	Enzymatic catalysis	Aspartame
10000	L-Phenylalanine	Fermentation	Aspartame
15000	L-Threonine	Fermentation	Feed additive
10000	Glycine	Chemical synthesis	Food additive, sweetener
3000	L-Cysteine	Reduction of cystine	Food additive, pharmaceutical
1000	L-Arginine	Fermentation, extraction	Pharmaceutical
500	L-Leucine	Fermentation, extraction	Pharmaceutical
500	L-Valine	Fermentation, extraction	Pesticides, pharmaceutical
300	L-Tryptophan	Whole cell process	Pharmaceutical
300	L-Isoleucine	Fermentation, extraction	Pharmaceutical

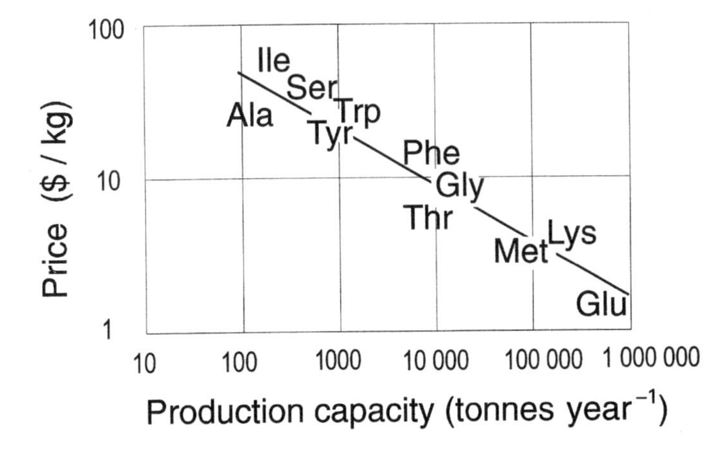

Fig. 13.5 The amino acids with the largest market are the cheapest.

current worldwide demand for the most relevant amino acids are given in Table 13.1. L-Glutamate continues to occupy the top position followed by L-lysine together with D,L-methionine, while the other amino acids trail behind at a considerable distance.

There is a close interaction between the prices of the amino acids and the dynamics of the market. More efficient fermentation technology can provide cheaper products and hence boost demand. This in turn will lead to production on a larger scale with a further reduction of costs. However, since the supply of some amino acids, e.g. L-lysine, as a feed additive is directly competitive with soybean meal (the natural L-lysine source) there are considerable fluctuations in the amino acid demand depending on the crop yields. The amino acids produced in the largest quantities are also the cheapest (Fig. 13.5). The low prices in turn dictate the location of the production plants. The main factors governing the location of production plants are the price of the carbon source and the local market. Large L-glutamate production plants are spread all over the world, with a significant presence in the Far East, e.g. Thailand and Indonesia. For L-lysine the situation is different. Since one-third of the world market is in North America and there is convenient access to maize as a feedstock material for the fermentation process, about one-third of the L-lysine production capacity is located there. In almost all cases, the companies producing L-lysine are associated with the maize milling industry, either as producers, in joint ventures or as suppliers of cheap sugar. This illustrates the fact that the commercial production of amino acids is a vigorously growing and changing field with many global interactions.

13.3 Production methods and tools

Some amino acids are chemically synthesised, such as glycine, which has no stereochemical centre, or D, L-methionine. This latter sulphur-containing amino acid can be added to feed as a racemic mixture, since animals contain a D-amino acid oxidase which, together with a transaminase activity, converts D-methionine to the nutritively effective L-form. The classical procedure of amino acid isolation from acid hydrolysates of proteins is still in use for selected amino acids with a low market volume, e.g. L-cysteine (Table 13.1). Other methods in use are those of precursor conversion with bacteria, or enzymatic synthesis. However, for L-amino acids required in large volumes, fermentation production with bacteria is the method of choice.

Classical strain development

However, bacteria do not normally excrete amino acids in significant amounts because regulatory mechanisms control the amino acid synthesis in an economical way. Therefore, mutants have to be generated which over-synthesise the respective amino acid. A great number of amino-acid-producing bacteria have been derived by mutagenesis and screening programmes. This has involved the consecutive application of:

- undirected mutagenesis;
- selection for a specific phenotype;
- selection of the mutant with the best amino acid accumulation.

Taking the best resulting strain, the entire procedure was repeated over several additional rounds to increase the productivity each time, and, eventually, resulted in an industrial producer (see Table 13.2 as an example). Due to this optimisation over several decades, together with the accompanying process adaptation, excellent high-performance strains are now available. They certainly carry a variety of unknown mutations also decisive for their production properties, as will become evident from the examples described below.

Application of recombinant techniques

In conjunction with this classical technique for strain development, recombinant DNA techniques are also applied. They serve

- to rapidly develop new producers by increasing limiting enzyme activities;
- to analyse mechanisms of flux control;
- to combine this knowledge with classically obtained strains for their further development.

Intracellular flux analysis

An exciting new approach in strain development combining both the genetic and classical procedure is the reliable quantification of the carbon fluxes in the living cell. A great deal of progress has been made here recently in developing to a high level of sophistication the old isotope labelling technique. In particular, with ^{13}C-NMR spectroscopy the intracellular fluxes were quantified to extreme high resolution. For instance, in *C. glutamicum* it has even been possible to quantify the exchange flux rates as are present in the pentose phosphate pathway. The method is described in detail in Chapter 2 of this book. Such flux identifications are of major assistance in selecting the reactions in the central metabolism to be modified by genetic engineering.

Functional genomics

Another tool whose potential is only now being exploited is the genome analysis of producer strains. The availability of the entire sequence of the chromosomes from *C. glutamicum* and *E. coli* opens up exciting possibilities to compare mutants and to uncover new mutations essential for high overproduction of metabolites. For instance, RNA analysis using chip technology will make it possible to detect whether a specific gene is altered in its expression for producers of different efficiency. New mutations and genes might thus be discovered which are not directly concerned with carbon fluxes, but rather with total cell control, or are involved in energy metabolism. Chip technology will also make it possible to use genome analysis as a tool to qualify individual fermentations, thus resulting in still further improvements and consolidations of the production processes.

13.4 L-Glutamate

As already mentioned, L-glutamate was the first amino acid to be produced. The very successful production still exclusively uses the original bacterium *C. glutamicum*. As metabolic pathways *C. glutamicum* uses glycolysis, the pentose phosphate pathway and the citric acid cycle to generate precursor metabolites and reduced pyridine nucleotides. However, this bacterium displays a special feature in the **anaplerotic reactions** of the citric acid cycle (Fig. 13.6). Since L-glutamate is directly derived from α-ketoglutarate, a high capability for replenishing the citric acid cycle is, of course, a prerequisite for high glutamate production. It was originally assumed that only the phospho*enol*pyruvate carboxylase is present as a carboxylating enzyme within the anaplerotic reactions. However, molecular research in close conjunction with

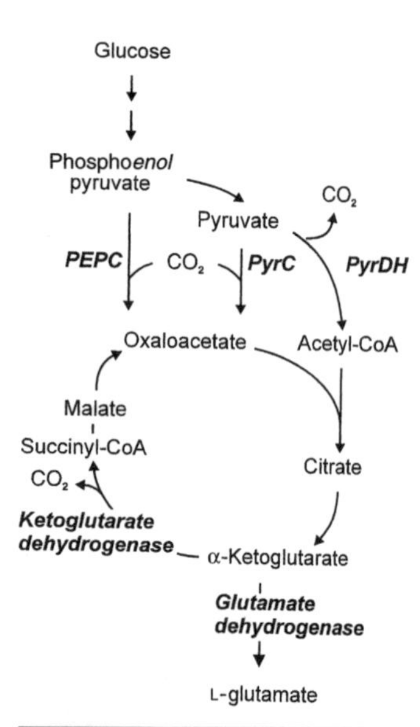

Fig. 13.6 Sketch of main reactions in *C. glutamicum* connected with the citric acid cycle and of relevance for L-glutamate production. Abbreviations: PyrDH, pyruvate dehydrogenase; PyrC, pyruvate carboxylase; PEPC; phospho*enol*pyruvate carboxylase.

^{13}C-labelling studies and flux analysis showed that an additional carboxylating reaction must be present. The pursuit of this enzyme activity resulted in the detection of pyruvate carboxylase activity, PyrC, and the cloning of its gene. This carboxylase was not detected by the original enzyme measurements since it is very unstable in crude extracts. Its detection requires an *in situ* enzyme assay using carefully permeabilised cells. Therefore, *C. glutamicum* has the pyruvate dehydrogenase (PyrDH) shuffling acetyl-CoA into the citric acid cycle but two enzymes supplying oxaloacetate: pyruvate carboxylase (PyrC) together with a phospho*enol*pyruvate carboxylase (PEPC) (Fig. 13.6). The successful cloning of both genes together with mutant studies showed that both carboxylases can basically replace each other to ensure conversion of glucose-derived C3-units to oxaloacetate. This is different from *E. coli*, which has exclusively the phospho*enol*pyruvate carboxylase serving this purpose, or *Bacillus subtilis*, where only the pyruvate carboxylase is present. Since *C. glutamicum* possesses both enzymes, it has an enormous flexibility for replenishing citric acid cycle intermediates upon their withdrawal.

The reductive amination of α-ketoglutarate to yield L-glutamate is catalysed by glutamate dehydrogenase. The enzyme is a multimer, each subunit having a molecular weight of 49 100. It has a high specific activity of 1.8 mmol $min^{-1} \cdot$mg protein, and L-glutamate is present in the cell in a rather high concentration of about 150 mM. In the case of other amino acids, in contrast, the intracellular concentrations are usually below 10 mM. The high concentration serves to ensure the supply of L-glutamate directly required for cell synthesis and also for the supply of amino groups via transaminase reactions for a variety of cellular reactions. As much as 70% of the amino groups in cell material stems from L-glutamate.

13.4.1 Production strains

For the biotechnological production of L-glutamate the intracellularly synthesised amino acid must be released from the cell. This is, of course, usually not the case since the charged L-glutamate is retained by the cytoplasmic membrane, otherwise the cell would not be viable. However, as shown by the special circumstances in discovering *C. glutamicum*, L-glutamate is already excreted when biotin is limiting. This striking fact is based on two essential characteristics:

- a carrier is present mediating the active excretion of L-glutamate;
- the lipid environment of this carrier triggers its activity.

A specific carrier is required since otherwise, in addition to the charged L-glutamate, other metabolites and ions would also leak from the cell. Moreover, only an active export enables the energy-dependent 'uphill' transport of L-glutamate from inside the cell (0.15 M) towards the very high concentrations obtained in fermentation broths (more than 1 M). However, for practical purposes, the triggering of active export by the appropriate molecular environment of the cytoplasmic membrane is important. The switches for tuning this environment and thus eliciting glutamate export are surprisingly diverse: (i) growth under biotin limitation, (ii) addition of local anaesthetics, (iii) addition of penicillin, (iv) addition of surfactants, (v) use of oleic acid auxotrophs, and (vi) use of glycerol auxotrophs. All of these means trigger L-glutamate excretion. Although, overall, there are as yet no completely conclusive ideas on the molecular changes thus caused, nevertheless in the classical biotin effect part of the causal link to glutamate excretion is well understood. Biotin is a cofactor of the acetyl-CoA carboxylase. With limited supply, the activity of this enzyme is thus decreased and consequently the fatty acid synthesis is diminished. This leads to a decreased availability of phospholipids and a greatly decreased lipid to protein ratio in the membrane as well as a change in the degree of saturation of the fatty acids. Under biotin limitation the phospholipid content is drastically decreased from 32 to 17 nmol mg^{-1} dry weight, and the content of the unsaturated oleic acid increased relative to the saturated palmitic acid by 45%. This represents a severe alteration of the physical state of the membrane which thus dramatically alters L-glutamate efflux. The membrane composition is also affected in oleic acid and glycerol auxotrophic mutants. The use of such mutants enables the production of monosodium glutamate from substrates which may be rich in biotin as well.

Apart from the export process and high glutamate dehydrogenase activity, another key reaction is that of α-ketoglutarate dehydrogenase (Fig. 13.6). This enzyme has a weak activity in *C. glutamicum* and it is also unstable. Therefore, under those conditions that result in glutamate efflux, the activity of this enzyme is also diminished. Exposing the cell to either penicillin, surfactants or biotin-limitation reduces the α-ketoglutarate dehydrogenase activity up to a residual activity of only 10%, whereas the activity of the glutamate dehydrogenase is hardly affected. The competing α-ketoglutarate dehydrogenase activity is therefore lowered, thus preventing an excess conversion of α-ketoglutarate to succinyl-CoA, and therefore favouring its conversion to L-glutamate.

13.4.2 Production process

The most relevant factors influencing L-glutamate formation are the ammonium concentration, the dissolved O_2 concentration and the pH. Although, in total, a large amount of ammonium is neccessary for sugar conversion to L-glutamate, a high concentration is inhibitory to growth as well to the production of L-glutamate. Therefore, ammonium is added in a low concentration at the beginning of the fermentation and is then added continuously during the course of the fermentation. The oxygen concentration is controlled, since under conditions of insufficient oxygen, the production of L-glutamate is poor and lactic acid as well as succinic acid accumulates, whereas with an excess oxygen supply the amount of α-ketoglutarate as a by-product accumulates. A flow diagram of the process is shown in Fig. 13.7.

For the actual fermentation the production strains are grown in fermenters as large as 500 m^3 (Fig. 13.8). After precultivation, the onset of L-glutamate excretion is controlled by the addition of surfactants like polyoxyethylene sorbitan monopalmitate (Tween 40). Yields of 60–70%

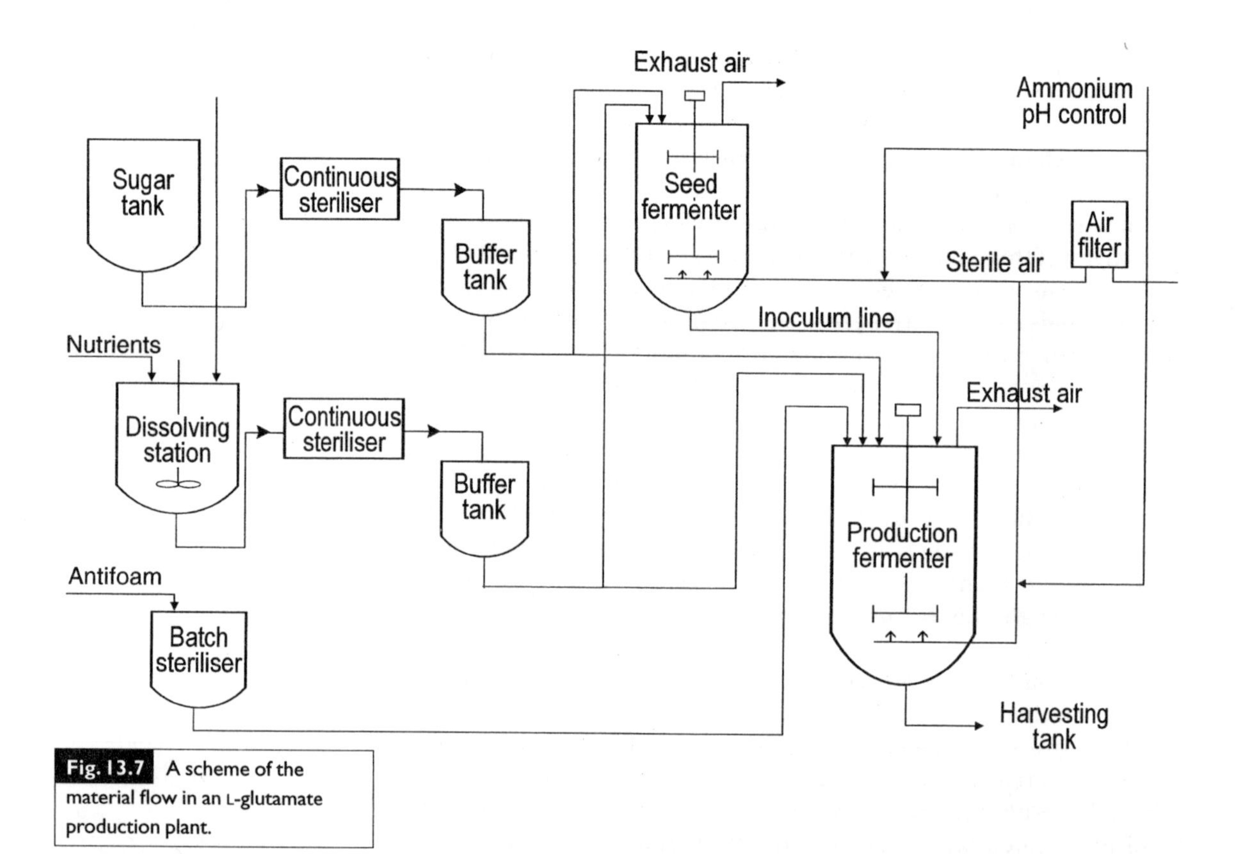

Fig. 13.7 A scheme of the material flow in an L-glutamate production plant.

Fig. 13.8 Amino acid production plant of Kyowa Hakko in Japan showing on the right seven large fermenters each 240 m^3 in size, suitable for L-glutamate production.

L-glutamate, based on the glucose used, have been reported. At the end of the fermentation the broth contains L- glutamate in the form of its ammonium salt. In a typical downstream process, the cells are separated and the broth is passed through a basic anion exchange resin. L-Glutamate anions will be bound to the resin and ammonia will be released. This ammonia can be recovered via distillation and reused in the fermentation. Elution is performed with NaOH to directly form MSG in the solution and to regenerate the basic anion exchanger. From the eluates, MSG may be crystallised directly followed by further conditioning steps like decolorisation and sieving to yield a food-grade quality.

13.5 L-Lysine

The second amino acid made exclusively with *C. glutamicum*, or its subspecies *lactofermentum* and *flavum*, is L-lysine. The carbons of L-lysine are derived in the central metabolism from pyruvate and oxaloacetate (Fig. 13.9). In contrast to the special situation with L-glutamate, where practically only a single reaction represents the synthesis pathway, L-lysine is synthesised via a long pathway. Moreover, the first two steps of L-lysine synthesis are shared with that of the other members of the aspartate family of amino acids: L-methionine, L-threonine and L-isoleucine.

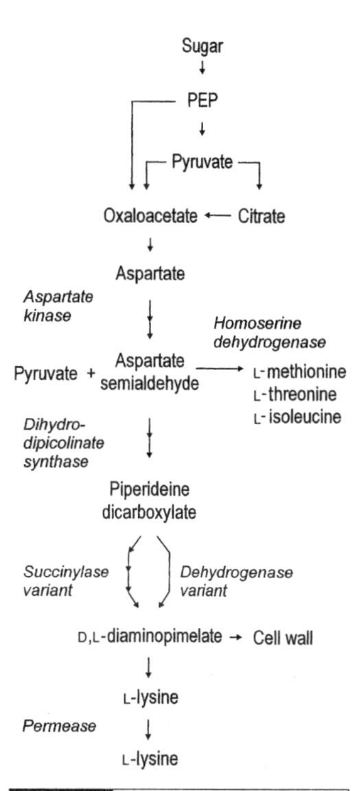

Fig. 13.9 L-Lysine biosynthesis in *C. glutamicum* with the reactions supplying oxaloacetate and pyruvate as precursors. PEP, phosphoenolpyruvate.

The kinase initiating lysine synthesis is feedback-inhibited by lysine plus threonine

The first reaction initiating L-lysine synthesis is catalysed by aspartate kinase. As is typical of an enzyme at the start of a lengthy synthesis pathway, aspartate kinase is controlled in its catalytic activity. The enzyme is inactive when L-lysine plus L-threonine together are present in excess, thus providing a feedback signal (see Chapter 2) concerning the availability of these two major metabolites of the aspartate family of amino acids. The kinase has an interesting structure. It consists of 2 α-subunits of 421 amino acid residues each, and 2 β-subunits of 171 amino acid residues. An exciting discovery was that the amino acid sequence of the β-subunit is identical to that in the carboxyterminal part of the α-subunit. The molecular basis is that the gene for the smaller β-subunit, *lysC*β, is an in-frame constituent part of the larger α-subunit (Fig. 13.10). Thus two promoters are present at this locus: one driving *lysC*α expression together with that of the downstream gene, *asd*, and one driving *lysC*β and *asd* expression. The regulatory features of the kinase reside in the β-subunit. Thus specifically altering the β-subunit structure, or those of both subunits together in their carboxyterminal part, results in a kinase which is no longer feedback regulated by L-lysine plus L-threonine. With such an insensitive kinase, *C. glutamicum* already excretes some L-lysine, showing the rather simple type of flux control in this organism.

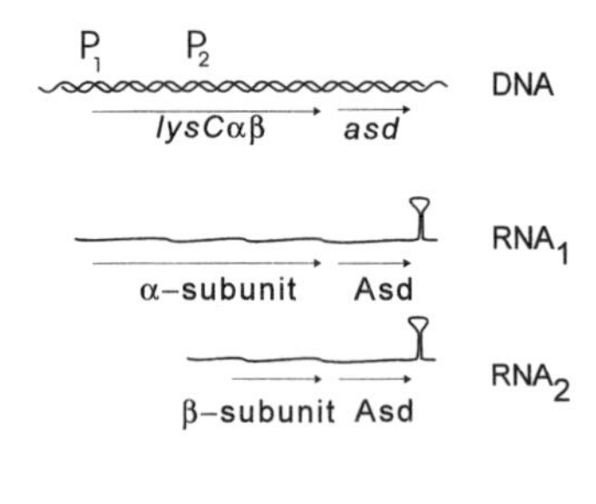

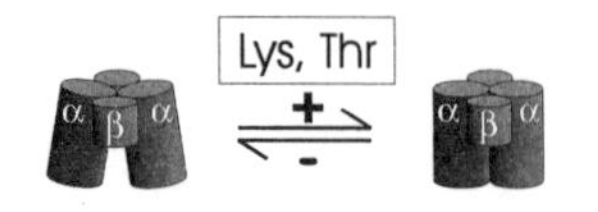

Active kinase

Inactive kinase

Fig. 13.10 The *lysCasd* operon of *C. glutamicum* and allosteric control of the kinase. The second promoter within *lysC* results in formation of the β subunit constituting the regulatory subunit of the kinase protein of $\alpha_2\beta_2$ structure.

The synthase limits flux

A further important step of flux control within lysine biosynthesis is at the level of aspartate semialdehyde distribution. The dihydrodipicolinate synthase activity competes with the homoserine dehydrogenase for the aspartate semialdehyde (Fig. 13.9). In *C. glutamicum*, the synthase is not regulated in its catalytic activity as is the corresponding enzyme in *E. coli*, for example. Instead, in *C. glutamicum* it is the amount of the protein which directly controls the flux. This is thus different from the kinase where the catalytic activity is regulated by L-lysine and thereby controls the flux at a constant amount of protein. Graded overexpression of the synthase gene, *dapA*, has shown that with an increasing amount of synthase a graded flux increase towards L-lysine is the result. Surprisingly, *dapA* overexpression also has a second consequence: the flux of aspartate semialdehyde into the branch leading to homoserine is already diminished with just two *dapA* copies. Due to the shortage of the homoserine-derived amino acids, this results in a weak growth limitation which is advantageous for L-lysine formation, since now more intermediates of the central metabolism are used for lysine synthesis instead for cell proliferation.

Lysine synthesis is split which ensures proper cell wall formation

A remarkable feature of *C. glutamicum* is its split pathway of L-lysine synthesis. At the level of piperideine-2,6-dicarboxylate, flux is possible either via the 4-step succinylase variant or the 1-step dehydrogenase variant (Fig. 13.9). In contrast, *E. coli*, for example, has only the succinylase variant and *Bacillus macerans* only the dehydrogenase variant. The flux distribution via both pathways has been quantified in a study using NMR spectroscopy and [1-^{13}C]glucose as the substrate. Surprisingly, the flux distribution is variable (Fig. 13.11). Whereas at the start of the cultivation about three-quarters of the L-lysine is made via the dehydrogenase variant, at the end the newly synthesised L-lysine is almost exclusively made via the succinylase route. There is a mechanistic reason for this. As kinetic characterisations have shown, the dehydrogenase has a weak affinity towards its substrate, ammonium, with a K_m of 28 mM. Thus at low ammonium concentrations, as are present at the end of the fermentation, the dehydrogenase cannot contribute to L-lysine formation. Instead, flux via the succinylase variant is favoured, where after succinylation of piperideine-2,6-dicarboxylate, a transaminase incorporates the second amino group into the final L-lysine molecule.

Dehydrogenase variant

Succinylase variant

Variant use

Lysine

Fermentation time

Fig. 13.11 At the beginning of the L-lysine fermentation use prevails of the dehydrogenase variant over that of the succinylase variant, whereas at the end the succinylase variant is used almost exclusively. Variant use is in percent.

The key to understanding this luxurious pathway construction is provided by the amino acid D,L-diaminopimelate. This amino acid is required for the synthesis of the activated muramyl peptide L-Ala-γ, D-Glu-D,L-Dap, which is one of the linking units in the peptidoglycan of the cell wall. Upon inactivation of the succinylase variant, a radical change to the cell morphology becomes apparent with low nitrogen supply. The cells are elongated, and furthermore less resistant to mechanical stress. If either the succinylase or the dehydrogenase variant is inactivated, L-lysine accumulation is reduced to 40%. Thus

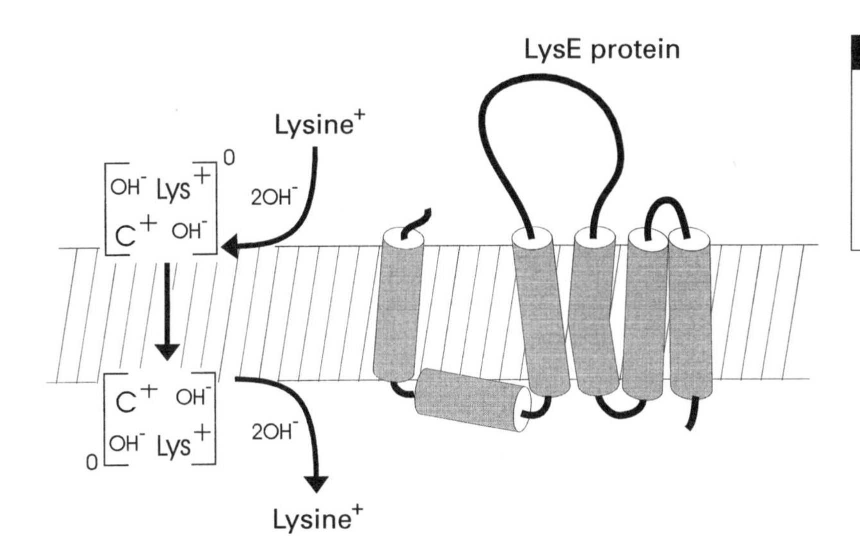

Fig. 13.12 Topology of the L-lysine exporter showing its five membrane spanning helices and the additional hydrophobic segment. The formally distinct steps of the translocation process driven by the membrane potential are included.

both variants together ensure the proper supply of the crucial linking unit D,L-diaminopimelate, as well as a high throughput for L-lysine formation. The split pathway in *C. glutamicum* is an example of an important principle in microbial physiology: pathway variants are generally not redundant but evolved to provide key metabolites under different environmental conditions.

Export of L-lysine

Amino acid transport has long been investigated in bacteria but, principally, this is only their import. In contrast, the molecular basis for amino acid export was completely unknown until 1996 since a specific export process appeared nonsensical. The breakthrough was achieved by the cloning of the lysine export carrier from *C. glutamicum*, which at one blow enabled amazing discoveries concerning the nature and relevance of a new type of exporter. The L-lysine carrier, LysE, is a comparatively small membrane protein of 25.4 Da. It has the transmembrane spanning helices typical of carriers, but only five of them (Fig. 13.12). A sixth hydrophobic segment is located between helix one and three and may dip into the membrane or be surface localised. Several distinct steps are involved in the translocation mechanism, which probably requires the dimerisation of LysE. These are: (i) the loading of the negatively charged carrier with its substrate L-lysine together with two hydroxyl ions, (ii) substrate translocation via the membrane, (iii) the release of L-lysine and the accompanying ions at the outside of the membrane, and finally, (iv) the reorientation of the carrier. The driving force for the entire translocation process is the membrane potential, $\Delta\Psi$, required for the reorientation of the carrier.

Access to the lysine-exporter gene, *lysE*, has also made it possible to solve the puzzle as to why *C. glutamicum* has such an exporter at all. In a *lysE* deletion mutant supplied with glucose and 1 mM of the dipeptide, lysyl-alanine, an extraordinarily high intracellular L-lysine concentration of more than 1 M accumulates, abolishing growth of the mutant.

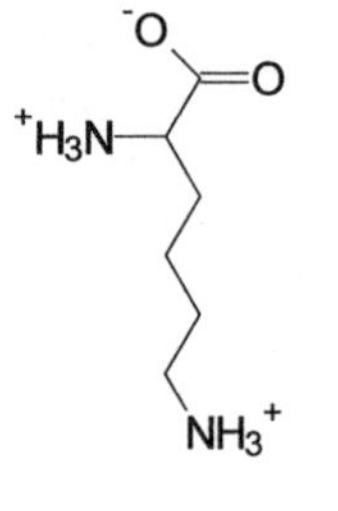

Lysine

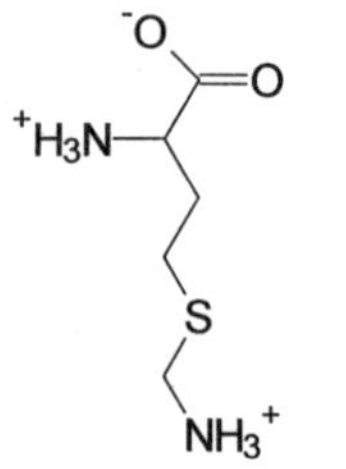

Aminoethyl-
cysteine

Fig. 13.13 Aminoethyl cysteine is a sulphur-containing analogue of L-lysine for generating mutants deregulated in L-lysine synthesis.

Thus, the exporter serves as a valve to excrete any excess intracellular L-lysine that may arise in the natural environment in the presence of peptides. As in the case of other bacteria, too, *C. glutamicum* has active peptide-uptake systems as well as hydrolysing enzymes giving access to the amino acids as valuable building blocks. However, *C. glutamicum* has no L-lysine-degrading activities and therefore must prevent any piling up of L-lysine. This also happens in the lysine producer strains where the biosynthesis pathway is mutated. As genome projects have now shown, homologous structures of the L-lysine carrier LysE are present in various Gram-negative and Gram-positive bacteria. Therefore, this type of intracellular amino acid control by an exporter is expected to be present in other bacteria, too. Since the LysE structure is not shared with other translocators, LysE also represents a new superfamily of translocators, which is probably related to its new function.

13.5.1 Production strains

L-Lysine producer strains have been derived over the decades by mutagenesis to give strains excreting more than 170 g L-lysine per litre. It is clear that these strains carry a long list of phenotypic characters to achieve this massive flux directioning (Table 13.2). Typically, the strains are resistant or sensitive to some analogue of lysine. A typical feature of some L-lysine producers is their resistance to the lysine analogue S-(2-aminoethyl)-L-cysteine (Fig. 13.13). In these mutants, the aspartate kinase (see Fig. 13.9) is mutated so that it is no longer inhibited by L-lysine. Dozens of other chemicals structurally related to L-lysine, such as γ-methyl-L-lysine or α-chlorocaprolactam, have been used in screenings to obtain improved producers. Fluoropyruvate has also been used to identify strains that are sensitive to it as these have decreased pyruvate dehydrogenase activities resulting in a diminished oxidation of pyruvate via the citric acid cycle (Fig. 13.9). This is also the case in strains with decreased citrate synthase activity. In another lineage of strains, over-producers were derived from mutants with diminished

Table 13.2 A genealogy of strains obtained by classical mutagenesis and screening, showing the yield improvement obtained and some phenotypic characters known

Strain	Character	Yield of L-Lysine (%)
AJ 1511	Wild type	0
AJ 3445	AEC^r	16
AJ 3424	AEC^r Ala$^-$	33
AJ 3796	AEC^r Ala$^-$ CCL^r	39
AJ 3990	AEC^r Ala$^-$ CCL^r ML^r	43
AJ 1204	AEC^r Ala$^-$ CCL^r ML^r FP^s	50

Notes:
AEC^r: Resistant to S-(β-aminoethyl)-L-cysteine; Ala$^-$: L-alanine-requiring; CCL^r: resistant to α-chlorocaprolactam; ML^r: resistant to γ-methyl-L-lysine; FP^s: sensitive to β-fluoropyruvate.

homoserine dehydrogenase activity to lower the availability of L-threonine inside the cell (Fig. 13.9). In this way, inhibition of the kinase activity was abolished and, at the same time, a favourable growth limitation was introduced.

13.5.2 Production process

The most common carbon sources for L-lysine fermentation and also other amino acids are molasses (cane or sugar beet molasses), high test molasses (inverted cane molasses) or sucrose and starch hydrolysates. In contrast to *E. coli*, the wild type of *C. glutamicum* can utilise both glucose and sucrose. There are also production technologies available based on acetic acid or ethanol as feedstocks. In the past, molasses was mostly used for production since it is a relatively cheap carbon source. However, the utilisation of molasses has severe disadvantages:

- waste is exported from the sugar company to the fermentation plant and causes additional costs there;
- the seasonal availability of molasses causes ageing effects in its quality during storage.

Therefore, there is a clear tendency away from molasses towards refined carbon sources such as hydrolysed starches. Profitable nitrogen sources are ammonium sulphate and ammonia (gaseous or ammonia water). The growth factors required are provided from plant protein hydrolysates, cornsteep liquor or by the addition of the defined compounds. A typical lysine fermentation is shown in Fig. 13.14. After consumption of the initial sugar, the substrates are added continuously and L-lysine accumulates up to 170 g l^{-1}. Ammonium sulphate provides the counterion to neutralise the accumulating basic amino acid. Therefore, L-lysine is present in the fermentation broth as its sulphate. As a convention in the literature, lysine is usually given as lysine.HCl. Due to the high sugar cost, the conversion yield is a very important criterion for the entire production process. Technical processes have been published with a yield of 45–50 g Lys.HCl per 100 g carbon source.

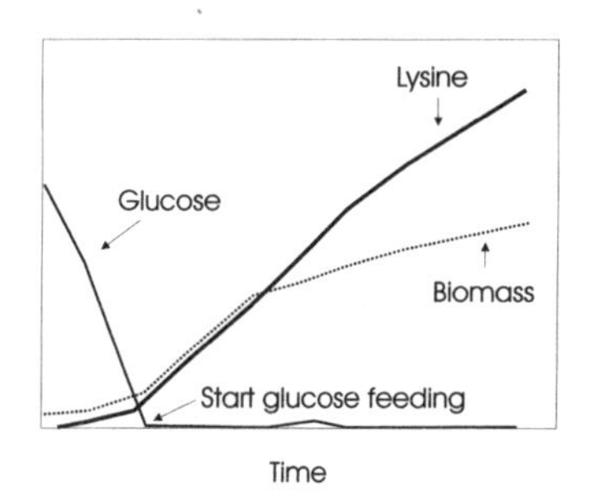

Fig. 13.14 Time course of L-lysine accumulation in a production plant. There are three phases of growth and L-lysine accumulation.

For the recovery of L-lysine, several basically different processes have been developed. Three processes are currently in use to supply L-lysine in a form suitable for feed purposes:

- A crystalline preparation containing 98.5% L-lysine.HCl. It can be made by ion exchange chromatography, evaporation and crystallisation. Also direct spray-drying of the ion exchange eluate is possible.
- An alkaline solution of concentrated L-lysine containing 50% L-lysine. It is obtained by biomass separation, evaporation and filtration.
- A granulated lysine sulphate preparation consisting of 47% L-lysine. It consists of the entire fermentation broth conditioned by spray-drying and granulation.

These processes differ in investment costs, losses during downstreaming, amount of waste volume, and user friendliness. All this, together with the fermentation itself, decides the success of the entire production process.

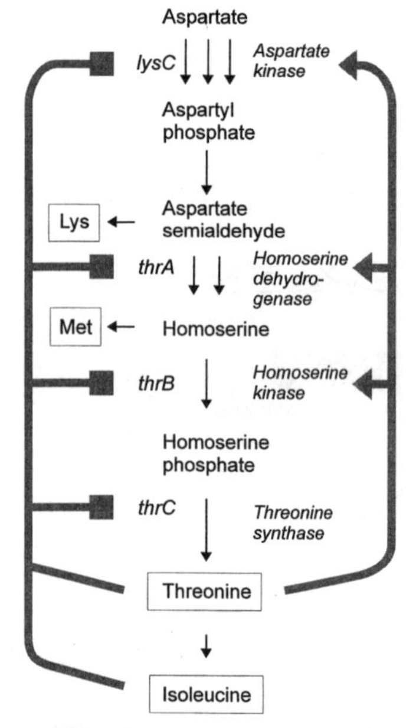

Fig. 13.15 L-Threonine synthesis in *E. coli*. The thin arrows indicate individual enzyme activities, and the genes *thrABC* constitute an operon. Only the regulation of genes and enzymes by L-threonine and L-isoleucine is shown, where the square ends indicate gene repression, and the arrowhead ends enzyme activity inhibition.

13.6 L-Threonine

The commercial production of L-threonine is possible with either *E. coli* or *C. glutamicum* mutants. However, the production figures of selected *E. coli* strains are superior. The synthesis of L-threonine proceeds via a short pathway comprising only five steps (Fig. 13.15). As already mentioned, the first steps are shared with that of L-lysine and L-methionine synthesis. Furthermore, L-threonine is also an intermediate in the L-isoleucine synthesis. This naturally requires special metabolic regulation. In *C. glutamicum* this was solved in such a way that the sole aspartate kinase present was only inhibited by the joint presence of L-lysine and L-threonine. In the case of *E. coli*, however, three isoenzymes are present each of which is separately inhibited by a different end-product: one by L-threonine, one by L-lysine and one by L-methionine. There are furthermore two homoserine dehydrogenase activities: one is inhibited by L-threonine, and one by L-methionine. Additionally, the corresponding genes are grouped into transcriptional units, thereby ensuring a balanced synthesis of the appropriate amino acid at the level of gene expression. The relevant operon for L-threonine synthesis in *E. coli* is *thrABC*. It encodes three polypeptides, with *thrA* encoding an apparently fused polypeptide with kinase plus dehydrogenase activity. Therefore, four enzyme activities of the five steps required to convert L-aspartate to L-threonine are encoded by *thrABC*. A strong expression control of this operon is provided by a transcription attenuation mechanism. The corresponding leader peptide at the beginning of the transcription unit is Thr-Thr-Ile-Thr-Thr-Thr-Ile-Thr-Ile-Thr-Thr, serving to sense the availability of L-threonine and L-isoleucine. When the corresponding tRNAs are uncharged, the leader peptide formation does not occur, and transcription of the operon is increased at least ten-fold.

13.6.1 Producer strains

Based on this regulation there is a clear focus on two major targets for the design of a producer strain: the prevention of L-isoleucine formation, and stable high-level expression of *thrABC*. Therefore, in one of the first steps of strain development, chromosomal mutations were introduced to give an isoleucine leaky strain (Fig. 13.16). The isoleucine mutation is a very specific and important one. L-Isoleucine is required for growth only at low L-threonine concentrations but, at high concentrations of L-threonine, growth is independent of L-isoleucine. The mutation therefore has several advantageous consequences. In the first place, it prevents an excess formation of the undesired byproduct L-isoleucine. Additionally, it prevents the L-isoleucine-dependent premature termination of the *thrABC* transcription due to limiting $tRNA^{Ile}$. A high transcription rate is, of course, required to have high specific enzyme activities.

Another consequence of the isoleucine mutation is more subtle. It relates to the stability of the plasmid-containing producer strain in the various precultivation steps. Starting from a single clone, a preculture is inoculated for each production run and is then enlarged in several stages. This means that the clone is fermented for about 25 generations so that there is a great danger of the plasmid containing the *thrABC* operon being lost. This would of course be a complete disaster if it happened in the final production stage. In the presence of the isoleucine leaky mutation, however, cells that have lost the plasmid now are clearly disadvantaged when not supplied with L-isoleucine. Their further proliferation is halted, thereby stabilising a culture where almost all the cells that are growing contain the plasmid. Further engineering during strain evolution involved the introduction of resistance to L-threonine and L-homoserine. Subsequently, *tdh*, which encodes threonine dehydrogenase, was inactivated thus preventing threonine degradation. To obtain very high activities of the *thrABC*-encoding enzymes, the operon was cloned from a strain whose kinase and dehydrogenase activities are resistant to L-threonine inhibition. In addition, the transcription attenuator region was deleted. In fermentations the operon engineered in this way was successfully used with pBR322 as a vector, but a further improvement was obtained by replacing this plasmid by a pRS1010 derivative, resulting in an even more stable high-level expression.

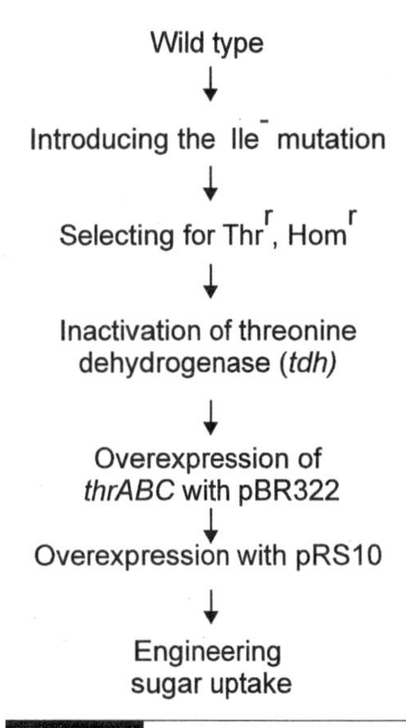

Fig. 13.16 Relevant steps in the development of an *E. coli* strain suitable for L-threonine production involving undirected mutagenesis, gene inactivation and use of different plasmids.

Substrate uptake

Since the cost of the sugar source has a decisive influence on the price of the amino acid produced it is essential to be able to switch between glucose and sucrose as substrates. However, only a few of the *E. coli* strains can use sucrose. Two different sucrose-utilising systems of *E. coli* are available to engineer sugar utilisation in L-threonine producing strains (Fig. 13.17). One of them is represented by the *scr* regulon, where

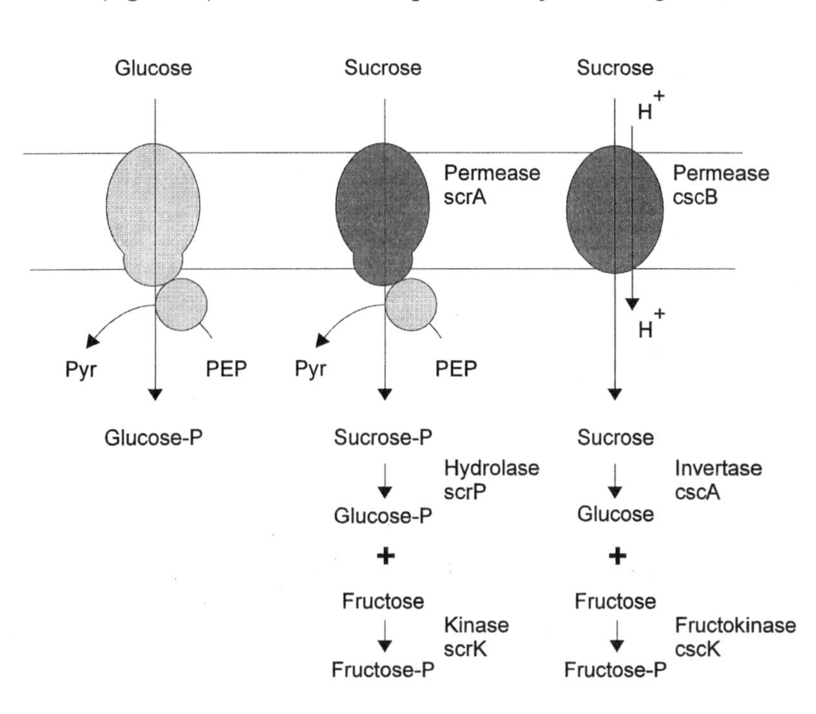

Fig. 13.17 Mechanisms of sugar uptake and phosphorylation in *E. coli*. Translocation is coupled by phosphorylation, as is the case for the phosphotransferase system (left and middle), or occurs in symport with protons without phosphorylation (right). The phosphotransferase translocating sucrose (middle) shares one of the phosphoryl transfer domains with a component of the phosphotransferase translocating glucose. Pyr, pyruvate; PEP, phosphoenolpyruvate.

the actual translocator consists of a phosphoenolpyruvate: sugar phosphotransferase system (PTS). Introduction of the *scr* genes into a glucose-utilising *E. coli* strain results in the uptake and phosphorylation of sucrose. Due to subsequent hydrolase and fructokinase activities the sugar is then channelled into the central metabolism. An alternative sucrose utilisation system is provided by the *csc* regulon of some *E. coli* strains. In this case, sucrose is translocated by the *cscB* encoded translocator in symport with protons. Using transposition the sucrose-utilisation capability of the *csc* regulon was introduced into a glucose-utilising strain. Although originally without uptake of sucrose, this strain now imported sucrose at a rate of 9 pmol min^{-1}·mg dry wt. With the plasmid-encoded regulon, the rate obtained was 43 pmol min^{-1}·mg cell dry wt, which was almost identical to that of the strain from which the *csc* regulon had been isolated.

13.6.2 Production process

The fermentation of the engineered L-threonine producer is in a simple mineral salts medium with either glucose or sucrose as the substrate with addition of a small amount of a complex medium component like yeast extract. After the inoculation and consumption of the initially provided sugar, continuous feeding of sugar begins. Additionally, ammonia has to be fed in the form of gas or as NH_4OH which is regulated via pH control. Thus the feeding strategy in the case of L-threonine fementation is quite easy compared to L-lysine fermentation where the accumulation of the basic product requires the feeding of sulphate as the counter-ion. At the end of the fermentation, L-threonine is present in concentrations of about 85 g l^{-1} with a conversion yield of up to 60% based on the carbon source used. Such fermentations with high yields show quite low byproduct levels. This is an advantage for downstream processing. Crystallisation of L-threonine is easy due to its low solubility (about 90 g l^{-1} in water) and the low salt concentration present. A process is described where the cells are initially coagulated by a heat- or pH-treatment step, followed by filtration. Subsequently, the broth is concentrated and crystallisation initiated by cooling. The separation and drying of the crystals leads to an isolation yield of 80 to 90% with the L-threonine having a purity of more than 90%. A recrystallisation step may be required for high-purity L-threonine.

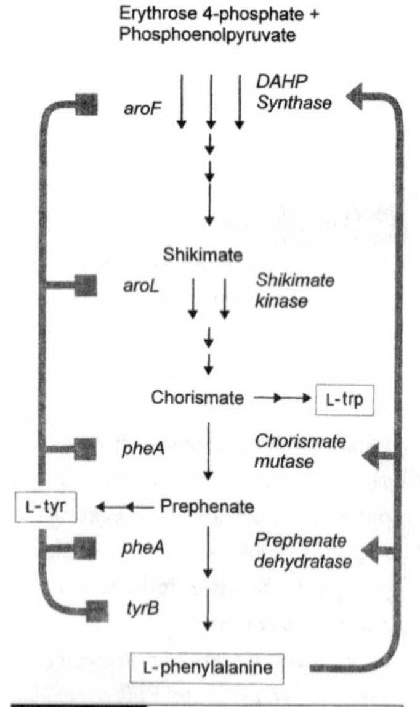

Fig. 13.18 Simplified pathway of L-phenylalanine synthesis and the relevant regulation by L-phenylalanine and L-tyrosine (L-tyr) with feedback control of enzyme activity (arrowhead ends) and gene repression (square ends).

13.7 L-Phenylalanine

L-Phenylalanine can be produced with *E. coli* or *C. glutamicum*. The pathway for L-phenylalanine synthesis is shared in part with that of L-tyrosine and L-tryptophan. These three aromatic amino acids have in common the condensation of erythrose 4-phosphate and phospho*enol*-pyruvate to deoxy*arabino*heptulosonate phosphate (DAHP) with further conversion in six steps up to chorismate. L-Phenylalanine is then finally made in three further steps (Fig. 13.18). There are three DAHP synthase enzymes in *E. coli* encoded by *aroF*, *aroG* and *aroH*. These enzymes play a key role in flux control. Their regulation of catalytic activity, in each case by one of the three aromatic amino acids, recalls the specific regulation of aspartate kinase in the synthesis of threonine. About 80% of the total DAHP-synthase activity is contributed by the *aroG*-encoded enzyme. Increased flux towards L-phenylalanine can be obtained by over-expression of either *aroF* or *aroG* encoding feedback-resistant enzymes. Furthermore, *pheA* overexpression is essential. This gene encodes the bifunctional corismate mutase-prephenate dehydratase. A second chorismate activity is present as a bifunctional chorismate mutase-prephenate dehydrogenase. The *pheA*-encoded enzyme activities are inhibited by L-phenylalanine and *pheA* expression is dependent on the level of $tRNA^{Phe}$.

13.7.1 Production strains

Producer strains have a DAHP activity that is resistant to feedback inhibition and which is encoded either by *aroF* or *aroG* and a feedback-resistant corismate mutase-prephenate dehydratase. As a rule, the producers are L-tyrosine auxotrophic mutants. There are very good reasons for this, one of which is that the enzymes of the common pathway from DAHP to prephenate are no longer regulated by L-tyrosine and enzyme activities are no longer feedback-inhibited. Another reason is that in this way tyrosine accumulation is prevented, which would otherwise undoubtedly result as a byproduct since there are only two additional steps from prephenate to L-tyrosine. An essential aspect is that due to the auxotrophy, a beneficial growth limitation is possible by appropriate tyrosine feeding (see below). In some *E. coli* strains, the temperature-sensitive cI_{857} repressor of bacteriophage λ has been used together with the λP_L promoter to enable inducible expression of the key genes *pheA* and *aroF*. This enables extremely high enzyme activities to be adjusted solely in the actual production runs thus eliminating the inherent problems of strain stability due to the resulting high metabolite concentrations or side activities of the enzymes. It enables the pre-cultivation steps up to the seed fermenter to be performed with low expression of the key genes but in the actual large production fermenter the genes are now induced to a high level of expression.

13.7.2 Production process

As with the other amino acids, effective L-phenylalanine production is the joint result of engineering the cellular metabolism and control of the production process. Control is necessary for two reasons. First, the carbon flux has to be optimally distributed between the four major products of glucose conversion, which are L-phenylalanine, biomass, acetic acid and CO_2. The second reason is that the cellular physiology is not constant during the course of fermentation, which correspondingly requires an adaptation of fermentation control during the process. Figure 13.19 shows the typical time curve of L-phenylalanine production. The major problem is that *E. coli* tends to produce acetic acid which has a strong negative effect on process efficiency. To prevent this, researchers have developed an ingenious sugar-feeding strategy, which

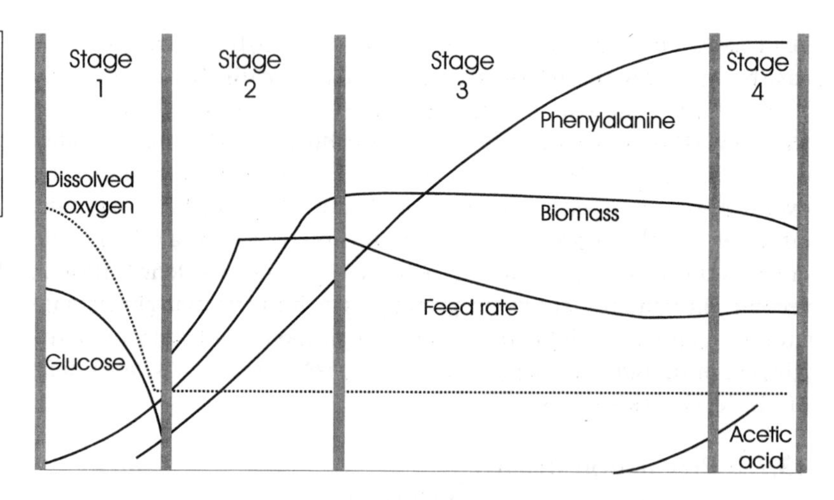

Fig. 13.19 The four stages of L-phenylalanine production characterised by different physiology requiring different process control regimes to give the highest yields in shortest times.

first collects on-line data and fluxes such as oxygen concentration, sugar consumption and biomass concentrations. These are then counterbalanced during the process to control the optimum sugar concentration. The feeding of sugar starts when the cells enter Stage 2 of the fermentation where the glucose initially provided has almost been consumed. The trick is to prevent too high a glucose concentration occurring since this would result in acetic acid formation and, at the same time, to prevent too low a glucose concentration since this would result in an excess of CO_2 evolution. Thus the feeding rate is a compromise where the process is run at the highest possible feeding rate which still provides a sufficiently strong limitation to prevent acetic acid excretion. When the L-tyrosine initially present has been consumed, the cells proceed to Stage 3. As already mentioned, almost all L-phenylalanine producers cannot synthesise tyrosine. The L-tyrosine concentration selected at the start of the culture therefore fixes the minimum amount of biomass necessary to efficiently metabolise the predetermined amount of glucose. In Stage 3, the metabolic capacity of the cells decreases which brings about a consequent decrease of the glucose feeding rate. At the end of Stage 3, acetic acid excretion begins and the cells enter Stage 4 where no further L-phenylalanine accumulation occurs and the process is eventually terminated. This example of amino acid production shows that by the sophisticated application of feeding strategies with adaptive control a very high L-phenylalanine concentration can be achieved with a high yield within 2.5 days. Values of 50.8 g phenylalanine per litre with a yield of 27.5% of carbon used have been reported.

13.8 L-Tryptophan

L-Tryptophan is a high-price amino acid which still has a rather low market volume. Effective production processes are available with mutants of different bacteria, including *Bacillus subtilis*. However, cellular synthesis is no longer performed due to originally not realised impurities in the final product used for medical purposes. These impurities arose during the isolation of L-tryptophan from a chemical reaction with traces of acetaldehyde at low pH. An alternative process is the enzymatic synthesis of L-tryptophan from precursors. The current enzymatic production process uses the activity of the biosynthetic tryptophan synthase (Fig. 13.20). This enzyme catalyses the last step in the tryptophan synthesis, which consists in fact of two partial reactions:

$$\text{Indole 3-glycerol phosphate} \rightarrow \text{indole} + \text{glyceraldehyde 3-phosphate}$$

$$\text{Indole} + \text{L-serine} \rightarrow \text{L-tryptophan} + H_2O$$

These separate reactions are catalysed by separate subunits of the enzyme: α and β. The enzyme of *E. coli* is an $\alpha_2\beta_2$ tetramer, which can be dissociated into two α subunits and a β_2 subunit. The α subunit catalyses the cleavage of indole 3-glycerol phosphate, whereas the β_2 subunit catalyses the condensation of L-serine with indole to form L-tryptophan. Each β subunit contains one molecule of covalently bound pyridoxal phosphate, forming a Schiff's base with L-serine. This enzyme-bound aminoacrylate is attacked when indole is provided from the α subunit. But how does indole get to the β subunit? The problem is that indole is very hydrophobic so that with free diffusion it can pass through the cell membrane and be lost. The crystal structure of the synthase revealed the ingenious solution for solving this problem. To prevent a loss of indole it is channelled within the enzyme protein. There is a 25 Å long tunnel from the α subunit, where indole is formed, to the β subunit where, as the enzyme-bound aminoacrylate, L-serine is ready to accept the indole. Furthermore, within the native tetramer both partial reactions are coordinated. Only when L-serine, as aminoacrylate, is ready to accept indole, does indole 3-glycerol phosphate conversion occur at the β subunit. Tryptophan synthase is thus an example of how an enzyme complex is used as a sophisticated device to handle a reactive and diffusible intermediate within the cell.

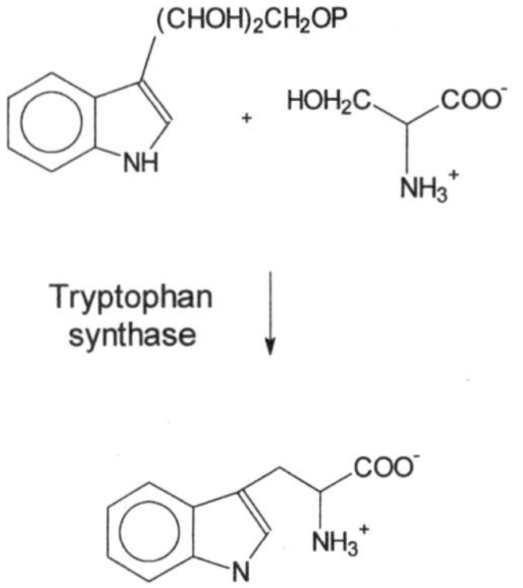

Fig. 13.20 The tryptophan synthase uses *in vivo* indole 3-glycerol phosphate plus L-serine, and in the production process indole plus L-serine.

13.8.1 Production from precursors

The process of L-tryptophan production with this enzyme is based on *E. coli* cells which have a high tryptophan synthase activity. The α, and β subunits encoding genes *trpA* and *trpB*, respectively, are located on the *trpEDCBA* operon which is regulated by repression and attenuation. In the *E. coli* mutant used, the repressor of that operon has been deleted as is part of the attenuator region together with the first structural genes of the operon. In the resulting strain, about 10% of the total protein is tryptophan synthase with an excess of the β subunit. Although indole is not the true substrate of the enzyme (see Fig. 13.20), with a sufficiently high concentration the enzyme will react with it. Indole is available from the petrochemical industry as a comparably cheap educt, whereas the second educt, L-serine, is recovered from molasses during sugar refinement using ion exclusion chromatography, and further

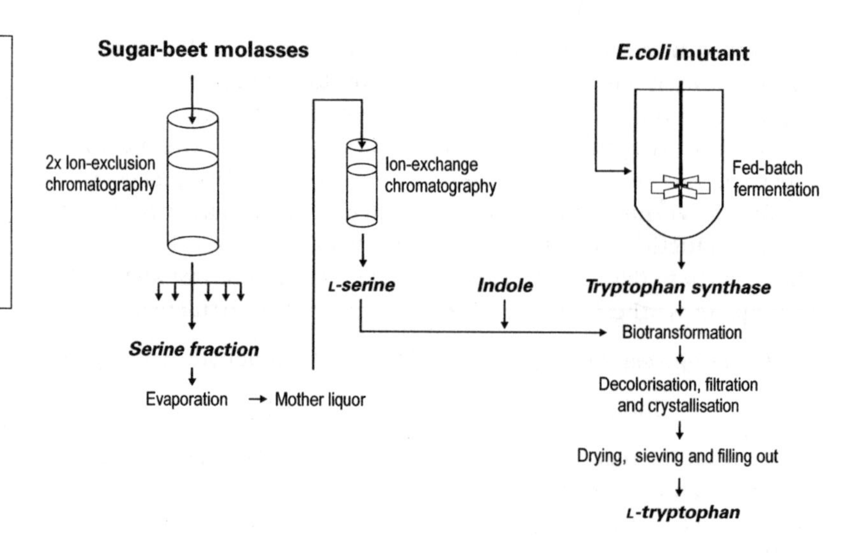

Fig. 13.21 Production plant to fractionate molasses by ion-exclusion chromatography, with isolation of L-serine. An *E. coli* mutant overexpressing tryptophan synthase is pregrown, and subsequently mixed with L-serine plus indole to convert these substrates to L-tryptophan.

purification steps (Fig. 13.21). The resulting L-serine is fed to the previously cultivated *E. coli* cells, and indole is added continuously at a concentration adjusted to 10 mM, which is controlled on-line. This type of process ensures an almost quantitative conversion of indole to yield L-tryptophan with a space-time yield of about 75 g per litre and day. Further processing of the L-tryptophan solution can be taken from Fig. 13.21 leading to a pyrogen-free pharmaceutical product of the highest quality.

13.9 L-Aspartate

L-Aspartic acid is widely used as a food additive and in pharmaceuticals. Demand increased rapidly with the introduction of aspartame as an artificial sweetener. This is a dipeptide consisting of L-aspartate and L-phenylalanine which is about 200-fold sweeter than sugar and was successfully introduced into the market as a low-calorie sweetener. Although L-aspartate was originally produced fermentatively, it is currently produced exclusively using aspartase due to the high productivities and the cost effectiveness of the process. In fact, the use of aspartase to make L-aspartate represents one of the highest productivities known for an enzyme used in biotechnology. The method developed allows reuse of the enzyme to the extent that over 220000 kg of product can be produced per kg of enzyme.

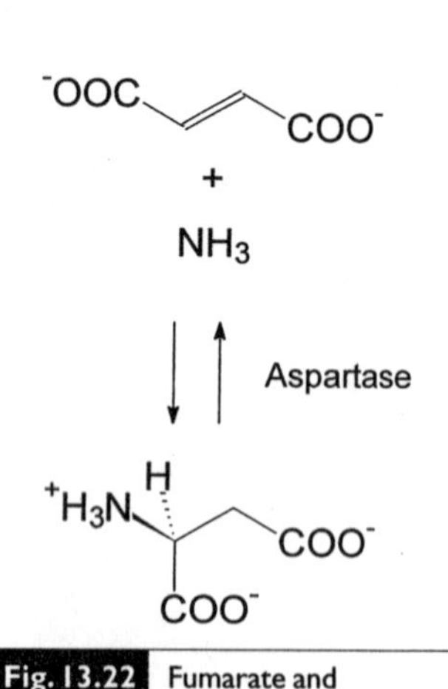

Fig. 13.22 Fumarate and ammonium serve as substrates for the aspartase.

Aspartase catalyses the interconversion between L-aspartate and fumarate plus ammonia (Fig. 13.22). The reaction favours the amination reaction. The enzyme of *E. coli* is a tetramer with a molecular weight of 196000 which has an absolute requirement for divalent metal ions. A severe disadvantage at the beginning of the work by the Tanabe Seiyaku company, which now successfully uses aspartase, was the instability of the enzyme. After incubation of the enzyme in solution for just half an hour at 50 °C, activity was no longer detectable. Nevertheless, a residual

Table 13.3 Comparison of immobilised *E. coli* cells used for production of L-aspartate

Immobilisation method	Aspartase activity (U/g cells)	Half-life (days)	Relative productivity (%)[a]
Polyacrylamide	18850	120	100
Carrageenan	56340	70	174
Carrageenan (GA)[b]	37460	240	397
Carrageenan (GA + HA)[b]	49400	680	1498

Notes:
[a] Considers the initial activity, decay constant and operation period.
[b] GA = glutaraldehyde, HA = hexamethylene diamine.

activity of 10% is present when the enzyme is immobilised in polyacrylamide. Such a physical confinement of cells in space turned out to be the method of choice. Table 13.3 shows that with the natural polymer κ-carrageenan, resulting from a screening of different polymers, and use of appropriate cross-linking exceptional improvements are obtained in the relative productivity as well as in the stability of the catalyst. The final material has a half-life of almost two years. This represents almost unimaginable progress in comparison to the initial situation where enzyme in free solution only had a half-life measured in minutes. An initial disadvantage of the original cells used, was their fumarase activity which results in the partial conversion of fumarate to L-malic acid. To solve this problem a heat treatment step of the cells is used which eliminates the fumarase activity almost completely. Using such conditioned cells and starting with 1 M ammonium fumarate, the final product solution contains 987 mM L-aspartate, 10.7 mM non-reacted fumarate and only trace quantities of L-malic acid of 1.9 mM.

For the production process the immobilised cells are packed into a column designed as a multistage system. The stages introduced, consisting of horizontal tubes, serve two purposes. On the one hand, they allow effective cooling to prevent decay of the catalytic activity since the aspartase reaction is exergonic. About 6 kcal heat mol^{-1} substrate evolves in the actual large-scale production process which is very close to that calculated from the standard free energy change of the aspartase reaction of 4 kcal mol^{-1}. On the other hand, the flow properties of the column are increased. Any compacting of the bed over time is prevented, and the preferred plug-flow characteristics are obtained. With such a column, flow rates of two column volumes per hour are possible. The continuous process enables full automation and control to achieve an optimum throughput with the highest product quality. Yet another advantage of such a controlled continuous process is its reduced waste production. A typical volumetric activity is about 200 mmol h^{-1} g cells. Assuming a 1000 litre column, the yield of L-aspartate is 3.4 tonnes per day which is 100 tonnes per month. The final product is eventually purified by crystallisation.

13.10 Outlook

Although amino acids are now among the classical products in biotechnology, their constant development means that processes must be improved, new processes established and our understanding of the exceptional capabilities of producer strains deepened. Just one example of molecular research is the recent discovery of the L-lysine export carrier, which opens up an entirely new field in the metabolism of amino acids in bacteria in general. Moreover, much information has been gathered from strain development in conjunction with fermentation technology, with the new science of metabolic engineering at the interface between them. In fact, amino acid production is an outstanding example of the integration of many different techniques. In this way, the early Japanese activities on the taste of kelp laid the foundation for the continuing very successful and flourishing production of amino acids.

13.11 Acknowledgements

We would like to thank the following for providing material for this article: R. Faurie, Amino GmbH; N. Kato, Kyoto University; Y. Kawahara, Ajinomoto Ltd.; W. Leuchtenberger, G. Thierbach, Degussa AG; S. Rhee, NHI Bethesda; T. Shibasaki, Kyowa Hakko Kogyo; T. Tosa, Tanabe Seiyaku.

13.12 Further reading

Chibata, I., Tosa, T. and Shibatani T. (1992). The industrial production of optically active compounds by immobilized biocatalysts. In *Chirality in Industry* (Collins, A.N., Sheldrake, G. N. and Crosby. J., eds.), John Wiley & Sons, London.

Eggeling, L., Morbach, S. and Sahm, H. (1997). The fruits of molecular physiology: Engineering the L-isoleucine biosynthesis pathway in *Corynebacterium glutamicum*. *J. Biotechnol.* **56**, 167–182.

Eggeling, L. and Sahm, H. (1999). L-Glutamate and L-lysine: traditional products with impetuous developments. *Appl. Microbiol. Biotechnol.* **52**, 146–153.

Hodgson, J. (1994.) Bulk amino acid fermentation: Technology and commodity trading. *Bio/Technology* **12**, 152–155.

Jetten, M. S. M., Follettie, M. T. and Sinskey, A. J. (1994). Metabolic engineering of *Corynebacterium glutamicum*. *New York Acad. Sciences* **721**, 12–29.

Katsumata, R. and Ikeda, M. (1993). Hyperproduction of tryptophan in *Corynebacterium glutamicum* by pathway engineering. *Bio/Technology* **11**, 801–806.

Kiss, R. D. and Stephanopoulos, G. (1991). Metabolic activity control of the L-lysine fermentation by restrained growth fed-batch strategies. *Biotechnol. Prog.* **7**, 501–509.

Konstantinov, K. B., Nishio, N., Seki, T. and Yoshida, T. (1990). Physiologically motivated strategies for control of the fed-batch cultivation of recombinant *Escherichia coli* for phenylalanine production. *J. Ferment. Bioeng.* **71**, 350–355.

Krämer, R. (1994.) Secretion of amino acids by bacteria: Physiology and mechanism. *FEMS Microbiol. Rev.* **13**, 75–79.

Leuchtenberger, W. (1996). Amino acids, technical production and use. In *Products of Primary Metabolism* (Rehm, H. J. and Reed G, eds.). *Biotechnology* **6**, 455–502.

Li, K., Mikola, M. R., Draths, K. M., Worden, R. M. and Frost, J. W. (1999). Fed-batch fermenter synthesis of 3-dehydroshikimic acid using *Escherichia coli*. *Biotechnol. Bioeng.* **64**, 61–73.

Peters-Wendisch, P., Kreutzer, C., Kalinowski, J., Pátek, M., Sahm, H. and Eikmanns, B. J. (1998). Pyruvate carboxylase from *Corynebacterium glutamicum*: Characterization, expression and inactivation of the pyc gene. *Microbiology, UK* **134**, 915–927.

Schilling, B. M., Pfefferle, W., Bachmann, B., Leuchtenberger, W. and Deckwer, W. D. (1999). A special reactor design for investigations of mixing time effects in a scaled-down industrial L-lysine fed-batch fermentation process. *Biotechnol. Bioeng.* **64**, 599–606.

Vrljic, M., Sahm, H. and Eggeling, L. (1996). A new type of transporter with a new type of cellular function: L-lysine export in *Corynebacterium glutamicum*. *Mol. Microbiol.* **22**, 815–826.

Chapter 14

Organic acids

Christian P. Kubicek

14.1 Introduction

Various organic acids are accumulated by several eukaryotic and prokaryotic micro-organisms. In anaerobic bacteria, their formation is usually a means by which these organisms regenerate NADH, and their accumulation therefore strictly parallels growth (e.g. lactic acid, propionic acid etc.; see Chapter 13). In aerobic bacteria and fungi, in contrast, the accumulation of organic acids is the result of incomplete substrate oxidation and is usually initiated by an imbalance in some essential nutrients, e.g. mineral ions. Despite the completely different physiological prerequisites for the formation of these products, no distinction will be made between these two types of products in this chapter. The organic acids described below are those which are manufactured in large volumes (see Table 14.1), and marketed as relatively pure chemicals or their salts.

Table 14.1 Annual production of major organic acids treated in this chapter

	Kilotonnes annum^{-1}
Citric acid	400
Gluconic acid	60
Lactic acid	50
L-Ascorbic acid	60

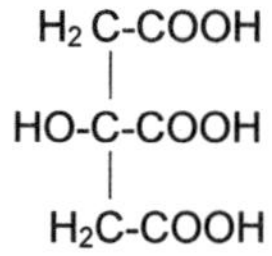

$$\begin{array}{l} H_2C\text{-}COOH \\ \quad | \\ HO\text{-}C\text{-}COOH \\ \quad | \\ H_2C\text{-}COOH \end{array}$$

Fig. 14.1 Citric acid.

14.2 Citric acid

Citric acid (2-hydroxy-propane-1,2,3-tricarboxylic acid; Fig. 14.1) was first discovered as a constituent of lemons but is today known as an intermediate of the ubiquitous tricarboxylic acid cycle (see page 26) and therefore occurs in almost every living organism. Originally produced from lemons by an Italian cartel, the discovery of its accumulation by *Aspergillus niger* (then named *Citromyces*) in the early 1920s led to a rapid development of a fermentation process which, 15 years later, accounted for more than 95% of the world's production of citric acid.

14.2.1 Microbial strains and biochemical pathways of citric acid accumulation

Most of today's citric acid is produced by *A. niger*. Industrial strains of this fungus producing citric acid are among the most secretly kept organisms in biotechnology and this precludes also the knowledge of the strategy used for their isolation during strain selection and improvement. Several mutant isolation procedures, on the other hand, have been reported by academic laboratories, which include tolerance against high sugar concentrations, 2-desoxyglucose, respiratory chain inhibitors, fluoroacetate, low pH and others, but the significance of these strategies to the industrial know-how has not been revealed. Other obvious strategies have been focused towards reduction or elimination of by-product formation, such as oxalic acid and gluconic acid (see below).

In addition to *Aspergillus*, several yeasts have been described which form large amounts of citric acid from n-alkanes and also – albeit in lower yields – from glucose. These include *Candida catenula* (former *C. brumptii*), *C. guilliermondii*, *Yarrowia lipolytica*, and *C. tropicalis*. A disadvantage in the use of these yeasts is their by-production of isocitric acid in amounts of up to 50% of the citric acid found. Mutant selection has therefore frequently sought to select for mutants with very low aconitase activity (see Figs 2.9 and 14.2), using monofluoroacetate resistance as a selection criterion.

The biochemical pathways of citric acid formation involve glycolytic catabolism of glucose to two moles of pyruvate, and their subsequent conversion to the precursors of citrate, oxaloacetate and pyruvate (Fig. 14.2). A key in this process is the use of one mole of pyruvate and the CO_2 released during the formation of acetyl-CoA to form oxaloacetate. The importance of this step becomes obvious by a simple calculation: if the oxaloacetate required for biosynthesis of citrate would have to be formed by one turn of the tricarboxylic acid cycle, two moles of CO_2 would thereby be lost, and consequently only two-thirds of the carbon of glucose accumulate as citric acid, i.e. 0.70 kg kg^{-1} sugar. Practical yields, however, are much higher, yet are perfectly consistent with the synthesis of oxaloacetate by an anaplerotic CO_2 fixation (i.e. a reaction, normally destined to balance the carbon needed for biosynthetic purposes; see page 25) by pyruvate carboxylase. In addition, the pyruvate

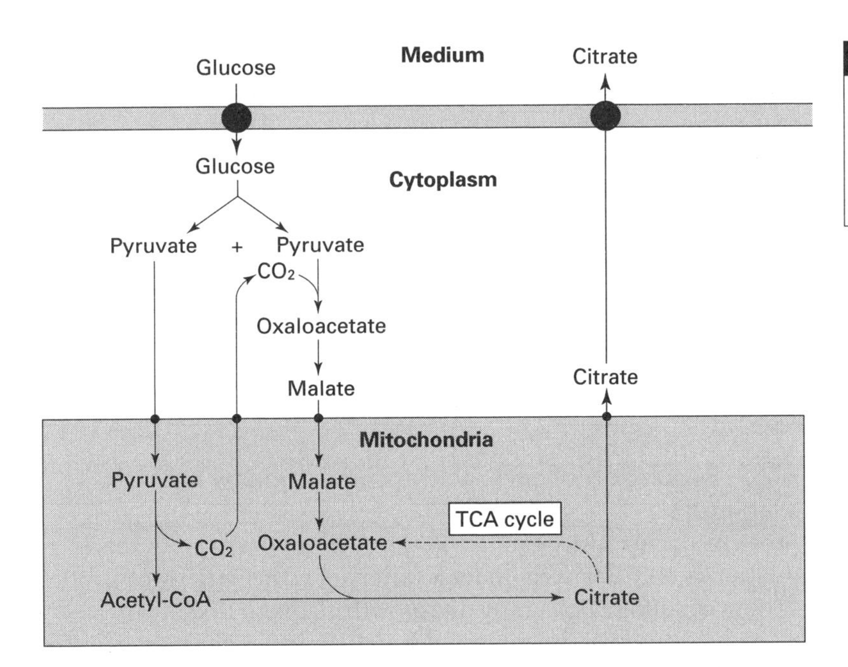

Fig. 14.2 Simplified metabolic scheme of citric acid biosynthesis. Side reactions and intermediates not relevant to citric acid biosynthesis have been omitted. TCA cycle, tricarboxylic acid cycle.

carboxylase reaction has a further important implication in citric acid biosynthesis: unlike in several other eukaryotes, pyruvate carboxylase of *A. niger* is localised in the cytosol, and the oxaloacetate formed is therefore converted further to malate by cytosolic malate dehydrogenase (Fig. 14.2), thereby also regenerating 50% of the glycolytically produced NADH. This provision of cytosolic malate as an 'end-product' of glycolysis is of utmost importance to citric acid overflow because it is the co-substrate of the mitochondrial tricarboxylic acid carrier in eukaryotes.

While the biochemical pathway for citric acid biosynthesis, as shown in Fig. 14.2, is experimentally well supported, the reason for accumulation of citric acid in molar yields of up to 90% of the consumed glucose is still not fully understood. In the past, it was thought that inactivation of an enzyme degrading citrate (e.g. aconitase or isocitrate dehydrogenase) would be the key to the accumulation of citric acid but since then solid evidence for the presence of an intact citric acid cycle during citric acid fermentation has been presented and hence these explanations have been abandoned. More likely, fine regulation of one or more of the enzymes degrading citrate by metabolites may be relevant for citric acid accumulation. However, equally likely, citrate accumulation may be the result of enhanced (deregulated) biosynthesis rather than inhibited degradation. A more detailed description of the various mechanisms proposed and the evidence presented in favour of them can be taken from some of the reviews and papers given in the Further reading list at the end of this chapter.

Table 14.2 Optimal conditions for citric acid production

Sugar concentration	120–250 g l^{-1}
Trace metal ion limitation	Mn < 10^{-8} M
	Zn < 10^{-6}–10^{-7} M
	Fe < 10^{-4} M
Dissolved oxygen tension	> 140 mbar
pH	1.6–2.2
Phosphate concentration	0.2–1.0 g l^{-1}
Ammonium salts	> 2.0 g l^{-1}
Time	160–240 h

14.2.2 Regulation of citric acid accumulation by nutrient parameters

While citric acid can accumulate in extremely high amounts, this accumulation is only observed under a variety of rather strictly controlled nutrient conditions. In fact, during growth of *A. niger* in standard media for the cultivation of fungi, little if any citric acid is accumulated. The conditions required for optimum yields vary with the type of fermentation (see below), and are most critical in the submerged fermentation process. Optimal conditions are given in Table 14.2, and are explained below.

Sugar type and concentration

The type and concentration of the carbon source is the most crucial parameter for successful citric acid production as it applies both to the submerged as well as to the surface production process. Only sugars which are rapidly catabolised by the fungus – such as sucrose, maltose or glucose – allow high yields as well as high rates of acid accumulation. In the currently used industrial processes, beet and cane molasses predominate as carbon source raw materials. However, both beet and cane molasses are very variable in quality both from season to season and from refinery to refinery. Several components present in molasses have been identified to affect citric acid production but, as the whole composition is so complex, as yet no general strategy for quality assessment can be claimed. Their selection therefore still depends on small-scale performance. A variety of alternative raw materials have been proven applicable on a laboratory scale as well (Table 14.3). Among them, the use of glucose syrup was previously more or less restricted to the USA because of fiscal and other restrictions in Europe. This attitude has been changing, however, in view of the increased availablity (and pressure for use) of various waste materials containing glucose polymers.

Concentration of the carbon source used for citric acid production is very high (100 to over 200 g l^{-1}), which seems logical when a bulk product is produced economically. However, the carbon source concentration has also been shown to have an influence on the regulation of citrate overproduction, as the yields from glucose (g g^{-1}) significantly decrease when the carbon concentration decreases below 100 g l^{-1}

Table 14.3 Raw materials which have been reported as carbon sources for citric acid production

Blackstrap molasses
Cane molasses
Bagasse
Starch
Date syrup
Apple pomace
Carob sugar
Cotton waste
Whey permeate
Brewery waste
Sweet potato pulp
Pineapple waste water
Banana extract

(Fig. 14.3). Only very little citric acid is produced at sugar concentrations below 50 g l^{-1}.

With respect to the biochemical basis for the relationship between citric acid accumulation and sugar concentration, it was shown that a high sugar concentration induces an additional glucose transport system. It is believed that the increased uptake of glucose under conditions of high sugar supply will counteract the inhibition of hexokinase by trehalose 6-phosphate. Support of this theory is obtained from the fact that *A. niger* strains in which the gene encoding trehalose 6-phosphate synthase has been knocked out, now accumulate citric acid at an increased rate even at lower sugar concentrations.

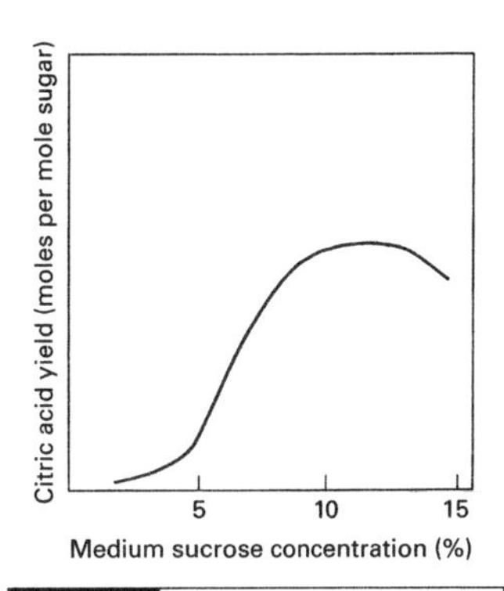

Fig. 14.3 Effect of sugar concentration of the molar citric acid yield Y_{ps} (moles citric acid produced per mole of sugar (calculated as glucose) consumed).

Trace metal ions

The effect of trace metal nutrition has been known for a long time and had been the key to the establishment of successful fermentation processes, although the effect is much more pronounced in the submerged fermentation. While all usual trace metal ions (Fe, Zn, Cu, Mn, Co) are essential for *A. niger* growth, some of them – particularly Mn^{2+}, Fe^{3+} and Zn^{2+} – have to be present in the medium at growth-limiting concentrations to give high citric acid yields. The effect of manganese ions is particularly striking, as even concentrations as high as 2 μg l^{-1} will decrease acid accumulation by about 20%. The concentration of metal ions below which citric acid is accumulated in high amounts is not absolute, however, but depends on their relative proportion to other nutrients, particularly phosphate.

Since these concentrations of metal ions, which affect citric acid production, are easily introduced into the medium by the high concentrations of the carbon source, all carbon sources to be used in citric acid fermentation have to be purified free of metal ions. This can be done in various ways, e.g. by precipitation or cation exchange treatment. The latter is usually performed only with glucose syrups. Purification of

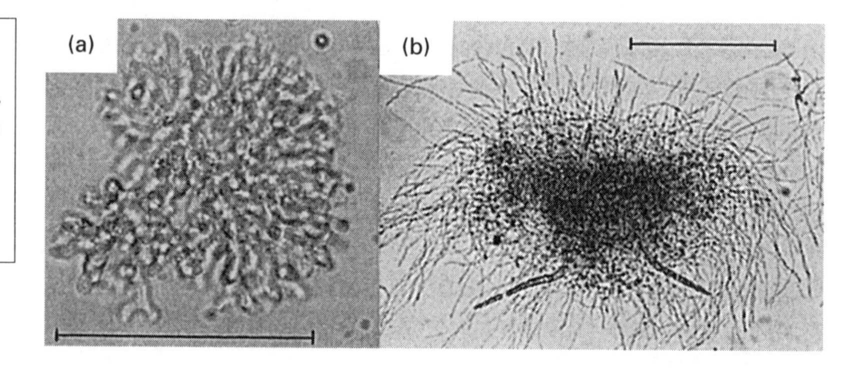

Fig. 14.4 Mycelial pellet of *Aspergillus niger*, grown under (a) manganese deficient conditions, and under (b) manganese sufficient (0.1 mM) conditions. Marker bars indicate 50 (a) and 250 (b) μM (from Roehr et al., 1996a; with permission).

industrial carbon sources such as sugar beet or sugar cane molasses is even more essential, and is mostly carried out by complexation with ferrocyanide and subsequent precipitation, which also seems to have a beneficial effect on the citric acid-forming metabolism of *A. niger*. Alternatively, the effect of trace metals can be antagonised either by the addition of copper, which blocks manganese transport into the mycelia, or by the addition of lower alcohols or of lipids which may facilitate citric acid export from the cells.

Several different hypotheses have been offered to explain the biochemical basis of this requirement for trace metal ion limitation but no single convincing explanation can yet be offered. The influence of manganese ions has been most thoroughly studied. The effect seems to be a multiple one, as it has been reported that a limiting concentration of Mn^{2+} increases the flux of carbon through glycolysis, alters the composition of the *A. niger* plasma membrane, and impairs protein turnover including that of a component of the standard respiratory chain and hence leads to impaired respiration. A further striking effect of manganese deficiency on several fungi, including *Aspergillus* spp., is its effect on the morphology of the fungus: manganese-deficient grown mycelia are strongly vacuolated, highly branched, contain strongly enthickened cell walls and exhibit a bulbous appearance (Fig. 14.4). The attached Further reading list provides more detailed information on the existing literature in this area.

The influence of other metal ions on the accumulation of citric acids by *Aspergillus* spp. is even less clear: some workers have claimed a particularly strong influence of Fe^{3+} which is, however, not supported by others. Iron limitation has repeatedly been claimed to lead to an inactivation of aconitase, the enzyme catalysing further degradation of citric acid within the tricarboxylic acid cycle and which contains covalently-bound Fe. However, this assumption has now been clearly refuted.

pH

Citric acid accumulation has been reported to accumulate in significant amounts only when the pH is below 2.5. Because of the pK values for citric acid, a pH of 1.8 is automatically reached when certain amounts of it accumulate in the medium in the absence of any other buffering agent, and hence there is no problem with this point. However, some carbon sources used (e.g. sugar beet molasses) contain a significant amount of several amino acids (particularly glutamate) which strongly buffer the medium between pH 4 and 5. The reason for the requirement of a low pH is not clear at the moment, but may be related to the formation of glucose oxidase, as gluconic acid accumulates at the expense of citric acid if the pH is above 4. Glucose oxidase is induced by high concentrations of glucose and strong aeration in the presence of low concentrations of other nutrients, i.e. conditions which are otherwise typical for citric acid fermentation and will thus inevitably be formed during the starting phase of citric acid fermentation and convert a significant amount of glucose into gluconic acid. However, due to the extracellular location of the enzyme, it is directly susceptible to the external pH and will be inactivated once the pH decreases below 3.5. Not all strains of *A. niger* show equally efficient induction of glucose oxidase under fermentation conditions and reports on the effect of the starting pH on the fermentation yield are therefore variable. Also, some strains accumulate oxalic acid at a pH >6, which must be avoided because of its toxicity. Its formation has been attributed to the hydrolysis of oxaloacetate but a possible involvement of the glyoxylic acid cycle under certain conditions has still not been completely ruled out.

Other explanations for the effect of pH have been proposed: one explanation suggests that citrate efflux from the cells may occur by diffusion driven by a gradient and only citrate^{3-} may be transported. If this assumption is correct, the low pH would be responsible for the citrate gradient necessary for transport and consequently less citrate can be secreted at higher pH values. Another explanation has discussed that the effect of pH may be related to the energetics of citrate biosynthesis by *A. niger*, as the exclusive operation of citric acid accumulation (as occurs during the idiophase of fermentation) yields 1 ATP and 3 NADH, whose turnover is limited by the absence of biosynthetic processes of equivalent capacity. Hence, while part of the NADH pool can be reoxidised by the alternative, salicylhydroxamic acid (SHAM) sensitive respiratory pathway described below, at a low external pH ATP will be consumed by the plasma-membrane-bound ATPase for maintainance of the pH gradient between the cytosol and the extracellular medium (see also page 31).

Dissolved O_2 tension

Accumulation of high amounts of citric acid is dependent on strong aeration. A dissolved O_2 tension higher than that required for vegetative growth of *A. niger* is essential, and even sparging with pure O_2 is in use. Sudden interruptions in the air supply on the other hand (as may occur with power breakdowns) cause an irreversible impairment of citric acid production without any harmful effect on mycelial growth. The biochemical basis of the high dissolved O_2 tension appears to be the induction of an alternative respiratory pathway which is required for reoxidation of the glycolytically produced NADH. Even short interruptions have been shown to impair the activity of the alternative oxidase. The role of the standard and alternative respiratory pathways in citric acid accumulation has been studied in some detail and the assembly of the proton-pumping NADH:ubiquinone oxidoreductase has been shown to be impaired during citric acid accumulation, which may be the reason for the importance of the activity of the alternative pathway.

Nitrogen
Nitrogen sources used in media for citric acid production have included ammonia salts, nitrates and the potential ammonia source, urea. No one material has been shown to be definitely superior to another, as long as it was guaranteed that the compounds did not lead to unfavourable changes in pH. Advantages sometimes observed may merely be a measure of the purity of the compunds used. It should be noted that the effect of nitrogen sources is mainly observed in chemically defined media, as no further nitrogen is necessary when beet molasses are used as carbon source.

Phosphate
The concentration of the phosphate source is usually not critical either. However, an appropriate balance of nitrogen, phosphate and trace metals appears to be important for the accumulation of citric acid in batch cultures. Thus, under special conditions such as in continuous culture, nitrogen must be limiting for attaining highest citric acid yields. On the other hand, several authors have described that the exogenous addition of ammonium ions during citric acid fermentation even stimulates citrate production.

14.2.3 Production processes for citric acid

Basically there are two different types of fermentations carried out for the production of citric acid, e.g. **the surface process** and **the submerged process** (see Fig. 14.5). In addition, some citric acid is also produced by solid state fermentations, particularly in less developed rural areas such as some East Asian countries. Citric acid production by yeast is exclusively done by submerged cultivation.

Surface fermentation is the older and more labour-intensive version of citric acid fermentation, yet it is still in use, even by some major producers of citric acid. The main reasons for this are the lower power requirements and the higher reproducibility of the process due to its lower susceptibility to interference by trace metal ions and variations in the dissolved O_2 tension. The fermentation is usually carried out in aluminium trays, filled with nutrient medium to a depth of between 50 and 200 cm. Spores are distributed over the surface of the trays, and sterile air (serving both as an oxygen supply as well as a cooling aid) is passed over them. The mycelium develops as a coherent felt, becoming progressively more convoluted. A final yield of 0.7–0.9 g g^{-1} supplied sugar is obtained within a period of 7 to 15 days.

The submerged fermentation process is desirable because of its higher efficacy due to higher susceptibility to automatisation. Yet the severe influence of trace metal ions and other impurities present in the carbohydrate raw materials and its disturbance by variations in O_2

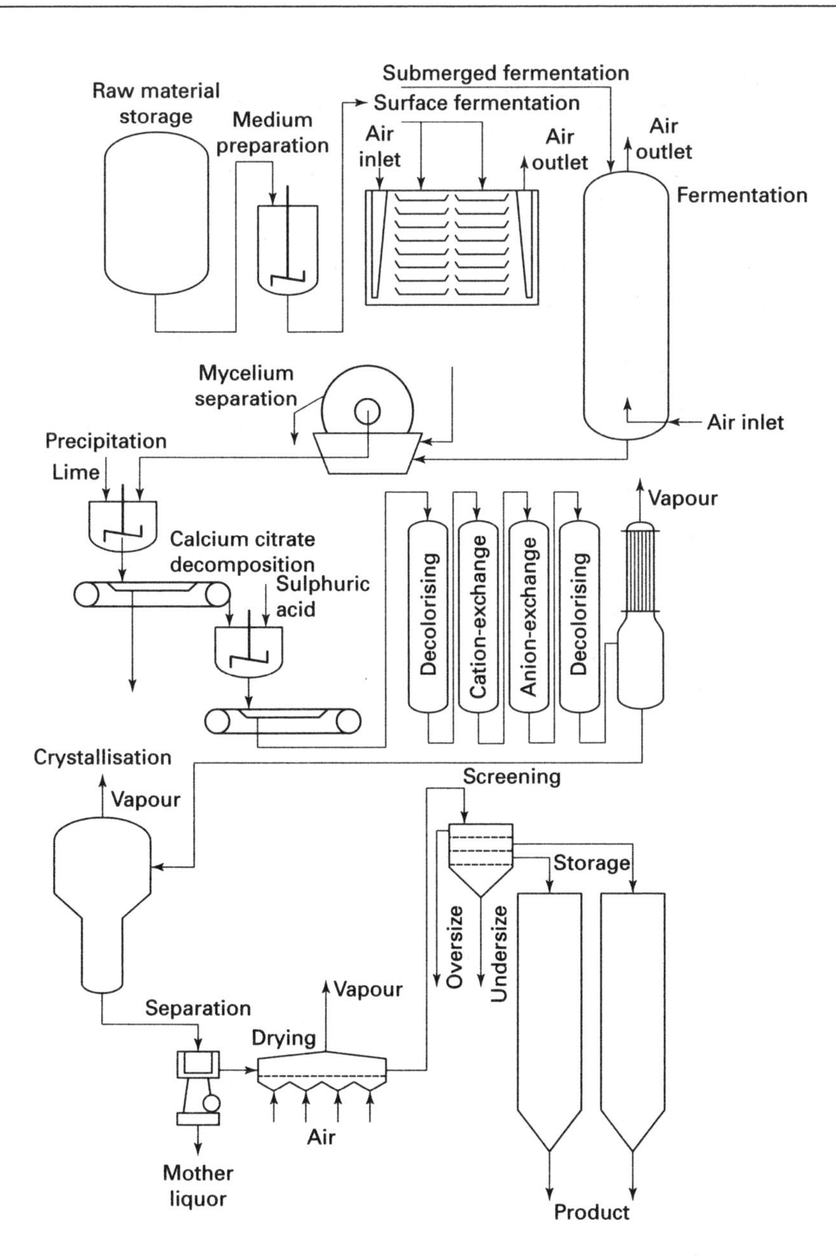

Fig. 14.5 Flow-sheet of citric acid manufacture by surface or submerged process (from Roehr *et al.*, 1992; with permission)

supply make it more difficult to manage, particularly since the quality of the carbohydrate source is variable. There are two types of fermenters in use: stirred tanks and aerated tower fermenters. Both types are constructed of high-grade stainless steel and contain facilities for cooling. Sparging with O_2 occurs from the base.

One of the most prominent features of submerged fermentation is the mycelial development which shows a characteristic pattern: the germinating spores form stubby, forked and bulbous hyphae, which aggregate to small (0.2–0.5 mm) pellets, which have a firm, smooth surface and sediment quickly when harvested (see Fig. 14.4). This striking morphology has been shown to be critical for attaining high yields by submerged fermentation and is dependent on an appropriate nutrient

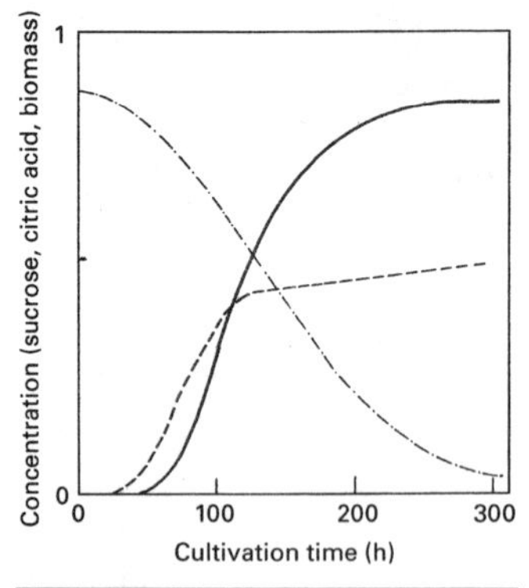

Fig. 14.6 Time course of a typical industrial citric acid fermentation showing citric acid monohydrate (—), biomass (- - -), and sugar (-·-·-). Typically, in 250–280 hours, 8–12 g l^{-1} biomass dry wt and 110–115 g l^{-1} of citric acid are obtained from 140 g l^{-1} sucrose.

composition. It is therefore a convenient indicator for the progress of fermentation, e.g. by procedures involving microscopy. A final yield of 0.8–0.9 kg kg^{-1} is obtained after 7 to 10 days (Fig. 14.6).

The Japanese wheat bran process accounts for about 20% of the annual citric acid production in Japan. Similar processes, frequently on a relatively small scale, are also carried out in China and South East Asia. The process uses solids from potato starch processing or wheat bran, adjusted to a pH of 4–5, and with a water content of 70–80%. Addition of several materials such as α-amylase or the filter cake of a glutamic acid fermentation have proven beneficial. After 5–8 days, the koji is harvested and placed in percolators, and the citric acid is extracted with water. Further purification occurs by the same procedures as for the surface or submerged fermentation (see below).

Citric acid production by yeast is carried out with either n-alkanes or glucose syrup as carbon sources. Various fractions of straight-chain paraffins (about C_9 to C_{20}) are preferred substrates. The pH should be kept above 5 and the P/C-ratio of the medium should be between 10^{-4} to 2×10^{-3}. However, as the world oil crisis in 1973/1974 almost entirely ended the exploitation of this process, it will not be dealt here in more detail.

Recovery of citric acid from the surface process usually starts with filtration of the culture broth and thorough washing of the mycelial cake, which may trap up to 15% of the citric acid produced. Filtration of the mycelium from the submerged process often requires the use of filter aids due to the by-production of a slimy heteropolysaccharide. In several cases, lime at pH <3 is added to the broth to precipitate any oxalic acid. Recovery of citric acid from the broth is then generally accomplished by three basic procedures: (1) precipitation, (2) extraction and (3) ion exchange adsorption. Several workers have also proposed solvent extraction, which makes use of various aliphatic alcohols, ketones, amines or phosphines. Obviously, the extractants require approval by the respective food and drug authorities. Crystallisation of citric acid is finally performed in vacuum crystallisers. Citric acid monohydrate, the main commercial product in Europe, is formed at below 36 °C, and citric acid anhydride is formed at higher temperatures.

14.2.4 Applications of citric acid

Due to its pleasant taste, low toxicity and excellent palatability, citric acid is widely used in industry for the preparation of food and sugar confectionery (21% of total production) and beverages (45%). Other major applications are in the pharmaceutical and detergent/cleaning industry (8 and 19%, respectively). It is also able to complex heavy metal ions, such as iron and copper, and therefore is applied in the stabilisation of oils and fats or ascorbic acid against metal ion-catalysed oxidation. In addition, citric acid esters of a wide range of alcohols are known and can be employed as non-toxic plasticisers. Finally, some of its salts have commercial importance, e.g. trisodium citrate as a blood preservative which prevents blood clotting by complexing calcium, or as a stabiliser of emulsions in the manufacture of cheese.

Today, citric acid is produced in bulk amounts with an estimated worldwide production of 400 000 tonnes per year, most of which is produced by fermentation with the fungus *A. niger*. The bulk of production occurs in Western Europe (41%) and North America (28%).

14.3 Gluconic acid

D-Glucono-δ-lactone, the simplest of the direct dehydrogenation products of D-glucose, and its free form – gluconic acid – are produced by a large variety of bacteria and fungi. The equilibrium of the lactone and the free acid in solution is dependent on pH and temperature.

14.3.1 Biology and biochemistry of gluconic acid accumulation

Microbial accumulation of gluconic acid was first observed in cultures of acetic acid bacteria, and a bacterial parasite of olive trees, *Pseudomonas savastanoi*. With regard to fungi, gluconic acid formation by *A. niger* was observed in 1922. Subsequently, gluconic acid has been shown to be produced by several prokaryotic as well as eukaryotic micro-organisms, such as members of the bacterial genera *Pseudomonas*, *Vibrio*, *Acetobacter* and *Gluconobacter*, as well as species of the fungal genera *Aspergillus*, *Penicillium* and *Gliocladium*.

Bacterial gluconic acid formation mainly occurs by a membrane-bound D-glucose dehydrogenase, which uses PQQ (pyrroloquinoline quinone) as a coenzyme (Fig. 14.7a), and converts extracellular glucose into extracellular gluconic acid. Another enzyme, an intracellular NADP-dependent glucose dehydrogenase, does not seem to be involved in gluconic acid accumulation. Gluconic acid is not usually an end-product, but will normally be transported into the cell and be further catabolised via the reactions of the pentose phosphate pathway. However, the pentose phosphate pathway is repressed by extracellular glucose concentrations > 15 mM and a pH below 3.5 (the latter also prevents the formation of 2-oxogluconate), and gluconic acid is therefore accumulated when these conditions are applied.

Fungal gluconic acid formation is catalysed by the enzyme glucose oxidase. The enzyme is extracellular, i.e. partially cell-wall bound in *Penicillium* spp., but secreted into the medium by *Aspergillus* spp. In

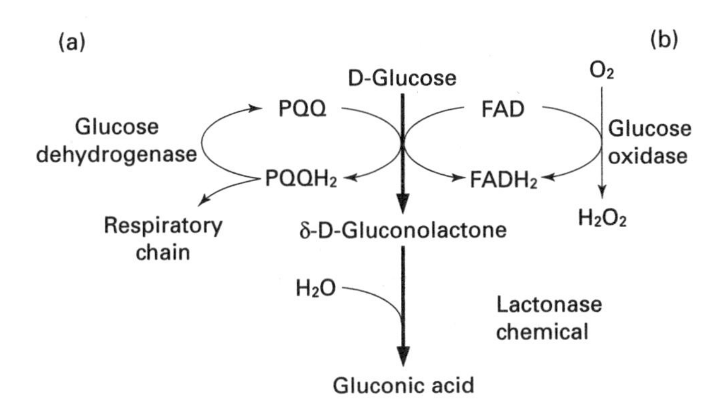

Fig. 14.7 Enzymic reactions leading to gluconic acid formation in (a) *G. suboxidans* and (b) *A. niger*.

addition, *A. niger* also produces a lactonase, and thus its product is almost exclusively gluconic acid. Glucose oxidase is a tetrameric, glycosylated flavoprotein, which uses O_2 in its reaction (Fig. 14.7b). The enzyme is most actively induced by high glucose concentrations, high aeration and at a pH above 4. It is inactivated below pH 3.0 (see section 14.2.1). Physiologically, glucose oxidase formation may be involved in the antagonistic reaction of *A. niger* against other micro-organisms, resulting in glucose withdrawal and formation of hydrogen peroxide. To protect itself against the arising hydrogen peroxide, *A. niger* also secretes multiple forms of catalase.

14.3.2 Fermentation processes for production of gluconic acid

Several processes for the production of gluconic acid have been developed, all of which use either *A. niger* or *G. oxidans* as producer organisms.

Gluconic acid production with *A. niger* was developed in the 1930s and is traditionally achieved by the calcium gluconate process. This name stems from the use of calcium carbonate for neutralisation of the fermentation broth; unless carried out, the decrease in pH would inactivate glucose oxidase and hence stop gluconic acid accumulation. The production medium contains up to 120–150 g l^{-1} glucose (most frequently derived from corn); further increases in the glucose concentration are hampered by the limited solubility of calcium gluconate, which would precipitate on the mycelia and inhibit O_2 – and substrate – uptake by the fungus. In the 1950s, the solubility of calcium gluconate was increased by the addition of boric acid to the fermentation solutions. However, the borogluconate formed was found to be deleterious to the blood vessels of animals, and the product withdrawn from the market. Other components of the nutrient medium – particularly salts to supply phosphorus and nitrogen – are added in limiting amounts in order to restrict growth of the fungus. Application of increased oxygen pressure has been shown to be advantageous, which is easily understandable by considering the stoichiometry of the reaction (see Fig. 14.7b). Fermentations with almost quantitative yields (corresponding to >90% on a molar basis) are usually completed in less than 24 h.

Sodium gluconate has been used as a superior alternative to the calcium gluconate process, as it enables the fermentation of even higher glucose concentrations (up to 350 g l^{-1}). In this process, the pH is maintained close to pH 6.5 by the addition of NaOH. In other respects, the process is similar to the calcium gluconate process. This process has been employed for the development of continuous fermentations in Japan, which claimed the conversion of 35% (w/v) glucose solutions with 95% yield.

Several different bacterial gluconic acid fermentation processes have been described but only few of them are actually performed on an industrial scale. As already mentioned, a high glucose concentration (>15%, w/v) and a pH below 3.5 are necessary for high yields. Several workers have also shown the possibility to use immobilised cells for gluconic acid production.

Methods for product recovery are similar for both fungal and bacterial fermentations but depend on the type of carbon source used and the method of broth neutralisation. Calcium gluconate is precipitated from hypersaturated solutions in the cold and is subsequently released by adding stoichiometric amounts of sulphuric acid. By repetition of this step, the clear liquid is concentrated to a 50% (w/v) solution of gluconic acid. Sodium gluconate is precipitated by concentration to a 45% (w/v) solution and raising the pH to 7.5. Today, sodium gluconate is the main manufactured form of gluconic acid, and hence free gluconic acid and δ-gluconolactone are prepared from it by ion exchange. As gluconic acid and its lactone are in a pH- and temperature-dependent equilibrium, either or both can be prepared by appropriate adjustment of these two conditions.

14.3.3 Commercial applications of gluconic acid

Gluconic acid is characterised by an extremely low toxicity, low corrosivity and the ability to form water-soluble complexes with a variety of di- and trivalent metal ions. Gluconic acid is thus exceptionally well-suited for use in removing calcareous and rust deposits from metals or other surfaces, including milk or beer scale on galvanised iron or stainless steel. Because of its physiological properties it is used as an additive in the food, beverage and pharmaceutical industries, where it is the preferred carrier used in calcium and iron therapy. In several food-directed applications, gluconic acid 1,5-lactone is advantageous over gluconic acid or gluconate because it enables acidic conditions to be reached gradually over a longer period, e.g. in the preparation of pickled goods, curing fresh sausages or leavening during baking. Mixtures of gelatin and sodium gluconate are used as sizing agents in the paper industry. Textile manufacturers employ gluconate for desizing polyester or polyamide fabrics. Concrete manufacturers use 0.02–0.2 wt% of sodium gluconate to produce concrete highly resistant to frost and cracking. According to recent estimates, its annual worldwide production is >60 000 tonnes.

14.4 Lactic acid

Lactic acid (Fig. 14.8) was first isolated from sour milk in 1798, and subsequently shown to occur in two isomeric forms, i.e. L(+) and D(−) isomers, and as a racemic mixture of these. The capital letter prefixed to the names indicate configuration in relation to isomers of glyceraldehyde, and the (+) and (−) symbols indicate the direction of rotation of a plane of polarised light. The mixture of isomers is called DL-lactic acid.

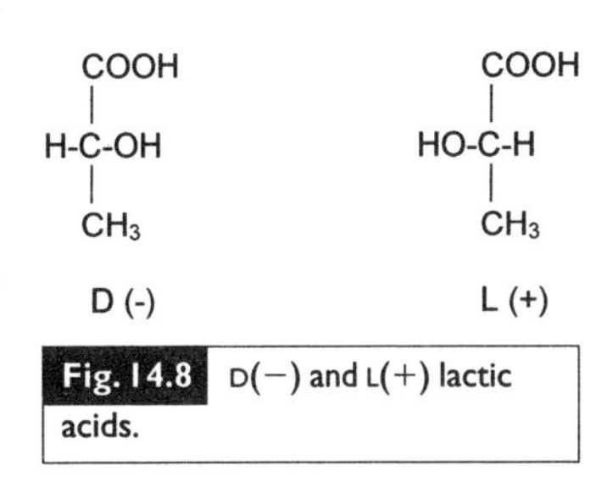

Fig. 14.8 D(−) and L(+) lactic acids.

14.4.1 Production organisms and biochemical pathways

Lactic acid was the first organic acid to be manufactured industrially by 5fermentation (around 1880 in Massachusetts, USA). The biology and biochemistry of lactic acid bacteria have been extensively reviewed. Traditionally, they are functionally classified into hetero- and homofermentative bacteria, each of which in turn can be divided according to their coccoid or rod-shaped form. Application of molecular genetic techniques to determine the relatedness of food-associated lactic acid bacteria has resulted in significant changes in their taxonomic classification. The lactic acid bacteria assciated with foods now include species of the genera *Carnobacterium, Enterococcus, Lactobacillus, Lactococcus, Leuconostoc, Oenococcus, Pediococcus, Streptococcus, Tetragenococcus, Vagococcus* and *Weisella*. The genus *Lactobacillus* remains heterogeneous with over 60 species, of which one-third are heterofermentative. Heterofermentative lactic acid bacteria are involved in most of the typical fermentations leading to food or feed preservation and transformation, whereas the homofermentative bacteria are used for bulk lactic acid production. Generally, strains operating at a higher temperature (45–62 °C) are preferred to the latter, as this reduces the power requirements needed for medium sterilisation. *Lactobacillus* spp. (e.g. *L. delbrueckii*) are used with glucose as the carbon source, whereas *L. delbrueckii* spp. *bulgaricus* and *L. helvetii* are used with lactose-containing media (whey). *L. delbrueckii* spp. *lactis* can ferment maltose, whereas *L. amylophilus* can even ferment starch.

Most lactic acid-producing micro-organisms produce only one isomer of lactic acid; however, some bacteria, which unfortunately can occur as infections during lactic acid fermentations, are known to contain racemates and are thus able to convert one isomeric form into the other.

In addition to lactic acid bacteria, other micro-organisms can produce lactic acid, e.g. *Rhizopus nigricans* and *Bacillus coagulans*. These organisms are not used for commercial purposes, however.

The biochemical pathway for lactic acid formation by homofermentative lactic acid bacteria or fungi occurs by catabolism of glucose or other hexoses via the glycolytic hexose bisphosphate pathway, and subsequent regeneration of the gained NADH by reduction of pyruvate (see page 35). Consequently, 2 mol lactic acid can theoretically be formed from 1 mol hexose, resulting in a theoretical yield of 1 kg lactic acid kg^{-1} hexose.

14.4.2 Lactic acid production

Although the molecular genetics of lactic acid bacteria are well advanced, strain selection is still carried out in traditional ways. Besides high yields of lactic acid, industrial strains are selected particularly for acid tolerance and phage insensitivity.

Raw materials used should meet certain criteria of purity as this strongly aids the final purification procedure of lactic acid, but this depends on the quality of the brand to be manufactured. As lactic acid has a very low selling price, appropriate selection of the carbon source is an important point. Materials frequently used include glucose syrups (e.g. derived from starch hydrolysis), maltose-containing materials, sucrose (e.g. from molasses) or lactose (whey).

Lactic acid is classically produced as its calcium salt. Most fermentation protocols in use today are only slight modifications of those developed in the early 1950s. They are carried out in reactor volumes up to 100 m^3, using the carbon source between 120 and 180 g l^{-1}, and appropriate concentrations of nitrogen- and phosphate-containing salts and micronutrients. As lactic acid bacteria display complex nutrient requirements for B-vitamins and some amino acids, appropriate supplements (crude vegetable materials, such as malt sprouts) have to be added. Fermentations are run at $>45\,°C$ with gentle stirring (lactic acid bacteria are anaerobic organisms and the introduction of O_2 therefore has to be avoided). The pH is maintained between 5.5 and 6.0 by the addition of sterile calcium carbonate. As an alternative to neutralisation with calcium carbonate, ammonia can be used, which also aids in the recovery of lactic acid by esterification (see below), but this results in a more expensive process. Due to the corrosive properties of lactic acid, wood or concrete were used as materials for the construction of the fermenters in the past. Today, however, stainless steel is used in the majority of cases, particularly at larger production volumes. Conversion yields of 85–95% of the theoretical maximum are usually obtained after 4–6 days.

Process variants using continuous cultivation or immobilised cells have been described in the research literature, but as yet industrial applications have not been realised.

Several techniques to purify lactic acid have been developed which are necessary to suffice the different purity requirements. It is very important that the residual sugar concentration has dropped to below 0.1% (w/v) when higher purity lactic acid is to be obtained. A standard procedure for recovery from rather pure nutrient media is given in Fig. 14.9. Broths from the fermentation of lower quality raw materials require even more extensive purification steps, including prepurification by filtering the hot calcium lactate solution, and its repeated recrystallisation. Alternatives used are solvent extraction (e.g. using isopropyl ether, isobutanol or trialkyl tertiary amines in organic solvents), or esterification with methanol and subsequent distillation.

14.4.3 Applications

Lactic acid is a highly hygroscopic, syrupy liquid which is technically available in various grades, i.e. technical grade, food grade, pharmacopoeia grade and plastic grade. The properties of these grades and their respective applications are given in Table 14.4. Recent estimates of the current market volume of lactic acid are around 50000 tonnes per annum, 70% of which is from fermentation, and the remainder from chemical manufacture.

14.5 Other acids

In addition to citric acid, gluconic acid and lactic acid, a number of other acids are commercially produced by fermentation in minor amounts.

Fig. 14.9 Scheme for recovery of lactic acid from fermentation broths.

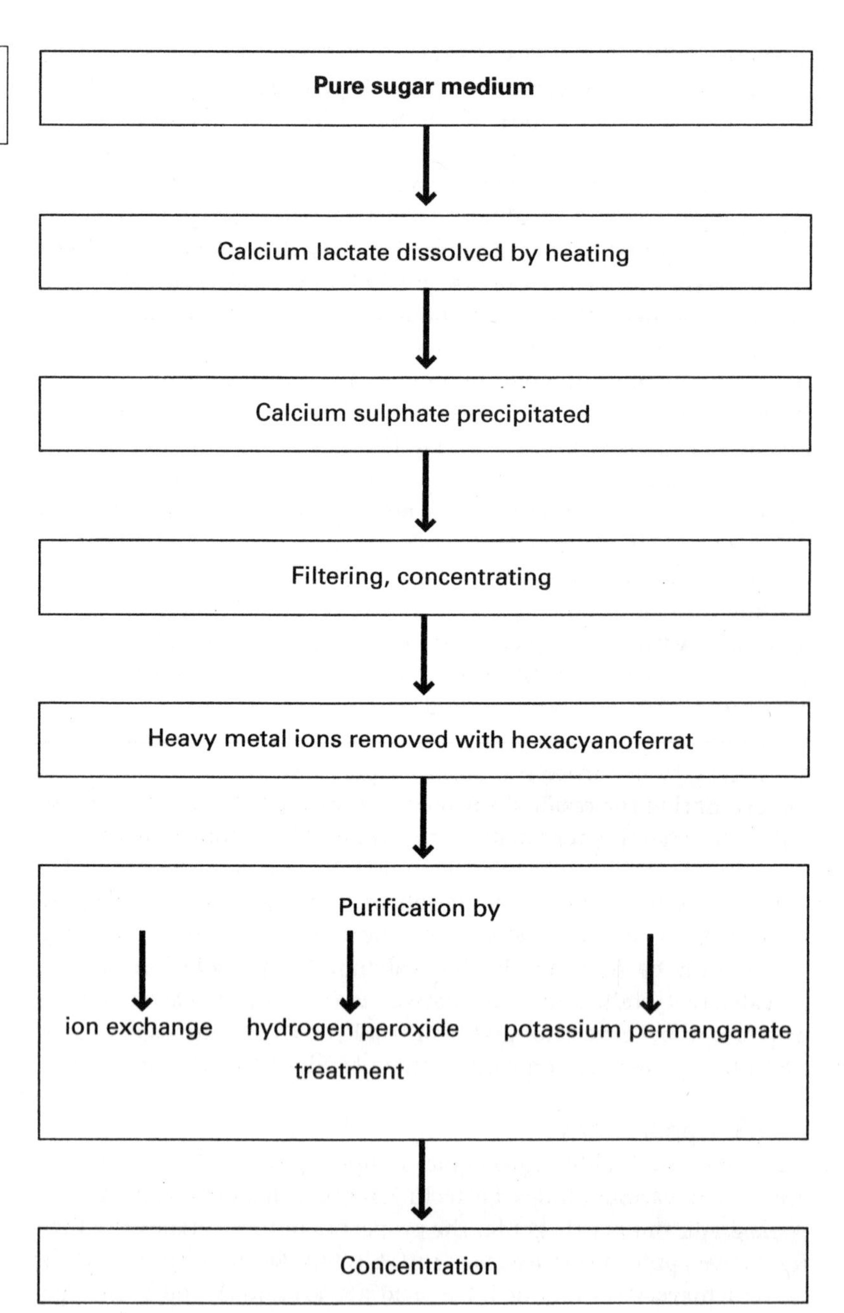

14.5.1 Itaconic acid

Itaconic acid (Fig. 14.10) was originally known as a product of pyrolytic distillation of citric acid. In the 1940s, it was found that this acid could be produced by *Aspergillus terreus* in fermentation. Chemically, it is a structurally substituted methacrylic acid, and its use therefore is mainly in the manufacturing of styrene butadiene copolymers, where it has to compete with similar petrochemistry-derived products.

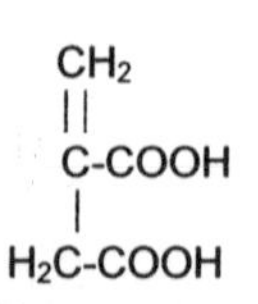

Fig. 14.10 Itaconic acid.

Commercially, itaconic acid is produced by strains of *A. terreus* or *A.*

Table 14.4 Commercial grades of lactic acid and their uses

Quality	Property	Application
Technical grade	Light brown colour Iron free 20–80% lactic acid	Deliming hides, textile industry, ester manufacture
Food grade	Colourless, odourless >80% lactic acid	Food additive, acidulant, production of sour flour and dough
Pharmacopoeia grade	Colourless, odourless >90% lactic acid <0.1% ash	Intestine treatment, hygienic preparations, metal ion lactates
Plastic grade	Colourless <0.01% ash	Lacquers, varnishes, biodegradable polymers

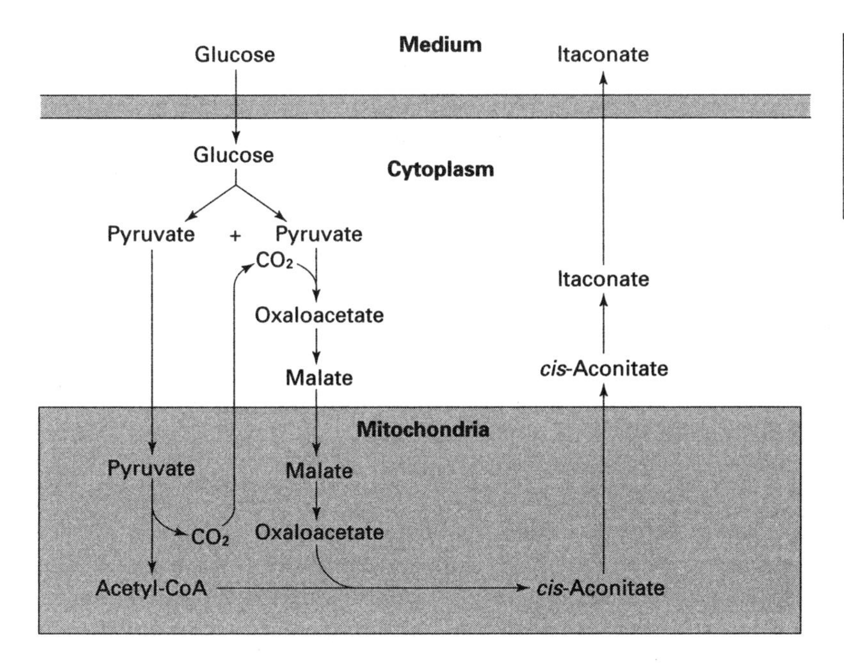

Fig. 14.11 Simplified metabolic scheme of itaconic acid biosynthesis. Side reactions and intermediates not relevant to itaconic acid biosynthesis have been omitted.

itaconicus. The biochemistry of its formation was controversial for some time but has now been established to occur by reactions similar to that involved in the accumulation of citric acid, e.g. carbon catabolism via the glycolytic pathway and anaplerotic formation of oxaloacetate by CO_2 fixation (Fig. 14.11). In addition – and in contrast to *A. niger* – *A. terreus* contains an additional enzyme, aconitate decarboxylase, which forms itaconate from *cis*-aconitate. As this reaction is localised in the cytosol, it has been implied that *A. terreus* transports *cis*-aconitate, rather than citrate, in exchange with malate out of the mitochondria (Fig. 14.11). During fermentation, itaconic acid formation is also accompanied by varying amounts of succinic, citramalic and itatartaric acid. Data currently available suggest that these are not degradation products of itaconic acid but rather are formed by other pathways.

The fermentation production of itaconic acid is largely similar to that of citric acid, i.e it requires an excess of an easily metabolisable carbon source (glucose syrup, crude starch hydrolysates, molasses), and a limitation in metal ions by the aid of complexation and/or precipitation with hexacyanoferrat or addition of copper (see Section 14.2.2). However, the effect of pH is different: several workers reported that the pH has to be maintained between 2.8 and 3.1, and lower pH values favour the formation of itatartaric acid. Yields of 85% (w/w) of the theoretical maximum have been reported to be obtained within 5 days of cultivation at rather high temperatures (39–42 °C).

Recovery is usually performed by evaporation, active carbon treatment and crystallisation/recrystallisation. Itaconic acid is sold in two grades: refined, which is a pale tan to white crystalline solid, and the industrial grade which is darker in colour. The main potential of utilisation of itaconic acid is the manufacture of styrene butadiene co-polymers, and for lattices and paint emulsions.

14.5.2 L-Ascorbic acid (vitamin C)

Ascorbic acid is the official IUPAC designation for vitamin C. It was discovered in 1928 by Szent-Györgi. Its most significant characteristic is its reversible oxidation to dehydro-L-ascorbic acid (Fig. 14.12), with which it forms a redox system. A number of enzymes are stimulated by ascorbic acid, notably Fe^{2+}-containing dioxygenases and Cu^{2+}-containing monooxygenases. One of the best known symptoms of ascorbic acid deficiency – scurvy – can be explained by the malfunctioning of these oxidases required for collagen biosynthesis. However, ascorbic acid also protects the body against formation of carcinogenic nitrosamines and oxygen radicals, and has essential functions in iron uptake. These properties, together with its nutritional qualities and low toxicity, are the main reasons for the numerous applications of vitamin C in the food and pharmaceutical industries.

Commercially, ascorbic acid is mainly produced by a combination of synthetic organic chemical steps and biotransformation, known as Reichstein synthesis. Its basic principle is reduction of C-1 of D-glucose, and oxidation at C-5 and C-6, while simultaneously preserving the chirality at C-2 and C-3. The classical scheme is shown in Fig. 14.13. The microbially catalysed step is the oxidation of D-sorbitol to L-sorbose, which is carried out by *Acetobacter xylinum*. Large scale fermentations occur at 30–35 °C and pH 4–6. All steps of Reichstein's synthesis generally have a yield >90%, and thus the final yield of ascorbic acid is about

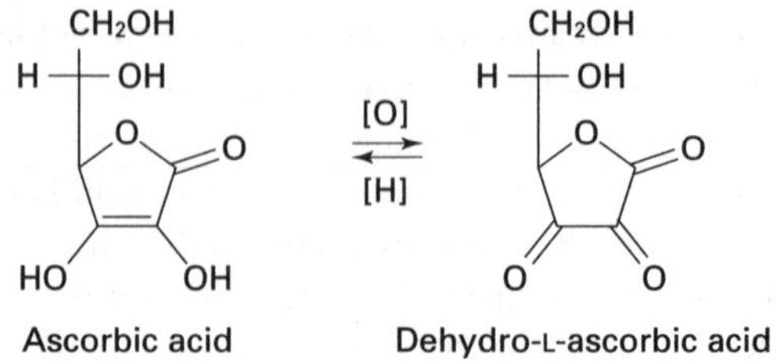

Fig. 14.12 L-Ascorbic acid in equilibrium with dehydro-L-ascorbic acid. [O] means oxidation, [H] reduction.

Ascorbic acid ⇌ Dehydro-L-ascorbic acid

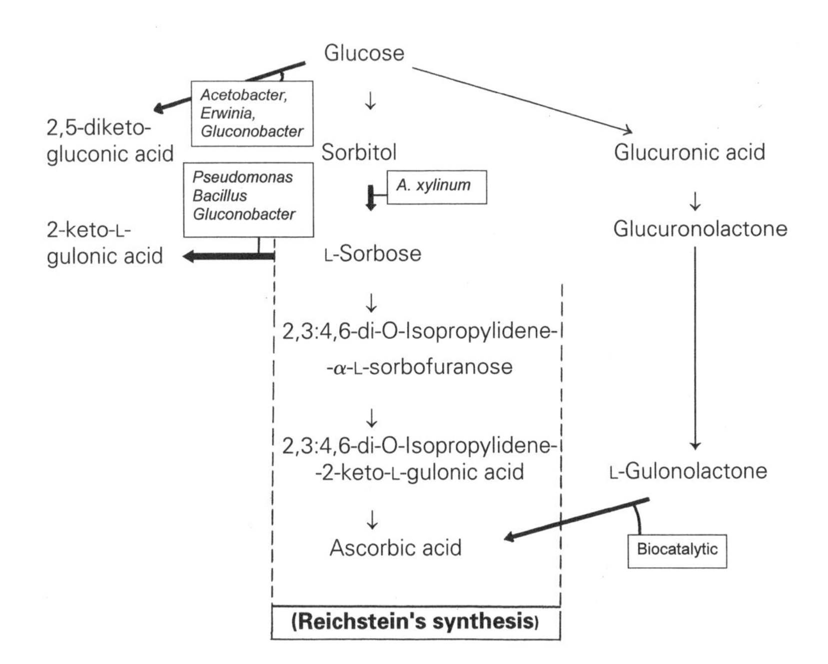

Fig. 14.13 Semi-synthetic pathways to L-ascorbic acid. Arrows printed in bold indicate steps, which can be carried out fermentatively or biocatalytically. The respective micro-organisms are boxed.

60%. Estimates of current industrial production are 60 000 tonnes per annum, the bulk of which is produced as free ascorbic acid (see p. 426).

There have been several attempts to produce ascorbic acid directly by fermentation but none of these has so far advanced to a commercial process. Micro-algae of the genus *Chlorella* can directly form L-ascorbic acid from glucose, although at very low yield. Their ascorbic acid-enriched biomass is currently used as an aquaculture fish feed or additive.

As a consequence, no alternative to the Reichstein synthesis has been established yet; however, there have been several trials to reduce the number of organic chemical synthetic steps by microbially producing more appropriate starting substances. The most successful ones are shown in Fig. 14.13: one possibility is the fermentative production of 2-keto-L-gulonic acid from L-sorbose with *Bacillus megaterium* or *Pseudogluconobacter saccharoketogenes*. Yields are in the range of 75–90%, and will therefore, once scaled up, enable a cheaper production route to L-ascorbic acid.

Another very short, mostly biocatalytic pathway to ascorbic acid would be possible via L-gulonolactone, which can be directly converted to ascorbic acid by L-gulonolactone dehydrogenase. While L-gulonolactone is readily obtained by chemical hydrogenation of D-glucuronolactone, the latter can be obtained from glucose or starch only in low yields.

14.5.3 Other acids

A small number of other, tricarboxylic acid cycle-related acids can be produced in commercially attractive amounts, none of which however is so far received industrial practice. They are shown in Table 14.5. Some of these, such as fumaric acid, have in the past been produced by

Table 14.5 Other organic acids, for which microbial production is possible

Acid	Producer	Potential application
Tartaric acid	*Gluconobacter oxydans*	Beverages, drug uses
Fumaric acid	*Rhizopus nigricans*	Polyester manufacture
	R. arrhizus	L-Aspartate manufacture
Malic acid	*Aspergillus wentii*	Beverages, flavour
trans-2,3-Epoxysuccinic acid	*Paecilomyces* spp.	β-Lactam precursor
	A. fumigatus	
	A. clavatus	
Succinic acid	*A. niger*	
Kojic acid	*A. oryzae*	Cosmetics, insecticides
Gallic acid	*A. wentii*	Blue pigments

fermentation on an industrial scale, but are currently unable to compete with the chemical productions.

14.6 Further reading

Kascak, K., Kominek, J. and Roehr, M. (1996). Lactic acid. In *Biotechnology, 2nd edition. Vol. 6: Products of Primary Metabolism* (H. J. Rehm and G. Reed, eds.; M. Roehr, volume editor), pp. 294–306. Verlag Chemie, Weinheim.

Roehr, M., Kubicek, C.P. and Kominek, J. (1992). Industrial acids and other small molecules. In *Aspergillus: Biology and Industrial Applications* (J. W. Bennett, and M.A. Klich, eds.), pp. 91–131. Butterworth-Heinemann, Reading, MA.

Roehr, M., Kubicek, C. P. and Kominek, J. (1996). Citric acid. In *Biotechnology, 2nd edition. Vol. 6: Products of Primary Metabolism* (H. J. Rehm and G. Reed, eds.; M. Roehr, volume editor), pp. 308–345. Verlag Chemie, Weinheim.

Roehr, M., Kubicek, C. P. and Kominek, J. (1996). Further organic acids. In *Biotechnology, 2nd edition. Vol. 6: Products of Primary Metabolism* (H. J. Rehm and G. Reed, eds.; M. Roehr, volume editor), pp. 364–379. Verlag Chemie, Weinheim.

Roehr, M., Kubicek, C. P. and Kominek, J. (1996). Gluconic acid. In *Biotechnology, 2nd edition. Vol. 6: Products of Primary Metabolism* (H. J. Rehm and G. Reed, eds.; M. Roehr, volume editor), pp. 347–362. Verlag Chemie, Weinheim.

Stiles, M. E. and Holzapfel, W. H. (1997). Lactic acid bacteria in food and their current taxonomy. *Int. J. Food Microbiol.* **36**, 1–29.

Chapter 15

Microbial polyhydroxyalkanoates, polysaccharides and lipids

Alistair J. Anderson and James P. Wynn

15.1 Introduction

When micro-organisms are provided with surplus glucose, or another source of carbon and energy, they may produce one or more intracellular storage compounds. Some yeasts and other fungi accumulate large amounts of oil, or 'lipid', whereas bacteria more commonly accumulate polyhydroxyalkanoates. These storage compounds are both hydrophobic and can be seen as distinct inclusions within the cells. Glycogen and trehalose are other well known examples of microbial storage compounds. Some micro-organisms excrete large amounts of polysaccharides into their growth medium, and may do so in addition to accumulating intracellular reserve compounds. The synthesis of these intracellular and extracellular products is promoted when growth is restricted by the availability of an essential nutrient other than the carbon source, and can usually occur in the absence of growth. This chapter focuses on both well established industrial products and developing areas of commercial interest.

15.2 Microbial polyhydroxyalkanoates

15.2.1 Introduction

Polyhydroxyalkanoates, or PHA, are intracellular carbon and energy reserve compounds produced by many bacteria. They can be isolated by treatment of the bacteria with chloroform. PHA are biodegradable plastics. They are broken down in soil and water by bacteria and fungi in the

Fig. 15.1 PHA granules in cells of *Ralstonia eutropha*.

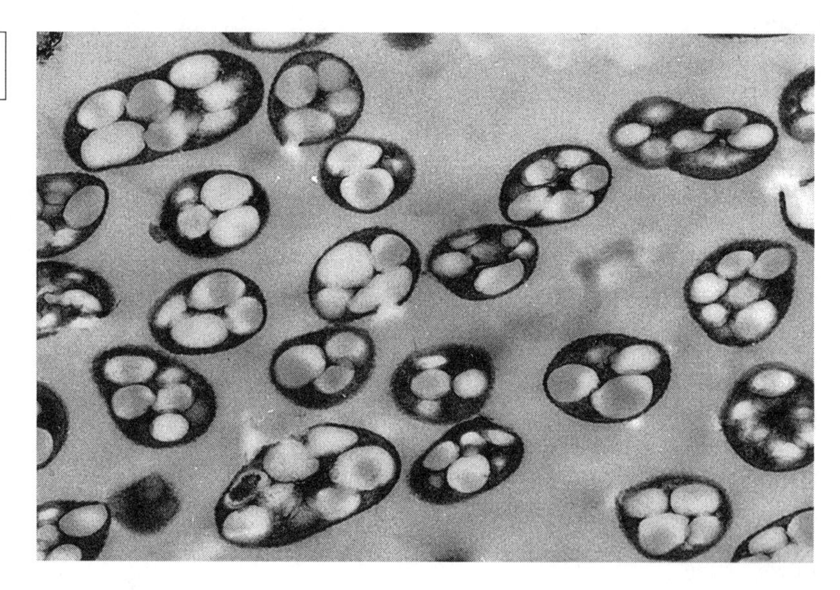

environment, in the same way that plant and animal waste is degraded. Their biodegradability and the fact that they can be produced from renewable resources make PHA of considerable commercial interest for use as packaging plastics. Conventional plastics are produced from our finite reserves of coal and oil. Moreover, these man-made materials are not biodegradable. They will survive for hundreds of years in landfill sites. A variety of other biodegradable polyesters are available, but these are made by chemical synthesis rather than by fermentation. They include poly(lactic acid), poly(glycolic acid) and poly(ε-caprolactone).

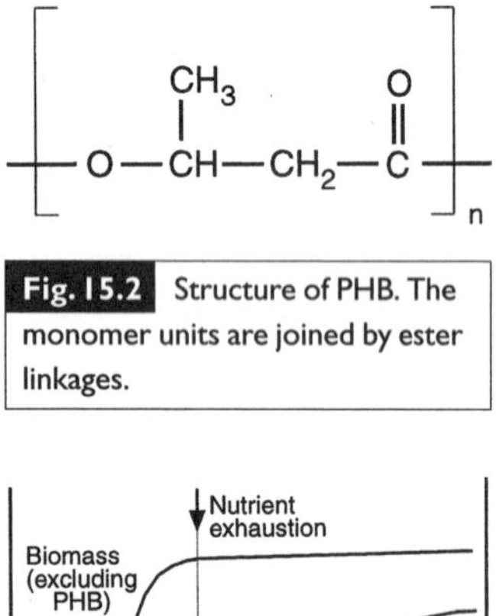

$$\left[-O-\overset{\displaystyle CH_3}{\overset{|}{CH}}-CH_2-\overset{\displaystyle O}{\overset{\|}{C}}- \right]_n$$

Fig. 15.2 Structure of PHB. The monomer units are joined by ester linkages.

15.2.2 PHA as lipid reserve materials in bacteria

PHA are analogous in function to the oils and fats produced by yeasts and other fungi. They accumulate as granules within the cells. The granules can be seen by phase-contrast light microscopy and, more clearly, by electron microscopy (Fig. 15.1). Some bacteria can accumulate massive amounts of PHA (e.g. 80% of their dry biomass) and, not surprisingly, the cells become swollen.

The most common PHA is polyhydroxybutyrate, which is usually known as PHB. It is a polyester composed of 3-hydroxybutyrate (3HB) repeating units (Fig. 15.2). As a result of its high molecular weight, even large amounts of PHB have little effect on the osmotic pressure within the cell. Oxidation of PHB to carbon dioxide and water yields a large amount of energy. For these reasons, PHB is an ideal carbon and energy reserve for bacteria.

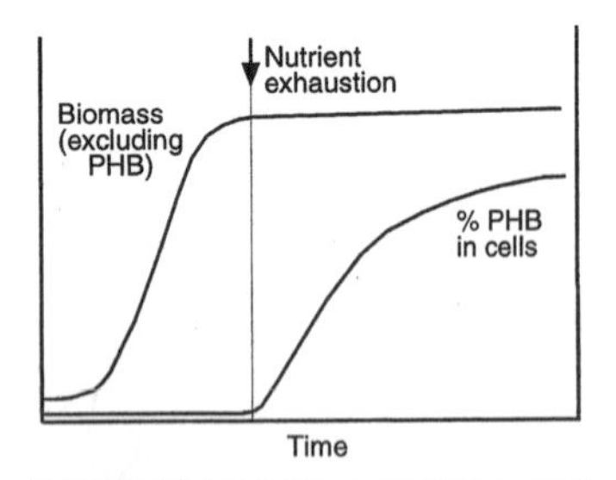

Fig. 15.3 PHB accumulation in batch culture. Rapid polymer synthesis commences at the time of cessation of growth due to nutrient exhaustion.

PHB and other PHA are produced under **nutrient limitation**, providing that surplus carbon source remains available. In batch culture, PHB is produced following cessation of growth due to exhaustion of an essential nutrient such as the N, P or S source (Fig. 15.3). Thus there is a growth phase and a polymer accumulation phase. PHB is also produced when growth of aerobic bacteria is restricted by the availability of oxygen. Most bacteria accumulate only a small amount of PHB during the growth phase, and the reasons for this are discussed below.

PHB can also be produced in chemostat culture, in which growth is restricted by the supply of one essential nutrient. The bacteria are thus subjected to **continuous nutrient limitation** and this allows PHB to be produced in actively growing bacteria. The amount of PHB produced in chemostat culture decreases at high growth rates because metabolism of the carbon source to support biosynthesis and energy generation take priority over PHB synthesis.

Bacteria require energy even when they are not growing, for example to maintain concentration and pH gradients across their cytoplasmic membrane. Degradation of PHB (or other intracellular reserve materials such as glycogen) can satisfy this **maintenance energy** requirement and so aid survival. Degradation of PHB and other PHA generally requires different enzymes from those used in biosynthesis. It is generally assumed that all PHA-producing bacteria are able to degrade their PHA but this is not established. It is certainly possible to produce recombinant strains that can produce PHB but lack the ability to degrade it.

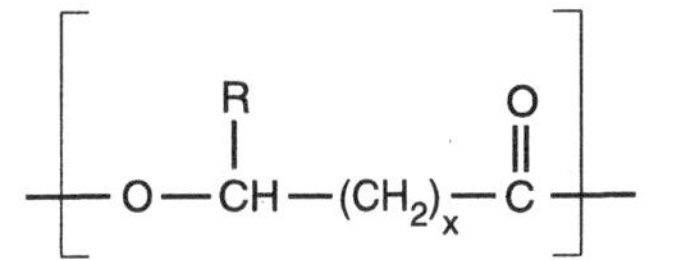

Fig. 15.4 General structure of monomer units in PHA. The most common monomers found are 3-hydroxyacids (x = 1) with a simple alkyl side chain, R. Side chains that are branched or contain an aromatic ring or halogen are also known.

15.2.3 PHA composition and properties

PHA are linear polyesters composed of hydroxyacid monomers (Fig. 15.4). 3-Hydroxyacid monomers are most common and 3-hydroxyacids with carbon chain lengths from C_3–C_{14} have been found in the range of PHA produced by bacteria. With the exception of PHB which is a homopolymer, most PHA contain two or more different monomers and are referred to as heteropolymers. The polymers generally consist of a random sequence of the constituent monomers rather than having different monomers in separate chains. In addition to 3-hydroxyacids, various other hydroxyacid monomers have been found in PHA. The composition of PHA is dependent both on the organism and the carbon source(s) available during polymer accumulation and research has shown that, collectively, bacteria are capable of incorporating over a hundred different hydroxyacid monomers into PHA. Some examples of bacterial PHA are given in Table 15.1.

In many examples, PHA monomers are produced from 'precursor substrates' that have a related chemical structure. For example, *Pseudomonas oleovorans* produces PHA containing mainly 3-hydroxyoctanoate monomers (Fig. 15.5) from octanoic acid or octane, and *Ralstonia eutropha*, formerly known as *Alcaligenes eutrophus*, synthesises 4-hydroxybutyrate and 3-hydroxypropionate monomers from 4-hydroxybutyric acid and 3-hydroxypropionic acid, respectively.

The physical properties of PHA depend on its constituent monomers. For example, PHB is hard and inflexible, whereas polymers of longer side chains (Fig. 15.5) are soft rubbers. In some cases it is easy to control the composition of PHA, and hence its physical properties can easily be controlled by varying the composition of the medium. For example, *Ralstonia eutropha* produces PHB from glucose and poly (3-hydroxybutyrate-*co*-3-hydroxyvalerate) (abbreviated PHB/V) from

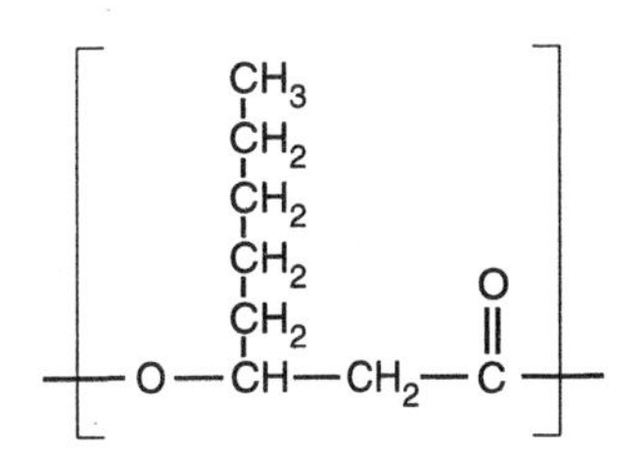

Fig. 15.5 PHA composed mainly of hydroxyacids with long side chains, such as 3-hydroxyoctanoate (shown here), are soft rubbers.

Table 15.1 Composition of PHA produced from various carbon sources by a range of bacteria

		Composition of PHA										
		3-Hydroxyacid monomers										Other monomers
Organism	Carbon source(s)	C_3	C_4	C_5	C_6	C_7	C_8	C_9	C_{10}	C_{11}	C_{12}	
Ralstonia eutropha	Glucose		●									
R. eutropha	Glucose + propionic acid		●	○								
R. eutropha	Glucose + 4-hydroxybutyric acid		●									○ (4HB)
Comamonas acidovorans	Glucose + 4-hydroxybutyric acid		○									● (4HB)
Alcaligenes latus	Sucrose + 3-hydroxypropionic acid	○	●									
Pseudomonas oleovorans	Octanoic acid				○		●		○			
P. oleovorans	Nonanoic acid				○	○	○	●	○	○		
P. aeruginosa	Gluconic acid				○		○		●		○	
Rhodococcus ruber	Glucose		○	●								

● principal monomer present in PHA
○ other monomers – 4HB 4-hydroxybutyrate

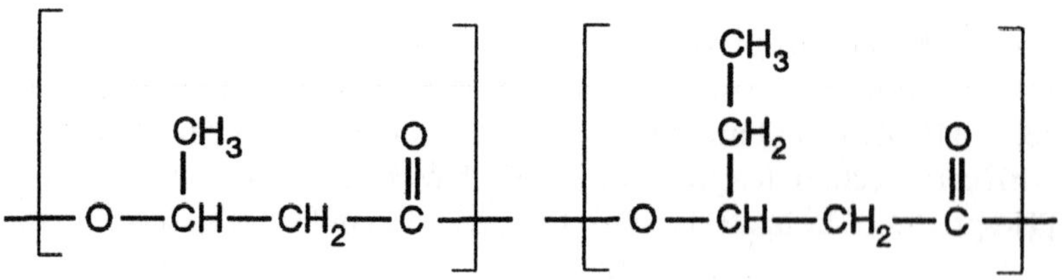

Fig. 15.6 Structure of PHB/V. It consists of a random sequence of 3-hydroxybutyrate and 3-hydroxyvalerate monomers, and is therefore described as a random copolymer. Most PHA contain two or more different monomers in the polymer chain.

glucose plus propionic acid. PHB/V (Fig. 15.6) is a copolymer of 3HB and 3HV monomers, and its composition can be controlled by varying the concentrations of glucose and propionic acid in the medium during the polymer accumulation phase. PHB is hard and brittle, but the incorporation of a small proportion of 3-hydroxyvalerate (3HV) monomers into the polymer chain results in a stronger and more flexible plastic. This is exploited in the commercial production of PHB/V (Section 15.2.8).

In some cases, bacteria can produce PHA monomers that are not related to the structure of the carbon sources provided. For example, fluorescent pseudomonads produce PHA containing 3-hydroxydecanoate from many carbon sources and some *Rhodococcus* and *Nocardia* species produce PHB/V (Fig. 15.6) containing a high proportion of 3HV monomers, again from a variety of carbon sources.

15.2.4 Biosynthesis of PHB

Of all the PHA, the biosynthesis of PHB has been studied in greatest detail. In most bacteria, PHB is synthesised from acetyl-CoA in three steps (Fig. 15.7). 3-Ketothiolase (encoded by gene *phbA*) catalyses the condensation of two molecules of acetyl-CoA to produce acetoacetyl-CoA, which is then reduced by an NADPH-dependent acetoacyl-CoA reductase (*phbB*) to yield *R*-3-hydroxybutryl-CoA. Addition of 3-hydroxybutyrate (3HB) to the growing PHB chain involves PHA synthase (*phbC*), an enzyme associated with the membrane surrounding PHB granules. In *Ralstonia eutropha*, the genes for these enzymes are organised in an operon: *phbCAB*. The genes have been cloned and expressed in other bacteria and also in plants (see below).

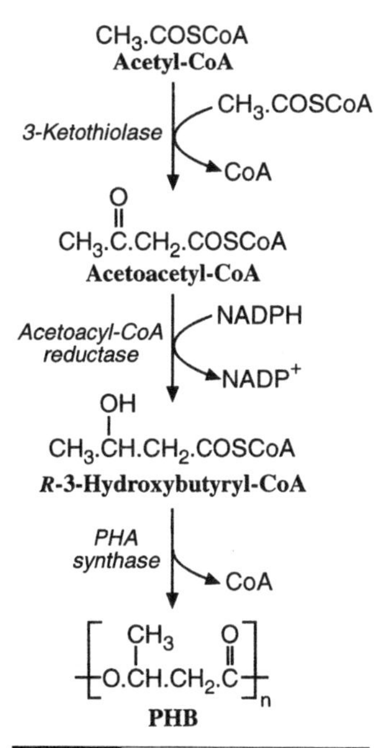

Fig. 15.7 Biosynthesis of PHB from glucose in *Ralstonia eutropha*.

15.2.5 Regulation of PHB metabolism

The enzymes for PHB biosynthesis are constitutive – they are present even during unrestricted growth. This allows immediate PHB synthesis as soon as growth becomes restricted by the availability of an essential nutrient. In natural environments, where the supply of nutrients can change rapidly, it is clearly advantageous to micro-organisms to be able to accumulate a carbon and energy reserve under conditions where growth is not possible.

The biosynthesis and degradation of PHB in *Ralstonia eutropha* are shown in Fig. 15.8. During the growth phase, 3-ketothiolase is inhibited by free coenzyme A and little PHB synthesis can occur. When growth ceases, or is restricted, metabolism of acetyl-CoA via the TCA cycle decreases as a result of inhibition of citrate synthase by NADH. The decrease in concentration of coenzyme A relieves the inhibition of 3-ketothiolase and the surplus acetyl-CoA is channelled into the production of PHB. The synthesis of PHB is also governed by the availability of reducing power, in the form of NADPH, which is required in the subsequent reaction.

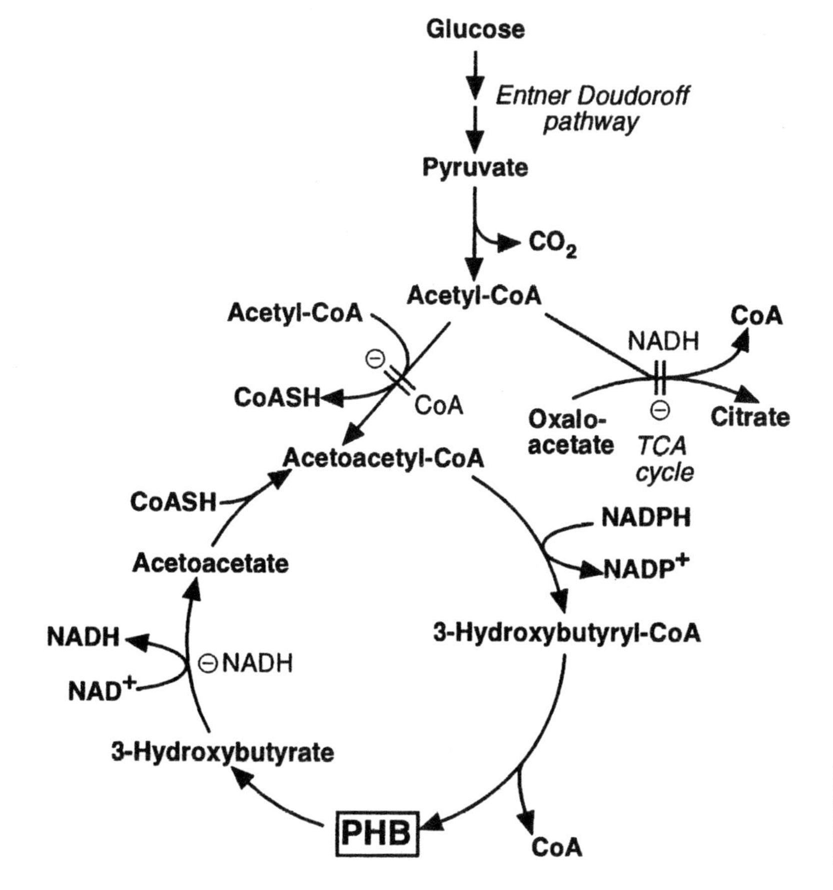

Fig. 15.8 Regulation of PHB synthesis and degradation in *Ralstonia eutropha*.

PHB biosynthesis and degradation form a cyclic process (Fig. 15.8). When the organism is deprived of a carbon and energy source, PHB can be broken down to acetyl-CoA and metabolised via the TCA cycle. The degradation of PHB has been studied in less detail than biosynthesis. PHB depolymerase is, like PHA synthase, associated with the granule. NADH is an inhibitor of the subsequent oxidation reaction, catalysed by 3-hydroxybutyrate dehydrogenase. This regulation will help to prevent **futile cycling** or simultaneous synthesis and degradation of PHB.

15.2.6 Biosynthesis of PHB/V

Certain strains of *Ralstonia eutropha* can produce PHB/V from glucose plus propionic acid, and the 3HV monomer is made exclusively from the latter substrate. Propionic acid is first activated to yield propionyl-CoA:

$$\text{Propionic acid} + \text{CoA} + \text{ATP} \rightarrow \text{Propionyl-CoA} + \text{AMP} + \text{PPi}$$

The subsequent reactions to produce 3HV monomers involve the same three enzymes used in the production of 3HB monomers for the biosynthesis of PHB. In this case, 3-ketothiolase catalyses the condensation of propionyl-CoA with acetyl-CoA instead of two molecules of acetyl-CoA. The resulting polymer (Fig. 15.6) has 3HB and 3HV monomers in a random sequence. It does **not** contain PHB and PHV as separate polymers. The concentrations of propionic acid and glucose in the medium determine the availability of precursors of the 3HB and 3HV monomers and hence the composition of the polymer.

15.2.7 Biosynthesis of other PHA

Pseudomonas oleovorans produces PHA from organic acids and alkanes. For example, PHA produced from *n*-octanoic acid or *n*-octane contains a high proportion of 3-hydroxyoctanoate. The observation that this organism, when grown on a substrate with a carbon chain length of n, could produce 3-hydroxyacid monomers containing n-2 or $n+2$ carbon atoms has suggested the involvement of reactions of fatty acid degradation (β-oxidation, see Chapter 2) and synthesis, respectively. The reactions of fatty acid metabolism are now known to have an important role in the production of 3-hydroxyacids for use in PHA biosynthesis.

15.2.8 *Biopol* – a commercial biodegradable plastic made from PHA

Although the first published study on PHB, by Lemoigne of the Institut Pasteur, was in 1926, it was not until the late 1960s that the first patents for production and recovery of the polymer were published. In the 1980s, ICI plc in the UK succeeded in developing both a high-density fermentation process and downstream processing methods for the recovery of PHB and PHB/V without the need for costly solvent extraction. *Biopol* was the trade name used for the range of polymers manufactured by ICI. These materials were the subject of extensive research before *Biopol* was used for the commercial manufacture of biodegradable plastic articles. The first commercial product made from *Biopol* was a shampoo bottle sold by Wella in Germany. The production of *Biopol* was continued by Zeneca plc (an offshoot of the original ICI company) and the process was subsequently acquired by Monsanto in the USA. The greatest demand for biodegradable plastics was for use in packaging and disposable products, but *Biopol* was too expensive to compete with conventional non-biodegradable plastics, and production ceased in 1998. Other companies have maintained an interest in PHA but none have, as yet, launched any products aimed, like *Biopol*, at the high volume market.

A glucose-utilising strain of the Gram-negative soil bacterium *Ralstonia eutropha* was selected by ICI for the production of *Biopol*. The fermentation was run as a two-stage process (Fig. 15.9); a biomass production phase and a polymer accumulation phase, are employed. The organism is grown in a simple glucose/salts medium and when the phosphate source is depleted at the end of the growth phase, glucose and propionic acid are fed to the culture until approximately 80% of the dry weight of the bacteria consists of PHA. Phosphate was chosen as the growth-limiting nutrient because it is a relatively expensive nutrient compared, for example, with ammonia and because surplus phosphate would substantially increase the cost of effluent treatment of waste from the process. At concentrations in excess of 1 g l^{-1}, propionic acid inhibits polymer accumulation and its concentration must therefore be monitored carefully during the polymer accumulation phase. Downstream processing involves cell rupture and solubilisation of components other than PHA. The polymer is washed and recovered by centrifugation.

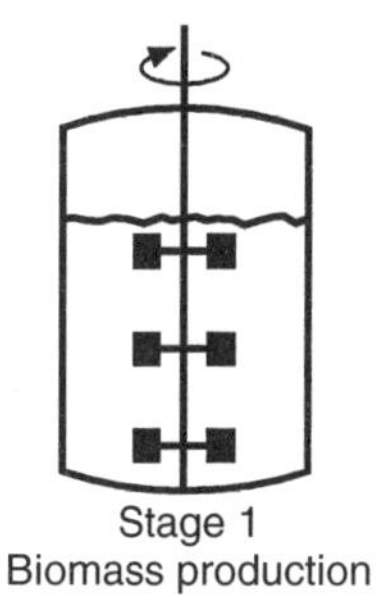

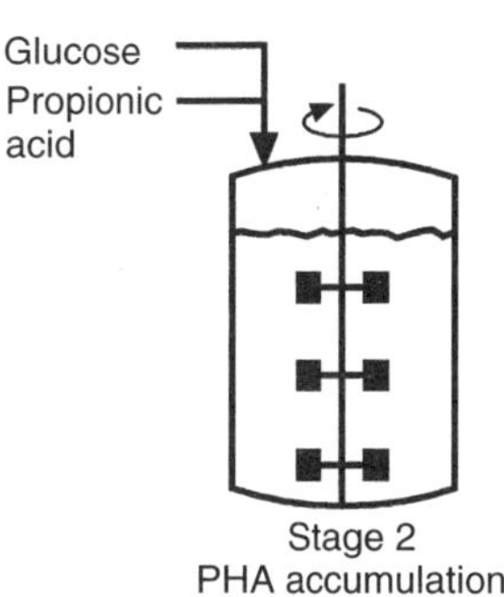

Fig. 15.9 *Biopol* is produced by a two-stage process, using a simple, defined medium containing glucose as the principal carbon source. When growth ceases due to exhaustion of phosphate (end of Stage 1), propionic acid and additional glucose are supplied to the culture (Stage 2). The temperature, pH and dissolved oxygen concentration of the stirred tank reactor are controlled throughout the fermentation.

The proportions of 3HB and 3HV monomers in *Biopol* (Fig. 15.10) are determined by the relative amounts of glucose and propionic acid fed to the culture. It is possible to produce PHB/V containing up to 30% 3HV units but, for many applications, polymers containing around 10% 3HV possess the required characteristics. Most conventional plastics include plasticisers to improve their flexibility and a biodegradable plasticiser was used in *Biopol* formulations.

Biopol can be blow-moulded to produce bottles and other items. Careful temperature control is essential to minimise thermal degradation, but standard industrial equipment for processing conventional plastics can be used. *Biopol* has food contact approval and can be used to coat card with a thin, waterproof layer for use in, for example, paper cups and food trays.

15.2.9 Medical applications of PHB

PHB is biocompatible and can be implanted in the body without producing an immune response, which is the cause of 'rejection' of most foreign materials by the body. It is slowly degraded to 3-hydroxybutyric acid, which is already present in the bloodstream, but the rate of degradation of PHB is too low for the plastic to be useful for degradable sutures and other implants intended to be degraded reasonably quickly. It can, however, be used for more durable implants, such as bone plates, and for wound dressings.

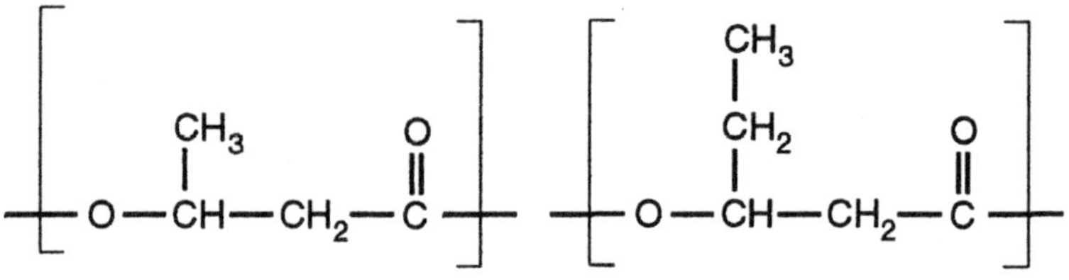

$$\left[-O-CH(CH_3)-CH_2-C(=O)- \right] \quad \left[-O-CH(CH_2CH_3)-CH_2-C(=O)- \right]$$

3HB monomer

3HV monomer
(0-30 mol%)

Fig. 15.10 *Biopol* is based on the co-polymer PHB/V. The polymer becomes more flexible as the proportion of 3HV monomers is increased. Since 3HV monomers are produced only from propionic acid, the composition of the polymer can be controlled by varying the amounts of glucose and propionic acid fed to the culture during the PHA accumulation phase.

PHB/V and other PHA are not, at present, used in medical applications because the fate of degradation products is uncertain.

15.2.10 Production of PHA by recombinant bacteria

It is possible to express the genes encoding the enzymes required for PHA biosynthesis in *Escherichia coli* (Fig. 15.11) and other bacteria that do not normally produce PHA. In principle it would be possible to produce any PHA in any organism but there are various practical problems. The key genes required for biosynthesis of a polymer must first be identified and cloned. The host organism is particularly important for large-scale production of PHA because it determines the range of substrates that can be used. Productivity, maximum polymer content and the ease of downstream processing are other important considerations. Thus Gram-positive bacteria are unsuitable as hosts because their thicker and stronger cell wall limits the amount of polymer that they can accumulate and makes product recovery difficult.

Escherichia coli was an obvious host for PHA production in a recombinant organism because it is well characterised with regard to its genetics, metabolism and the range of substrates that it can utilise. Recombinant strains of *Escherichia coli* are already used in the biotechnology industry to produce a range of other products, so the technology

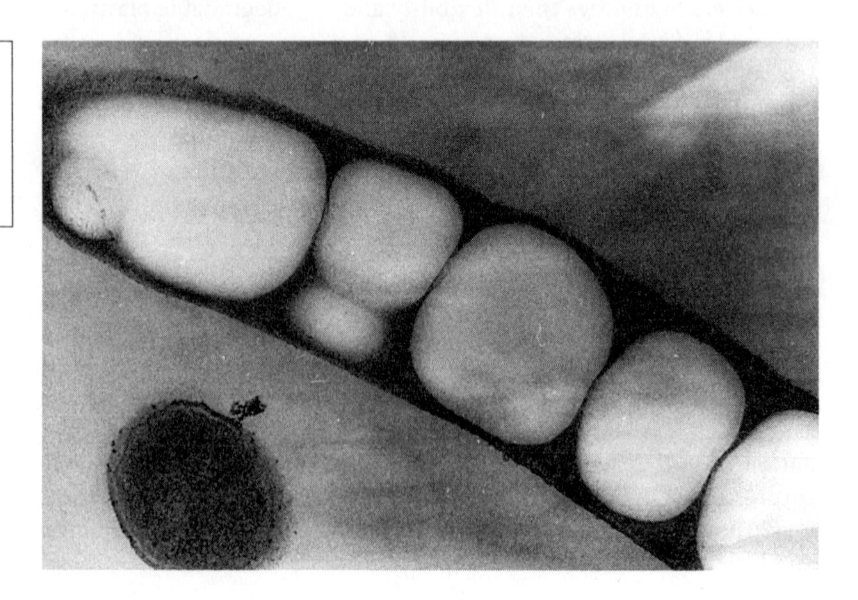

Fig. 15.11 PHB granules in a recombinant strain of *Escherichia coli*. The genes encoding the enzymes of PHB biosynthesis in *Ralstonia eutropha* have been transferred to this strain.

for operation of high density culture and recovery of intracellular products is well established. *Escherichia coli* also grows more rapidly than *Ralstonia eutropha*, the best studied PHA producing organism, and produces at least as much polymer. Not all recombinant strains are stable and their ability to produce PHA may decline during cultivation. Some of the *Escherichia coli* strains require a more complex and, therefore, more costly medium than that used for *Ralstonia eutropha*, and none can yet match the culture density possible with the latter organism. Most work with recombinant strains of *Escherichia coli* has focused on the production of PHB, but PHB/V and other co-polymers have also been produced.

15.2.11 Plants as prospective sources of PHA

Increased environmental awareness in most developed countries has not been sufficient to encourage manufacturers to use PHA or other biodegradable plastics, or even to achieve the recycling of more than a small fraction of packaging plastics. The fundamental problem with PHA produced by fermentation is that it can cost up to ten times as much as common petrochemical plastics.

The use of crop plants to produce PHA is potentially attractive because the costs of production of fermentation substrates and the fermentation process are replaced by simple and relatively inexpensive cultivation of plants, making use of photosynthesis to provide the carbon and energy required for polymer synthesis. The first transgenic plant (*Arabidopsis thaliana*) harbouring bacterial genes for PHB synthesis was reported in 1992 but the quantity of polymer present in the leaves was only 0.1%, although this has since been improved over a hundredfold. *Arabidopsis* was chosen because it is a model organism widely used for genetic manipulation studies. Oil-seed plants are particularly attractive candidates for PHB production because they already make large quantities of acetyl-CoA for oil biosynthesis. Eukaryotic systems are much more complex than bacteria and there are problems to overcome before plant systems are viable alternatives to fermentation. These include the problem of stunting as a result of the influence of polymer synthesis on normal metabolism and growth and the need to produce polymers other than PHB, which is of very limited commercial use. Thus fermentation is likely to remain the only real option for commercial production of PHA for a few more years.

15.3 Microbial polysaccharides

15.3.1 Introduction

Many micro-organisms produce substantial amounts of polysaccharide when surplus carbon source is available. Some of these polysaccharides accumulate within the cell and act as storage compounds, glycogen being a well-known example. Other polysaccharides, known as **exopolysaccharides (EPS)**, are excreted by the cell and are generally the microbial polysaccharides of commercial interest. They may remain

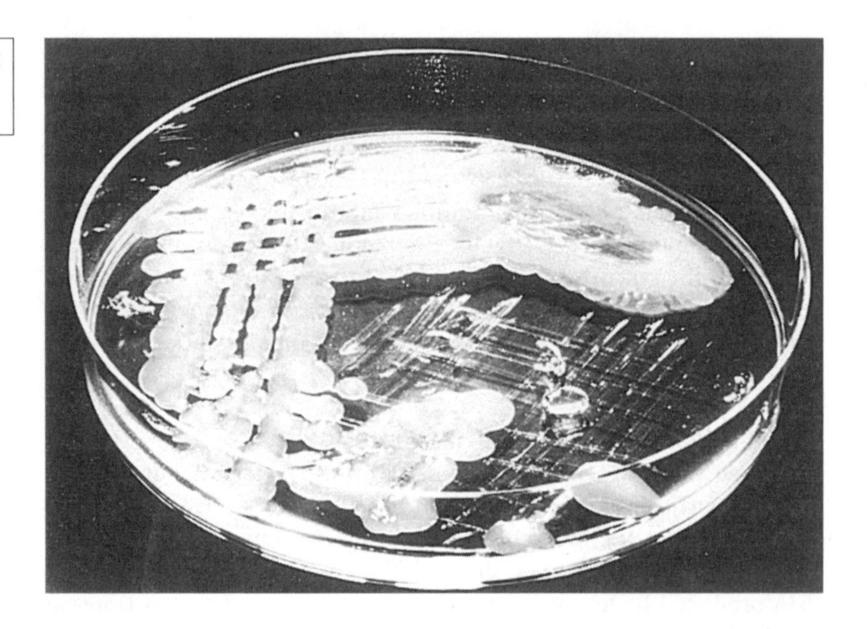

Fig. 15.12 An alginate-producing strain of *Pseudomonas mendocina* growing on agar.

associated with the cell, as a capsule or slime, or simply dissolved in the medium. This depends on various factors, including the chemical structure of the polysaccharide, and how vigorously the culture is agitated. On solid media, large slimy colonies may be produced (Fig. 15.12).

While some microbial exopolysaccharides, or **gums** as they are generally known in industry, are well established as commercial products, they must compete with plant polysaccharides, some of which are manufactured on a vast scale and at a low price. Production of microbial exopolysaccharides by fermentation can continue throughout the year, unlike production of plant polysaccharides, and fermentation, if carefully controlled, can yield a very consistent and reliable product. However, fermentation is a relatively costly process, which is not ideally suited to the manufacture of cheap products, even at high volume.

15.3.2 General properties

Microbial polysaccharides are, like plant and seaweed polysaccharides, of value because they can be used to modify the rheology (i.e. flow characteristics) of solutions. They increase viscosity and are commonly used as thickening, gelling and suspending agents.

Some polysaccharides, such as dextran and scleroglucan, are **neutral** and lack ionisable groups. Others, such as xanthan and gellan, are **acidic**. Acidic polysaccharides, which are of greater industrial importance, are polyelectrolytes, and possess carboxyl groups from uronic acids, such as glucuronic acid (Fig 15.13) and/or pyruvate residues.

The conformation (shape) of polysaccharide molecules in solution is affected by the ionic strength (salt concentration), pH and the concentration of the polysaccharide. The acidic polysaccharides are generally more affected by the presence of cations in solution.

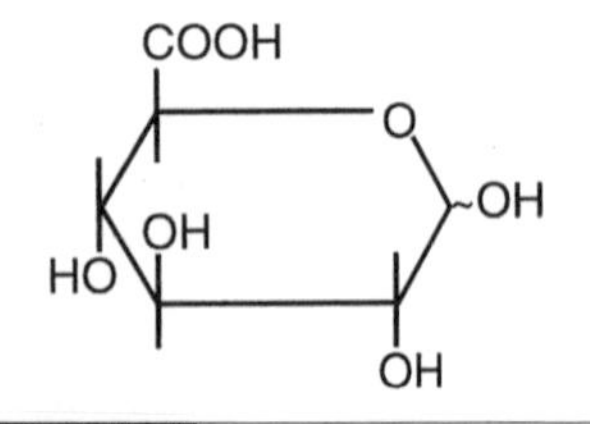

Fig. 15.13 Structure of glucuronic acid, which is commonly found in microbial exopolysaccharides.

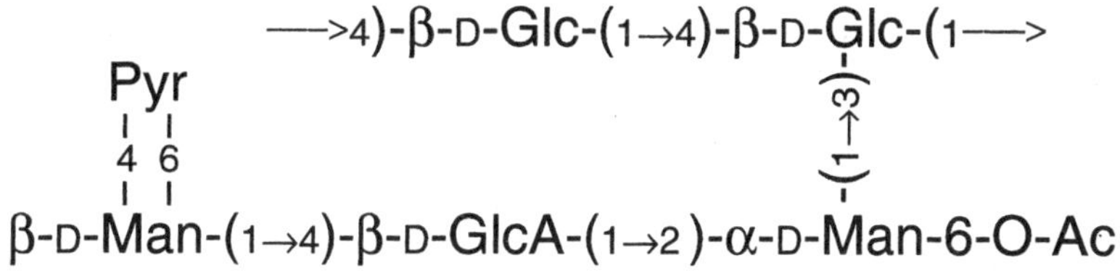

```
             —>4)-β-D-Glc-(1→4)-β-D-Glc-(1—>
                                   |
Pyr                              (1→3)
 | |                               |
 4 6                               |
 | |                               |
β-D-Man-(1→4)-β-D-GlcA-(1→2)-α-D-Man-6-O-Ac
```

Fig. 15.14 The structure of xanthan. The extent of acetylation of the mannose unit adjacent to the backbone is commonly 30%, but can be significantly lower or higher.

Divalent cations can cross-link polysaccharide chains to produce a strong gel.

15.3.3 Xanthan

Xanthan is produced by the Gram-negative bacterium, *Xanthomonas campestris*. It is the best-studied and most widely used exopolysaccharide. Xanthan is a large polymer, having an M_r in excess of 10^6 daltons. It is a branched polymer with a β-(1→4) linked glucan (i.e. polymer of glucose) backbone with a trisaccharide side chain on alternate glucose residues (Fig. 15.14). The pyruvate and acetate content depend on the bacterial strain, culture conditions and processing of the polymer. These substituents do not have a great influence on the properties of the polymer.

Xanthan is a polyelectrolyte due to the glucuronic acid residues in the side chains. Despite being an acidic polysaccharide, the viscosity of xanthan is relatively independent of the salt concentration.

Xanthan is the most important commercial microbial polysaccharide, and current production is around 20000 tonnes each year. Kelco, now part of Monsanto, is the principal manufacturer. Xanthan was first used in 1967 and approved for food use in the United States in 1969. It is widely used for stabilisation, suspension, gelling and viscosity control in the food industry. These properties are also exploited for water-based paints and a wide variety of other domestic and industrial applications. Crude xanthan is employed as a suspending and lubricating agent in drilling muds used by the oil industry.

15.3.4 Dextran

Dextran (Fig. 15.15) is an α-glucan containing various linkages, depending on the producing organism. It is produced by a wide variety of Gram-positive and Gram-negative bacteria, including *Leuconostoc mesenteroides* and *Streptococcus* species.

Unlike most exopolysaccharides, which are synthesised within the cell, dextran is produced from sucrose by an extracellular enzyme, dextransucrase, which acts on sucrose polymerising the glucose units and liberating free fructose into the medium.

The properties of dextrans are manipulated by hydrolysis of the solvent-precipitated polymer using *exo-* or *endo-*dextranases or mild acid treatment, to generate a product with the desired molecular weight range. Dextran was the first commercial microbial polysaccharide and has been manufactured by Pharmacia for almost 50 years. It was first used as a blood plasma extender. Dextrans now have many clinical

—>6)-α-D-Glc-(1—>
—>2)-α-D-Glc-(1—>
—>3)-α-D-Glc-(1—>
—>4)-α-D-Glc-(1—>

Fig. 15.15 Structure of dextran. The predominant likage is α-(1→6).

->3)-β-D-Glc*-(1→4)-β-D-GlcA-(1→4)-β-D-Glc-(1→4)-α-L-Rha-(1->

Fig. 15.16 Structure of gellan. *This glucose carries *O*-acetyl and glyceryl residues in the native polymer.

applications, including the prevention of thrombosis and use in wound dressings to absorb fluid. Sephadex remains a well-known gel filtration medium and dextrans now have many other laboratory applications. Dextrans are also used in foodstuffs.

15.3.5 Gellan

Gellan (Fig. 15.16) is a linear heteropolysaccharide whose repeating unit contains two glucose, one glucuronic acid and one rhamnose residue. Gellan is an acidic gel-forming polysaccharide produced by *Pseudomonas elodea*. It was developed by Kelco Inc, USA, as *Gelrite* by deacetylation of native gellan gum (by heating at pH 10), which is partially *O*-acetylated on one of the two glucose residues. The deacetylated product forms firm, brittle gels, which have the potential to replace agar and carrageenan. Gellan offers various advantages over agar for microbiological applications: it is resistant to enzymatic degradation and has a high gel strength at low concentration. Gel formation is influenced by temperature and the presence of cations and the polymer undergoes a coil to double helix transition upon gel formation.

Gellan is approved for use in food and is widely used, at low concentration, as a thickener.

15.3.6 Scleroglucan

Scleroglucan (Fig. 15.17) is a neutral polysaccharide with a 1→3-*β*-glucan backbone and branches consisting of a single glucose residue attached in an apparently regular sequence to every third glucose unit in the polymer chain.

Scleroglucan is a fungal exopolysaccharide, and is produced by various *Sclerotium* species. *Sclerotium rolfsii* and *Sclerotium glucanicum* are the most important species for commercial production of scleroglucan.

Scleroglucan is a soluble polysaccharide and is pseudoplastic over a broad pH and temperature range, and is unaffected by various salts. It is used to stabilise drilling muds, latex paints, printing inks and seed coatings.

->3)-β-D-Glc-(1→3)-β-D-Glc-(1→3)-β-D-Glc-(1->
(1→6)
β-D-Glc

Fig. 15.17 Structure of scleroglucan.

15.3.7 Curdlan

Curdlan (Fig. 15.18) is a 1→3-*β*-glucan produced as an exopolysaccharide by *Alcaligenes faecalis* var. *myxogenes*. Similar polysaccharides are produced by *Agrobacterium radiobacter* and *Agrobacterium rhizogenes*, and *Rhizobium trifolii*.

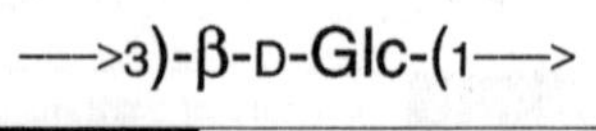

Fig. 15.18 Structure of curdlan.

—>6)-α-D-Glc-(1→4)-α-D-Glc-(1→4)-α-D-Glc-(1—>

Fig. 15.19 Structure of pullulan.

Unlike scleroglucan, curdlan is insoluble in water and forms a strong gel on heating above 55 °C and this gel formation is irreversible. Curdlan can be used as a gelling agent in cooked foods and as a support for immobilised enzymes. The properties of curdlan resemble those of the 1→3 -β-glucan, laminarin, which is found in many brown algae.

15.3.8 Pullulan

Pullulan (Fig. 15.19) is an α-glucan with a trisaccharide repeating unit. It is produced commercially using the fungus *Aureobasideum pullulans*. The fermentation is relatively slow (5 days) compared with the production of bacterial exopolysaccharides but 70% of the substrate (glucose) is converted to polysaccharide.

Pullulan forms strong, resilient films and fibres, and can be moulded. The films have a lower permeability to O_2 than cellophane or polypropylene and, being a natural product, the pullulan is biodegradable. Similar polymers are produced by some bacteria.

15.3.9 Alginate

Alginate is linear polymer composed of mannuronic and guluronic acids (Fig. 15.20). It is produced by the Gram-negative bacteria *Azotobacter vinelandii* and *Pseudomonas* species. The bacterial exopolysaccharide is similar to algal (seaweed) alginate, except that some of the mannuronic acid residues are *O*-acetylated.

The relative abundance of mannuronic and guluronic acids and the degree of acetylation depends on the organism and growth conditions. Polymers containing a high mannuronic acid content are elastic gels, whereas those with a high guluronic acid content adopt a different conformation and are strong, brittle gels. Alginates are not random co-polymers of mannuronic and guluronic acids, and regions containing a single monomer (i.e. -M-M-M-M-M-M- and -G-G-G-G-G-G-) may be present in the chain. These are known as block structures and also affect the shape and properties of the polymer.

Seaweed alginates are widely used in the food industry as thickening and gelling agents. Alginate beads provide a simple and effective method of immobilising cells and enzymes. The cell suspension or enzyme solution is mixed with a calcium salt and allowed to drip into a solution of alginate. The polysaccharide chains are crosslinked by interaction of the divalent cation with the carboxyl groups, forming a gel. Alginate is a useful matrix for immobilising cells, but may not retain enzymes efficiently.

Bacterial alginates are not used commercially because the producing strains are relatively unstable and they also excrete a degradative enzyme which decreases the molecular weight of the product. They have, however, considerable potential for commercial use because polymers with a wide range of properties can be produced by appropriate selection of the producing strain and fermentation conditions.

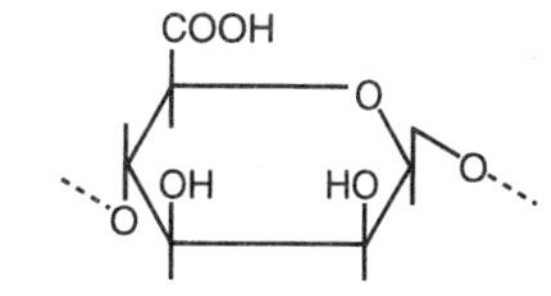

COOH
O
OH
HO
O
O
—>4)-β-D-Mannuronic acid-(1—>

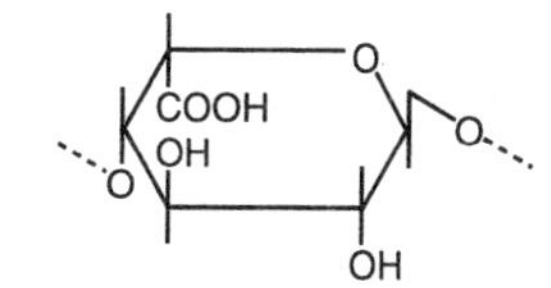

O
COOH
OH
O
O
OH
—>4)-α-L-Guluronic acid-(1—>

Fig. 15.20 Alginate is composed of mannuronic acid and guluronic acid. The proportions and sequence of these monomers depend on the source of the polymer.

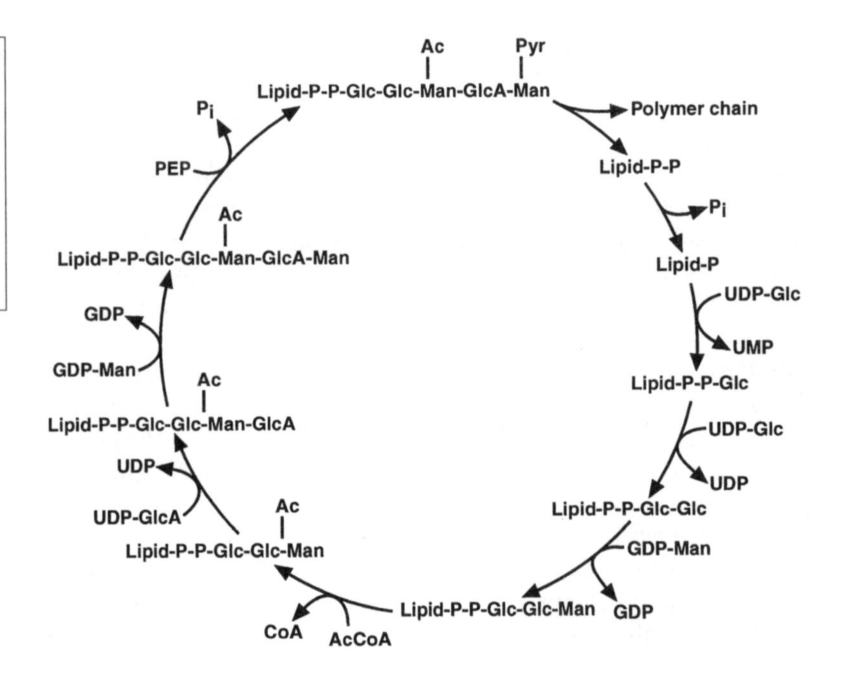

Fig. 15.21 Biosynthesis of xanthan in *Xanthomonas campestris*. Glc, glucose; Man, mannose; GlcA, glucuronic acid; UDP, uridine diphosphate; GDP, guanosine diphosphate; Ac, acetate; AcCoA, acetyl-CoA; Pyr, pyruvate; PEP, phosphoenolpyruvate; Lipid, lipid carrier (see Fig. 15.22).

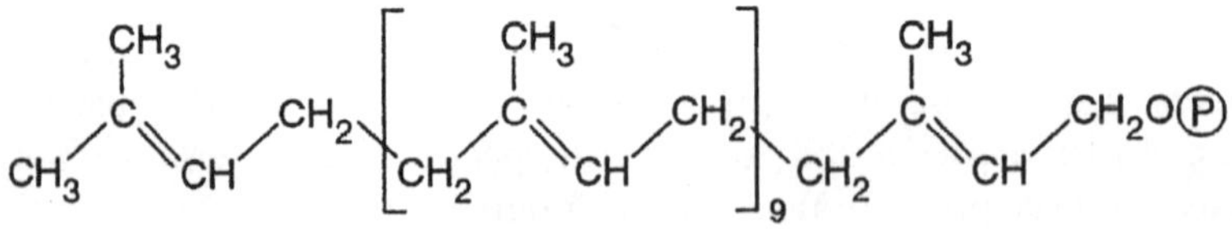

Fig. 15.22 Structure of the lipid carrier commonly involved in biosynthesis of microbial polysaccharides.

15.3.10 Biosynthesis of polysaccharides

The biosynthesis of xanthan, shown in Fig. 15.21. Each monomer is assembled on a lipid carrier (Fig. 15.22), anchored in the cytoplasmic membrane, prior to transfer to the growing polymer chain. The lipid carrier is similar, or the same as, the C_{55} isoprenyl phosphate used in the biosynthesis of peptidoglycan and lipopolysaccharides in bacterial cell walls.

In xanthan biosynthesis, sugar nucleotides, for example uridine diphosphate glucose (UDP-glucose), act as activated precursors, providing the energy for the formation of glycosidic bonds between adjacent monosaccharides.

The biosynthesis of most exopolysaccharides is essentially similar to that of xanthan and differences are beyond the scope of this chapter. Dextran synthesis is, however, quite different and is synthesised outside the cell. A single extracellular enzyme, dextran sucrase, cleaves the disaccharide sucrose to glucose and fructose, and polymerises the glucose units to form dextran.

15.3.11 Production of polysaccharides

Microbial polysaccharides are produced in batch culture in aerated stirred tank reactors. Polysaccharide synthesis generally commences during growth and continues after cessation of growth. Excretion of polysaccharide increases the viscosity of the culture. This limits the attainable polysaccharide concentration because it becomes increasingly difficult to achieve adequate mixing and O_2 transfer in the viscous cultures. Furthermore, the power required to stir viscous cultures is high and consequently the cost of heat removal to maintain the required temperature is increased.

Production of microbial polysaccharides is generally favoured by a high carbon/nitrogen ratio in the medium. The nitrogen source is the growth-limiting nutrient and its concentration is set to produce the required biomass concentration. Additional carbon source may be added after cessation of growth. Since cations can affect the rheological properties of polysaccharide solutions, care must be taken in optimising the concentrations of salts provided as nutrients in the medium.

Continuous culture is not used for production of microbial polysaccharides. At higher growth rates, which are desirable for high productivity, an increasing proportion of the carbon source is used to produce biomass rather than polysaccharide. Furthermore, some polysaccharide-producing micro-organisms are not stable in continuous culture and can be outgrown by variants that produce little polysaccharide. Strain stability is less of a problem in batch cultures which are of shorter duration.

The stages of xanthan production are summarised in Fig. 15.23.

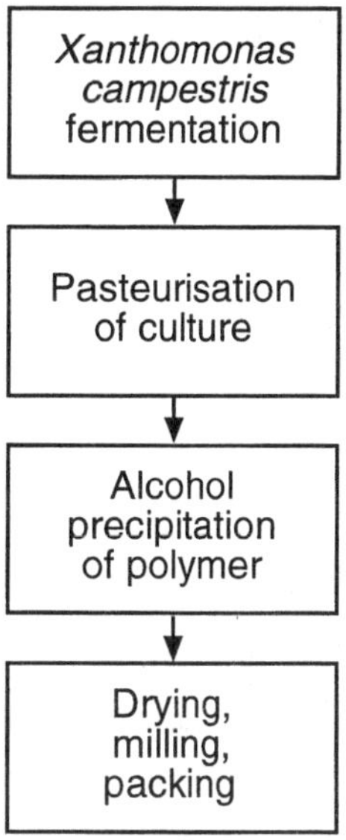

Xanthomonas campestris fermentation
↓
Pasteurisation of culture
↓
Alcohol precipitation of polymer
↓
Drying, milling, packing

Fig. 15.23 Production of xanthan.

15.4 Microbial lipids

15.4.1 Structure of lipids

What are lipids? Put simply the lipids this chapter will be primarily concerned with (i.e. triacylglycerols) are composed of three fatty acids attached to a three carbon (glycerol) backbone (see Fig 15.24). Although all triacylglycerols share this common structure their physical properties vary enormously, from hard waxy solids at room temperature (**fats**) to translucent liquids (**oils**). It is the structure of the fatty acid molecules (more correctly fatty acyl chains) attached to the glycerol backbone that accounts for the properties of lipids.

The oils that this chapter will focus on are the so-called **single cell oils**. Single cell oils are oils derived from microbial sources, produced on a commercial basis, and which are destined for human consumption.

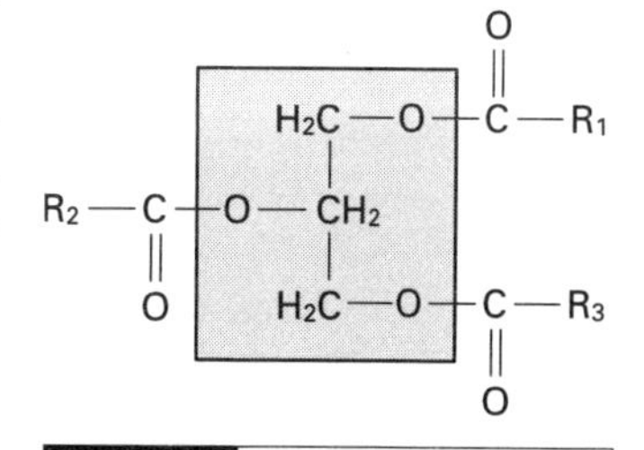

Fig. 15.24 Structure of a triacylglyceride molecule: the glyceryl backbone is inside the shaded box. Attached to this backbone are three fatty acyl residues containing aliphatic chains R_1, R_2 and R_3 respectively, which may all be identical or all different.

15.4.2 Fatty acid nomenclature

The naming of fatty acids can appear confusing as in most cases a single fatty acid can be assigned any one of three names, depending upon the personal preference of the author. The three names can be thought of as (i) a systematic name, (ii) a trivial name and (iii) a numerical designation. The three different names for some common fatty acids and the fatty acids that have been developed as single cell oils are shown in Fig 15.25. The systematic names, although precise, are often long and confusing to those unfamiliar with lipid chemistry. As a result these names

Fig. 15.25 Fatty acid structure and nomenclature.

Trivial name	Molecular structure	Systematic name	Numeric designation
Palmitic acid	$CH_3(CH_2)_{14}COOH$	Hexadecanoic acid	16:0
γ-Linolenic acid	$CH_3(CH_2)_4(CH{=}CH.CH_2)_3(CH_2)_3COOH$	All *cis*-6, 9, 12-octatrienoic acid	18:3(n-6)
Arachidonic acid	$CH_3(CH_2)_4(CH{=}CH.CH_2)_4(CH_2)_2COOH$	All *cis*-5, 8, 11, 14-eicosatetraenoic acid	20:4(n-6)
DHA	$CH_3CH_2(CH{=}CH.CH_2)_6CH_2COOH$	All *cis*-4, 7, 10, 13, 16, 19-docosahexaenoic acid	22:6(n-3)

Fig. 15.26 Overall scheme of modifications made to fatty acids after *de novo* synthesis. **Elongases** serve to increase the fatty acid chain length by addition of a C_2 unit (acetyl-CoA). **Desaturases**, indicated by Δ, introduce a double bond between two adjacent C atoms. Only the position of the first C atom is given and this is indicated by the number: thus Δ9 means that the bond from C atom 9 to C atom 10 is now a double bond. The fatty acid has become **unsaturated**.

16:0
↓ elongase
18:0 —Δ9→ 18:1(Δ9) —Δ12→ 18:2(Δ9,12) —Δ15→ 18:3(Δ9,12,15)

n-9 series	n-6 series	n-3 series
18:1(Δ9)	18:2(Δ9,12)	18:3(Δ9,12,15)
↓ Δ6	↓ Δ6	↓ Δ6
18:2(Δ6,9)	18:3(Δ6,9,12)	18:4(Δ6,9,12,15)
↓ elongase	↓ elongase	↓ elongase
20:2(Δ8,11)	20:3(Δ8,11,14)	20:4(Δ8,11,14,17)
↓ Δ5	↓ Δ5	↓ Δ5
20:3(Δ5,8,11)	20:4(Δ5,8,11,14)	20:5(Δ5,8,11,14,17)
		↓ elongase
		22:5(Δ7,10,13,16,19)
		↓ Δ4
		22:6(Δ4,7,10,13,16,19)

are seldom used. In contrast the trivial names are still in common usage, both in scientific and non-scientific circles. Trivial names have the disadvantage of giving no direct information about the chemical structure of the fatty acid. The numeric designation has the benefit of simplicity, whilst explicitly denoting the structure of a fatty acid. In this designation the number before the colon (see Fig. 15.25) denotes the number of carbons in the acyl chain whilst the number after the colon indicates the number of double bonds in the fatty acid. The n-3, n-6 or n-9 in brackets informs the reader which series the fatty acid belongs to (see Fig. 15.26) and indicates the position of the **last double bond** relative to the terminal methyl group. Once the position of the last double bond is fixed the position of all the remaining unsaturations can be deduced (see Section 15.4.4). The numerical designation gives the exact structure of any straight chain polyunsaturated fatty acid in a concise and easily understood format.

15.4.3 Fatty acids: the building blocks of lipids

Fatty acids can be considered the primary building blocks of lipids in the same way that amino acids are the building blocks of proteins. Fatty acids are usually aliphatic long chain carboxylic acids [the group of compounds that includes formic acid and ethanoic (acetic) acid as its first two members]. Although some organisms (particularly bacteria) produce fatty acids that have branched chains or even contain cyclopropane rings, those of current biotechnological importance are simple straight chained fatty acids (Fig. 15.25). Fatty acids synthesised by eukaryotic micro-organisms (fungi, yeast, algae), which are the only sources of commercially important single cell oils to date, normally contain 16 to 24 carbon atoms and can possess up to six double bonds. The presence of double bonds in a fatty acid structure, referred to as **unsaturation** or **desaturation** (and therefore to the fatty acid being an unsaturated fatty acid), introduces a 'kink' into the molecule and stops the fatty acid packing closely to its neighbouring fatty acids in the lipid. This effect on the packing of fatty acids causes unsaturated fatty acids to have lower melting points than saturated fatty acids (fatty acids with no double bonds). The more double bonds introduced into a fatty acid the lower its melting point. The introduction of double bonds into fatty acids containing 18 carbons demonstrates this, whilst the saturated fatty acid, stearic acid (18:0), has a melting point of 65 °C, the mono-unsaturated oleic acid (18:1(n-9)) has a melting point of 13 °C and the diunsaturated linoleic acid ((18:2(n-6)) has a melting point of 5 °C. As a result, **polyunsaturated fatty acids (PUFAs)** are more fluid at room temperature than diunsaturated fatty acids which are in turn more fluid than mono-unsaturated and saturated fatty acids. In general terms, unsaturated fatty acids are liquid at room temperature and lipids that are rich in these fatty acids are oils, whereas saturated fatty acids are solid, and lipids rich in saturated fatty acids are fats.

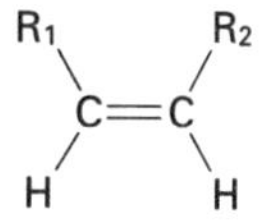

15.4.4 Unsaturated fatty acids

When fatty acids are synthesised *de novo* from acetyl-CoA and malonyl-CoA by the enzyme complex **fatty acid synthase**, they are saturated. Double bonds are added to saturated fatty acids after synthesis by enzymes called **fatty acid desaturases**. When double bonds are inserted into fatty acids they are introduced in specific conformation. In theory, as the introduction of a double bond 'locks' the fatty acid structure, the double bonds could be introduced in either a *cis* or *trans* form (Fig. 15.27), however, in nature the double bonds in fatty acids are almost exclusively in the *cis* form. The position of the double bonds and the order of their insertion is also highly ordered (Fig. 15.26). The first double bond is inserted between carbons nine and ten in the acyl chain (carbons numbered from the carboxylic acid group). Subsequent double bonds can be inserted at a number of sites giving rise to the n-3, n-6 and n-9 series of fatty acids. In polyunsaturated fatty acids the double bonds are always methylene interupted, so that double bonds have a saturated carbon between them (i.e. -CH = CH-CH_2-CH = CH-).

cis double bond

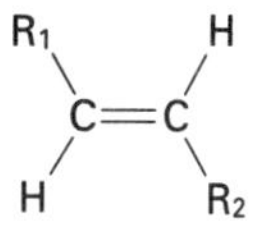

trans double bond

Fig. 15.27 Structure of *cis* and *trans* double bonds. Double bonds 'lock' the fatty acid structure and lead to the existence of *cis* and *trans* isomers. Fatty acids in biological systems are almost exclusively the *cis* isomer. R_1 and R_2 represent acyl chains, in a fatty acid molecule one will possess the terminal methyl (CH_3) group whilst the other will possess the carboxylic acid (COOH) group.

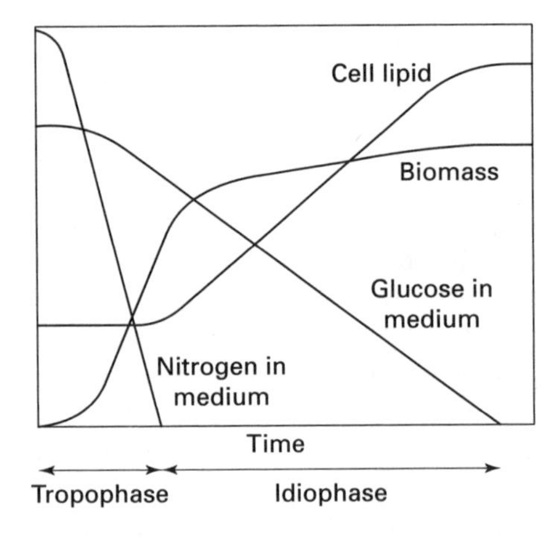

Fig. 15.28 A schematic representation of the timing of lipid accumulation in oleaginous micro-organisms during a batch cultivation.

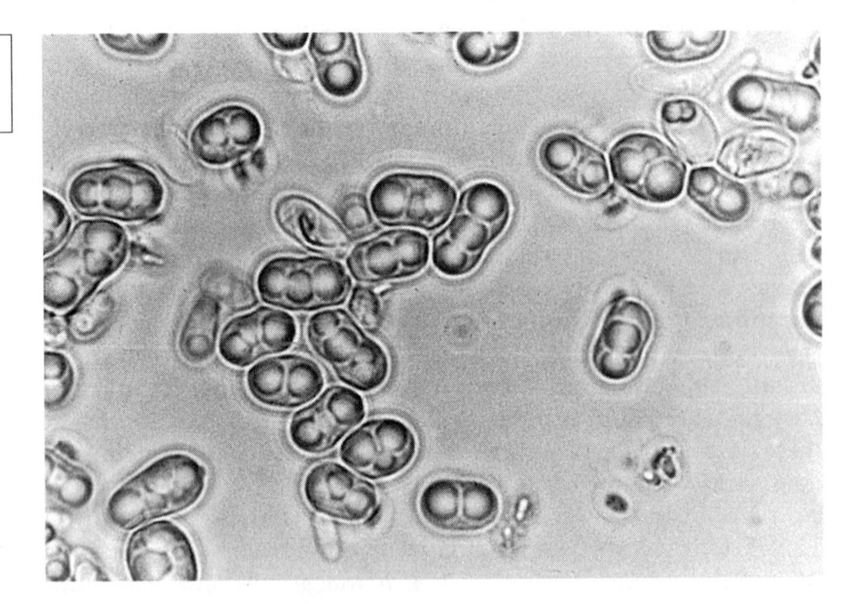

Fig. 15.29 Oleaginous yeast *Apiotrichium curvatum*. Cells are packed with oil droplets.

15.4.5 The cellular role of lipid

Triacylglycerols are generally storage compounds, accumulated in eukaryotic cells under conditions of carbon excess when growth has ceased due to the exhaustion of some other essential nutrient, usually nitrogen (Fig. 15.28). When produced in substantive amounts, as in some micro-organisms (termed **oleaginous**), the accumulated oil coalesces to form an oil droplet(s) which can occupy a significant portion of the cell volume (see Fig. 15.29). Stored triacylglycerols act as a carbon store to maintain essential metabolic processes in the event of subsequent carbon starvation. Another suggested function, albeit restricted to marine micro-organisms, is that the accumulation of large lipid droplets in the cytosol acts as an aid to buoyancy.

Of greater metabolic significance is another class of lipids, the **phospholipids**. Phospholipids differ from triacylglycerols in that instead of three fatty acyl chains attached to the glycerol backbone they only have two. The third position, on the glycerol backbone, is occupied by (as the name suggests) a phospho-group. The phospho-group can be phosphate itself or more commonly contains a hydroxyl-containing compound linked to the phosphate group by an ester linkage (see Fig. 15.30). Regardless of the exact nature of the 'head group' the effect of the polar phospho-group gives all phospholipids their crucial amphiphilic nature. By **amphiphilic** it is meant that phospholipids have separate portions of their structure that have very different solubility in water. The fatty acyl chains are non-polar and hydrophobic (insoluble in water) whereas the polar phospho-group is hydrophilic (water soluble). Due to their amphiphilic nature, when phospholipid molecules are suspended in water they organise themselves to form micelles with the polar (phospho-containing) portion of the molecule on the outside, in contact with water, and the fatty acyl chains inside the micelle forming a hydrophobic core. It is this property of phospholipids that allows them to play such a major role in the structural integrity of all biological membranes. Phospholipids form the lipid bilayer of cell membranes (see Fig. 15.31).

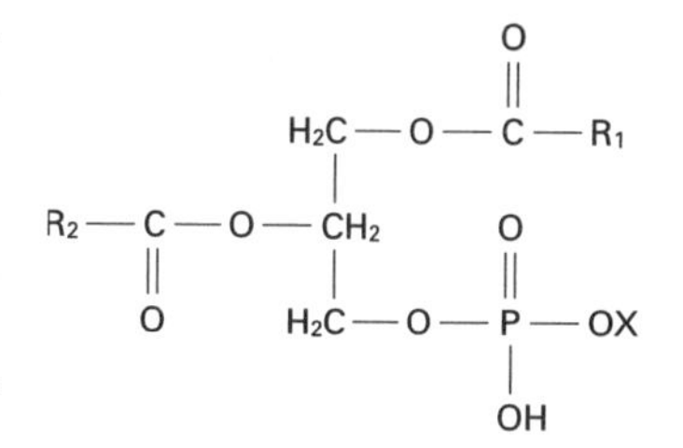

Fig. 15.30 Structure of a phospholipid molecule: X can be H, ethanolamine, serine, inositol, choline, glycerol etc. When the attached group is H the molecule is phosphatidic acid, the others are named phosphatidyl X (i.e. phosphatidylethanolamine etc.).

Just as the nature of the fatty acids in triacylglycerols determines their physical properties so the same applies to phospholipids and the biological membranes of which they form such an important part. Alteration of the fatty acyl profile of the phospholipid component of cell membranes plays a key role in the regulation of membrane fluidity, a crucial parameter in the maintenance of many membrane associated processes (photosynthesis, the electron transport chain etc.).

In mammals certain key fatty acids, notably arachidonic acid [20:4(n-6)] and eicosapentaeonic acid [20:5)n-3)], are precursors for the synthesis of a group of short lived but potent cellular signal molecules, the **eicosanoids**. The eicosanoids synthesised from 20:4(n-6) and 20:5(n-3) have antagonistic effects on such diverse cellular processes as gastrointestinal function and the immune system.

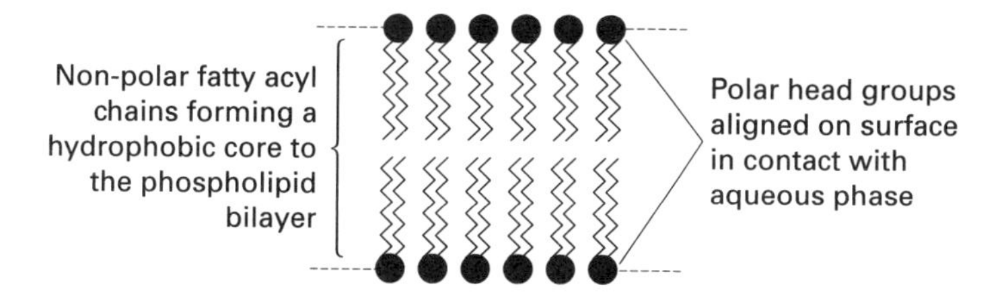

Fig. 15.31 Self-organisation of phospholipid molecules into a phospholipid bilayer.

15.4.6 The biochemistry of oleaginicity

The synthesis of fatty acids and their incorporation into triacylglycerols is a process that differs little between different cell types. This process is well documented (see Further reading) and will not be described in detail here. In general fatty acids are synthesised by the large multi-enzyme protein, **fatty acid synthase**, according to the overall reaction:

$$\text{acetyl-CoA} + 7\ \text{malonyl-CoA} + 14\ \text{NADPH} \rightarrow \text{palmitoyl-CoA (16:0-CoA)} + 7\ \text{CoA-SH} + 7\ CO_2 + 14\ NADP^+$$

Fig. 15.32 Schematic representation of biochemistry underlying oleaginicity in micro-organisms. Enzymes: 1, pyruvate dehydrogenase; 2, citrate synthase; 3, aconitase; 4, NADH isocitrate dehydrogenase; 5, AMP deaminase; 6, ATP:citrate lyase; 7, acetyl-CoA carboxylase; 8, fatty acid synthase; 9, malate dehydrogenase; 10, malic enzyme

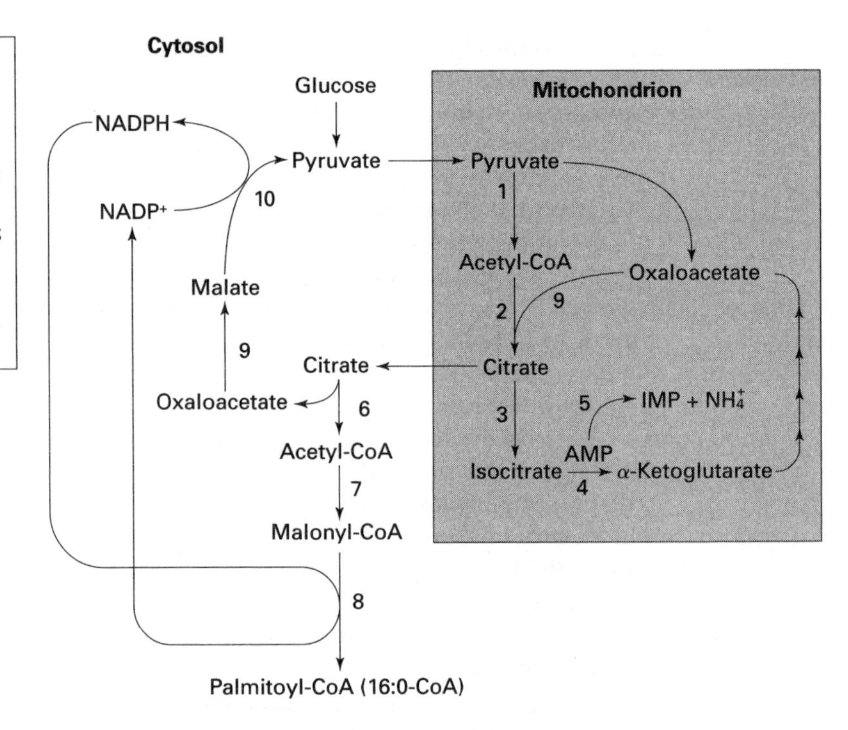

Malonyl-CoA is itself derived from acetyl-CoA, which is carboxylated by the enzyme **acetyl-CoA carboxylase**, in the first committed enzyme of lipid biosynthesis.

Although acetyl-CoA carboxylase is widely regarded as a key regulatory site in lipogenesis, it now appears more likely that the major regulation of lipid biosynthesis is imposed by the supply of the primary substrates (i.e. acetyl-CoA and NADPH). In this regard two enzymes, **ATP:citrate lyase** and **malic enzyme**, appear crucial. ATP:citrate lyase is involved in the production of cytosolic acetyl-CoA for lipid biosynthesis whilst malic enzyme is crucial for the production of NADPH for fatty acid synthesis (see Fig. 15.32).

In eukaryotes (most prokaryotes do not accumulate significant amounts of triacylglycerol, instead they tend to accumulate PHB/As, see Section 15.2) acetyl-CoA is generated from pyruvate in the mitochondria. Acetyl-CoA is too large to cross biological membranes so transport mechanisms exist to export it out of the mitochondria and into the cytosol for anabolic pathways, including lipogenesis. The major pathway for the export of acetyl-CoA, from the mitochondria involves citrate. Citrate is formed in the mitochondria by the combination of acetyl-CoA and oxaloacetate as part of the citric acid cycle (see Chapter 2). Citrate is normally converted into isocitrate and then α-ketoglutarate respectively (see Fig. 15.32). However, under conditions of nitrogen depletion (which triggers lipogenesis), a nitrogen-scavenging enzyme, AMP deaminase, is activated and causes the cellular levels of AMP to fall. In oleaginous micro-organisms, isocitrate dehydrogenase activity is dependent on AMP as a cofactor, so the decrease in AMP concentration restricts the conversion of isocitrate to α-ketoglutarate. As a result of this metabolic 'bottle-neck' the intra-mitochondrial concentration of citrate (rather than isocitrate, as a result of the reversibility of aconitase) increases and stimulates the export of citrate into the cytosol. In the cytosol, citrate is cleaved by ATP:citrate lyase to produce acetyl-CoA for lipid synthesis and oxaloacetate which is converted to malate. The malate is decarboxylated to yield pyruvate and NADPH the latter which can then be utilised by fatty acid synthase as the necessary reducing power needed for fatty acid biosynthesis. Decarboxylation of malate (and hence the generation of NADPH) is catalysed by malic enzyme. Although malic enzyme is not the only NADPH-generating system in oleaginous micro-organisms this enzyme appears to be intimately associated with the lipogenic pathway as a deficiency of malic enzyme inhibits lipid accumulation.

15.4.7 Nutritional importance of polyunsaturated fatty acids (PUFAs)

Animals (including man) are capable of synthesising the saturated fatty acid, palmitic acid (16:0), and of carrying out a wide range of subsequent modifications including desaturations and elongation reactions (see Fig. 15.26). Nevertheless, a dietary intake of PUFAs is important for maintaining human health. This is because animals are incapable of carrying out $\Delta 12$ and $\Delta 15$ desaturations (see Fig. 15.26) whilst fatty acids possessing desaturations in these positions [i.e. the **essential fatty acids** linoleic acid, 18:2(n-6) and α-linolenic acid, 18:3 (n-3)] are required by the body.

The need for dietary PUFAs by animals is due to the conversion of certain fatty acids into the eicosanoids (see Section 15.4.5), which play key physiological roles in the control of a wide range of bodily functions, including processes such as homeostasis and blood clotting.

Furthermore, a range of human diseases appear to be the result of an inability to desaturate fatty acids, though other factors may also be involved. Disorders such as rheumatoid arthritis, multiple sclerosis, schizophrenia and pre-menstrual syndrome have been all reported to fall into this category, and their symptoms can, in some cases, be eleviated by an increased dietary intake of PUFAs.

At the moment the dietary importance of polyunsaturated fatty acids is becoming appreciated both by governments and the general public. In particular the importance of PUFAs in the development of new born babies is generating more interest. Two polyunsaturated fatty acids, arachidonic acid [20:4(n-6)] and DHA [22:6(n-3)] in particular have been strongly implicated in the development of brain and eye function in new-born infants.

15.4.8 Micro-organisms as 'oil factories'

Commercial oils (used in cosmetics and foodstuffs) are usually obtained from plants or animals. If an oil from a microbial source is to be produced on a commercial scale it must compete with oils from these 'traditional' sources. As fermentation technology is expensive single cell oils can only compete with the most expensive speciality oils. These are

Table 15.2 The fatty acid profiles of lipid from a number of eukaryotic micro-organisms demonstrating the range and variation of fatty acids produced by different organisms. *Apiotrichum curvatum* is a yeast, *Mortierella alpina* is a fungus, *Crypthecodinium cohnii* is a marine alga and *Thraustochytrium aureum* is a marine fungoid protist

Organism	Relative % (w/w) of fatty acid in total cell lipid									
	16:0	18:0	18:1	18:2	18:3(n-6)	18:3(n-3)	20:4(n-6)	20:5(n-6)	22:6(n-3)	Others
Apiotrichum curvatum	17	12	55	8	–	–	–	–	–	8
Mortierella alpina[a]	16	14	14	10	–	4	35	–	–	7
Crypthecodinium cohnii[a]	25	2	12	–	–	–	–	–	40	21
Thraustochytrium aureum[a]	10	10	30	5	2	–	–	8	15	20

Note:
[a] Organism used for commercial production of single cell oil.

inevitably the very long chain polyunsaturated fatty acid rich oils destined for human consumption for which no convenient plant or animal source currently exists.

Although plants produce a number of unsaturated fatty acids including the essential fatty acids, linoleic acid 18:2 (common sources being sunflower and rape seed oils) and α-linolenic acid 18:3(n-6) (common sources being flax and linseed oils), they do not produce very long chain polyunsaturated fatty acids (> 18 carbons long). In comparison, animals (including fish) are a source of a number of very long chain, highly unsaturated fatty acids [up to 22:6(n-3)]. These fatty acids tend to be present in animal or fish oils in relatively small quantities, however, making processing the oil to enrich the desired fatty acid difficult and expensive. Furthermore, fatty acids from animal sources are unacceptable to a significant portion of society on moral and/or religious grounds. The possible transfer of disease causing agents (prions) in animal oils and the risk of pollutants persisting in fish oils is also a potential problem. Likewise, plant oils have the potential to contain residues of the pesticides and herbicides used in the cultivation of the oil seed plants.

Eukaryotic micro-organisms have the advantage over 'traditional' sources in that they not only produce a wide variety of PUFAs (the fatty acid profiles of some selected micro-organisms are shown in Table 15.2) but also some species accumulate large quantities of single PUFAs in their cell lipids which simplifies oil processing. Microbially derived oils also present no problems for the consumer on ethical or religious grounds and, moreover can be essentially guaranteed to be devoid of unwanted and potentially harmful contaminants.

Although expensive, the use of fermentation technology in the production of single cell oils has the advantage that a high degree of control over the process can be maintained. As a result the quantity and quality of oil produced can be controlled far more precisely than is possible with oils from animal and plant sources.

15.4.9 Current applications for single cell oils

To date only three PUFAs have been produced commercially using micro-organisms. The first single cell oil was rich in γ-linolenic acid [18:3(n-6)] produced using the fungus *Mucor circinelloides*, a process developed by the Lipid Research Group at the University of Hull. This oil was produced in the UK between 1985 and 1990, by J & E Sturge Ltd at Selby in Yorkshire, in competition with the 'traditional' oil of evening primrose. The process was discontinued in the face of a decreasing price of 18:3(n-6) when alternative agricultural sources came onto the market in the form of starflower oil and blackcurrant seed oil. The other single cell oils, rich in either 20:4(n-6) or 22:6(n-3) continue to be produced commercially in the absence of any real competition from oils from traditional sources.

Arachidonic acid [20:4(n-6)], is produced using the common soil fungus, *Mortierella alpina*, which has the ability to accumulate up to 50% (w/w) of its dry weight as lipid of which as much as 40% can be 20:4(n-6). Processes using this organism have been developed by DSM–Gist in the Netherlands and Zeneca-Roche in the UK although the future of these processes (beyond 2000) is uncertain.

Docosahexaenoic acid [DHA, 22:6n-3], is produced commercially by Martek Biosciences, Maryland, USA and Omega-Tech (in collaboration with Monsanto), Boulder, Colorado, USA both utilising marine algae. Although fish oils are a potential source of 22:6(n-3) they cannot be used to obtain this fatty acid for inclusion in baby milk formula because fish oil also contains another fatty acid 20:5(n-3) which cannot be separated from 22:6(n-3) using conventional oil processing, and which should not be given to infants. The marine micro-organisms employed to produce 22:6(n-3), devoid of contaminating 20:5(n-3), on a commercial scale are the heterotrophic (i.e. non photosynthetic) marine algae *Crypthecodinium cohnii* (Martek Bioscience) and *Schizochytrium* (Omega-Tech).

15.4.10 Future applications

If current data on the nutritional benefits of long chain PUFAs (particularly for babies) are confirmed then the future for single cell oils looks assured in the short term at least. As more research is carried out it is also likely that nutritional benefits of other PUFAs will emerge (as well as additional benefits of those presently in production).

In the long term, plant geneticists claim, single cell oils will become redundant as the genes for desired PUFAs will be transferred from microbial sources into plants. Production of PUFAs by transgenic plants will allow very large quantities to be obtained more cheaply than can be produced using fermentation processes. This, however, relies upon public acceptance of genetically engineered food products, which at present looks problematic (at least in Europe), especially for inclusion in baby milk formula! It may be that the general public will be more willing to accept nutritional supplements from naturally occurring microbial sources than from transgenic plants.

15.4.11 Functional foods

A recent development is the concept of 'functional foods'; foodstuffs that contain ingredients that are included specifically to give the product a (supposedly) defined health benefit to the consumer. Functional foods containing PUFAs 20:4(n-6) and 22:6(n-3) are being produced by Omega-Tech, in the US. The basis of this approach is to feed farm animals diets supplemented with microbial sources of desired PUFAs to produce animal-derived food stuffs (meat, dairy products and eggs) that are enriched with these fatty acids. This approach may overcome the reluctance of some sections of the community to ingest PUFAs from microbial sources but has the same drawbacks as the use of oils from animal sources (see Section 15.4.8).

15.5 Further reading

Carlson, S. E. (1995). The role of PUFA in infant nutrition. *Inform* **6**, 940–946.

Doi, Y. (1990). *Microbial Polyesters.* VCH, Weinheim.

Gill, I. & Valivety, R. (1997). Polyunsaturated fatty acids, part 1: Occurrence, biological activities and applications. *Trends Biotechnol.* **15**, 401–409.

Madison, L. L. and Huisman, G. W. (1999). Metabolic engineering of poly(3-hydroxyalkanoates): From DNA to plastic. *Microbiol. Mol. Biol. Rev.* **63**, 21–53.

Ratledge, C. (1997). Microbial lipids. In *Biotechnology, Vol. 7* (Rehm, H-J. and Reed, G. eds.), pp. 135–197. VCH, Weinheim.

Stryer, L. (1995). Fatty acid metabolism. In *Biochemistry, 4th Edition*, pp. 603–628. W. H. Freeman & Co, New York.

Sutherland, I. W. (1990). *Biotechnology of Microbial Exopolysaccharides.* Cambridge University Press, Cambridge.

Sutherland, I. W. (1998). Novel and established applications of microbial polysaccharides. *Trends Biotechnol.* **16**, 41–46.

Tombs, M. and Harding, S. E. (1998). *An Introduction to Polysaccharide Biotechnology.* Taylor & Francis, London.

Chapter 16

Antibiotics

David A. Lowe

16.1 Introduction

Antibiotics have changed the world we live in. Their wide-scale introduction in the middle of the 20th century led to new standards of health for billions of people. Many of the life-threatening infections of previous centuries are now conveniently cured by oral medicine. Penicillin was the first major antibiotic from a microbial source to be commercialised. Its acceptance and success led to the search and identification of thousands of novel antibiotics, many of which are now available for therapeutic use. Antibiotics also have applications as feed additives, growth stimulants, pesticides and wider agricultural uses. The discovery of major antibiotics, such as penicillin, cephalosporin, streptomycin, tetracycline and erythromycins, and their subsequent development, have been well documented. Their commercial development over the past 50 years serves as an excellent example of how the applied research has contributed to producing low cost commodities that support therapeutic products with annual sales in the multi-billions of dollars.

Table 16.1 Annual sales of therapeutic antibiotics (1991)

	$ million	
Antibiotic	World	US
Cephalosporins	7300	2500
Penicillins	2750	1000
Macrolides	1000	250
Aminoglycosides	800	110
Tetracyclines	500	100

In 1991 world sales of therapeutic antibiotics were estimated at 15 billion dollars, approximately 10% of the total world pharmaceutical market. Eighty per cent of these sales were comprised of five antibiotics groups (Table 16.1). Sales in 1999 have now doubled for most categories, probably tripled for the macrolides (see Section 16.3). A sixth category, that of the glycopeptides (e.g. vancomycin), have current annual sales of over one billion dollars due to their effectiveness against methicillin-resistant micro-organisms.

Manufactured quantities of antibiotics are difficult to assess. World fermentation of penicillins is in excess of 65000 tonnes, the largest volume category of therapeutic antibiotics. The volume of antibiotics produced for veterinary use approaches that for therapeutics, however their monetary value is less than 10%.

Large pharmaceutical companies are totally international both in their manufacturing sites and sales forces. Intermediates for an antibiotic can be made in one country, the final product assembled in another and packaged in a third.

As world sales of antibiotics continue to increase, their production costs decrease. These current trends in what are relatively established manufacturing processes are due to the continued development efforts of the industrial scientists and engineers particularly in the area of biotechnology. What in the 1950s to the 1980s was the working domain of the chemist, microbiologist and chemical engineer, has now in the 1990s encompassed the skills of the molecular biologist, enzymologist, protein chemist and the biochemical engineer. Biotechnology continues to contribute dramatic changes to the ways antibiotics are manufactured through fermentation yield improvements, recovery processes and final product purity.

All these skills have been developed in individual companies as trade secrets. Each company has its own unique strain and its own unique fermentation and extraction technologies to maximise the genetic properties of these strains. Due to the proprietary nature of this type of development actual processes cannot be described and only generalised consensus processes discussed. However, from the use of published literature, patents and review articles, many common approaches and processes can be used as illustrations to describe a generalised manufacturing process.

Antibiotics sold today are made either by total chemical synthesis or by a combination of microbial fermentation and subsequent chemical modification. The choice is one of simple economics. The microbial fermentation produces the basic active molecule at relatively low cost, and, through chemical modification, the therapeutic effects of the molecule can be increased, e.g. by increasing stability to low pH or temperature, widening the spectrum of activity, altering tissue distribution, increasing absorption and decreasing excretion.

This chapter will discuss the biotechnology involved with the manufacture of the five major antibiotic groups: penicillins, cephalosporins, aminoglycosides, tetracyclines and macrolides.

In the development of all these antibiotics there are many common approaches. These will be summarised first. Specific examples associated individual antibiotics will be addressed in later sections.

16.2 Biosynthesis

Knowledge of the biosynthetic pathway of the antibiotic is not necessary for the early empirical development of the fermentation process. However in order to progress in a rational approach some knowledge of the biosynthesis is essential. This is particularly important for investigations into the genetic and enzymic regulation. For most of the important antibiotics the synthetic pathways are known, together with their relevant enzymes and gene locations, and this detailed knowledge has had a significant impact on the development of improved strains and the optimisation of productive fermentations.

16.3 Strain improvement

The increase in the production of a specific microbial product such as an antibiotic has several important consequences. Higher concentrations of the antibiotic increase the volumetric productivity (output per fermenter), increase the extraction efficiency, decrease the proportion of unwanted products and make purification easier and, most importantly, reduce the cost of the product.

Strain improvement programmes involve the forced creation of mutations in the DNA material of the micro-organism generally using ultraviolet radiation or a chemical mutagen such as nitrosoguanidine (NTG). The latter gives better results as it has a higher mutagenic effect compared to kill rate, however, it has to be handled carefully due its carcinogenic nature. With either treatment, the protocols are optimised to produce a kill range of 60–90%, which gives the highest percentage chance of single point mutations.

Initially, improved strains can be selected empirically by choosing surviving colonies with minor morphological changes or altered colour production, however well-established strain improvement programmes

Table 16.2 Selection environments for strain improvement

Resistance	Possible effect
Analogues of amino acids, sugars involved in biosynthesis	Remove feedback control
Antifungal agents e.g. nystatin	Altered cell wall composition increased, permeability
Toxic metals e.g. Cu, Cd, Hg	Increase in thiols, glutathione
Toxic metals Fe, Mn	Improved sporulation
Selenomethionine, ethionine	Increase in sulphate metabolism
Selenide, methyl selenide	Improved cysteine synthesis
Deoxyglucose	Reduced glucose regulated feedback
Carbon dioxide	Tolerate high levels CO_2 in fermentation
High phosphate	Reduced phosphate regulation
High salt	Tolerate high-salt raw materials
Nitrophenol, azide	Improved oxidative phosphorylation
Polypropylene glycol	Tolerate high levels of antifoam
Water miscible solvents	Improved downstream processing
Peroxide	Higher catalase activity
Others	
Improved tolerance to low oxygen	Improved metabolism at low dissolved oxygen
Sensitivity to chromate, selenate	Increased sulphate uptake
Selection of auxotrophs	Redirection of metabolism

use many selective approaches designed around the known biochemistry of the biosynthesis of the antibiotic and the metabolism of the micro-organism (Table 16.2).

Mutation programmes have continued for over 50 years and have yielded large productivity increases and cost reductions (Fig. 16.1).

The highest percentage gains occurred in the earlier years and now current strain selection is subject to the law of diminishing returns. However, even though increases of less than 5% are difficult to achieve and detect analytically, they are desirable in terms of cost reduction and volumetric capacity increase. To recognise improvements of 5% or less needs the careful design of shake-flask fermentations, with the appropriate controls and replicates, and the exacting skills of the analyst in sample preparation and analyte measurement. Today's screening programmes rely on the use of miniaturisation, automation, and high throughput screening to process the large numbers necessary for the recognition of superior strains. Automation also decreases procedural variabilities seen in sample preparation and dilution. Replication at the shake-flask, re-test stage is necessary to minimise the inherent variability of the biological process to provide the confidence in recognising mutants with only small percentage increases in titres. Replication decreases the number of different cultures that can be handled by the system, however, good screening programmes usually have an excess of analytical capacity to meet these challenges.

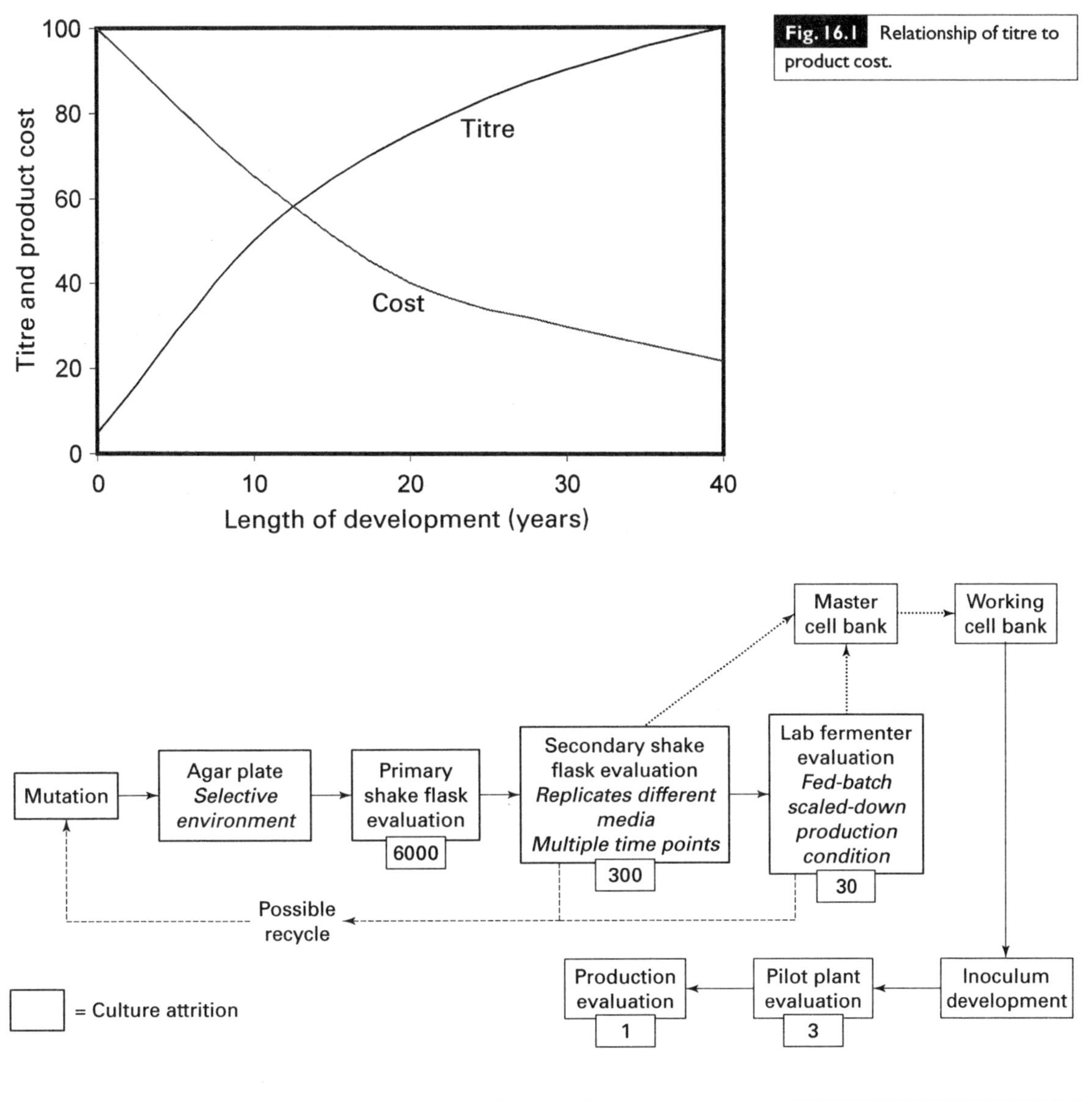

Fig. 16.1 Relationship of titre to product cost.

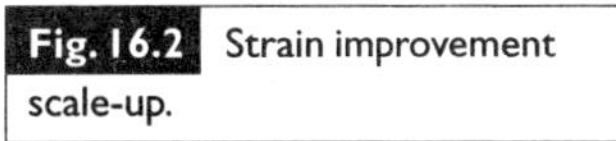

Fig. 16.2 Strain improvement scale-up.

Rapid recycling of the mutation cycles is often used to compensate for the selection of new strains with only minor improvements. Thus several minor changes can be built up into a culture at the laboratory level before further evaluation in the pilot plant. Good cultures typically show their superiority across different media culture conditions and through scale-up (Fig. 16.2).

Historically strain improvement programmes have been carried out as in-house projects, which have the benefit of control and rapid integration into scale-up and downstream process. However there is an increased trend nowadays to contract such work to third parties: companies that are specialised in the multiple skills of culture mutation, selection, automated fermentation and analysis using high throughput screening. Such companies are always a cost- effective option.

16.4 Genetic engineering

Conventional strain selection programmes will remain central to culture improvement. Biotechnology, however, has opened up other possibilities. Together with the elucidation of biochemical pathways and the isolation of biosynthetic enzymes, genetic engineering techniques can now be used to express selected enzymes both in recombinant *Escherichia coli* and in the producing micro-organism. Using kinetic studies and the measurement of metabolic pools, presumptive rate-limiting enzyme steps can be identified in the biosynthesis of secondary metabolites such as antibiotics. Pharmaceutical research groups have attempted to strengthen these weak links by the introduction of additional copies of the genes encoding these rate-limiting enzymes. In other approaches attempts have been made to add new enzymes to produce new metabolites.

16.5 Analysis

The development of modern analytical instruments, i.e. nuclear magnetic resonance (NMR), mass spectroscopy (MS), high performance liquid chromatography (HPLC), capillary electrophoresis (CE), gas-liquid chromatography (GLC) with their associated automated injection systems, multi-faceted detection and data handling, have played an important role in the pharmaceutical industry. The reproducibility, ease of automation, increase in sensitivity and accuracy of multiple analyte detection have facilitated the selection of improved strains and the optimisation of fermentation and recovery parameters, and provided confidence in the final quality of the product.

16.6 Culture preservation and aseptic propagation

Attention has to be given to the correct preservation and consistent propagation of high producing strains. Today's cultures with their long mutation history, increased copy number of certain genes and possible recombinant status do have questionable stability. Repeated slant-to-slant transfer of high yielding strains can produce sub-populations with lower productivity with the appearance of wild type morphology. Storage in liquid N_2 is the most convenient way of long-term culture preservation. Stock cultures are typically maintained through a master cell bank hierarchy, where each master frozen culture, from a stock of many such cultures, is used to make a large number of working stock cultures. In this way, there is always a common lineage to start cellular propagation.

Preparation of a new master cell line is carried out through single cell or spore reisolation and each lot rigorously evaluated both in shake-flask and pilot plant fermentations to confirm superiority and stability before the culture is used in large scale manufacturing. Considerable

Table 16.3 Differences between shake-flask and stirred tank fermentations

Shake-flask	Stirred tank
10–50 ml in 125–500 ml flasks	10–100 000 litre vessels
Batch only	Batch and feed possible
Limited controls: temperature	Continuous controls: pH, temperature, dissolved oxygen, pressure
Slow metabolising carbohydrate: lactose, starch	Readily metabolised carbohydrate: glucose
High initial salts: ammonium, precursors, stimulators added in large shots	Ammonium salts or ammonia added precursors, stimulators added continuously
Ambient pressure	Pressure at two atmospheres possible
Buffers needed to control pH: phosphate or calcium carbonate	No buffer necessary
In-process sampling difficult	In-process sampling easy and often necessary for feedback control
Volume decrease by evaporation	Volume increase by sugar feed
Solid growth on side walls	Very uniform growth
Antifoam not needed	Antifoam often required
Agitation limited: shaker speed, radial throw, baffled flasks	Wide variety of impellers and baffles to optimise mixing
No control of dissolved oxygen	Dissolved oxygen controlled by aeration, pressure, water addition

care is taken to maintain aseptic conditions throughout the build-up of culture volumes. This is especially critical at the seed stage where the cultures are growing fast and scheduled tank transfers occur before full status of asepsis is known. The presence of contaminants late in the fermentation cycle, and even in post-harvest work up, is of concern due to the possibility of introducing minor impurities into the final bulk material. Impurity profiling by gradient HPLC and MS is now standard practice in the evaluation of new strains, new media components and major engineering changes. It is also routinely performed on the final isolated purified product.

16.7 Scale-up

Shake-flask media and conditions are selected to provide environments as close as possible to the stirred-tank large-scale fermentations. This is not always possible and many compromises have to be taken (Table 16.3).

A good relationship between shake-flask performance, pilot plant and large-scale fermentations can only be established after years of careful comparison. Potential titre increases of 5% or less are not only difficult to assess in shake-flask experiments but also difficult to assess at the pilot

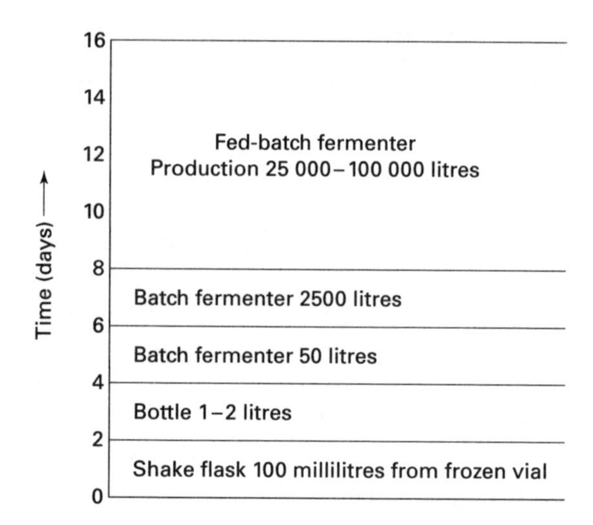

Fig. 16.3 Culture build-up train.

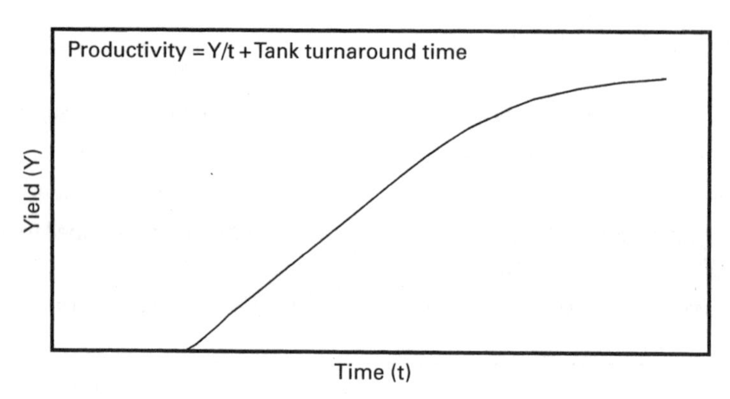

Fig. 16.4 Stirred tank productivities.

plant stage where resources are limited and evaluations expensive. It is always desirable to have new cultures that easily fit into the existing fermentation protocols without further development work. However new cultures often have properties that need further development to express their full potential. Here the interdisciplinary skills of the bioengineers, microbiologists and biochemists can prove to be rewarding.

16.8 Fermentation

Large-scale antibiotic fermentations are optimised for fast culture growth, early production rates and maximum productivity. High productivity plants are characterised by their ability to maximise the use of all vessels. A train of vessels of increasing size allows for the rapid build-up of cell mass, each typically 1–3 day fermentations, with inoculum transfers at 5–10% (v/v) into the next stage. The final stage is run in conditions to maximise productivity in terms of amount of antibiotic per unit fermenter volume per time period (Fig. 16.3). Down time (or turn-around time) for production fermenters is usually kept at a minimum by use of separate continuous media sterilisation, rapid harvesting and tank cleaning, sterilisation and inoculation (Fig. 16.4).

Table 16.4 Media for cell growth and final production

Stage	Components	Range % (w/v)
Shake-flask/tank seed	Glucose/sucrose/starch	3.0–5.0
	Corn steep liquor	3.0–5.0
	Calcium carbonate	0.5–1.0
	Phosphate	0.1–0.5
	Ammonium sulphate	0.1–1.0
	Urea	0.1–0.5
	Oil	0.1–0.5
Shake-flask production	Glucose	0.2–1.0
	Starch	0.5–5.0
	Lactose	5.0–8.0
	Corn steep liquor	5.0–8.0
	Pharmamedia	1.0–5.0
	Soy flour	1.0–5.0
	Oil	0.5–5.0
	Ammonium sulphate	0.5–1.0
	Calcium carbonate	0.5–1.0
	Phosphate	0.1–1.0
	MOPS/MES buffers	0.1–1.0
Tank production	Glucose	0.5–1.0
	Starch	0.5–5.0
	Corn steep liquor	5.0–8.0
	Pharmamedia	1.0–5.0
	Soy flour	1.0–5.0
	Ammonium sulphate	0.5–1.0 and fed periodically
	Phosphate	0.1–1.0
	Oil	1.0 and fed continuously
	Corn syrup	Fed continuously

Media for cell mass build-up are designed to provide fast growth in a batch mode with minimal changes in pH. Individual medium components do not have to be greater than 3–5% and provide a readily available carbohydrate, such as glucose or sucrose, and a soluble form of nitrogen such as corn steep liquor or yeast extract. Calcium carbonate or phosphates can be added if buffering is required which is often the case due to organic acids that can be produced by the rapid metabolism of sugars. Ammonium sulphate can be used to provide additional nitrogen (Table 16.4).

Media for the production stage are proprietary and have been developed and fine-tuned over the years. They are a compromise between cost and performance. The most suitable media are those that use inexpensive raw materials in combinations that can maximise productivity. Final stage fermentations are fed-batch, which gives the bioengineer the ability to optimise the fermentation to provide the fine balance

Table 16.5 Complex nitrogen source raw materials
Beef blood
Casein hydrolysate
Cotton seed flour (Pharmamedia)
Cottonseed meal
Corn germ meal
Corn gluten meal
Corn steep liquor
Corn steep liquor (solid)
Distillers solubles
Fish meal
Fish solubles
Lard water solids
Linseed meal
Meat and bone meal
Peanut meal
Rape seed meal
Soybean meal
Soybean flour
Soybean protein concentrate
Whey solids
Whey permeate
Whole yeast, brewer's
Whole yeast, torula
Yeast extract

between controlled cell growth and maximum biosynthesis. A fed-batch fermenter can be controlled in a number of ways: physically, e.g. by temperature, aeration, agitation, pH; or biochemically, e.g. by the addition of nutrients, precursors, inducers.

Raw materials for use in the initial batch have to provide both immediate utilisable soluble nutrients as well as longer lasting and therefore less soluble sources. Initial carbon sources are the least critical as they are easily added in a soluble form during the fermentation. Nitrogen sources are more critical as they serve as a main nutrient source throughout the fermentation. Ideal nitrogen sources are derived from agricultural sources, however questions of quality and variability can arise both with seasons and between seasons (Table 16.5). This presents an on-going concern for maintaining reproducible fermentations. To alleviate this situation several different raw materials can be used to prevent excess variation.

In some of today's highly productive fermentations there is no clear separation of the primary (tropophasic) and secondary (idiophasic) stages. This lack of division generally depends upon the state of the fermentation technology. In batch type fermentations, clear primary and secondary stages can be seen, however with the use of continuous feed these differences are not always apparent (Fig. 16.5). To obtain

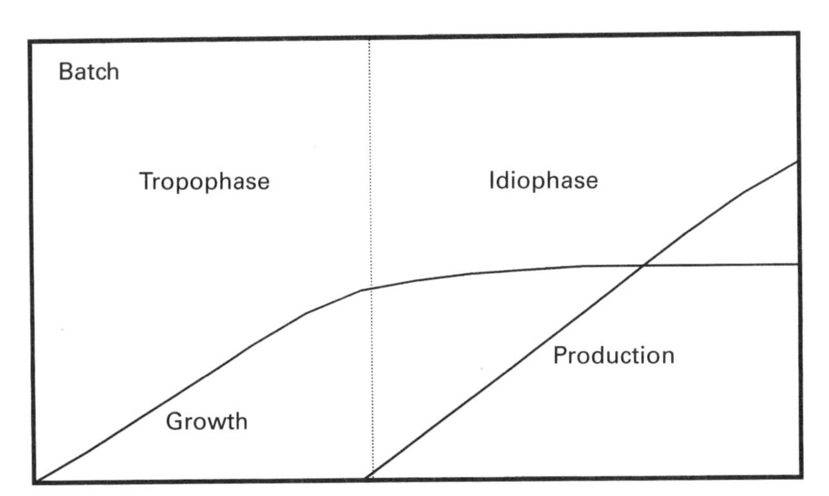

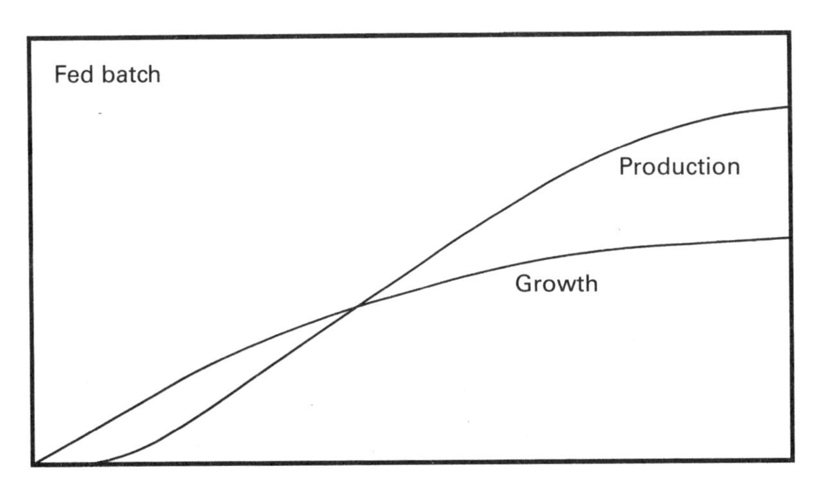

Fig. 16.5 Comparison of batch and fed-batch.

maximum production rates, conditions are created that can provide rapid, early, antibiotic production with continued cell growth.

Supplemented raw materials are soluble and rapidly utilised. Suitable carbohydrates are sucrose, glucose or enzyme-hydrolysed corn syrups. Other carbon sources can be used (Table 16.6). If necessary they can be supplemented with soluble nitrogen from corn steep liquor. The diligent feeding of a soluble, readily utilised carbohydrate such as glucose can prevent catabolic repression (see Chapter 2), as the concentration of the sugar will always be very low.

Oil, as triacylglycerol, can be lard oil, soy oil, palm oil, peanut oil or rape seed oil, the final choice often dictated by local availability. Oil addition has the additional benefit of controlling excessive foaming and air hold-up. Antifoams, such as silicone-based products or polypropylene glycol, can be used to supplement or replace oil feeding. It is important that antifoam addition is available on an as-needed basis and not simply batched into the starting medium due to the toxic nature of some antifoams. The metabolism of the proteinaceous nutrients from the complex raw materials can create foaming, often at unpredicted

Table 16.6 Carbon source raw materials

Beet and cane molasses
Glucose
Citric acid
Corn syrup incompletely hydrolysed
Corn syrup fully hydrolysed
Dextrins
Ethanol
Glycerol
Maltose syrup
Methanol
Starch
Lactose
Cottonseed oil
Lard oil
Methyl oleate
Palm oil
Palm kernel oil
Peanut oil
Rape oil (Canola)
Soy oil
Tallow

times, thus it is important to have automated feed-back control for effective antifoam addition to provide sufficient control without excess usage of these agents. Excess use can cause processing difficulties on downstream recovery. Control of foaming and the minimisation of air hold-up are important factors in obtaining the maximum volumetric output from a fermenter. Typically, the final harvest volumes should be in the range 80–85% of the total fermenter capacity.

The added volume of soluble nutrient feed can vary depending upon its concentration (typically 30–65%). At lower sugar concentrations early partial harvests may be necessary to decrease the increase in broth volume caused by the high volume of feed addition. This addition of dilute solutions has the added benefit of lowering the viscosity of the broth, typically a problem with filamentous cultures. Early, partial harvests, produce large volumes of dilute antibiotic for product recovery. With correct handling however, such protocols can be very productive as the maximum production rate of the fermentation can be maintained for long periods.

The pH of the broth can be controlled to within 0.1 pH units by the addition of acid (sulphuric) or base (ammonia or caustic). Often ammonia gas can be added through the air input. The pH can also be controlled by using the culture's own metabolism of sugar. Excess feeding of sugar in some conditions will produce acetic acid, which will lower the pH. Conversely, a cutback in the sugar feed-rate can raise the pH.

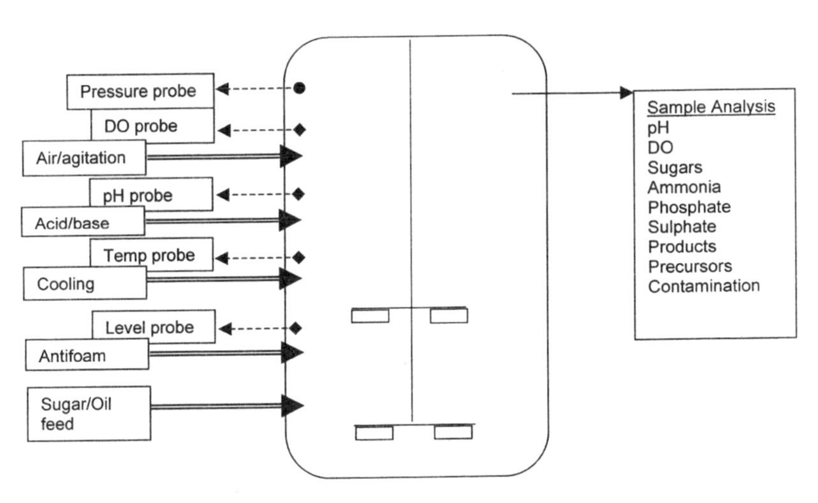

Fig. 16.6 Fermentation control parameters.

Dissolved O_2 DO levels are critical for maintaining the maximum rate of antibiotic production and culture viability. As O_2 supplementation is too costly, as well as a safety concern, ambient air is used as the source of O_2. A fine balance has to be established between aeration and the agitation necessary to distribute the O_2 into the liquid phase, the back pressure in the tank to increase oxygen solubility, the volume expansion of the fermentation broth, and the compounding of several of these effects on the dissolved CO_2 levels. Dissolved O_2 levels should be maintained higher than 20% saturation at 1.5–2 atmospheres pressure throughout the fermentation, at air flow rates high enough to sweep out as much CO_2 as possible, build up of which can be detrimental to some fermentations (Fig. 16.6).

Monitoring of the fermentation can be performed by the use of multiple pH and dissolved O_2 probes, pressure and temperature measurements. However, for more detailed analysis, broth samples are taken at convenient intervals depending on the length of the fermentation and critical nature of the measurement. Samples are used to check the correct performance of pH probes and for a variety of chemical measurements, e.g. concentration of the product, ammonia, sulphate and other factors that may be considered important to control the fermentation, i.e. sugar, amino acids, organic acids, degradation products or intermediates. These analyses can be performed rapidly by shift workers, and the results can be available within several hours if necessary. Additional samples are taken for microbiological analysis to assess the aseptic nature of the fermentation.

16.9 Penicillins

16.9.1 Therapeutic penicillins and antibiotics derived from penicillin

Penicillin, and its related β-lactam cephalosporin, are bactericidal antibiotics. They inhibit the formation of peptide cross-linkages in the final stages of bacterial cell wall synthesis. Penicillin G and V are active

Fig. 16.7 Penicillins.

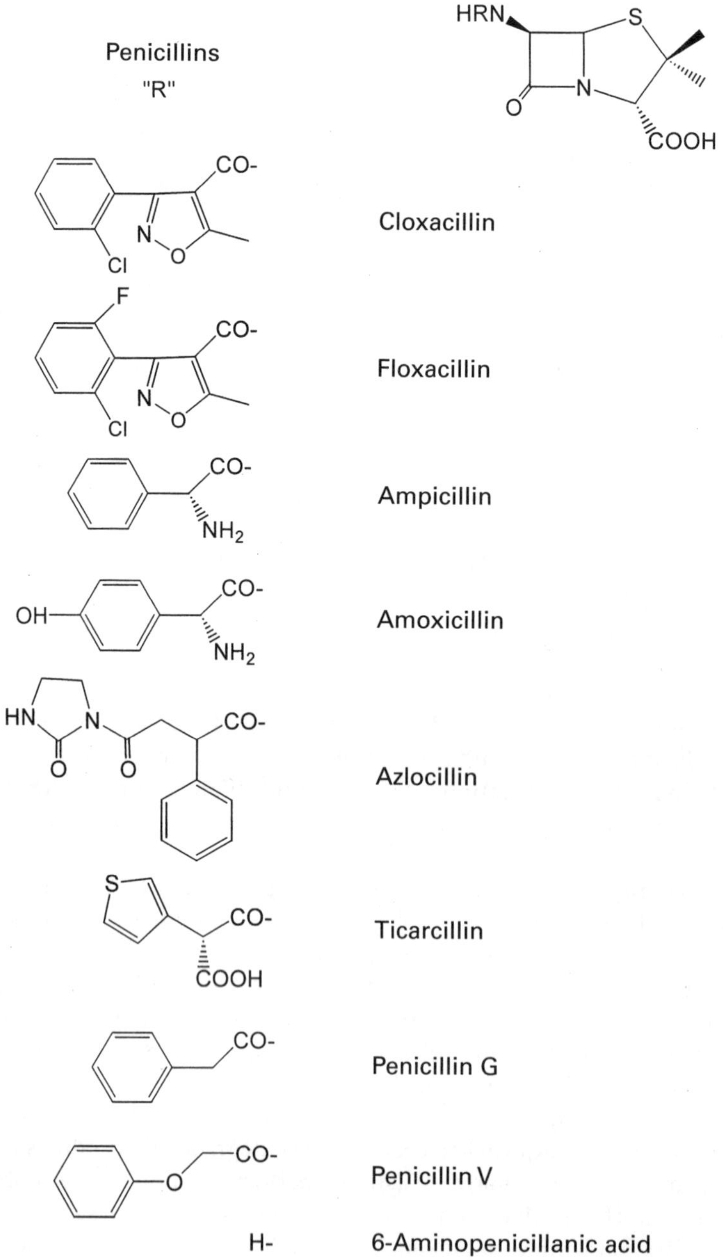

Penicillins "R"

HRN — COOH (penicillin nucleus)

R	Name
3-(2-chlorophenyl)-5-methylisoxazol-4-yl CO-	Cloxacillin
3-(2-chloro-6-fluorophenyl)-5-methylisoxazol-4-yl CO-	Floxacillin
Ph–CH(NH_2)–CO-	Ampicillin
OH–C$_6$H$_4$–CH(NH_2)–CO-	Amoxicillin
(2-oxoimidazolidin-1-yl)–CO–CH$_2$–CH(Ph)–CO-	Azlocillin
3-thienyl–CH(COOH)–CO-	Ticarcillin
Ph–CH$_2$–CO-	Penicillin G
Ph–O–CH$_2$–CO-	Penicillin V
H-	6-Aminopenicillanic acid

against Gram-positive cocci but are readily inactivated by hydrolysis by β-lactamase-producing cultures and are therefore ineffective against *Staphylococcus aureus.* Cloxacillin and floxacillin are resistant to β-lactamase and are used against *Staph. aureus.* The broad spectrum ampicillin and amoxicillin extend activity against Gram-negative bacteria such as *Haemophilus influenzae, E. coli* and *Proteus mirabilis.* Amoxicillin is used in combination with clavulanic acid, a potent β-lactamase inhibitor, to extend the use of this antibiotic. Azlocillin and ticarcillin are used to combat severe pseudomonal infections (Fig. 16.7).

Penicillin G and V are fermented products from the fungus, *Penicillium chrysogenum*. The bulk of penicillin G and V, however, is now used as starting material for the production of the active β-lactam nucleus, 6-aminopenicillinanic acid (6-APA).

Penicillin G can also be ring-expanded chemically to the cephalosporin nucleus which, after enzyme hydrolysis, yields the active nucleus 7-aminodesacetoxycephalosporanic acid (7-ADCA).

Both of these nuclei are important bulk products, and are used for the chemical synthesis of the semi-synthetic penicillins mentioned above, and in the case of 7-ADCA, for the synthesis of oral, broad spectrum cephalosporins, such as cefadroxil and cephalexin. The cephalosporin nucleus provides resistance to β-lactamase hydrolysis and these oral cephalosporins are particularly effective against urinary tract infections. The β-lactam nucleus can also be modified further to form the basis of antibiotics such as cefaclor and cephprozil (Fig. 16.8).

Cefadroxil and amoxicillin can now be made enzymically by using the reversible (synthetic) catalytic property of penicillin amidase. The penicillin G amidase from *E. coli* has been crystallised and its three-dimensional structure determined. Changing amino acids at the active site by site-directed mutagenesis has produced enzymes with improved ability to work in the synthetic direction.

16.9.2 Biochemistry and fermentation

Penicillin strain improvement programmes have been in existence for over 50 years. By using convention mutation and selection the original titres of less than 0.1 mg ml^{-1} have been increased 400-fold. Further gains have been realised by media modification and engineering developments (Fig. 16.9).

The biosynthesis of this molecule together with the biosynthetic enzymes and associated genes are well characterised (Fig. 16.10). Rate limiting steps in the biosynthesis have been identified and attempts made to increase the production of limiting enzymes by recombinant technology. For example extra genes coding for cyclase (*pcb*C) and acyl transferase (*pen*DE) have been inserted into *P. chrysogenum*.

Addition of the precursor molecules, phenylacetic acid or phenoxyacetic acid, to fermentations of *P. chrysogenum* produce either penicillin G or penicillin V, respectively. For optimum production the culture is grown on a batch medium of corn steep liquor or soy flour plus minerals, and fed carbohydrate as a corn syrup throughout the cycle. In addition, the precursor and ammonium sulphate are fed to maintain critical concentrations of these components needed for the biosynthesis of the penicillin (Fig. 16.11).

16.9.3 Recovery

Penicillin is recovered by solvent extraction at an acidic pH at temperatures below 10 °C to minimise both chemical and enzymic penicillin breakdown. Solvents of choice are n-butyl acetate, or methylisobutyl ketone. Solvent extraction can be done on the whole broth itself or on clear filtrates. Depending on the nature of the mycelium the solids can

Fig. 16.8 Cephalosporins.

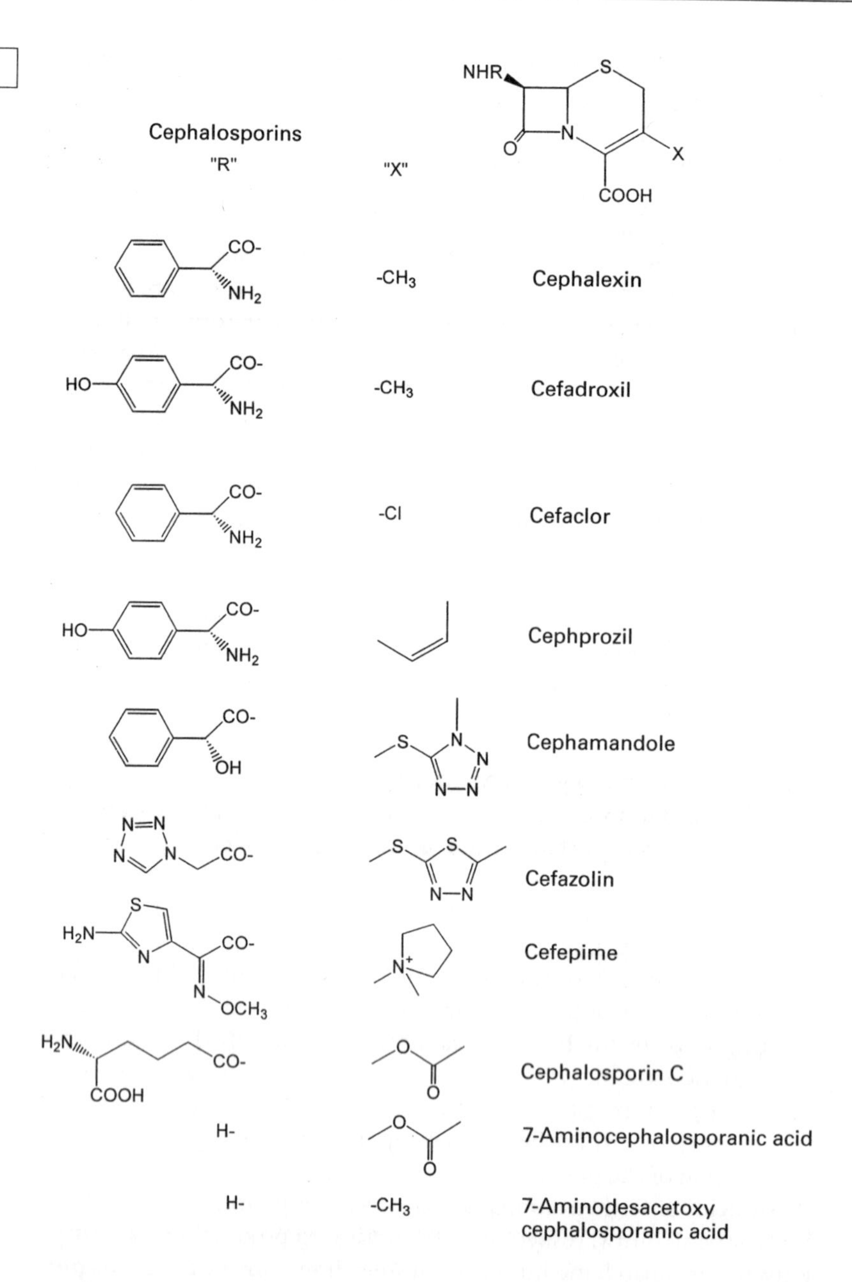

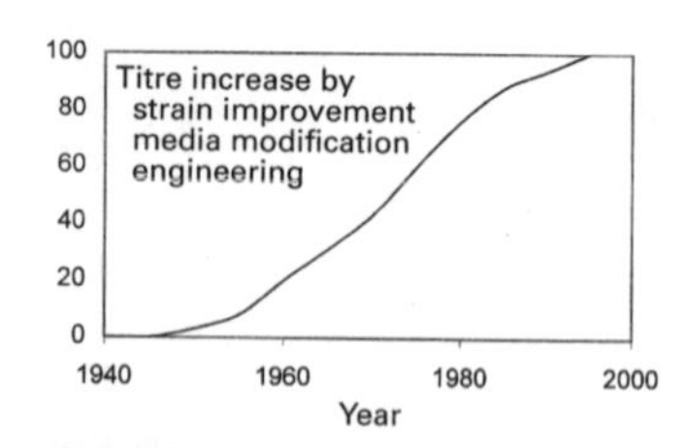

Fig. 16.9 Penicillin fermentation improvements.

be removed by string filters, pre-coated diatomaceous filters or by ultrafiltration. The nature of the removal of the mycelial mass can have implications on the choice of final waste treatment. Mycelium can be treated, dried and used as a soil conditioner. Otherwise it has to be added back to the residual liquid waste for conventional anaerobic/aerobic digestion.

The penicillin-rich solvent can be treated with activated carbon to remove pigments and other impurities, and the penicillin recovered as the potassium or sodium salt by adding potassium or sodium acetate to the solvent. Further impurities can be removed by washing the recovered salts with a dry solvent such as isopropanol or n-butanol.

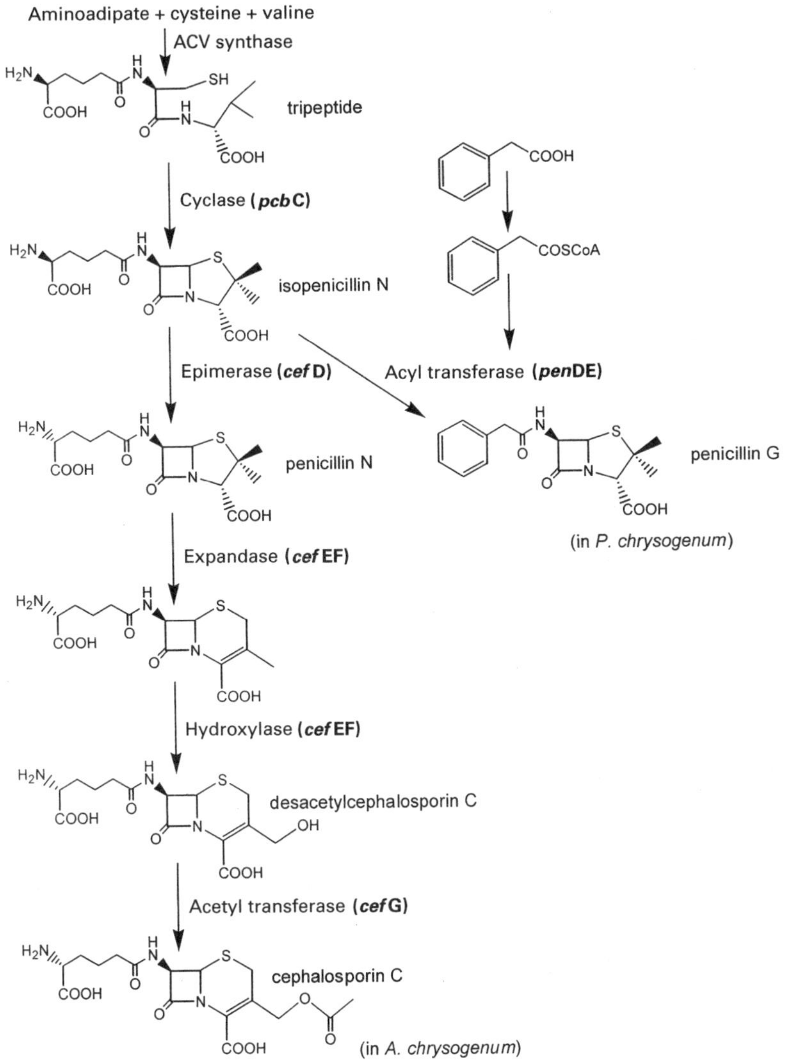

Aminoadipate + cysteine + valine

ACV synthase

tripeptide

Cyclase (***pcb*C**)

isopenicillin N

Epimerase (***cef*D**)

Acyl transferase (***pen*DE**)

penicillin N

penicillin G

(in *P. chrysogenum*)

Expandase (***cef*EF**)

Hydroxylase (***cef*EF**)

desacetylcephalosporin C

Acetyl transferase (***cef*G**)

cephalosporin C

(in *A. chrysogenum*)

Fig. 16.10 Biosynthesis, enzymes and genes involved in β-lactam production.

Penicillin V is stable to acid and can be precipitated directly from clear filtrates at a pH of 2. Impurities in the penicillin can be removed by dissolution of the acid in organic solvent to allow treatment with activated carbon. Direct precipitation reduces the use of organic solvents, which can have positive cost and environmental impacts.

As most of the penicillin nowadays is used to make 6-APA, technology is being developed to bypass the precipitation of the penicillin salt and hence eliminate mother liquor losses. This is the 'direct process' and typically employs the back-extraction of the solvent-based penicillin into an aqueous solution which, after the removal of residual solvents, is used directly for enzyme hydrolysis by the appropriate enzyme system.

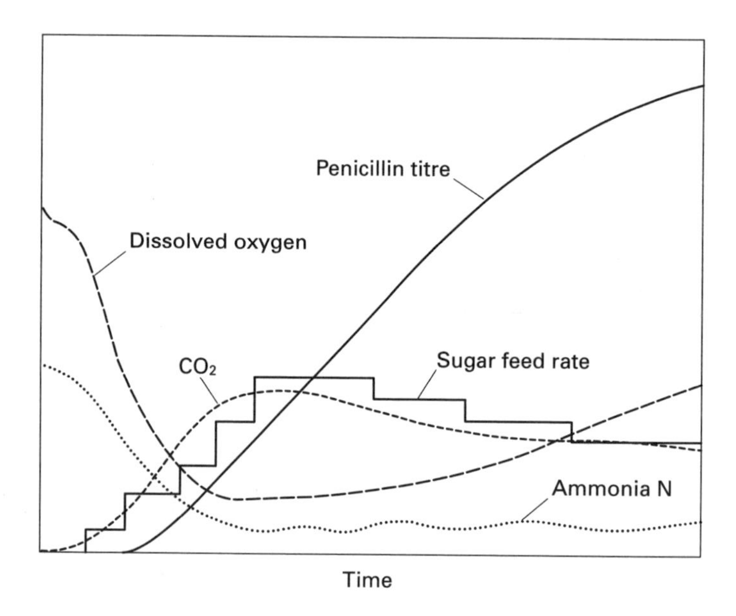

Fig. 16.11 Penicillin fermentation profile.

Table 16.7 Benefits of enzyme hydrolysis of penicillin

Elimination of chlorinated and alcohol solvents
Elimination of solvent recovery and solvent/odour release
Elimination of hazardous chemical reagents
Elimination of liquid nitrogen for cooling
Elimination of hazardous and toxic waste products and their disposal
All aqueous reactions, neutral pH and ambient temperatures
Good control and monitoring of reaction through pH measurement and adjustment
Quick removal of soluble reaction products from immobilised catalyst
Re-use of immobilised enzyme catalyst
Easy recovery of side chain for re-use
Pleasant working environment for all personnel
Improved product quality, less impurities
Improved yields and manufacturing capacity
Decreased cost of manufacture

16.9.4 Production of 6-aminopenicillanic acid

Over the last 10 years the industry has switched from chemical hydrolysis of penicillins to enzyme hydrolysis to decrease cost and attain environmental benefits (Table 16.7).

Specific, immobilised penicillin amidases have been developed for penicillin G and penicillin V hydrolysis. Immobilised enzyme can be made in-house or purchased from third parties. From the thermodynamic equilibrium of 6-APA and the side chain, hydrolysis is somewhat greater for penicillin V than penicillin G. However penicillin G is a more versatile product due to its application in ring expansion, which partially explains its fermentation volume dominance over penicillin V.

The final choice between either process is often directed by the company's own historical development success.

In conventional splitting technology, the penicillin salt is used at 12–15% (w/v) for enzymic hydrolysis by the appropriate immobilised penicillin amidase system. This yields mixtures of 6-APA and the precursor acid. During the hydrolysis the pH is maintained between 7–8 by the addition of base, either caustic or ammonium hydroxide. The product 6-APA can be recovered by precipitation at pH 4 in the presence of a water immiscible-solvent for the convenient removal of the precursor acid. In operations that have both penicillin fermentation and splitting processes, the recovered precursor can be conveniently recycled.

16.10 Cephalosporins

16.10.1 Therapeutic cephalosporins

Cephalosporins were developed to overcome the allergic problems associated with penicillins. They can, however, be modified chemically at two sites: the 7-amino and the 3-methylene, to produce a variety of very effective antibiotics, notably cephamandole, cefazolin and cefepime (Fig. 16.8).

Cephalosporins are made from cephalosporin C, a fermented product of *Acremonium chrysogenum* which, after extraction and purification, is hydrolysed, either enzymically or chemically, to the active nucleus, 7-aminocephalosporanic acid (7-ACA), which serves as substrate for the chemical synthesis of injectable, semi-synthetic cephalosporins. Cephalosporins with a 7-α-methoxy group (cephamycins) are produced by several *Streptomyces* spp., and they serve as the precursor of cefoxitin and others.

16.10.2 Biosynthesis and fermentation

The early part of the biosynthesis pathway is shared with penicillin. The cephalosporin molecule is derived from penicillin by a ring expansion of penicillin N. This enzyme, expandase, and other subsequent synthesis enzymes are not found in *P. chrysogenum*. As in the case of penicillins, yield improvements have been sought by the insertion of extra rate-limiting enzymes. Genes for cyclase (*pcb*C), expandase and hydroxylase (*cef*EF) and acetyltransferase (*cef*G) have been enriched in *A. chrysogenum* in attempts to both increase the production of cephalosporin C and reduce the production of unwanted intermediates e.g. penicillin N and desacetoxycephalosporin C (DAOC).

High-producing cultures of *A. chrysogenum* are fermented using corn steep liquor and soy flour-based media with continuous feeding of both corn syrup and triacylglycerols (soy, rape or lard oil). Methionine is used both as a source of sulphur and as an inducer of morphological changes. Cephalosporin C is unstable and degrades chemically to desacetylcephalosporin C (DAC) and a thiazole-4-carboxylate. Cephalosporin C is also hydrolysed to DAC by esterases released by the fungus. These, however, can be inhibited by the use of phosphates.

16.10.3 Production of 7-aminocephalosporanic acid

Cephalosporin C is recovered from broth filtrates by a variety of hydrophobic and ion exchange resins. The column chromatography is designed to separate cephalosporin C from related intermediates and breakdown products. The rich fractions are either treated with zinc acetate to precipitate the low solubility zinc salt, or with sodium or potassium acetate followed by a water-miscible solvent to precipitate the salt complex.

Isolated cephalosporin C is efficiently hydrolysed chemically to 7-ACA. Unfortunately, the process uses similar reactants and solvents used in the chemical hydrolysis process for penicillin, with the familiar drawbacks of hazardous material handling, solvent use and negative environmental issues. The switch to enzyme hydrolysis has proved to be difficult due to the inability to identify enzymes to directly hydrolyse off the side chain, the unnatural D-amino acid D-α-aminoadipate. Indirect enzyme systems, though, have been developed which rely on the sequential use of two enzymes (Fig. 16.12).

The first enzyme, a D-aminoacid oxidase, removes the chirality of the side chain by oxidative deamination to produce a keto acid which, in the presence of the co-produced peroxide, is conveniently decarboxylated to the glutaryl side chain. The yeast, *Trigonopsis variabilis*, is a suitable source of this enzyme. The second enzyme was discovered in *Pseudomonas* sp. and can directly hydrolyse the glutaryl side chain to produce 7-ACA. In a similar manner to penicillin hydrolysis these two enzymes are now available from recombinant sources and have been immobilised. Most of the major industrial producers of 7-ACA are now switching over to the enzyme process.

16.11 New β-lactam technologies

There is interest in developing alternative ways to make the cephalosporin intermediates, 7-ADCA and 7-ACA, using the *P. chrysogenum* fermentation. The availability of biosynthetic genes has been used to this purpose to design new biosynthetic pathways.

The expandase enzymes have a strict substrate preference for penicillin N-like molecules and will not expand penicillin G-like molecules. It has been demonstrated, however, that adipic acid can serve as the precursor to adipyl-penicillin in *P. chrysogenum*. On the insertion of the gene, *cefE* (expandase from *Streptomyces clavuligerus*), into *P. chrysogenum*, the transformants produced adipyl-6-APA, and adipyl-7-ADCA. Transformants with the genes *cefEF* and *cefG* (acetyltransferase) produced adipyl-7-ACA in addition to the above (Fig. 16.13). The adipyl derivatives do have the advantage of being solvent-extractable and their hydrolysis has been demonstrated using glutaryl amidases from *Pseudomonas* sp., enzymes known to have some affinity for the adipyl side chain. Similarly directed synthesis has been carried out using carboxymethylthiopropionate, a molecule of similar structure to adipic acid.

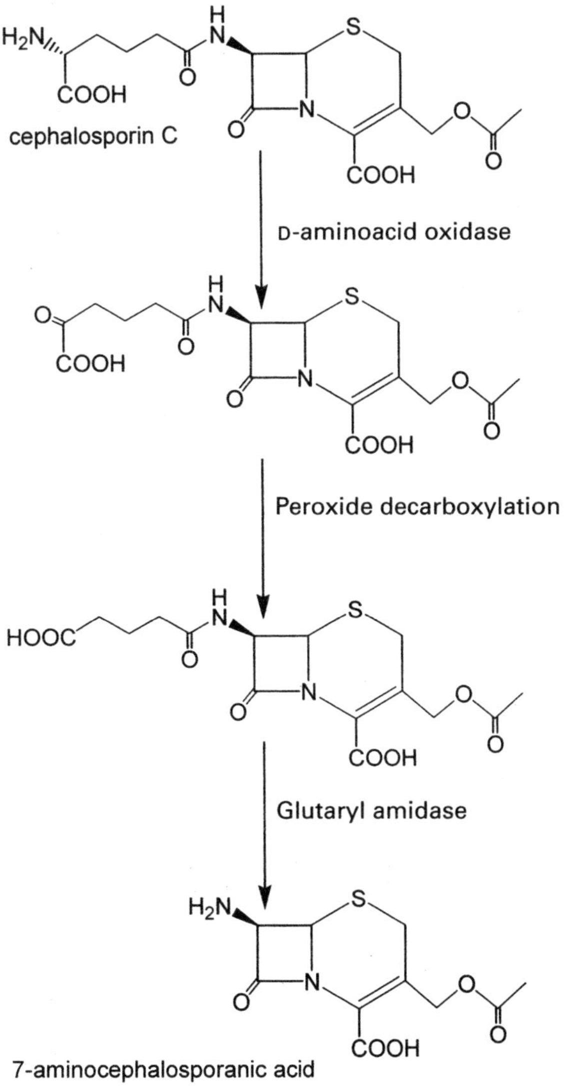

Fig. 16.12 Enzymic hydrolysis of cephalosporin.

cephalosporin C

D-aminoacid oxidase

Peroxide decarboxylation

Glutaryl amidase

7-aminocephalosporanic acid

The direct fermentation of 7-ACA has already been demonstrated by the insertion of genes for D-aminoacid oxidase from *Fusarium solani* and for glutaryl amidase from *Pseudomonas diminuta* into *A. chrysogenum.*

A more challenging objective would be to produce the 7-ADCA nucleus directly by the expandase working directly on 6-APA or isopenicillin N. Again, strict substrate preferences do not permit this with the current natural enzymes.

Direct 6-APA fermentations have been developed. The instability, however, of this product, especially in the presence of CO_2, and its poor extractability were major barriers to commercialisation. On the other hand, 7-ADCA is stable in solution, does not react with CO_2 and has very low solubility at pH 4. This could conceivably be an easier compound to recover.

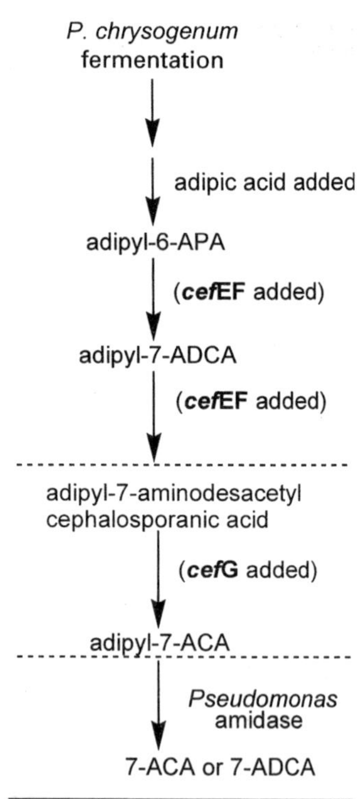

Fig. 16.13 Adipic acid precursored β-lactams.

16.12 Aminoglycosides

Streptomycin was the first aminoglycoside used for antibiotic therapy. Its activity against *Mycobacterium tuberculosis* initiated the widespread introduction of antibiotic treatment to combat tuberculosis. Aminoglycosides are potent antibiotics and have activity against both Gram-positive and Gram-negative bacteria as well as against mycobacteria. Unfortunately they can have nephro-(kidney) and oto-toxicities (hearing), and care has to be taken in their use in treatment of serious infections.

Aminoglycosides are bactericidal and work by binding to the 30S ribosome subunit which prevents protein synthesis.

There are many aminoglycosides in medical use and are all derived from actinomyces spp. For example; streptomycin (*S. griseus*), gentamicin (*Micromonospora purpurea*), tobramycin (*S. tenebrarius*), kanamycin (*S. kanamyceticus*), sisomicin (*M. inyoesis*).

Some have been modified chemically to produce derivatives with resistance to clinical isolates with acquired resistance to earlier aminoglycoside types. Of particular interest is the use of the hydroxy-γ-aminobutyryl side-chain to give anti-pseudomonal activity to amikacin. This side chain occurs naturally in the aminoglycoside butirosin produced by *Bacillus circulans*. Netilmicin is chemically derived from sisomicin. Bacterial resistance occurs by enzymic modification, e.g. acylation, phosphylation or adenylation of the various amine and hydroxyl groups (Fig. 16.14).

All ring structures of these antibiotics are derived from glucose, synthesised separately and then assembled into the final molecule. Most of the biosynthetic enzymes and their associated genes have been identified. Many similarities in biosynthesis have been seen across the wide variety of aminoglycosides. In addition one culture can produce a variety of molecules, e.g. kanamycin A, B, C or gentamicin C_1, C_2, C_{1a}, C_{2a}, A. Recombinant DNA techniques have been used to produce hybrid aminoglycosides (mutasynthesis), and many novel structures have been produced, however, none has been found to be superior to existing structures. Strain improvement programmes have been successful in increasing fermentation titres to 15–20 mg ml^{-1}. Additional challenges have been to either reduce the production of unwanted products e.g. kanamycin C, or maintain the required ratios, e.g. gentamicin C_1, C_2, C_{1a}.

Large-scale fermentations of aminoglycosides have several similar features. Use of soy products is common, e.g. soy flour or soy meal. Antibiotic synthesis is sensitive to feedback repression by glucose, ammonia and phosphate. For these reasons ammonium and phosphate salts are not used in the starting batch. Nitrogen is obtained from the slow metabolism of the soy proteins and the necessary phosphate is obtained from organic sources such as phytic acid. Starch is commonly used in the starting batch as streptomyces have poor amylase activities and the enzymic release of glucose is slow and rate limiting. Alternatively, corn syrups can be fed at pre-determined rates.

Fig. 16.14 Aminoglycosides.

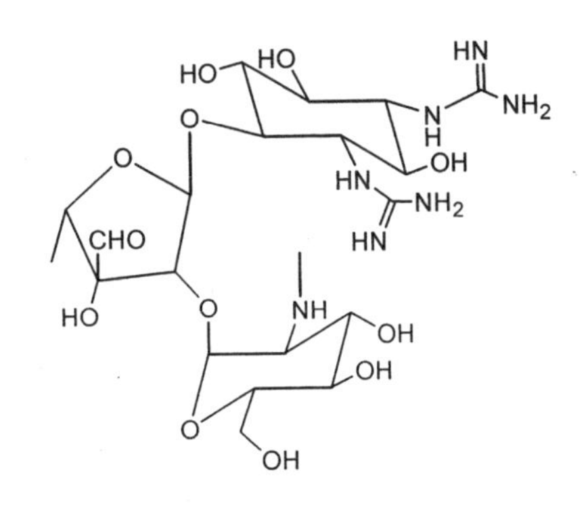

Streptomycin

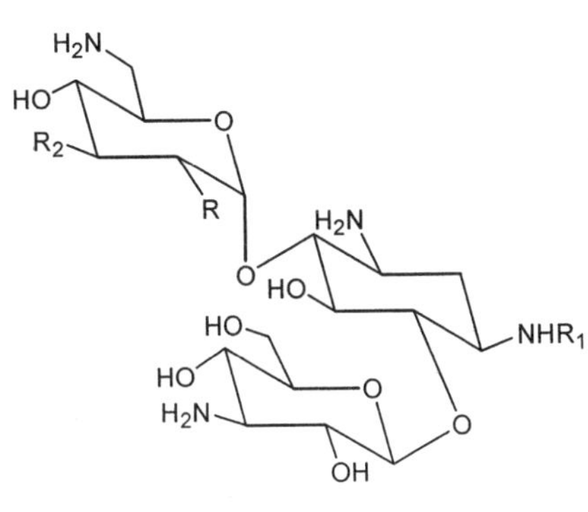

Kanamycin A
R = OH
R_1 = H
R_2 = OH

Tobramycin
R = NH_2
R_1 = H
R_2 = H

Amikacin
R = OH
R_1 =
R_2 = OH

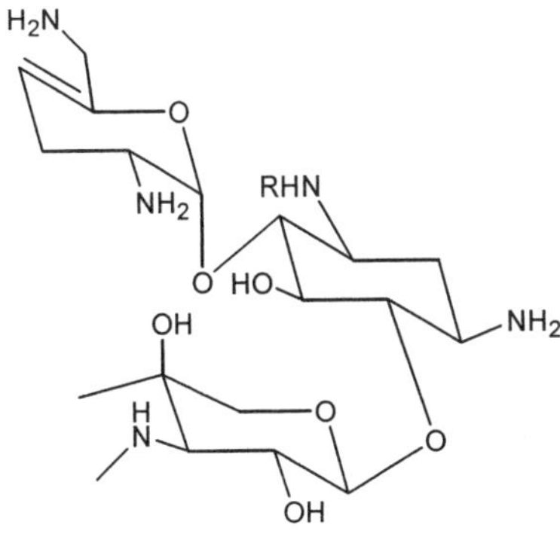

Sisomicin
R = H

Netilimicin
R = CH_2CH_3

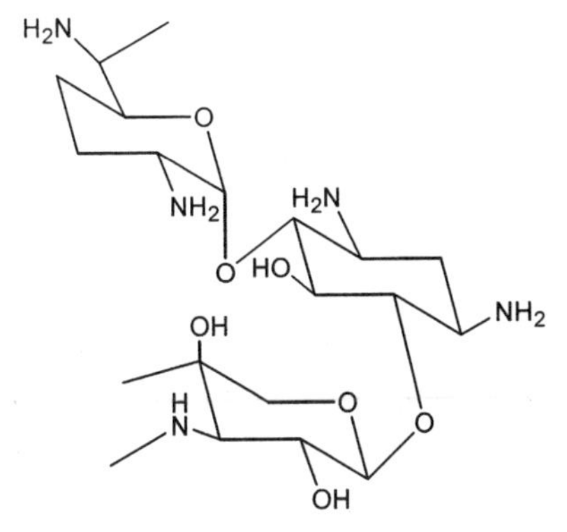

Gentamicin C1

Fig. 16.15 Tetracyclines.

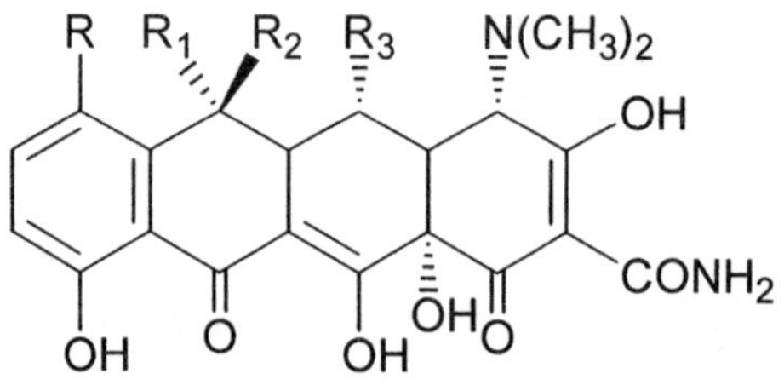

	R	R_1	R_2	R_3
Chlorotetracycline	Cl	CH_3	OH	H
Oxytetracycline	H	CH_3	OH	OH
Tetracycline	H	CH_3	OH	H
Doxycycline	H	CH_3	H	OH
Minocycline	$N(CH_3)_2$	H	H	H

Due to the general basic nature of aminoglycosides, they are generally recovered by a combination of resin column treatments, e.g. weak cationic IRC 50, non-ionic XAD, or alumina. Activated carbon treatment is often necessary and the final product can be precipitated as the sulphate salt.

16.12.1 Tetracyclines

Tetracyclines were the first group of antibiotics recognised to have broad spectrum activity. They act by preventing protein synthesis at the 30S ribosome interaction with tRNA. They are used for urinary tract infections, chronic bronchitis, rickettsial and chlamydial infections. They also have broad applications in veterinary use, despite the concern and known relationship of widespread use with resistance build-up. Novel applications include activity against *Helicobacter pylori* to combat stomach ulcers and as a prophylactic against malaria. Chlorotetracycline and tetracycline are produced by *S. aureofaciens*, and oxytetracycline by *S. rimosus*. Chlorotetracycline production is stimulated by chloride ions and tetracycline by bromide ions. The chlorination gene can be deleted making the bacterium produce only tetracycline. Tetracyclines have been modified chemically to produce products with improved activity and stability. These include doxycycline and minocycline (Fig. 16.15)

The biosynthesis and genetics of tetracyclines have been well described. The starting polyketide chain is first cyclised into the four ring structure that is then sequentially modified in a specific order. From cloning studies of the biosynthetic enzymes the phenomenon of gene clustering was first recognised. Knowledge of the genetics of the producing organism has been a great asset to strain improvement. Tetracycline resistance genes have been identified and mapped, and have played an important role in the build up of product resistance in the producing culture (which itself has 30S ribosomes).

Little detail exists for the industrial fermentation of tetracyclines. In common with other streptomycete fermentations, soy flour, peanut meal or corn steep liquor are the main supply of nitrogen in the initial batch medium. Corn syrups are used as carbon feeds throughout the fermentation to maintain a balanced control of growth and product synthesis. Ammonium and phosphate have to be maintained at low concentrations to achieve successful fermentations. Various methods have been described for the recovery of tetracyclines. The antibiotic can be extracted into n-butanol or methylisobutyl ketone at acid or alkaline conditions, or in the presence of quaternary ammonium compounds, or adsorbed on to active carbon for subsequent selective elution.

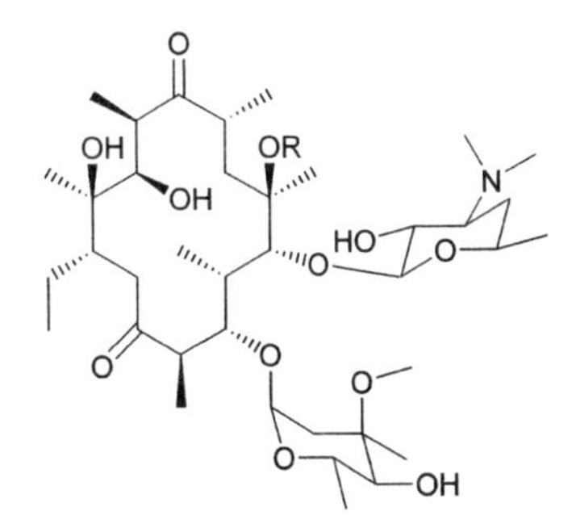

Fig. 16.16 Macrolides.

16.13 Macrolides

Macrolides are a diverse class of antibiotics, produced by actinomyces. Macrolides with antibacterial properties have in common a 12, 14 or 16 carbon macrocyclic lactone ring, substituted with sugar molecules. Larger ring macrolides, the polyene macrolides, can have lactone rings of 26–38 carbons. These polyenes are mainly antifungal, e.g. nystatin and amphotericin. The non-polyene macrolides are bacteriostatic. They inhibit protein synthesis by reversibly binding to the 50S portion of the ribosome.

Erythromycin and clarithromycin (chemical derivative of erythromycin) are the most prescribed macrolides (Fig. 16.16). They have a similar activity spectrum to the penicillins and are used by penicillin-sensitive people to combat Gram-positive bacteria, and, in addition, are used against *Mycoplasma, Campylobacter, Bordetella* and *Legionella*. Clarithromycin is currently prescribed to combat *Helicobacter pylori*.

Erythromycin is a 14 carbon macrolide produced by *S. erythreus*. Other 14 carbon macrolides include oleandomycin from *S. antibioticus*, pikromycin from *S. felleus*, megalomicin from *Micromonospora inositola*. Tylosin is a 16 carbon macrolide produced by *S. fradiae* and is produced industrially for animal use.

The general biosynthetic pathways and their associated enzymes and genes have been identified for many of the macrolides. Acetate, propionate and butyrate are the building blocks of the lactone ring and glucose is the precursor of the sugar units. Many commonalities have been recognised and considerable research has been focused on the creation of hybrid macrolides using recombinant techniques.

16.14 Economics

Based on the quantities of carbon and nitrogen required to obtain maximum broth titres, the biosynthesis of secondary metabolites such as antibiotics is inefficient. To produce a metabolite at a 3% concentration in the fermentation broth typically requires the utilisation of 20–25% carbohydrate and 3–5% protein.

Raw materials themselves contribute 30–45% of the final cost of the recovered antibiotic, with utilities at 10–20%, fixed costs, i.e. plant overheads, at 20–30%. Recovery costs can be 20–40% at a recovery yield of 85% plus. Fixed costs vary depending on the quantity of product produced and on the scale of operations.

Commercialisation of antibiotics has produced a competitive worldwide market. The difference between manufacturing cost and sale price is dependent upon a variety of fluctuating factors. Main factors are the annual volume of the operation, quality of technology available, local cost of the manufactured product, establishment of long-term contracts, and currency fluctuations between the major developed countries. Manufacturing costs are continually being lowered through technical development, improved efficiencies, increases in production volume and increases in market share. For a best case situation, a company should produce sufficient material to support its own internal captive demands for further processing to more expensive products. It should also have long term contracts to supply third party sales preferably to more than one customer, and have an active sales force to sell any remaining capacity to other third parties. This is not always the case as over the last 10 years several large pharmaceutical companies in the US and Europe have stopped making penicillin. This has been due to a need to switch to patented products with greater profit margins and a reluctance to compete in the commodity market place.

In contrast, several emerging Third World countries have now entered the field encouraged by their lower labour costs, the easy availability of producing strains of antibiotics, and the need to earn export dollars. The quality of the product can sometimes be an issue especially for export of the antibiotic to developed countries. The ready availability of modern manufacturing technologies, coupled with good quality assurance, will soon change this situation.

16.15 Good Manufacturing Practices

Over half of all antibiotics manufactured today are for human use. Their extensive use necessitates that the consumer should have confidence that the product is safe, consistent, clean and pose no additional adverse health conditions. Government authorities have established a philosophy and guidelines to ensure that products for human consumption are made under well controlled conditions. These are referred to as Good Manufacturing Practices. Regulations and guidelines are in place to ensure that correct procedures are followed throughout the many stages of product manufacture. Rigorous toxicology tests and detailed clinical trials have to be performed before a product can be considered for full-scale manufacture. Raw materials have to meet certain pre-established quality criteria and consistency. All manufacturing and pilot processes have to be detailed as standard operating procedures. All analytical procedures have to be validated to ensure that they always give true results under a wide variety of conditions. These procedures have been established to ensure that the product manufactured is of a consistent high quality. Any changes to a manufacturing process could result in differences in the final product so it is extremely important to adhere to all established operating procedures. Quality checks are always performed at suitable stages in the manufacturing process.

The purity of intermediates and final products cannot be expected to be 100%. However specifications have to be set to ensure product uniformity. The product should be of as highest purity possible at an established acceptance level. The presence and identity of all impurities should be known and should not exceed set limits. The toxicity of these impurities should be known. Analytical procedures should be in place to ensure the recognition and identification of any new impurities. To follow Good Manufacturing Practices, all procedures have to be documented and working copies available for operators to follow. Instructional sheets have to be signed at the completion of each stage and the records checked by management and retained as the batch record. These records are available to any inspections by Government regulatory bodies. Strict adherence to these policies will satisfy the regulatory authorities, and will ensure confidence in the general public that their medicines are safe.

16.16 Further reading

Elander, R. P. (1989). Bioprocess technology in industrial fungi. In *Fermentation Process Development of Industrial Organisms.* (J. O. Neway, ed.), pp. 169–219. Marcel Dekker, New York.

Hersbach, G. J. M., Van Der Beek, C. P. and Van Dick, P. W. M. (1984). The penicillins: properties, biosynthesis, and fermentation. In *Biotechnology of Industrial Antibiotics* (E. J. Vandamme, ed.), pp. 45–140. Marcel Dekker, New York.

Lowe, D. A. (1986). Manufacture of penicillins. In *Beta-Lactam Antibiotics for Clinical Use* (S. F. Queener, J. A. Webber and S. W. Queener, eds.), pp. 117–161. Marcel Dekker, New York.

Paradkar, A. S., Jensen, S. E. and Mosher, R. H. (1997). Comparative genetics and molecular biology of beta-lactam biosynthesis. In *Biotechnology of Antibiotics, 2nd Edition* (W. R. Strohl, ed.), pp. 241–277. Marcel Dekker, New York.

Queener, S. and Schwartz, R. W. (1979). Penicillins: biosynthetic and semi-synthetic. In *Economic Microbiology, Vol. 3* (A. H. Rose, ed.), pp. 35–122. Academic Press, London.

Smith, A. (1985). Cephalosporins. In *Comprehensive Biotechnology, Vol. 3* (M. Moo-Young, ed.), pp. 163–185. Pergamon Press, New York.

Strohl, W. R. (1997). Industrial antibiotics: today and the future. In *Biotechnology of Antibiotics, 2nd Edition* (W. R. Strohl, ed.), pp. 1–47. Marcel Dekker, New York.

Vandamme, E. J. (1984). Antibiotic search and production: an overview. In *Biotechnology of Industrial Antibiotics* (E. J. Vandamme, ed.), pp. 3–31. Marcel Dekker, New York.

Chapter 17

Baker's yeast

Sven-Olof Enfors

Nomenclature

Symbol	Description	Unit
C_E	Ethanol carbon concentration	(kg m^{-3})
C_S	Sugar carbon concentration	(kg m^{-3})
C_X	Cell carbon concentration	(kg m^{-3})
DOT	Dissolved oxygen tension	(% air saturation)
DOT*	DOT in equilibrium with gas	(% air saturation)
F	Substrate flow rate	(m^3 h^{-1})
H	Conversion constant	(% air sat. kg^{-1}·m^3)
K_La	Oxygen transfer coefficient	(h^{-1})
K_S	Saturation constant	(kg m^{-3})
q_{Ec}	Specific rate of ethanol consumption	(h^{-1})
q_{Ep}	Specific rate of ethanol production	(h^{-1})
q_m	Maintenance coefficient	(h^{-1})
q_O	Specific rate of oxygen consumption	(h^{-1})
q_{Os}	Specific rate of oxygen consumption for sugar oxidation	(h^{-1})
q_S	Specific rate of sugar consumption	(h^{-1})
q_{San}	Specific rate of sugar to anabolism	(h^{-1})
q_{Scrit}	Specific rate of ethanol consumption when overflow metabolism sets in	(h^{-1})
q_{Sen}	Specific rate of sugar to aerobic energy metabolism	(h^{-1})
q_{Smax}	Maximum q_S	(h^{-1})
S	Sugar concentration	(kg m^{-3})
t	Time	(h)
V	Volume of medium	(m^3)
X	Biomass concentration	(kg m^{-3})

Y_{em}	Yield coefficient exclusive maintenance	(kg kg^{-1})
Y_{OS}	Coefficient of oxygen per sugar	(kg kg^{-1})
Y_{XN}	Yield coefficient of cells per ethanol	(kg kg^{-1})
Y_{XO}	Yield coefficient of cells per oxygen	(kg kg^{-1})
Y_{XS}	Yield coefficient of cells per sugar	(kg kg^{-1})
μ	Specific growth rate	(h^{-1})
μ_{crit}	μ when overflow metabolism sets in	(h^{-1})

17.1 Introduction

The use of micro-organisms to raise the dough for bread making is one of the oldest examples of man's employment of micro-organisms. Bread of this type is known in Egypt from at least 3000 BC, when a slave's payment was settled in units of bread and beer. This bread was probably raised by a mixture of yeast and lactic acid bacteria, both of which were ingredients in the beer mash and foam that was used as starter culture for the bread production. The yeast used for today's baking, *Saccharomyces cerevisiae*, does not grow during dough raising conditions and it must therefore be supplied from external sources. Until the middle of the nineteenth century this yeast was obtained from the distillers and breweries, though the baker did not know what kind of active agents was contained in the fermentation foam that was used.

During the nineteenth century the bottom fermenting *S. carlbergensis* (now regarded as a synonym of *S. cerevisiae*) gradually replaced the top fermenting *S. cerevisiae* in large parts of Europe and, since the bottom fermenting yeast was less suitable for bread making a shortage of yeast was encountered. In 1846, the so-called **Vienna process** was introduced for production of dedicated yeast starters for bread making. From this time, most bread has been produced with specially propagated baker's yeast. The process was mainly an anaerobic fermentation of barley mash even though some aeration was applied, and thus the biomass yield was low and the economy was based on the concomitant ethanol production. At this time, the science of microbiology quickly developed and it became clear that it was a micro-organism that was responsible for the fermentation, in fermentation of alcoholic beverages as well as in bread dough raising. This resulted in the introduction of the **Air Process** in Denmark in 1877, based on Pasteur's finding that aeration inhibits ethanol production and promotes yeast growth.

In 1917 two patents were filed almost at the same time: by Zak in Denmark and Hayduck in Germany. Their inventions described a process for production of baker's yeast by intermittent feeding of the sugar, rather than adding all of it from the beginning. This was the so-called **Zulaufverfahren** or, in English, **fed-batch technique**. Although this chapter concerns the production of baker's yeast, its secondary function is to illustrate the technology that has become dominant for most industrial scale, aerobic fermentation processes, namely the fed-batch culture. This technique offers a means, not only to control overflow metabolism, but also to control catabolite repression, substrate inhibition and O_2 and cooling demand. In baker's yeast production the

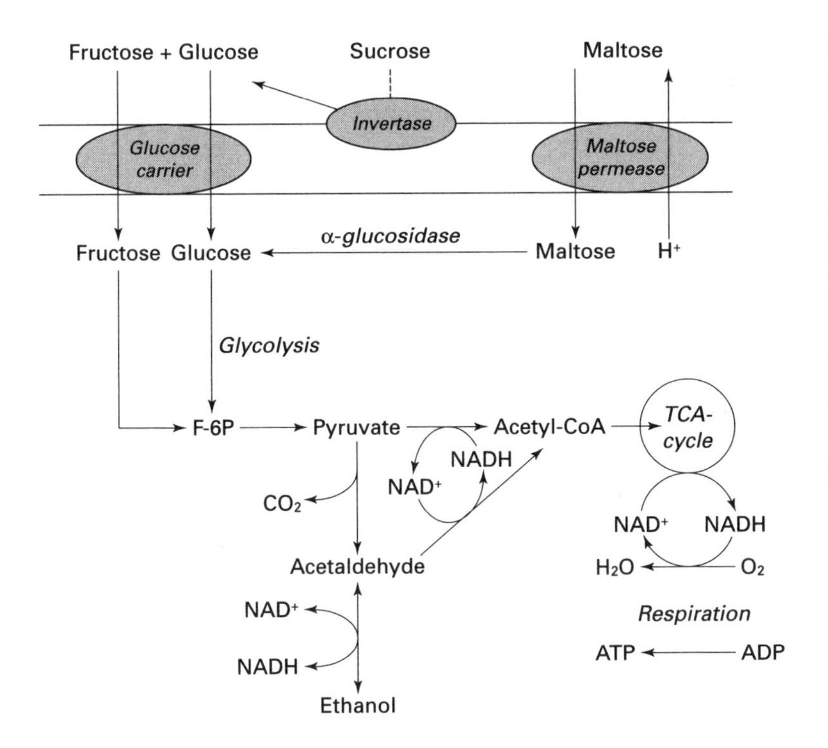

Fig. 17.1 Uptake and energy metabolism of the main sugars utilised by baker's yeast. Sucrose is hydrolysed by invertase to glucose plus fructose which are then taken up by the cell. Maltose is first transported into the cell and then hydrolysed in the cytoplasm to glucose. An important quality aspect of baker's yeast is that the maltose consumption genes are subjected to glucose repression and the enzymes involved are very unstable which means that the ability to utilise maltose is variable.

fed-batch technique is employed to control the sugar concentration; in other fermentation processes it may be other substrates or micronutrients to keep them at concentrations which will permit optimal metabolic activity in the cultivated micro-organisms. The uptake and energy metabolism of the main sugars utilised by baker's yeast is shown in Fig.17.1.

Baker's yeast is composed of living cells of aerobically grown *S. cerevisiae*. The commercial producers use various strains of this species. They differ from the strains of *S. cerevisiae* used for beer production mainly in their pattern of utilisation of medium components. The product is either delivered as a dried powder (dry yeast) with about 95% dry weight or as a cake with about 25–29% dry weight, containing only washed cells and residual water. The yeast is used to raise the dough in the baking process and to give special texture and taste to the bread. Dough raising is caused by the production of CO_2 during alcoholic fermentation of sugars available in the dough. These sugars are mainly maltose and glucose, produced from the flour starch by the α-amylase activity in the flour, or sucrose if added by the baker.

The main reaction of the dough raising can be considered as anaerobic fermentation of hexose to CO_2 and ethanol:

$$C_6H_{12}O_6 \rightarrow 2\,C_2H_5OH + 2\,CO_2 \tag{17.1}$$

The carbon dioxide is entrapped in the dough and causes its expansion. The ethanol, even though it evaporates in the oven, contributes to formation of esters. However, there are many other, less well characterised, properties of the yeast that are important for the bread quality, as evident from the difference between yeast fermented bread and bread produced with baking powder, that also evolves CO_2. Thus, baker's yeast should be considered as a package of enzymes rather than just biomass. The composition of this enzyme package is subject to optimisation by strain development and control of the fermentation process.

17.2 Medium for baker's yeast production

The stoichiometry for production of baker's yeast can be summarised as

$$200 \text{ g glucose} + 10 \text{ g } NH_3 + 100 \text{ g } O_2 + 7.5 \text{ g salts} \rightarrow 100 \text{ g biomass} + 140 \text{ g } CO_2 + 70 \text{ g } H_2O \quad (17.2)$$

This results in the following approximate yield coefficients:
$Y_{XS} = 0.5 \text{ kg kg}^{-1}$
$Y_{XO} = 1.0 \text{ kg kg}^{-1}$
$Y_{XN} = 0.1 \text{ kg kg}^{-1}$.
The production is an aerobic fed-batch process on a medium of molasses, ammonia or ammonium salts, phosphates, vitamins and antifoam. Which specific vitamins and additional salts have to be included in the medium depends on the strain, the quality of the sugar source (molasses) and the quality of the water. *S. cerevisiae* has a demand for many components, as evident from the complexity of a defined medium for its growth (see Table 17.1).

For commercial production, however, the molasses and the process water furnish most of these components. Molasses of both sugar cane and sugar beet can be used for baker's yeast production. The sugar content of the commercial molasses is 45–50%. A major difference between the two types of molasses is that sugar beet molasses contains mainly sucrose and little biotin, while in sugar cane molasses the sucrose to a large extent has been hydrolysed to glucose plus fructose, and it is also richer in biotin. Furthermore, molasses contains other fermentable sugars and amino acids that are utilised by the cells. A problem with the beet molasses is that 0.5 to 3% of the sugar is raffinose, a trisaccharide (fructose-glucose-galactose) that is only partially hydrolysed by baker's yeast that does not have α-galactosidase activity. This results in a substantial effluent of melibiose (glu-gal). Brewer's yeast, on the other hand, often has α-galactosidase activity, and cloning the gene coding for this enzyme into baker's yeast is therefore an obvious possibility to improve the yield and decrease the biological oxygen demand of the effluents arising in baker's yeast production.

Ethanol can be used by *S. cerevisiae* under aerobic, but not under anaerobic, conditions. In the baker's yeast process, some ethanol is initially produced and the process is controlled in such a way that this ethanol is later on utilised as energy source, as explained in (Fig. 17.2).

The main source of nitrogen is ammonia, but most of the amino acids of the molasses are also consumed and contribute to the total nitrogen supply. This uptake of amino acids and other organic compounds from the molasses is important for environmental reasons, since the remaining organic compounds in the medium contribute to

Table 17.1 Defined and commercial media for production of about 50 g baker's yeast per litre. The glucose and molasses are fed continuously with a solution containing about 300 g sugar per litre

Defined medium		Commercial medium	
Main medium (g l^{-1})			
Glucose (total feed)	100	Molasses (total feed)	340
$(NH)_4SO_4$	10	NH_3 aq.	0.3
KH_2PO_4	5	H_3PO_4(85%)	2
$MgSO_4{\cdot}7H_2O$	8	$MgSO_4{\cdot}7H_2O$	1.5
Trace elements (mg l^{-1})			
EDTA-Na_2	60		
$CaCl_2{\cdot}2H_2O$	18		
$FeSO_4{\cdot}7H_2O$	12		
$MnCl_2{\cdot}2H_2O$	4		
$CuSO_4{\cdot}5H_2O$	1.2		
$ZnSO_4{\cdot}7H_2O$	16		
$CoCl_2{\cdot}6H_22O$	1.2		
$Na_2MoO_4{\cdot}2H_2O$	1.6		
H_3BO_3	4		
KI	0.4		
Vitamins (mg l^{-1})			
Thiamine·HCl	4		
Pyridoxine·HCl	4		
Nicotinic acid	4		
D-biotin	0.2	D-biotin	0.1
Ca D-pantothenate	4		
Meso-inositol	100		

the considerable biological oxygen demand of the effluent process water. The ammonia consumption results in pH-decrease since one proton is liberated per NH_4^+ ion that is consumed. This is compensated for by pH regulation in which aqueous NH_4OH is used.

17.3 Aerobic ethanol formation and consumption

The main sugar sources in the production medium, glucose and fructose, enter the glycolytic pathway (EMP – see Chapter 2) as visualised in Fig. 17.1. The fate of the sugar can be divided into three main reactions:

- complete oxidation to CO_2;
- partial oxidation to ethanol;
- assimilation into the biomass.

Under certain conditions, the ethanol may be consumed by the cells and used as a carbon and/or energy source.

It is an old observation that *S. cerevisiae*, and some other yeast species, produce ethanol from sugar, not only when they are grown anaerobically

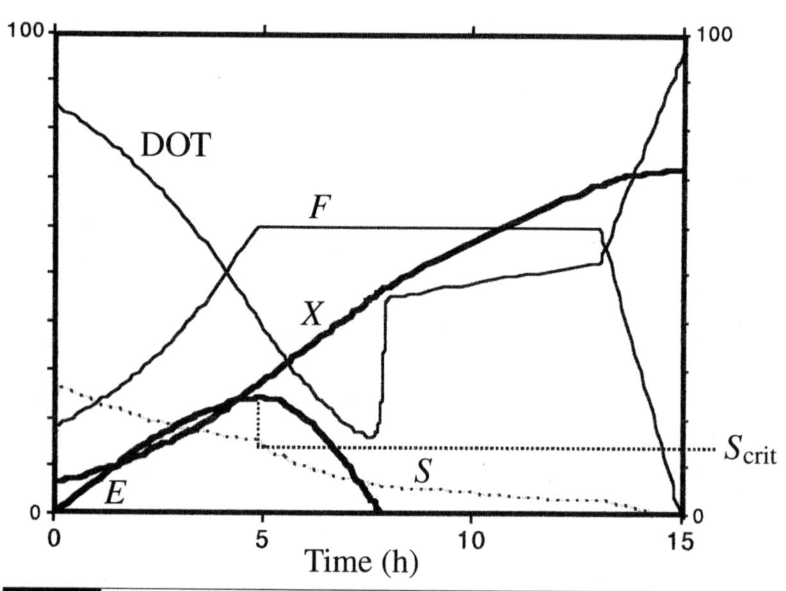

Fig. 17.2 Simulation of a baker's yeast fermentation process. Initially the molasses inflow rate (F) is increased exponentially to provide a given specific growth rate which results in a sugar concentration (S) slightly larger than the critical value (S_{crit}) and thus leading to some ethanol (E) being produced. After the feed rate has reached a pre-set value, the flow rate is kept constant to avoid O_2 limitation. During the constant feeding phase the sugar concentration declines and soon falls below the critical value. The ethanol peak, about 1 g l^{-1}, coincides with the time when the sugar concentration is equal to the critical value (S_{crit}). The DOT continues to decrease as long as ethanol is present but rapidly increases when the ethanol is exhausted and stabilises at a concentration corresponding to the feed rate. During the last hours, both the molasses feed rate (F) and the ammonia feed rate (not shown) are decreased to force the cells to stop dividing and acquire starvation resistance for future storing.

but also in aerobic cultures at high sugar concentration. This aerobic ethanol production is called **overflow metabolism**. Although not fully understood, this phenomenon is related to a critical, strain dependent, rate of sugar uptake. At sugar uptake rates below the critical rate, all sugar that is not assimilated into biomass is fully oxidised to CO_2 by pyruvate dehydrogenase and in the tricarboxylic acid (TCA) cycle (see Chapter 2). This oxidation generates reduced co-enzymes, that must be re-oxidised by the respiratory chain, which ultimately requires O_2. However, at glucose concentrations above about 30–40 mg l^{-1} the sugar is metabolised faster than the critical rate and the surplus of pyruvate that cannot be oxidised aerobically is reduced to ethanol instead of entering the TCA cycle. In this way, the NADH that was produced in the oxidation of this glucose to pyruvate is re-oxidised to NAD^+ (see Fig. 17.1).

Under certain conditions ethanol can also be used by *S. cerevisiae*. Ethanol then diffuses back into the cell where it is oxidised by NAD^+ to acetate and this is converted to acetyl-CoA, which enters the TCA cycle. The switch from production of ethanol to its consumption differs between a batch culture and a fed-batch culture. In a batch culture on sugar, ethanol is accumulated due to the overflow metabolism until the

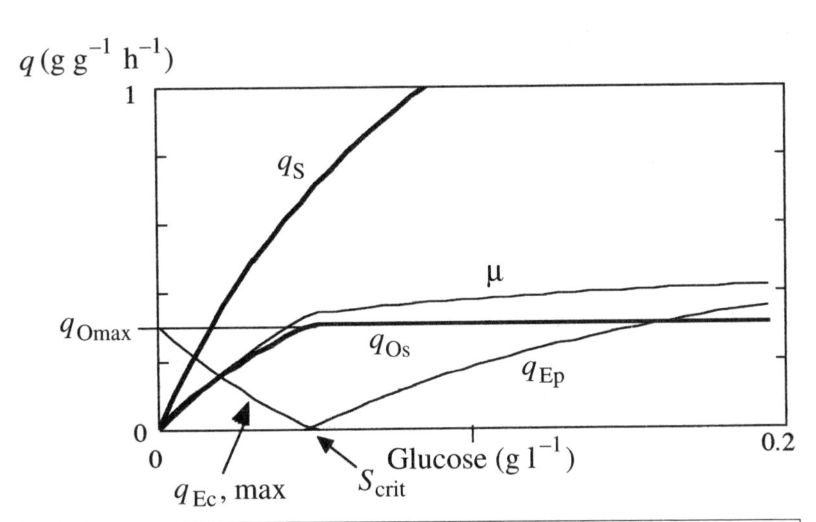

Fig. 17.3 The bottle-neck model illustrated as Monod plots of the specific rates of sugar (q_S) and oxygen (q_{Os}) consumption, growth (μ) and ethanol production (q_{Ep}) and consumption (q_{Ec}). As long as the sugar concentration increases at concentrations below the critical value, the specific oxygen consumption rate increases proportionally, but when the sugar concentration is further increased, no further increase in the specific oxygen consumption rate is observed, resulting in a maximum oxygen consumption rate (q_{Omax}). When the sugar concentration declines below the critical value and if ethanol is present, the ethanol is consumed at a rate that saturates the respiration, keeping the oxygen consumption rate at the maximum until the ethanol is exhausted.

sugar concentration is below the critical value. Shortly afterwards the growth abruptly ceases due to lack of sugar. Growth on ethanol alone requires a biosynthetic pathway to C_3–C_6 substances, which is provided by the so-called gluconeogenesis and the glyoxylate shunt (see Chapter 2).

Several enzymes of these pathways are repressed by glucose catabolite repression and their induction takes some time which results in a so-called diauxic lag of about one hour between growth on glucose and growth on ethanol in batch culture. This lag is not observed in the fed-batch process since glucose is continuously supplied and no gluconeogenesis is required. Instead, the acetyl-CoA generated from ethanol can be used in the TCA cycle and in biosynthetic processes based on acetyl-CoA.

In a fed-batch process with constant feed, the sugar concentration declines slowly, as illustrated in Fig. 17.2. As long as the sugar concentration is higher than the critical value, ethanol is produced but, when the concentration declines below the critical value, ethanol is consumed resulting in an ethanol concentration peak. The corresponding growth rate is then about 0.25 h^{-1} (strain dependent value). This model is also called the **bottle-neck model**, since it can be explained in terms of a critical maximum reaction rate (= bottle-neck) somewhere in the metabolism from pyruvate to the respiratory chain. This model is illustrated in Fig. 17.3.

17.4 The fed-batch technique used to control ethanol production

It is difficult to reach high concentrations of biomass of *S. cerevisiae* in a batch process since that would require a high initial sugar concentration and, due to the overflow metabolism this would result in inhibitory high ethanol concentrations. Furthermore, the high growth rate would result in a high oxygen consumption rate. As industrial bioreactors have quite modest O_2 transfer capacities in the range of 100 mmol $l^{-1}\cdot h$, high concentrations of biomass cannot be obtained without controlling the oxygen consumption rate within the range of the bioreactor capacity. Both these goals, the metabolic control of the overflow metabolism and the oxygen consumption rate control can be achieved through the fed-batch technique.

There are no simple analytical solutions of the mass balance equations of a fed-batch process as in the case of the chemostat. Instead, numerical solutions are applied from given initial conditions such as concentrations of biomass and substrate. A simple fed-batch model, not including the overflow metabolism, is summarised below.

The specific uptake rate for the carbon/energy source with concentration S is given by the Monod model

$$q_s = q_{s,max} \frac{S}{S + K_s} \tag{17.3}$$

This substrate is channelled into two main metabolic fluxes used for anabolism and aerobic energy metabolism, respectively. The flux used for anabolism can be estimated from a carbon mass balance as

$$q_{San} = \frac{C_X}{C_S}(q_S - q_m)Y_{em} \tag{17.4}$$

and the flux to aerobic energy metabolism is obtained as the rest:

$$q_{Sen} = q_S - q_{San} \tag{17.5}$$

The growth rate is obtained by subtracting the maintenance demand from the total substrate uptake and multiplication by the yield coefficient exclusive maintenance:

$$\mu = (q_S - q_m)Y_{em} \tag{17.6}$$

Since almost all oxygen consumption is derived from the respiration when cells grow on common carbon/energy substrates, the oxygen consumption rate is obtained from the flux to aerobic energy metabolism:

$$q_O = q_{Sen} Y_{O/S} \tag{17.7}$$

These specific rates are then inserted into the corresponding mass balance equations:

$$\frac{dS}{dt} = \frac{F}{V}(S_i - S) - q_S X \tag{17.8}$$

$$\frac{dX}{dt} = -\frac{F}{V}\cdot X + \mu X \tag{17.9}$$

$$\frac{dDOT}{dt} = K_L a\,(DOT^* - DOT) - q_O XH \tag{17.10}$$

A fed-batch process is often started at the end of a batch process when the dissolved O_2 concentration is approaching a limiting value. Then a constant feed rate, corresponding to a desired consumption rate, is applied. In the baker's yeast process, however, the principle of exponential feed is used to replace the batch phase, since too much ethanol would be produced during a batch phase. An exponential feed profile for control of overflow metabolism can be derived by simple means.

Assume that you want the cells to grow at a specific growth rate below μ_{max}, e.g. at μ_{crit}, which is the highest growth rate without ethanol production. The corresponding specific substrate uptake rate is obtained from the specific growth rate if the biomass yield coefficient is known:

$$q_{crit} = \frac{\mu_{crit}}{Y_{XS}} \tag{17.11}$$

Note that the yield coefficient is decreased if growth rates above the critical value for ethanol formation are selected. The corresponding sugar concentration, S, is then obtained by rearrangement of the Monod model:

$$S = \frac{q_{S,crit} K_S}{q_{S,max} - q_{S,crit}} \tag{17.12}$$

where K_S is the saturation constant for sugar uptake.

This initial substrate concentration can be kept constant by application of an exponential feed rate, $F(t)$, calculated from:

$$F(t) = F_0\, e^{\mu_{crit} t} \tag{17.13}$$

The initial feed rate, F_0, is estimated from:

$$F_0 = \frac{\mu_{crit}}{S_i Y_{xs}} (XV)_0 \tag{17.14}$$

where $(XV)_0$ is the initial biomass (kg) and S_i is the inlet feed sugar concentration (kg m^{-3}).

In practice, the coefficients in this model are not constant and therefore the concentration of sugar may not be absolutely steady. A lag phase with respect to sugar uptake is often observed and then the sugar concentration may initially rise, but this is later on compensated by an increased growth rate, so eventually the concentration stabilises at the control value as is visualised in Fig. 17.2.

Exponential feed of sugar gives exponential growth of biomass and thus also exponential increase of the total O_2 consumption rate. The exponential feed must therefore be switched to a constant feed before O_2 becomes limiting. This is illustrated by a simulation in Fig. 17.2. Note that the O_2 consumption rate does not decline until all ethanol has been consumed.

17.5 Industrial process control

The process lasts for about 15 hours and 50–60 g dry cells per litre will be reached. The inoculum size is about 10% (v/v). The initial rate of sugar feed must be low to avoid accumulation of sugar and excessive ethanol formation as the critical value above which ethanol is produced by overflow metabolism is only about 100 mg l^{-1} sugar (present as glucose and fructose at a ratio about 1:3 due to the rapid hydrolysis of sucrose). To use a constant feed rate then, corresponding to the critical consumption rate of the cells, would make the biomass productivity too low. Therefore the principle of exponential feed is applied to increase the sugar feed rate at the same rate as the biomass is increasing.

In the baker's yeast process, a specific growth rate of approximately 0.25 h^{-1} is used during the exponential phase, since a higher growth rate would give too much ethanol by overflow metabolism even though the productivity would be higher. If the feed flow profile is selected to give a growth rate at or below μ_{crit} no ethanol is formed and the biomass yield would be the maximal, but the productivity of biomass would be lower. Thus, selection of the feed profile is a typical optimisation case between high yield/low productivity and low yield/high productivity. In practice, a flow profile that gives an initial phase with some ethanol production, in the range of a few grams per litre, is chosen with the total cell mass, hence the loss in yield being low. When the flow rate is held constant the initially formed ethanol is consumed. This is illustrated in Fig. 17.3.

The constant feed phase can in principle be extended to achieve very high cell concentrations. In practice, however, the process is stopped at about 50–60 g dry cells per litre, since accumulation of non-metabolisable compounds from the molasses reaches inhibitory concentrations that raise the maintenance energy requirement of the cells, which means that the net biomass yield per molasses declines if the process is extended. The termination of a baker's yeast process includes critical procedures that aim at maturing the yeast to give it suitable qualities. The most important of these qualities are the gassing power, i.e. the rate of CO_2 production in the dough, the storage stability, i.e. the rate by which the gassing power declines during the storing, and the drying resistance for yeast intended for drying. It is also of utmost importance to keep a constant gassing power from batch to batch due to the mechanised bread production.

The maturation methods include cessation of the nitrogen and molasses feed according to company proprietary profiles. When the ammonia feeding is decreased and stopped while molasses feeding is continued, the cells are forced to terminate their cell cycle. This, and the subsequent step-down of the molasses feeding, also contributes to a decrease of the biological oxygen demand of the remaining broth. During this process the cell composition changes considerably with respect to its content of proteins and carbohydrates. In particular the concentration of trehalose which increases during the maturation phase, is considered important for the storage quality. About 1% of the cell's content of glycogen + trehalose is degraded per day during storage at 4 °C. Trehalose has further been suggested to function as a protective agent that increases the drying and freezing resistance of the yeast cells. Enzyme activities are also controlled by the feed profiles and nitrogen to sugar ratio during the process. There is a positive correlation between high protein concentration and high gassing power but a negative correlation between protein concentration and storage stability. Thus, control of the protein concentration is one means by which the manufacturer can adjust these properties of the product.

17.6 Process outline

The flow scheme of a baker's yeast plant is illustrated in Fig. 17.4. Each production company has its own, special, strain of *S. cerevisiae*, preserved under strictly aseptic conditions. From this strain, a first inoculum may be produced and then stored in frozen state. This inoculum is then propagated step-wisely to larger volumes; first in an anaerobic fermentation and later, when the yield becomes important, in aerated fermenters. The production stage mostly uses non-agitated fermenters of the air-lift or bubble column type. Typical size may be 100–200 m^3 and several fermenters may be used in a plant.

The medium is not fully sterilised. The process is run at pH between 4 and 5 and the large inoculum and short fermentation time, help to prevent any contaminating organisms from taking over. However, the molasses, which may contain large quantities of micro-organisms, is sterilised by continuous high-temperature/short-time sterilisation after dilution and standardisation of the sugar concentration. Water and the dissolved medium components are heated by steam under normal pressure.

The air for the process is filtered but not with a sterilisation filter. Thus, the baker's yeast product is not a monoculture but contains also some lactic acid bacteria and may occasionally also contain 'wild yeast' from the environment. It is sometimes thought that these lactic acid bacteria influence the leavening process. Since they constitute less than 1% of the total biomass in a typical baker's yeast product, it is not likely, however, that they have a measurable effect unless the leavening time is extended.

The fermentation time needed to complete one fed-batch depends on the ratio selected between the starting concentration of the yeast and its final concentration, and on the feed rate which is limited by the cooling or aeration capacity of the reactor. A typical value can be 15 hours to produce a broth containing about 50–60 g dry cells per litre. The cells are then washed with continuous centrifugal separators connected for counter-current operation. This concentrated yeast cream is then further de-watered in a rotating vacuum filter which provides the

Fig. 17.4 Flow scheme for production of baker's yeast

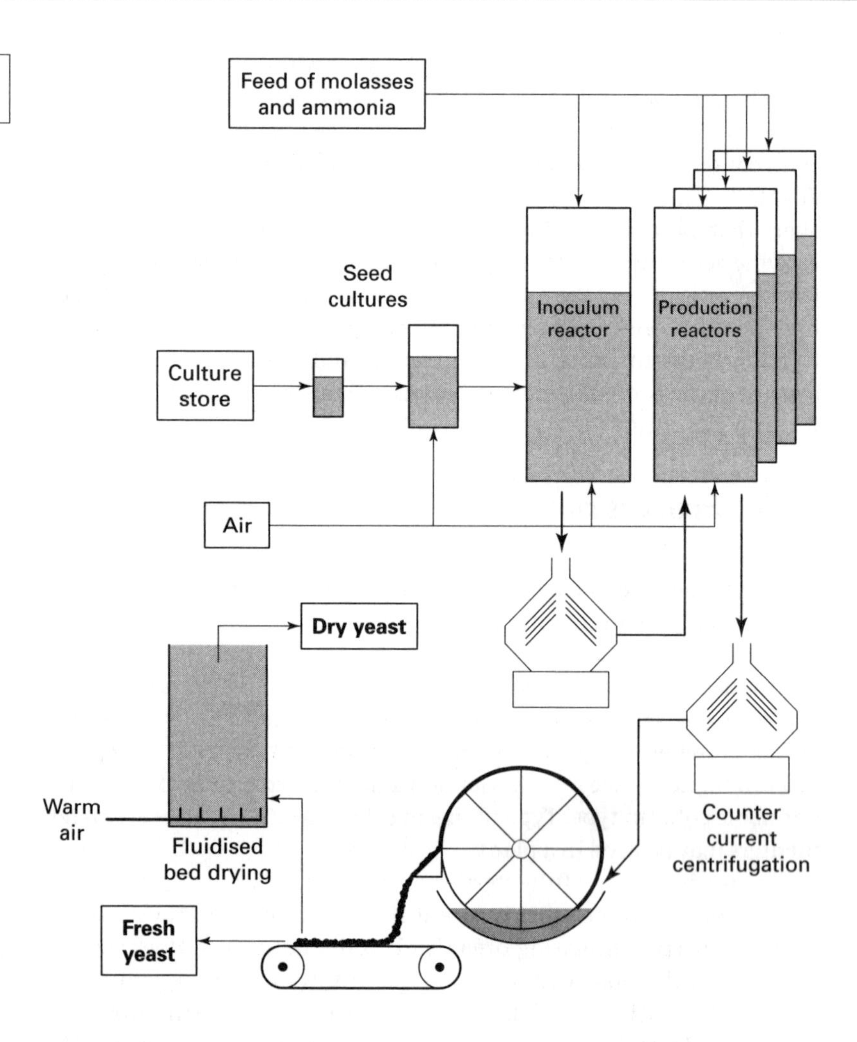

common yeast cake with about 27% dry weight that is used commercially as fresh yeast. Alternatively, the vacuum-filtered yeast cake is subjected to further drying in fluidised beds to produce dry yeast powder.

The fresh yeast has a shelf life of about one month when stored at refrigerator temperature (about 4 °C). The shelf life is not limited to this value for hygiene reasons, but rather because the baking quality declines. Some important quality parameters are listed in Box 17.1. Other important analyses to control the production are concentrations of protein and carbohydrate including trehalose and glycogen.

It may appear that the baker's yeast process would be a perfect candidate for continuous production according to the chemostat principle. Efforts have been made to develop a continuous process but have failed. There are two major difficulties. Firstly, the baker's yeast process is not a completely aseptic process but contains lactic acid bacteria, some other yeasts and endospores, none of which can develop during the short batch process but they will develop in a continuous process. Thus a continuous process must be absolutely aseptic which increases the investment and running costs considerably. Secondly, the maturation

Box 17.1 Standard quality parameters of baker's yeast

Gassing power, i.e. the rate of CO_2 production under baking conditions

Osmotolerance, i.e. the corresponding behaviour of the yeast in sweet dough

Storage stability, i.e. the rate of declination of the gassing power during normal cold storage for about 4 weeks

Colour

Smell

Rheology of the de-watered product

Microbial infections

procedures, in which both nitrogen and the molasses feeding are stepped down according to specific schemes, would require a multi-step continuous process which also increases the production costs. Thus, the fed-batch technique, in which the proportions between the supplies of nitrogen, sugar and O_2 can be easily controlled, is superior from the quality control point of view.

17.7 Further reading

deJong-Gubbels, P., Vanrolleghem, P., Heijnen, S., van Deijken, J. P. and Pronk, J. P. (1995). Regulation of carbon metabolism in chemostat cultures of *Saccharomyces cerevisiae* grown on mixtures of glucose and ethanol. *Yeast* **11**, 407–418.

Pham, H., Larsson, G. and Enfors, S.-O. (1998). Growth and energy metabolism in aerobic fed-batch cultures of *Saccharomyces cerevisiae:* simulation and model verification. *Biotechnol. Bioeng.* **60**, 474–482.

Rose, A. H. and Harrison, J. S. (eds.) (1989). *The Yeasts Vol. 3, 2nd Edition*. Academic Press, London.

Rose, A. H. and Vijayalakshimi, G. (1989). Baker's yeasts. In *The Yeasts, Vol. 5, 2nd Edition* (A. H. Rose and J. S. Harrison, eds.), pp. 357–397. Academic Press, London.

Sonnleitner, B. and Kappeli, O. (1986). Growth of *Saccharomyces cerevisiae* is controlled by its limited respiration capacity: Formulation and verification of a hypothesis. *Biotechnol. Bioeng.* **28**, 927–937.

Chapter 18

Production of enzymes

David A. Lowe

18.1 Introduction

Enzymes have been used both directly and indirectly by mankind for thousands of years. Their initial discovery and use was through serendipitous observations, adoption and continued adaptation. In general the enzymes themselves were expressed through the use of live microorganisms, for example in the leavening of bread, fermentation of fruit juices, bating of leather, or from crude tissue extracts, for example the conversion of milk to cheese. With hindsight these early discoveries are marvellous examples of the observant and creative nature of our early ancestors. Today, enzymes have many applications in a wide variety of areas of which the general consumer is unaware.

The world market for enzymes is over $1.5 billion and is anticipated to double by the year 2008. There has been a 12% annual increase in the volume of enzymes manufactured in the last 10 years. Approximately 400 companies are currently involved in the manufacture of enzymes, 14 of which can be considered to be major producers (Table 18.1).

In addition, over the last five years several new companies have emerged with interesting new technologies for enzyme isolation and production. Sixty per cent of enzyme production occurs in Europe, with 15% in the US and 15% in Japan. In terms of dollar usage, the US and Europe each

Table 18.1 Major bulk enzyme manufacturers and suppliers

Amano Pharmaceutical Co., Japan
Biocatalysts Ltd, Wales
Enzyme Development Corp., USA
Danisco Cultor, Finland
DSM-Gist, The Netherlands
Meito Sankyo Co., Japan
Nagase Biochemicals, Japan
Novo Nordisk, Denmark
Rhône-Poulenc, England
Rohm GmbH, Germany
Sankyo Co., Japan
Shin Nihon Chemical Co., Japan
Solvay Enzymes GmbH, Germany
Yakult Biochemical Co., Japan

Table 18.2 Total bulk enzyme distribution by value

Food	45%
Detergents	34%
Textiles	11%
Leather	3%
Pulp/paper	1%
Others	6%

Table 18.3 Food enzyme type distribution by value

Rennet	25%
Glucoamylase	20%
Alpha-amylase	16%
Glucose isomerase	15%
Papain	3%
Trypsin	3%
Other proteases	2%
Phytase	2%
Pectinase	2%
Others	12%
	Total value = US$ 1.5 billion

consume 30% of world output. Approximately 75% of industrial enzymes are used for hydrolysis and de-polymerisation of complex natural substances with proteases dominating due to their use in the detergent and dairy industries. Food applications are the largest category (Table 18.2).

These applications include conversions of starch to glucose and fructose, cheese, wine, beer, baking, flavouring, fruit juices and animal feeds (Table 18.3).

Production of enzymes has greatly expanded since the 1960s due to the widespread introduction of fermentation technology and more recently from the introduction of genetic engineering. Recombinant micro-organisms are now becoming the dominant source for enzymes for a wide variety of types. This trend will increase in the future due to the ease of genetic manipulation and the wide variety of enzymes available from micro-organisms found in diverse and extreme environments. Many microbial enzymes have been found and developed to replace existing enzymes from animal and plant origin.

18.1.1 Commercial considerations

Currently, enzymes are produced from a wide range of biological sources. An approximate breakdown of the sources for bulk enzyme source is as follows: filamentous fungi, 60%; bacteria, 24%; animal, 6%; plant, 4%; yeast, 4%; streptomyces, 2%.

Three general approaches can be taken to increase market share of bulk enzyme sales:

- develop less expensive manufacturing processes to reduce final cost and the sale prices;
- identify and develop new enzymes from new sources and seek new applications;
- find new uses for existing enzymes.

Many producers produce more than one enzyme from the same source. This is because market changes can produce a shortage in one type of enzyme and over-production of other enzymes. These imbalances can produce price variations. Bulk enzyme producers usually make more than one grade of enzyme with the high quality food grade dominating due to the high production volumes for this type of product.

Manufacturing processes for enzymes vary a great deal and are governed by the required quality, application, cost and market volume. Bulk commodity enzymes have an inherent low value and therefore necessitate a low cost manufacture with minimal processing. At the other extreme, high cost research and diagnostic enzymes have an expected high quality associated with them and inherent high manufacturing cost. Enzymes that are sold in annual volumes of over 10 000 tonnes typically cost $5–30 per kilogram. Speciality enzymes with limited annual volumes of less than one tonne can cost over $50 000 per kilogram, and some therapeutic enzymes can cost over $5000 per gram. At each extreme the total annual value of the enzyme product would range from $5 to $50 million.

There are clear differences between companies involved in fine and bulk enzyme manufacture, and companies will specialise in either but not in both as manufacture of either type involve radically different approaches, as indicated in Table 18.4.

18.2 Enzymes from animal and plant sources

The early sources of enzymes were animal or plant material. Initially their use was local. However, with the advent of slaughter houses and

Table 18.4 | Differences between the manufacture of bulk and fine chemicals

Bulk	Fine
Low unit cost, priced by weight	High unit cost, priced by enzyme units
Low purity, less than 10% protein	High purity, greater than 90% enzyme
Presence of other enzymes	Absence of other enzymes, or their activities quantified
Spray dried or concentrated liquid	Lyophilised solid, frozen solid or ammonium sulphate suspension
Absence of preservatives	Variety of preservatives used
Limited variety available ~200	Wide variety available ~2000
Purification limited to batch steps	Wide variety of purification techniques, especially column chromatography
Sample available gratis	No free samples
Minimum order 1–2 kg	Wide range of quantities

Table 18.5 | Enzymes from plant sources

Source	Enzyme	Application
Jack bean (*Canavalia ensiformis*)	Urease	Diagnostic
Papaya (*Carica papaya*)	Papain	baking, dairy, tanning, meat tenderiser, beer haze
Fig (*Ficus carica*)	Ficin	Meat tenderiser
Pineapple (*Ananas comosus*)	Bromelain	Baking (gluten complex reduction)
Horseradish (*Armoracia rusticana*)	Peroxidase	Diagnostic
Almond (*Amygdalus communis*)	β-glucosidase	Research
Wheat (*Triticum aestivum*)	Esterase	Ester hydrolysis and synthesis
Barley (*Hordeum vulgare*)	β-amylase	Baking, maltose syrups
Soybean (*Glycine max*)	β-amylase	Baking, maltose syrups

organised agriculture, increases in scale were possible. There still are many efficient enzyme manufacturing processes reliant on these sources. Animal tissue and organs provide excellent sources for some lipases, esterases and proteases, the most notable of which are the rennets, pepsin, trypsin and chymosin. Hen eggs continue to be a good source for lysozyme. Cultivated plants serve as excellent sources of proteases, for example bromelain from pineapple, papain from papaya and ficin from figs (Table 18.5.)

Animal sources tend to be more variable than plant sources (Table 18.6). There can be significant differences between the type, breed, age, condition and life history of animals prior to slaughter. Agricultural plants, on the other hand, can be cultivated specifically for the production of enzymes and some uniformity of product ensured. However all crops are at the mercy of the weather and seasonal, climatic and occasionally political events. Crude tissue from animal and plant sources has to be stabilised, usually by freezing or drying, prior to transporta-

Table 18.6 Enzymes from animal sources

Source	Enzyme
Calf, bovine, kid, lamb, porcine	Diastase (amylase), pre-gastric esterase, lipase, pepsin, trypsin, phytase, chymosin (rennin), phospholipase
Hen eggs	Lysozyme
Human urine	Urokinase

tion to enable the collection of quantities large enough for economical batch handling.

There are obvious limitations to relying solely on animal secondary products and agricultural materials. Often materials are not always available consistently and quantities can be limited. These limitations have been adequately addressed by the development of microbial enzymes with similar activities and the practice of large-scale microbial fermentations.

18.3 Enzymes from microbial sources

Micro-organisms are the most convenient source of enzymes. The number and diversity of enzymes is proportional to the number and diversity of micro-organisms. Microbes have been collected from environmental extremes such as hot springs, the arctic, rain forest and deserts. Each biological species has associated specific microbes and therefore the potential spectrum of enzyme activities is large. Genetic engineering techniques have enabled the enzyme industry to increase the fermentation productivity of these enzymes by many orders of magnitude. Even the properties of these enzymes can be altered and improved by protein engineering.

Many enzymes are naturally repressed and can only be expressed under certain culture conditions. These enzymes can be both intracellular, typically in the case of *E. coli*, and extracellular as in *Bacillus* species. Commercial strains of non-recombinant bacilli and aspergilli are known to produce enzymes up to a concentration of 20 kg m^{-3}. Recombinant cultures can also produce enzymes at these productivities.

18.3.1 Species-specific enzymes

A number of *Aspergillus* strains are prolific producers of many types of enzymes. Due to well-developed fermentation processes, *A. niger* is also widely used as a host for expression of recombinant enzymes (see also Chapter 5). This species is known to produce over 40 different commercial enzymes, the most common of which are shown in Table 18.7.

Similar enzymes can also be produced from different micro-organisms, for example there are seven different microbial sources for glucose isomerase and eight for α-amylase (Table 18.8).

Table 18.7 Bulk enzymes produced by *Aspergillus niger*

Lipase	Pectin esterase
Amyloglucosidase	Cellulase
Pentosanase	Catalase
Protease	α-galactosidase
α-amylase	Inulinase
Phospholipase	β-glucanase
Phytase	Galactomannase
Glucose oxidase	Arabinase
Pectinase	

Table 18.8 Common sources of α-amylase and glucose isomerase

α-amylase	Glucose isomerase
Malted cereals	*Actinoplanes missouriensis*
Animal pancreas	*Bacillus coagulans*
Aspergillus oryzae	*Mycobacterium arborescens*
Aspergillus niger	*Streptomyces murins*
Bacillus amyloliquefaciens	*Streptomyces olivaceous*
Bacillus licheniformis	*Streptomyces olivochromogens*
Bacillus stearothermophilus	*Streptomyces phoenics*
Bacillus subtilis	*Endomyces* spp.
Rhizopus oryzea	

18.4 Large-scale production

The cultivation of micro-organisms is economical on a large scale due to the use of inexpensive media and short fermentation cycles. The physiological state of the micro-organism in a fermentation process can be well controlled and the uniformity of each batch ensured. The harvest can be conveniently scheduled to fit in with the downstream processing. The choice of enzyme to be fermented is easy to schedule and different enzyme production campaigns can be planned and adjusted to meet sales demands. Enzyme productivity can be increased many-fold by both conventional strain improvement and fermentation process development and, with the additional use of genetic engineering, several orders of magnitude of improvement can be realised within a relatively short period of time (one to two years). Recombinant DNA techniques have also opened up opportunities for the mass production of enzymes from other microbial cultures, which conventionally were fastidious growers, required expensive media or inducers or were potential pathogens. Enzymes from extremophiles (growing at the extremes of temperature, salinity, pressure, alkalinity) can now be conveniently grown in mesophilic cultures, yet produce enzymes with the benefits of temperature resistance or high salt tolerance.

18.4.1 Recombinant *E. coli* fermentation

Enzymes originating from prokaryote sources can be conveniently produced on a large scale at high productivities in recombinant *E. coli* hosts. These fermentations can be carried out at a 3000 to 60 000 litre scale, and do not require complex inoculum build up. A typical fermentation protocol could be as follows:

The *E. coli* host would harbour a plasmid with the DNA coding for the required enzyme, together with a suitable antibiotic resistance marker such as ampicillin or neomycin, and an inducer such as TAC (codon specific to lactose or isopropylthiogalactose induction). Satisfactory high production can be achieved by a fed-batch fermentation where the initial batch medium contains components for initial growth e.g. glucose (2%), yeast extract (1%), phosphate (1%) and other salts together with the chosen antibiotic. After the initial growth has been established, further nutrients are fed at pre-determined rates to provide a readily available supply of carbon and nitrogen. Carbon is conveniently supplied as glucose, and nitrogen can be either of a complex nature e.g. yeast extract, casein hydrolysate or corn steep liquor (all providing amino acids), or as a simple ammonium salt or urea. Growth on amino acid mixtures provides faster growth and enzyme production, however the use of simpler, more defined nitrogen, although it may require a longer fermentation period, can support similar high cell growth and enzyme productivities. Use of defined media is a benefit to downstream processing particularly when ultrafiltration and ion exchange purification are involved. In either case the chosen antibiotic is included in the feed in order to prevent any shedding of the plasmid. Enzyme production is conveniently induced by addition of isopropylthiogalactoside to 30–300 mg l^{-1}, or lactose to 1–10 g l^{-1}. Timing and frequency of addition of the inducer are very important for maximum enzyme expression. Optical densities over 200 OD600nm (50 g dry cell weight l^{-1}) and enzyme expression at 10 g l^{-1} (30% of total protein) are desirable fermentation targets. Fermentations are normally run for 2 days at temperatures lower than the organism's optimum growth temperature, i.e. 24–28 °C.

18.4.2 Fungal fermentation

Fungal hosts, such as *Aspergillus* and *Fusarium*, and the methylotropic yeast, *Pichia*, are suitable for the production of glycoylated enzymes i.e. originating from fungal or animal sources. The enzyme DNA is integrated directly into the chromosomal DNA together with a promoter system.

The recombinant hosts can be fermented in a similar manner to non-recombinant cultures and high cell density fermentations can easily be obtained using conventional bioengineering approaches. *Aspergillus* or *Fusarium* hosts can be grown on inexpensive raw materials such as soy flour or Pharmamedia with the controlled feeding of corn syrups. Yeasts, such as *Saccharomyces* and *Pichia*, can be grown on yeast extract or protein hydrolysates with corn syrup feeding. Cultures can be grown to high cell mass using defined media, thus providing alternate less expensive nutirents which can have benefits in downstream processing. Fermentations are typically for periods of 4–8 days. Enzyme expression in excess of 10 g l^{-1} have been reported at the industrial scale. Enzymes are usually produced extracellularly which is a benefit as mechanical cell breakage is not needed.

18.4.3 Microbial enzymes replacing plant enzymes

Some industrial enzymes continue to be extracted from bovine sources which pose the danger of contamination with bovine spongiform encephalopathy (BSE). For example rennin, obtained from the stomachs of newborn calves, continues to be used for cheese making. It remains to be seen whether this presents any health risk to consumers, however, recombinant calf rennin can now be produced by microbial fermentations. Alternatively, with the convenient isolation of enzymes from microbial sources, and the use of appropriate screening technology, many new animal- and plant-like enzymes are now available. For instance, *Mucor* protease enzymes have in many instances replaced rennin (chymosin) from calf stomach; *Aspergillus* and *Bacillus* proteases have replaced the faecal enzymes used in leather bating, and a wide variety of microbial amylases have been developed to replace or supplement the plant amylases in malted barley and wheat. In addition some enzymes with mammalian origins, e.g. tissue plasminogen activator, can be produced in recombinant microbial hosts provided protein chain refolding can be attained. Glycosylated mammalian enzymes can be expressed in eukaryotic organisms such as *Saccharomyces* and *Aspergillus.* Here the host can glycosylate the enzyme to provide full activity even though the glycosylation sugars may be different from the mammalian source (also see Chapter 5). Enzymes can be produced directly by mammalian cell culture, however the manufacturing cost will be very high at $1000–$5000 per gram, a cost which can only be justified by therapeutic enzymes such as tissue plasminogen activator and related proteins.

18.5 Biochemical fundamentals

Microbial enzyme fermentations can be improved by classical techniques similar to those used for productivity improvements in antibiotic fermentations. Improvements both at the genetic level and at the process level have led to the development of fermentations capable of producing enzyme protein up to 20 kg m^{-3}. An understanding of the genetic regulation of enzyme synthesis (see Chapter 2) has been very helpful in selecting for improved strains and optimising the fermentation processes. There are many important factors that can influence the production of enzymes.

18.5.1 Induction

Enzyme synthesis is normally repressed, a condition that helps conserve energy from unnecessary protein synthesis, i.e. the enzyme will only be produced in the presence of an inducer, normally its substrate. The level of induction can be very strong (a more than 1000-fold increase over non-induced conditions) and acts by interfering with the controlling repressor. Many catabolic enzymes are inducible. For example:

- sucrose is needed for invertase production;
- starch for amylase production;
- galactosides for β-galactosidase production.

In some instances a product or intermediate can act as an inducer:

- phenylacetate induces penicillin G amidase;
- fatty acids induce lipase;
- xylobiose induces xylanases.

Product induction is common in the synthesis of extracellular enzymes required for the hydrolysis of large polymers that otherwise would not have the ability to enter the cell and cause the induction. Co-enzymes can act as inducers, i.e. pyruvate decarboxylase is induced by thiamine. In addition to being effective in enzyme production, induction can be useful in controlling the timing of enzyme production in the fermenter, i.e. a late rapid induction for an enzyme that is unstable under fermentation conditions. However, in practice, induction does often necessitate the handling of expensive inducer compounds, which have to be sterilised and added at specified times to established fermentations. To avoid these problems regulatory mutants can be produced in which the inducer dependence has been eliminated and are thus called constitutive mutants.

18.5.2 Feedback repression

Enzyme synthesis is also controlled by feedback repression. This occurs particularly in enzymes involved in the biosynthesis of small molecules where the accumulation of the final product can cause the repression of the synthesis of particular enzymes, normally the first enzyme in the biosynthesis route. Mutants lacking feedback repression can be obtained by selecting for cultures resistant to the toxic effects of an analogue of the product or intermediate. These survivors have lost the feedback sensitivity towards the product and its toxic mimic. Similar mutants can be obtained by isolating nutritional auxotrophs where the culture cannot make the final product, but instead depends on the addition of this compound for normal growth. The controlled feeding of this nutrient will limit the intracellular concentrations to below feedback repression levels.

18.5.3 Nutrient repression

Enzyme synthesis can also be controlled by nutrient repression typically by carbon, nitrogen, phosphate or sulphate. These mechanisms exist to conserve the production of unnecessary enzymes. Thus the cell only produces enzymes for the assimilation of the most easily metabolised or most readily available form of nutrient. The best known example is the control caused by the presence of glucose where this carbohydrate can effectively shut down the production of enzymes involved in the metabolism of other related and non-related compounds. Glucose catabolite repression can be very strong and can often suppress the inducer effect (see Chapter 2). Other carbon sources, such as lactate, pyruvate, succinate and citrate, are also effective repressors in some micro-organisms. Citrate can even repress the metabolism of glucose in some bacteria.

Glucose repression can be of major concern in the large-scale cultivation of cell mass as it is the most cost-effective raw material for fermentations. The effects of glucose can often be reduced by the limited feeding of the carbohydrate in such a manner that maintains practically zero concentrations of glucose in solution (the fermentation procedure for this is described in Chapter 7). If this approach fails alternative carbon sources can be used.

Glucose catabolite repression can also be solved genetically by selection of mutants resistant to this phenomenon. Mutants can easily be selected from media containing glucose and the substrate of the required enzyme, e.g. a glucose/aspartate mixture will select for aspartase producers with no glucose repression provided the aspartate is the only source of nitrogen. Penicillin G amidase production in *E. coli* has been increased many-fold by selecting for mutants capable of growth on an amide as sole nitrogen source in the presence of glucose. The resulting hyper-producers are constitutive and not subject to glucose repression. The glucose analogue, 2-deoxyglucose, is also a similar and an effective way of selecting for mutants with no glucose repression.

NH_4^+ is a much utilised source of nitrogen. It can however have strong control over the metabolism of many amino acids and other complex sources of nitrogen. This can often present problems in large-scale fermentations where both ammonium salts, amino acids and proteins are used as nitrogen sources. Ammonium salts are very inexpensive sources of nitrogen compared to amino acid mixtures, but the presence of the latter can often lead to rapid vigorous microbial growth. Some mutants resistant to nitrogen source repression can be selected by using the ammonium anti-metabolite, methylammonium.

18.6 Genetic engineering

The ready availability of recombinant techniques over the past 10 years has had a profound effect on the microbial production of enzymes. The host *E. coli* is very suitable for high enzyme expression provided the enzyme is not glycosylated. *Bacillus* species are suitable for the production of extracellular non-glycosylated enzymes. *Aspergillus* species are also very prolific producers of extracellular proteins and in addition can produce glycosylated enzymes.

Several properties should be considered in the construction of recombinant bacteria suitable for large-scale enzyme production. The choice of host is critical in that it should have a known lineage, have vigorous growth, have no auxotrophy and possess no enzyme systems that would be undesirable in the final enzyme product, i.e. no β-lactamase activity in a host used for the production of penicillin G amidase. The plasmid construct should be as simple as possible.

After synthesis the recombinant enzyme can either accumulate in the cytoplasm or be transported to the periplasm. Over-production of this foreign protein in the cytoplasm can sometimes cause the formation of inclusion bodies, which generally contain the recombinant protein in an unfolded, or wrongly processed, state. This is often undesirable as it is difficult, though not impossible, to extract and refold these proteins. Extraction from inclusion bodies is only cost-effective for the production of high cost proteins and enzymes such as insulin and tissue plasminogen activator. It is often more desirable to have the protein transported into the periplasm. For this to occur, signal or secretion sequences are required. These are located at the amino terminal end of the enzyme (upstream) and can be conveniently removed by signal peptidases. Excretion into the extracellular medium in *E. coli* is limited to certain proteins. For the production of extracelluar enzymes, *Bacillus* spp., yeasts and *Aspergillus* spp. are the hosts of choice.

Plasmids should be present in high copy numbers in the host and, in order to do this, some selection pressure is necessary as a host cell without plasmid would outgrow a host with plasmid. The most convenient way of ensuring plasmid retention is the placement of an antibiotic-resistant gene on the plasmid and maintenance of the relevant antibiotic in the growth medium. Ideal systems are kanamycin phosphotransferase, ampicillin β-lactamase, and chloramphenicol acetyltransferase. Antibiotic concentrations of above 30 mg l^{-1} are adequate to maintain the presence of the plasmid. Expression of recombinant enzymes in fungi such as *A. niger* generally involves the integration of the gene and a fungal promoter together with an associated selection mechanism (antibiotic resistance or a wild type gene to overcome a host auxotrophic factor) directly into the chromosomal DNA. Once a stable construct has been selected it is no longer necessary to have the selective agent present (unlike the necessary plasmid retention in *E. coli*). Often multiple copies of the gene can be inserted. Constitutive promoters, such as phosphoglycerate kinase and glyceraldehyde phosphate dehydrogenase, are very powerful and are a popular choice. Promoter genes are inserted upstream of the desired enzyme gene for optimum production, e.g. a starch promoter can be used as a convenient signal for α-amylase production, or phenoxyacetic acid promoter can be used for penicillin V amidase production. Often the inducer is unrelated to the produced enzyme, e.g. methanol is used to stimulate the alcohol dehydrogenase promoter in *Pichia*, which can be coupled to many different enzymes. The inducer of choice has to be maintained above a critical level, however the timing of addition is not critical.

18.6.1 Site-directed mutagenesis

Enzymes which have known amino acid sequences and three dimension structures can be altered by site-directed mutagenesis (see Chapter 4). Amino acids in the active site or other important areas can be identified as targets for changing to other amino acids. This can be performed easily by changing the specific trinucleotide codons in the enzyme gene and expressing the altered enzyme by conventional cloning. Once the beneficial change has been selected the process can be repeated to create further amino acid changes in the enzyme. The new enzymes are easy to scale-up using the same production procedures developed for the parent wild type as these amino acid changes have no effect on the growth of the host or protein expression. Well-known such amino acid changes are increased stability, changed substrate preference, resistance to oxidation, tolerance to solvents and alkali, and changes in chiral activity.

18.7 Recovery of enzymes

Enzyme recovery and purification are as important to the economics of production as the fermentation stages. The main challenge in the recovery steps is to minimise losses in enzyme activity. Many of the steps employ conventional recovery and purification units that are described in detail in Chapter 9. The following text will concentrate on enzyme-specific issues in the recovery and purification of enzymes.

18.7.1 Recovery of extracellular enzymes

Extracellular enzymes are relatively easy to recover and purify. They can represent a major portion of the total extracellular protein and often simple cell removal and concentration of the active solution can yield enzyme preparations directly suitable for some applications. Relatively clean enzyme preparations can be obtained by cultures growing on simple defined media. The recovery and purification steps are the same as those used for intracellular enzymes, once the cells have been broken (see below).

18.7.2 Recovery of intracellular enzymes

For preparation of intracellular enzymes from animal or plant sources, the tissue has to be disrupted to release the enzyme. Detergents or surface-active agents may have to be used to dissociate enzymes that are membrane bound. Tissue drying can be a convenient method of stabilising and disrupting animal and plant tissue. Freeze-drying is the least disruptive and avoids protein degradation, although it is prohibitively expensive on a large scale. Tissue can be air- or vacuum-dried, or subjected to water-miscible solvent precipitation. Enzyme extraction from dried materials can be as simple as rehydration in an appropriate buffered solution. Simple freezing can break some tissue although this is not a convenient method for scale-up. Some plant and animal materials require tissue homogenisation where the tissue is shredded and blended by a mechanical means to break open the cells. Once the enzyme has been solubilised, the residual cell debris can be conveniently removed by filtration. Low speed centrifugation can also be used. In any tissue with fat present, the removal of the fat layer can be problematic for centrifugation. Fatty material can be easily removed by solvent extraction; acetone precipitation is a convenient way to remove protein from lipid materials. As an alternative a water-immiscible solvent, such as hexane, can be used.

The breakage of microbial cells presents greater problems than with animal or plant cells. The cells are often tougher and sizes range from 0.2–10 microns, often necessitating specialised breakage. Many enzymes from yeast and Gram-negative bacteria can be extracted by the use of water immiscible solvents, such as toluene or chloroform, which can cause disruption of the cell membranes resulting in enzyme leakage. Water-soluble solvents, such as ethanol and 2-propanol, can be used to extract enzymes from the periplasmic space, but the use of these solvents can provide handling and safety issues on a large scale. Suitable detergents can be used in extracting small enzymes (less than 70 000 MWt). In some cases where enzymes are stable in the presence of high pH, alkali can be used to lyse bacterial cells. Cell walls of bacteria can also be lysed by the use of lysozyme, however the cost of this enzyme can be prohibitive. Similarly, yeasts can be lysed by β-glucanases. Autolysis can be effective in yeasts, however, long periods of time are required and it is difficult to control or optimise the process.

Microbial cells can be broken effectively by physical means. Sonication is convenient and effective at the laboratory scale. This process, however, cannot be scaled up. High pressure extrusion equipment such as the French Press is very effective at treating small quantities, i.e. 10–100 ml at the laboratory scale. This process of cell breakage uses high pressure to extrude material through a small orifice to impact abruptly onto a plate, thus causing very high sheer. This principle has been scaled up, e.g. Gaulin, Microfluidics, to equipment capable of breaking cell cultures at 100–1000 litres per hour. This is currently the method of choice in the industry.

Many types of cells can be broken by rapid agitation or blending with small glass or ceramic beads. At the laboratory scale this principle is available as a 'bead beater'. At a larger scale ball mills are available. With the appropriate choice of bead size and retaining mesh screens a ball mill can be a very effective, continuous process for cell disruption.

18.8 Isolation of soluble enzymes

It is critical to separate the soluble enzyme quickly and efficiently from the remaining cell debris. Chilling, use of an appropriate buffer, and the presence of enzyme protectants such as mercaptoethanol may be necessary to stabilise some enzyme preparations. Often protein inhibitors have to be added to reduce the destructive effect of proteases. Soluble enzymes can be collected by membrane filtration or by centrifugation – the latter has been used since the turn of the century. It is relatively easy to perform at a laboratory scale, where high centrifugal forces can easily be obtained. The same high **g** forces cannot be attained on large-scale equipment. Large centrifuges are often constructed from expensive titanium alloys to withstand the high **g** forces, but even then only moderately high **g** forces can be obtained. To combat this problem, protein- and nucleoprotein-precipitants, such as polyethyleneimine, can be added to the cell homogenates to flocculate unwanted materials and aid the settling times.

Filtration, with the use of filtration aids, such as a diatomaceous earth, can be a convenient way to obtain soluble enzymes. Scale-up is easy. Additives can be used to speed up the filtration. Often several precipitating agents can be used together, in order to reduce subsequent processing steps. With washing, product recoveries of 95% can be obtained. Disposal of the filter aid can be an issue. Filter aids can be used for the collection of the initial whole cells provided a cell breakage technology is used that does not rely on mechanical homogenisers. For example, the cells in the filter aid mix can be lysed chemically, enzymically or physically by simple agitation due to the abrasive nature of diatomaceous earth. The released soluble enzymes can be conveniently collected by simple filtration.

Ultrafiltration is a very versatile technology, easily scaled up and, with the correct choice of membrane porosity (molecular weight cut-off), enzymes can be collected selectively according to their molecular weight range. Plate and frame, and hollow fibres have been used for 20–30 years. Ceramic membranes are now becoming more popular due to their ease of cleaning and sterilisation, as they can tolerate high temperatures under high alkaline and detergent conditions.

Often the soluble enzyme solution requires no further treatment except concentration, either under reduced pressure or by membrane filtration, to produce a concentrated protein solution (10–50% solids). Stabilisers such as ammonium sulphate can be added if needed. Excipients such as lactose, dextrins can be added to provide bulk and act as stabilisers. Other additives are added to reduce dust formation.

18.9 Enzyme purification

Enzyme purification may be necessary but only if the extra cost is justified by the enzyme's application. The scale of the purification process will dictate the choice of separation technology, as some are difficult to practise on large scale. The same separation medium can often be used in different techniques, e.g. ion exchange chromatography is best practised as gradient elution from columns at the smaller scale, and by batch adsorption/elution at the larger scale.

18.9.1 Enzyme precipitation

It is often desirable to have an initial batch precipitation stage in order to reduce the total volume of the enzyme solution. Simple precipitation can be carried out, sequentially or in combination, with ammonium sulphate, sodium sulphate, polyethyleneimine and polyallylamine, or organic solvents such as isopropanol, ethanol and acetone. Streptomycin sulphate, polyethyleneimine and other polyamines can precipitate acidic nucleic acid and nucleoproteins. Ammonium sulphate and organic solvents can selectively precipitate the enzyme of choice with 80–90% activity recovery. Simple precipitation can have the additional benefit of protease removal. The main benefit is the ability to reduce the total operational volume, often by a factor of 20. This will significantly reduce the volume of resins and the associated downstream handling used for further processing. Changes in pH and temperature can also result in the selective precipitation and removal of unwanted enzyme proteins.

Liquid–liquid partition with polyethylene glycol, dextrans or polyamines can produce separate liquid phases into which desired enzymes can be concentrated. These treatments involve the use of large quantities of additives, and recovery of the polymers is desirable from a cost point of view, although recycling can reduce the initial selectivity and yield a cruder separation. These methods are easily applied to whole broths where the enzyme is soluble, or to extracellular enzymes, as the removal of the biomass is avoided.

18.9.2 Separation by chromatography

Chromatographic separation techniques commonly used, include ion exchange, hydrophobic interaction, size exclusion, affinity and dye ligand chromatography. Ion exchange chromatography is the most widely used and can encompass the use of strong and weak anion and cation exchange, and phosphate interaction. Weakly bound ions are more suitable for enzyme adsorption as stronger ionic charges can denature enzymes due to extremes of local pH. Enzyme proteins are generally adsorbed at low ionic strengths or at pH values where their total ionic charge is strong enough to interact with the opposite resin charge. Protein elution is easily accomplished by changing the pH or by increasing the ionic strength of the eluting solution. Such resins can be used either as a means to actively adsorb and selectively elute the desired enzyme, or by simple adsorption of unwanted protein materials.

Hydrophobic interaction chromatography uses the surface hydrophobicity of the enzyme as the selection property. Protein solutions are loaded on the resin at high salt concentrations and enzyme separations occur when a decreasing salt gradient is applied to the resin. This type of separation is useful as a follow-up to ammonium sulphate precipitation as salt removal is not needed prior to this chromatographic step. Hydrophobic resins are an order of magnitude more expansive than ion exchange resins.

By using these adsorption and desorption techniques the volume of the enzyme solutions can be greatly decreased and enzymes of up to 90–95% purity produced. This quality is sufficient for most specific enzyme applications. The chromatographic techniques can be chosen to remove other contaminating enzymes that would otherwise interfere with the use of the desired enzyme. Further clean up can be achieved using size exclusion chromatography. This, however, is difficult to perform on a large scale. It is often not necessary to purify an enzyme to 99%-plus purity. If greater than 95% purity is required further adsorption/desorption chromatography can be used taking finer and more selective cuts, and by using shallower eluting gradients. Such techniques do not give high activity yields, a factor that has to be taken into consideration when making very pure enzyme preparations for sale. Enzyme preparations of high purity are generally lyophilised for storage and transport. Polyhydroxy cryoprotectants, buffers, reducing agents and anti-microbial agents can be included if necessary. Alternatively enzymes can be stabilised as ammonium sulphate suspensions.

18.10 Immobilised enzymes

Over 95% of enzymes are sold in soluble forms, the majority of which are used directly on a single use basis, particularly in the food and beverage industries. For certain applications it is more desirable to have the enzyme in an immobilised form so that it can be re-used and easily removed from the reaction mixture. This is particularly important in the pharmaceutical industry where an enzyme step can have the benefits of chiral resolution and environmental positive processing. The easy removal of the enzyme catalyst ensures better product recovery and minimum carry over of residual enzyme protein. Generally, bulk enzyme manufacturers do not manufacture immobilised enzymes. It is more cost-effective for these catalysts to be manufactured in-house using proprietary supports and immobilisation techniques.

18.10.1 Immobilisation techniques

There is a wide range of techniques for enzyme immobilisation. The correct choice depends upon the final application of the catalyst; whether the enzyme step is in an aqueous or solvent medium, at the beginning or at the end of a synthesis or hydrolysis process and the type of reactor. Adsorption onto an inert support is the simplest technique, however, such supports do not permit permanent adsorption when used in aqueous environments without further modification. Lipases can be effectively adsorbed onto hydrophobic supports for repeated use in non-aqueous lipid systems. Adsorbed enzymes can be permanently fixed by cross-linking reactions with glutaraldehyde treatment, which can make them re-usable in aqueous solutions. Cross-linking with glutaraldehyde onto inert supports is a common technique. The choice of inert support will dictate the final physical form of the immobilised catalyst, which in turn must be compatible with the reactor choice. Supports such as cellulose or diatomaceous earth produce a product that is suitable for stirred reactor use and recovery by filtration. Enzymes adsorbed onto beaded supports are more suitable for column reactors or stirred reactors where simple settling is sufficient to remove the final reactants from the support catalyst.

18.10.2 Immobilisation supports

Many inert and activated supports are available for enzyme immobilisation. Inert supports, such as resins, porous glass beads, cellulose, gelatin, alginates or chitosans, require chemical activation or derivatisation prior to immobilisation; treatments that can add substantial costs to the final process. However some activated supports can produce high enzyme loading with a high retention of activity. Many chemically activated supports, predominately agarose-based, are available for laboratory research. Their use at the large scale is prohibitive due to very high costs, with the possible exception of antibody-immunological applications. Chemically activated supports, such as Eupergit, are available at the large scale, and are priced competitively for some commercial applications.

Enzymes can also be immobilised at high activity recoveries by simple entrapment where the enzyme is mixed with a soluble material that can be easily solidified. Such reactions include cellulose acetate precipitation from an organic solvent to an aqueous change, acid to alkali changes with chitosan polymers, temperature reduction with gelatine, sodium to calcium ion change with alginates, chemical polymerisation with polyacrylamides. The resulting entrapped enzyme is in a fibre or beaded form. Enzymes in these forms have been used industrially for many years for the resolution of amino acids and chiral separations.

18.11 Legislative and safety aspects

Although enzymes are natural products and their past use in food processing generally proves them safe for consumer use, there is a community and government awareness of the potential hazards associated with enzymes. This perception arose from the rapid introduction of enzymes as detergent additives in the 1960s, which had the unfortunate negative consequences of allergic reactions on inhalation, and skin rashes. The collapse of the detergent enzyme business left manufacturers with excess fermentation capacity, which stimulated diversification and development of other enzymes, which can now be regarded as non-traditional. Further diversification occurred with the introduction of genetic engineering techniques where large quantities of enzymes from a wide variety of sources could be easily manufactured. In response to government regulatory concerns, the industry has been proactive and has reduced and eliminated the exposure of workers and consumers to their products. Many enzymes are now sold as concentrated solutions, or as pelleted solids to avoid dust formation.

Many enzymes are GRAS (generally regarded as safe) listed provided they are manufactured under Good Manufacturing Practices (GMP) and many are now made using recombinant hosts such as *Bacillus*, *E. coli*, yeasts and *Aspergillus*. If not GRAS, these micro-organisms are listed under the hazard group 1; most unlikely to cause human disease, the lowest category of microbial culture apart from those actually cultivated for human consumption, i.e. yeast.

Enzymes from animal and plant sources do not require toxicology studies. Enzymes from non-pathogenic micro-organisms are usually subjected to short-term toxicology, however, those from less well-known micro-organisms do require extensive testing at least in rodents. With the cultivation of any micro-organism there is always the potential danger of the inadvertent production of a toxic metabolite. The number of known mycotoxins, exotoxins and endotoxins continues to grow annually. The quantification of risk assessment associated with this potential is very difficult. However with controlled fermentations and standardised extraction and recovery procedures, all operated under GMP, these risks can be minimised. In most food use, enzymes are only added at low levels, typically much lower than a variety of other components of biological origin.

In non-food applications such as chemical synthesis and hydrolysis, most enzymes are used in intermediary steps and the resulting products can be subjected to purification and detailed quality and impurity analysis. In the rare case where the enzyme is used as the last step, analytical procedures must ensure that any residual enzyme protein levels meet pre-determined minimum levels in the final product. All bulk enzyme products sold today come with a detailed Material Safety Data Sheet which outlines all potential dangers and describes recommended handling procedures for using the enzyme.

18.12 Further reading

Arnold, F. H. and Volkov, A. A. (1999). Directed evolution of biocatalysts. *Curr. Opin. Chem. Biol.* 3, 54–59.

Atkinson, B. and Mavituna, F. (eds.) (1991). *Biochemical Engineering and Biotechnology Handbook, 2nd Edition*. Stockton Press, New York.

Bullock, C. (1995). Immobilised Enzymes. *Science Prog.* **78**, 119–134.

Darbyshire, J. (1981). Large scale enzyme extraction and recovery. In *Topics in Enzyme and Fermentation Biotechnology, Vol. 5*. (A. Wiseman, ed.), pp. 147–186. John Wiley, New York.

Demain, A. L. (1990). Regulation and exploitation of enzyme biosynthesis. In *Microbial Enzymes and Biotechnology, 2nd Edition* (W. M. Fogarty and C. T. Kelly, eds.), pp. 331–368. Elsevier, London.

Godfrey, T. and West, S. (eds.) (1996). *Industrial Enzymology, 2nd Edition*. Macmillan Press, London.

Lowe, D. A. (1992). Fungal enzymes. In *Handbook of Applied Mycology, Vol. 4* (D. K. Arora, *et al.*, eds.) pp. 681–706. Marcel Dekker, New York.

Patel, R. N. (ed.) (1999). *Stereoselective Biocatalysis*. Marcel Dekker, New York.

Tanaka, A., Tosa, T. and Kobayashi, T. (eds.) (1993). *Industrial Application of Immobilized Enzymes*. Marcel Dekker, New York.

Zaks, A. and Dodds, D. R. (1998). Biotransformations in the discovery and development of pharmaceuticals. *Curr. Opin. Drug Discovery Develop.* **1**, 290–303.

Chapter 19

Synthesis of chemicals using enzymes

Thorleif Anthonsen

19.1 Introduction

19.1.1 Use of industrial enzymes

Enzymic processes have a long tradition in human history. In ancient times brewing of beer and making of wine are examples. In more recent years, enzyme catalysis has played important roles in production of fine chemicals and drugs such as vitamin C, amino acids, antibiotics and steroids. The best known examples to the public are use of enzymes in detergents. In this case, enzymes are contained in the marketed product itself. The first 'bio-detergents' appeared in the mid-1960s and they contained proteases to dissolve protein stains on clothes. At present a detergent also contains lipases and amylases to dissolve fat and starch stains. In order to restore colour of cotton that has been washed several times, cellulases are added. Cellulases are also used to give jeans the so-called 'stone-wash' look. Enzymes are also extensively used in other industries such as pulp and paper industry, textile industry, leather industry and for baking. Furthermore, large quantities of high-fructose corn syrup sweetener is produced from corn starch using hydrolytic enzymes and glucose isomerase in order to isomerise glucose to fructose which is sweeter than glucose. Enzymes are also used in dairy industry and for the production of wine, fruit juice, beer and alcohol. In order to transform low value fats and oils into more valuable ones, lipases are used as transesterification catalysts. In particular due to need of enantiopure pharmaceuticals and building blocks for organic synthesis, the use of enzyme catalysis in fine chemicals and pharmaceutical industry is increasing.

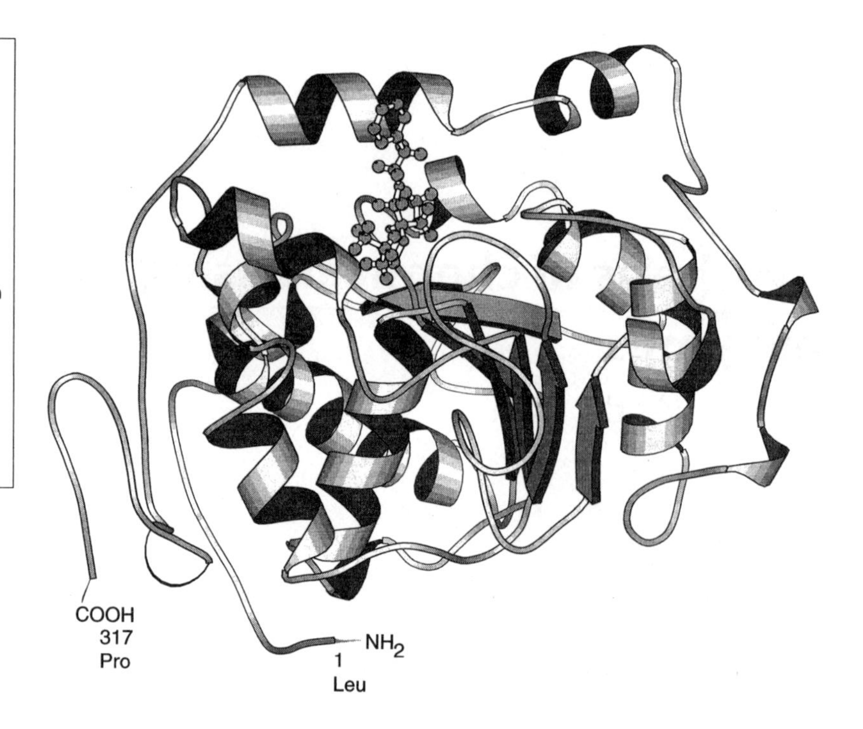

Fig. 19.1 Structure of lipase B from *Candida antarctica* (CALB) as displayed by Molscript. This lipase has the α/β-hydrolase fold (a specific sequence of α-helices and β-strands) characteristic of lipases of the serine-hydrolase group. This lipase is made up of 317 amino acids comprising 4625 atoms which gives a molecular weight of 33 273. CALB is a relatively small protein – most enzymes have molecular weights twice or three times as high. It contains 10 α-helices which may be seen in the diagram.

19.1.2 What are enzymes?

Enzymes are proteins, i.e. they are made of amino acids held together by amide bonds, called peptide bonds. An amino group in one amino acid is united with a carboxylic acid in another and in this way a long peptide chain is formed. The **primary structure** of a protein is characterised by the type and order of amino acids in the peptide chain. Parts of this chain may then be shaped into helices and sheets due to hydrogen bonds between electronegative groups like carbonyl and hydrogen atoms attached to O or N atoms. Although much weaker than the covalent bonds forming the primary structure, several H-bonds provide the strength of the **secondary structure**. When this molecule is folded into a bundle kept together by covalent -S-S- bonds, formed from two cysteine residues in different parts of the chain, H-bonds and van der Waal's forces, the catalytically active **tertiary structure** is formed. Catalysis takes place in a limited area, called the **active site** of the enzyme. Some enzymes need co-factors, such as small organic molecules or metal ions, for their activity. These co-factors are regenerated in the cells and the ultimate reduced or oxidised compounds are sugars, O_2 and other simple molecules (see Section 19.4). Figure 19.1 shows the three-dimensional structure of the backbone of lipase B from *Candida antarctica* (CALB), an enzyme that in a natural environment can hydrolyse the glycerol ester bonds which build up fat.

19.1.3 Why are enzymes of interest as catalysts in synthesis?

Why use enzyme catalysis in the laboratory or in process industry? There are unique advantages that enzymes offer which are difficult to obtain by conventional catalysis. First of all, they have great selectivity and specificity. No matter how simple the enzyme-catalysed chemical reaction is, this may be on three levels: *chemoselectivity*, *regioselectivity* and *stereo selectivity* and *stereospecificity*. Chemoselectivity is the ability of the enzyme to direct the catalytic action to a specific functional group in the molecule so as to distinguish between OH or NH. When the substrate contains several functional groups of the same kind, as seen in carbohydrates, the enzyme is able to catalyse a regioselective reaction of one particular OH-group. The stereochemical properties of enzymes are extremely attractive in organic synthesis. Enzyme catalysis may be used for production of enantiopure chiral molecules either by enantioselective asymmetric synthesis or to resolve racemic mixtures (see Section 19.3).

The stereochemical properties of enzymes are important but so is the fact that enzymes work under mild conditions. The latter is becoming more and more important as greater demands are made on chemical process industry concerning environmental aspects.

19.1.4 Classification of enzymes

Enzymes are the tools of biocatalysis and are classified and numbered by the Enzyme Commission (EC), International Union of Biochemistry and Molecular Biology. They are divided into six classes according to the chemical reactions they catalyse: (i) oxido-reductases, (ii) transferases, (iii) hydrolases, (iv) lyases, (v) isomerases and (vi) ligases.

The classes that are currently most used by chemists are oxido-reductases, hydrolases and aldolases, the latter belonging to the lyases. However, it still remains to be seen which class of enzymes will have most success in the chemical and pharmaceutical industries.

19.1.5 Importance of enantiopure compounds

No matter if a pair of enantiomers (i.e. molecules that are non-superimposable mirror images of each other) have exactly the same chemical and physical properties, such as melting point, boiling point and spectra and even show the same reactivity in an achiral (see Section 19.3) environment, they are, in principle, totally different compounds when they interact with chiral molecules. It is well known that some enantiomers may have different odours and tastes. For example, *(S)*-carvone tastes of caraway while the *(R)*-enantiomer tastes of spearmint. A useful metaphor for interaction of receptors with the wrong enantiomer may be trying to fit the left hand glove on to the right hand. The effect of different enantiomers may be particularly significant for drugs. Hence drugs that are chiral must be administered as single enantiomers.

19.1.6 How do enzymes work?

Whilst a typical chemical industry process may take place under high pressure and temperature and at a pH far from neutral, this is certainly not the case for a biological process. As catalysts, enzymes make a chemical process go faster but they do not influence the equilibrium position of a chemical reaction. The rate of an enzyme-catalysed reaction is

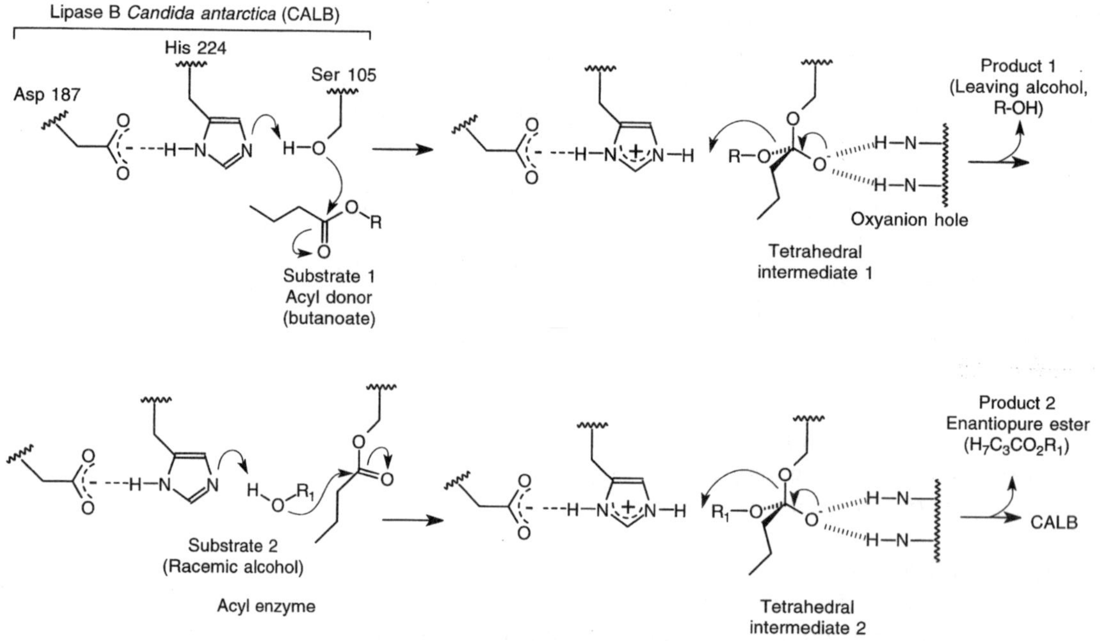

Fig. 19.2 Detailed mechanism of transesterification of a racemic mixture of a secondary alcohol with a butanoic acyl donor following a ping-pong bi-bi mechanism. Substrate 1 (acyl donor) enters the enzyme, forms an acyl enzyme via tetrahedral intermediate 1 and expels product 1 (the leaving alcohol from the acyl donor). Then substrate 2 (the enantiomers of the alcohol to be resolved) reacts with the acyl enzyme to form another tetrahedral intermediate 2. Subsequently product 2 (the enantiomers of the produced esters) is liberated, leaving the enzyme in its original form. In a kinetic resolution one of the enantiomeric alcohols reacts faster than the other to form an excess of one enantiomer of the esters (ideally enantiopure) The success of the resolution is expressed by the enantiomeric ratio *E*, which depends on the difference in free energy of activation of the two diastereomeric transition states formed which in turn is related to the two tetrahedral intermediates.

proportional to the rate constant (k) and the concentrations of the reactants. It is the free energy of activation, $\Delta G^{\ddagger}$, that decides magnitude of the rate constant of a reaction. If a reaction includes several steps, it is the step with the largest $\Delta G^{\ddagger}$ that is the rate determining step of the reaction. An enzyme-catalysed reaction follows a different mechanism from that of a reaction catalysed in a non-enzymic manner. The difference in rate of catalysed and uncatalysed reaction depends on their difference in free energy of activation ($\Delta\Delta G^{\ddagger}$). A relatively well understood reaction is hydrolysis of either a peptide or carboxylic ester bond catalysed by a serine hydrolase, such as trypsin, chymotrypsin or lipase B from *Candida antarctica*, as shown in Fig. 19.2.

19.2 Hydrolytic enzymes

Hydrolytic enzymes (Class 3) are the most commonly used enzymes in organic chemistry. There are several reasons for this. Firstly, they are easy to use because they do not need co-factors like the oxidoreductases. Secondly, there is a large number of hydrolytic enzymes available because of their industrial interest. Detergent enzymes com-

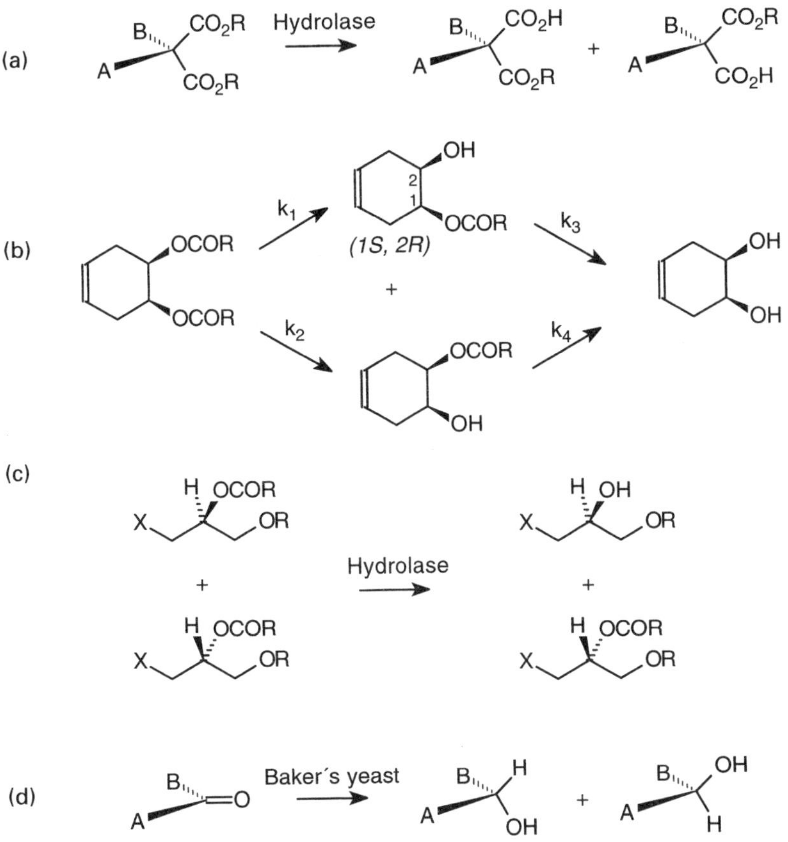

(a) (b) (c) (d)

Hydrolase

k_1 k_2 k_3 k_4

(1S, 2R)

Hydrolase

Baker's yeast

Fig. 19.3 Production of enantiopure compounds using hydrolytic enzymes (a, b and c) and oxido-reductases (d). In (a) a prochiral diester is hydrolysed to yield unequal amounts of enantiomeric carboxylic acids; in (b) a *meso*-diester is hydrolysed to yield predominance (in theory 100%) of one enantiomer of the monoester. If $k_1 > k_2$ the *(1S,2R)*-enantiomer is formed to the greatest extent. Due to the preference of the enzyme, $k_4 > k_3$ and the lower monoester *(1R,2S)* will be consumed fastest. Hence both steps will lead to an increase of the upper enantiomer at the monoester stage. If the reaction proceeds to completion, however, the result will be another *meso*-compound, a diol. The first step is asymmetric synthesis, the second step is resolution. In example (c) a racemic secondary ester is resolved by hydrolysis. One monoester is hydrolysed faster than the other and this leads to kinetic resolution. Maximum yield of each enantiomer is 50%. In the last example (d) a prochiral ketone is reduced enantioselectively to give unequal amounts of secondary alcohols. (a), (b) and (d) are examples of asymmetric synthesis which in theory may give 100% yield of one enantiomer.

prise proteases, cellulases, amylases and lipases. Even if hydrolytic enzymes catalyse a chemically simple reaction, many important features of catalysis are still contained such as chemo-, regio- and stereoselectivity and specificity.

19.2.1 Hydrolysis

Hydrolytic enzymes will hydrolyse ester, amide and glycoside bonds. They have different substrate specificity although many hydrolases can accept a wide range of substrates. Examples of carboxylic ester hydrolysis are given in Fig. 19.3. Carboxylic ester hydrolases (EC 3.1.1) comprise

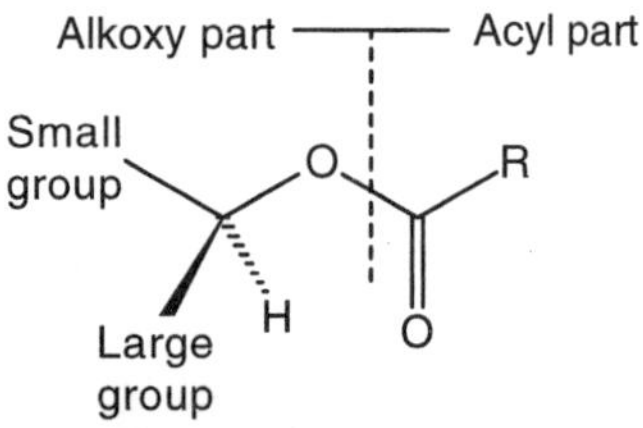

Fig. 19.4 Carboxyl esterases and carboxyl lipases both act on carboxylic esters. They differ in the structure of esters they hydrolyse. Lipases work best on substrates with relatively uncomplicated (not branched) acyl parts. Esterases on the other hand are not that specific and may accept bulky acyl groups (R). Moreover, the structure of the alkoxy part is not so critical. A general observation, however, is that if the stereocentre is far away from the centre of reaction (ester oxygen), the catalysis is not stereospecific. Hydrolysis of esters of secondary alcohols are more stereospecific. The shown enantiomer will usually be the faster reacting.

several enzymes of which carboxylesterase (EC 3.1.1.1) and triacylglycerol lipase (EC 3.1.1.3) have been frequently used as catalysts in biocatalysis. They both act on carboxylic esters but they differ in the type of esters they hydrolyse (Fig. 19.4). With the currently available esterases and lipases it may be summarised that:

- both esterases and lipases give hydrolysis at mild conditions (neutral pH, ambient temperature);
- lipases do not hydrolyse esters with 'bulky' acyl groups;
- hydrolysis of esters of secondary alcohols is more stereospecific.

Of course, the complete picture is more complicated.

19.2.2 Hydrolytic enzymes in organic solvents

There are several reasons for choosing organic solvent as a medium for catalysis by hydrolytic enzymes;

- better solubility of substrate and product;
- better stability of enzyme (most deactivating processes need water to occur);
- simpler removal of solvent (most organic solvents have lower boiling point than water);
- shift of equilibrium since water is not present (synthesis takes place instead of hydrolysis);
- easy removal of enzyme after reaction since it is not dissolved.

Generally, enzyme catalysis depends on the medium. Properties of the medium, which solvent, co-solvents, the amount of water in the system as expressed by the water activity (a_w) are important parameters. Use of organic solvents in kinetic resolutions is discussed in Section 19.3.3.

19.2.3 Ester synthesis and transesterification

It is possible to catalyse the formation of esters from acid and alcohol by a hydrolase. However, the water formed in the reaction ($RCO_2H + R'OH = RCO_2R' + H_2O$) creates a problem for the equilibrium of the reaction and, moreover, the enzyme gradually associates with the formed water and becomes inactive. Therefore, transesterification is a much more frequently used procedure. This is discussed in Section 19.3.3.

19.3 Chiral building blocks for synthesis

Chiral building blocks for synthesis of complicated organic molecules can be provided by three basically different methods:

- chemical transformation of enantiopure natural products;
- asymmetric synthesis from prochiral substrates;
- resolution of racemic mixtures.

Enzymes as chiral catalysts play a role in all three methods. In nature, enzymes catalyse production of chiral compounds. Enzymes may also catalyse asymmetric synthesis, as well as resolve racemates. Which method is chosen in different cases depends on several factors, like price of starting materials, number of synthetic steps, available production technology, know-how etc.

19.3.1 Asymmetric synthesis

Asymmetric synthesis is the term used when a prochiral substrate or a *meso*-substrate is converted into an unequal amount of chiral product. A **prochiral** compound is a compound that may be converted into a chiral compound in one step. A *meso*-compound has stereocentres but they are organised in such a way that the compound as a whole is achiral. The product of an asymmetric synthesis is characterised by the **enantiomeric excess,** *ee*. For instance if the product mixture contains 95% of one enantiomer and 5% of the other, $ee = 90\%$. A **racemic mixture** which contains 50% of each enantiomer, has $ee = 0$. The theoretical yield of 100% of one single enantiomer may be obtained if all of the starting material is converted into one single isomer. If the two possible products are enantiomers the reaction is **enantioselective**. Typical examples of enzyme-catalysed enantioselective asymmetric synthesis is reduction of a non-symmetrical ketone (Fig. 19.3d) or hydrolysis of a prochiral diester (Fig. 19.3a). The starting material may also be a *meso* compound as in Fig. 19.3b.

In asymmetric synthesis, the enantiomeric excess of the product will be constant throughout the reaction and it will depend only on the $\Delta\Delta G^{\ddagger}$ of the two possible courses of reaction. Reaction between the prochiral substrate and the enzyme leads to two diastereomeric transition states with different energy. The difference in free energy of activation is related to the ratio of the two rate constants of reactions. If the measured *ee* of the product is 90%, the relative rate constant for formation of the isomers is $95/5 = 19$ which corresponds to $\Delta\Delta G^{\ddagger} = 7.3$ kJ mole^{-1} ($\Delta\Delta G^{\ddagger} = -RT\ln K = -8.31441 \times 10^{-3} \times 298 \times 19$) This is a small number as compared to the total $\Delta G^{\ddagger}$ for the reaction which may be in the order of 60–80 kJ mole^{-1}.

19.3.2 Resolution by hydrolysis: irreversible reactions

A racemic mixture (racemate) of a desired building block may be produced by conventional organic synthetic methods. For instance, racemic amino acids can be obtained by the Strecker synthesis from an aldehyde, ammonia and hydrogen cyanide (see Fig. 19.6). Classical resolution has been performed by formation of salts using enantiopure amines. If the amine for instance has *(R)*-configuration while the racemic amino acid is *(R)* and *(S)*, *(RR)* and *(RS)* salts will be formed. They are diastereomers and they are chemically different also in symmetric environments. They can be separated by standard methods such as crystallisation. In a series of synthetic steps and when resolution is one step, it is of utmost importance that the correct chirality is introduced at an early stage. When a racemate is subject to enzyme catalysis, one enantiomer reacts faster than the other and this leads to kinetic resolution (Fig. 19.3c). The important parameter of kinetic resolution process is the enantiomeric ratio, the *E*-value. This is the ratio of the specificity constants (k_{cat}/K_M) of the enzyme for the two enantiomers *R* and *S*. An *E*-value of 50 means that one enantiomer reacts 50 times faster than the other.

$$E = \frac{\left(\frac{k_{cat}}{K_m}\right)_R}{\left(\frac{k_{cat}}{K_M}\right)_S}$$

In a resolution process there are two products called **product** and **remaining substrate.** Both products can reach very high enantiomeric excess provided the *E*-value is high. The enantiomeric excesses are termed ee_p and ee_s, respectively, and their values depend on the degree of conversion *c*. Both ee_p, ee_s and the conversion, *c*, have values between 0 and 1, but they are sometimes dealt with in percentages. In hydrolysis, which is an irreversible reaction since water is present at 55 M, the *E*-value may be calculated from either ee_p or ee_s and *c* at one single measurement according to:

$$E = \frac{\ln[1 - c(1 + ee_p)]}{\ln[1 - c(1 - ee_p)]} \qquad E = \frac{\ln[(1 - c)(1 - ee_s)]}{\ln[(1 - c)(1 + ee_s)]}$$

The degree of conversion under most circumstances (equal amounts of enantiomers at the beginning of the reaction, no side reactions) is related to ee_p and ee_s by:

$$c = \frac{ee_s}{ee_s + ee_p}$$

Hence, another expression may be used to calculate E:

$$E = \frac{\ln\frac{[ee_p(1 - ee_s)]}{(ee_p + ee_s)}}{\ln\frac{[ee_p(1 + ee_s)]}{(ee_p + ee_s)}}$$

The advantage of this expression is that it does not involve *c* which may be difficult to measure accurately. As opposed to ee_p and ee_s, which are relative quantities, *c* is an absolute quantity. The most accurate way, however, is to use a computer program to fit many measured data points from several conversions to calculated curves for different *E*-values.

In a resolution *ee* of the substrate fraction is zero when the reaction starts. Provided the enantiomeric ratio *E* is high, the product fraction will have high *ee*. For instance if $E = 19$ (95:5), ee_p will be 90% at the start of the reaction. As the reaction proceeds, the concentrations of the enantiomers change and also ee_p and ee_s. The relationship between ee_p, ee_s and *c* for three different values of *E* is shown in Fig. 19.5. Ideally, if *E* is very high ($E > 100$) both ee_p and ee_s will be close to 100% at 50% conversion and the reaction virtually stops. Even reactions with moderate *E*-values can give the remaining substrate with very high *ee* providing that yield can be sacrificed. A resolution that proceeds with $E = 12$ will have ee_s of 100% at 75% conversion. However, half of the theoretical 50% yield is lost.

19.3.3 Resolution in organic solvents: reversible reactions

Hydrolytic enzymes may be used in organic solvents. Since water is not present, instead of hydrolysis, ester synthesis takes place. Either esterifi-

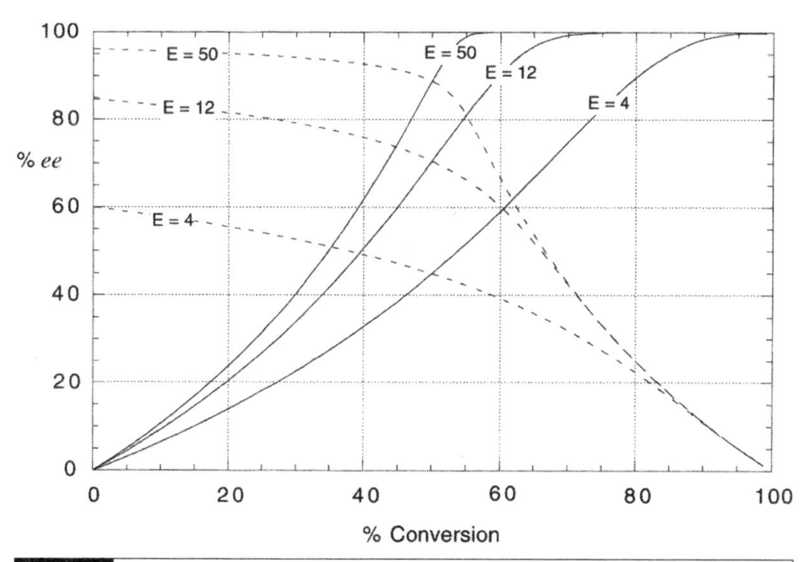

Fig. 19.5 Enantiomeric excess of product (ee_p, full lines) and remaining substrate (ee_s, stippled lines) vs. degree of conversion calculated for three different values of the enantiomeric ratio *E* for an irreversible resolution. Ideally, if *E* is very high ($E > 100$) both ee_p and ee_s will be close to 100% at 50% conversion and the reaction virtually stops. The ee vs. conversion curves for the different *E*-values infer that ee_p is at its maximum in the beginning of the reaction while ee_s reaches maximum at a later stage. This has an important consequence; *no matter how low the enantiomeric ratio is, it is always possible to obtain a very high* ee_s *provided a reduced yield is tolerable*. It is clear that even for a low *E*-value such as 12, an ee_s close to 100% may be achieved if 25% of yield can be sacrificed. This difference between resolution and asymmetric synthesis is very important. For this reason it may be easier to obtain the remaining substrate with higher ee.

cation or better transesterification in non-aqueous media ($RCOOR_1 + R_2OH = RCOOR_2 + R_1OH$) is performed. A starting ester is needed: the acyl donor, $RCOOR_1$. It reacts with the enzyme to form the acyl enzyme which in turn reacts with the racemic alcohol, the acyl acceptor R_2OH (see Fig. 19.2). Since the enzyme shows the same stereopreference no matter hydrolysis or transesterification, either the ester or the alcohol may be separated as the remaining substrate, but with the same configuration. If the *(S)*-ester is the remaining substrate in hydrolysis the *(S)*-alcohol will be the remaining substrate of transesterification (Fig. 19.6).

The mathematical expressions presented in Section 19.3.2 are restricted to irreversible reactions. Hydrolysis, due to the large excess of one reactant, water at 55 M, is for practical purposes irreversible. In a transesterification the concentration of the leaving alcohol, R_1OH (from the acyl donor), will accumulate and eventually the reverse reaction will become important. This will lead to depressed enantiomeric excesses for the product and the substrate as the reaction proceeds. For reversible reactions, the equilibrium constant, K_{eq}, also has to be taken into account.

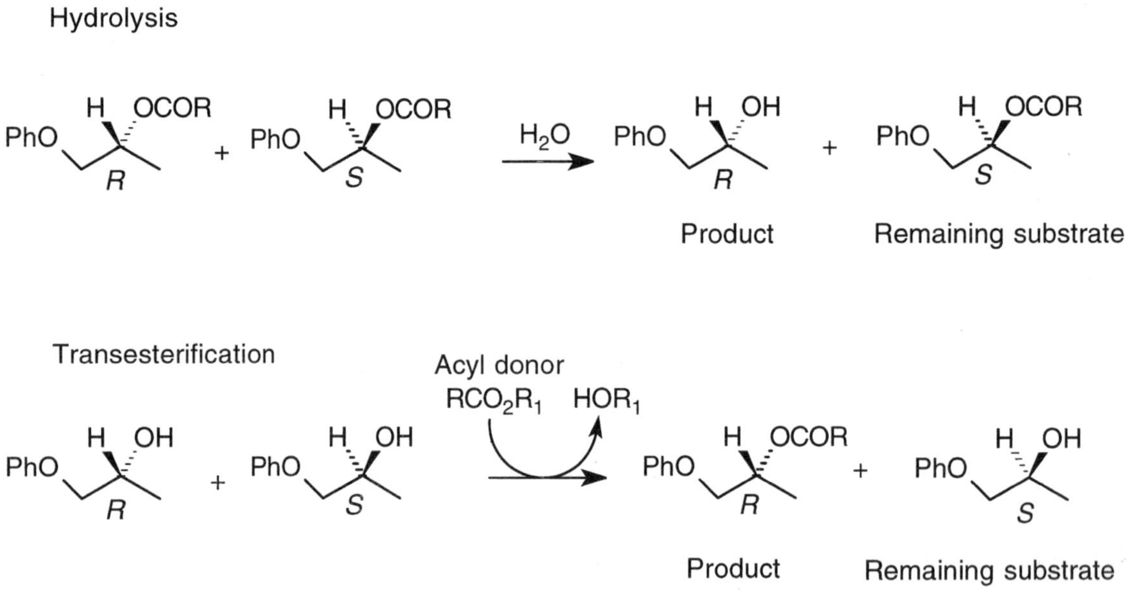

Hydrolysis

PhO–CH(H, OCOR)–CH3 (R) + PhO–CH(H, OCOR)–CH3 (S) —H_2O→ PhO–CH(H, OH)–CH3 (R) + PhO–CH(H, OCOR)–CH3 (S)

Product | Remaining substrate

Transesterification

PhO–CH(H, OH)–CH3 (R) + PhO–CH(H, OH)–CH3 (S) —Acyl donor RCO_2R_1 → HOR_1→ PhO–CH(H, OCOR)–CH3 (R) + PhO–CH(H, OH)–CH3 (S)

Product | Remaining substrate

Fig. 19.6 Hydrolysis of a racemic secondary ester or transesterification of the corresponding secondary alcohol (1-phenoxy-2-propanol) with a butanoic acyl donor and CALB as catalyst both yield the same enantiomer as product. The product of hydrolysis is the *(R)*-alcohol while the product of transesterification is the *(R)*-ester.

Computer programs for ping-pong bi-bi kinetics, which use *ee*-values measured at several degrees of conversion, are available. If both enantiomers can be provided in pure forms, it is also possible to determine E and K_{eq} from initial rate measurements.

19.3.4 Problems with reversibility

A problem with transesterification is that the reaction will become reversible and the equilibrium constant will become important. The enantiomer that reacts fastest in the forward direction will also react fastest in the reverse direction. The effect of this is clearly inferred in Fig. 19.7 in which the ee_p and ee_s are calculated for a resolution with $E = 50$ and for three different equilibrium constants. As is inferred the effect of reversibility is particularly dramatic for the substrate fraction. When the equilibrium constant is low, for example $K_{eq} = 0.5$, the enantiomeric excess reaches a maximum value at around 45% conversion and then decreases. The point where decrease of *ee* occurs may be shifted towards higher conversion when the acyl donor/substrate alcohol is increased. Another way would be to change the nature of the alkoxy group of the acyl donor. When the pK_a value of the corresponding leaving alcohol is decreased, the reaction becomes less reversible. Completely irreversible conditions are obtained when vinyl esters are used as acyl donors. After reaction the expelled vinyl alcohol immediately tautomerises to the corresponding aldehyde or ketone which provide irreversibility.

19.3.5 Determination of enantiomeric excess

The enantiomeric excess values may be determined in different ways:

- derivatisation to diastereomers which may be separated by chromatography or NMR;
- chiral solvation and detection of solvated diastereomeric species by NMR;

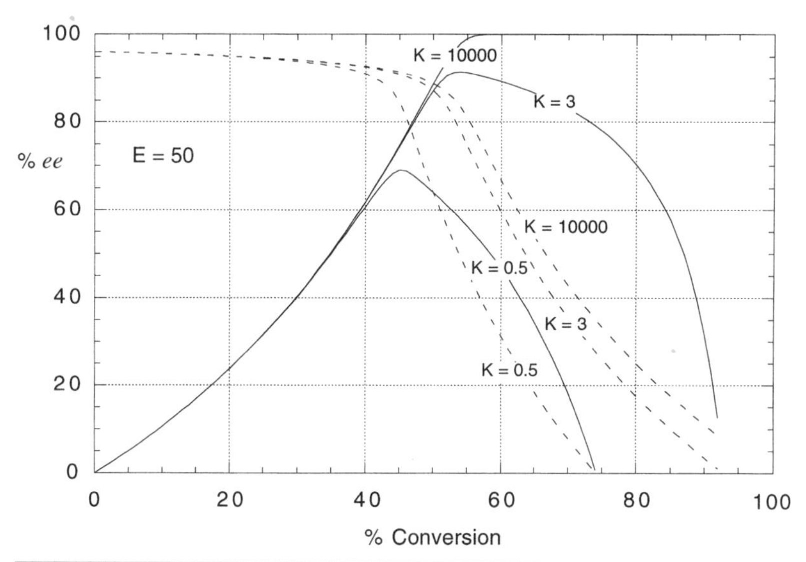

Fig. 19.7 Enantiomeric excess of product (ee_p, full lines) and remaining substrate (ee_s, stippled lines) vs. degree of conversion calculated for $E = 50$ and three different equilibrium constants, 10 000, 3 and 0.5. With a large K_{eq} the reaction is irreversible and the progress curves look like the examples of Fig. 19. 5. For reactions with smaller K_{eq} values a dramatic effect is observed for ee_s. The curve reaches a maximum, as the reaction progresses further, ee_s is reduced and the curve never reaches 100% as it always does in the irreversible case. The effect of reversibility is not as dramatic on ee_p. The curve dips down at an earlier degree of conversion when K_{eq} is lowered. An obvious way to proceed is to push the reaction towards the product side by increasing the concentration of the reactants.

- direct separation on chiral stationary phases of GLC or HPLC.

The latter methods are by far the most accurate and easiest to perform provided suitable columns are available.

19.4 Reductions and oxidations

Oxido-reductases (Class 1) are responsible for reductions and oxidations in nature. As opposed to hydrolases, their action depends on co-factors mainly NAD^+/NADH or $NADP^+$/NADPH. When one mole of substrate is reduced, one mole of co-factor is oxidised and vice versa with the enzyme remaining unchanged. Since co-factors are extremely expensive, they have to be regenerated in order to assure an economically feasible process (see Section 19.4.4). The alternative is to use whole cells for the reduction-oxidation processes when the cells will take care of the regeneration process. For example, if growing cells of baker's yeast are reducing a ketone, it is the sugar in the medium that is oxidised. The first sub-class of the oxido-reductases are the dehydrogenases which act on primary or secondary alcohols or hemiacetals. They are mostly used for reduction of ketones and aldehydes.

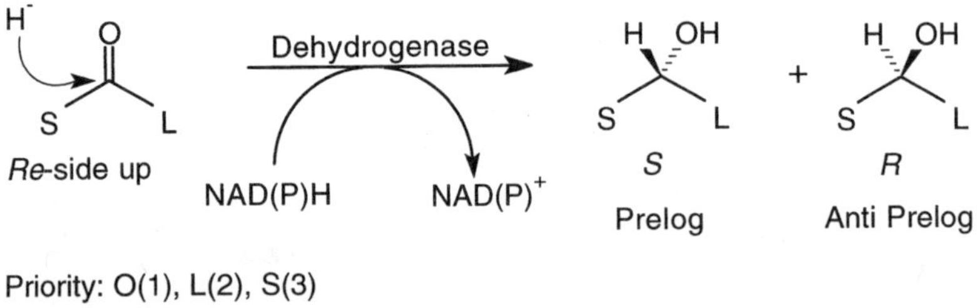

H⁻ O S L *Re*-side up
Dehydrogenase
NAD(P)H NAD(P)$^+$
H OH S L *S* Prelog + H OH S L *R* Anti Prelog

Priority: O(1), L(2), S(3)

Fig. 19.8 Enantioselective reduction of non-symmetrically substituted ketones by dehydrogenases yields secondary alcohols. There are numerous examples of such reactions using the commercially available horse liver alcohol dehydrogenase (HLADH) or yeast alcohol dehydrogenase (YADH). Both of which follows the so-called Prelog´s rule for stereoselectivity which states that the hydride is delivered from the *re*-side of the keto group. Most commonly this leads to predominance of the *(S)*-alcohol. Using whole cells for reductions eliminates the need for regeneration of NADH. Most widely used is baker´s yeast (*Saccharomyces cerevisiae*) which also in most cases follows Prelog´s rule. Baker´s yeast may also reduce selectively carbon-carbon double bonds in certain cases when the double bond is activated.

19.4.1 Reductions

Enantioselective reduction of non-symmetrically substituted ketones by dehydrogenases yields secondary alcohols. This reaction is important since it is an asymmetric synthesis capable of giving 100% product. The stereochemical course of the reaction is said to be either according to Prelog or anti-Prelog rules, see Fig. 19.8.

19.4.2 Oxidations

Oxidation of secondary or primary alcohols by dehydrogenases is usually not performed biocatalytically. The reaction destroys a stereocentre, it is not favoured thermodynamically and product inhibition is a problem. Only in cases where it is necessary to discern between several hydroxy groups in a molecule is it attractive.

19.4.3 Hydroxylation of carbon centres, mono-oxygenases

In addition to the dehydrogenases, the oxygenases is a useful group of oxido-reductases from a synthetic viewpoint. Oxygenases are very much involved in important reactions. The mono-oxygenases insert one of the two oxygen atoms from O_2 into the substrate while di-oxygenases insert both.

The mono-oxygenases which catalyse a series of oxidations such as hydroxylations, epoxidations, heteroatom oxidations and Baeyer–Villiger oxidations, as shown in Fig. 19.9, depend on NADH or NADPH and additional co-factors, usually Fe or Cu. A particularly important reaction is the direct incorporation of O_2 into non-activated carbon centres. Examples are synthesis of important steroidal drugs by microbial 11α-hydroxylation of progesterone and 7β-hydroxylation of lithiocholic acid.

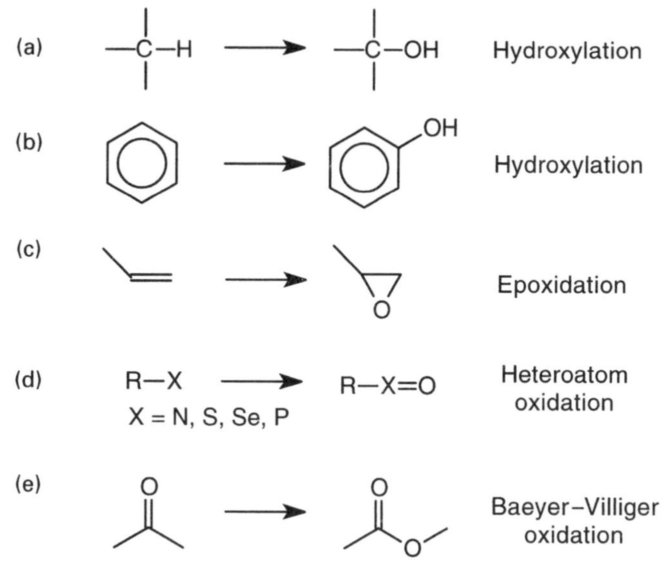

(a) Hydroxylation (b) Hydroxylation (c) Epoxidation (d) R—X → R—X=O, X = N, S, Se, P, Heteroatom oxidation (e) Baeyer–Villiger oxidation

Fig. 19.9 Reactions catalysed by mono-oxygenases, (a) hydroxylation of carbon centres, (b) aromatic hydroxylation, (c) epoxidation of alkenes, (d) heteroatom oxidation and (e) Baeyer–Villiger oxidation of a ketone.

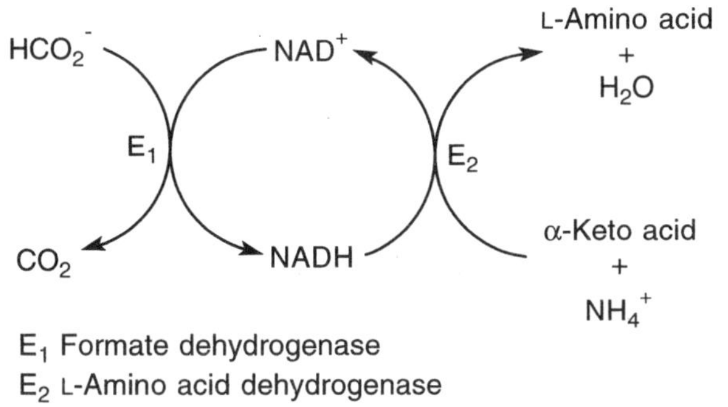

HCO_2^-, NAD^+, L-Amino acid + H_2O, E_1, E_2, CO_2, NADH, α-Keto acid + NH_4^+

E_1 Formate dehydrogenase
E_2 L-Amino acid dehydrogenase

Fig. 19.10 Formate dehydrogenase system for regeneration of NADH. The process for production of L-amino acid by reductive amination from an α-keto acid is shown on the right. In this process NADH is consumed. The NAD^+ produced is regenerated by conversion of formate to carbon dioxide catalysed by formate dehydrogenase.

19.4.4 Methods for regeneration of co-factors

When using pure enzymes for a redox process it is of utmost importance to regenerate the co-factor. The efficiency of such a recycling reaction is measured by the **total turnover number**, TTN, which is the number of cycles obtained before the co-factor is destroyed. TTN may vary from 1000 on a laboratory scale to 100 000 on a technical scale.

There are several ways of regenerating nicotinamide co-factors, which are the most commonly involved in redox processes. To regenerate the reduced form, NADH, from NAD^+ there are basically four different ways:

- non-enzymatic reduction using sodium dithionite, $Na_2S_2O_4$;
- electrochemical or photochemical regeneration;
- coupled substrate process using the same enzyme;
- coupled enzyme process using two different enzymes.

The enzymatic methods give the highest turnover numbers. In particular the last method has had some technical success. In the formate dehydrogenase system, cheap formate is converted into CO_2 and NADH is regenerated, as shown in Fig. 19.10. This system has been successfully applied in production of L-amino acids from α-keto acids in a membrane reactor (see section 19.6.1). By reductive amination catalysed by L-leucine dehydrogenase, α-ketoisocaproic acid was converted into L-leucine and the TTN for the co-enzyme was 80 000. In order to keep the low molecular weight co-factor on the same side of the membrane as the enzyme, its molecular weight is increased by covalently binding it to polyethylene glycol.

19.5 Use of enzymes in sugar chemistry

Carbohydrates, such as trioses, tetroses, pentoses, hexoses etc., are important natural molecules. Their biological significance for cell-cell interaction is increasingly understood. The simple carbohydrates are the building blocks of oligo- and polysaccharides. Biocatalysis is important for synthesis of simple carbohydrates as well as for oligosaccharides.

19.5.1 Derivatisation of sugars

The most characteristic feature of sugars, chemically speaking, is their more or less chemically equivalent hydroxy groups. Performing selective synthetic transformations on sugar molecules is therefore always a matter of protection and deprotection steps. Primary hydroxy groups may be tritylated, vicinal diols may be protected as acetonides, other OH-groups may be acetylated etc. After the wanted reaction has been carried out, the protecting groups have to be removed. This often leads to long and tedious routes for synthesis of specific carbohydrates, however, sugar chemists have for decades taken this discipline to perfection. Nevertheless, tempted by the aspect of developing methods not needing protection and deprotection steps, enzyme catalysis has been exploited. Regioselective acylations have been performed, however, with modest success. Sugars which are typically hydrophilic molecules, have been dissolved in solvents like pyridine and dimethyl formamide and acylated using hydrolytic enzymes. The selectivity obtained has been mostly similar to the selectivity that is easily obtained by more classical methods. A drawback is the necessity to use polar, high boiling solvents, such as those mentioned above, which are difficult to get rid of. Due to low solubility of carbohydrates in organic solvents and lack of selectivity of reactions, it may be concluded that this strategy for use of hydrolytic enzymes in the carbohydrate field will only have limited value.

19.5.2 Synthesis of sugars from small molecules

In several recent applications of enzyme catalysis, the substrates on which the enzymes act are not the kind of substrates that are 'natural' to the enzyme. However, enzyme catalysed synthesis of hexoses in the laboratory depends solely on enzymes acting on natural or near natural substrates. The relevant enzymes belong to the class of lyases specifically called aldolases (EC 4.1.2 aldehyde-lyases) since they catalyse an

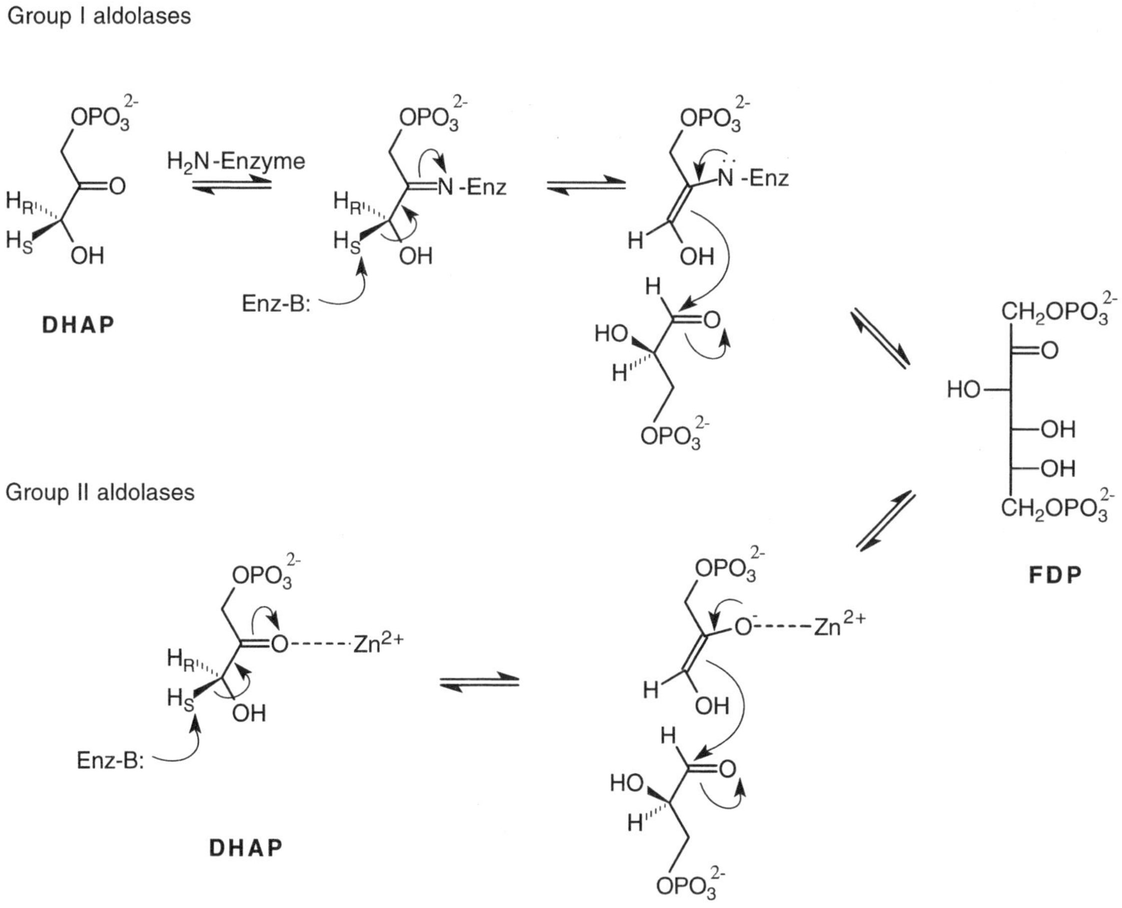

Group I aldolases

DHAP

H_2N-Enzyme

Enz-B:

FDP

Group II aldolases

DHAP

aldol type of C-C bond forming aldol addition reaction. The aldolases most commonly join two C-3 units, called *donor* and *acceptor*, and two new stereocentres are formed with great stereoselectivity.

Aldolases may be divided into two groups according to their mechanism of action and occurrence in nature. Aldolases from animals and higher plants (group I) use an amino group in the enzyme to form a Schiff's base intermediate to activate the aldol donors. Those from lower organisms, bacteria and fungi, group II, use a metal ion, usually Zn^{2+}, in the active site to form an enolate intermediate. The two mechanisms, exemplified by fructose-1,6-diphosphate aldolase, a very important aldolase in synthesis and breakdown of sugars, are shown in Fig. 19.11. The most frequently used FDP-aldolase for synthesis of specific carbohydrates is rabbit muscle aldolase (RAMA). It belongs to a group of aldolases for which dihydroxyacetone phosphate (DHAP) is the sole acceptable donor. (Group (a) in Fig. 19.12) However, several aldehydes and ketones may function as acceptors. The different aldolases are commonly divided into four groups requiring in addition to DHAP as donor, pyruvate (b), acetaldehyde (c) and glycine (d).

Since the range of substrates of aldolases are rather limited they are primarily useful for synthesis of carbohydrate-like compounds. Their

Fig. 19.11 The two mechanisms of aldolases. Group I enzymes from animals and higher plants use an amino group in the enzyme to form a Schiff's base intermediate to activate the aldol donors. Group II enzymes from lower organisms, use a metal ion, usually Zn^{2+} in the active site to form an enolate intermediate. The two mechanisms are examplified by fructose-1,6-diphosphate aldolase, a very important aldolase in synthesis and breakdown of sugars.

Fig. 19.12 The four groups ((a), (b), (c), (d)) of aldolases according to their donor dependence.

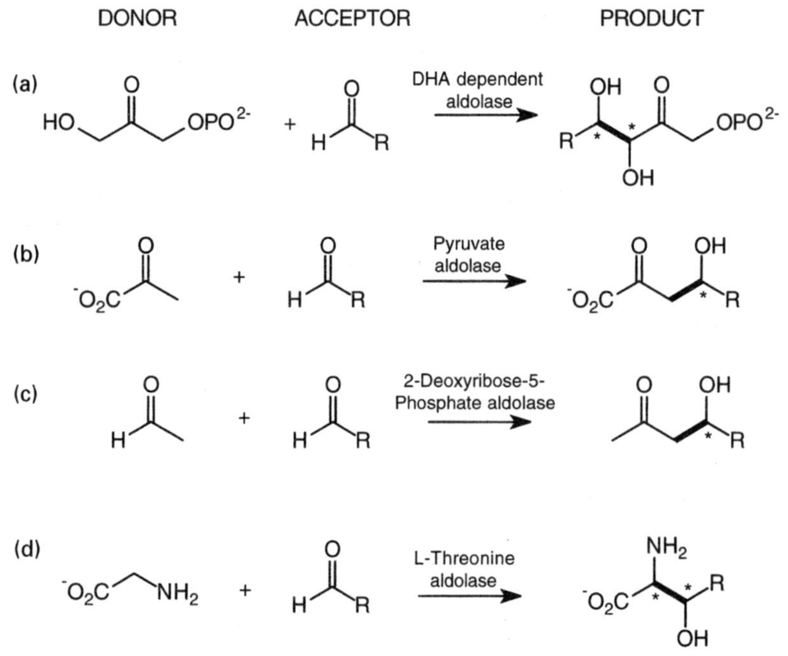

utility cannot be compared to the wide range of aldol reactions known in organic chemical synthesis. Thus it may be concluded that aldolases may be used for synthesis of:

- rare carbohydrates;
- isotopically labelled carbohydrates;
- carbohydrates with unusual heteroatoms

19.5.3 Synthesis of oligosaccharides

The enzymes involved in breakdown and build up of oligo- and polysaccharides in nature are either glycosidases or glycosyl transferases. The first type is a hydrolytic enzyme and mainly catalyses breakdown of oligo- or polysaccharides in nature. They have been found to have only limited use as catalysts for build up of oligosaccharides mainly due to low solubility of carbohydrates in organic media which are necessary to reverse hydrolysis. On the other hand, in recent years there has been tremendous progress in the use of glycosyl transferases. This is due to a combination of challenging biological problems, production of relevant enzymes by molecular biological techniques and skilled organic chemists. Glycosyltransferases are divided into two groups according to which activated donors they use for transfer of monosaccharides. The Leloir glycosyltransferases utilise eight nucleoside mono- or diphosphate sugars, UDP-Glc, UDP-GlcNAc, UDP-Gal, UDP-GalNAc, GDP-Man, GDP-Fuc, UDP-GlcUA and CMP-NeuAc. The Non-Leloir glycosyltransferases utilise glycosylphosphates as activated donors.

Carbohydrate-mediated cell adhesion is an important event which can be initiated by tissue injury or infection and is involved in metastasis. One such adhesion process is the interaction between the glycoprotein E-selectin and oligosaccharides on the surface of neutrophils

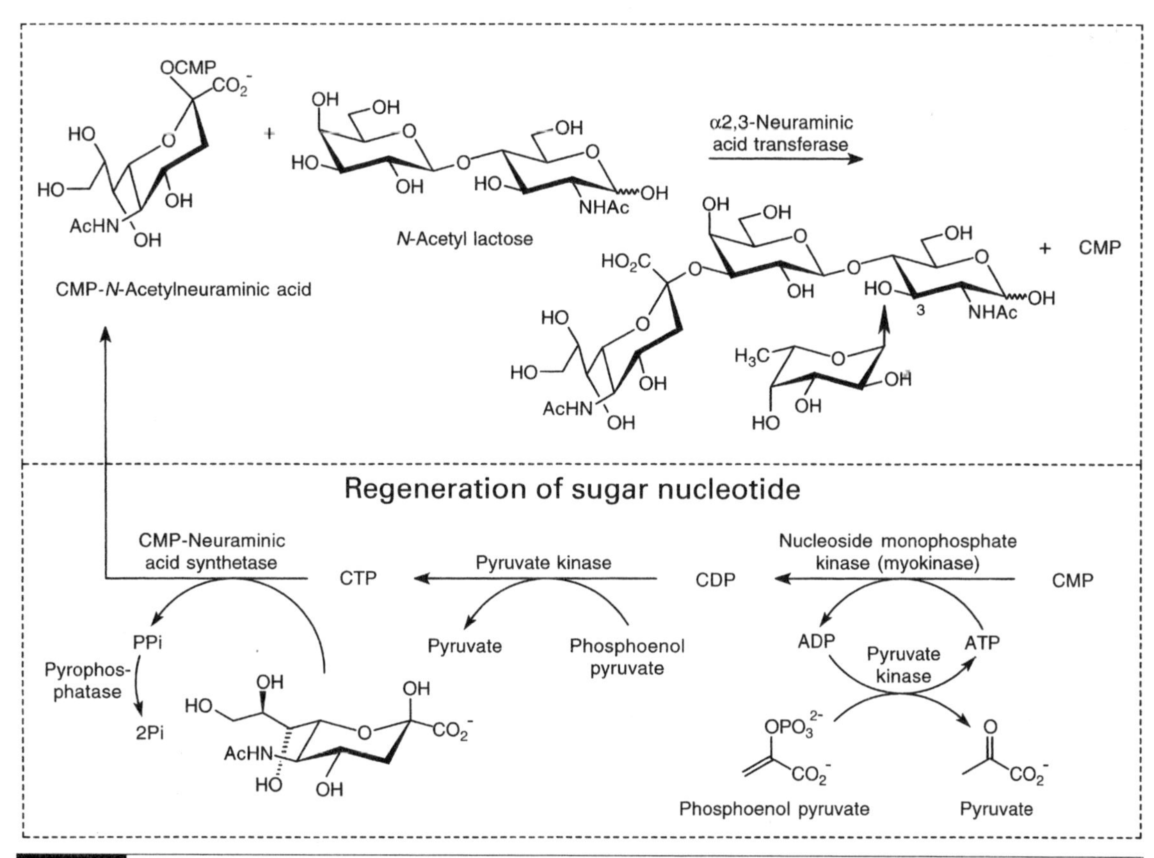

Fig. 19.13 The synthesis of sialyl Lewis X (SLe^x) comprises three transferase catalysed steps. In the first N-acetyl lactose is formed using galactosyl transferase and UDP galactose. Sialylation (N-acetylneuraminic acid) in the second step by cytidyl monophosphate (CMP)-N-acetylneuraminic-N-acetylneuraminic acid and α2,3-neuraminic acid transferase is shown in the figure, upper box. The formed trisaccharide is fucosylated at the Glc 3-position in the next step using fucosyl transferase and GDP fucose (see arrowed fucosyl unit). Regeneration of the sugar nucleotide is shown in the lower box. CMP is converted into CTP in two steps using two different kinases. In the final step CMP-N-acetylneuraminic acid is synthesised from CTP and N-acetylneuraminic acid (sialic acid) using the appropriate synthetase. The formed pyrophosphate is converted into inorganic phosphate. Altogether five different enzymes are involved in the process.

(white blood cells). The ligand that E-selectin recognises is the tetrasaccharide sialyl Lewis X (SLe^x). Since SLe^x competes with white blood cells for binding to E-selectin, thus inhibiting the adhesion process, it may useful as an anti-inflammatory and anticancer agent.

Non-enzymic synthesis of SLe^x involves a large number of protection and deprotection steps which are not suited for large scale production. However, enzymic processes using transferases have been developed with great success. The crucial factor in order to succeed is regeneration of the activated monosaccharides. Synthesis of SLe^x and related oligosaccharides have been performed on a large scale (kilograms) using this technology. The synthesis comprises three transferase catalysed steps. In the first, *N*-acetyl lactose is formed using galactosyl transferase and UDP-galactose. Sialylation (*N*-acetylneuraminic acid) in the second step by CMP-*N*-acetylneuraminic acid and α2,3-neuraminic acid transferase is shown in Fig. 19.13. The formed trisaccharide is fucosylated in the Glc

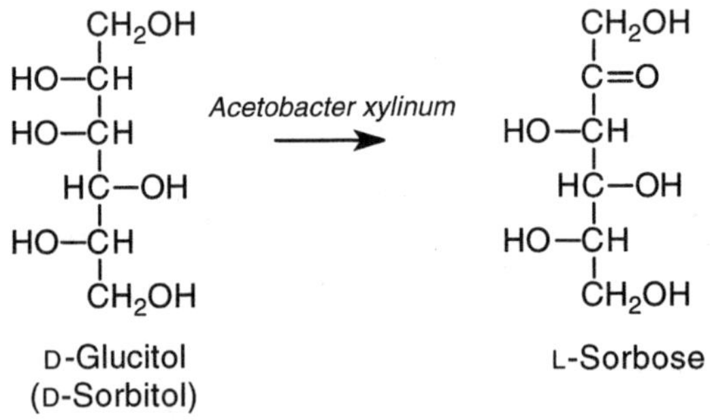

Fig. 19.14 An important step in the production of vitamin C (ascorbic acid) from glucose is the regioselective oxidation of D-glucitol to yield L-sorbose by *Acetobacter xylinum*. This biocatalytic process which was developed by Reichstein and Gössner in 1934 is responsible for the manufacture of about 50 000 tonnes of ascorbic acid each year.

Acetobacter xylinum

D-Glucitol (D-Sorbitol) → L-Sorbose

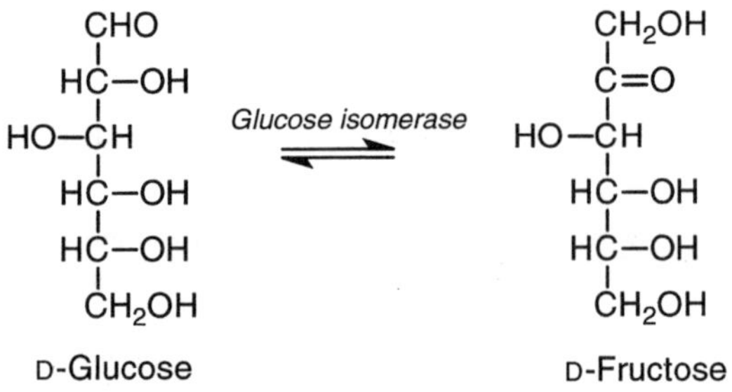

Fig. 19.15 In the production of the sweetener high fructose corn syrup (HFCS), which is produced from corn starch, glucose isomerase is used to convert glucose into an equilibrium mixture consisting of fructose: glucose: oligo saccharides in the ratio 42:51:7. The mixture has improved sweetness since fructose is nearly three times as sweet as glucose.

Glucose isomerase

D-Glucose ⇌ D-Fructose

3-position in the next step using fucosyl transferase and GDP–fucose. Regeneration of CMP-*N*-acetylneuraminic acid is shown in the lower box. As inferred it involves a series of enzymes. If all of the enzymes are available, the process itself is not complicated.

19.5.4 Other enzyme-catalysed reactions on carbohydrates

Some biocatalytic reactions with sugars have tremendous industrial importance. The crucial step in the production of vitamin C (ascorbic acid) from glucose is the regioselective oxidation of D-glucitol to L-sorbose by *Acetobacter xylinum* (Fig. 19.14). This biocatalytic process, developed by Reichstein and Gössner in 1934, is responsible for the manufacture of 60 000 tonnes of ascorbic acid each year (see pp. 322–323).

The sweetener, high-fructose corn syrup (HFCS), is produced from corn starch which is hydrolysed by α-amylase and amyloglucosidase to give a refined glucose syrup. Glucose isomerase is used to convert glucose into fructose (equilibrium mixture, fructose : glucose : oligo saccharides, 42:51:7) since fructose is nearly three times as sweet as glucose (Fig. 19.15). About 8 million tonnes per year of glucose is treated in this way.

19.6 Use of enzymes to make amino acids and peptides

19.6.1 Production of amino acids

Amino acids may be produced by biocatalysis either by asymmetric synthesis or resolution, see Fig. 19.16.

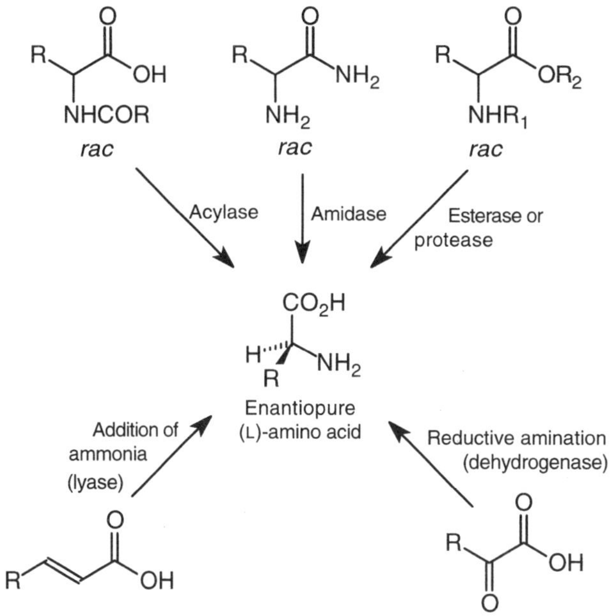

Fig. 19.16 Biocatalytic synthesis of amino acids either by asymmetric synthesis (lower part of figure) or resolution (upper part of figure). Addition of ammonia to the double bond of an α,β-unsaturated carboxylic acid using a lyase such as fumarase is one example of the former. For instance L-phenylalanine has been produced from cinnamic acid in high yield. The most frequently used method, however, is by resolution of an α-amino acid obtained for instance by the Strecker synthesis using hydrolytic enzymes (upper part of the figure). Asymmetric synthesis gives predominantly the natural L-form of the amino acid. By resolution both the L and the D-form are obtainable, however, since the L-form has higher economic interests, processes are designed to give only this enantiomer. There are basically three ways to use hydrolytic enzymes in resolution processes, an alkyl ester at the carboxylic group may be hydrolysed by an esterase or protease, an amide at the carboxylic group may be hydrolysed by an amidase or an acyl group at the amino group may be hydrolysed by an acylase. The latter process is exploited commercially in a hollow fibre membrane reactor where the unreacted D-isomer is separated, racemised and recycled. When ester hydrolysis is performed under conditions of **dynamic resolution**, the substrate is continuously racemised and the enantiopure product may be formed in 100% yield.

19.6.2 Peptide synthesis

Formation of an amide bond (peptide bond) will take place if Substrate 2 in Fig. 19.2 is an amine and not an alcohol. If Substrate 2 is an amino acid (acid protected), reactions can be continued to form oligo peptides and the process will be a kinetically controlled aminolysis. If Substrate 1 is an amino acid (amino protected) it will be reversed hydrolysis, if it is a protected amide or peptide it will be transpeptidation. Both of the latter methods are thermodynamically controlled. However, synthesis of peptides using biocatalytic methods (esterase, lipase or protease) is only of limited importance for two reasons. Synthesis by either of the above mentioned biocatalytic methods will take place in low water media and low solubility of peptides with more than two or three amino

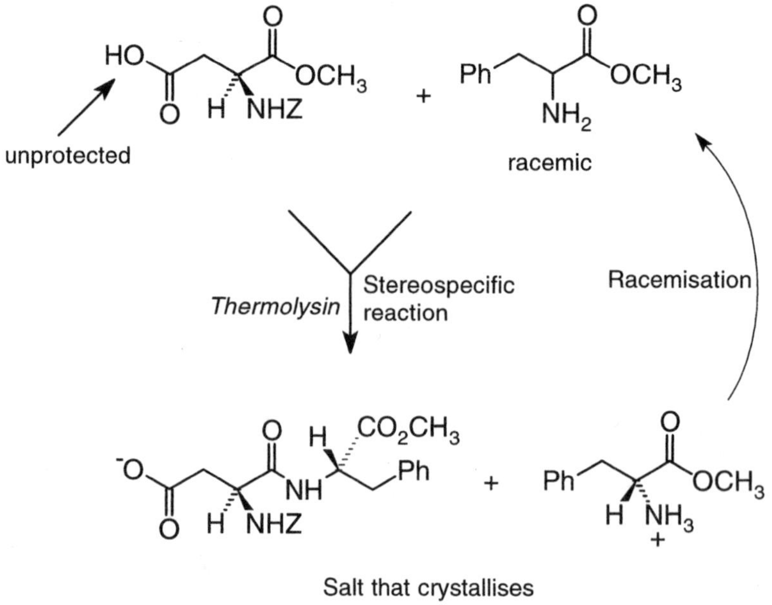

Fig. 19.17 Synthesis of the low calorie sweetener, Aspartame, which is a methyl ester of a dipeptide, Asp-Phe-OMe involves a biocatalytic step. In the Tosoh process aspartic acid amino protected by benzyloxycarbonyl group is reacted with two moles of racemic phenylalanine methyl ester catalysed by the protease thermolysin. The extra mole of ester makes the dipeptide precipitate and after liberation of the product, the extra molecule of phenylalanine is racemised and recycled.

acids limits their value. Secondly, there are well developed non-biocatalytic methods for peptide synthesis. For small quantities the automated Merrifield method works well. Nevertheless, one process for synthesis of the low calorie sweetener, Aspartame, which is a methyl ester of a dipeptide (Asp-Phe-OMe), involves a biocatalytic step (the Tosoh process). Aspartic acid amino protected by benzyloxycarbonyl group, is reacted with two moles of racemic phenylalanine methyl ester catalysed by the protease thermolysin. The extra mole of ester makes the dipeptide precipitate (Fig. 19.17).

19.7 Further reading

Bornscheuer, U. T. and Kazlauskas, R. J. (1999) *Hydrolases in Organic Synthesis*. Wiley-VCH, Weinheim.

Faber, K. (2000). *Biotransformations in Organic Chemistry, 4th Edition*. Springer-Verlag, Berlin.

Fersht, A. (1998). *Structure and Mechanism in Protein Science. A Guide to Enzyme Catalysis and Protein Folding*. W. H. Freeman & Co, New York.

Palmer, T. (1995). *Understanding Enzymes, 4th Edition*. Ellis Horwood.

Roberts, S. M., Turner, N. J., Willets, A. and Turner, M. K. (1995). *Introduction to Biocatalysis using Whole Enzymes and Micro-organisms*. Cambridge University Press, Cambridge.

Chapter 20

Recombinant proteins of high value

Georg-B. Kresse

20.1 Applications of high-value proteins

Proteins used in industrial enzyme technology, e.g. detergent proteinases or enzymes applied in the food industry, are in most cases rather crude preparations and usually mixtures of different enzymes. In contrast, there are a number of commercial applications where highly purified (and therefore high-value) proteins are needed. Examples are:

- **Analytical enzymes and antibodies**. For use in medical diagnostics, food analysis, as well as biochemical and molecular biological analysis (see Section 20.2).
- Enzymes used as **tools in genetic engineering technology**. Gene technology has become possible through the availability of highly purified enzymes such as restriction endonucleases, DNA or RNA polymerases, nucleases and modifying enzymes. Similarly, glycohydrolases and glycosyl transferases are used increasingly in glycobiotechnology in order to modify the sugar residues of glycoproteins (see Chapters 4 and 5).
- **Therapeutic proteins**. Growth factors, antibodies or enzymes are used as the active drug ingredients for the treatment of diseases (see Section 20.3).

Furthermore, proteins with proven or supposed biological relevance in pathomechanisms are needed as **targets for the search of new ligands** (agonists or antagonists), inhibitors and for X-ray or NMR structural analysis in order to design novel interacting compounds by structure-based molecular modelling. This requires the production of these proteins on a relatively small scale (10 to 100 mg) but often with high purity depending on the intended use.

20.2 Analytical enzymes

Enzymes are highly specific both in the reaction catalysed as well as in their choice of substrates. Indeed, enzymes are, besides antibodies, the most specific reagents known. The use of enzymes in analysis, therefore, offers a number of advantages compared to chemical reagents. The reactants may either become chemically transformed in the presence of an enzyme (if they are substrates), or they may modulate the enzymatic activity in a manner related to their concentration (if they act as activators or inhibitors). Enzymes also serve as 'markers' in assay techniques based on non-enzymatic interactions, such as antigen-antibody binding (see Chapter 23) or DNA-oligonucleotide hybridisation.

The advances in recombinant DNA technology have made it possible to clone any gene of interest and to manipulate bacterial, fungal, insect, or mammalian cells to overproduce the desired protein. Many enzymes have been expressed in manipulated micro-organisms at levels 10 to 100 times higher than in the natural host cell. This has allowed not only better economics in enzyme production, but also significant reductions of the environmental burden because of the decrease in fermentation volumes (and, thereby, in waste formation). Furthermore, the techniques of unspecific and, more recently, site-directed mutagenesis have opened the way to improve relevant properties of enzymes for analytical applications, e.g. stability under the assay conditions, pH optimum, or solubility.

20.2.1 Enzymes in diagnostic assays

End-point assays

In this case, the compound to be determined (the analyte) takes part as the substrate in an enzyme-catalysed reaction and is converted with the simultaneous and stoichiometric production of a detectable signal (e.g., a positive or negative change in absorbance). The reaction is allowed to proceed until completion, and the result can easily be calculated from known physical constants, e.g. the molar absorption coefficient in the case of light-absorbing substances. The specificity of the assay system depends on the substrate specificity of the enzyme employed. An example of this is the determination of uric acid using urate oxidase (see Fig. 20.1).

The same principles will, of course, apply if it is not the analyte conversion itself but the stoichiometrically linked conversion of a co-substrate or co-factor that is detected. This is the case in dehydrogenase-catalysed assays, which make use of the absorption of the reduced coenzyme NADH or NADPH at 340 nm for detection. An overview of important enzymes used in diagnostic assays is given in Table 20.1. Most of the enzymes used in enzymatic analysis are derived from bacteria or yeast, and are produced using bacterial (in most cases *E. coli*) or yeast expression systems due to the lower production costs.

In many cases, none of the reactants or products of an enzymatic

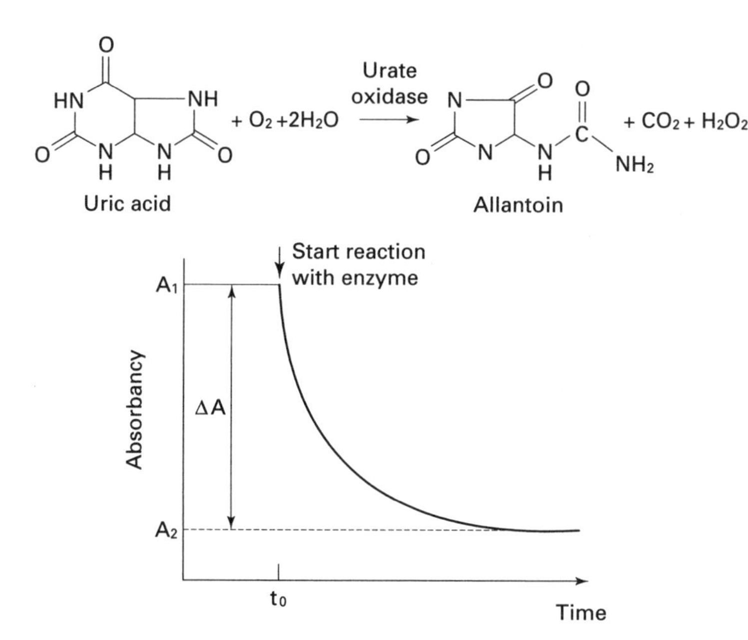

Fig. 20.1 Determination of substrate concentrations by an enzymatic end-point assay. The assay of uric acid is used as an example. Uric acid absorbs UV light (λ_{max}: 293 nm) whereas the product of its enzymatic oxidation, allantoin, shows no absorption at this wavelength. To determine the concentration of uric acid in a sample, the initial absorbance at 293 nm is measured, then the enzyme, uricase (urate oxidase), is added. Oxidation of uric acid (with O_2 as the oxidant) occurs until all the substrate has been converted to allantoin. From the difference of the final and the initial absorbance values ($\Delta A = A_1 - A_2$), the uric acid concentration can be calculated using Beer's law.

Table 20.1 Examples of enzymes important in diagnostics

Enzyme	Source (original)	Used for the assay of
Cholesterol oxidase	*Nocardia erythropolis* or *Brevibacterium* sp.	Cholesterol
Creatinase	*Pseudomonas* sp.	Creatine, creatinine
Creatininase	*Pseudomonas* sp.	Creatinine
β-Galactosidase	*Escherichia coli*	Sodium ions; immunoassay marker enzyme
Glucose oxidase	*Aspergillus niger*	Glucose
Glucose-6-phosphate dehydrogenase	*Leuconostoc mesenteroides*	Glucose (indicator enzyme)
α-Glucosidase	Yeast or *Bacillus* sp.	α-Amylase activity
Glycerol-3-phosphate oxidase	*Aerococcus viridans*	Triacylglycerols
Hexokinase	Yeast	Glucose and other hexoses
Peroxidase	Horseradish	Indicator enzyme and immunoassay marker enzyme
Pyruvate oxidase	*Pediococcus* sp.	Pyruvate; transaminase activity
Sarcosine oxidase	*Pseudomonas* sp., *Bacillus* sp.	Creatinine
Urate oxidase (uricase)	*Arthrobacter protophormiae*	Uric acid
Urease	*Klebsiella aerogenes*	Urea

Notes:
All listed enzymes are manufactured from recombinant expression systems and commercially available.

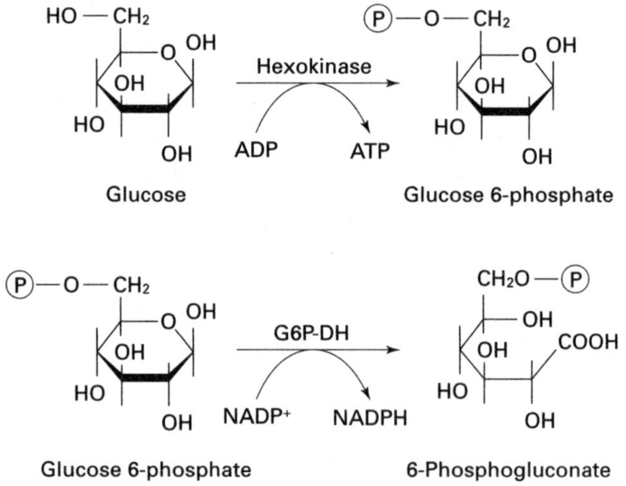

Fig. 20.2 An example of a coupled enzymatic assay system using an indicator enzyme: glucose assay with hexokinase and glucose-6-phosphate dehydrogenase. The determination of glucose in blood or food materials comprises the phosphorylation of glucose catalysed by yeast hexokinase as an auxiliary reaction which cannot be detected directly. This is coupled to the oxidation of glucose 6-phosphate to 6-phosphogluconate catalysed by glucose-6-phosphate dehydrogenase (G6P-DH). In this second 'indicator' reaction, 1 mol of $NADP^+$ is reduced per mol of glucose 6-phosphate which in turn is stoichiometrically equivalent to the glucose present in the original sample, to give NADPH which can be determined spectrophotometrically at 340 nm. Hexokinase is an unspecific enzyme which would phosphorylate many hexoses. Since however G6P-DH is strictly specific for glucose 6-phosphate and would not accept other sugar phosphates, the reaction system is specific for the assay of glucose also in the presence of other carbohydrates.

reaction lend themselves readily to physical or chemical measurement because no detectable signal is produced. In these cases, the primary (called 'auxiliary') reaction usually is coupled to a stoichiometrically linked 'indicator' reaction (mostly also enzyme-catalysed), with one of the products of the second reaction being easily detectable. In **coupled systems**, it is sufficient to employ an enzyme with narrow substrate specificity for only one of the reaction steps, because the specificity of the whole reaction sequence will depend on the most specific enzyme. This means that in practice, general (i.e. unspecific) indicator reactions can be coupled without individual optimisation to various specific auxiliary reactions. To reach the endpoint in an appropriate reaction time, enzymes with low K_m values for the analyte substrates are preferred to ensure that the reaction is rapid even at the low substrate concentrations reached when complete conversion is approached.

Horseradish peroxidase is a well-known example of an indicator enzyme used in a large number of commercial oxidase-based coupled assays but, in most cases, this enzyme is obtained from a non-recombinant source. An example of a coupled glucose assay which uses recombinant enzymes as the indicator enzyme is shown in Fig. 20.2.

Kinetic assays

If the substrate concentration [S] is much smaller than the Michaelis–Menten constant K_m of the enzyme, it follows from the Michaelis–Menten equation that the observed reaction rate becomes linearly proportional to [S]:

$$v = \frac{V \times [S]}{K_m + [S]}$$

where v = observed reaction rate; V = theoretical maximal rate with fixed amount of enzyme; [S] = substrate concentration; K_m = Michaelis–Menten constant.

By monitoring the reaction kinetics, either by the rate of disappearance of substrate $-d[S]/dt$, or the rate of formation of product, $d[P]/dt$, catalysed by a fixed amount of enzyme, the analyte concentrations can be determined by measuring the reaction rate. Kinetic assays allow a drastic reduction of the time required for analysis and are less sensitive to interferences than end-point assays. However, for rate measurements the reaction conditions, such as temperature, pH, amount of enzyme, etc., must be held constant. Therefore, kinetic assays are best performed with automated analysers rather than carried out manually.

Substrate concentration determination by rate measurement is only possible if the initial substrate concentration is low compared to the K_m value of the enzyme. Therefore, enzymes with high K_m values (i.e. low substrate affinity) are used in kinetic assays to increase the dynamic concentration range of the assay.

In contrast, when $[S] \gg K_m$, it follows from the Michaelis–Menten equation that the observed reaction rate now becomes independent of the substrate concentration (zero-order kinetics). Under these conditions, the enzyme activity cannot be used to determine the substrate concentration. However, in zero-order reactions a direct correlation exists between the reaction rates v and V, which is proportional to the total enzyme activity present in the assay mixture. Therefore, if a substance exerts an activating or inhibiting effect on the catalyst in a concentration-dependent manner, it will be possible to determine its concentration indirectly by kinetic measurement of the enzyme activity under substrate saturation conditions. This principle has been used to design kinetic substrate concentration assays for a number of effectors, e.g. heparin (based on inhibition of thrombin), insecticides (inhibition of acetylcholinesterase), theophyllin (inhibition of alkaline phosphatase), and several ions able to enhance the activity of apoenzymes, e.g. Na^+, K^+, Mg^{2+}, Cu^{2+} or Cl^-.

Test strip systems and biosensors

In order to facilitate handling of the reagents used for determination of analyte concentration or enzyme activity, attempts have been made to use immobilised enzymes for analytical purposes. An early and still useful approach is to bind the enzyme non-covalently to paper. This has been used in the design of the test strips for use in biological samples such as urine or blood. In biosensors, the enzyme (usually an

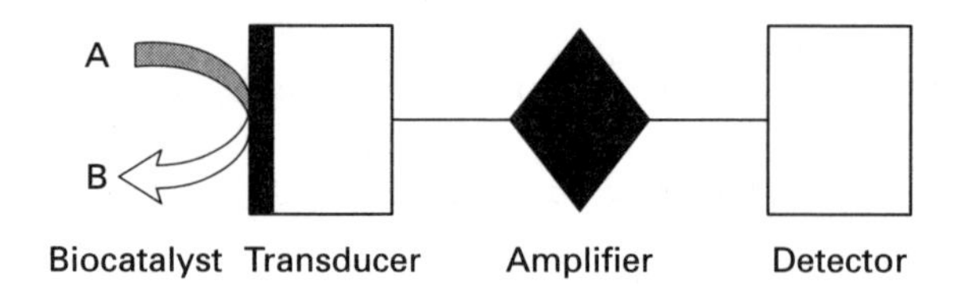

Fig. 20.3 The general concept of a biosensor. The biocatalyst converts the substrate into product with a concurrent change in a physicochemical parameter (e.g. heat, electron transfer, light, ion or proton flow, etc.) which is concerted into an electrical signal by the transducer, amplified, and processed by a detector.

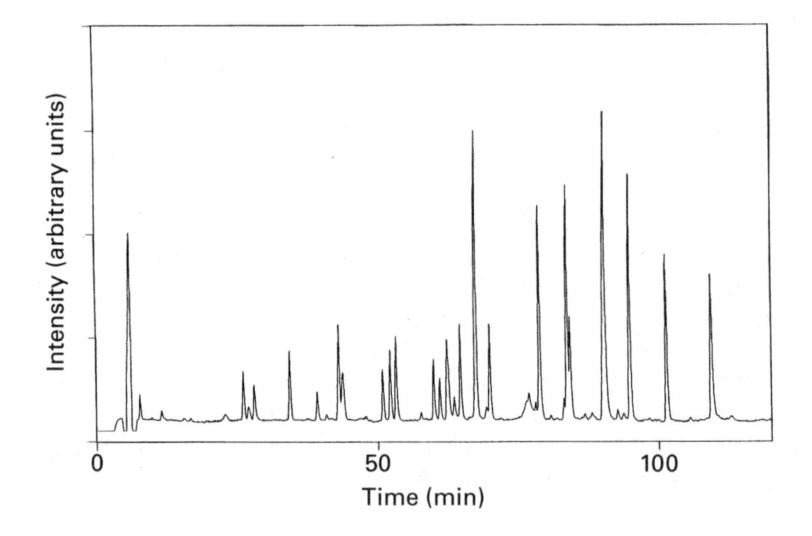

Fig. 20.4 Peptide map of a therapeutic protein. The protein (Reteplase) was digested with highly purified trypsin, and the resulting peptide fragments were separated by high-performance liquid chromatography. Each peak represents one peptide fragment. Because of the cleavage specificity of trypsin, all peptides (except the *C*-terminal fragment) are expected to have lysine or arginine residues at their *C*-terminus. The individual peptides can be analysed, e.g. by mass spectroscopy, and sequenced. (Illustration courtesy of Dr M. Wozny, Penzberg, Germany.)

oxidoreductase) serves as the 'specifier' mediating analyte recognition. This biological interaction is then transformed through a 'transducer' component into an electrical signal which can be amplified and processed electronically (see Fig. 20.3). A number of commercial test strip and biosensor systems for use in clinical chemistry have been developed.

20.2.2 Enzymes as tools in biochemical analysis

Purified enzymes are widely used for analytical problems outside medical diagnostics and food analysis, e.g. in molecular biology and gene technology, and in protein and glycoconjugate analysis.

In **molecular biology**, enzymes are used for molecular analysis of chromosome and genome structure, for the characterisation of genetic defects on the DNA level, for taxonomy of viruses and other organisms by correlation of characteristic fragment patterns, as well as for elucidating phylogenetic relationships. In addition, DNA and RNA modifying enzymes as well as polymerases (e.g. *Taq* DNA polymerase) are essential tools for all cloning techniques as described in Chapters 4 and 5.

Highly purified enzymes are also important tools in **analysis of protein structure and modification** (including glycosylation). Specific proteases are used for protein fragmentation and fingerprinting as well as for C-terminal protein sequencing (an example is given in Fig. 20.4).

Table 20.2 Some enzymes used in biochemical analysis

Enzyme (source)	Use
Proteases Trypsin (bovine) Chymotrypsin (bovine) Endoproteinase Lys-C from *Lysobacter enzymogenes* Endoproteinase Glu-C (V-8 protease) from *Staphylococcus aureus V8*	Protein fragmentation for sequence analysis, peptide fingerprinting, limited proteolysis of enzymes or receptors to study structure-function relationships
Carboxypeptidases A, B, C, Y	C-terminal protein sequencing
Restriction proteases Factor Xa (bovine or human) Enterokinase (bovine) IgA protease from *Neisseria gonorrhoe*	Processing of recombinant fusion proteins
Glycosidases Endoglycosidase D and *O*-Glycosidase from *Diplococcus pneumoniae* Endoglycosidase F and *N*-Glycosidase F from *Flavobacterium meningosepticum* Endoglycosidase H from *Streptomyces plicatus* Many exoglycosidases	Carbohydrate and glycoprotein analysis

Similarly, a set of glycohydrolases is used to analyse the carbohydrate residue structures of glycoproteins. A number of enzymes used for this purpose are listed in Table 20.2.

20.2.3 Special requirements for analytical enzymes

Enzymes to be used for analytical applications have to satisfy a number of quality criteria concerning:

- **Specificity** – absence of side activities towards other substances which may be present in the sample or the reaction mixture.
- **Purity** – absence of contaminating activities or other contaminants interfering with the analytical and detection systems (and not necessarily purity with respect to the absence of other inactive proteins unless the enzyme is used in protein analysis).
- **Stability** – in the reaction mixture during the reaction as well as in long-term storage.
- **Kinetic properties** – suitable K_m and k_{cat} values and no inhibition by substances present in the sample.
- **pH optimum** – suitable for the required experimental conditions.
- **Solubility and surface properties** – no interference by adsorption or aggregation effects.
- **Cost.**

The criteria are dependent on each other. Therefore the choice and quality of an enzyme must be optimised in each case with regard to the particular analytical application.

20.3 Therapeutic proteins

Proteins are part of numerous traditional medicines, such as snake and bee venom and enzyme preparations but, in most cases, these are ill-defined mixtures. Furthermore, naturally occurring proteins obtained from animals, plants or micro-organisms as well as from the human body (blood, urine, placenta or adenohypophysis) have long been used as drug ingredients. Typical examples are porcine insulin, blood coagulation factors VIII and IX from human blood fractionation, pancreatin as a digestive aid, or the Ancrod and Batroxobin proteases obtained from snake venoms. However, foreign proteins are immunogenic for humans and will give rise to immune responses if injected into the bloodstream. This may lead to rapid inactivation and may prevent repeated application of the protein drug. Furthermore, the isolation of therapeutic proteins from human (and animal) fluids and tissues poses a potential risk of infecting the patients with other diseases, for example with HIV.

The concentration of physiologically active proteins in human body fluids or organs is very low, necessitating the processing of much material for low amounts of product. Chemical synthesis of proteins, although feasible in principle, is not normally an economically viable production method for protein drugs. So, only through the advent of gene technology has it become possible to produce human proteins in large amounts and high purity. In many cases, recombinant human proteins allowed, for the first time, a rational therapy with the body's own substances based on knowledge of the causes of disease and pathobiological mechanisms. Therapeutic proteins such as hormones, growth and differentiation factors, act in signalling while others function as biocatalysts or inhibitors. They are used for substitution, amplification, or inhibition of physiological processes.

Although biopharmaceutical sales worldwide made up just 4% of world prescription drug sales in 1994, it has grown tremendously over the past decade. At present more than 60 recombinant proteins are used in therapy (a selection is given in Table 20.3), and more than 200 others are in development. Erythropoietin (EPO), insulin, somatotropin (human growth hormone), granulocyte-colony stimulating factor (G-CSF), and alpha-interferon are among the most successful drugs and are placed in the 'top twenty' on the list of worldwide sales of pharmaceuticals. Analysts expect that in the near future, more than 100 pharmaceuticals based on gene technology will be on the market.

20.3.1 Choice of expression system

Bacteria, yeast, insect and mammalian cells are the most commonly used hosts for heterologous protein expression. A brief overview of the main advantages and disadvantages of the various host systems referring especially to the use of the expression systems for therapeutic proteins will be given below.

Table 20.3 A selection of recombinant protein drugs

Protein	Diseases treated (selection)	Year of first approval
Insulin	Diabetes mellitus	1982
Human growth hormone (hGH)	Short stature (hGH deficiency)	1985
Interferon-alpha	Hepatitis, cancers, genital warts	1986
Hepatitis B vaccine	Hepatitis B prevention	1986
Plasminogen activators (alteplase, reteplase)	Acute myocardial infarction	1987/1996
Erythropoietin	Anaemia of chronic renal failure	1989
Interferon-gamma	Chronic granulomatous disease	1990
Granulocyte colony stimulating factor (G-CSF)	Neutropenia, bone marrow transplantation	1991
Granulocyte–macrophage colony stimulating factor (GM-CSF)	Bone marrow transplantation	1991
Interleukin-2	Renal cell carcinoma	1992
Coagulation factor VIII	Haemophilia A	1992
Interferon-beta	Relapsing multiple sclerosis	1993
DNase	Cystic fibrosis	1994
Glucocerebrosidase	Gaucher's disease	1994
Coagulation factor IX	Christmas disease	1997
Interleukin-10	Prevention of thrombocytopenia	1997

Notes:
Many of the proteins listed are commercialised as various brands by the same or different pharmaceutical companies.

Mammalian cells

In mammalian host cells, such as Chinese hamster ovary (CHO) or baby hamster kidney (BHK) cells, high-level expression (from 10 to more than 100 picogram per day per cell) of the recombinant protein can be achieved. Usually, the proteins are secreted into the fermentation medium in properly folded, active form and in most cases, glycosylation and other post-translational modifications occur in a more or less 'human-like' manner, although minor differences may exist. However, development of a stable cell line that expresses the therapeutic protein at a high level may take many months, and the manufacturing costs are high. Therefore, the developed technology using CHO, or BHK, cells is today the systems of choice for large-scale production of modified, e.g. glycosylated, therapeutic proteins, especially if correct protein modification is crucial for the therapeutic effect.

Human cells

One way to ensure the identity of the recombinant protein product with the original human protein is to use human cell lines as the expression system. Instead of cloning a gene or DNA sequence coding for the desired protein into the host cell, it is an interesting alternative to manipulate not the gene itself, but its promoter, in order to activate expression of the endogenous human gene. This 'gene activation' technology may eventually become a commercially advantageous way for the production of many therapeutic proteins.

Insect cells

The gene coding for a recombinant protein can be inserted into the genome of the baculovirus, *Autographa californica*. It will very efficiently infect insect cells, and uses their protein synthesis machinery to produce large amounts of protein (up to 500 mg of protein per litre of culture medium) within two to three days after infection. Therefore, this system is very suitable to obtain protein rapidly for feasibility studies. However, post-translational processing differs from the mammalian cell systems, and the system is not very suitable for scaling-up.

Yeast

Yeast and other fungi are eukaryotic micro-organisms that are routinely cultivated on a large scale. Recombinant proteins are usually located inside the yeast cell, however, it is also possible to attach a leader sequence in order to induce protein secretion (see Chapter 5). Whereas intracellular heterologous proteins may accumulate to g l^{-1} level, secreted proteins usually reach titres of about 10 to 100 mg l^{-1}. When expressed in yeast, human proteins are correctly folded and disulphide-bridged, but glycosylation differs significantly from the mammalian pattern. In most cases, the well-known baker's yeast (*Saccharomyces cerevisiae*) is used, but for the production of large amounts of recombinant proteins, *Pichia pastoris* is the system of choice in many cases.

Bacteria

Bacterial host systems offer fast development, high efficiency and relatively inexpensive production as the main advantages. Recombinant proteins can be accumulated intracellularly or secreted into the periplasmic space (*Escherichia coli*) or into the fermentation medium (*Bacillus* species). However, post-translational processing of complex eukaryotic proteins, such as by glycosylation or disulphide bond formation, does not occur in the bacterial cell. Protein expressed in bacteria may also differ from their natural form concerning their amino terminus where an *N*-formylmethionine residue may be present. A correct *N*-terminus can however be obtained when the recombinant protein is expressed as a fusion construct with an *N*-terminal extension that is then cleaved off with a suitable protease. In many cases, the sequence of this fusion tail is chosen so that it can be exploited to facilitate purification, e.g. poly(His) tails, which mediate binding to metal chelate chromatography materials.

In addition, upon high-level expression in bacterial cells, whether cytosolic or periplasmic, many eukaryotic proteins are accumulated as misfolded, aggregated and insoluble particles (termed '**inclusion bodies**') that have to be refolded *in vitro* into their native conformations (see also Section 20.3.2). If this refolding process can be designed to offer a fair yield and good economics, the bacterial (especially *E. coli*) host system is very well suited for the production of all the therapeutic proteins that do not require post-translational processing for *in vivo* bioactivity.

Transgenic animals and plants

Transgenic manipulation means that a gene from one species (the **transgene**) is introduced into the germ line of another species, either plant or animal. Milk, blood as well as urine have been proposed for transgenic protein production, and a number of different proteins have already been produced in this way; some proteins, e.g. milk-produced antithrombin-III and α_1-antitrypsin, are currently in clinical trials. Expression levels up to 35 g l^{-1} milk have been reported, suggesting that transgenic dairy animals may provide a cost-effective route to the large-scale manufacture of biotherapeutics. However, development times are long because the gestation period and the onset of sexual maturity of the animal are rate limiting, and there are still a number of concerns with respect to the consistency of protein production from different animals. Nevertheless, transgenic technology represents a real challenge for biotechnology.

Production from transgenic plants is potentially a more economically attractive system for large-scale production of recombinant proteins, offering advantages in the low cost of growing plants on large acreage, the availability of natural protein-storage organs, and the established practices for harvesting, transporting, storing and processing. At present, the main disadvantages are low accumulation levels of recombinant proteins, insufficient information on post-translational events and limited knowledge of relevant downstream processing technology.

20.3.2 Protein folding from inclusion bodies

Protein folding *in vitro* has often been compared to the task of unboiling an egg; to reform the biologically active, native protein conformation from insoluble and inactive aggregates. This 'naturation' process is usually done in several steps, as shown in Fig. 20.5.

In unfolded proteins, hydrophobic regions that would be buried within the native globular protein structure are exposed. These parts of the polypeptide chain tend to induce unspecific aggregation, thereby decreasing the refolding yield, see Fig. 20.6.

Kinetically, aggregation is a bimolecular reaction and therefore concentration dependent. As a consequence, protein naturation usually has to be performed at high dilution, and the resulting low protein concentrations and large reaction volumes lead to unfavourable economics for large-scale production. However, it has been demonstrated that a correctly folded (thus, hydrophilic) protein does not interfere with folding of further portions of the same, still unfolded protein. Therefore, if one starts refolding at a low protein concentration and adds further portions of unfolded protein continuously or discontinuously to the same mixture only after the initial amount has already

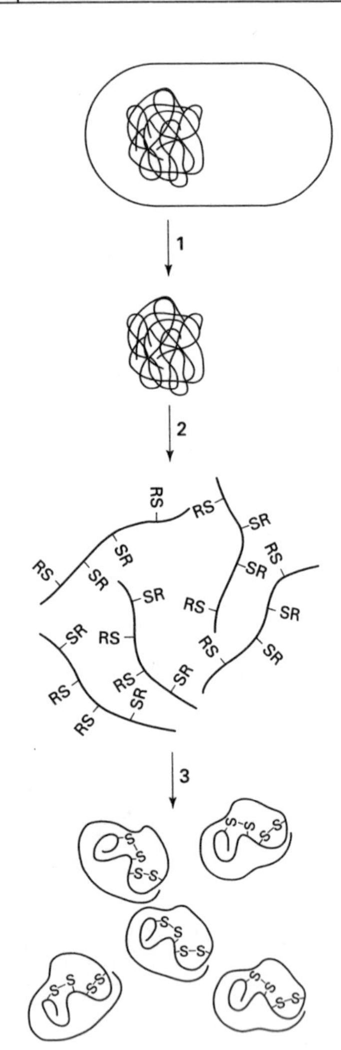

Fig. 20.5 Protein 'naturation' from inclusion bodies. Recombinant proteins overexpressed in bacterial cells often are formed as insoluble and misfolded 'inclusion bodies' (1). After cell lysis, the inclusion bodies are collected by centrifugation, washed with buffer to remove soluble cell components (which may already lead to ≥ 90% purity of the desired protein), and are then dissolved in a concentrated solution of a strong denaturant, e.g. 6–8 M urea or 5–6 M guanidinium.HCl (2). If the recombinant protein contains cysteine residues, a redox buffer system such as a mixture of reduced and oxidised glutathione (GSH/GSSG) at a slightly alkaline pH is added. In some cases, it has proven advantageous to modify the sulphydryl groups reversibly, e.g. by formation of mixed disulphides with glutathione, to increase solubility of the denatured polypeptide chain (R). The protein is then allowed to refold by slow removal of the denaturing agent, usually by dilution or dialysis, with concomitant formation of the correct disulphide bonds (3). In many cases, the addition of additives such as arginine, tris(hydroxymethyl)amino methane, or alkylurea derivatives has been shown to improve the refolding yields considerably. [Figure modified from: Marston, F.A.O. (1986). *Biochem. J.* **240**, 1–12, © The Biochemical Society (with permission).]

found its correct conformation, the total protein concentration can be increased stepwise up to economically attractive levels. This process of **'pulse naturation'** is commercially used in the production of a plasminogen activator (see Section 20.3.5).

20.3.3 Application, delivery and targeting of therapeutic proteins

Because of their typical substance class properties, proteins generally would by no means be considered 'ideal' therapeutic agents for reasons related to stability and application:

Stability

Proteins are polypeptides and therefore labile against heat, extreme pH values and biological degradation. This may lead to limited shelf-life as well as short half-lives in the human body, e.g. due to proteolysis in the stomach and intestine, and to receptor mediated clearance from the blood followed by proteolytic degradation in the liver. In addition to the risk of proteolytic cleavage of the polypeptide chain, amino acid side chains of proteins may also be modified during storage, e.g. by oxidation or isopeptide bond formation. Whereas shelf-life can be improved by suitable additives such as sugars or amino acids, attempts to prolong the biological *in vivo* half-life of protein drugs have met limited success until now. Microencapsulation into biodegradable polymers, mostly polylactide-polyglycolide co-polymers, appears to offer an interesting approach in this respect.

Ways of application

The molecular surface of soluble proteins is hydrophilic. Therefore, as a rule, proteins cannot pass through biological membranes and will not enter into tissue through the intestinal wall or into human cells from the bloodstream. Proteolytic degradation in the stomach could be prevented by encapsulation, giving the protein resistance against the acid

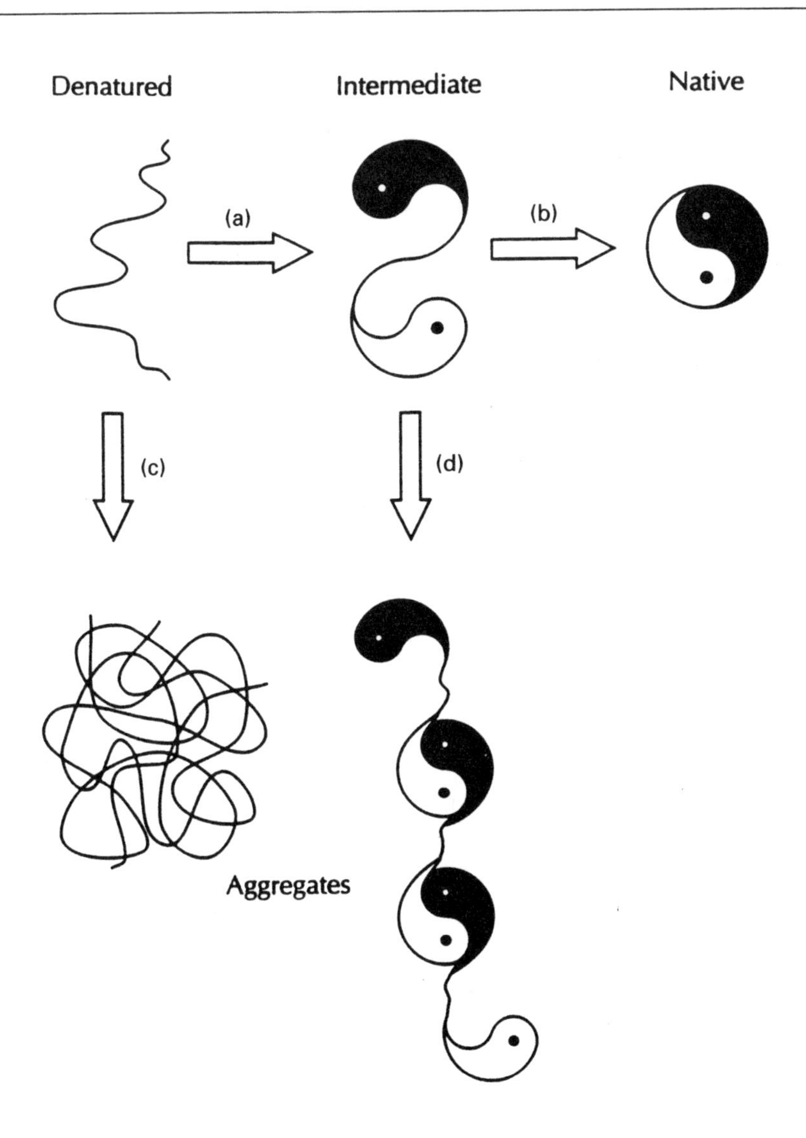

Fig. 20.6 Kinetic competition between protein folding and association. Steps (a) and (b) are productive first-order folding steps whereas steps (c) and (d) are unproductive second or higher order association processes. (Illustration courtesy of Dr R. Rudolph, Halle, Germany.)

pH and proteases. Therefore, oral application of protein drugs would not result in sufficient bioavailability unless the protein is intended to act in the oral cavity itself (as for example lysozyme, used to inhibit bacterial infections in the mouth cavity) or in the gastrointestinal tract (e.g. lipases and amylases, used to support food digestion). Protein therapeutics therefore cannot be given orally but have to be injected or infused into the bloodstream.

Immunogenicity of foreign proteins

Proteins that are foreign to the human body are immunogenic. When injected into the bloodstream, they may induce the formation of antibodies and cellular immune response. Furthermore, proteins obtained from natural sources may contain immunogenic contaminants. This may prevent repeated or prolonged application of the same protein drug. (The immunogenicity is desired when proteins are used as vaccines.)

One way to decrease immunogenicity of proteins is chemical coupling to water-soluble polymers, especially to polyethylene glycol. Such 'pegylated' proteins are in use as therapeutics, for example PEG-adenosine deaminase (PEG-ADA) for treatment of ADA deficiency (SCID – severe combined immunodeficiency disease) by substitution of the missing enzyme, as well as PEG-asparaginase in tumour therapy. It is, however, not easy to ensure product homogeneity after chemical modification, and of course production costs are increased by the additional chemical modification step. On the other hand recombinant human proteins are expected not to be immunogenic. Depending on the expression system used, however, proteins may differ from original human proteins in their post-translational modification (e.g. glycosylation, processing of *N*-terminus, etc.).

20.3.4 First-generation therapeutic proteins

The first-generation of recombinant therapeutic proteins are protein drugs made with the aid of gene technology. These are identical to natural human proteins. A few typical examples are listed below.

Insulin

This is a pancreatic hormone which has been used for treatment of type I diabetes since 1922 because of its effect in lowering blood glucose levels. Insulin consists of two polypeptide chains connected by disulphide bonds. The A chain has 21 amino acid residues and the B chain has 30 amino acid residues. Insulin biosynthesis involves proteolytic processing from the single-chain precursor molecule proinsulin, with release of a connecting (C-)peptide, as illustrated in Fig. 20.7.

During the first decades of insulin therapy, bovine or porcine insulin had to be used. In these animal proteins, there are some amino acid sequence differences from human insulin that may lead to formation of insulin antibodies during long-term application. In the 1970s it became possible to replace the alanine residue, B30, of porcine insulin with a threonine residue by protease-catalysed semisynthesis and, thus, insulin identical to the human molecule could now be produced. However, due to the growing population of patients needing insulin (about 1 in 1000), there were concerns that the supply of porcine insulin might become limited and the porcine or semisynthetic human material has been replaced by recombinant production of human insulin. Several strategies have been developed to produce recombinant insulin. In the original process described by Genentech, Inc. and Eli Lilly, the A and B chains are expressed separately in *E. coli* as fusion proteins with tryptophan synthease or β-galactosidase and, after processing by cleavage with cyanogen bromide, the two chains are connected by chemical reoxidation. In an alternative process, the physiological biosynthetic intermediate proinsulin (Fig. 20.7) or analogues with shortened connecting peptide sequences are expressed in *E. coli* or yeast, and the connecting peptide is removed enzymatically.

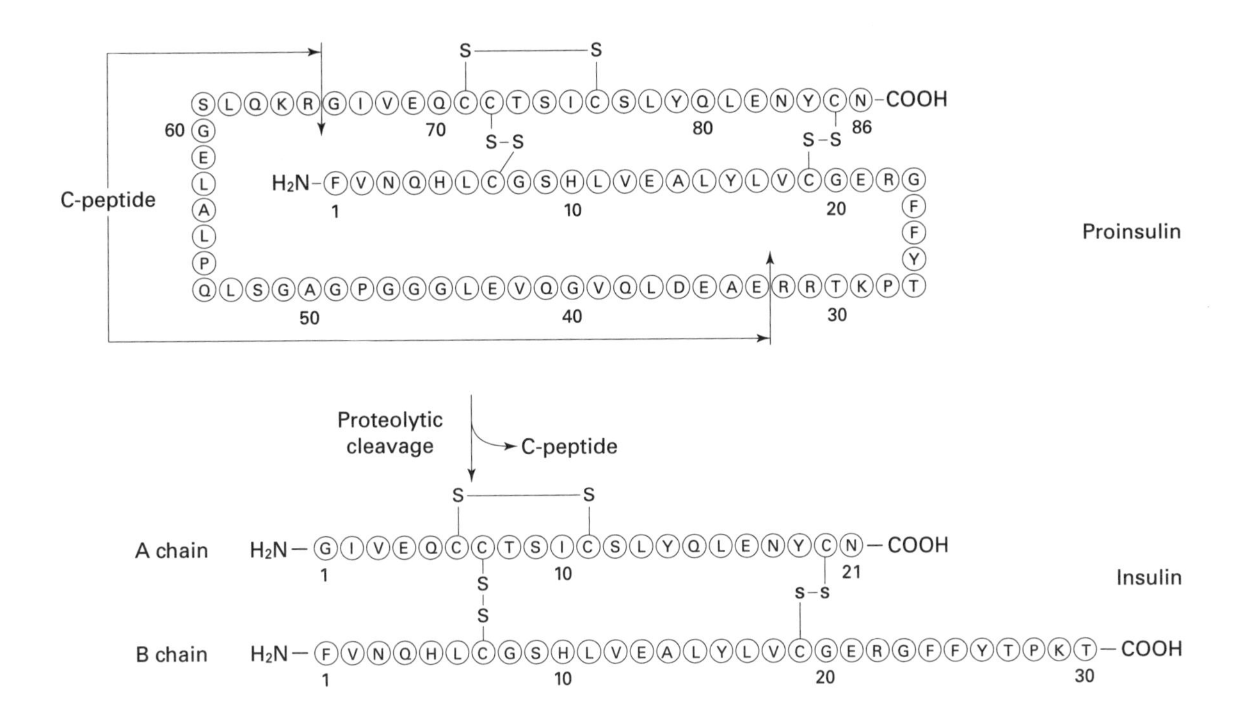

Fig. 20.7 Biosynthesis and amino acid sequence of human proinsulin and insulin. Proteolytic cleavage removes the connecting *C*-peptide from the single-chain precursor, proinsulin, to release the active two-chain, disulphide-linked insulin. In the analogue 'insulin lispro', the order of residues B28 and B29 is changed (Lys-Pro instead of Pro-Lys). B30 is the only amino acid residue which is different between human (B30 = Thr) and porcine (B30 = Ala) insulin.

Erythropoietin

Erythropoietin (Epoietin alpha and beta, EPO) is a glycoprotein of 165 amino acid residues. It is formed in the foetal liver and in the kidneys of adults. The EPO hormone belongs to the haematopoietic growth factors and induces the formation of erythrocytes from precursor cells (termed BFU-E und CFU-E) in the bone marrow. Recombinant erythropoietin has to be produced in mammalian cell systems due to the necessity of glycosylation (Chinese hamster ovary (CHO) cells are used in the commercial processes), and is used therapeutically mainly in renal anaemia, but also in other indications, e.g. in tumour anaemia.

Granulocyte-colony stimulating factor (G-CSF)

G-CSF belongs, as EPO, to the class of haematopoietic growth factors. G-CSF stimulates proliferation and differentiation of neutrophil precursor cells to mature granulocytes. It is therefore used as an adjunct in chemotherapy of cancer to treat neutropenia caused by the destruction of white blood cells by the cytotoxic agent. Furthermore, G-CSF is also used in the treatment of myelosuppression after bone marrow transplantation, chronic neutropenia, acute leukaemia, aplastic anaemia, as well as to mobilise haematopoietic precursor cells from peripheral blood.

G-CSF is a glycoprotein containing 174 amino acid residues. Products have been launched which contain either the glycosylated molecule produced from recombinant CHO cells (Lenograstim) or alternatively an unglycosylated, but therapeutically equally effective, form produced from recombinant *E. coli* (Filgrastim) which additionally possesses an *N*-terminal methionine residue.

Among the first-generation of recombinant therapeutic proteins, there are also various **antibodies**, enzymes, such as **glucocerebrosidase**, used for treatment of glucocerebrosidase deficiency (Gaucher's disease) as well as coagulation factors, **Factor VIIa, Factor VIII** and **Factor IX**, used for substitution therapy of haemophilia. **Tissue plasminogen activator** (t-PA) is described below. Additional examples are listed in Table 20.3.

20.3.5 Second-generation therapeutic proteins (muteins)

DNA sequences coding for proteins can nowadays be modified by site-directed mutagenesis so that the amino acid sequence of recombinant proteins can be designed as desired. This is known as **protein engineering**. Mutated proteins obtained in this way are called **muteins**. The changes may be restricted to isolated amino acid residues (point mutations), but may also involve the deletion or insertion of larger sequence regions, or newly introduced connection of originally unrelated sequences (protein fusions). These technologies may offer a strategic approach to modify protein properties in a rational way, such as stability, solubility, substrate or receptor binding specificity, or pharmacokinetics. Muteins obtained by rational design have been described as the second generation of therapeutic proteins.

However, present knowledge of structure-function relationships in proteins is far from complete. Only in some simple cases has it been possible to predict the effects of sequence changes on observable protein properties but in most cases, the assumptions leading to the introduction of mutations have to be verified empirically. Nevertheless, in several cases recombinant proteins have been successfully designed for use as therapeutic agents. Protein engineering has been used in these cases for quite diverse purposes.

Insulin lispro

After an insulin injection, the plasma concentration of insulin rises so slowly that the injection should be done at least 15 minutes before a meal. Similarly, the plasma level insulin also decreases more slowly than physiologically required, so there is the danger of hyperinsulinaemia. The slower increase in concentration is due to the time needed for dissociation of insulin hexamer to the pharmacologically active dimers and monomers. To accelerate this process, a large number of insulin muteins have been constructed which still are biologically active but show faster dissociation of hexamers in solution. One of these fast-acting insulin analogues is *insulin lispro* where, in analogy to the naturally occurring insulin homologue insulin-like growth factor-I (IGF-I), the order of the amino acid residues B28 and B29 was changed, see Fig. 20.7. It was reported that insulin lispro reaches pharmacologically efficient levels faster and therefore could be injected immediately before a meal.

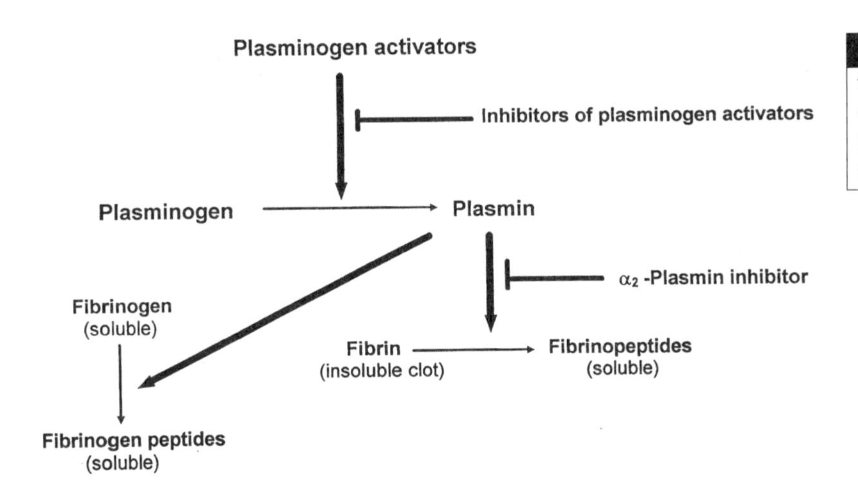

Fig. 20.8 General scheme of fibrinolysis. Thick arrows (⬇) designate catalytic activation by proteolysis which is under control of plasma inhibitors (⊥).

Tissue plasminogen activators

Acute myocardial infarction (AMI) is the principal cause of deaths in most Western hemisphere countries. One approach to improve treatment of AMI is the use of thrombolytic enzymes. Plasminogen activators catalyse the proteolytic processing of the inactive proenzyme plasminogen, which circulates in the bloodstream, into the active protease plasmin. Plasmin is able to cleave the insoluble fibrin of blood clots into soluble fibrin fragment peptides so that the clot is dissolved and the blood vessel is opened. The reaction scheme is outlined in Fig. 20.8.

Plasminogen activators (e.g. Alteplase and Reteplase, a mutein with an increasd *in vivo* half-life) are used increasingly as thrombolytic agents in the treatment of AMI, and are also used in studies on related diseases such as stroke or deep vein thrombosis.

Other second generation recombinant protein drugs

In the near future, much progress is expected in the field of **recombinant immunotoxins** used in experimental treatment of various cancers. These are chemical conjugates or recombinant fusion proteins constructed from a cell-binding part (mostly the antigen binding parts of an antibody), a translocation domain mediating transfer through the cell membrane, and a cytotoxic portion, e.g. protein domains from bacterial toxins (such as *Diphtheria* toxin or *Pseudomonas* exotoxin) or a chemical cytotoxic agent. The idea is that the toxic agent should be targeted to the selected cancer cell population by the antibody domain directed against specific surface antigens of these cells, which then should be killed after internalisation of the toxin part. In clinical experiments, encouraging results have been obtained with immunotoxins targeting the transferrin or interleukin-2 receptors or the erbB2 or Lewis-Y antigens. However, problems concerning in particular the immunogenicity of the bacterial toxin sequences still have to be solved.

20.4 Regulatory aspects of therapeutic proteins

20.4.1 Development and approval risk

The toxicity of proteins is usually less severe than with chemically synthesised substances since they have fewer undesirable side-effects associated with them, due to their specific physiological roles. As natural substances, proteins are neither carcinogenic nor teratogenic. If the principle of action has been identified, development of recombinant human proteins into therapeutic agents should carry less risk than the development of new low-molecular weight drugs. This is true once efficacy has been demonstrated, i.e. in the later clinical development phases, where the main part of development costs arises. Furthermore, after completion of clinical development, innovative protein therapeutics will usually be approved faster for market launch due to internationally agreed common quality standards.

In the United States, approval of biopharmaceuticals is regulated by the Center for Biologics Evaluation and Research (CBER) of the Federal Drug Administration (FDA). They have defined guidelines for approval of therapeutic proteins as 'well characterised biotechnology pharmaceuticals'. In Europe, the approval procedure for biotherapeutics has been centralised through the European Medicines Evaluation Agency (EMEA) which follows similar guidelines.

20.4.2 Safety

In contrast to proteins isolated from human or animal, including transgenic, sources or from pathogenic organisms, e.g. vaccines obtained from bacteria or viruses, highly purified and carefully analysed recombinant proteins do not bear the risk of contamination with allergenic substances, pathogenic viruses, e.g. HIV, or prions from cattle or humans causing new variant Creutzfeldt–Jakob disease. For this reason, products such as coagulation factors (formerly produced from human blood or plasma), human growth hormone (in the past obtained from adenohypophysis extracts), or hepatitis B vaccines are today manufactured from recombinant systems.

20.5 Outlook to the future of protein therapies

Considering the general advantages and disadvantages of protein therapeutics, it can be concluded that they are not equally attractive in all therapeutic areas and indications when compared with competing approaches such as low-molecular weight chemical substances on the one hand, and gene therapy on the other. Protein drugs would be especially useful in the following cases:

- In indications where no alternative therapy is available, particularly for potentially live-threatening diseases such as acute myocardial infarction, cancer or viral infections.
- For substitution therapy if essential human proteins are missing or inactive, e.g. in ADA deficiency or in coagulation factor deficiencies.

- To modulate the regulation of biological processes such as metabolism, cell growth, wound healing, etc. or to influence the immune system by proteins acting as hormones, growth factors, or cytokines (e.g. insulin, erythropoietin, G-CSF, somatotropin, interferons or interleukins). In these cases, protein–protein interactions have to be modulated. This may be more effective with therapeutic proteins as 'nature's own ligands' optimised in the course of evolution, than with small chemical substances.
- As vaccines, especially against viral infectious diseases.

Human proteins identical to the body's own substances have become available through the advent of gene technology. Besides the first-generation biotherapeutics, an increasing number of redesigned, second-generation protein muteins with improved properties are being introduced to the marketplace. Once the present problems of low transfection and expression efficiency have been solved, it may be possible to substitute defect genes, or add therapeutic genes, to human cells *in vivo* so that the patient's body itself will act as the manufacturing facility where the synthesis of therapeutic proteins occurs. In this sense, gene therapy may represent the future third-generation of therapeutic proteins, and may help to approach the final goal to cure, rather than treat, disease.

20.6 Further reading

Bergmeyer, H. U., Grassl, M. and Bergmeyer, J. (eds.) (1983–1986). *Methods of Enzymatic Analysis, Vol. 1–12*. VCH, Weinheim.

Brange, J. and Vølund, A. (1999). Insulin analogs with improved pharmakokinetic profiles. *Adv. Drug Delivery Rev.* **35**, 307–335.

Bristow, A. F. (1993). Recombinant-DNA-derived insulin analogues as potentially useful therapeutic agents *Trends Biotechnol.* **11**, 301–305.

Klegerman, M. E. and Groves, M. J. (1992). *Pharmaceutical Biotechnology: Fundamentals and Essentials.* Interpharm Press, Inc., Buffalo Grove, IL.

Kopetzki, E., Lehnert, K. and Buckel, P. (1994). Enzymes in Diagnostics: Achievements and Possibilities of Recombinant DNA Technology. *Clin. Chem.* **40**, 688–704.

Kresse, G.-B. (1995). Analytical uses of enzymes. In *Biotechnology, 2nd edition, Vol. 9* (H.-J. Rehm & G. Reed, eds.), pp. 138–163. Verlag Chemie, Weinheim.

Nicola, N.A. (1994). *Guidebook to Cytokines and Their Receptors*. Oxford University Press, Oxford.

Perham, R. N. *et al.* (1987). *Enzymes*. In *Ullmann's Encyclopedia of Industrial Chemistry, Vol. A9*, pp. 341–530. Also republished separately in: W. Gerhartz, ed. (1990) *Enzymes in Industry – Production and Applications*. Verlag Chemie, Weinheim.

Rouf, S.A., Moo-Young, M. and Chisti, Y. (1996). Tissue-type plasminogen activator: characteristics, applications and production technology. *Biotechnol. Adv.* **14**, 239–266.

Rudolph, R. and Lilie, H. (1996). In vitro folding of inclusion body proteins. FASEB J. **10**, 49–56.

Steinberg, F. M. and Raso, J. (1998). Biotech pharmaceuticals and biotherapy: an overview. *J. Pharm. Pharmaceut. Sci.* **1**, 48–59.

Chapter 21

Mammalian cell culture

N. Vriezen, J. P. van Dijken and L. Häggström

21.1 Introduction

The cultivation of mammalian cells *in vitro* (e.g. in a bioreactor) has evolved from an empirical art to a modern quantitative science since 1945. Media and cultivation conditions that can support viability and proliferation of a large number of different cell types from different organisms have been developed. Cell lines have been established from a range of mammals, such as humans, rats, mice, hamsters, cats, dogs, monkeys, sheep, cattle and horses, and from individual organs such as the lung, kidney, liver, skin, lymph nodes, muscles, ovaries, thymus and heart and also various types of cancer. Cell lines from reptiles, fish and insects have also been established.

The driving force for development of *in vitro* cultivation techniques for mammalian cells was the need for polio vaccine in the 1950s. Vaccine production is still a major application but, today, mammalian cell cultures are also used for toxicological and pharmaceutical research, thereby reducing the need for animal testing, and for production of artificial organs. For example, layers of cultured keratinocytes function as artificial skin and attempts are being made to construct artificial liver and kidney units. The most productive application of mammalian cell cultures has, however, so far been in the manufacture of proteins for diagnostic and therapeutic use.

21.2 Mammalian cell lines and their characteristics

Mammalian cells are normally part of an organ where they differentiate to perform specific functions. When transferred to *in vitro* conditions, some cell types will stay alive without multiplying, whilst others will multiply. The cell types most likely to multiply *in vitro* are those which will also do so in the body, such as cancer cells, epithelial cells and fibroblasts.

A normal diploid cell line has a finite life span. After a certain number of cell divisions, proliferation ceases and the culture dies eventually. Continuous (or immortalised) cell lines, such as cancer cells, have acquired the capacity to grow and multiply for an unlimited number of generations like micro-organisms. A number of other changes in physiology and metabolism also follow from the transformation of a normal cell line to a continuously growing one. For example, continuous cell lines are easier to handle in the laboratory as they are less dependent on serum and growth factors (see also Section 21.5), i.e. the normal proliferation control mechanisms are deregulated. Continuous cell lines have also increased growth rates, are less sensitive to environmental disturbances than their normal counterparts, and can be cultivated in suspension culture like micro-organisms. On the negative side are increased metabolic rates resulting in formation of inhibitory by-products (see Sections 21.6.4 and 21.6.5). Typically, cell lines used for protein production are grown continuously because of the advantages of using suspension culture, simpler media formulations and the gain in productivity. In cultured rodent cells, the transition from the normal state to a continuous cell line often occurs spontaneously, while in other cell types special modification procedures may be needed. Methods to create continuous cell lines include fusion with another cell line that is already 'immortal', infection with a virus, transfection with an oncogene or mutagenesis.

Like all eukaryotic cells, mammalian cells are bound to follow the cell cycle. The cell cycle phases, G1, S, G2 and M, denote gap 1, synthesis phase, gap 2 and mitosis, respectively. DNA is replicated during the S phase, while G1 prepares the cell for replication, and G2 for cell division. The time required for one round in the cell cycle is typically about 24 h. Normal mammalian cells that do not proliferate, are arrested in G1. However, continuous cell lines have lost the property to withdraw from the replication cycle. Therefore, if the conditions are unsuitable for growth, such cells may instead respond with cell death through a mechanism named **apoptosis** – programmed cell death. All mammalian cells contain the genetic programme for apoptosis. Apoptosis results in the packaging of the whole cell into small membrane-bounded, so-called, apoptotic bodies. In normal cells, DNA strand-breaks caused by UV radiation lead to apoptosis. Conditions that provoke apoptosis in cultures of continuous cell lines include, among others, a lack of growth factor or serum, nutrient limitation and mechanical stress.

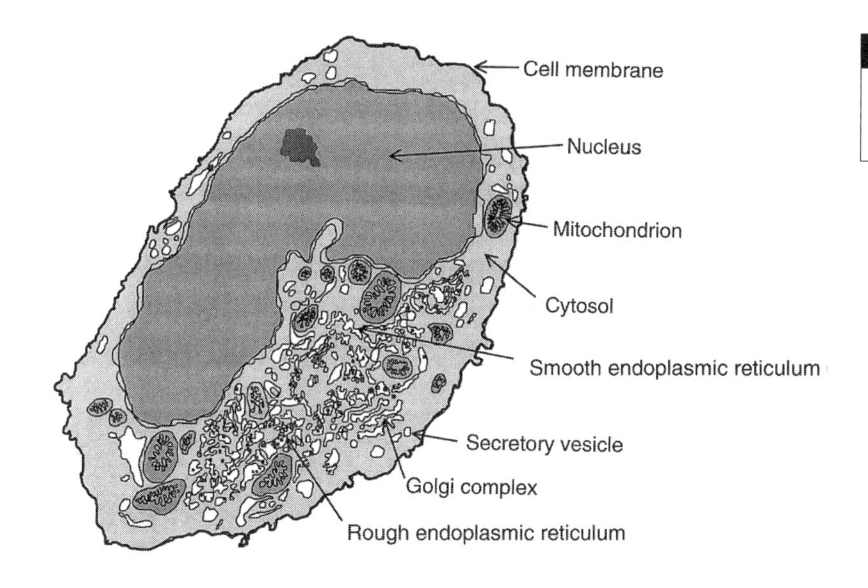

Fig. 21.1 Schematic representation of the structure of a myeloma cell. The cell diameter is approximately 12 μm.

Mammalian cells in culture act as unicellular organisms, i.e. they grow and multiply by division, as long as nutrients are available in sufficient amounts. In suspension culture, the cells assume a spherical shape with a diameter of 7 to 20 μm. The structure of a typical, undifferentiated eukaryotic cell is shown in Fig. 21.1. In contrast to microorganisms, the eukaryotic cell is complex, containing a variety of organelles. A further difference of technical significance between animal cells on one hand and microbial and plant cells on the other is the lack of a rigid cell wall, which makes mammalian cells vulnerable to changes in osmolarity, to shear forces, and to damage caused by air bubbles.

Examples of mammalian cells used for industrial production are hybridoma, myeloma, Chinese hamster ovary (CHO), and baby hamster kidney (BHK) cells. Fusion of lymphocytes and myeloma cells (lymphoid cancer cells) results in **artificially constructed hybridoma cells**, as first described by Köhler and Milstein. In the hybridoma cell, features of both parent cells are combined; the capacity for production of antibodies stems from the lymphocyte and the proliferative potential from the myeloma. Typical for hybridoma cells is the large potential to produce and secrete glycosylated proteins (see Section 21.4). Myeloma cells are being used as host cells for production of recombinant protein, in particular recombinant antibodies, as these cells, like the hybridomas, have the capacity for extensive secretion of proteins. CHO cells are the favoured industrial cells for production of recombinant proteins, for example, Factor VII and Factor VIII (see Section 21.3). Baby hamster kidney (BHK) cells have been used for a long time for vaccine production, an example being the veterinarian vaccine against foot and mouth disease.

Table 21.1 Examples of pharmaceutical proteins produced with mammalian cell cultures. The type of glycosylation is indicated

Pharmaceutical protein	Function	Type of glycosylation
Tissue plasminogen activator (tPA)	Fibrinolytic agent	*N*-linked
Erythropoietin (EPO)	Antianaemic agent (blood doping)	*N*- and *O*-linked
Factor VII, VIII, IX and X	Haemophilia, blood clotting agents	*N*- and *O*-linked
Follicle stimulating hormone (FSH), human chorionic gonadotrophin (hCG)	Infertility treatment	*N*- and *O*-linked
Interleukin-2	Anticancer, immunomodulator, HIV treatment	*O*-linked
Interferon-alpha (IFN-α)	Anticancer, immunomodulator	*N*- and *O*-linked
Interferon-beta (IFN-β)	Anticancer anti viral agent	*N*-linked
Interferon-gamma (IFN-γ)	Anticancer agent, immunomodulator	*N*-linked
Granulocyte colony stimulating factor (G-CSF)	Anticancer	*O*-linked
Monoclonal antibodies	Therapeutic and diagnostic	*N*-linked

21.3 Commercial products

Products made in bioprocesses with mammalian cells are mainly glycoproteins. Table 21.1 shows some representative examples. The complexity and costs of mammalian cell processes dictate that protein production with mammalian cells is economically viable only for high added-value products (>US\$ 10^6 kg^{-1}). Mammalian cell protein products are therefore mainly pharmaceutical products. **Monoclonal antibodies** (MAbs) (see also Chapter 23) are the best known mammalian cell culture products. The highly specific binding properties of the MAbs can be used in diagnostics (both medical and veterinary), imaging (cancer and heart disease), product purification (affinity chromatography) and as therapeutic agents. Other pharmaceutical proteins that are produced with mammalian cell cultures are aimed at treating cancer, heart diseases, blood diseases and hormonal disorders.

The products listed in Table 21.1 are all glycoproteins, i.e. a protein to which a sugar moiety has been added in a post translational process called **glycosylation**. Glycosylation of a protein occurs in the endoplasmic reticulum (ER) and Golgi complex of a eukaryotic cell and depends on the presence of specific enzymes: glycosyltransferases and glycosidases. Bacteria neither contain these organelles nor the enzymes and are therefore not able to perform this post translational modification. Yeasts and filamentous fungi (eukaryotes) are able to glycosylate proteins but do so differently from mammalian cells (see Chapter 5). The

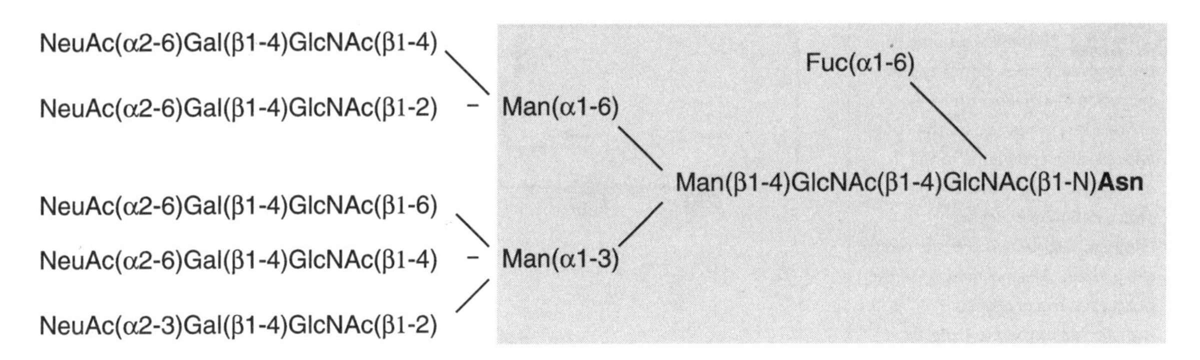

Fig. 21.2 Example of complex *N*-linked glycosylation. The grey area indicates the core unit of the five sugar residues found in *N*-linked glyco-structures that are attached to an asparaginyl residue. GlcNAc, *N*-acetylglucose amine; Fuc, fucose; Man, mannose; Gal, galactose; NeuAc, sialic acid.

presence and conformation of the sugar moiety of a glycoprotein is, in many cases, essential for a functional product as this functionality prolongs the half-life in the bloodstream, and diverts the protein to its specific location in the cell. If this is so, the production organism of choice is often a mammalian cell line. Pharmaceutical proteins that are not glycosylated or need not be glycosylated for proper function, like insulin or human growth hormone, human serum albumin and haemoglobin, may be produced more cost-effectively with bacteria, yeasts or filamentous fungi (see Chapter 20).

21.4 Protein glycosylation

Whereas protein synthesis is guided by DNA and RNA templates, the addition of sugar to a protein is a process without a template. Therefore a large variation in the oligosaccharide structures of glycoproteins can be found. Glycoproteins with the same amino acid sequence, but different oligosaccharide structures are called **glycoforms**. The oligosaccharide structures are covalently bound to the protein either at a nitrogen (*N*-glycosylation) or at an oxygen (*O*-glycosylation) atom. These two forms of glycosylation differ not only in the position where sugars are attached but also in the type and quantity of sugars added. In *N*-glycosylation, the glycan moiety is linked to the asparaginyl residue in a consensus sequence Asn – X – Thr/Ser, in which X represents any amino acid except for proline. The core of *N*-linked glycans consists of a pentasaccharide of three mannose and two *N*-acetylglucosamine molecules (Man3GlcNAc2), to which a variety of other sugars may be added in varying degrees of branching (Fig. 21.2). O-linked glycans are present on threonyl or seryl residues. No specific consensus sequence has yet been found, but the glycosylation is influenced by the local conformation of the peptide chain. The glycan structures in *O*-glycosylation are generally smaller and more variable than those in *N*-glycosylation.

The process of glycosylation is closely linked to the secretion of the product (Fig. 21.3). The protein moiety is made on the rough ER, where translation of the mRNA to the protein takes place. If *N*-glycosylation sites are present, a pre-made $Glu_3Man_9GlcNAc$ structure is transferred to the glycosylation site as the protein enters the ER and the terminal glucose units are trimmed off while the protein still remains in the ER.

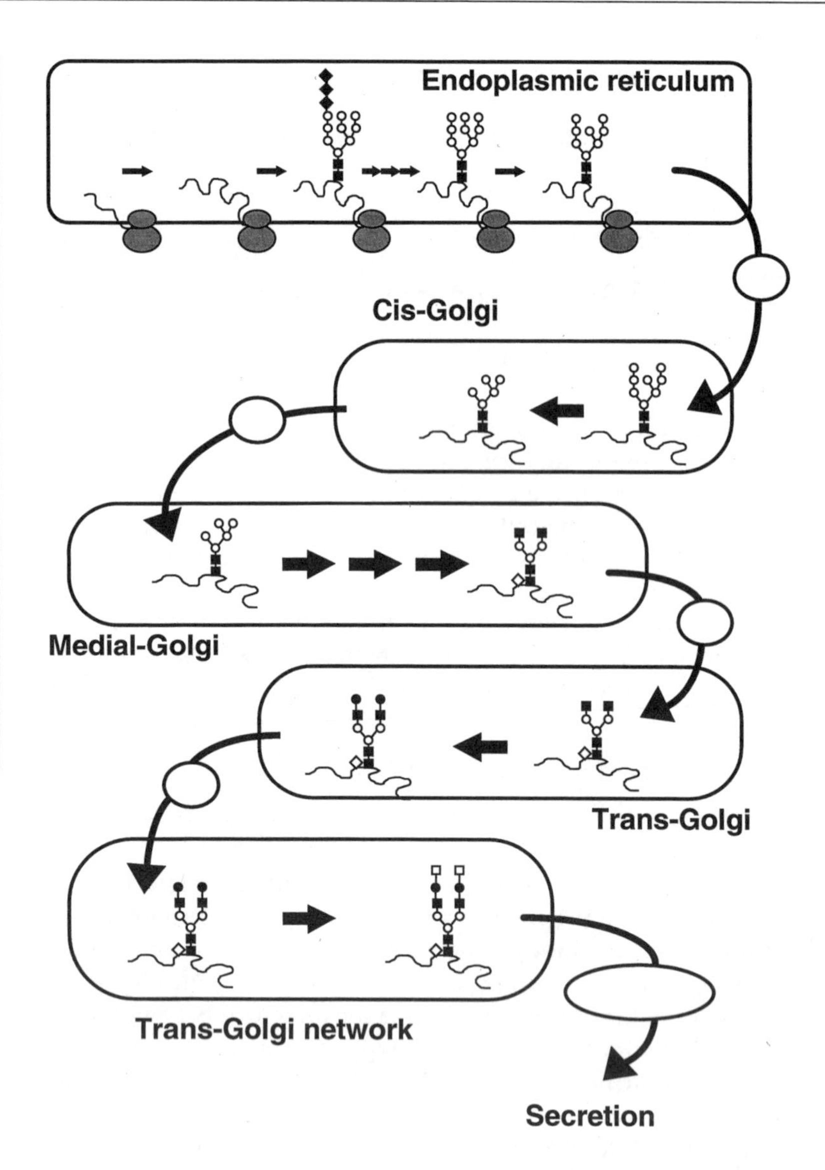

Fig. 21.3 Simplified scheme of the *N*-glycosylation pathway and secretion of a glycoprotein. A protein is synthesised on the rough endoplasmic reticulum. In the endoplasmic reticulum the *N*-glycosylation precursor $Glc_3Man_9GlcNAc_2$ is transferred to the protein. After trimming off the precursor, the protein is transferred to the cis-Golgi by vesicular transport. The glycostructure is trimmed and extended as the protein traverses the Golgi complex from cis- to trans-Golgi. Transport takes place by vesicles. When the glycostructure is complete the glycoprotein may be excreted by way of a secretion vesicle that fuses with the cell membrane. Symbols used: ○, mannose; ■, GlcNAc; ◆, glucose; ●, galactose; open diamonds, fucose; ❑, sialic acid.

The protein is then transported to the cis-Golgi compartment of the cell by vesicles that bud off from the ER. In the cis-Golgi, part of the mannose structure is trimmed off. The protein traverses the Golgi complex via vesicular transport. During progression through the medial and trans layers of the Golgi complex, the galactose and GlcNAc units of the complex type N-glycosylation are added. Finally, sialic acid units may be added in the trans-Golgi network. The completed glycoprotein is then transported to its destination. Secretory proteins are released from the cell by fusion of the final secretory vesicle with the cell membrane. *O*-type glycosylation also takes place during the trafficking of the protein through the Golgi complex; the location and reactions for this process are, however, less well known.

The specific glycosylation pattern differs between cell lines. For instance, CHO cells do not synthesise bisecting *N*-acetylglucosamine structures and mouse cell lines are known to sporadically generate terminal galactose units (Galα1$\rightarrow$3Gal) that are immunogenic in humans. Knowledge on the desired glycan structure is therefore beneficial in selecting a cell line for production. The glycan structures of glycoproteins may influence its key characteristics, essential for the activity of a pharmaceutical productprotein. For example, erythropoietin activity is totally dependent on the presence and structure of its glycan moieties while the biological activity of interferons and some interleukins does not depend on the presence of glycans.

The *in vivo* half-life of a glycoprotein is influenced by the amount of terminal sialic acid which protects the protein against clearance from the bloodstream by hepatocytes or macrophages. The glycan structure also influences physico-chemical properties such as the three dimensional structure, solubility, viscosity, thermal stability, pH stability and charge. The presence of the glycan can provide protection against proteolytic attack and antigenic sites can be presented or masked by the glycan.

21.5 Media for the cultivation of mammalian cells

Mammalian cells in the body of an organism receive nutrients from the blood circulation. Cell culture media for the *in vitro* propagation of mammalian cells must therefore supply nutrients similar to those present in the blood stream. Initial attempts to grow mammalian cells *in vitro* involved media derived from complex natural sources such as chick embryos, blood serum or clots and lymph fluids. Since about 1950, partly defined media, consisting of a great number of components, have been developed (Table 21.2). The basis for cell culture media is a balanced salt solution. These salt solutions were originally used to create a physiological pH and osmolarity, required for maintaining cell viability *in vitro*. To create conditions promoting proliferation, glucose, amino acids and vitamins were added to the salt solution, according to the requirements of the specific cell line. This development resulted in various of media formulations, each designed for a limited number of cell types. Many modifications of existing media have since been developed for new or more specialised applications. Four main groups of media can now be discerned:

- Eagle's medium and derivatives thereof, such as BME (Basal Medium Eagle's), EMEM (Eagle's Minimal Essential Medium), DMEM (Dulbecco's Modification of Eagle's Medium) and GMEM (Glasgow's Modification of Eagle's Medium).
- Media from Roswell Park Memorial Institute (RPMI), such as RPMI 1630 and RPMI 1640.
- Media designed for use with serum, such as Liebovitz, Trowell and Williams media.
- Media designed for a specific cell line for use without serum, such as CMRL 1060, Ham's F10 and F12, TC199, and IMDM (Iscove's Modification of Dulbecco's Modification of Eagle's Medium).

Table 21.2 Examples of medium composition (as mg l^{-1})

Component	Eagle's MEM	RPMI 1640	Ham's F12	IMDM
Amino acids				
L-alanine			8.91	25
L-arginine HCl	105	200	211	84
L-asparagine H2O		50	15.0	28.4
L-aspartic acid		20	13.3	30
L-cystine	24	50	24.0	70
L-glutamic acid		20	14.7	75
L-glutamine	292	300	146.2	584
glycine		10	7.51	30
L-histidine HCl H_2O	31	15	21.0	42
L-isoleucine	52	50	3.94	104.8
L-leucine	52	50	13.12	104.8
L-lysine	58	40[a]	36.54[a]	146.2
L-methionine	15	15	4.48	30
L-phenylalanine	32	15	4.96	66
L-proline		20	34.5	40
L-serine		30	10.51	42
L-threonine	48	20	11.91	95
L-tryptophan	10	5	2.042	16
L-tyrosine	36	20	5.43	84
L-valine	46	20	11.7	93.6
glutathione (red)		1		
L-hydroxyproline		20		
Vitamins				
D-biotin		0.2	0.007	0.013
Ca D-panthothenate	1	0.25	0.26	4
choline chloride	1	3.0	13.96	4
folic acid	1	1.0	1.32	4
i-inositol	2		18.02	7.2
nicotinamide	1	35	0.037	4
p-aminobenzoic acid		1.0		
pyridoxine HCl		1	0.062	
pyridoxal HCl	1			4
riboflavin	0.1	0.2	0.038	0.4
thiamine HCl	1	1.0	0.34	4
vitamin B12		0.005	1.36	0.013
Inorganic salts				
$CaCl_2 \cdot 2H_2O$	200		44.1	218
$CaNO_3 \cdot 4H_2O$		100		
$CuSO_4 \cdot 5H_2O$			0.0025	
$FeSO_4 \cdot 7H_2O$			0.83	
KCl	400	400	223	330
KNO_3				0.076
$MgSO_4 \cdot 7H_2O$	220	100	133	200
NaCl	6800	6000	7599	4505

Table 21.2 (*cont.*)

Component	Eagle's MEM	RPMI 1640	Ham's F12	IMDM
$NaHCO_3$	2000	2000	1176	3024
$Na_2HPO_4 \cdot 7H_2O$		1512	268	
$NaH_2PO_4 \cdot 2H_2O$	150			141
Other components				
D-glucose	1000	2000	1801	4500
HEPES				5962
phenol red		5.0	1.2	15
sodium pyruvate			110	110
sodium selenite				0.017
BSA				400
transferrin				1.0
soybean lipid				100
lipoic acid			0.21	
linoleic acid			0.084	
hypoxanthine			4.08	
putrescine·2 HCl			0.16	

Notes:
[a] HCl salt used.

Eagle, a pioneer in this field with many important articles published during the 1950s and 1960s, determined which amino acids were essential for mammalian cells in culture (i.e. amino acids that cannot be synthesised by the cells themselves in amounts adequate for growth). EMEM is based on the results of this investigation. Yet, many media contain all amino acids, whether essential or not. The amino acid, glutamine, is provided in higher concentrations than the other amino acids, as glutamine is also used by most cells as a metabolisable energy and carbon source.

Addition of 5 to 20% (v/v) blood serum to the medium is still required for many applications. Serum supplies growth factors, trace elements and lipids and enhances the buffer capacity, chelates heavy metals, and protects against proteolytic activity, shear forces and bubble damage. The use of serum also allows a single medium formulation to be used for many cell lines, reducing the need for extensive medium optimisation for every cell line. Disadvantages of using serum include a dependency on its supply, the lack of reproducibility due to variation in quality between batches, complications in downstream processing including protein purification and the risk of contamination of the product with virion or prion particles. The last factor has gained much attention since the outbreak of bovine spongiform encephalopathy (BSE). Serum-free media based on the existing nutrient media, but fortified with growth factors from defined sources (e.g. recombinant proteins produced by micro-organisms) and other supplements, have been developed. The most common additions are insulin (growth factor), transferrin (carrier of Fe^{3+}), selenium (trace element), fatty acids, dexamethason (an artificial glucocorticoid with growth promoting activity in certain cell types) and bovine serum albumin (protects cells from bubble damage and is a carrier of lipids). Trace elements such as zinc, molybdenum and nickel are added to some media. Serum-free media for a variety of cell lines are now commercially available.

21.6 Metabolism

Most cultured mammalian cells use both **glucose** and **glutamine** as sources of energy and anabolic precursors. This provides mammalian cells with a certain flexibility. A limited supply of glucose can be compensated for by an increased consumption of glutamine and vice versa. As glutamine also is a nitrogen source for mammalian cells, glutamine limitation can lead to an increased consumption of other amino acids to compensate for the lower nitrogen intake.

21.6.1 Metabolic routes for glucose and glutamine

Glucose is mainly metabolised via the glycolytic pathway to pyruvate (Fig. 21.4) – see also Chapter 2. The main fate of glucose-derived pyruvate is reduction to lactate. Lactate is excreted and accumulates in the culture medium. Alternatively, pyruvate may be converted to acetyl-CoA which enters the tricarboxylic acid cycle (TCA cycle). A small fraction (4–8%) of consumed glucose goes via the pentose phosphate pathway (PPP) which supplies ribose-5-phosphate for the synthesis of nucleotides, as well as reducing equivalents (NADPH) for biosynthesis. PPP is the most important part of the sugar metabolism for mammalian cells as shown by the following example. If glucose is exchanged for fructose, very little sugar is consumed by both normal and continuous cell lines as compared to the corresponding amount of glucose. The consumed fructose, which supports growth to the same extent as glucose, is almost completely metabolised via the PPP and no lactate is produced. During such conditions cells derive energy from the metabolism of glutamine.

Glutamine can be catabolised in a number of ways. The first step is always the deamidation of glutamine to glutamate (Fig. 21.4). Most of the glutamine consumed by the cell is deamidated by glutaminase, a reaction that yields glutamate and releases an ammonium ion. A smaller amount is metabolised via transamidation reactions in which the amide group is transferred to a precursor metabolite which is used in biosynthesis of, for example, purines, pyrimidines and amino sugar moieties incorporated into glycan structures. The second step in the glutamine metabolism is the conversion of glutamate to α-ketoglutarate. This conversion can either occur via glutamate dehydrogenase, a reaction that liberates another ammonium ion, or via transamination with pyruvate or oxaloacetate as amino-group acceptors resulting in the formation of alanine or aspartate. Alanine, in particular, is easily excreted from the cell and accumulates in the culture medium. The transamina-

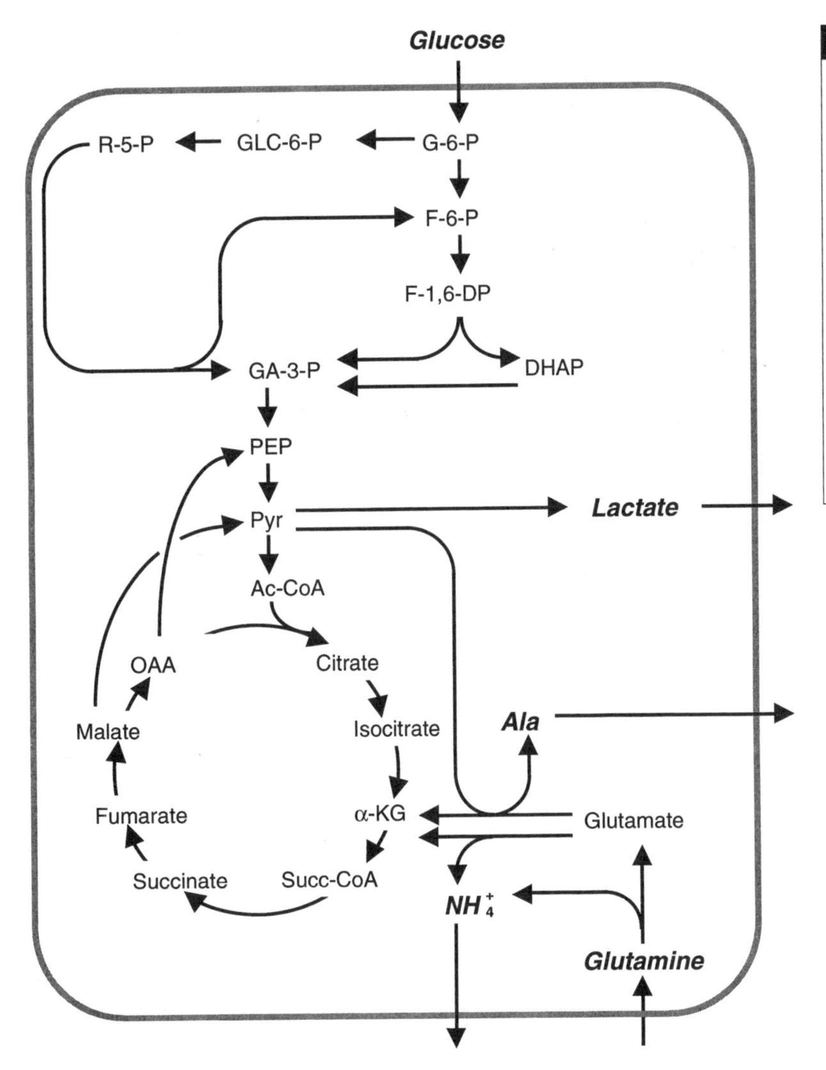

Fig. 21.4 Schematic representation of glucose and glutamine catabolism in mammalian cells. Compounds in bold print are consumed from or excreted into the medium. Abbreviations used: G-6-P, glucose-6-phosphate; GLC-6-P, 6-phosphogluconate; R-5-P, ribose-5-phosphate; DHAP, dihydroxyacetone phosphate; PEP, phospho*enol*pyruvate; Pyr, pyruvate; Lac, lactate; Ac-CoA, acetyl-CoA; α-KG, α-ketoglutarate; Succ-CoA, succinyl-CoA; OAA, oxaloacetate; Ala, alanine.

tion pathway is the dominant route of glutamine metabolism in rapidly growing cultured mammalian cells.

Both glucose-derived and glutamine-derived carbon enter the TCA cycle, as acetyl-CoA and *α*-ketoglutarate, respectively (Fig. 21.4). In a complete turn of the cycle, one molecule of acetyl-CoA is oxidised to CO_2 and water. However, fluxes, such as that of *α*-ketoglutarate, entering the TCA cycle must be balanced by fluxes leaving the cycle, otherwise TCA cycle intermediates would accumulate. To some extent, this balance can be met by anabolic fluxes that consume TCA cycle intermediates. Balancing the influx and efflux of the TCA cycle can also be accomplished by a catabolic process. Malate or oxaloacetate may leave the cycle by being converted into pyruvate and/or phospho*enol*pyruvate which, in turn, can re-enter the cycle as acetyl-CoA. Another possibility is that pyruvate so formed, or oxaloacetate, undergoes transamination with glutamate, resulting in alanine and aspartate production. These amino acids are, in fact, end-products of glutamine metabolism. Thus, the metabolic routes for glucose and glutamine partially overlap with pyruvate being formed from glutamine as well as from glucose. Oxaloacetate is formed from glutamine via reactions of the TCA cycle and, in normal cells, also by carboxylation of glucose-derived pyruvate via pyruvate carboxylase. However, no flux occurs via this enzyme in continuous cell lines. Energy, in the form of ATP and reducing equivalents, is generated from both glucose and glutamine.

21.6.2 Stoichiometry of metabolism and energy yields

A description of mammalian cell metabolism as a balance between catabolism and anabolism is complicated by the use of both glucose and glutamine as energy and carbon sources and by the utilisation of other amino acids for both anabolic and catabolic purposes. Moreover, the diverse metabolic fates of glucose and glutamine have different consequences for the amounts of by-products and energy produced. Glucose may be fermented to lactate, yielding a theoretical maximum of 2 mol ATP and 2 mol lactate per mol glucose. Alternatively, glucose may be completely oxidised in the TCA cycle, yielding 6 CO_2, 10 NADH, 2 ATP, 2 GTP and 2 $FADH_2$, which at a P/O ratio of 3 for NADH and 2 for $FADH_2$ translates to give a maximum of 38 ATP. Complete oxidation of glutamine generates 5 CO_2, 2 NH_4^+ and 27 ATP. The transamination pathway of the glutamine metabolism yields 1 NH_4^+ and 9 ATP, together with alanine and 2 CO_2, or aspartate and 1 CO_2, depending on the type of transamination reaction.

21.6.3 Metabolic compartmentation

Like all eukaryotes, mammalian cells are compartmented into organelles (Fig. 21.1). Different metabolic routes are located in different compartments. Some reactions have a unique location: glycolysis including lactate dehydrogenase are cytoplasmic while the TCA cycle and certain enzymes of the glutamine metabolism (glutaminase, glutamate dehydrogenase) are mitochondrial. Metabolic intermediates may need to be transported from one compartment to another to be available for particular routes. This is especially important for the metabolism of reducing equivalents, NADH. Large amounts of NADH formed in the cytosol need to be reoxidised. Although all the shuttle systems responsible for transporting NADH into mitochondria (for subsequent oxidation in the respiratory system) likely are present in continuous cell lines, they are obviously not keeping up with glycolytic NADH production as judged by the extensive formation of lactate, which is a means of regenerating NAD^+ via cytosolic reduction of pyruvate.

The transamination reactions with glutamate as amino group donor, leading to alanine and/or aspartate formation as end products of the glutamine metabolism, can occur both in mitochondria and in the cytosol. While transamination of oxaloacetate to aspartate mainly is a mitochondrial event, transamination of pyruvate to alanine takes place both in the cytosol and in mitochondria. Furthermore, alanine formation from glutamine requires that a four-carbon compound of the TCA cycle is converted to a three-carbon glycolytic intermediate. The involved enzymes (phospho*enol*pyruvate carboxykinase and malic enzyme) are both cytosolic and mitochondrial implicating that either substrates or products may need to be transported between the compartments. The compartmentation of metabolism and the kinetics of metabolite transport between the compartments probably plays an important role in the physiology of mammalian cells. However, the factors that determine the localisation of reactions that can take place in more than one compartment remains to be established.

21.6.4 Inhibition by by-products

The metabolic by-products, lactate and ammonia/ammonium ions, are inhibitory to mammalian cells. Alanine is also a major metabolic by-product in many cell lines but is not believed to harm the cells. Ammonia/ammonium, even at 1 to 5 mM, which is easily reached in cell cultures, may slow down the growth rates of mammalian cells. Differences in ammonia/ammonium ion tolerance have been observed between growth phases of batch and continuous cultures. The mechanisms of ammonia/ammonium ion inhibition include interference with electrochemical gradients, changes in intracellular pH, apoptosis (cell death) and a futile cycle of ammonia/ammonium ions which increases the demand for maintenance energy. Another serious effect of ammonia/ammonium ions is the disturbance of the glycosylation pattern of the product. The major negative effect of lactate is caused by the decrease in pH that follows from its excretion to the culture medium. In processes with automatic pH control, as for example in a bioreactor, accumulation of lactate is not considered a problem as it normally does not reach inhibitory levels (ca. 60 mM).

21.6.5 Causes of overflow metabolism

Mammalian cells may excrete up to 90% of the consumed glucose as lactate in situations where excess glucose is present. This occurs even in completely aerobic conditions. Similarly, feeding cells excess amounts of glutamine leads to the accumulation of ammonia/ammonium ions and alanine and/or aspartate. An interesting question here is why the cells carry out this apparently wasteful type of metabolism? However, it must be remembered that most fast-growing, cultured mammalian cells are continuous cell lines, i.e. they carry mutations that de-regulate the proliferation control. Other, concomitant genetic changes, such as an increased capacity for glucose and amino acid consumption, affect the energy metabolism of the cells. Rapidly proliferating normal cells, e.g. cells from the immune system or intestinal cells, have the same type of metabolism. Therefore, it has been suggested that this feature, i.e. a high flux in the major metabolic pathways, is necessary for supporting a high growth rate, not in terms of quantity of energy or precursors but for increasing the sensitivity of the pathways to arising demands for precursor metabolites.

Limiting the supply of glucose, as for example in a chemostat or a fed batch culture, decreases total lactate production and the apparent yield of lactate from glucose ($Y'_{lac/glc}$) and increases the cellular yield coefficient for glucose (Y_{xs}). This indicates that under glucose limitation a larger part of the glucose consumed is used for oxidation and biosynthesis, similar to observations with micro-organisms. Limiting the amount of glutamine fed to a culture likewise decreases the amounts of ammonium and amino acids that are formed. If both glutamine and glucose are kept limiting (as with a double-limited, fed-batch culture) the production of lactate and ammonia/ammonium ions can be decreased simultaneously. Thus, limiting one or both of the two major substrates forces the cellular metabolism to become more efficient. Hence, understanding the interactions between glucose and glutamine catabolism and amino acid metabolism is essential for rational design of feed and control strategies in production processes.

21.7 Large-scale cultivation of mammalian cells

21.7.1 General conditions

Many mammalian cell lines can be cultivated in suspension culture in the same way as micro-organisms. However, some cell types, typically normal diploid cells, are anchorage dependent, requiring a surface to grow on. These cells may be grown on the inside surface of plastic or glass bottles or on the surface of microcarriers. **Microcarriers** are small solid spherical particles (diameter 100–200 μm) which can be suspended in liquid culture medium. **Porous microcarrier beads** are an alternative to obtain a high surface area to volume ratio. Cells growing in these carriers are protected against shear damage, but as they grow into layers, diffusion limitations of nutrients and (by-)products will develop.

Mammalian cells originate from the body of an organism and are consequently adapted to an environment that is kept in homeostasis. An artificial culture environment should therefore maintain its pH, dissolved O_2 and temperature within narrow limits. The pH optimum (pH 6.7 to 7.9) and tolerance (0.05 to 0.9 pH units) are dependent on the cell line. The range for the optimal dissolved O_2 concentrations is usually quite large; concentrations between 20–80% being appropriate. In general, growth may be negatively affected below 20% of air saturation; above 80% the O_2 concentration becomes toxic. As mammalian cells lack the rigid cell wall that bacteria have, they are sensitive to shear forces. Shear occurs not only because of stirring but also as a result of sparging. Cells attached to air bubbles are exposed to enormous forces when these bubbles leave the bulk liquid at the surface and burst due to decompression. Cells in sparged cultures can be protected from shear forces by using a medium with a high viscosity. This can be achieved through high cell densities ($>10^7$ cells ml^{-1}), addition of extra serum or components like Pluronic PF68. The main effect of surface active agents, such as Pluronic, is via the coating of rising air bubbles. Cells do not attach to bubbles coated with Pluronic to the same extent as to naked bubbles and thereby do not follow the bubble to the surface and to the deadly bursting zone.

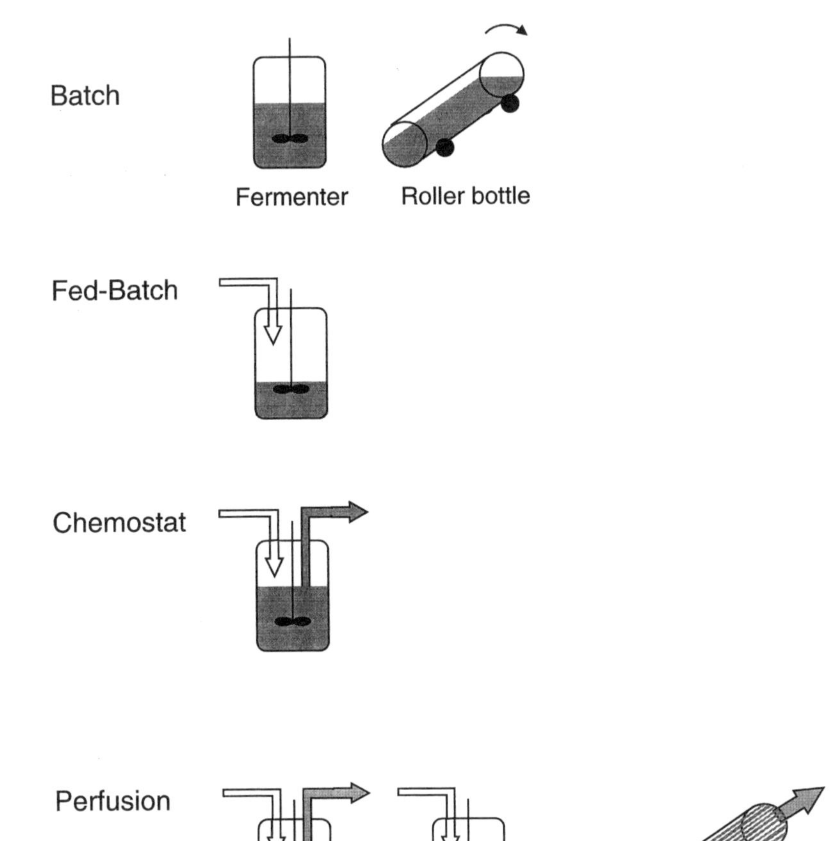

Fig. 21.5 Cultivation methods for mammalian cells. Open arrows indicate a flow of medium, thick black arrows represent a flow of culture fluid with biomass, light grey indicates culture fluid from which biomass has been removed.

Like micro-organisms, mammalian cells can be cultivated in batch, fed-batch or continuous mode. Continuous processes can be run as a chemostat, or as a perfusion culture. Perfusion systems are a specific mode of continuous cultivation in which the biomass is retained in the reactor whilst cell-free culture liquid is removed.

21.7.2 Batch

In batch cultures, the inoculum of cells is added to the total volume of medium to be used (Fig. 21.5). During growth, cells deplete the nutrients in the medium and excrete by-products (Fig. 21.6). Growth stops when a substrate is depleted or a by-product has reached inhibitory levels. However, in many cases it is not obvious why growth ceases.

Mammalian cells are routinely maintained in the laboratory by successive sub-cultures in stationary flat-bottomed plastic flasks, called T-flasks or Roux bottles, containing 10–100 ml medium, with a large surface-to-volume ratio. Anchorage-dependent cells will attach to the bottom of the flask so that further passages require that cells are

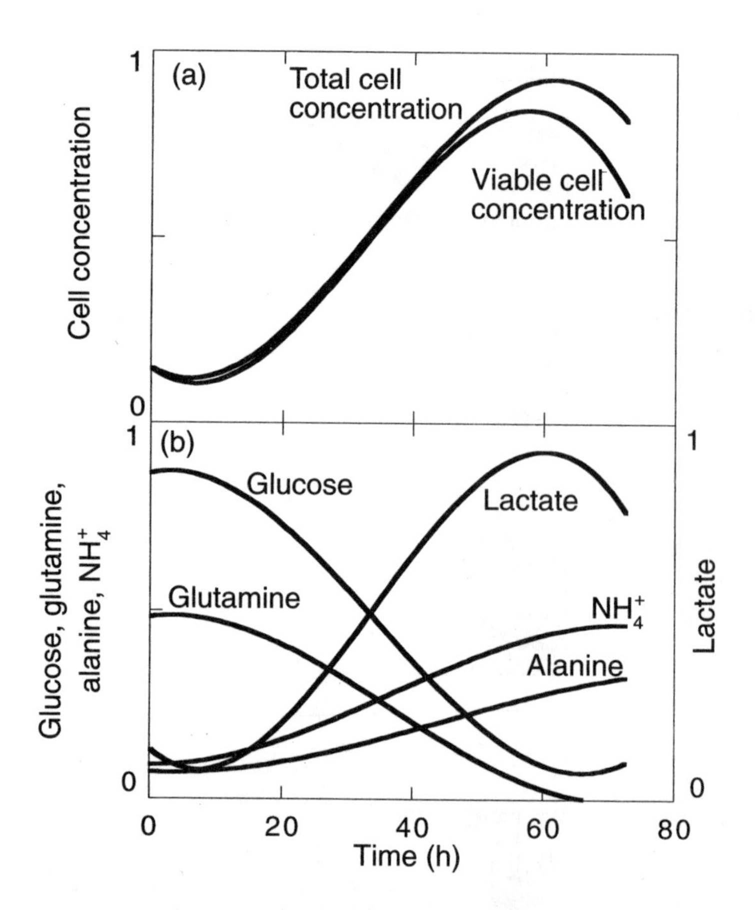

Fig. 21.6 Batch culture of hybridoma cells. (a) Increase in total and viable cell concentration. The inoculum cell density is about 0.2×10^6 cells ml^{-1} and the final cell density about 1.5×10^6 cells ml^{-1}. (b) Concentration profiles of glucose, glutamine, lactate, alanine and ammonium ions. The initial concentration of glucose is typically 5–25 mM and of glutamine 2–6 mM. The final concentration of lactate is typically 1.7 times the initial glucose concentration. Alanine and ammonium ions amount to 2–4 mM, depending on the initial glutamine concentration.

detached by using trypsin, a protease that 'dissolves' bridging proteins. Suspension cells will attach more loosely and can often be removed by shaking the flask. It is essential that the inoculum size is not too small. About 2×10^5 cells ml^{-1}, or more, are often used together with some spent medium which may contain secreted factors that stimulate the cells' own growth. Alternatively, the spent medium must be removed by centrifugation to dispose of inhibitory by-products.

For large-scale cultures of mammalian cells, 200 litres or less is often sufficient to satisfy the demand for high-value therapeutic proteins. Even so, scaling-up requires several intermediate steps, the first being the transfer of cells from stationary culture to shake flasks or spinner flasks. The spinner flask, equipped with a magnetically driven impeller hanging down from the lid without touching the bottom, was originally developed to provide gentle stirring of microcarrier cultures, but is now used for suspension cultures as well. The scale-up factor from stationary cultures, or shake flasks without pH control, is not more than five, often less, meaning that an inoculum volume of at least 20% must be used. In bioreactors, where higher cell densities are obtained, the scale-up factor can be up to 10 (i.e. the inoculum is 10% v/v or less).

In a bioreactor, a typical batch culture of hybridoma cells lasts 3 to 5 days and reaches a cell density of $2–5 \times 10^6$ cells ml^{-1} (corresponding to ca. 1 g dry wt cells l^{-1}). The maximum specific growth rate (μ) of hybridoma and myeloma cells is about 0.05 h^{-1}. The amount of monoclonal antibodies produced in a batch culture of hybridoma cells ranges from 10 to 100 mg of protein l^{-1}. A large-scale, batch production of monoclonal antibodies has been described with a 1 m^3 stirred tank reactor, in which cell densities of up to 5×10^6 cells ml^{-1} were obtained over 3.5 days. Early commercial production with anchorage-dependent cells was often performed in roller bottles (Fig. 21.5). Roller bottles are kept in constant motion by rotation and anchorage-dependent cells grow on the bottle surface. Typically a surface of 750–1500 cm^2 with 200–500 ml medium will yield $1–2 \times 10^8$ cells. A larger surface area is obtained by the use of microcarriers in stirred tank reactors.

21.7.3 Fed-batch

A fed-batch culture is, in the strict sense, controlled in the same way as a chemostat, i.e. the cellular growth rate is restricted by the dilution rate and the growth limiting substrate. The reasons for using the (substrate-limited) fed-batch technique in microbial processes is that O_2 limitation and overflow metabolism are avoided, resulting in much higher cell densities than batch cultures. Although a glucose- and glutamine-limited fed-batch culture also solves the problem of overflow metabolism in mammalian cells, it is not enough to bring about a substantial increase in cell density. By feeding a balanced mixture of nutrients, both the cell density and the product titre can be improved more than 10-fold as compared to batch cultivation. Fed-batch cultivations can last up to a month. Processes up to 15 m^3 have been described. Cell densities of around 1 to 1.4×10^7 viable cells ml^{-1} have been reported for fed-batch processes (see also Section 21.7.6).

21.7.4 Chemostat

Continuous, chemostat cultivation is characterised by the continuous addition of fresh medium and withdrawal of culture fluid, keeping the culture volume constant (Fig. 21.5), as described elsewhere in this volume (see also Chapter 6). In steady-state microbial cultures the relationship between the dilution rate (D) and the specific growth rate (μ) is expressed as $\mu = D$. However, in mammalian cell cultures, the viability of the culture must be taken into account. Proliferating cells not only replace the viable cells in the effluent stream, but also the cells that die within the culture. This leads to a steady state description for the growth rate:

$$\mu = D\,(N_t \times N_v^{-1}),$$

in which N_t is the total cell concentration (viable plus dead cells) and N_v is the viable cell concentration. From this relationship it follows that μ is greater than D when cell death occurs in the system. Generally, in microbial chemostat cultures a single nutrient is growth-limiting and the concentration of the growth-limiting substrate in the feed medium dictates the maximal biomass concentration in the culture. Further, the specific consumption rates of other nutrients are independent of the concentration of the limiting substrate in the feed medium.

In mammalian cell cultures, which contain multiple carbon and nitrogen sources, it is difficult to establish steady-state growth limited by a single nutrient. Although one of the energy sources, glucose or glutamine, can limit the biomass yield in a steady state culture, the specific consumption rates of other nutrients may, nevertheless, depend on the reservoir concentration of the energy source, or on the concentration of the individual nutrient. Growth of mammalian cells in chemostat cultures fed with complex media is therefore likely to result in multiple nutrient limitation.

Many aspects of mammalian cell physiology and medium optimisation, such as the influence of μ on product formation and the effects of dissolved O_2 concentration, pH, glucose and glutamine concentration, and amino acid and vitamin concentrations on growth and product formation, have been investigated using chemostat cultures. Chemostat production processes with up to $2\ m^3$ reactor volume have been described. Chemostat cultivation for production purposes has some disadvantages. The long duration of a culture, at least five weeks, creates a marked increase in contamination risk, and the time needed to re-establish a steady-state culture after contamination has occurred is longer than for re-starting fed-batch or batch processes. Moreover, validation of a process based on a continuous cultivation has to include proof that the cell line used is stable over the cultivation period.

21.7.5 Perfusion

In **perfusion cultures**, biomass is accumulated as the cells are retained within the reactor via a retention device, while fresh medium is introduced and spent medium removed. In this way, cell densities up to 3×10^7 cells ml^{-1} and product titres an order of magnitude higher than in batch cultures can be achieved. Devices to separate cells from the culture fluid can be placed inside or outside the reactor (Fig. 21.5). The latter option has the disadvantage that a substantial part of the culture is not in the controlled environment of the culture vessel itself. Several perfusion systems can be distinguished, based on the method used to separate cells and medium. **Spin-filter devices** make use of a rotating cage of wire mesh with pores of 5 to 75 μm. Spin-filters are prone to fouling, leading to a diminished flow rate through the filter and ultimately to a total clogging of the filter mesh. Alternatively, **membrane filters** (hollow fibres) can be used for separation of cells from the culture fluid. Fouling of such filters can occur too but may be remediated by back-flushing. Settling devices, utilising the slightly higher density of cells (compared to the medium) to separate cells from the culture fluid, have been developed. A specific device that uses gravity to keep cells in the reactor is the **acoustic filter**. This system uses static acoustic waves to concentrate cells in the effluent stream. Cells accumulate in the nodes of the wave and sediment back into the culture, against the up-flowing effluent stream. Finally, centrifugation as a means of cell retention has been applied to large-scale processes.

Hollow fibre culture systems can be considered a special type of perfusion culture in which the cells are physically separated from the medium flow (Fig. 21.5). Cells are grown in the extra capillary space of the unit, while fresh medium is fed through a large number of hollow membrane fibres, that pass through the unit. Cell densities of up to 10^8 cells per ml of extra-capillary space can be achieved and the effluent medium from this space contains a high product concentration. However, concentration gradients of nutrients and (by-)products are formed over the fibres. These gradients limit the possibilities of scaling-up hollow fibre units to large production reactors. Nevertheless, hollow fibre units are easy to use and have been successfully applied to commercial production processes (see Section 21.7.6).

21.7.6 Product quality and quantity

A product purified from a mammalian cell culture may not be 100% biologically active depending on variations in the glycosylation pattern or on proteolytic degradation. Both these parameters are influenced by the environmental conditions. The glycosylation pattern changes in response to many factors such as the mode of cultivation, the growth phase of a batch culture, whether cells are grown on microcarriers or in suspension, the glucose concentration, the ammonium concentration, the availability of hormones in the medium, the presence of serum, the protein and lipid content of the medium, pH and the O_2 concentration. Thus, choosing the appropriate physiological conditions in a production process is important for obtaining the correct glycosylation of a pharmaceutical protein.

Not only the quality of mammalian cell products but also the overall productivity of mammalian cell cultures is influenced by many parameters such as pH, ammonia/ammonium ion and lactate concentrations, serum concentration, cultivation method, culture age, inoculum size and medium composition. Due to the complexity of mammalian cell physiology, in combination with different media and cultivation methods that are used, it is often difficult to single out the influence of one specific factor. However, a parameter that clearly has a major effect on the **specific productivity** of mammalian cell products is the growth rate. Both growth-associated and non-growth associated production kinetics occur.

The specific productivity may also be enhanced by compounds that are not normal components of cell culture media. Several mammalian cell lines show a higher specific productivity in media where the osmolarity is increased from the normal 330 mOsmol to above 400 mosmol. Although not completely understood, this effect is, however, dependent on the cell line and basal medium used. Interestingly, addition of butyric acid has been reported to enhance productivity in mammalian cells. This may depend on the ability of butyric acid to arrest cells in the G1 phase of the cell cycle. Consequently, for those products that exhibit non-growth associated production kinetics, growth arrest will lead to increased productivity.

The amount of product made by a culture can be expressed as the percentage of the total amount of protein produced. With non-growth associated production, a large fall in this percentage occurs with increasing growth rate. For example, the specific rate of protein production in a hybridoma cell line was reported as 1.5 mg (10^9 cells)$^{-1}$·h at a specific growth rate of 0.02 h^{-1}. The amount of product made corresponds to 28% of the total protein. The same cell line had a much lower specific production rate [0.2 mg (10^9 cells)$^{-1}$·h] at a growth rate of 0.058 h^{-1}, i.e. only 1% of the total protein production went towards the product during these conditions. On the other hand, in a myeloma cell line producing a recombinant antibody with growth-associated kinetics, an increase in the percentage of product protein from 18% to 29% was observed as the growth rate increased from 0.016 h^{-1} to 0.042 h^{-1}.

The protein production in mammalian cell cultures can be as high as in micro-organisms, as is evident from the following comparison. Filamentous fungi are generally regarded as very good producers of excreted proteins (see Chapter 4). For example, an *Aspergillus oryzae* strain with a growth rate of 0.09 h^{-1} and a protein content of 40% produces 0.4 g biomass protein per hour. The specific productivity of α-amylase is 0.15 g (g dry biomass)$^{-1}$·h. The amount of excreted product therefore amounts to 27% of the total protein production.

The type of mammalian cell process that has been most successful so far, with respect to product concentration and productivity, is monoclonal antibody production with hybridoma or myeloma cells. As shown above, the production potential of mammalian cells is not the limiting factor but rather it is the attainable biomass concentration. To meet this demand, fed-batch cultures and hollow fibre reactors have been used to obtain high cell density cultures of hybridoma and myeloma cells. Glucose and glutamine limitation has been combined with feeding of amino acids and serum, resulting in a total cell concentration of approximately 5×10^7 cells ml^{-1} (of which less than half was viable) over 550 h, and a final antibody concentration of 2.4 g l^{-1}, i.e. giving a volumetric productivity of 0.1 g l^{-1}·day. Commercial production of monoclonal antibodies in hollow fibre reactors can yield about 700 g product per month at about 2 g l^{-1}. Each run lasts for about three months but the first run is not productive since this time is required for building up the biomass in the extra capillary space. The productivity in this system is 0.3 g l^{-1}·day during the harvest period.

21.8 Genetic engineering of mammalian cells

Genetic modification of mammalian cells can be used to introduce the genetic information needed for production of a specific protein or to improve the characteristics of a production cell line. There are many methods that can be used to introduce foreign DNA into a mammalian cell, amongst others are: electroporation, lipofection in which the DNA is introduced via liposomes, micro-injection of the DNA directly into the cell, fusion of the mammalian cell with a bacterial protoplast containing the DNA or viral vector systems. A transfected cell line will express the introduced DNA stably only if it is integrated in the genome. In contrast to micro-organisms, like *S. cerevisiae* and *E.coli*, the integration of the introduced DNA is mostly non-homologous. The gene encoding a protein product may therefore be integrated into regions of the genome that are not favourable for efficient expression of the gene. Selection for the best producing transfectants is therefore always necessary.

Several **selectable markers** for mammalian cell lines are available. Dominant markers that can be used irrespective of the host cell line are mostly concerned with drug resistance. Recessive markers, that are used in combination with a specific host cell genetic background, can involve enzymes of the salvage pathways of the purine and pyrimidine metabolism, drug resistance or amino acid metabolism. The two most successful systems are the glutamine synthetase (GS) system and the dihydrofolate reductase (dhfr) system.

The enzyme glutamine synthetase catalyses the formation of glutamine from glutamate and ammonium ions. The **GS gene** can be used as a selectable marker in hybridoma and myeloma cells and other cells that do not possess GS. Stable transfected cells will express the GS gene and are therefore able to grow in glutamine-free media. As with the dhfr system (see below) the GS system can be used to amplify the product gene, a procedure also leading to amplification of the GS gene. The metabolic consequence of this situation would be that the cell actually overproduces glutamine.

The **dhfr system** is mostly used in combination with a *dhfr*$^-$ CHO cell line. A dhfr$^-$ cell line is unable to synthesise tetrahydrofolate which is an essential cofactor in the one-carbon metabolism. *Dhfr*$^-$ cell lines are only able to grow in media containing thymidine, glycine and hypoxanthine, precursors and building blocks necessary to overcome this deficiency. Stable, transfected cells that express the *dhfr* gene are capable of growth in unsupplemented medium. Methotrexate (MTX) can be used to amplify the *dhfr* gene. This folate analogue inhibits the *dhfr* gene product. By selecting for cells capable of growth in a medium with increasing concentrations of MTX, cells with an increased number of gene copies, and thereby with enhanced expression of the *dhfr* gene product, are obtained. An enhanced expression of the product protein is obtained at the same time. A disadvantage of the *dhfr* system is that MTX resistance can develop that is independent of *dhfr* expression.

The introduction of foreign genes into mammalian cells is quite common, while the deletion of specific genes is not. As mammalian cells show **heterologous recombination**, the opportunities for site-specific insertions and deletions are lacking as is possible in yeasts and *E. coli*. Mutations to prevent expression of genes can be made by less specific classical methods, like UV treatment of cells, combined with selection for the desired phenotype as has been done for the generation of glycosylation mutants. A more recent approach to gene 'knock-out' is the use of **antisense oligo nucleotides** that hybridise with a specific mRNA, thereby preventing its translation into mature protein.

Genetic modification of mammalian cells for cell line improvement is not yet wide-spread but is increasing in importance. Areas of interest are the prolongation of productive cell life, growth in serum-free media, the decrease of by-product formation and glycosylation characteristics. Apoptosis, that occurs in most mammalian cell cultures, can be influenced by introducing the *bcl2* gene, an anti-apoptotic gene. This prolongs cell life, and thereby the productive phase of a process. An example of decreasing by-product formation is the introduction of the GS gene. Cells with GS produce less ammonia/ammonium as they can be cultivated in media without glutamine. As a result of this, the production of MAbs in hybridoma cells is increased. CHO cell lines with glycosylation mutations have been developed with the aim of generating a less heterogeneous glycosylation of the product formed by these cells.

21.9 Further reading

Butler, M. (ed.) (1991). *Mammalian Cell Biotechnology. A Practical Approach.* Oxford University Press, New York.

Spier, R. E. (ed.) (2000). *The Encyclopedia of Cell Technology.* John Wiley, New York.

Chapter 22

Biotransformations

Joaquim M. S. Cabral

22.1 Introduction

Biotransformation deals with the use of biological catalysts to convert a substrate into a product in a limited number of enzymatic steps. The establishment of an efficient biotransformation process requires the extensive examination of factors affecting the development of optimal biocatalysts, reaction media and bioreactors (Fig. 22.1).

There are many opportunities for industrial use of biological catalysts for biotransformations. These include not only the traditional hydrolytic (e.g. starch and protein hydrolysis) and isomerisation (e.g. glucose conversion to fructose) reactions but, more recently, synthesis of chiral compounds, reversal of hydrolytic reactions, complex synthetic reactions such as aromatic hydroxylations and enzymatic group protection chemistry and degradation of toxic and environmentally harmful compounds.

Biological catalysts when compared with chemical catalysts have the advantages of their regioselectivity and stereospecificity which lead to single enantiomeric products with regulatory requisites for pharmaceutical, food and agricultural use. They are also energy effective catalysts working at moderate temperatures, pressures and pH values.

Biotransformations have been performed by a variety of biological catalysts, such as isolated enzymes, cells, immobilised enzymes and cells. The developments of recombinant DNA technology have led to improvements in the enzyme production in different host organisms giving the bioprocess engineer a greater choice of biocatalyst option.

The optimal biocatalyst must be selective, active and stable under

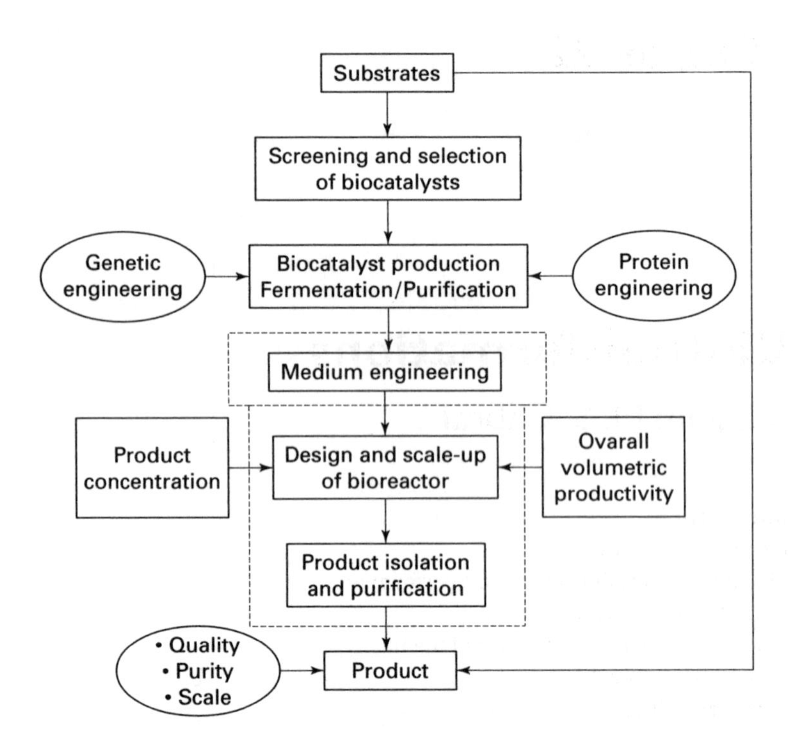

Fig. 22.1 Biotransformation process development.

operational conditions in the bioreactor, which may not be necessarily conventional in terms of composition, concentration, pressure and temperature. In particular it is necessary to evaluate the biocatalyst performance in non-conventional media (e.g. organic solvents and supercritical fluids).

A key issue is the availability of suitable biocatalysts. More rational screening and selection techniques are required to: (a) isolate biocatalysts, e.g. enzymes and cells, able to catalyse novel reactions of industrial interest, and (b) select and design catalysts suitable for industrial use with improved operational stabilities and kinetic properties. This requires a much greater understanding of the mechanisms of protein denaturation and decay of catalytic activities under process conditions and an evaluation of methods to maintain and improve biocatalyst stability, e.g. chemical modification, immobilisation and protein engineering.

In the optimisation of the overall process it is also important to enhance the predictability and performance of the biocatalyst in the reaction media in particular in multiphasic media for example involving a solid phase, e.g. immobilised biocatalyst, and one (aqueous) or two (aqueous and organic) liquid phases. It is very important to obtain reliable data and models on physical/chemical transport and interfacial phenomena. Medium engineering plays an important role in the definition of the optimal biocatalyst operation and to evaluate the effect of medium composition on the biocatalyst.

The optimal bioreactor should be simple, safe, well controlled, easy to design and flexible. The design of bioreactors requires knowledge of reaction kinetics as well as fluid dynamics, substrate dispersion and mass transfer. In addition for multiphase bioreactions, interfacial phenomena, substrate and product partitioning, and separation of two liquid phases should also be taken into account.

22.2 Biocatalyst selection

After selecting an appropriate starting material to be converted into the product, it is necessary to select the appropriate biocatalyst with suitable activity, selectivity and stability to work under the required operational conditions (temperature, salt concentration, pH, organic solvents, substrate and product concentrations). Several strategies can be followed to obtain the biocatalyst for the pertinent biotransformation: (a) screening for novel biocatalysts; (b) use of existing biocatalysts; and (c) genetic modification of existing biocatalysts.

22.2.1 Screening for novel biocatalysts

Selection of new micro-organisms with novel activities is still worthwhile taking into account the overwhelming biochemical diversity present in nature. To screen large numbers of organisms requires that cheap, simple, rapid and selective detection methods, preferably capable of some automation, should be available to facilitate this usually tedious process.

Selective selection methods for colonies on plates can be very useful, as shown for the isolation of micro-organisms able to hydroxylate L-tyrosine to L-DOPA, a drug used in the treatment of Parkinson's disease. The colonies which produce L-DOPA turn violet-black as a result of the reaction of L-DOPA with ferrous ions added to the agar plates.

Microbial selection has also been performed in the presence of high concentrations of the target compound. This approach was used to isolate benzoic acid-assimilating strains for the production of *cis,cis*-muconic acid from benzoic acid. Similar approaches have been followed to isolate nitrile-hydrolysing enzymes, such as nitrile hydratase, nitrilase and amidase, which have great potential as catalysts for producing high value amides and acids from the corresponding nitriles.

The resistance to organic solvents is often an important criterion in the selection of a suitable biocatalyst. *Pseudomonas* strains have been isolated with the ability to grow in the presence of toluene and aromatic and aliphatic hydrocarbons and long chain alcohols. These strains and their enzymatic activities are therefore important biocatalytic sources for the degradation of harmful compounds as well as for the synthesis of important chiral compounds.

22.2.2 Use of existing biocatalysts

A well-known way to accomplish a desired biotransformation is the use of existing biocatalysts (e.g. commercial enzymes) on natural and unnatural substrates. The substrate specificities of lipases and proteases are currently under intense investigations. The hydrolytic capacity of lipases is not restricted only to triacylglycerols. This type of enzyme is

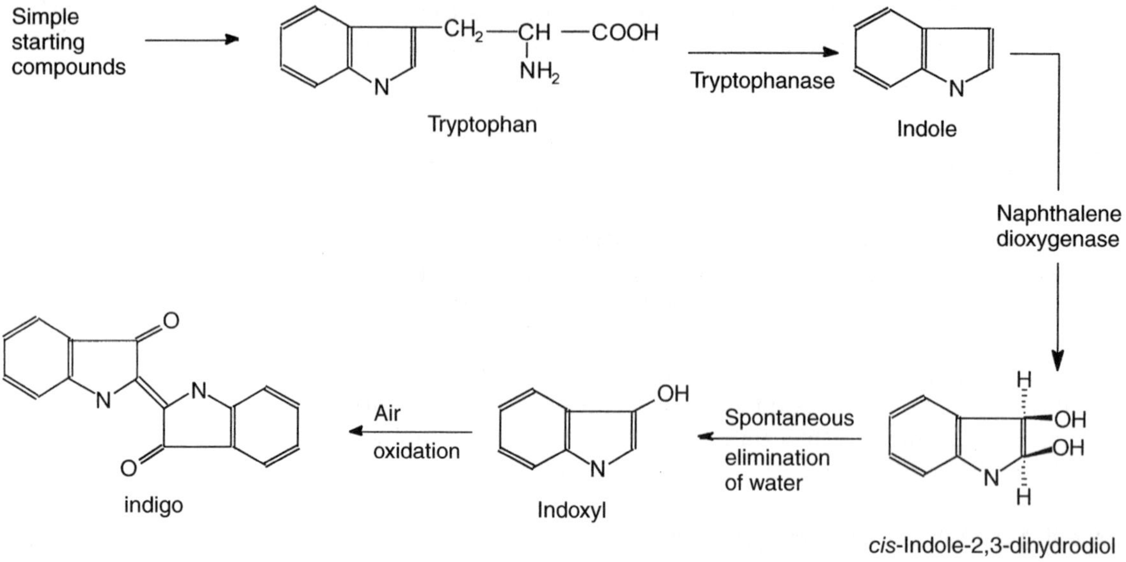

Fig. 22.2 Metabolic pathway engineering for indigo biosynthesis.

also able to hydrolyse mono-, di- and triacyl esters with different chain lengths of the various acyl groups.

The exploitation of existing enzymes under different reaction conditions could lead to the finding of a biocatalyst for the desired biotransformation. For example, lipases have been used to perform synthetic reactions in media under controlled water activity, e.g. esterification, inter-esterification and trans-esterification reactions. Methods to optimise the enantioselectivity of lipases have been reported, namely the non-covalent modification of lipase and the control of the surface tension of an emulsion.

22.2.3 Genetic modification of existing biocatalysts

A distinct way to obtain a biocatalyst is by *in vivo* (**metabolic pathway engineering**) and *in vitro* (**protein engineering**) construction of a novel biocatalyst. *In vivo* genetic engineering has been applied in large scale to obtain a recombinant organism with the desired enzymatic activity. Mutational events leading to the novel enzyme activities include transfer of genes, gene duplication, gene fusion, recombination between genes, deletion or insertion of gene segments, and one or more single site mutations, or combination of these activities. An example of this metabolic pathway engineering is the production of dyes, such as the indigo biosynthesis in *E. coli* (Fig. 22.2). By assembling, on a single operon, genes encoding for tryptophan formation, the gene specifying tryptophanase and a fragment of the NAH plasmid of a *Pseudomonas* encoding the naphthalene dioxygenase, a recombinant *E. coli* was obtained which was able to synthesise indigo from simple starting compounds.

Another approach is the use of protein engineering to modify an existing protein/enzyme or create *de novo* a protein of pre-specified properties. The protein engineering process can be viewed as an interactive cycle of several interconnected steps (protein engineering cycle). The aim of protein engineering has been to elucidate the structure–function relationship of proteins and to use this information to develop novel/modified proteins (enzymes) with improved characteristics for process applications. An elucidative example is the design of subtilisin mutants with altered properties (substrate specificity and pH activity profile) and improved thermal and oxidative stabilities. For example in subtilisin BPN′, two methionines, Met^{124} and Met^{222}, are especially susceptible to oxidation. To prevent the negative influence caused by the formation of methionine sulphoxide, Met can be replaced, using site-directed mutagenesis, by a non-oxidative amino acid, such as Ala, Ser or Leu, without losing more than 12–53% of the initial activity. The mutant Met^{222} – Ala^{222} is currently in use as a commercial detergent enzyme: 'Durazyme'.

22.3 Biocatalyst immobilisation and performance

22.3.1 Biocatalyst immobilisation

The immobilisation of biocatalysts for laboratory studies, analytical and medical applications and large-scale industrial processes is presently a widespread technique. Immobilisation can be defined as the confinement of a biocatalyst inside a bioreaction system, with retention of its catalytic activity and stability, and which can be used repeatedly and continuously. Table 22.1 lists some advantages and limitations which can arise from the use of immobilised biocatalysts.

The biocatalysts which can be immobilised range from purified enzymes to viable microbial cells, animal and plant tissues. Isolated enzymes can give high activities per unit mass or mole, high specificity and minimum side reactions. They are, however, often difficult and costly to prepare. In addition, they are frequently unstable and, in many cases, require parallel co-factor regenerating systems. Due to their relatively simple chemical nature, as compared to organelles or whole cells, isolated or partially purified enzymes are the biocatalysts most extensively studied in relation to immobilisation. Immobilised, purified enzymes find suitable applications in developing biosensors and preparing high added-value substances, such as chiral compounds. In more crude forms, immobilised enzymes are also used in large-scale applications in the carbohydrate, food and pharmaceutical industries.

Multi-enzyme systems, such as organelles, whole cells or cell tissues, have some clear advantages for immobilisation over isolated enzymes. They can be efficiently retained by mild, physical means, preserving, in adequate conditions, the enzyme-synthesising and co-factor regenerating capabilities and producing a suitable micro-environment for single and multiple enzymatic activities. However, the efficient use of immobilised cells relies on the control of metabolic and physiological alterations throughout the retention procedure and the subsquent catalytic process. The major large-scale utilisations of immobilised cell systems take advantage of the natural tendency of many microbial species to flocculate or to adhere to solid surfaces. Other applications are

Table 22.1 Advantages and limitations on the use of immobilised biocatalysts

General aspects		Specific aspects
Advantages		
Retention of the biocatalyst in the bioreactor		Possible biocatalyst re-use Product contamination avoided High dilution rates allowed without biocatalyst wash-out
High biocatalyst concentration		Increased volumetric productivity Rapid conversion of unstable substrates Minimised side-reactions
Control of biocatalyst micro-environment		Manipulation of biocatalyst activity and specificity Stabilisation of biocatalyst activity Protection of shear-sensitive biocatalysts
Facilitated separation of the biocatalyst from the product		Precise control of bioreaction time Minimisation of further product transformation
Limitations		
Increased costs of biocatalyst production		Increased requirements of materials and equipment Need for specific reactor configurations
Loss of biocatalyst activity during immobilisation	Biocatayst-related	Exposure to pH and temperature extremes Exposure to toxic reactants Exposure to high shear or mechanical strain
	Micro-environment related	Exclusion of macromolecular substrates Blocking of the enzymatic active site Local pH shifts Mass transfer limitations
Loss of biocatalyst activity during bioreactor operation	Leakage of biocatalyst	Matrix erosion or solubilisation Small support particles carried in the outflow Cell growth inside the matrix Broad pore-size range
	Matrix poisoning or fouling	Build-up of inhibitors in the micro-environment Retention of suspended solids Growth of contaminating species (biofilms) Need for a stricter control of feed composition
Empiricism		Need for case specific, multi-parameter optimisation Difficult process modelling and control

restricted to single-enzyme transformations with non-growing cells in the manufacture of pharmaceuticals and amino acids.

22.3.2 Methods for biocatalyst immobilisation

A wide range of basic immobilisation procedures with their specific variations has been described in a large number of reviews. Several classification schemes have also been proposed, one of which is given in Fig. 22.3.

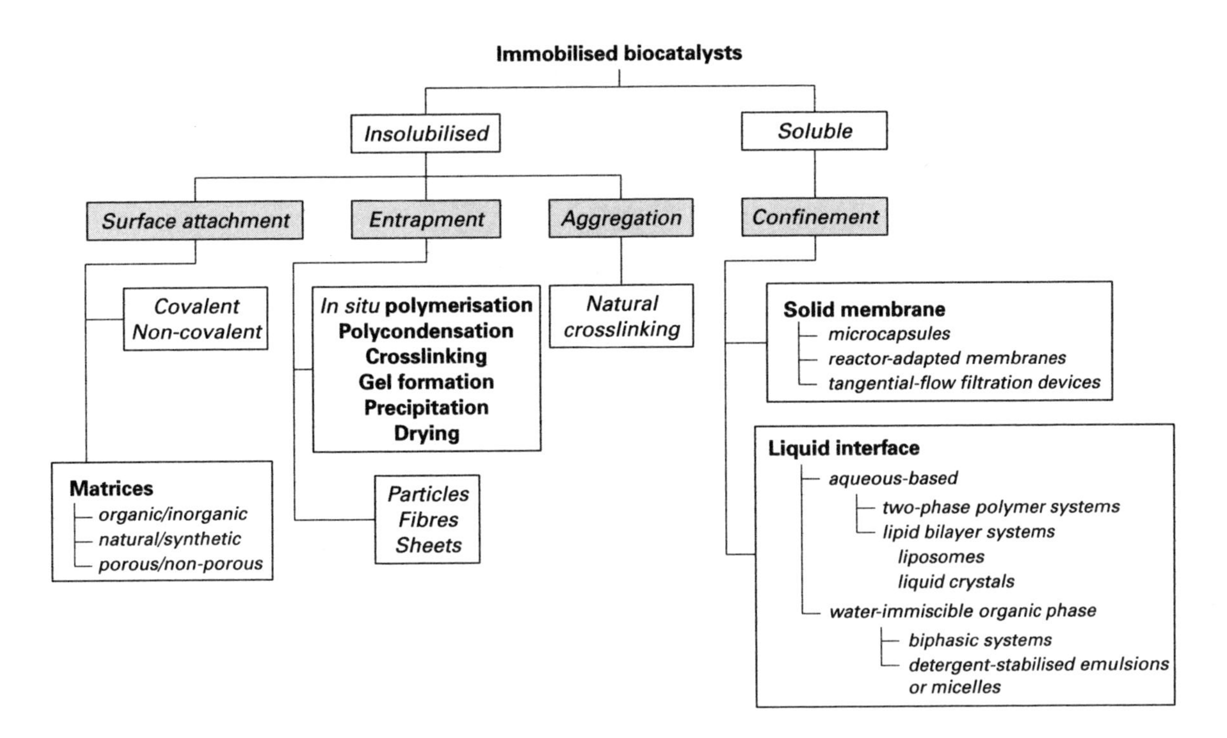

Fig. 22.3 General methods for biocatalyst immobilisation.

Cross-linking with bi-functional reagents

Both cells and enzymes can be covalently cross-linked with bi- or multifunctional reagents, such as aldehydes or amines. However the toxicity of these reagents limits their applicability to the immobilisation of non-viable cells and enzymes. This method produces three-dimensional, crosslinked enzyme aggregates, which are then insoluble in water. Glutaraldehyde has been the most extensively used crosslinking reagent which reacts with the lysyl residues of the enzyme forming a Schiff's base :

$$\text{Enzyme-NH}_2 + \text{OHC(CH}_2)_3\text{CHO} + \text{H}_2\text{N-Enzyme} \rightarrow$$
$$\text{Enzyme-N=CH(CH}_2)_3\text{CH=N-Enzyme}$$

The linkages formed between enzyme and glutaraldehyde are irreversible and survive extreme values of pH and temperature, which suggests that the aldimine bond is stabilised.

Biocatalyst crosslinking with glutaraldehyde is critically dependent on a delicate balance of factors such as the concentration of the biocatalyst and crosslinking reagent, pH and ionic strength of the aqueous solution, temperature and time of reaction. The most important advantage of this method is that only a single reagent is required and the reaction is easy to carry out. This method has been used successfully to immobilise industrial biocatalysts such as glucose isomerase and penicillin amidase.

Supported immobilisation methods

The available methods for biocatalyst immobilisation, involving solid supports, fall into two general categories: **surface attachment** and

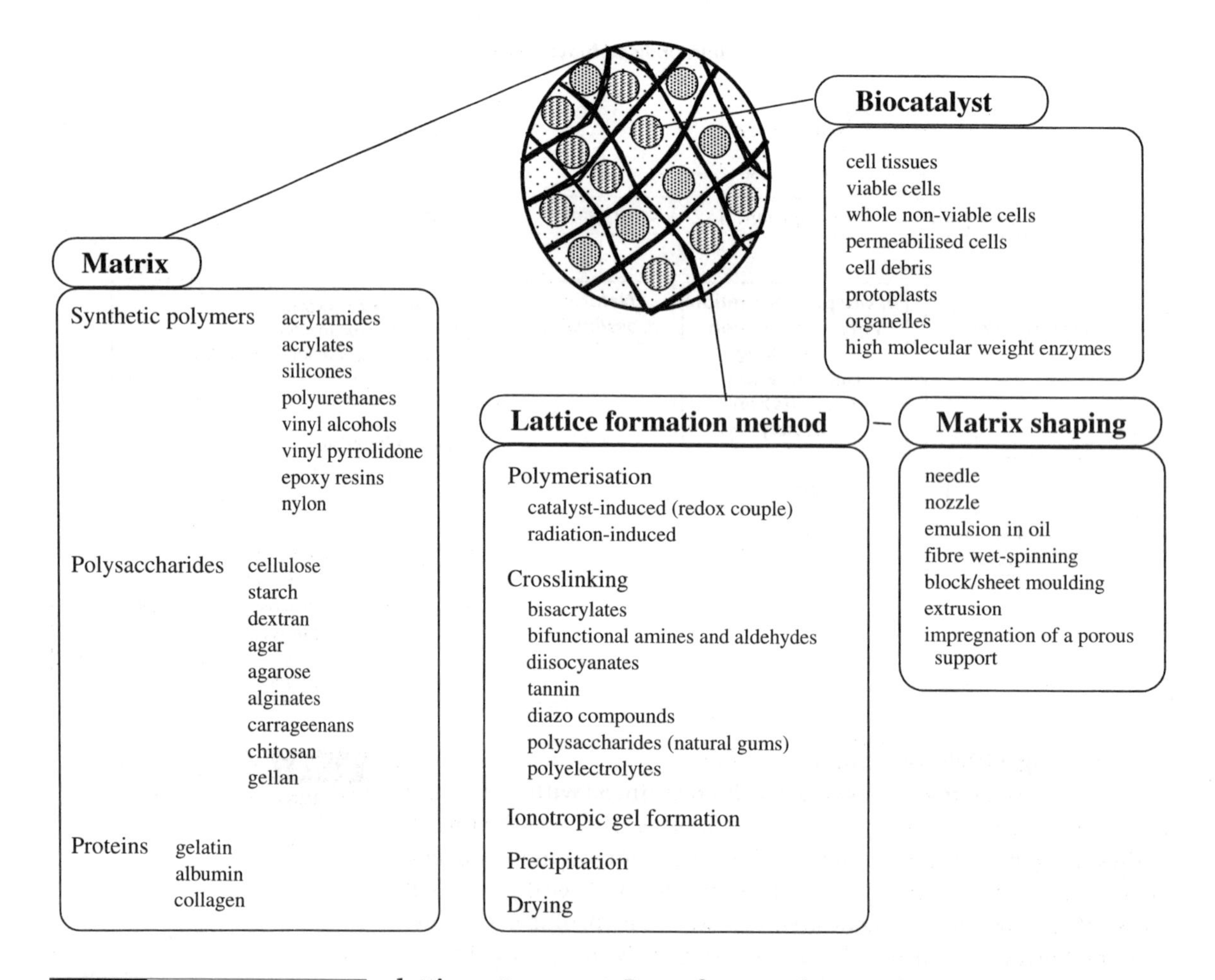

Fig. 22.4 Overview of surface attachment of biocatalysts.

lattice entrapment. By surface attachment, the enzyme, organelle or cell is bound to a solid interface through interactions which range from weak van der Waals forces to essentially irreversible, covalent bonding. The milder interactions can result from direct contact, in suitable conditions, between the biocatalyst and a natural, unmodified surface. However, the versatility and effectiveness of surface immobilisation have been greatly increased by introducing synthetic carriers and chemical modifications to natural and fabricated matrices. In lattice entrapment, a chemical or physical solidification process is induced in a solution containing the biocatalyst, ideally resulting in a water-insoluble lattice retaining the biocatalyst in its active or viable form. Mechanisms like polymerisation, thermal gelation or precipitation can be employed in this type of procedure. Figures 22.4 and 22.5 give a general overview of biocatalysts, supports and retention methods used in surface attachment and entrapment.

Supports for biocatalyst immobilisation

The development of a useful, support-immobilised biocatalyst necessarily involves a choice of a solid support. Ideally, this selection step should be based on established structural and activity data for the biocatalyst

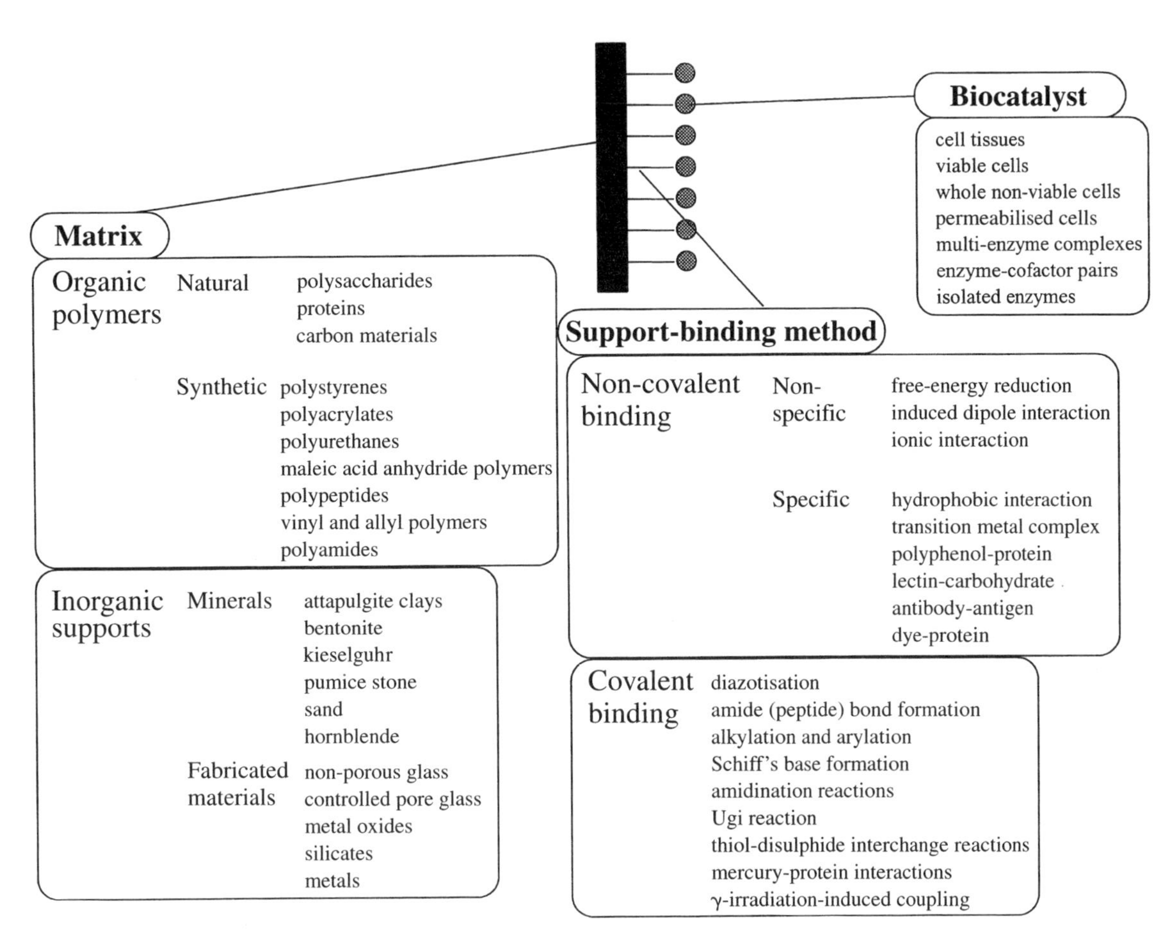

Fig. 22.5 Overview of lattice entrapment of biocatalysts.

and the general immobilisation method to be used, and process conditions. The important factors for examining a broad range of possible supports are summarised in Tables 22.2 and 22.3. Because several factors have usually to be considered when integrating biocatalyst immobilisation in a process, the optimal solution is frequently a compromise. In view of this, a support with a more flexible character is most often used. For example, a porous support can immobilise large biocatalyst loads, however to avoid diffusional limitations, this same support should be used as very small particles (see Section 22.3.3). In other cases, a single or few factors determine the choice of the support. Such is the case of systems where the aim is to preserve biocatalytic activity in the presence of aggressive components in the reaction medium, such as toxic species, organic solvents or strong inhibitors. With these systems, a porous matrix entrapping the biocatalyst and excluding the inhibitor is often the only efficient choice, regardless of the diffusional hindrances slowing down the reaction. Here, the possibility of changing the shape, porosity or hydrophobicity of the support can be advantageous in fine-tuning substrate or non-substrate size exclusion and external and internal mass transfer rates.

Organic polymers are the most widely employed supports for biocatalyst immobilisation (for examples, see Figs. 22.4 and 22.5). This preference

Table 22.2 Important aspects in the chemical nature of potential supports for biocatalyst immobilisation

	Chemical nature / Origin			
	Organic		Inorganic	
	Natural	Synthetic	Mineral	Fabricated
Availability of reactive functional groups				
Usable with large variety of biocatalysts	++	+++	+	+
Wide range of techniques for surface activation	++	+++	+	+
Commercially available pre-activated supports	+++	++	–	+
Usable with lattice formation techniques	+++	+++	–	–
Possibility of adjusting hydrophilic/hydrophobic character	+	+++	+	+
Sensitivity to physical, chemical and microbial agents				
Resistance to changes in reaction medium composition (pH, ionic strength, organic media)	+	++	+++	+++
Resistance to high temperatures	–	+	+++	+++
Resistance to large hydrostatic or hydrodynamic pressures	–	++	+++	+++
Regenerability	–	+	+++	++
Low cost / availability	+++	+	+++	+
Obtainable support morphology				
Usable diameter or thickness ranges	++	+++	+	+++
Available porosity ranges	+	++	++	++
Obtainable shapes				
Sphere	++	++	–	+++
Fibre	++	+++	–	+++
Sheet / membrane	++	+++	–	+++

Notes:
–, inadequate; +, poor; ++, fair; +++, good.

derives from the adaptability of these supports to nearly all kinds of surface-binding or entrapment techniques and to the broad variety of their chemical and physical characteristics. Their major drawbacks come from insufficient mechanical and chemical resistances, which limit both their use under harsher conditions and their regenerability.

Adsorption and ionic binding

Adsorption on a support is the oldest and simplest retention method for biocatalysts, involving no previous modification of the solid surface and relaying on weak interactions of van der Waals, electrostatic, hydrophobic or hydrogen-bond types. This is a low-cost procedure largely retaining the native conformation of an immobilised enzyme and its intrinsic catalytic activity. The general application of physical adsorption methods is, however, severely limited by the reversible nature of the biocatalyst-support bond, which is critically dependent on process conditions, such as temperature, pH, ionic strength and dielectric constant.

Table 22.3 Important aspects of the morphology of potential supports for biocatalyst immobilisation

Characteristics	Morphology			
	Porous			Non-porous
Total surface area available per unit weight	+++			+
Low incidence of diffusional limitations	+			+++
Attainable biocatalyst load	+++			+
Biocatalyst protection from external aggressions	++			+
Usable with macromolecular substrates	+			+++
	Fabricated			
	Mineral	Inorganic	Gel	
Pore size uniformity	+	+++	++	not applicable
Pore size stability	++	+++	+	not applicable
Low cost	+++	−	++	+++

Notes:
−, inadequate; +, poor; ++, fair; +++, good.

This factor makes it difficult to operate large scale bioreactors without significant biocatalyst leakage which then result in loss of productivity and product contamination. On the other hand, reversible adsorption methods allow straightforward regeneration of the supports. After adsorption, the enzyme may be cross-linked, however, this limits the possibility of reusing the support.

A growing field for applying physically adsorbed enzymes is non-aqueous biocatalysis. By using an organic solvent in which the protein is insoluble, an enzyme bound to a carrier by simple adsorption undergoes virtually no desorption in prolonged processes. A lipase from *Rhizomucor miehei* was adsorbed on a polyacrylate support with a very high activity retention (90%) and used in hydrolytic and synthetic reactions. An industrial example is the use of a *Rhizopus* lipase adsorbed onto Celite for the continuous production of cocoa butter-like fats in organic media.

A slightly stronger biocatalyst-support linkage can be achieved by the use of ionic supports. Their advantages and limitations are mostly the same as those of physical adsorption. The stability of the established ionic bonds is particularly sensitive to pH and ionic strength. Glucose isomerase from *Streptomyces rubiginous* was adsorbed onto an anion-exchange resin consisting of DEAE-cellulose agglomerated with polystyrene and TiO_2. This process was industrially implemented for the isomerisation of glucose into fructose.

Covalent coupling of enzymes

Probably the most throughly investigated approach to enzyme immobilisation involves covalent binding of amino acid residues in the protein to reactive groups in the support. In principle, the wide variety of surface activation and coupling reactions available makes it a generally applicable method. However, the high cost of materials, the often complicated procedures and the almost unavoidable loss of part of the catalytic activity, restrict practical applications of covalent surface immobilisation to specific cases with outstanding advantages.

Covalent coupling of enzymes to supports produces highly stable conjugates with no protein leakage over a wide range of operational conditions. Trypsin, penicillin acylase and lipases immobilised by multipoint covalent attachment onto CNBr-activated agarose yielded derivatives much more stable (300- to 50 000-fold) than their free counterparts. Ideally, the binding reaction should not interfere with the amino acids residues at the active site of the enzyme and should not significantly distort the native protein conformation nor alter its flexibility, either as a result of single or multipoint linkages. To meet these requirements, methods include multi-stage activation steps. Although ten different amino acids residues from enzymes can, in principle, be used for covalent coupling, most procedures are targeted at amino, thiol, phenolic and hydroxyl groups. Some of the basic coupling reactions are presented in Fig. 22.6.

Lattice entrapment

The immobilisation of biocatalysts within the lattices of solid matrices aims to take the advantage of the size difference between the substrates, or products, and the biocatalyst so as to achieve total retention of the latter while the former move freely between the bulk medium and the catalytic site. In lattice entrapment, the biocatalyst is not generally subject to strong binding forces; no structure distortion or active site blocking take place. Some deactivation can, nevertheless, occur during the immobilisation process due to pH and temperature changes and contact with aggressive monomers or solvents. Matrices are formed in the presence of the biocatalyst by *in situ* polymerisation, starting from the appropriate monomers, or by solidifying polymer solutions through ionic (alginates) or covalent (polyurethanes) crosslinking, cooling (agarose, gelatin), drying or induced (carrageenan) precipitation (Fig. 22.7). Critical parameters in such processes are always related to optimising pore size and its influence on support rigidity and substrate diffusion within the lattice.

Several attempts have been made to entrap isolated enzymes, however, due to enzyme leakage, this immobilisation method is mainly used for whole cell immobilisation. Several industrial examples include the use of entrapped cells for the production of amino acids (L-aspartic acid, L-isoleucine), L-malic acid, hydroquinone and acrylamide.

Immobilised soluble enzyme and suspended cell methods

All the methods of biocatalyst immobilisation described so far involve the modification of the biocatalyst (enzyme) or its micro-environment, with subsquent alteration of its kinetics and catalytic properties. In order to use a biocatalyst in its native state continuously over a long

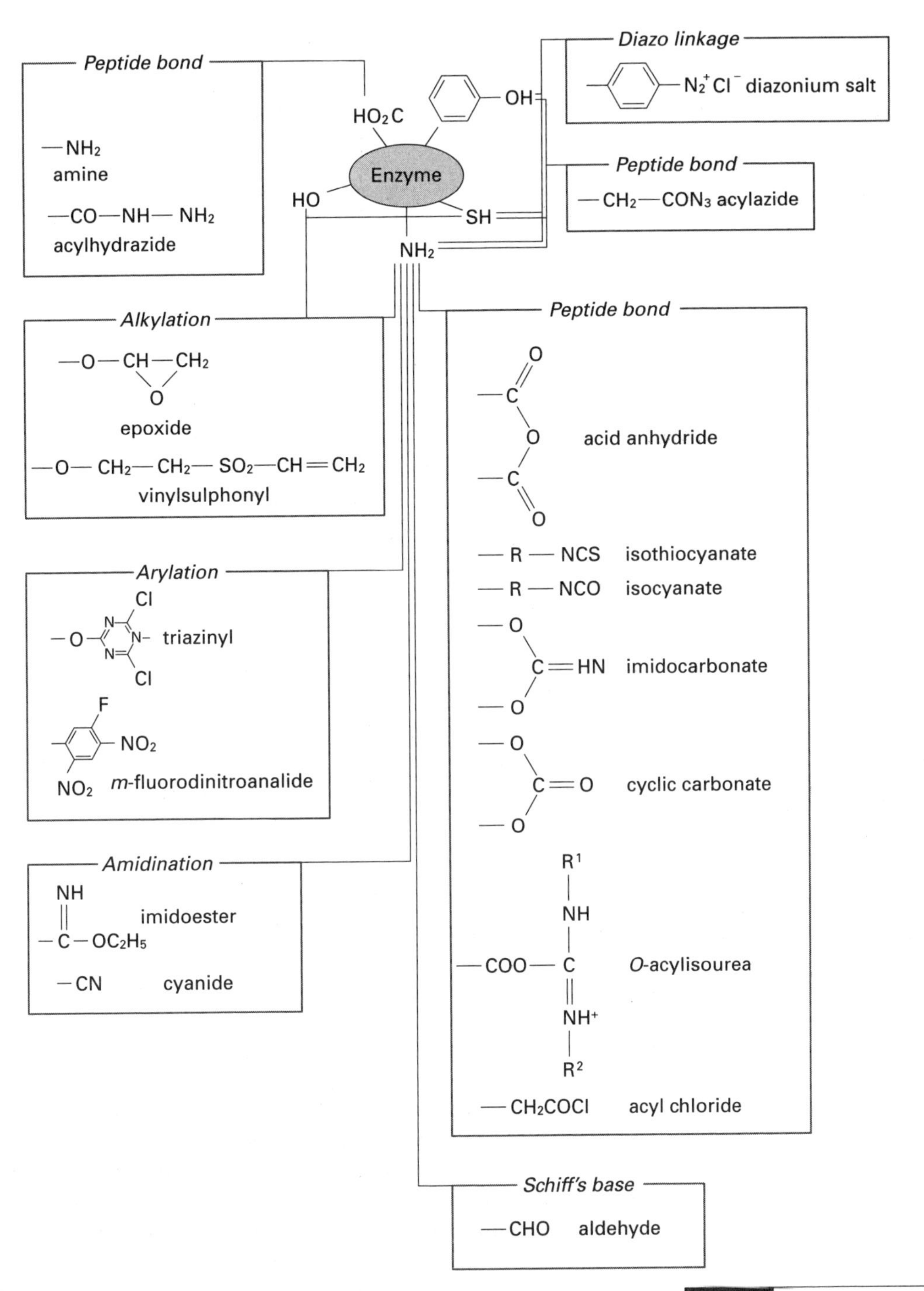

Fig. 22.6 Examples of covalent coupling of enzymes to active groups on supports.

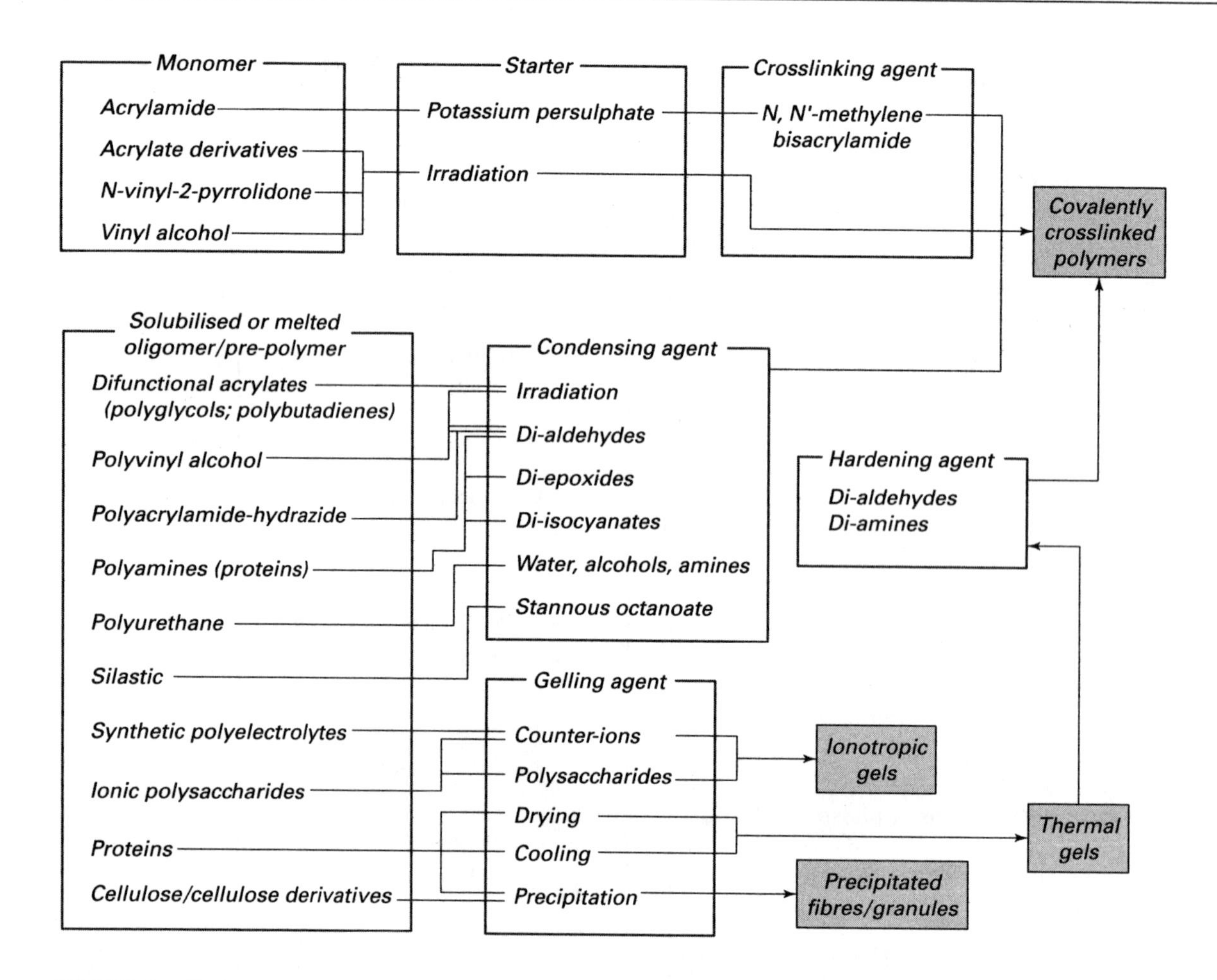

Fig. 22.7 Examples of lattice-formation strategies for entrapping biocatalysts.

period of time, biocatalysts have been confined within semipermeable membranes in the form of hollow fibres or flat sheet ultrafiltration membrane reactors (Fig. 22.8). The membrane retains the biocatalyst but is permeable to the products and sometimes to the substrates. This method offers several advantages relative to other immobilisation methods. Chemical modification of the biocatalyst is not necessary and the biocatalyst retain its kinetic properties.

This method is particularly suited for conversion of high molecular weight or insoluble substrates, such as starch, cellulose and proteins, as it allows the intimate contact of the biocatalyst with the substrate achieving an efficient conversion of the substrates. However some disadvantages are inherent in the method: the possible decrease in the reaction rate as a result of the permeability resistance of the membrane; and the adsorption of the biocatalyst and/or substrates and products on the membrane surface. This type of immobilisation has found applications on the modification of fats and oils (e.g. olive oil, palm oil) by lipases and dipeptide synthesis (acetylphenylalanine-leucinamide) by proteases in organic media.

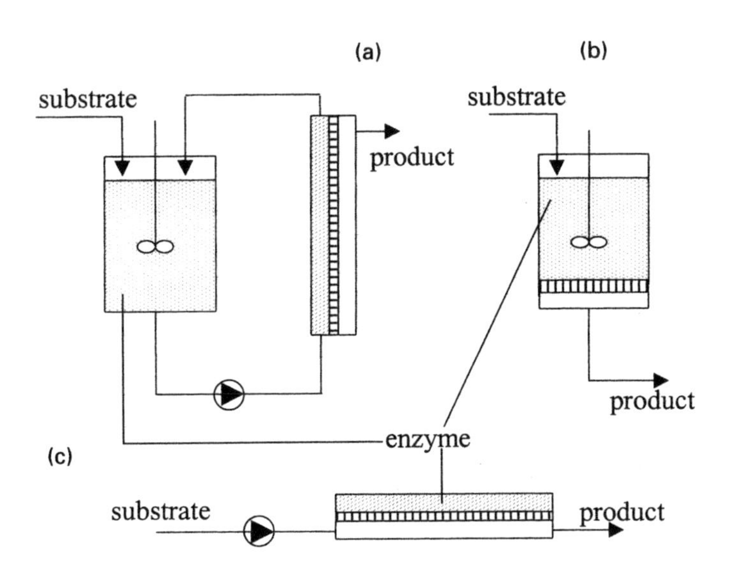

Fig. 22.8 Enzyme membrane reactors: (a) continuous stirred tank reactor with recirculation; (b) dead-end cell; and (c) tubular.

Immobilisation of multi-enzyme systems and cells

One of the strong disadvantages of single enzyme systems, free or immobilised, is their limitation to single step transformations. This limitation is particularly acute with thermodynamically unfavourable conversions and those requiring the regeneration of enzyme co-factors, such as redox reactions or phosphorylations. Among the possible solutions is coupling the intended enzymatic reaction to a chemical one or to a second enzymatic reaction. The latter alternative has been investigated for several NAD(P)- or ATP-dependent systems, using a pair of enzymes and two substrates with the co-factor shuttling between them. When two or more biotransformations are carried out simultaneously in the same vessel – to regenerate co-factors, to shift thermodynamic equilibria, or to favour process economics – the intermediate products should be rapidly converted. In this context, co-immobilisation of the involved enzymes is likely to minimise the diffusion paths of the intermediates between active sites, thus accelerating the potential rate-limiting steps. However, such systems suffer from severe problems related to the correct relative positioning of the immobilised enzymes and co-factor which are extremely difficult to achieve in practice. An example of this type of immobilisation method is the synthesis of L-tertiary leucine (Fig. 22.9), a chiral intermediate for chemicals.

The immobilisation of the multienzyme systems contained in whole cells or cell particles can lead to marked improvements in a process when compared to using free cells or immobilised single or multiple enzymes. However, a clear distinction should be made between the cases in which cell viability is indispensable and those employing whole cells or cell parts as non-viable, crude preparations of single-activity biocatalysts.

In the first situation, one deals with sensitive catalytic forms which require very mild immobilisation and close control of operating conditions in order that cell viability is preserved. This situation corresponds,

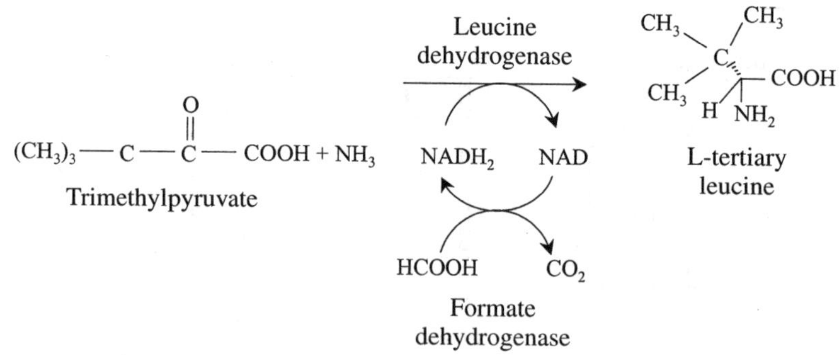

Fig. 22.9 Synthesis of L-tertiary leucine from trimethylpyruvate and ammonia.

$(CH_3)_3-C(=O)-C-COOH + NH_3$ (Trimethylpyruvate) → Leucine dehydrogenase ($NADH_2$ → NAD; HCOOH → CO_2, Formate dehydrogenase) → L-tertiary leucine

for example, to the immobilised cell fermentations and the culture of anchorage-dependent mammalian cells.

Immobilised, non-viable cell systems are sometimes preferred to immobilised single enzymes to avoid costly purification processes, or to increase catalytic stability and to retain lattice-entrapped enzymes more efficiently without the need for tight control of matrix porosities. With this type of biocatalyst, entrapment or attachment procedures designed for enzymes can be safely used, though lower productivities per unit weight of biocatalyst are to be expected. In addition, diffusional resistances are enhanced by the cell membrane, and permeabilising treatments with heat, surfactants or solvents are often required. Such treatments can also be required to inactivate contaminating enzyme activities in the cells. On the whole, single-activity, immobilised cell preparations are convenient for industrial applications, where cost reduction is necessary and process control as well as sanitisation procedures are feasible.

One important and industrial example is the immobilisation of glucose isomerase, being an expensive intracellular enzyme, the producing cells were successfully immobilised and used continuously at industrial scale in packed bed reactors. A heat treatment was included which degraded other enzymes and thus avoided substrate (glucose) and product (fructose) degradation. Other examples are listed in Table 22.4.

22.3.3 Effect of immobilisation on the enzyme kinetics and properties

Although enzyme immobilisation can be very useful, immobilisation may also change the kinetics and other properties of the enzyme, usually with a decrease of enzyme-specific activity. This may be ascribed to several factors: (i) conformational and steric effects; (ii) partitioning effects; and (iii) mass transfer or diffusional effects.

Conformational and steric effects

The decrease of specific activity of enzymes, which occurs on their binding either to solid supports or upon inter-molecular cross-linking, is usually attributed to conformational changes in the tertiary structure of the enzymes. For instance, covalent bonds between the enzyme

Table 22.4 Examples of immobilised whole cells for single- or two-enzyme conversions of industrial interest

Microbial biocatalyst	Immobilisation method	Application
Escherichia coli	Entrapment	L-Tryptophan production from indole and DL-serine
Bacteria and yeasts	Entrapment	Biosensors
Escherichia coli	Entrapment	L-Aspartic acid production from fumaric acid and ammonia
Arthrobacter simplex	Entrapment	Prednisolone production from hydrocortisone
Rhodococcus rhodocrous *Pseudomonas chlororaphis*	Entrapment	Acrylamide production from acrylonitrile
Saccharomyces cerevisiae	Surface attachment	Sucrose hydrolysis
Humicola sp.	Entrapment	Conversion of Rifamycin B to Rifamycin S
Zymomonas mobilis	Entrapment	Sorbitol and gluconic acid production from glucose and fructose
Pseudomonas AM1	Entrapment	L-Serine production from glycine and methanol

and the matrix can stretch the enzyme molecule and thus the three-dimensional structure at the active site. Denaturation of the enzyme can arise by the action of reagents used in entrapment methods.

The specific activity decrease may also be attributed to steric hindrance resulting in limits on the accessibility of the substrate. In these two cases, the decrease in enzyme activity can be minimised or prevented by choosing suitable conditions for immobilisation. Thus the active centre of the enzyme can be protected with a specific inhibitor, substrate, or product, and the shielding effect of the support that causes steric hindrances can be reduced by the introduction of 'spacers' that keep the enzyme at a definite and certain distance from the support. By introducing bi-functional componds, such as 1,6 diaminohexane, as 'spacers', the specific activity of covalent coupled glucoamylase on porous silica was increased by a factor of 10-fold.

In addition to their influence on the enzyme activity, any physical or chemical matrix–enzyme interactions may additionally modify the selectivity and stability of the bound enzyme from that which it normally possesses in free solution.

Partition effects

In the support binding method, when the support is charged or has a hydrophobic character, the kinetic behaviour of the immobilised enzyme may differ from that of the free enzyme even in the absence of mass transfer effects. This difference is commonly attributed to partition effects that cause different concentrations of charge species, substrates, products, hydrogen ions, hydroxyl ions and so on, in the micro-environment of the immobilised enzyme and in the domain of the bulk solution, owing to electrostatic with fixed charges on the support.

The main consequences of these partition effects is a shift in the optimum pH, with a displacement of the pH-activity profile of the immobilised enzyme towards more alkaline or acidic pH values for negatively or positively charge carriers, respectively. For example, chymotrypsin immobilised on a polyanionic support – ethylene/maleic anhydride co-polymer – shifted 1 pH unit to the alkaline side, while immobilised on a polycationic support, polyornithine, a shift of 1.5 pH units to the acid side occurred.

By similar considerations, the partitioning of charged compounds, substrate or product, between a charged enzyme particle and the bulk solution can also be evaluated. For a positively charged substrate, when using a negatively charged immobilised enzyme, a higher concentration of substrate is obtained in the local environment or microenvironment than in the bulk solution, and a higher value of relative activity is obtained than with a neutrally charged matrix. However, when effects other than partitioning are present, it is possible to have no shift of the enzyme's pH optimum on charged supports.

Mass transfer effects

When an enzyme is immobilised on or within a solid matrix, mass transfer effects may exist because the substrate must diffuse from the bulk solution to the active site of the immobilised enzyme. If the enzyme is attached to non-porous supports there are only external mass transfer effects on the catalytically active outer surface; in the reaction solution, being surrounded by a stagnant film, substrate and product are transported across the Nernst layer by diffusion. The driving force for this diffusion is the concentration difference between the surface and the bulk concentration of substrate and product.

When an enzyme is immobilised within a porous support, in addition to possible external mass transfer effects, there could also be resistance to the internal diffusion of the substrate (as it must diffuse through the pores in order to reach the enzyme) and resistance of product for its diffusion into the bulk solution. Consequently, a substrate concentration gradient is established within the pores, resulting in a concentration decreasing with increased distance (in depth) from the surface of immobilised enzyme preparation. A corresponding product concentration gradient is obtained in the opposite direction.

Unlike external diffusion, internal mass transfer proceeds in parallel with the enzyme reaction and takes into account the depletion of substrate within the pores with increasing distance from the surface of the enzyme support. The rate of reaction will also decrease, for the same reason. The overall reaction is dependent on the substrate concentration and the distance from the outside support surface.

Miscellaneous effects

Other properties of the enzyme can change upon immobilisation. The substrate specificity alters, particularly when using a substrate of high molecular weight by the effect of steric hindrance and diffusional resistances. The kinetics constants K_m and V_m of the immobilised enzyme are

Table 22.5 Classification of enzyme reactors

Mode of operation	Flow pattern	Type of reactor
Batch	Well mixed	Batch stirred tank reactor (BSTR)
	Plug flow	Total recycle reactor
Continuous	Well mixed	Continuous stirred tank reactor (CSTR)
		CSTR with ultrafiltration membrane
	Plug flow	Packed bed reactor (PBR)
		Fluidised bed reactor (FBR)
		Tubular reactor (other)
		Hollow fibre reactor

different from the free enzyme as a consequence of conformational changes of the immobilised form, which affect the affinity between enzyme and substrate. The increase of activity energy for some immobilised enzymes may be attributed to diffusional resistances, mainly in porous supports.

22.4 Immobilised enzyme reactors

22.4.1 Classification of enzyme reactors

Among the applications of immobilised enzymes, their utilisation in industry is perhaps the most important and consequently the most frequently discussed. The use of immobilised enzymes in industrial processes is performed in basic chemical reactors. A classification of enzyme reactor based on the mode of operation and the flow characteristics of substrate and product is presented in Table 22.5. The configurations of the different reactor types are shown in Fig. 22.10.

22.4.2 Batch reactors

Batch reactors are most commonly used when soluble enzymes are used as catalysts. The soluble enzymes are not generally separated from the products and consequently are not recovered for re-use.

Since one of the main goals of immobilising an enzyme is to permit its re-use, the application of immobilised enzymes in batch reactors requires a separation (or an additional separation) to recover the enzyme preparation. During this recovery process, appreciable loss of immobilised enzyme material may occur as well as loss of enzyme activity. Traditionally, the stirred tank reactor has been used for batchwise work. Composed of a reactor and a stirrer, it is the simplest type of reactor that allows good mixing and relative ease of temperature and pH control. However, some matrices, such as inorganic supports, are broken by shearing in such vessels, and alternative designs have therefore been attempted. A possible laboratory alternative is the **basket reactor**, in which the catalyst is retained within a 'basket' either forming the impeller 'blades' or the baffles of the tank reactor.

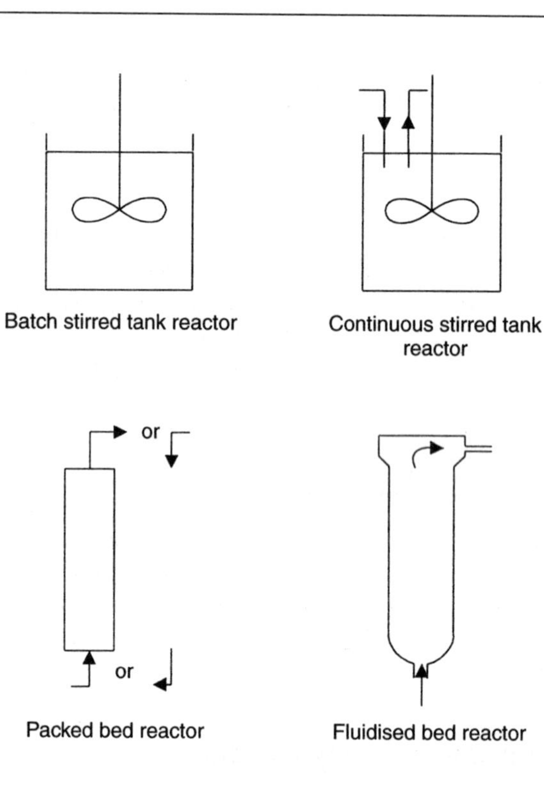

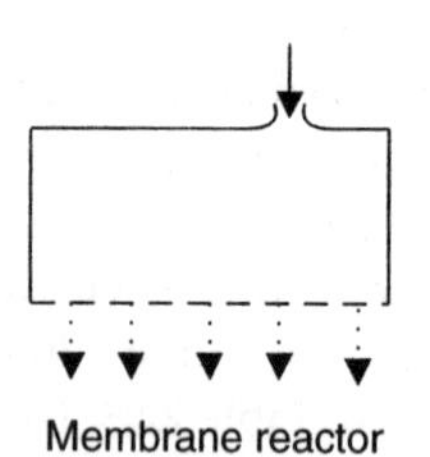

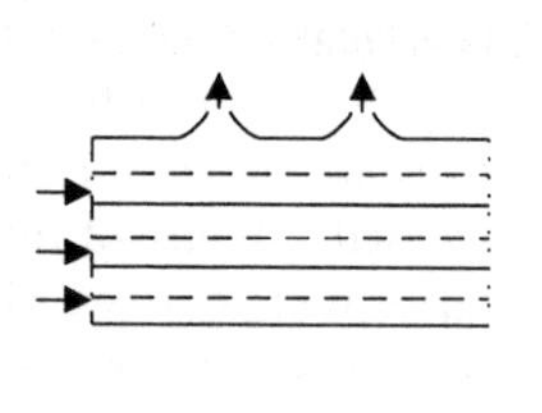

Fig. 22.10 Examples of immobilised biocatalyst reactors.

Another alternative is to change the flow pattern, using a plug flow type of reactor: the total recycle reactor or batch recirculation reactor, which may be a packed bed or fluidised bed reactor, or even a coated tubular reactor. This type of reactor may be useful where a single pass gives inadequate conversions. However, it has found greatest application in the laboratory for the acquisition of kinetic data, when the recycle rate is adjusted so that the conversion in the reactor is low and it can be considered as a differential reactor. One advantage of this type of reactor is that the external mass transfer effects can be reduced by the operational high fluid velocities.

22.4.3 Continuous reactors

The continuous operation of immobilised enzymes has some advantages when compared with batch processes, such as ease of automatic control, ease of operation, and quality control of products. Continuous reactors can be divided into two basic types: the **continuous feed, stirred tank reactor** (CSTR) and the **plug flow reactor** (PFR).

In the ideal CSTR the degree of conversion is independent of the position in the vessel, as a complete mixing is obtained with stirring and the conditions within the CSTR are the same as the outlet stream, that is, low substrate and high product concentrations. With the ideal PFR, the conversion degree is dependent on the length of the reactor as no mixing device at all exists and the conditions within the reactor are never uniform.

While a nearly ideal CSTR is readily obtained (since it is only necessary to have good stirring to obtain complete mixing), an ideal PFR is very difficult to obtain. Several adverse factors to obtaining an ideal PFR often occur, such as temperature and velocity gradients normal to the flow direction and axial dispersion of substrate.

Several considerations influence the type of continuous reactor to be chosen for a particular application. One of the most important criteria is based on kinetic considerations. For Michaelis–Menten kinetics, the PFR is preferable to the CSTR as the CSTR requires more enzyme to obtain the same degree of conversion as a PFR. If product inhibition occurs, this problem is accentuated, as in a CSTR high product concentration is always in direct contact with all of the catalyst. There is only one situation where a CSTR may be kinetically more favourable than a PFR, namely, when substrate inhibition occurs.

The form and characteristics of the immobilised enzyme preparations also influence the choice of reactor type, and operational requirements are still another factor to be taken into account. Thus, when pH control is necessary, for instance with penicillin acylase, the CSTR or batch stirred tank reactor is more suitable than PFR reactors. Due to possible disintegration of support through mechanical shearing, only durable preparations of immobilised enzyme should be used in a CSTR. With very small immobilised enzyme particles, problems such as high pressure drop and plugging arise from the utilisation of this catalyst in packed bed reactors (the most used type of PFR). To overcome these problems, a fluidised bed reactor, which provides a degree of mixing intermediate to the CSTR and the ideal PFR, can be used with low pressure drop.

Reactant characteristics can also influence the choice of reactor. Insoluble substrates and products and highly viscous fluids are preferably processed in fluidised bed reactors or CSTR, where no plugging of the reactor is likely to occur, as would be the case in a packed bed reactor.

As can be deduced from this outline, there are no simple rules for choosing reactor type and the different factors mentioned must be analysed individually for a specific case.

22.5 Biocatalysis in non-conventional media

Water as an essential reaction medium for biocatalysts has been advocated for many years as one of the major advantages of biotransformations. However, this so-called advantage has proved to be one of the severest limitations for broadening the scope of applications of

Table 22.6 Characteristics of two-liquid phase bioconversion

Potential advantages
High substrate and product solubilities
Reduction in substrate and product inhibition
Facilitated recovery of product and biocatalyst
High gas solubility in organic solvents
Shift of reaction equilibrium
Potential disadvantages
Biocatalyst denaturation and/or inhibition by organic solvent
Increasing complexity of the reaction

biocatalysts, especially when the reactants are poorly soluble in water. Non-conventional media that have been used include organic solvents, some gases and supercritical fluids. The scopes and limitations of these different systems are described below.

22.5.1 Biocatalysis in organic media

The first examples of biocatalyst/enzyme use in organic solvents for the conversion of hydrophobic compounds were presented over 20 years ago. Several examples show the synthesis of peptides (e.g. $AcPheAlaNH_2$) from amino acids catalysed by proteases, the production of the sweetener, Aspartame (L-aspartyl-L-phenylalanine methyl ester), catalysed by thermolysin using ethyl acetate as solvent, and the use of lipases on esterification, transesterification and inter-esterification reactions and the resolution of enantiomers.

The introduction of an organic solvent in the reaction system has several advantages (Table 22.6). The organic solvent increases the solubility of the poorly water-soluble or insoluble compounds, thereby increasing the volumetric productivity of the reaction system. Another important advantage is that the equilibrium of a hydrolytic reaction can be shifted in favour of the product, this being extracted into the organic phase, therefore biocatalyst and product recovery will be facilitated (extractive bioconversions). High product yields may also be achieved by decreasing possible substrate or product inhibition and prevention of unwanted side-reactions. In spite of these advantages of using organic solvents, limitations also exist. The biocatalyst may be denaturated or inhibited by the solvent and in addition, the introduction of an organic solvent leads to an increased complexity of the reaction process system.

These reaction systems were first applied to enzymatic conversions before their use was extended to whole-cell systems. Recently, the discovery of bacterial strains which are able to grow in the presence of organic solvents has made this field even more promising, opening further possibilities for the understanding of the mechanisms underlying the tolerance or toxicity responses of micro-organisms to organic solvents.

Selection of solvent

Two of the most important technical criteria for solvent selection are high product recovery and biocompatibility, although other characteristics like chemical and thermal stability, low tendency to form emulsions with water media, non-biodegradability, non-hazardous nature and low market price are desirable.

Whereas the other desirable solvent attributes are relatively mild conditions, the requirement of biocompatibility is a particularly restrictive criterion. Several attempts have been made to associate the toxicity of different solvents to some of their physico-chemical properties. The parameters used to classify solvents in terms of biocompatibility have been related to the polarity of the solvent. Laane and coworkers from Wageningen Agriculture University have described a correlation between bioactivity and the logarithm of the partition coefficient of the solvent in the octanol/water two-phase system (log P_{oct}), known as **the Hansch parameter**. Log P_{oct} denotes hydrophobicity, which is not exactly the same as polarity, but it shows a much better correlation with the biocatalytic rates than other models based on solvent polarity. The Hansch parameter has currently been used in the pharmaceutical and medical fields as a part of drug activity studies and can be determined experimentally or calculated by Rekker's hydrophobic fragmental constant approach. Many attempts have been made to explain the empirical correlation between log P_{oct} and the activity retention of cellular biocatalysts but so far the mechanisms of solvent-caused toxicity are poorly understood.

Laane and coworkers observed that a correlation exists between log P_{oct} and the epoxidising reaction activity of immobilised cells and the gas-producing activity of anaerobic cells in various water-saturated organic solvents. When plotting the cellular activity retention against log P_{oct} sigmoidal curves are obtained (Fig. 22.11). A solvent with a log P_{oct} value lower than the inflection point is usually toxic and one with a log P_{oct} value higher than the inflection point is biocompatible. The inflection point of these curves depends on the micro-organism studied. In general, solvents having a log P_{oct} lower than 2 are relatively polar solvents not suitable for biocatalytic systems, and biological activities vary in solvents having a log P_{oct} between 2 and 4, being high in apolar solvents having log P_{oct} values above 4 (Table 22.7).

Similar sigmoidal shapes were observed for the effect of solvents on micro-organisms, however different inflection points were obtained for different micro-organisms which could be due to differences in the characteristics of their cellular membranes. It has also been observed that increasing the agitation rate caused the log P_{oct} curve to shift to the right. A good correlation between the metabolic activity of *Arthrobacter*, *Acinetobacter*, *Nocardia* and *Pseudomonas* and the log P_{oct} of the solvent was found, however the transition between toxic and non-toxic solvents was observed in the log P_{oct} range from 3 to 5. Tramper and coworkers have investigated the relationship between the metabolic activity of cells exposed to organic solvents at 10% (v/v) concentrations and their log P_{oct} values for different homologous series of solvents. They found that the

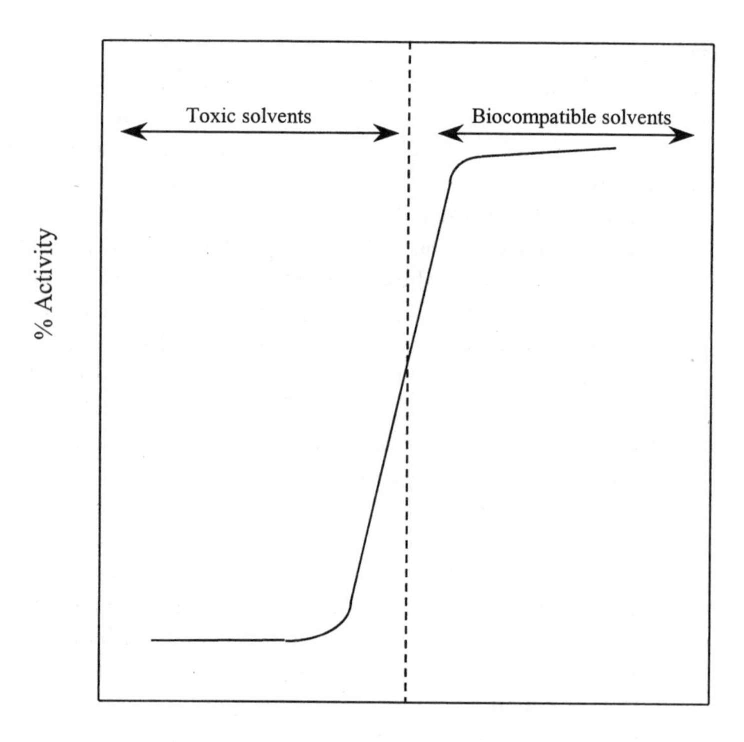

Fig. 22.11 Retention of enzyme activity versus log P. P, partition coefficient

log P_{oct} value, above which all solvents are non-toxic, is different for different homologous series: e.g. *Arthrobacter* and *Nocardia* tolerate alkanols with log P_{oct} above 4 but are only able to tolerate phthalates having a log P_{oct} value higher than 5.

The influence of the characteristics of the cell membrane on the solvent tolerance of the micro-organisms has been evaluated for Gram-negative and Gram-positive bacteria. Gram-negative bacteria, and particularly *Pseudomonas*, are in general more tolerant than Gram-positive bacteria. The difference in solvent tolerance is probably due to the presence of the outer membrane of Gram-negative bacteria. This membrane contains major structural proteins, lipoproteins and lipopolysaccharides. A percentage of the total lipoproteins of the outer membrane is bound covalently to the peptidoglycan layer. This linkage might be expected to protect cells from environmental stress. Plant cells appear to be even more sensitive to the presence of organic solvents as shown for cell suspensions of *Morinda citrifolia* which had a biocompatibility limiting log P_{oct} value of 5. Thus the transition range from toxic to non-toxic clearly depends on the type of cellular biocatalyst.

Other criteria for correlation with solvent biocompatibility have recently been described, especially for the effect of solvents on enzyme stability. These are the 'three-dimensional solubility parameter' or the 'denaturing capacity' used to predict, for example, the concentration of the organic solvent at which half-inactivation is observed.

Table 22.7 Biocompatible organic solvents

Solvents	Hansch parameter log P
Alcohols	
Decanol	4.0
Undecanol	4.5
Dodecanol	5.0
Oleyl alcohol	7.0
Ethers	
Diphenyl ether	4.3
Carboxylic acids	
Oleic acid	7.9
Esters	
Pentyl benzoate	4.2
Ethyl decanoate	4.9
Butyl oleate	9.8
Dibutylphthalate	5.4
Dipentylphthalate	6.5
Dihexylphthalate	7.5
Dioctylphthalate	9.6
Didecylphthalate	11.7
Hydrocarbons	
Heptane	4.0
Octane	4.5
Nonane	5.1
Decane	5.6
Undecane	6.1
Dodecane	6.6
Tetradecane	8.8
Hexadecane	9.6

Classification of organic reaction systems

Biocatalysts can be used in different ways in combination with organic solvents: (a) homogeneous mixture of water and water-miscible solvent; (b) aqueous/organic two-liquid phase systems; (c) microheterogeneous systems (micro-emulsions and reversed micelles); (d) enzyme powder and immobilised biocatalysts suspended in solvent without aqueous phase; and (e) covalently modified enzymes dissolved in organic solvent (Fig. 22.12a–g).

Homogeneous mixture of water and water-miscible solvent

An easy way to increase the solubility of a hydrophobic substrate is to add a water-miscible organic solvent, such as methanol, acetone, ethyl acetate, dimethyl formamide, dimethylsulphoxide etc., to the reaction medium. These systems have the advantages of generally not presenting

Fig. 22.12 Classification of biocatalysis in organic media systems. (a) Water-in-oil biphasic emulsion; (b) oil-in-water biphasic emulsion; (c) enzyme immobilised in a porous support in a biphasic system; (d) enzyme in a reversed micelle; (e) enzyme modified with polyethylene glycol and solubilised in organic media; (f) immobilised enzyme suspended in organic media; and (g) enzyme powder suspended in organic media.

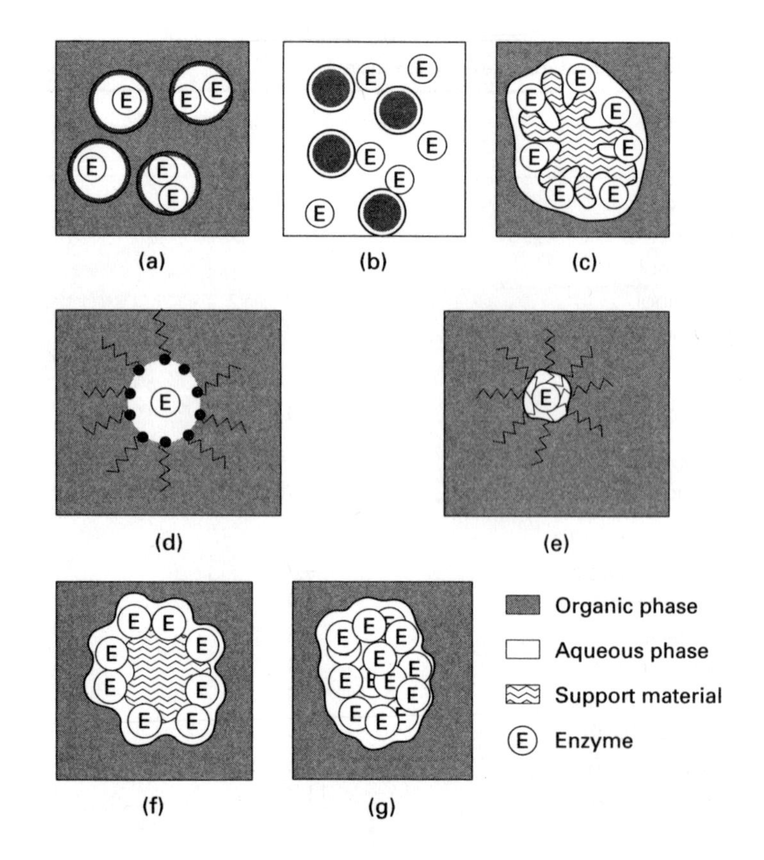

mass transfer limitations as they are homogeneous systems. However, the biocatalyst in the presence of these systems usually has poor operational stability, particularly if a high concentration of solvent is needed. This results from the fact that water-miscible solvents are polar compounds with log P values lower than 2, being considered as toxic solvents. For example, ribonuclease dissolved in increasing concentrations of 2-chloroethanol undergoes a transition from the native state to an unfolded form.

Aqueous/organic two-phase systems

Two-liquid (aqueous/organic) phase systems are useful when reactants of poor water solubility have to be employed. The organic solvent may be the substrate itself (e.g. olive oil) to be converted or may serve as a reservoir (a hydrocarbon) for substrate(s) and/or product(s) (Figs. 22.12a,b,c and 22.13). These systems can also be used to confine (immobilise) the biocatalyst physically in the aqueous phase whilst the organic phase is being renewed. In these systems, it is important that the interfacial area is large enough to improve mass transfer. The partitioning of the substrate(s) and product(s) in the two-liquid phases can be controlled by choosing a suitable solvent and, in certain cases, such as those involving ionic species (e.g. organic acids), the pH (which should be lower than the pK_a of those ionic species in order to get the unprotonated compound, which is the one readily to be extracted) of the water phase. In the latter case, the pH selected should also be compatible with the enzymatic activity. The partitioning of substrate(s) and product(s) is particularly suited when one or both of these compounds is an inhibitor of the enzymatic activity. Its accumulation in the organic phase will alleviate this inhibition. The overall displacement of reaction equilibrium by extraction of the products is also an advantage of this type of systems. The phase ratio can be varied over a wide range leading to an optimisation of reactor capacity. Applications of two-liquid phase systems include: olive oil hydrolysis by lipases, peptide synthesis, extractive ethanol fermentations by coupling an extractant (e.g. dodecanol) with fermentation step, steroid transformations (e.g. prednisolone dehydrogenation by adsorbed and gel-entrapped *Arthrobacter simplex* cells, etc.).

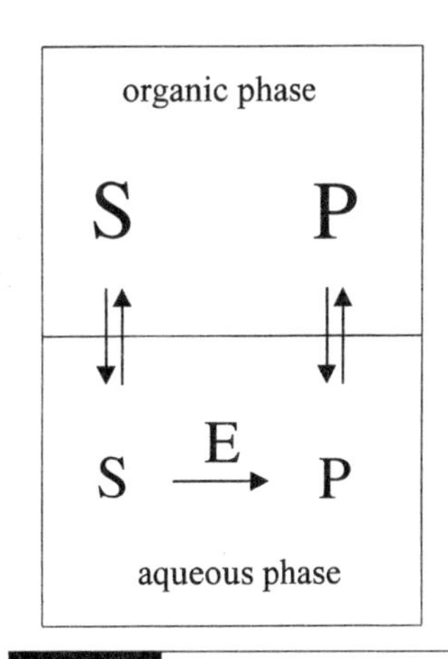

Fig. 22.13 Schematic presentation of an enzymatic conversion in a two-phase system. S, substrate; P, product; E, enzyme.

Micro-heterogeneous systems

Micro-emulsions and **reversed micelles** represent a special case of two-liquid phase systems in which the aqueous phase is no longer macroscopically distinguishable from the continuous organic phase. These systems usually contain surfactants to stabilise the distribution of water and its contents in the continuous organic phase.

Reversed micelles are aggregates formed by surfactants in apolar solvents. Surfactants are amphipathic molecules that possess both a hydrophilic and a hydrophobic part (Fig. 22.12d). The hydrophobic tails of the surfactant molecules are in contact with the apolar bulk solution; the polar head groups are turned towards the interior of the aggregate forming a polar core. This core can solubilise water (water pool) and host macro-molecules such as proteins. The group of amphipathic molecules used in the formation of reversed micelles in hydrocarbon solvents include both natural membrane lipids and artificial surfactants (Table 22.8).

The amount of water solubilised in the reversed micellar systems is commonly referred to as w_0, the molar ratio of surfactant to water ($w_0 = H_2O$/surfactant). This is an extremely important parameter, since it will determine the number of surfactant molecules per micelle, the availability of water molecules for protein hydration and biocatalysis, and is the main factor affecting the micelle size.

The formation of reversed micelles depends largely on the energy change due to dipole-dipole interactions between the polar head groups of the surfactant molecules. The solubilisation properties of surfactants are often expressed by a three or four component phase diagram, the reversed micelles being identified by the regions of optical transparency. Most of the work performed with reversed micelles in biological systems uses AOT (sodium dioctyl sulphosuccinate), an anionic surfactant. AOT forms stable micellar aggregates in organic solvents, the most used being isooctane. The maximum amount of solubilised water for an AOT/iso-octane/H_2O system is around $w_0 = 60$. Above this value the transparent reversed micelle solution becomes a turbid emulsion and phase separation occurs.

Biocatalysis in reversed micelles was first reported in 1978 since when several major studies and biocatalytic applications have been

Table 22.8 Example of surfactant-forming reversed micelles

Surfactant	Solvent
Sodium dioctyl sulphosuccinate (AOT)	n-Hydrocarbon (C_6–C_{10})
	Iso-octane
	Cyclohexane
	Carbon tetrachloride
	Benzene
Cetyltrimethylammonium bromide (CTAB)	Hexanol / iso-octane
	Hexanol / octane
	Chloroform / octane
Methyltrioctylammonium chloride (TOMAC)	Cyclohexane
Brij 60	Octane
Triton X	Hexanol / cyclohexane
Phosphatidylcholine	Benzene
	Heptane
Phosphatidylethanolamine	Benzene
	Heptane

described in the literature. Examples include the controlled hydrolysis of lipids and vegetable oils, the esterification and trans-esterification reactions catalysed by lipases, and peptides synthesis by proteases. The optimum water level can be controlled by the w_0 value, this value being greater than 10 for synthesis reactions. Currently important aspects of research on these systems include the stabilisation of biocatalysts and the development of appropriate reactors to accomplish enzyme retention, product separation and avoid product contamination with the surfactant molecules.

Very low water systems

For many biotransformations it is advantageous to decrease considerably the amount of water in the reaction medium. The first studies in this area were performed in 1966 and since then extensive research has been carried out. It is believed that retention of enzyme activity is due to the minimum essential water necessary to maintain the protein structure and the enzymatic function. The selection of the organic solvent is critical as hydrophilic solvents strip the essential hydration shell from the enzyme molecule. The amount of water bound to the enzyme decreases dramatically with the increasing hydrophilicity of the solvent. For example the reactivity of α-chymotrypsin in octane is 10 000-fold higher than that in pyridine.

The amount of water in the reaction medium can be measured in several ways. The most common way is to measure the water concentration (in% v/v or $mol\,l^{-1}$). This parameter, however, does not describe the reaction conditions for the enzyme. A better way to characterise the degree of hydration of the reaction medium is to use the thermodynamic water activity, a_w, as parameter. This parameter, which is a measure of the amount of water in the system, determines directly the effects of water on the chemical equilibrium. During the reaction the water activity may change, especially if water is formed (esterification reactions) or consumed (hydrolysis). Therefore it is important to keep the water activity constant by controlling its concentration during the reaction, for example by adding salt hydrates directly to the reaction medium which give a constant water activity suitable for the enzymatic conversion. The salt hydrates act as a 'buffer' of the water activity.

Very low water systems involve the use of enzymes as solid suspensions, powders (Fig. 22.12g) or immobilised onto a support, in organic solvents (Fig. 22.12f); and covalently modified enzymes soluble in organic media (Fig. 22.12e).

When enzyme powders are used as biocatalysts, problems can occur due to aggregation of the enzyme molecules. A solution to solve these problems is the immobilisation of the enzyme on to an appropriate solid support. The support should be judiciously selected taking into account its surface properties, namely its capacity to attract water (**aquaphilicity**), as the water in the reaction medium will partition between the enzyme, the support and the reaction medium affecting the enzyme micro-environment.

A very interesting feature of enzyme suspensions in organic media is the phenomenon of '**pH memory**'. It has been observed that lyophilised lipases and proteases and several oxido-reductases display pH optima in organic media identical to optima in aqueous solutions. This phenomenon is attributed to the right ionisation state of the enzyme for catalysis. Enzymes suspended in organic solvents have been observed to exhibit altered substrate specificity compared to that in aqueous media. This has been attributed to steric hindrance caused by the lack of conformational mobility of the enzyme (lipases and proteases) in organic solvents and to the substrate's inability to displace water from the hydrophobic binding pockets of the enzyme molecules in organic media.

Since dehydration drastically decreases the conformational mobility of enzymes, it has been observed, as expected, that the thermal denaturation process is slowed down considerably for enzymes suspended in organic solvents.

Suspensions of solid enzymes in organic solvents have been used for a number of biotechnological applications, namely the transesterification of fats and oils by lipases, which is an industrial process for upgrading triacylglycerols.

In spite of the advantages of solid enzyme suspensions in organic solvents, these systems are limited by mass tranfer. To overcome these diffusional limitations, enzymes have been covalently modified, namely using polyethylene glycol for this purpose, to make them soluble in organic solvents. As the enzyme becomes soluble in the reaction there are no diffusional limitations, however in order to re-use the enzyme, it has to be recovered by precipitation from the reaction mixture with a non-polar solvent such as hexane. Other drawbacks of this type of enzyme are inactivation of the enzyme that may occur during the derivatisation procedure and that the enzyme preparations are soluble only in a limited number of solvents, like aromatic and chlorinated hydrocarbons.

22.5.2 Gas-phase reaction media

It is possible to perform biotransformations with a solid biocatalyst preparation and gaseous substrates. The oxidation of ethanol by alcohol oxidase was a pioneer work. Further applications have involved the use of lipases to catalyse transesterifications reactions between short chain alcohols and esters. As the substrates are in the gaseous phase, relatively high temperatures are used, consequently it is advantageous to use thermostable enzymes for these applications. In gas phase reactions, as with using organic media, the water content of the enzyme preparation should be well controlled to achieve the best performance both in terms of activity and stability.

22.5.3 Supercritical fluids as bioreaction media

Enzymatic reactions in supercritical and near critical fluids require pressurised systems. Such systems permit high mass transfer rates and easy separation of reaction products. Due to its non-toxic character and its relatively low critical temperature (31 °C) CO_2 has been the most used fluid. In some aspects, supercritical fluids have properties resembling those of non-polar solvents being adequate for biotransformations of hydrophobic compounds. The oxidation of cholesterol by cholesterol oxidase and the stereoselective hydrolysis of racemic glycidyl butyrate by immobilised *Rhizomucor miehei* lipase yielding the homochiral R(-)glycidyl butyrate have been achieved in the presence of supercritical CO_2. The solubility of these compounds may also be improved by the addition of small amounts of co-solvents known as entrainers. For example methanol (3.5% mol) was used as an entrainer to enhance the solubility of cholesterol in supercritical carbon dioxide. The major drawback of this reaction system is the high energy and equipment costs due to the use of high pressures.

22.6 Concluding remarks

This chapter described the use and performance of biocatalysts in biotransformations relevant to industrial, analytical and biomedical applications and environment bioremediation.

From a process point of view, there are advantages and limitations for both chemical and biochemical routes. Part of the limitations of the biocatalytic route has been solved through the new developments in the areas of biology, chemistry and process engineering. The recent advances in recombinant DNA technology, metabolic engineering, fermentation and biocatalysis in non-conventional media have also broaden the applications of biocatalysts to synthetic and oxidative/reductive biotransformations. It is also important to emphasise the integration of both processes and disciplines (engineering, biology and chemistry) which is a key feature for the development of competitive biocatalytic routes. New applications of biocatalysts (native or modified) in the fields of chemical synthesis, analysis (biosensors), biomedical and environment are foreseen.

22.7 Further reading

Blanch, H.W. and Clark, D. S. (eds.) (1991). *Applied Biocatalysis, Vol. 1.* Marcel Dekker Inc., New York.

Cabral, J. M. S., Best, D., Boross, L. and Tramper, J. (eds.) (1994). *Applied Biocatalysis.* Harwood Academic Press, Switzerland.

Kelly, D. R. (vol. ed.) (1998). *Biotechnology Series.* (H.-J. Rehm and G. Reed, eds.), *Vol. 8a, Biotransformations I, 2nd Edition.* Wiley–VCH Verlag GmbH, Weinheim.

Lilly, M. D. (1992). The design and operation of biotransformation processes. In *Recent Advances in Biotechnology* (F. Vardar-Sukan and S.S. Sukan, eds.), pp. 47–68. Kluwer Academic, Amsterdam.

Straathof, A.J.J. and Adlercreutz, P. (eds.) (2000). *Applied Biocatalysis, 2nd Edition.* Harwood Academic Press, Switzerland.

Tanaka, A., Tosa, T. and Kobayashi, T. (eds.) (1993). *Industrial Applications of Immobilized Biocatalysis.* Marcel Dekker, Inc., New York.

Tramper, J., Vermüe, M., Beeftink, H.H . and van Stocksar, U. (eds.) (1992). *Biocatalysis in non-conventional media.* Elsevier, Amsterdam.

Wells, J. A. and Estell, D. A. (1988). Subtilisin: an enzyme designed to be engineered. *TIBS* **13**, 291–297.

Wingard,L. B., Katchalski-Katzir, E. and Goldstein, L. (eds.) (1976). *Immobilized Enzyme Principles.* Academic Press, New York.

Chapter 23

Immunochemical applications

Mike Clark

Glossary

Adjuvants Substances which when mixed with an antigen will make them more immunogenic, i.e. they enhance the immune response. Adjuvants cause inflammation and irritation and help to activate cells of the immune system.

Affinity The measured binding constant of an antibody for its antigen at equilibrium.

Allo-immunisation Immunisation of an animal with cells or tissues derived from another animal of the same species where there are allelic differences in their genes.

Antibody Adaptive proteins in the plasma of an immune individual with binding specificity for antigens (cf. immunoglobulin).

Antigen A molecule or complex of molecules which is recognised by an antibody (immunoglobulin) by binding to the antibody's variable or V-regions.

APC An antigen presenting cell is a specialised cell (dendritic cells and macrophages) which can ingest, degrade and then present on its cell surface, fragments of pathogens and other antigens, to other cells of the immune system (e.g. B-cells and T-cells).

Autoimmune Immunity to molecules (antigens) within an animal's own body which can lead to a disease e.g. rheumatoid arthritis, or some forms of diabetes.

Avidity Antibodies frequently interact with antigen using multiple antigen binding sites and thus they have a functional affinity termed avidity which is a complex function of the individual binding affinities.

B-cells A subset of white cells (lymphocytes) in the blood which produce antibodies.

CDR (1,2,& 3) The three complementarity determining regions of the immunoglobulin variable region domains which form the major interaction with antigen. Structurally the complementarity determining regions form the loops at one end of the globular domain.

Chimaeric antibody Recombinant DNA technology allows artificial antibodies to be prepared in which domains from one antibody are substituted by domains from another antibody or protein.

Class The major type or classification of an immunoglobulin e.g. IgM, IgG, IgA or IgE.

Complement An auto-catalytic enzyme cascade found in the plasma which can be triggered by antibody-antigen complexes and which can lead to antigen destruction and removal.

D-segment The diversity segment, a gene segment found in immunoglobulin heavy chains which is re-arranged between the V-segment and the J-segment.

Effector functions Immune functions triggered through specific binding of antibody to antigen. These include complement in the plasma and Fc receptors on many different cell types.

Epitope The epitope is a single antibody binding site on an antigen. Any given antigen may bind different antibodies through different epitopes.

ELISA An enzyme-linked, immunoadsorbent assay is a commonly used assay system in which antibody is covalently linked to an enzyme so that conversion of a substrate can be used to quantify the amount of antibody bound.

Fab The antigen binding proteolytic fragment of an immunoglobulin.

$F(ab')_2$ A proteolytic fragment of an immunoglobulin in which the two antigen binding 'Fab' fragments are still attached at the hinge.

Fc The crystallisable proteolytic fragment of an immunoglobulin. This fragment also contains the sequences needed for interacting with and triggering effector functions.

Fc receptor A protein molecular complex expressed on a cell which is able to specifically bind to and recognise sequences within the Fc fragment of an immunoglobulin.

FR (1,2,3 & 4) The framework regions are four partially conserved (less variable) regions of sequence within the immunoglobulin variable region domains. Structurally the framework regions form the conserved anti-parallel β-strands of the protein domains.

Fv fragments The minimal component of an immunoglobulin still capable of binding to antigen. It consists of the heavy- and light-chain variable-domains.

Hapten A hapten is a small molecule which can be recognised by antibodies but which is not immunogenic in itself. They thus must be coupled covalently to carrier proteins in order to use them for immunisation.

Humanised antibodies In order to reduce the immunogenicity of monoclonal antibodies in human patients many of the rodent-derived sequences are substituted with homologous human sequences. This can also be done for the framework regions within the variable-region domains to give a 'fully humanised' or 'reshaped' antibody.

Hybridoma cells In order to produce long-term cell lines secreting a single specific antibody, B-cells from the spleens of immunised animals are fused with myeloma cells adapted to growth in cell culture. These hybrid cell lines made with myeloma cells are called hybridomas.

Immunoadhesins Fusion proteins, generated using recombinant DNA technology, in which cellular adhesion molecules are made as a chimaeric hybrid molecule with an immunoglobulin Fc region.
Immune complex A complex of antibodies bound to their antigens.
Immunogenic A form of an antigen which is capable of generating an immune response when injected or administered to an animal.
Immunoglobulin A globulin fraction of plasma which contains the specific immune proteins termed antibodies.
Immunoprecipitation The use of antibodies to remove an antigen from solution through the formation of an insoluble or immobilised immune complex.
Immunosuppress To lower, or suppress, the ability of an animal to make an active immune response. This may be desired, and can be achieved using drugs, or antibodies, specific for regulatory cells of the immune system. It can also occur as an unwanted effect in some diseases such as in AIDS resulting from HIV infection.
J-chain The J-chain is a 'joining' protein subunit which is found covalently associated, through disulphide bonds, with multimeric immunoglobulins such as IgA dimers and IgM pentamers. The J-chain should not be confused with the similar sounding 'J-segment' (see below).
J-segment J-segment is the junctional or joining DNA segment which is rearranged with the V-segment for immunoglobulin light chains, or the V and the D-segments for immunoglobulin heavy chains, to give a fully formed immunoglobulin variable (V-) region.
MHC class I and class II Major histocompatibility locus class-I and class-II molecules are the molecules on a cell surface used to present peptide fragments of an antigen to the T-cell receptor.
Monoclonal antibody This term is applied to an antibody produced from a clonal cell line in tissue culture. It is a well defined antibody of predictable characteristics unlike the complex mixtures of antibodies found in an animal's plasma.
Polyclonal antisera This term is used to distinguish the inherently heterogeneous mixture of antibodies found in the sera derived from an immunised animal, from the laboratory prepared monoclonal antibodies.
Sub-class A sub-classification of an immunoglobulin within a given class e.g. IgG1, IgG2, IgG3 and IgG4 are all IgG subclass antibodies.
ScFv A single chain Fv fragment is an artificial genetic construct in which a polypeptide linker has been inserted between the N-terminus of one variable-region domain and the C-terminus of the other variable-region domain.
Specificity The specificity of an antibody is the ability to show a level of discrimination in binding avidities between different antigens. It is thus in a sense a relative term i.e. the antibody is specific for 'antigen A' but not for 'antigen B'. This would then be termed an *anti-A antibody.*
T-cells A subset of white cells (lymphocytes) in the blood which either directly kill infected cells or help other cells such as B-cells to respond to an antigen.
V-region Variable region of an immunoglobulin.
V_H The variable region domain of the immunoglobulin heavy chain.
V_L The variable region domain of the immunoglobulin light chain.
V-segment A gene segment which is re-arranged to give a functional immunoglobulin variable region.
Xeno-immunisation Immunisation of one species with cells or tissues of a different species.

23.1 Introduction

This chapter will discuss immunochemical applications in basic biotechnology and thus will mainly concentrate on the derivation and applications of **antibodies** otherwise known as **immunoglobulins** (Ig). These proteins are so named because of the way in which they were first discovered. They were first identified as a particular globulin protein fraction of the blood, which was called the gammaglobulin and, because it was then recognised that this protein fraction was a major specific component of the immune response made to infection, they were also called immunoglobulins. The term 'antibody' refers to the fact that they recognise or are specific ('anti-') for 'foreign bodies'. Antibodies are important within biotechnology because of the ease with which it is possible to exploit the immune system's ability to generate a diverse population of immunoglobulins with specificity for binding to a huge range of different molecular structures, the 'foreign bodies' recognised by antibodies which we call **antigens**.

In order to describe these applications it is necessary that the reader has some basic knowledge of the immunobiology of antibody production and of immunoglobulin structure and function. A very brief and simplified overview will be given here but it should be noted that the immune system has evolved of necessity to be highly complex in organisation and the interested reader is recommended to look at more detailed and fuller explanations given in the many widely available immunology textbooks (see the Further reading list).

23.2 Antibody structure and functions

Antibodies are proteins made as part of the humoral immune response to immunogenic substances and infectious agents. They serve as key adapter molecules within the immune system enabling the host's inherited **effector functions** to recognise the many unpredictable, diverse and varied antigen structures which might be encountered during an animal's lifetime. These **effector functions** are inherited mechanisms of inactivating or killing infectious pathogens and then causing their breakdown and removal from the body. However, these **effector systems** do not have the ability to easily recognise the infectious agents in all of their many diverse forms. This recognition, or targeting, of the effector systems is, in part, dependent upon the antibody's ability to interface between antigens on the infectious agent and also the body's effector systems. The effector systems are inherited within the germ line genes of an individual but the antibody specificities are derived by complex somatic re-arrangements of the genes encoding immunoglobulins within the so-called B-cells (a sub-population of the white blood cells or lymphocytes). This means that even two identical twins, or two mice from the same laboratory strain, will have different immunoglobulin sequences expressed at any one time.

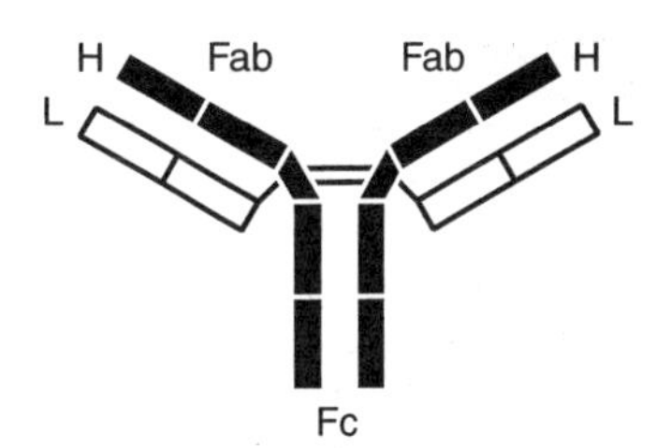

Fig. 23.1 The basic IgG immunogloblin structure of two heavy chains (black) and two light chains (white). The two heavy chains are disulphide-bonded together and each light chain is disulphide-bonded to a heavy chain. The antibody also has two antigen binding Fab regions and a single Fc region.

The basic schematic representation of an antibody is the familiar Y-shaped structure of an IgG with two identical **Fab** (antigen binding) arms and a single **Fc** (crystallisable) region joined by a more flexible **hinge region** (see Fig. 23.1). Again, these terms come about from the original protein chemistry in which the whole molecule was fragmented by cleavage with proteolytic enzymes and different properties were then assigned to the different isolated fragments. This basic molecular structure (or subunit) is made up of two identical heavy (H) chains and two identical light (L) chains, based upon their molecular size, and each chain contains repeated immunoglobulin type globular domains with a conserved structure. In protein structural terms the domains have anti-parallel strands which loop back on themselves to form β-sheets and these sheets are then rolled up into a barrel-like structure (see Fig. 23.2). Light chains have two of these globular domains whereas heavy chains have four or more (depending upon their '**class**', see below). The heavy and light chains come in several different forms which give rise to the concept of immunoglobulin classes and sub-classes, and, for example in man (and most other mammals), we have heavy chain types μ, γ, ε and α giving, respectively, the classes of antibody IgM, IgG, IgE and IgA. Each of these classes can have light chains of either the κ or the λ type. The proportion of immunoglobulins in the plasma with each light chain type varies between species with man having a κ:λ ratio of approximately 60:40 whereas mouse has a ratio of about 90:10. In man, the IgG class has four sub-classes called IgG1, IgG2, IgG3 and IgG4 using γ1, γ2, γ3 and γ4 chains whilst the IgA class has two sub-classes IgA1 and IgA2 using α1 and α2 chains. Some of these classes of immunoglobulin are secreted into plasma in the form of more complex oligomerised structures of subunits often associated with a molecule called J-chain. Thus IgM is a pentamer of five identical protein subunits, and IgA is frequently found as dimers and trimers of identical protein subunits, again associated with J-chain (see Fig. 23.3).

It is the heavy chain which is largely responsible for the 'effector functions' (antigen destruction and removal) triggered through interactions with cells of the immune system by *ligation* (binding to and cross-linking) of cell surface receptors (called **Fc receptors** because they require the Fc fragment of the antibody) or, alternatively, through activation of the **complement cascade** and the binding to **complement receptors**. *Complement* is another family of proteins found in the blood and which are involved in immune reactions. The components of complement are mainly specific proteolytic enzymes whose substrates are themselves other complement components which are activated by proteolysis. This gives rise to a classical biochemical amplification of an initial small activation step. Once activated some of these complement components also rapidly form covalent chemical bonds with antigen, thus marking them for clearance by complement receptors of the immune system, whilst others are able to create pores in cell or viral membranes of infectious organisms and thus kill the cells or viruses.

Each of the different immunoglobulin (antibody) classes and also sub-classes exhibit a different pattern of effector functions some of

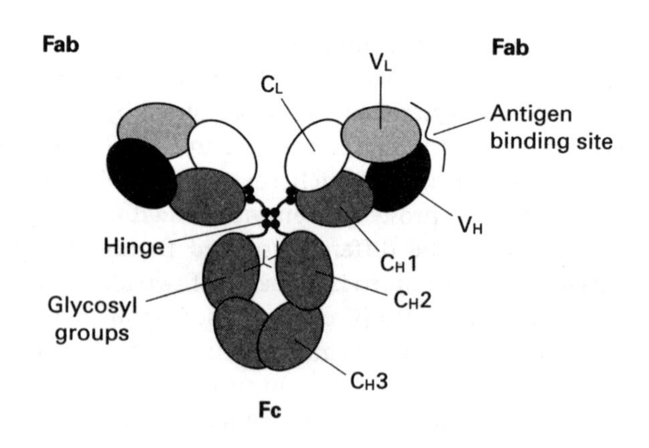

Fig. 23.2 An alternative schematic of an IgG structure. Each globular domain of the molecule is illustrated as an ellipse. The heavy chain domains are shown in darker shades and the light chain domains in lighter shades. The heavy and light chain variable domains V_H and V_L are also indicated along with the position of the antigen binding site at the extremeties of each Fab. Each C_H2 domain is glycosylated and the carbohydrate sits in the space between the two heavy chains. Disulphide bridges between the chains are indicated as black dots within the flexible hinge region.

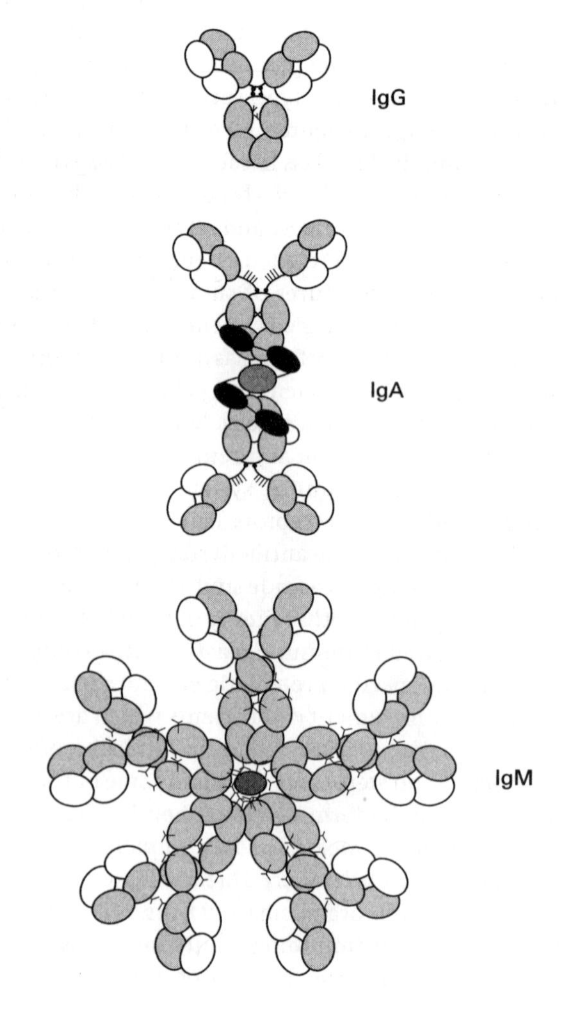

Fig. 23.3 The origins of immunoglobulins. Different classes of immunoglobulin are built up from the same basic structure. The top shows IgG which can be compared with Fig. 23.2. In secretory IgA, two subunits of IgA, each of which is similar to IgG, are joined together by covalent disulphide bridges and through a subunit known as a J-chain (dark grey). During transport and secretion of IgA across the gut lining, a second set of subunits, which are derived from the transport receptor, also become associated. These extra subunits found only on secreted IgA are known as 'secretory components' (black). IgM has a pentameric structure of five subunits covalently bonded together by disulphide bonds and associated like IgA with a single subunit of J-chain (dark grey).

which may be more appropriate in dealing with certain types of antigens or infectious agents. Each of the different classes of immunoglobulin has their own class of Fc effector functions. Thus there are well characterised Fc receptors for the IgG class (FcγRI, FcγRII, FcγRIII), the IgA class (FcαR) and the IgE class (FcεRI and FcεRII) which have different cellular distributions, affinities for Fc type and sub-class, and which also mediate different signals and thus trigger different effector functions. In addition for some of these classes, there are **transport receptors** which enable, for example, IgA to be secreted into the gut, the urinary tract, the respiratory tract, into tears and saliva and also into milk and colostrum.

The receptor for transport of IgA is the poly Ig receptor and, during the transport of the IgA, the receptor is cleaved leaving a fragment of the receptor termed secretory component (because it was initially characterised on secreted but not plasma IgA) associated with secreted IgA (see Fig. 23.3). In humans, IgG is actively transported across the placenta during the late stages of pregnancy to provide the neonate with a primary immune defence whilst in other animals, such as rodents, the IgG is transported across the gut from colostrum during the first few hours after birth. The receptor which transports IgG is called FcRn (the neonatal Fc receptor). FcRn is also responsible for protecting IgG from catabolism and is thus responsible for extending the plasma half-life of IgG from days to weeks. FcRn achieves this by binding at low pH to the Fc region of IgG which is within intracellular endosomal vesicles containing proteins destined for degradation. The bound IgG is then recycled back to the plasma before it is degraded, whereupon it is released under the neutral pH conditions encountered at the plasma membrane. This has considerable importance and consequences with regard to pharmaceutical applications of IgG antibodies *in vivo*. A long antibody half-life in the plasma means that a smaller amount of antibody is needed at less frequent intervals to maintain a required plasma concentration. The extent of the half-life is a function of the specific binding of receptor FcRn to the Fc region of IgG and is thus lost in fragments of antibodies such as Fab fragments. Obviously, because FcRn is an IgG-specific receptor, these properties of placental transfer and extended half-life are unique to this class of immunoglobulins. The name FcRn had originally been applied to just the form of receptor found in the gut of neonatal rats, but in a much earlier series of papers, published by Professor Brambell in the mid 1960s, the existence of both forms of receptor had been predicted, and thus some now refer to both functions of the receptor under the unified name of FcRB.

As already mentioned, the antigen specificity of the antibody is a property of the Fab fragment of the molecule. Specificity is the result of variation in parts of the sequences of the Fab. The *N*-terminal domain of both the heavy and light chain is called the variable region (V_H and V_L domains). Sequence analysis of amino acids of large numbers of variable regions for both V_H and V_L has allowed three small regions of hypervariability to be defined within four more conserved framework regions (FR1, FR2, FR3 and FR4). In the three-dimensional structures, the

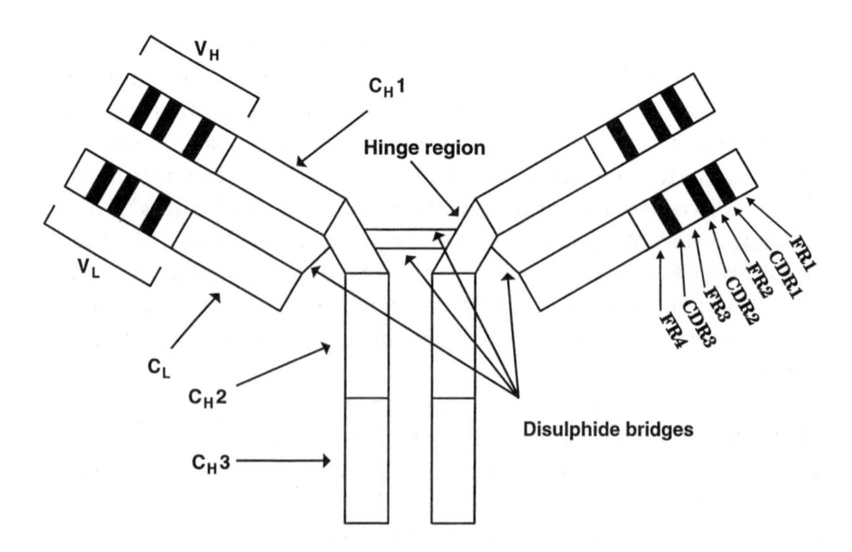

Fig. 23.4 Regions of the variable domains of IgG. The sequences of these domains are classified as either framework region sequences (FR1, FR2, FR3 and FR4), or complementarity determining regions (CDR1, CDR2 and CDR3). Framework region sequences are sequences which go to make up the conserved beta-pleated strands which form the globular barrel shape of the domain structure. The complementarity determining regions form the variable loop structures which make the antigen binding sites. There are three CDR loops from each heavy chain and three CDR loops from each light chain. These six loops act together to form the antigen binding site of the antibody.

hypervariable regions form loops which combine together to form the principle antigen binding surfaces and thus these sequences have also been named the **complementarity determining regions** or CDRs (CDR1, CDR2 and CDR3) (see Fig. 23.4).

In terms of the genetics of immunoglobulin expression, the unique sequences of each different antibody are the direct result of somatic re-arrangements of different gene segments during B-cell development (see Fig. 23.5). In the case of the heavy chain, three segments, V, D and J are rearranged and in the case of light chains, two segments, V and J, are rearranged. These gene rearrangements lead to expression of a surface immunoglobulin receptor and this is followed by selection of individual B-cell clones based on their binding to antigen. Further differentiation can result in somatic mutations of the V-region sequences and/or further somatic cell rearrangements to bring the constant region segments for different heavy chain sub-classes adjacent to the variable domain encoding gene segments.

23.3 Antibody protein fragments

Various protein fragments of antibodies can be produced individually and separately from the other protein components which may be of

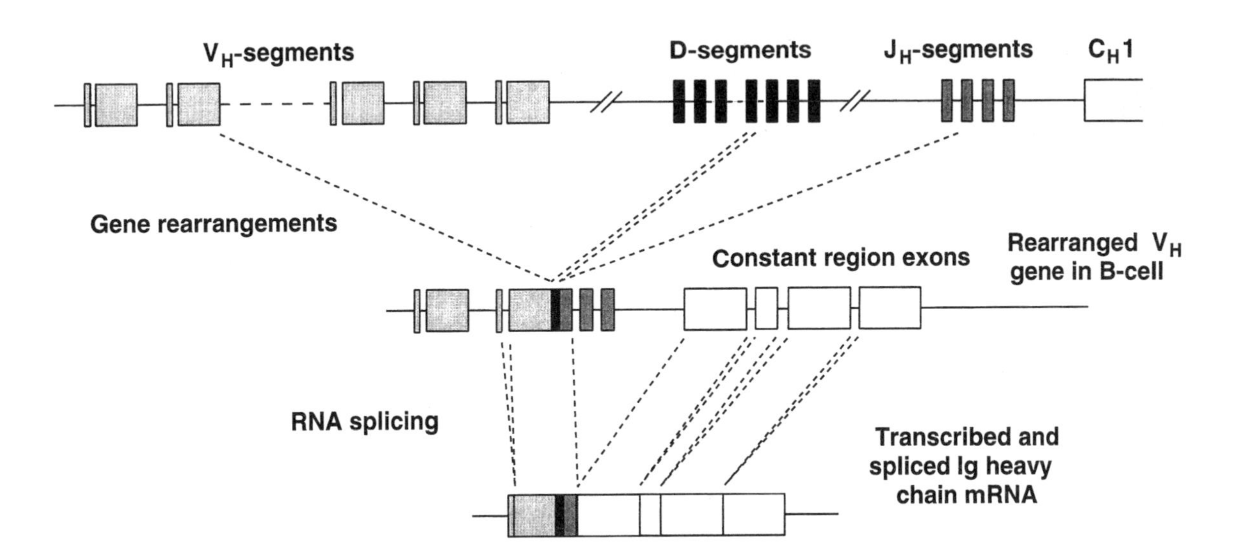

Fig. 23.5 Germline genomic gene re-arrangements. During B-cell differentiation the genomic immunoglobulin segment sequences are rearranged to give a single functional immunoglobulin heavy chain encoding gene and a single functional light chain encoding gene in each clone of B-cells. These rearranged genes still include introns which must be spliced out of the transcribed RNA. The V-D-J segment rearrangements of the immunoglobulin heavy chain encoding genes are shown. Immunoglobulin light chains have just V- and J- segments and no D-segments.

practical use in different circumstances (see Fig. 23.6). These fragments can conveniently be derived by enzyme proteolysis. In general the Fab region is relatively resistant to proteolysis whereas the Fc region, and, particularly, the hinge region are comparatively susceptible. Depending upon which protease is used and the particular antibody isotype under examination (and the animal species from which the antibody is derived), the proteolytic cleavage may occur at the hinge region. If it does, and the cleavage site is on the *C*-terminal side of the inter-chain disulphide bridges, then $F(ab')_2$ fragments are generated; if cleavage is on the *N*-terminal side, Fab fragments are generated. Alternatively, mildly reducing conditions can be used to separate the F(ab')2 fragment into two F(ab') fragments.

Protein fragments can also be expressed using recombinant DNA techniques. Other recombinant products, such as the so-called **Fv fragments**, contain only the V-region domains which are not covalently associated and could be considered as the smallest unit of antibody which should still be capable of antigen binding with the original single-site affinity. In order to stabilise this association of the V-regions of the recombinant heavy and light chains, a gene segment encoding an artificial linker from the *C*-terminus of one domain to the N-terminus of the other can be introduced and the whole fusion protein expressed, by a suitable cell, as a single chain Fv (**ScFv**).

All the small fragments of antibodies may be of use, both *in vitro* and *in vivo*, because they are still capable of binding to antigen but have lost the ability to bind to Fc receptors and to activate the complement cascade. Their smaller size can in certain situations improve their diffusion and penetration properties particularly, for example, when they are used, for staining of tissue sections *in vitro*, or for targeting of cellular antigens *in vivo*. However, for IgG the loss of the Fc in addition to the smaller size of the fragments will, of course, result in a considerable decrease in the plasma half-life as discussed above.

It should also be noted that post-translational modifications of an

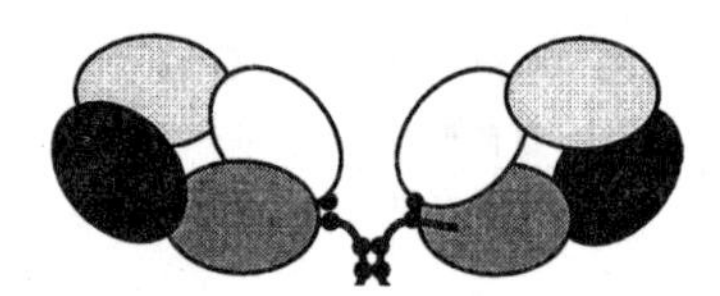
F(ab')$_2$ Fragment

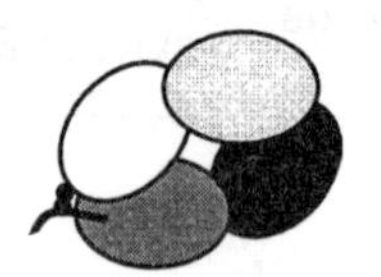
Fab Fragment

F_V Fragment

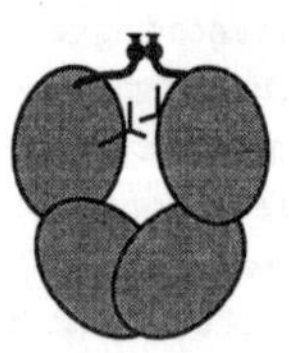
Fc Fragment

Fig. 23.6 Functional sub-fragments of IgG molecule. These can be generated either through limited proteolysis or through expression of recombinant genes.

antibody may be critical for its functions. Antibodies of different classes and sub-classes show conserved sites for both *N*- and *O*-linked sugars. In IgG, the conserved *N*-linked glycosylation of the C_H2 region is essential for many of the molecule's effector functions (i.e. binding to some Fc receptors and also activation of complement is dependent upon the correct glycosylation). Similarly, the intra- and inter-chain disulphide bridges are important for the overall structure and function of the antibody. Thus the manner in which antibodies are produced, as well as the particular methods of purification are important issues to consider. This is the case particularly with recombinantly produced immunoglobulins, for example where the inability of bacteria to glycosylate or reliably assemble and disulphide bond complex proteins must be taken into account.

23.4 Antibody affinity

The concept of the **affinity**, or more correctly the **avidity**, of an antibody for antigen is important. For many uses, both *in vivo* and *in vitro*, the affinity of the antibody for antigen is an important factor in determining not only the utility but also the commercial success of a product. Strictly, the affinity of an antibody for its antigen (association constant or K_a expressed in units of M^{-1}) is a measure of the ratio of the concentrations of bound antibody-antigen complex to free antibody and free antigen at a thermodynamic equilibrium. It assumes that the interaction with bound antigen is of single valency, which is more likely to be the case only for very simple antigens or for antibody Fab or F_v fragments. In the past, antibody affinities were often determined by equilibrium dialysis or by measuring radiolabelled antibody binding to antigens under conditions near to equilibrium. It is quite common today, however, to carry out direct determination of percentage of antibody association and disassociation using techniques such as plasma resonance. However, it is worth remembering that a good approximation to the affinity of an antibody for an antigen can be estimated by measuring the concentration of antibody needed to give half maximal binding to the antigen. This gives the dissociation constant, K_d, expressed in units of M which is, in fact, the reciprocal of the association constant K_a (i.e. $K_d = 1/K_a$).

It must be remembered that antibodies usually have two or more identical binding sites for an antigen. Often the interaction of a bivalent (e.g. whole IgG with two Fab regions) or multivalent (e.g. IgM with ten Fab regions) antibody with multivalent antigen (e.g. a cell surface or antigen immobilised on a solid surface) is the critical parameter in determining the strength of interaction between antibody and antigen. This functional affinity of an antibody is referred to as its **avidity**. The affinity or avidity of an antibody for antigen is also related to the ratio of the rates of the forward reaction for formation of the complex to back reaction for decay of the complex (Eqn 23.1).

$$[Ab] + [Ag] \underset{k_{back}}{\overset{k_{forward}}{\rightleftharpoons}} [AbAg] \tag{23.1}$$

$$K_a = \frac{1}{K_d} = \frac{[AbAg]}{[Ab].[Ag]}$$

Two antibodies can have a similar affinity for antigen measured at equilibrium but one may have a much slower on-rate ($k_{forward}$) and, of course, a proportionally slower off-rate (k_{back}). For many uses, the antibody will not be used under conditions of thermodynamic equilibrium: for example, when using an antibody to affinity purify an antigen or when using antibodies in immunometric assays. In such situations, the antibody is usually in excess and a faster rate of the forward reaction ($k_{forward}$) may then be desirable. In a different example, such as the use of radiolabelled antibodies for the radio-imaging of tumours *in vivo,* the antibody needs first to circulate through the body and then to diffuse and penetrate through the tissues before it even has a chance to interact with the antigen. The stability of antibody on the tumour once it is bound (affinity and off-rate) as well as the diffusion rates of the antibody in tissues (a product of the antibody or fragment size) are both factors which determine the suitability of one antibody versus another.

23.5 Antibody specificity

The **specificity** of an antibody is another important concept and one which is often highly confused with the concept of affinity. In a practical sense, the specificity of an antibody for its antigen is only, in part, related to its affinity or avidity. It is highly likely that an antibody will have a spectrum of affinities for a range of different antigens. Sometimes these antigens may be completely unrelated whilst more often they may share related structural features (for example many different complex carbohydrate structures share features in common). Different antibodies to the same antigen may therefore show different functional cross-reactions on other antigens. Clearly, if the intended use of the antibody is to discriminate between different antigens in a complex mixture then the cross-reactions of the antibody are as critical as the avidity for the correct antigen. For an antibody which is to be used in a situation where the 'alternative' antigens are not likely to be encountered, for example in the affinity purification of an antigen product from a batch culture process, any cross-reactions may be considered as irrelevant. In the use of antibodies for *in vivo* therapy or diagnostics, there are so many different tissue antigens that unexpected cross-reactions of the antibody on tissues other than the intended target may frequently be a complicating factor in the development of an antibody-based product. The observation of the cross-reaction of an antibody on a second antigen, is of course, related to the avidity of the antibody for that antigen and the sensitivity of the assay

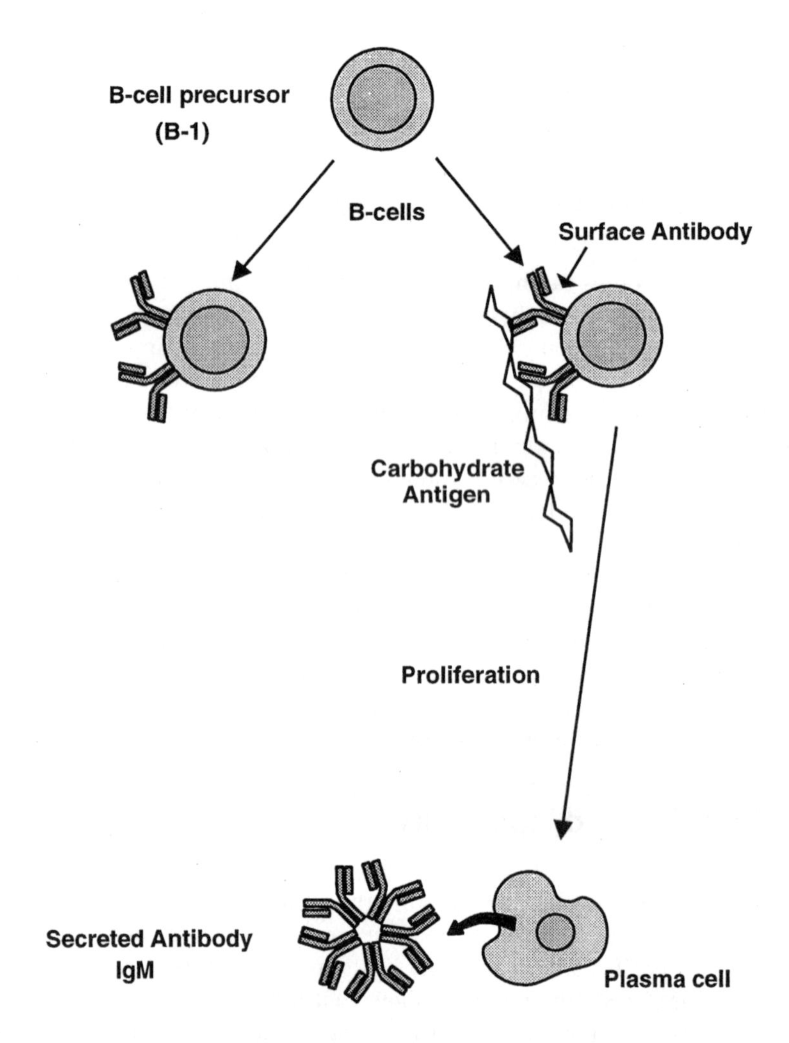

Fig. 23.7 The T-cell-independent B-cell response. Some key steps in the production of secreted IgM by a so-called T-cell-independent B-cell response are illustrated. The critical feature is that the antigen is usually a multimeric repeating structure (e.g. bacterial carbohydrate) and is capable of cross-linking the surface antibody on the B-cells which have specificity for this antigen. These B-cells are then activated by this event and go on to differentiate into plasma cells secreting the IgM.

being used to measure the interaction. This can lead to the situation where an apparent improvement in the sensitivity of an assay leads to a deterioration in the specificity of the assay.

23.6 Immunisation and production of polyclonal antisera

The earliest but still a widely used way to exploit the immune system is to immunise an animal with an **immunogenic** form of the substance or a pathogen of interest (perhaps repeatedly over several months) and then, some weeks after the final immunisation, to collect the blood plasma or serum and use it whole or fractionated. It is necessary to understand some of the complexity of the immune response in order to appreciate some of the problems associated with derivation of antisera to different types of antigen. Figures 23.7 and 23.8 show a highly schematic and simplified view of two different types of B-cell response to antigen, T-cell independent (Fig. 23.7) and T-cell dependent (Fig. 23.8)

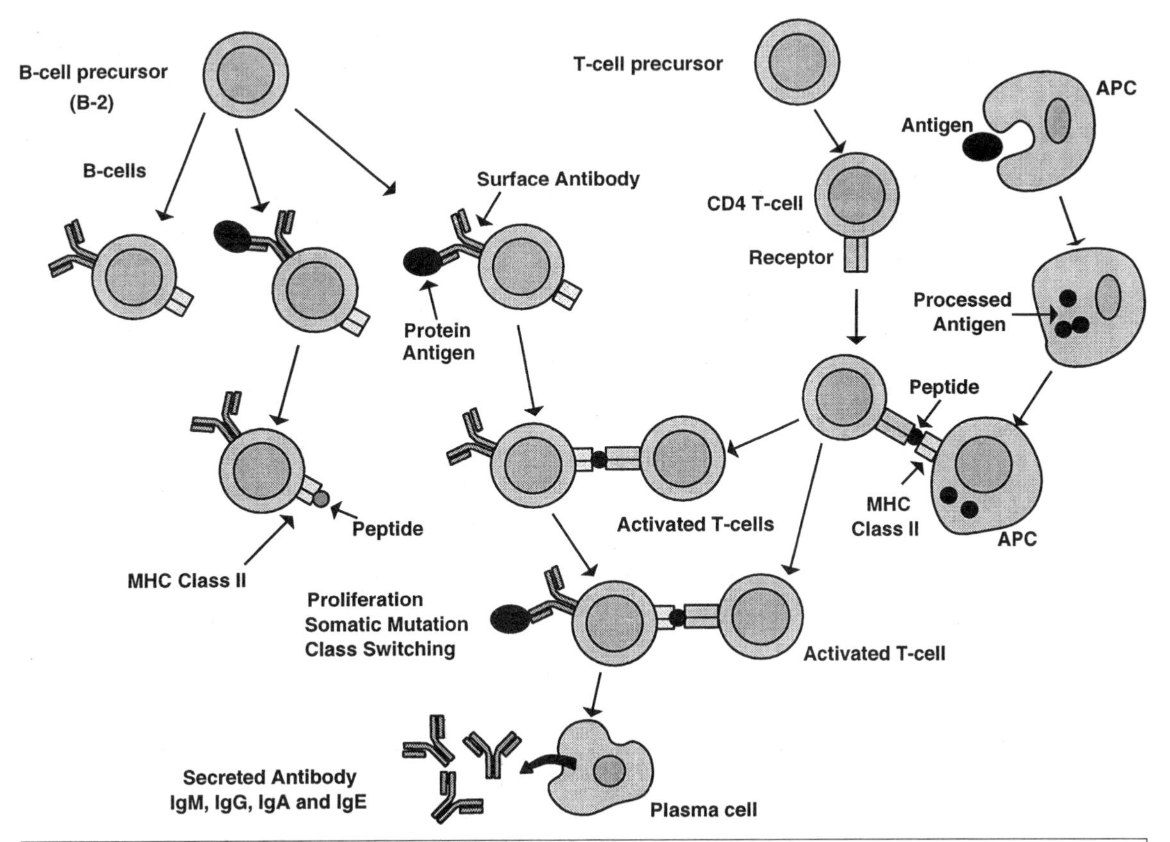

Fig. 23.8 The T-cell dependent B-cell response. Antibody production resulting from the T-cell dependent B-cell response involves antigen recognition and co-operation between a number of different cell types including T helper-cells (which express the CD4 molecule, a co-receptor for MHC Class II) and so-called antigen presenting cells or APCs (macrophages and dendritic cells). Antigen presenting cells are called this because they present on their cell surface a complex of the MHC Class II molecule containing, within a binding groove peptides from the antigen. These peptides are derived through proteolysis, from antigen molecules which have been endocytosed. Before these antigen-specific B-cells can differentiate antibody secreting plasma cells they must be helped by an activated, CD4 positive, T-helper cell. For this activation the CD4 positive T-helper cells must first see 'processed' antigen presented by macrophages or dendritic cells (APC).

responses to antigen. The T-cell independent, B-cell response is largely the result of triggering the surface immunoglobulin on the B-cells by cross-linking with an antigen of a highly repetitive structure. Such antigens include carbohydrates, glycolipids, phospholipids and nucleic acids. The B-cells are driven into proliferation and differentiate into plasma cells which secrete large amounts of immunoglobulin (mainly of the IgM class). Immune responses of this type include the human anti-blood group A and anti-blood group B responses which are thought to be triggered by exposure to bacterial carbohydrates and which then cross-react with the blood group antigens from other individuals. This is an excellent example of natural cross-reactions of antibodies because, except for individuals who have been transfused with mismatched blood or women following pregnancy, a majority of individuals with such antibodies are unlikely to have encountered blood cells of these other blood groups. Anti-blood group A and B antibodies, as well as being principally of the IgM class, are also generally of low affinity but, because of the valency of IgM (five subunits and thus ten possible binding sites per molecule) and the repetitive structures within the antigen, they may interact with a high avidity.

In contrast, immune responses to T-cell dependent antigens seem more complex and involve several steps whereby different cell types are required to interact in an antigen specific way thus allowing for complex regulation (see Fig. 23.5). Proteins are taken up by specialist **antigen-presenting cells** (APCs) and are broken down into peptides. Some of these peptides are capable of binding to MHC Class II molecules and are presented as a complex on the APCs cell surface. CD4-positive, Class II restricted T-cells are able to bind the MHC peptide complex and can be activated by the APCs. B-cells are also capable of taking up antigen through their specific receptor, which is the membrane-bound surface immunoglobulin, and as a consequence they too can present peptides in the context of MHC Class II. If an activated CD4 T-cell interacts with such an antigen-presenting B-cell it is able to help the B-cell by providing signals which activate the B-cell to divide, differentiate and secrete its antibody. During several such rounds of specific T- and B-cell co-operation the B-cell may switch to produce other antibodies such as IgG, IgA and IgE and it may also undergo somatic mutation and be selected for higher affinity binding to the antigen. Thus, in general, T-cell dependent B-cell antigens should be proteins or protein-associated.

There is an important feature which is common to both T-cell independent and T-cell dependent B-cells and that is the concept of self-tolerance. In general, the immune system has checks and controls which act to minimise the chances of an immunoglobulin recognising a self-antigen being made in quantity. Such auto-reactive B-cells are generally eliminated. In extreme situations a breakdown of tolerance can occur and, in such cases, pathology can result from this so called **auto-immune response**.

Self-tolerance means that it is again, in general, easier to generate antibody responses to antigens which are unrelated to any self antigens within the animal being immunised. For example, there are more likely to be many differences if human-derived proteins are used to immunise a mouse (**xeno-immunisation**) than if mouse-derived proteins are used to immunise a different strain of mouse (**allo-immunisation**). The different regions on the antigen recognised by antibodies are called **antigenic epitopes** and thus a xeno-immunisation is likely to raise antibodies to more epitopes of an antigen than an allo-immunisation. This may be important and will be discussed later because simultaneous recognition of an antigen by several antibodies to different epitopes may result in apparently improved affinity (avidity) and specificity of reaction.

Another important factor of immunisation is that some antigens are more immunogenic than others. Partly this may relate to self-tolerance but it is also now thought that an important component of an immune response is the activation of the immune system by danger

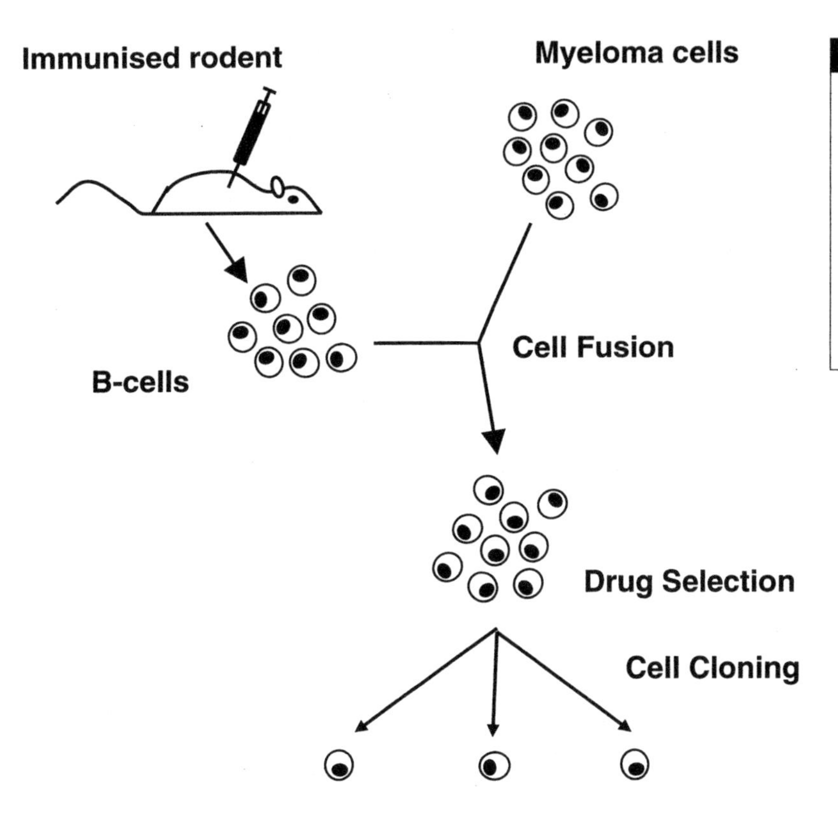

Fig. 23.9 Production of monoclonal antibodies. This involves fusion of spleen cells, from immunised rodents, with myeloma cells adapted to cell culture. The resulting hybridoma cells are selected for growth in media which is toxic to the parental myeloma cells and they are then cloned. Each clone produces a single monoclonal antibody.

signals. Thus an antigen can be combined with other substances, such as mineral oils and components derived from micro-organisms, which can act as **adjuvants** and activate the immune system and improve antigen processing and presentation by the APCs.

Following production, the antisera can be used in very many systems as a specific tool for detection of antigen. The immunoglobulin fraction of the antisera can be purified and then used in several assay and detection systems. For example, it can be labelled by covalent conjugation with fluorescent dyes and used in microscopy or flow cytometry for detecting antigen binding on or in cells. Equally, the antibody could be labelled with an enzyme and used in histology or in an enzyme-linked, immunosorbent assay (ELISA) (see Section 23.10.2) again for detection of the specific antigen.

23.7 Monoclonal antibodies

Conventionally, cell lines secreting monoclonal antibodies have been derived by taking immune B-cells, which have a limited capacity to proliferate *in vitro*, and then immortalising them by somatic cell fusion with a suitable tissue culture cell line (see Fig. 23.9). For reasons that are most likely to be related to the complex interactions of regulatory genes (e.g. transcription factors) encoded on different chromosomes in different species, this technique has proved to be most successful for a limited range of species, particularly for monoclonal antibodies of the IgM and IgG classes derived from rat and mice, although other species such as sheep, hamsters and human have been used.

In the methods used for cell fusion and subsequent selection of hybridomas a large number of variations exist. These are well documented in text books and reviews devoted to the methodology. Although Kohler and Milstein used the *Sendai* virus to induce cell fusion in their earlier experiments, this virus has been replaced almost universally with polyethylene glycol (PEG) or electrofusion techniques. For mouse and rat hybridomas, the efficiencies of these procedures are all high and typically several hundreds to thousands of individual hybridoma clones can be obtained from one animal spleen.

Using somatic cell fusion or cell transformation, either alone or in combination, it has proved very difficult to make human monoclonal antibodies. The time and effort expended is far greater than that needed to produce the equivalent mouse or rat monoclonal antibodies. It is this difficulty in production of human monoclonal antibodies which has driven the strategies for the rescue of human antibodies by phage display or alternatively to 'humanise' or 'reshape' rodent antibodies using recombinant DNA technology for so-called 'antibody engineering'.

23.8 Antibody engineering

It is now possible to genetically engineer and express a whole range of differing novel antibody constructs thus freeing biotechnologists from the constraints imposed by the natural biology of the immune system. It is the modular structure of antibody molecules which are composed of a collection of discrete globular domains, encoded by genes with a similar modular structure whereby each domain is coded in a separate exon, which makes the manipulation of immunoglobulin genes a relatively straightforward proposition (see Fig. 23.5). There are several obvious advantages of recombinant antibodies over conventionally derived monoclonal antibodies.

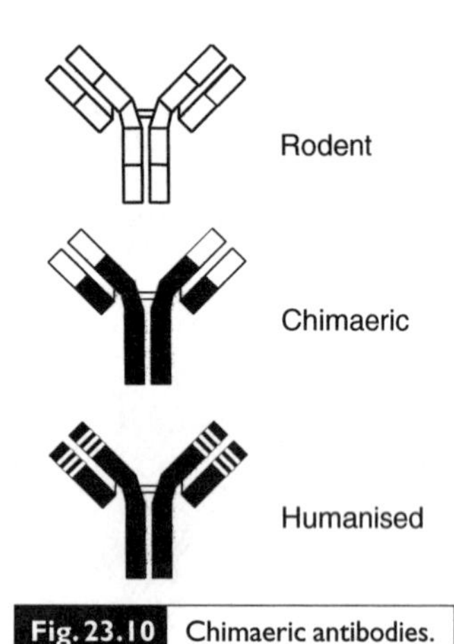

Fig. 23.10 Chimaeric antibodies. Through the use of recombinant DNA technology it is possible to engineer antibodies with novel properties. One simple step is to make chimaeric antibodies in which the variable regions from a rodent antibody (white) are combined with the constant regions of a human antibody (black). A step further in the technology is to combine just the complementarity determining region (CDR) encoding DNA sequences of a rodent antibody with framework region (FR) encoding DNA sequences of a human antibody, to give a fully humanised antibody.

It is technically feasible, through the use of appropriate cloning strategies, to isolate the genes encoding any antibody made from any immunised species and so future applications need not be restricted to the derivation of the mouse, rat and human antibody classes.

Often monoclonal antibodies can be derived with the correct specificity but they may exhibit the wrong effector functions because they are not of the desired species, class or sub-class of immunoglobulin. Obviously, using recombinant DNA technology any V-regions can be expressed in combination with any constant regions selected for desirable properties. These antibodies are called **chimaeric antibodies** (see Fig. 23.10). Thus variable regions from rodent antibodies specific for chosen antigens can be combined with constant regions encoding human immunoglobulin classes/sub-classes, the final product having potential *in vivo* therapeutic uses in man. It is also possible to introduce further mutations into the Fc regions to modify the properties to suit the proposed applications of the antibody, for example to remove the ability to bind to some Fc receptors or to activate complement.

Where *in vivo* therapeutic applications are concerned, rodent antibodies often are limited because they provoke an immune response in the patient to the antibody usually within a week of their first use. This precludes any further treatment beyond this time. As described above, useful rodent monoclonal antibodies can be partially 'humanised' by making chimeric antibodies by combining the rodent variable-regions with human constant-regions, thus introducing the effector mechanisms of the human whilst at the same time minimising the number of potential immunogenic epitopes. For immunotherapy, there are several key features that an antibody should have if it is to be successfully used. Obviously, the antibody must possess a desired specificity to bind to a relevant antigen, such as, for example an antigen expressed on a tumour cell surface, a viral antigen or perhaps a bacterial toxin. Once bound to that antigen, the antibody is then normally required to carry out a function. The antibody could be used for targeting of a radioimaging label or used for the destruction of a tumour cell or of a virus. Alternatively, the antibody might be used for neutralisation of a virus or toxin. The production of a chimaeric antibody and the selection of the most appropriate human immunogolobulin class/sub-class, allows for the retention or addition of desirable functions but, at the same time, reduces the 'foreignness' of the antibody to the patient (see Fig. 23.10).

As a further step in lessening the immunogenicity or 'foreignness', rodent monoclonal antibodies can also be fully 'humanised' or 'reshaped', to produce human antibodies that contain only those key residues from the rodent variable regions responsible for antigen binding combined with framework regions from human variable regions (see Fig. 23.10).

After manipulation of the antibody genes, they can be expressed in a number of different expression systems. Transfection of antibody genes cloned in suitable vectors into myeloma cells may result in expression of the antibody molecules which are processed and glycosylated in a manner which is characteristic of immunoglobulin produced by hybridoma cells and normal B-cells. It is also possible to express antibodies in other mammalian cell types such as Chinese Hamster Ovary (CHO) cells. Antibodies have also been expressed in a number of other eukaryotic and also prokaryotic expression systems including plant cells, yeast and bacteria. However there are certain problems which are encountered in some of these different expression systems chiefly with regard to glycosylation and to disulphide bonding which preclude expression of complete molecules or complicate the purification of the antibody product. Natural antibodies have multiple domains per chain and multiple chains per molecule and these chains have intra-domain disulphide bonds as well as inter-chain disulphide bonds. Thus the cells used for antibody expression must be capable of correctly assembling the molecule. In addition for most of the IgG antibody effector functions the appropriate *N*-linked glycosylation is required. Bacteria are thus only really appropriate for the expression of smaller fragments from antibodies such as Fab or Fv fragments. At present most commercial large-scale production of recombinant antibody molecules, particularly for therapeutic applications is carried out using either B-lymphoid cell lines or CHO cells.

23.9 Combinatorial and phage display libraries

Recent advances in molecular biology mean that mammalian genes can be rapidly cloned and expressed in bacteria, usually using phage vector systems (see Chapter 4). In **phage display**, genes encoding variable regions of immunoglobulins are cloned into the phage vectors (see Fig. 23.11). These modified bacteriophage vectors are then used to transform bacteria and, during assembly of the phage particle, the immunoglobulin variable regions are expressed on the surface of the phage particles. Thus, if each bacterium is infected by only one phage type all of the newly synthesised phage will carry the DNA which encodes the same antibody Fv or Fab fragments on the surface of the phage. For the system to work it is obvious that the phage which encodes the required antigen specific Fv or Fab must be separated from other phage and then propagated further. This is conveniently achieved by affinity selection of the phage on antigen (see Fig. 23.11). During several rounds of selection, phage with higher affinity can be selected and propagated. Finally, it is possible to re-isolate the genes from the phage particles and to express them in other systems. For example expression systems exist which produce soluble Fab or Fv fragments from the phage vector systems.

Antibody responses in a whole animal are transient, the antibodies appearing in reponse to immunisation or infection, and then disappearing over time once the antigen has been cleared from the body. Phage display provides the ability to rescue the antibody response from almost any immunised animal in the form of cloned genes encoding the individual heavy and light chain variable regions. Most strategies for cloning antibody genes in phage utilise random cloning of 'libraries' of the heavy and light chain sequences and then the expression of these libraries in randomised pairings of a single immunoglobulin heavy chain and a single immunoglobulin light chain in each individual phage. These **combinatorial libraries** are generated by cloning a repertoire of immunoglobulin heavy and light chains, usually by using the polymerase chain reaction, from mRNA isolated from tissue containing B-cells from an immune donor. It should be remembered however, that the combinations rescued after screening such a library are not necessarily representative of the combinations present in the native B-cells. This last point may be of importance because as described above, the B-cell repertoire found *in vivo* has been selected through a complex system of random gene re-arrangements followed by both positive and negative selection. The reason for this selection which involves T-cell antigen specific recognition and T-cell help is to restrict the immune

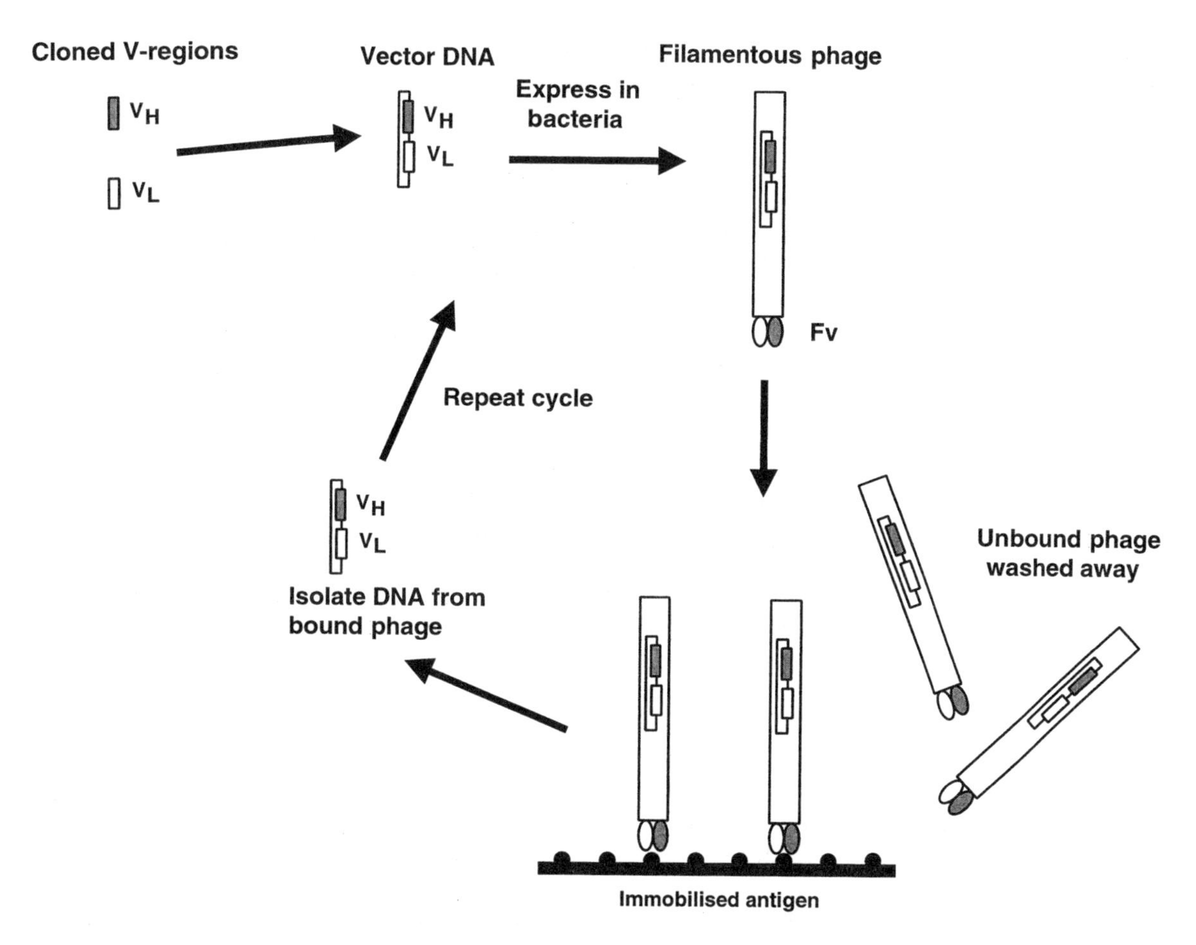

Fig. 23.11 Making a phage display library. Antibody fragments can be expressed on the surface of a bacteriophage in which the antibody variable regions are encoded within the DNA which is packaged inside the phage. Thus by selecting for antigen binding phage, it is possible to isolate the DNA which in turn encodes the antigen binding antibody V-regions. The cycle can be repeated to improve enrichment and to select for phage which bind with higher affinity.

response to 'foreign' antigens and to prevent cross-reactions to self-antigens. Thus, certain combinations of heavy and light chains may be generated and selected for in combinatorial libraries which would be selected against in a normal immune response. Phage display can also be used to mimic the immune response by generating an artificial, randomised library of synthetic genes with random complementarity determining sequences (see Section 23.2) (i.e. not derived by cloning genes from B-cells). Such libraries have been used successfully to screen for a number of different specificities.

A major disadvantage of phage display libraries is that the only antibody function being tested is antigen binding and this may not be the crucial function for the final application. Although the genes once isolated can be expressed along with any immunoglobulin constant regions, assays which are dependent upon the effector function cannot be used for the detection and isolation of the phage antibodies with appropriate specificity. Additionally, it is relatively easy to screen phage libraries on purified and homogeneous antigen preparations but it is very difficult to screen for specific binding to complex mixtures, such as cell surface antigens, where the required antigen may be a minor component of the mixture.

23.10 *In vitro* uses of recombinant and monoclonal antibodies

23.10.1 Affinity purification

Major uses of antibodies include roles in the purification of other molecules using affinity binding procedures, often in single step. This relies on the ability to derive antibodies and, in particular, monoclonal or recombinant antibodies, which have a unique and discriminating specificity for the chosen antigen. To raise useful antisera for affinity purification of an antigen it is usually necessary to have a highly purified antigen to start with. This is because the antisera will contain many different antibodies, i.e. it will be polyclonal, and thus the required antibodies must be affinity purified in some way. However, during the process of derivation of the monoclonal antibodies it is possible to work with impure mixtures of antigens and yet still obtain a useful reagent for the affinity purification of the antigen. This is because when the animal is immunised with an impure antigen there will be antibodies made against all of the different antigens present so the antisera from the animal will contain a complete mixture of antibodies. However, when the individual B-cell hybridomas are cloned in culture, all of these different antibody specificities are separated and the clones secreting antibody specific for a chosen antigen can be selected and large amounts of the antibody produced. Similarly, recombinant antibodies allow single, pure antibodies of defined specificity and affinity to be produced.

The affinity of an antibody for its antigen and its selectivity in binding can be exploited in techniques such as **immunoprecipitation**. The antibody is mixed with the antigen and it forms immune complexes. The basis for these immune complexes is that an antibody normally has at least two binding sites and so can, in theory, bind to at least two identical antigens. If the antigen, in turn, has more than one antigenic binding site (or epitopes) then the antigens and antibodies can form chains or higher order aggregates (immune complexes). Sometimes large immune complexes are formed and these then become insoluble and will form a precipitate. This insoluble immune precipitate can be separated away from the other antigens in solution by centrifugation and washing of the precipitate and, finally, it will contain a relatively pure mixture of the chosen antigen and its antibody.

There are several problems associated with immunoprecipitation reactions of this type. First, because they rely heavily on the valency of the antigen-antibody interactions in the immune complex, they do not work well using single monoclonal antibodies and they tend to work better with mixtures of monoclonal antibodies or with polyclonal antisera. Also, the immunoprecipitation reaction works best over a narrow concentration range where the antibody and antigen are said to be at equivalence. Either side of this range either the antigen or the antibody are in excess and only small soluble immune complexes are likely to be

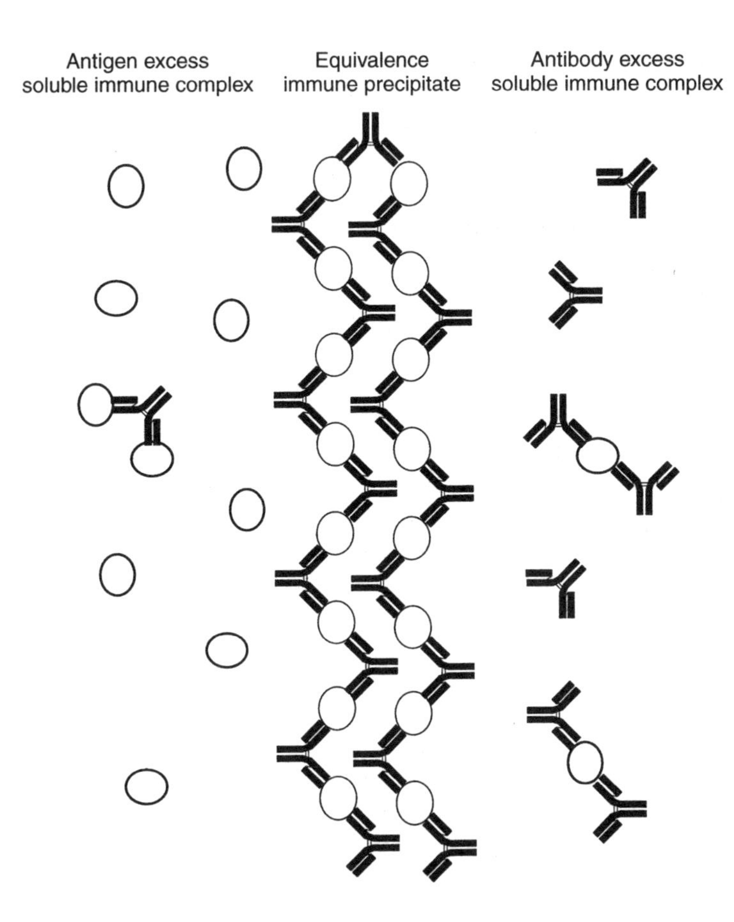

Fig. 23.12 Formation of immunoprecipitates. Antibodies and antigens can combine to form insoluble immune complexes. Thus the antigen is immunoprecipitated by the antibody. For this to occur, the antibody and antigen must be at appropriate concentrations otherwise small, soluble, immune complexes are formed.

formed. This principle is exploited in immunodiffusion reactions where antibody and antigen are allowed to diffuse towards each other in a semi-solid agarose layer. Immunoprecipitin lines can seen by eye where the points of equivalence have been reached (see Fig. 23.12). With appropriate standards and controls it is possible to adapt this technique to estimate the concentrations of antigen or antibody in mixtures and even to assess the purity of them.

An alternative strategy is to immobilise the antibody on to a solid matrix support such as on Sepharose beads using a covalent chemical reaction. The beads can then be packed into a column and solutions containing the antigen passed through (see Fig. 23.13). The antibody will remove the antigen from the rest of the mixture by a process of affinity chromatography. This process can be of direct use, for example in the removal of a contaminant such as a toxin from another protein, when an antibody or antisera specific for the toxin exists. If the antigen which is adsorbed to the antibody on the matrix is required, it is necessary to find conditions which disrupt the affinity of binding. Several methods are appropriate under different conditions. For some low affinity antibody-antigen interactions it may be achieved by competition with an alternative ligand. For higher affinity interactions, it is usually necessary to use partially denaturing conditions and chaotropic agents or extremes of pH. There is often a compromise which has to be taken

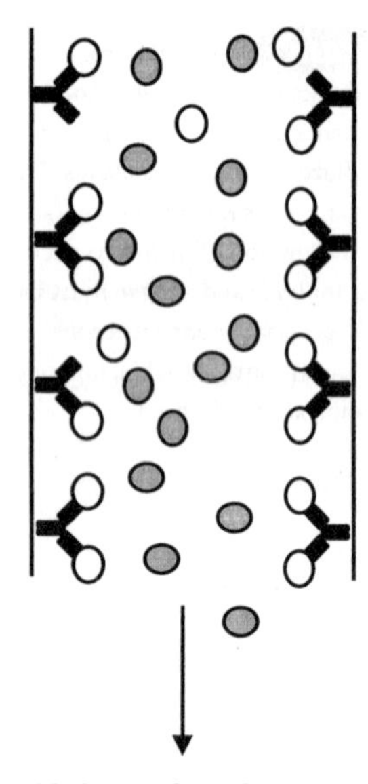

Fig. 23.13 Affinity chromatography using an immobilised antibody. Purification based on the antibody's affinity for antigen can be easily carried out using antibody immobilised on a column matrix. In the schematic shown the antibody is removing the white antigen from the grey antigen. Thus the fluid which flows through the column should be depleted of antigen to which the antibody is specific whereas the fraction which is at first bound and then later eluted in a subsequent step will be enriched for antigen. The elution of bound antigen is usually carried out using mildly denaturing conditions e.g. low pH buffers.

between the ease of elution of the antigen and the long-term stability of the antibody on the column (if it is to be re-used) or of the antigen (if it is required intact and functional).

Immunoprecipitation reactions are often used in experimental situations where the antigen mixture is radiolabelled and then run in gel electrophoresis. Immunoprecipitating the antigen, or purifying it on an antibody-affinity column, allows the individual radiolabelled components to be separated and identified by their reactivity with the antibody. As described above, affinity purification on antibody columns can be used either to remove contaminants from a mixture e.g. toxins or alternatively to purify an antigen out of a mixture. These affinity columns tend to work best when its antibody is in excess over the antigen, whereas direct precipitation relies on antibody-antigen equivalence. If the column is to be recycled and re-used, or the antigen is to be recovered intact, then it is important that the antibody-antigen affinity is not too great. However if it is important that, for example, all traces of an antigen, such as a toxin, are removed from a mixture, the antibody on the column must be in excess over antigen and not of too low an affinity.

It is possible to adapt the methods above to use indirect methods. Thus, molecules with an affinity for antibody such as Protein A or Protein G can be coupled to an affinity matrix and a mixture of antibody and antigen passed through. The binding of antibody to the column will indirectly adsorb the antigen. Similarly, antibodies can be chemically modified with chemical haptens (small molecules which can be recognised by antibodies but which must be coupled covalently to carrier proteins in order to make them immunogenic for use in immunisation), such as biotin, and then molecules, such as avidin or streptavidin, with affinity for biotin can be attached to the column matrix. Also anti-immunoglobulin antibodies (e.g. sheep antibodies specific for mouse IgG) can be used to affinity adsorb or immunoprecipitate soluble immune complexes.

23.10.2 Diagnostics

An important use of antibodies and, in particular, monoclonal antibodies is in diagnostic applications. The specificity of antibodies allows them to be used for the direct determination of the antigen even in complex mixtures. For example they can be used to determine concentrations of a single hormone in samples of human blood. With appropriate standards and controls, the detection methods can quantify the chosen antigens in the system usually by labelling the antibody with a marker which itself is quantitatively determined. Commonly used labels are radioisotopes, enzymes or fluorochromes. Appropriate detection systems are then used to detect these.

The **enzyme-linked immunosorbent assay** (or **ELISA**) is one of the most commonly used diagnostic techniques in use today. The basic principle is simple. An enzyme is coupled directly to an antibody, usually using a chemical cross-linking procedure. The amount of antibody bound to an antigen can then be determined indirectly by measuring the conversion of a substrate to a product by the enzyme. This is stoichiometric but also includes an amplification of the signal because one molecule of enzyme can convert many molecules of substrate over a given time. If coloured substrates are used or coloured products are formed a simple photometric adsorbancy measurement will quantify the enzyme reaction. It is also possible to make this measurement as a real time determination of the rate of the reaction.

There are many subtle variations on the basic ELISA system. In its simplest form, an antigen is adsorbed on to a solid surface, either non-specifically, or through an affinity ligand or covalent chemical bond. An enzyme-labelled antibody is then added in excess to the system and some of it binds to the immobilised antigen. Excess antibody is removed by washing and then substrate is added. The amount of enzyme, and hence amount of antibody–antigen complex in the system, is estimated from the amount of substrate it converts. More usually, ELISA involves a two site recognition with two different antibodies or an indirect detection (see Fig. 23.14). For example, one antibody may be immobilised on a solid matrix and used to capture ('affinity adsorb') the antigen. The amount of antigen captured can be determined by a second antibody coupled to enzyme which recognises a different site on the antigen and so does not compete with the first antibody. Indirect detection systems can employ multiple layers of anti-antibodies or of biotin-avidin layers giving even greater amplification in the system. Alternatively, the systems can be designed to determine the quantity of unknown antigen in the system through competition with binding a known amount of a labelled and pure form of the same antigen (a method originally widely employed in radioimmunoassays). The maximum binding of labelled antigen is seen when there is no competitor antigen in the test sample and the minimum binding of labelled antigen is seen when there is a huge excess of competitor antigen in the test sample.

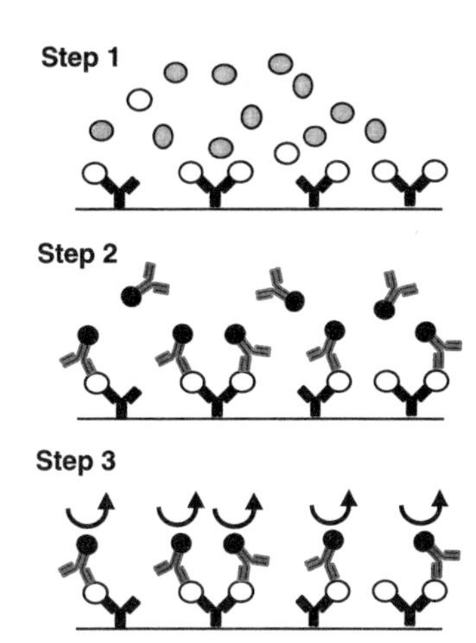

Fig. 23.14 A simple, two-site ELISA procedure. In the first step antibody which is immobilised onto a surface is used to capture the antigen from solution. The excess and therefore unbound antigen is then washed away. In a second step an enzyme-labelled antibody specific to a second site on the antigen is added. Again the excess labelled antibody which does not bind to the antigen is then washed away. Finally a substrate is added and the conversion of this by the enzyme is determined over a given time period. Usually a colour change resulting from formation of a coloured product is monitored using a spectrophotometer.

Enzyme-labelled antibodies are also employed in **immunocytochemistry**. Tissue sections or cell cytosmears are prepared on glass microscope slides. These are then incubated with antibodies specific for different tissue antigens and coupled with enzymes. After washing away excess antibody, the enzyme substrates are added. Substrates are chosen such that insoluble, coloured products are deposited in the section and these can be visualised in light microscopy and, along with suitable counter staining, may allow for very detailed classification of the cells stained. Using appropriate counter stains and, through co-localisation of test antibodies with known markers, it is possible to identify which parts of the cell, e.g. surface, cytoplasmic or nuclear staining, contain the antigen recognised by the antibody. Again, as described above for the ELISA system, the techniques can be modified to use multiple layers of antibody and anti-antibody in order to amplify the staining. As well as enzyme-labelled antibodies, it is possible to use antibodies coupled to fluorescent dyes (fluorophores) and to use them in fluorescent microscopy. Fluorescently conjugated antibodies can be used in the powerful technique of **confocal microscopy** which allows the precise localisation of the fluorescence on, or in, the cell to be visualised in a time-dependent way. In confocal microscopy images collected in a precise focal plane are digitised and then stored in a computer. These digitised images, which thus represent 'slices' through the cell, can then be built up into a three-dimensional representation of the intensity of fluorescence throughout the cell and a model can be displayed on a high resolution graphics monitor. If images are collected at intervals over a given time, then the computer can also be used to generate a time-lapse movie of the movement of fluorescence within the cell.

Fluorescent antibody cell sorting and analysis have developed in parallel with development of the monoclonal antibody technology. Again, the principles of the technique are simple. Monoclonal antibodies are labelled with a fluorochrome and used to stain cells. These cells are then passed at high velocity through a nozzle in a stream of liquid droplets such that the cells pass one at a time through a laser light beam which excites the fluorophore. Detectors then measure the fluorescence output from each cell on an individual basis. At the same time, other properties of the cells can be measured through their abilities to scatter the light beam (size and granularity of the cells). Also, different fluorophores with different emission spectra can be used to tag different antibodies. Thus, a very sophisticated analysis of cells even in a complex mixture can be carried out; for example human blood cells can be separated into their various types and sub-types. The whole classification of human (and now other animal) cell surface antigens using the 'CD', which stands for cluster designation, nomenclature has relied very heavily on the use of fluorescent cell analysis. The results from the analysis, carried out in many laboratories, on panels of monclonal antibodies are used to cluster these antibodies into groups with similar reactivities, and provided this clustering seems to be statistically robust, an international committee authorises the designation of a cluster with a new sequential number in the CD series (e.g. CD1, CD2, CD3 etc.). Although many people now commonly refer to the antigens by the CD name, the original designation was of groups of antibodies. Thus 'anti-CD1 antibodies', for example, is nonsensical in the purest sense since the cluster of antibodies is CD1, and the antigen is that which is recognised by the CD1 cluster of antibodies.

23.11 *In vivo* uses of recombinant and monoclonal antibodies

Again, the uses of antibodies *in vivo* rely heavily on their great specificity for antigen. It is sometimes easy to forget when dealing with uses of monoclonal antibodies *in vivo* that our own antibodies play a major role in our natural immune system in protecting us from infection by killing pathogens and by removing harmful antigens from our system. However, despite this obvious role for antibodies, there are, in fact, only a few therapies currently in use which exploit monoclonal antibodies for these properties. This is largely due to the commercial, practical and also ethical considerations involved in developing antibody therapies.

It is an enormously expensive and also a time-consuming undertaking to get even a single antibody through clinical trials and regulatory approval for widespread commercial sale and use. The situation is further complicated if accepted, existing treatments are already in use. Thus, polyclonal human IgG is manufactured and given (e.g. as a preparation called IVIG, intravenous IgG) for many disorders where passive immunity might be beneficial. Equally, polyclonal horse or sheep antisera against bacterial toxins (e.g. tetanus toxin), snake venoms or toxic drugs (e.g. digoxin) are a tried and tested treatment for acute poisoning. These antibody treatments work and could, in theory, be replaced with monoclonal antibodies but this may not be economically and practically viable. Clearly, there is a potential role for monoclonal antibody-based therapies wherever efficacy has already been demonstrated for polyclonal antisera. The obstacles are mainly commercial and regulatory issues. One example where these barriers might be overcome is the likely replacement of polyclonal human anti-RhD antibodies, used in the prevention of haemolytic disease of the newborn (HDN), with a monoclonal or mixture of monoclonal human antibodies specific for RhD antigen. The RhD antigen is not expressed on the red blood cells of a significant percentage of the population, and thus, mothers who are blood group RhD negative are able to make immune responses to their foetus's red blood cells if they are RhD positive as a result of inheriting this phenotype from their father. The mothers anti-RhD antibody can cross the placenta (if it is IgG) and cause red blood cell destruction in her unborn child. HDN does not affect the first born RhD-positive child because the mother's immune response takes time to develop IgG antibodies to the RhD antigen. However, during subsequent pregnancies with RhD antigen positive foetuses, the IgG antibodies develop more quickly (secondary immune response). It has been found that administering anti-RhD antibodies to the mother at the time of birth can suppress her immunisation by RhD antigens. In the example of RhD, concerns over the safe use (in the light of for example bovine spongiform encephalopathy (BSE), and new variant Creutzfeldt–Jakob disease, (nvCJD)) of pooled blood products, for treatment of young healthy women of child-bearing age, may play a significant role in pushing forward the switch to a monoclonal based product.

Thus, in contrast, many antibodies have been developed for use *in vivo* in situations where natural antibodies may not play any significant role, such as in attempts to eradicate tumour cells from the body. This is mainly because alternative treatments are not available making it logistically easier to try new antibody-based therapies. Requiring monoclonal antibodies to achieve what polyclonal antibodies are unable to do may be one reason for the large number of apparently unsuccessful antibody-based trials.

Imaging is an area where the combination of antibodies and modern computerised techniques can provide a detailed analytical tool for looking inside the human body in a non-invasive way. Thus, for example, radiolabelled antibodies which localise to a tumour can be detected with gamma cameras and, by using a series of moving detectors, a three-dimensional image can be built up as a computer model showing exactly where the labelled antibodies are sequestered. There are many problems with this technology: for example, for antibodies of moderate affinity only a fraction will become localised to antigen with the rest remaining unbound. Also, some antibody will be taken up non-specifically by some tissues or even specifically through Fc receptors and also receptors for carbohydrate. One way round this is to image two antibodies with different isotopes. One antibody will be chosen to be specific for antigen, for example a tumour-associated antigen, the other antibody will be a matched control but with no specificity for the antigen. The two images can be subtracted one from the other leaving just the image of the specific binding. In this way, the antibody appears to be more specific than it really is but it does provide a useful diagnostic tool in looking for tumour metastasis and similar malignancies.

Cancer therapy is one application where most people are familiar with the concept of antibodies as so-called 'magic bullets', a term used to describe them by the popular press and on broadcast news items. The idea is that the specificity of the antibody allows it to target tumour cells for destruction. The problems here are two-fold: first, it is necessary to identify a suitable specificity associated with the tumour; second, the antibody must be capable of delivering some kind of destructive effector mechanism to the tumour cells. Tumour-associated or tumour-specific antigens are not easy to identify and the examples where they can be found are usually such that a new antibody would have to be made for each patient. It is then often the case that tumour cells are resistant to killing by natural antibody-effector mechanisms, such as through complement or processes that are triggered by cross-linking of Fc receptors. There have been some examples where antibodies seem to be effective at least in a proportion of patients with some types of tumour e.g. leukaemias and lymphomas but there have been many failures in clinical trials.

As an alternative to natural effector mechanisms in tumour cell targeting and destruction some scientists have tried coupling other toxic agents to antibodies. Obviously, radioisotopes may deliver a lethal radiation dose to the tumour tissue providing there is a high enough degree of specific localisation of the antibody (a function of the antibodies' affinity, half-life and ease of tissue penetration). Others have tried coupling highly active toxins to antibodies, such as the plant toxins ricin, abrin and gelonin. These work very well against some target antigens and for some cell types but there are still problems with non-specific toxicity to the patient versus the degree of tumour cell kill. The other aspect is that the toxins seem to be highly immunogenic and, in the medium to long term, provoke a strong antiglobulin response to the antibody-toxin conjugate. As another alternative approach, enzymes can be coupled to antibodies which will convert non-toxic, pro-drugs to highly toxic but short-lived active drugs at the site of tumour localisation. I would argue this is reminiscent of how the natural complement system works. One problem with all these strategies is that the degree of tumour localisation is critical. It is undesirable to have too much antibody circulating round the body and triggering non-specific toxicity in other tissues. Ironically, natural effector mechanisms have evolved to work under precisely these conditions, i.e. an antibody excess. They rely mainly on a succession of low affinity, but higher avidity, steps to distinguish immune complexes from free antibody.

One area where antibody-based therapies have had some success is in specifically targeting cells involved in immune functions and thus creating a state of **immunosuppression**. Antibodies have been targeted at whole populations of cells, such as all lymphocytes, at specific lineages, such as T-lymphocytes, or at activation-antigens expressed only by smaller sub-populations of cells, e.g. those expressing certain cytokine receptors. Earlier strategies were aimed at killing these cells either using natural antibody effector mechanisms or through use of immunotoxins. More recently, there has been a shift in antibody therapy towards the use of non-depleting blocking antibodies. This comes from the realisation that cells respond to different signals and that the very nature of the signals, e.g. whether they are linked together or independent, can either result in cells which participate in an inflammatory reaction or, alternatively, regulatory cells which can attenuate a response. Through the use of antibodies which are able to block normal cellular processes, it is hoped that cells can be re-programmed in autoimmune reactions to stop reacting against self and, similarly, the immune system might be taught to accept foreign-grafted tissues.

These blocking functions of antibodies require that they still are able to bind to antigen but that they should not activate complement or trigger Fc receptors on effector cells. Such properties can be achieved by modifying sequences within the constant regions of antibodies known to be critical for individual antibody functions. As an additional step on from these strategies, chimaeric molecules are being constructed in which the genes encoding Fc regions of antibodies are combined with genes encoding domains of cytokine receptors or adhesion molecules to create **immunoadhesins**. These domains replace the antibody variable region but still provide a highly specific recognition of a ligand. The Fc region provides the whole molecule with a multiple valency and also a longer half-life.

As mentioned above the problems with use of antibodies *in vivo* is that the development time and clinical trials are procedures that last for many years. Thus, many of the antibodies in the final stages of clinical trials today are based on scientific ideas of perhaps ten or more years ago. Equally, it will be many years before some of the newest ideas in laboratory science today find their way into the next generation of clinical trials.

23.12 Further reading

Birch, J. R. and Lennox, E. S. (eds.) (1995). *Monoclonal Antibodies, Principles and Applications*. John Wiley, New York.

Capra, J. D. (ed.) (1997). *Antibody Engineering, Vol. 65, Chemical Immunology*. Karger, Basel.

Goldsby, R.A., Kindt, T.J. and Osborne, B.A. (2000). *Kuby Immunology, 4th Edition*. W.H. Freeman and Company, New York.

Janeway, C. A. and Travers, P. (1999). *Immunobiology, 4th Edition*. Churchill Livingston, Edinburgh.

Harris, W. J. and Adair, J. R. (eds.) (1997). *Antibody Therapeutics*. CRC Press, New York.

King, D. J. (1998). *Applications and Engineering of Monoclonal Antibodies*, Taylor and Francis, London.

Male, D., Cooke, A., Owen, M., Trowsdale, J. and Champion. B. (1996). *Advanced Immunology, 3rd Edition*. Mosby, London.

Chapter 24

Environmental applications

Philippe Vandevivere and Willy Verstraete

24.1 Introduction

Until recently, **sanitary engineering** monopolised environmental related industrial activities. Because sanitary engineering gradually developed as an offshoot of civil engineering during the past century, emphasis has been on conventional engineering techniques in which the 'bio' component is largely ignored and dealt with stochastically rather than mechanistically. Sanitary engineering is well established for:

- the catchment, treatment and distribution of drinking water;
- the treatment of waste water;
- the treatment and disposal of solid wastes (e.g. municipal);
- the treatment of industrial off-gases.

Many of the conventional technologies used in sanitary engineering are, however, perfect illustrations of Murphy's law in that they transform one problem into another often more intractable one, as when water pollutants are stripped into the air or concentrated and dumped in the soil. Environmental strategies have to be conceived with respect to the 'whole' of the environment in a long-term perspective. This integrated holistic approach requires a detailed knowledge of environmental biology and, more particularly, of the functioning of complex microbial communities. The new focus on the environment as a whole and on the detailed functioning of the 'bio' component has led to the development of new industrial activities, referred to as **environmental biotechnologies**. These must address formidable environmental problems now facing the world:

- acid rain and ozone depletion;
- enrichment of ground and surface waters with nutrients and recalcitrant pesticides;
- recovery of reusable products and energy from wastes;
- soil remediation;
- disposal of animal manures.

While industrial biotechnologists use well-defined micro-organisms to make products of predictable composition and quality such as lactic acid, beer or monosodium glutamate, environmental biotechnologists, on the other hand, start with poorly defined inocula and wait until desired phenomena occur. There is therefore a need to isolate, identify and characterise the micro-organisms which exist and interact in soils, activated sludges, anaerobic granules, etc. Only when it will become possible to re-assemble these micro-organisms and their functions in a predictable way will environmental biotechnology become more generally accepted.

New developments are concerned with the introduction of organisms and genes in mixed cultures. Practical application of these new developments is somewhat impeded by poor survival of introduced micro-organisms and regulatory constraints on deliberate introduction of modified organisms in the environment. The potential is, however, enormous as advances in molecular biology now make feasible the construction of novel genes and enzymes for the degradation of compounds which could not, until now, be biodegraded. These novel genes may become incorporated in the genomes of existing microbial communities, a process called 'horizontal gene transfer'. For example, broad host range plasmids specialised in the degradation of synthetic chemicals, can be introduced into soil microbial communities, thereby enhancing their degradative capabilities.

24.2 Treatment of waste water

24.2.1 Aerobic treatment by the activated sludge system

The most widely used process to purify waste water is via aerobic biodegradation with the activated sludge system (see Box 24.1 for definitions). The waste water flows through an aerated tank where the dissolved organic matter is mineralised, i.e. oxidised to carbon dioxide, nitrate and phosphate:

$$\text{dissolved organic matter} + O_2 \rightarrow \text{new biomass} + CO_2 + HNO_3 + H_3PO_4$$

This reaction is carried out mostly by bacteria which are aggregated in flocs, about 0.1 mm in diameter. After a reaction time of several hours (municipal sewage) up to several days (more concentrated industrial effluents), the **mixed liquor** flows through a settling tank where the flocs are separated by gravity from the clean effluent (Fig. 24.1). The concentration of flocs in the aerated tank should not exceed 4 g l^{-1} in order to ensure proper settling. The settled flocs (called the **sludge**) are partly

Box 24.1 Treatment of waste water – definitions

BOD_5: Biological oxygen demand (after 5 days of incubation) is a parameter which quantifies the concentration of biodegradable organic matter present in waste water. It is the amount of O_2 used by micro-organisms to degrade the organic matter as determined in a standardised laboratory test.
Mixed liquor: is the suspension of microbial flocs (tiny aggregates of micro-organisms) in the aeration tank of an activated sludge plant.
Sludge: refers to the microbial flocs in an activated sludge plant after these have been separated from the purified effluent via sedimentation in the settling tank.
Bulking sludge: refers to a sludge wherein overgrowth of filamentous micro-organisms prevents the sedimentation of the 'bulky' microbial flocs.
Nitrification: is the biochemical conversion of ammonium to nitrate carried out by autotrophic bacteria. It is an essential step during biological nitrogen removal in waste water treatment plants, following the mineralisation of organic nitrogen to ammonium.
Denitrification: refers to the biological reduction of nitrate to N_2. It occurs when O_2 is absent (anoxic conditions) and readily oxidisable organic compounds are present.
Anoxic: refers to a liquid wherein O_2 is absent but other oxidised species such as nitrate or ferric iron are present.
Anaerobic: refers to a liquid wherein oxidised species are absent, the redox potential is below zero, and where biochemical reactions such as fermentations, sulphate reduction and methanogenesis take place.

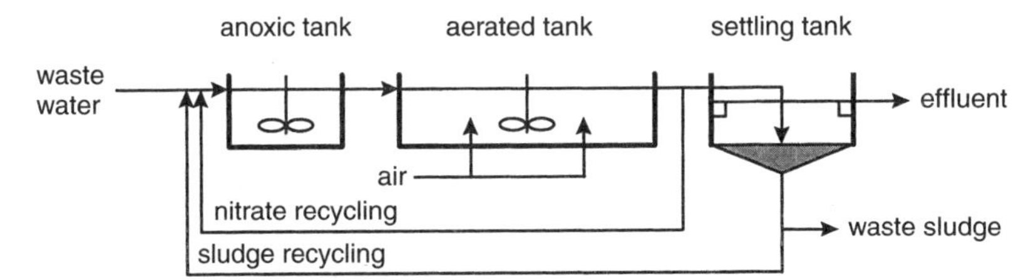

Fig. 24.1 Process flow diagram of an activated sludge plant with biological nitrogen removal. Since the first tank is anoxic (oxygen-free), micro-organisms use nitrates to oxidise the organic matter to carbon dioxide and ammonium, thereby reducing nitrate to dinitrogen gas (denitrification). In the subsequent aerated tank, residual organic matter is oxidised with oxygen as electron acceptor. Simultaneously, ammonium is oxidised to nitrate (nitrification) which is then recycled to the anoxic tank. The micro-organisms are separated from the clean effluent in the settling tank.

re-injected in the aerated tank and partly wasted. Good performance depends on the right choice of volumetric loading rate, which should lie in the range 0.5–1.5 g **BOD_5** per litre mixed liquor per day in order to ensure proper floc formation and obtain 90+% removal of dissolved organic matter. As one inhabitant equivalent produces 30 g BOD_5 per day on the average, with peak values of 100 g per day, aerated tanks are designed to have 100 litres mixed liquor for each inhabitant equivalent.

Table 24.1 Comparison of different processes for the treatment of waste water. Activated sludge systems provide good effluent quality but suffer several drawbacks. Membrane bio-reactors achieve the same or better effluent quality in much smaller installations and produce much less waste sludge. UASB reactors offer the additional advantage of small energy expenditure (no aeration) but are less efficient in terms of nutrients and BOD_5 removal

	Aerobic treatment		Anaerobic
	Activated sludge	MBR	UASB
Residual BOD	low	very low	high
Residual N, P	low	low	high
Sludge production	high	very low	very low
Energy	high	high	low
Floor area	large	very small	very small
Reliability	sludge bulking	robust	granule flotation

Notes:
MBR, membrane bio-reactor.
UASB, upflow anaerobic sludge blanket reactor.

The primary advantage of the activated sludge process, relative to other types of treatment, is a good effluent quality, with little BOD_5 (<20 mg l^{-1}) and little nutrients (<15 mg N l^{-1}) remaining after treatment. The process suffers however several drawbacks (Table 24.1). The biggest drawback is the large production of excess sludge since each kg of BOD_5 produces about 0.3 kg of excess sludge solids. This excess sludge is usually stabilised in anaerobic digesters, dehydrated, and finally disposed on agricultural land or landfilled. Disposal on land or in landfills is however becoming increasingly restricted in Europe and sludge disposal is becoming problematic.

Production of excess sludge can be somewhat lessened by including a carrier material in the aerated tank. In such reactors, the micro-organisms will not form suspended flocs as in the activated sludge system but rather form a film on the surface of the carrier material. The latter can be stones, in which case the aerated tank is called a trickling filter, or fine suspended sand particles, as in fluidised bed reactors. These variants of the activated sludge process produce less excess sludge because the attached micro-organisms remain longer in the aerated tank where autolysis takes place. Moreover, no settling tank is necessary. It was recently observed that excess sludge production in waste water treatment plants does occasionally drop drastically during periods of a few weeks. These periods correspond to the sporadic growth of tiny worms (*Nais elinguis*) which graze on the sludge flocs (Fig. 24.2). If the current attempts to ensure the continuous presence of these worms in mixed liquors succeed, it would provide a very elegant and simple solution to the problem of sludge disposal.

An important breakthrough in terms of excess sludge production was the **membrane bio-reactor** (MBR). In MBR, the settling tank is replaced by a micro-filtration unit which separates the micro-organisms

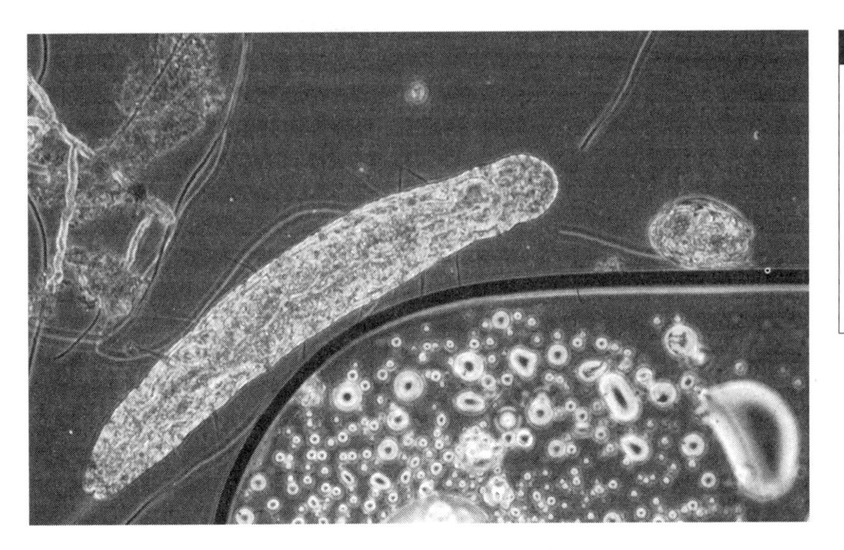

Fig. 24.2 Floc of micro-organisms as they occur in the mixed liquor of activated sludge tanks. Note the presence of a worm (*Nais elinguis*), which grazes the flocs actively. These worms could offer a very simple and elegant solution to the problem of sludge disposal (courtesy of Prof. Eikelboom).

from the treated effluent. In such a system, separated sludge can be recirculated almost indefinitely in the aerated tank and, under these circumstances, sludge age is very long and excess sludge production very low (<0.1 kg per kg BOD removed). A second major advantage of MBR is that a very high sludge concentration is attained (up to 30 g l^{-1}) which allows much larger volumetric loading rates to be used than in the activated sludge system. As a consequence, very compact MBR installations can be built on a small fraction of the space required by an activated sludge plant. This small footprint is very attractive to industries producing concentrated waste waters (BOD_5 > 2–3 g l^{-1}).

24.2.2 Anaerobic treatment of waste water

Until recently, anaerobic digestion was only applied for the stabilisation of concentrated organic slurries such as animal manures and waste sewage sludge. The consensus was that anaerobic was slow, did not remove much more than 50% of the organic load and, moreover, required high temperatures and was not reliable. This perception has changed drastically during the last two decades and anaerobic digestion is now an established-performance high-rate waste water treatment technology. Through the use of anaerobic granular sludge, very high biomass concentrations (50 g l^{-1}) can be attained in reactors, allowing very high volumetric loading rates to be used (20 g BOD_5 per litre reactor per day). New reactor designs that optimise the mass transfer of metabolites in the granular sludge now make it possible to treat waste water of almost any composition (0.3–100 g BOD_5 l^{-1}) over a temperature range of 10–55 °C in a reliable manner.

Anaerobic conversion of organic compounds to biogas is a stepwise process wherein different groups of bacteria operating sequentially effect full degradation of the substrates:

- hydrolytic acidogens: cleave polymers into short chain fatty acids;
- syntrophic acetogens: degrade the fatty acids into acetate and H_2;
- methanogens: transform acetate and H_2 into CH_4 and CO_2 (biogas).

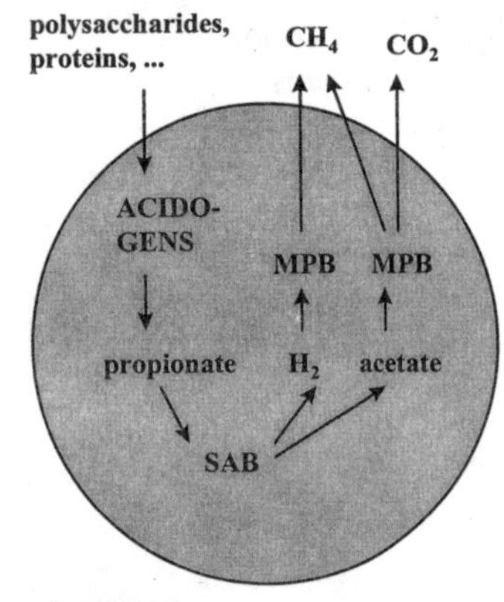

Fig. 24.3 Sequence of biochemical reactions taking place in an anaerobic sludge granule, e.g. in an upflow anaerobic sludge blanket reactor treating waste water. The sequence of reactions is thermodynamically favourable only in a narrow range of very low H_2 partial pressures. Growth closely together of acetogenic and methanogenic bacteria in a packed granule make the transfer of H_2 at low partial pressures much more efficient. SAB, syntrophic acetogenic bacteria; MPB, methane-producing bacteria.

The two latter groups are normally strictly dependent on one another (due to H_2 transfer) and are therefore referred to as the methanogenic association. Their metabolism is greatly enhanced by growing the anaerobic sludge in the form of densely packed granules which facilitate the transfer of H_2 and other intermediate degradation products (Fig. 24.3). The understanding of syntrophism, where several anaerobic micro-organisms can share the energy available in the bioconversion of a molecule to CH_4 and CO_2 and thus can achieve intermediate reactions which are endergonic under standard conditions, has been essential in the rather striking development of anaerobic digestion during the last decades. It has been postulated that the minimum energy quantum for life is about -21 kJ per mol product formed or substrate converted. Applying the concept of minimal energy to the fermentation of propionate to methane suggests that both syntrophs have to operate in a very narrow region of H_2 partial pressure, pH_2 (Fig. 24.4). The conversion of one mole propionate yields >21 kJ only when $pH_2 < 10^{-5.4}$ atm, while the minimum pH_2 value allowing the production of one mol methane to generate -21 kJ is also in the range 10^{-5} atm. Thus only when pH_2 lies around 10^{-5} atm is the sequential conversion of propionate to acetate and acetate to methane possible. Similar conclusions can be drawn for the conversion of butyrate to methane and also with formate instead of H_2 as intermediate. The understanding of the nature of the 'symbiosis' among syntrophic organisms is a challenging task and essential to the optimisation of anaerobic biotechnology.

Most anaerobic reactors treating waste waters are **upflow anaerobic sludge blanket**, or UASB, reactors (Fig. 24.5). The waste water enters the reactor at the bottom via a specially designed influent distribution system and subsequently flows through a sludge bed consisting of anaerobic bacteria growing in the form of granules which settle very well (60–80 m h^{-1}). The mixture of sludge, biogas and water is separated in the three phase separator situated in the top of the reactor.

The major advantages of anerobic waste water treatment over aerobic treatment are the small sludge production (0.1 kg per kg BOD_5), the low energy consumption since no aeration is required and the small floor area, typically 0.01 m^2 per inhabitant compared to 0.05 m^2 for activated sludge plants (Table 24.1). Moreover, energy is recovered in the form of biogas (0.35 l methane per g BOD_5).

The rate of BOD removal in anaerobic reactors drops markedly below 20 °C and the optimum temperature is about 35 °C. This explains why UASB reactors were first applied in tropical regions about 20 years ago. In temperate regions, UASB reactors have only been used to treat concentrated waste water (>2 g BOD_5 l^{-1}) since the large production of biogas can be used to warm up the reactor. A new reactor design has recently been developed which permits sufficiently high rates to be attained even at 10 °C. This so-called **expanded granulated sludge blanket** (EGSB) reactor maximises mass transfer rates of nutrients with more intensive hydraulic mixing and makes it possible to treat sewage anaerobically even in temperate regions.

The major disadvantage of anaerobic digestion is that only negligible portions of the nutrients (N, P) are removed, due to the small excess sludge production. It is therefore necessary to apply a post-treatment step in order to further remove these nutrients, e.g. via the sequence nitrification/denitrification. The aerobic post-treatment step is also necessary to remove the residual BOD_5 remaining in the UASB effluent, because anaerobic bacteria do not easily scavenge substrates present at less than 50 mg l^{-1} while aerobic bacteria can easily lower BOD under 10 mg l^{-1}. The fact that nitrification requires costly aeration and that denitrification requires oxidisable organic matter (which is degraded in the prior anaerobic step!) has spurred the search for alternative types of post-treatments. A very interesting alternative, currently under development, uses the Anammox reaction (anaerobic ammonium oxidation). It was found that NH_4^+ was oxidised anaerobically to N_2 in the presence of NO_2^- according to:

$$NH_4^+ + NO_2^- \rightarrow N_2 + 2H_2O \qquad \Delta G^{\circ\prime} = -358 \text{ kJ per mol of } NH_4^+$$

Thus by splitting the ammonium-laden anaerobic effluent into two sub-streams, nitrifying partially one sub-stream to nitrite, and mixing again the two streams in a reactor where the Anammox reaction would occur, much less aeration would be required for the nitrification and no oxidisable organic matter would have to be added. The full scale implementation of the Anammox reaction would open new doors for anaerobic digestion because it would enable a coherent sequence of organic carbon to methane and organic N via ammonium and nitrite to N_2. Using this scenario, even N-rich waste water could be treated anaerobically at low cost.

Direct anaerobic treatment of domestic sewage, either in the sewer or in low-capital anaerobic-aerobic combined plants, will only attract the interest of the environmental industry provided it offers adequate profit margins. Hence, the challenge is to locate in anaerobic sewage treatment opportunities for high-tech added-value engineering. Two possibilities are discussed below.

Development of engineered anaerobic granulated sludges (biocatalyst)

Certain organic compounds produced by the chemical industry (xenobiotics) are not degraded in either aerobic or anaerobic digesters but are degraded in a sequential anaerobic/aerobic treatment. Examples are organic compounds with halo, nitro, or azo substituents. It may take however several months and even up to a year in some cases before the sludge becomes adapted to these compounds. This time is needed for development or invasion of all species necessary to completely degrade the substrates, as complex xenobiotics often require more than one species to be completely mineralised. One option to accelerate the biodegradation of xenobiotics is to inoculate reactors with adequate bacterial strains. This was demonstrated with strains capable of dechlorinating chlorobenzoate or pentachlorophenol. The inocula were shown to have colonised the reactor in the long term and rapid breakdown of chlorobenzoate and pentachlorophenol could be

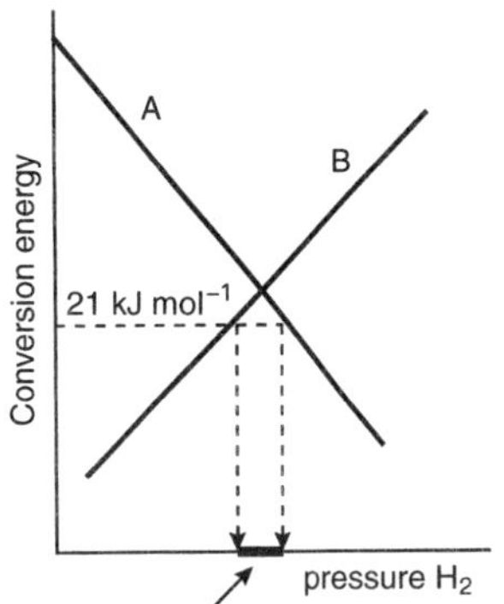

Fig. 24.4 Effect of H_2 partial pressure on the free energy of conversion of propionate by acetogens (reaction A) and of the subsequent transformation by methanogens of H_2 into methane (reaction B). Only in a very narrow range of H_2 partial pressure (around 10^{-5} atm) are both reactions thermodynamically favourable, i.e. they both yield >21 kJ mol^{-1} transformed.

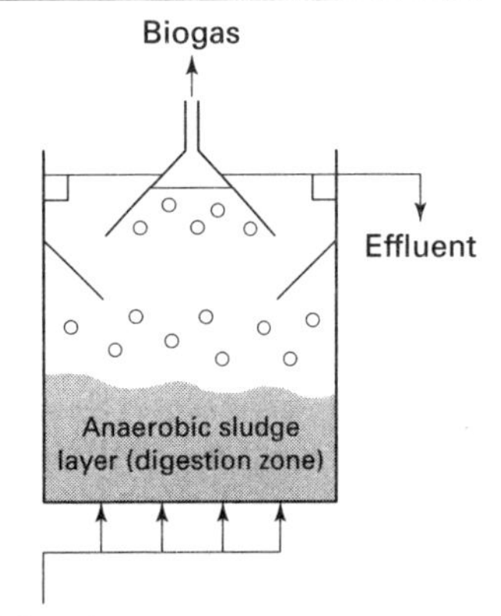

Fig. 24.5 Schematic diagram of the upflow anaerobic sludge blanket reactor (UASB) used extensively for the treatment of concentrated waste waters in temperate regions and also for the treatment of sewage (dilute waste water) in tropical regions.

obtained. Because adaptation of the association is probably also based on the proliferation of the right plasmids, there is clearly a need for better insight in genetic evolution, plasmid transfer and species interaction in anaerobic communities dealing with xenobiotics. Another potential benefit associated with the large-scale availability of specialised microbial consortia is 'biochemical re-routing', i.e. the induction of desirable biochemical pathways, as for example the degradation of malodorous primary amines, anaerobic ammonium oxidation or homoacetogenesis.

Development of performance-enhancing additives

Biomass retention through adequate granulation is of utmost importance in UASB technology, first in order to obtain a good effluent quality and second, in order to ensure a minimal cell residence time of 7 to 12 days which is required to avoid the wash-out of the slowest-growing anaerobic bacteria. One way to foster granular growth is to add polymers, clay or surfactants which have a physico-chemical effect on granule formation. Another way is to provide the adequate nutrients, e.g. sugars, that stimulate the growth of micro-organisms which cement anaerobic granules through the production of extracellular polymers. It appears therefore worthwhile, in order to make UASB technology more reliable, to develop bio-supportive additives able to maintain the granular sludge in a proper state in periods of start-up or low-quality input waste water.

24.2.3 Water recycling

In view of the steadily increasing shortage of water worldwide, the use of reclaimed waste water will be an issue of growing concern in the next decade. Since two-thirds of the world water consumption is used to irrigate cropland, there are several instances in developing countries where raw domestic sewage of very large cities is directly re-used to irrigate food crops. Such a closed loop system brings about the possibility of contaminating the food crops with pathogenic viruses or prions. It is a major challenge to work out cost-effective technologies to produce hygienically safe irrigation water without removing the fertilisers N and P. Anaerobic digestion might, in this respect, offer certain possibilities.

The second main consumer of water is industry, such as the food, metal, textile and paper sectors. These sectors are currently developing new treatment systems enabling them to recycle their waste waters in a closed loop system. Typically, a battery of modular processes are used that produce high-quality process water. The sequence usually combines biological treatments with final physico-chemical polishing treatments. For example, a potato chips factory uses a process train consisting of anaerobic and aerobic treatment, deep bed filtration, disinfection with ozone gas and reverse osmosis. Such a complex treatment system is necessary to achieve complete removal of carbohydrates, anti-sprout herbicides and micro-organisms.

The making of one tonne of steel requires 280 tonnes of water. Efforts to recycle this water in coke plants via activated sludge treatment were confronted by rapid sludge intoxication when more than 50% of the process water was re-used. This was due to the accumulation of highly toxic organic compounds, indicating the need for careful research on residual organics and even microbial products giving rise to abortive metabolism. A great many textile wet-processing plants are currently upgrading their waste water treatment systems in order to recycle water. Because of the greatly variable chemical composition of the liquid effluents, depending on the types of fabrics and dyes being processed, no two textile factories apply the same treatment scheme to treat their effluent (Fig. 24.6).

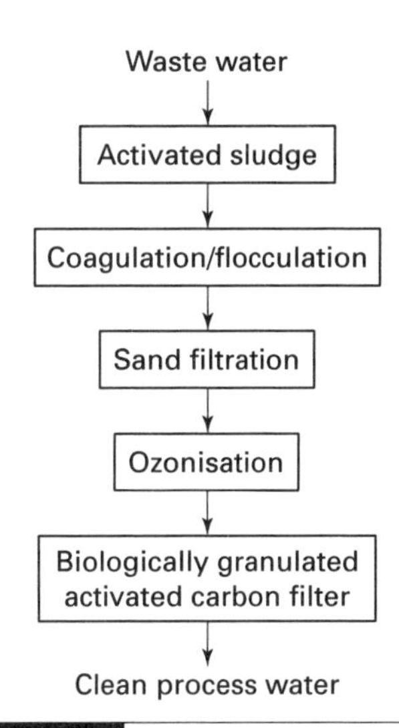

Fig. 24.6 This process flow diagram illustrates the state-of-the-art technology employed in the textile industry to convert large volumes of waste water into high-quality process water used for washing, scouring, bleaching, dyeing and printing. Biological treatments are combined with physico-chemical treatments in order to achieve the required purity. The final biofiltration step on activated carbon, combining physical sorption with *in situ* biodegradation, is necessary to remove toxic compounds produced during the ozonisation step.

24.2.4 Automatisation of waste water treament plants

At present, most biological waste treatment systems, even multi-million dollar plants, generally are operated on the basis of a few rudimentary physical parameters such as pH, dissolved oxygen (DO) or redox potential. DO probes are used in activated sludge plants to minimise energy expenditure to that just necessary to maintain a DO level around 2 mg l^{-1} in the aerated basin. Redox probes are used to monitor ammonium oxidation and nitrate removal in sequencing batch reactors. These control strategies fail however to ensure constant effluent quality because they do not detect variations in load, toxic shocks or process performance. The current control strategies must therefore be supplemented with dynamic mathematical models, i.e. models which can simulate and predict transient responses thus providing flexible automatic control strategies. The use of dynamic models requires the continuous input of data collected with on-line sensors.

On-line biomonitoring devices capable of quantifying the incoming load and effluent quality, and continuously transferring this information to the operation control system are currently being developed. One newly developed on-line biosensor measures the BOD of the incoming waste water and its potential toxicity toward different groups of micro-organisms present in the activated sludge (Fig. 24.7).

The development of other types of biosensors would help to ensure a more stable biological activity and therefore a more reliable treatment. For example, it was mentioned above that the occasional appearance of small worms in activated sludge plants was very beneficial to decrease sludge production (Fig. 24.2). These worms do however disappear as inexplicably as they appear and very variable process performance ensues. On-line biosensors capable of following and predicting the population size of these worms may help to maintain their profitable activity. The same strategy could be employed to stabilise the populations of other very valuable micro-organisms, such as bactivorous protozoa which are essential to obtain good quality effluents, or to show the development of detrimental micro-organisms, such as the filamentous bacteria which cause sludge bulking.

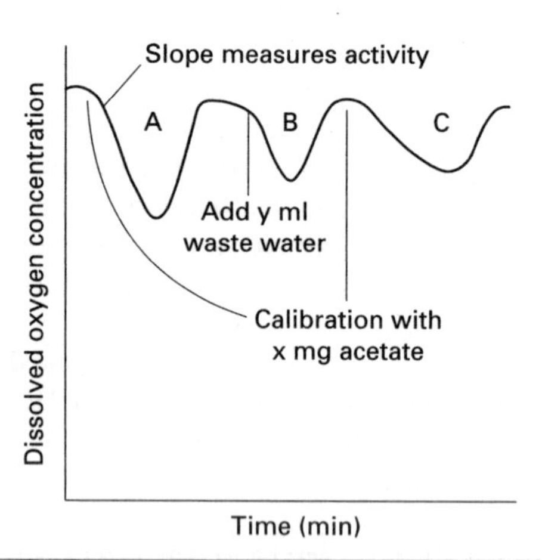

Fig. 24.7 Respirogram obtained with a biosensor used to measure on-line the BOD and potential toxicity of waste water before it enters a treatment plant. The addition of acetate to an aerated vessel containing activated sludge causes a temporary drop in the O_2 concentration (trough A). The initial slope of O_2 decline is an indication of the activity of acetate-utilising micro-organisms, while the surface area of trough A reflects the amount of BOD added. The latter can be used to measure the BOD_5 and the total BOD of a waste water sample by comparing the surface areas of troughs A and B. The BOD measurement is then used to adjust the flow rate to the plant. Trough C, obtained with a second pulse of acetate, indicates that the activity of acetate-utilising micro-organisms has decreased (smaller slope) due to the presence of a toxic compound in the waste water sample added in B. Possible remedial actions are (i) the addition of toxicant-neutralising additives in the main flow to the plant, e.g. powder activated carbon, (ii) the buffering and dilution of the toxic waste water with non-toxic effluent, or (iii) the addition of stored sludge in order to boost the microbial activity. Use of ammonium, or other substrates, in place of acetate allows the activity of other types of micro-organisms to be followed.

24.3 Digestion of organic slurries

Production of organic slurries, e.g. sewage sludge or animal manures, is increasing in many parts of the world causing the traditional disposal schemes, such as their application onto agricultural land, to become saturated. An increasing number of countries are even banning these disposal schemes due to contamination of ground water. More environmentally friendly treatment processes for organic slurries suffer high cost and/or poor efficiency.

A well-known treatment process for sewage sludge and animal manures is anaerobic digestion in completely-mixed anaerobic reactors. During this process, about 50% of the solids are converted to biogas, while the remainder is more or less stabilised. The performance, profitability and biogas output of anaerobic digesters can be increased by co-digesting animal manure or waste sewage sludge with 10–20% solid wastes from the agro- and food-industry such as slaughterhouse, pharmaceutical, kitchen, fermentation or municipal wastes. Many full-scale installations using this co-digestion approach have recently been built in several European countries.

Table 24.2 Design parameters for various types of anaerobic reactors. The loading rate, a measure of the process efficiency, is high with UASB reactors due to biomass retention in sludge granules and high in solid state reactors due to high biomass concentration. Completely mixed reactors share none of these advantages and are therefore less efficient (small loading rate and long hydraulic retention times).

	UASB reactor	Completely mixed reactor	Solid state reactor
Effluent treated	Waste water	Organic slurry	Solid wastes
Solid concentration in reactor ($g\ l^{-1}$)	<50	50–100	200–400
Loading rate (kg organics $m^{-3}\cdot$day)	10–30	2–5	20–40
Hydraulic retention time (days)	0.3–1	20–40	10–20
Solid retention time (days)	>20	20–40	10–20

Notes:
UASB, upflow anaerobic sludge blanket.

The completely mixed reactors treating organic slurries are operated at low volumetric loading rates, i.e. 2 to 5 kg organics $m^{-3}\cdot$day because the particulate organics must be solubilised before they can be subjected to anaerobic conversions (Table 24.2). The rate of solubilisation of particulate organics may be rather slow as in the case of waste activated sludge which takes 15 days to reach 90% hydrolysis. As a consequence, retention times of at least 20 days and up to 60 days or longer are used. Several new developments increase the performance of anaerobic digesters. For example, the hydraulic retention time in the reactor can be uncoupled from the solid retention time by filtering the treated effluent and re-injecting the solids in the reactor until the hydrolysis products pass through the membrane. This reactor design removes a greater proportion of solids due to the longer solid retention time and achieves this in a smaller (cheaper) reactor due to the smaller hydraulic retention time.

Improved performance can also be obtained by running the digestion at higher temperatures since the rate of hydrolysis of particulate matter increases with temperature. New insights in thermophilic digestion resulted in the construction in Denmark of several large-scale thermophilic digestors to treat farm manure. Being run at higher temperatures, these reactors yield a pathogen-free effluent, unlike the mesophilic digesters which often fail to meet the regulations in terms of faecal pathogens (Fig. 24.8). Several drawbacks have, in the past, kept thermophilic digestion from becoming popular, for example the difficulty of start-up and the sensitivity to certain stress factors such as NH_3 and H_2S. Bentonite clay can be used to remove NH_3 inhibition. H_2S, on the other hand, can be destroyed by injecting electron acceptors, e.g. oxygen or nitrate in the reactor.

Perhaps the major problem, at least for sewage sludge digesters, is to minimise the mass of N and P being recycled to the main plant flow via the so-called 'sludge water'. Indeed, more than 50% of the sludge N is hydrolysed during digestion and the resulting recycle load contains typically about 1 g $NH_4^+\ l^{-1}$ and may contribute 20% of the influent N load.

Fig. 24.8 Survival times of various pathogenic micro-organisms upon continuous exposure to different temperatures. Mesophilic reactors, treating animal manures or sewage sludge at 20–30 °C with retention times of one month do not eliminate the *Salmonella* completely. The thermophilic reactors, run at 55 °C, succeed in killing all pathogens after a few days' retention time.

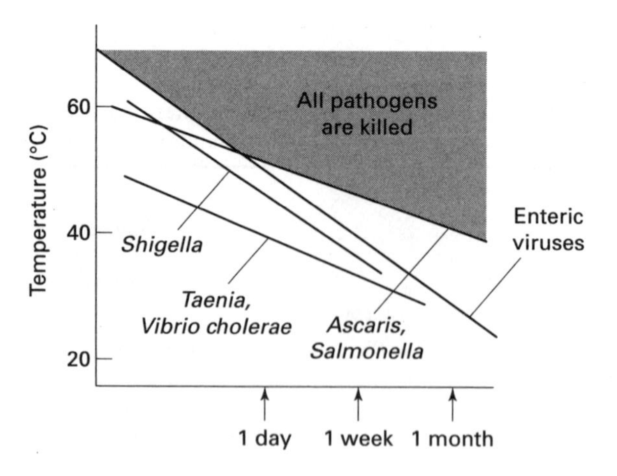

This extra nutrient load may cause problems in view of the new, more stringent, standards concerning the nutrient content of discharged effluents. This may also be the case for P as some investigators have found that up to 60% of the sludge-bound P may be released during anaerobic digestion. Various treatments have in the past been optimised to precipitate P chemically. The cost of these treatments have, however, prevented them from being used in practice. The pH-controlled precipitation with lime seems attractive because the high pH may also serve to remove the ammonium by stripping. The cost associated with the lime addition can be greatly reduced by pre-aerating the effluent in order to remove the buffering capacity associated with the alkalinity. This method can also be combined with the addition of Al or Fe salts, preferably from a cheap source such as Al/Fe-rich sludge from drinking water production plants. Still another method is to optimise the conditions for struvite ($MgNH_4PO_4$) precipitation through cooling and CO_2 stripping.

Current technologies such as aerobic or anaerobic stabilisation, land disposal and incineration of organic slurries could become much less costly provided more efficient and cheaper methods of dewatering sludges from 2–5% to 25–40% dry matter could be developed. A major challenge to environmental biotechnologists is to develop enzymes, products and treatments that will permit more satisfactory dewatering of surplus sludge microbial biomass. New developments are being applied commercially which rely on the heat production during aerobic post-treatment to evaporate the excess water. This 'biological drying' process requires less energy than the thermal drying techniques. One very sensitive problem, however, is the generation of bad odours which have caused the shut-down of several plants.

24.4 Treatment of solid wastes

Solid waste treatment is at present dominated by landfilling and incineration. Landfills are becoming less and less viewed as an option because they prevent the recycling of re-usable products (plastics, paper, construction materials . . .) and they are inefficient in terms of energy (biogas) recuperation. Moreover, landfill leachates and gas emissions pollute the environment. Likewise, incinerators do not allow material recovery though they may be designed to recover energy from waste. Incinerators suffer the drawbacks of high costs (ca. 100–250 euros per tonne municipal waste incinerated) and moreover require very sophisticated and costly flue gas purification systems to avoid environmental harm.

An elegant alternative for the treatment of municipal and industrial solid waste is currently making its way to the market place, the so-called **separation and composting plant**. These are very large and sophisticated plants, working at high capacities (100 000 to 300 000 tonnes of waste per year), and wherein a battery of physical separation units recover the following materials from rubbish:

- **sand** and **gravel** sold as construction material;
- **iron** sold to metallurgic industry;
- **aluminium** and other non-ferrous metals with high re-sale value;
- **cardboard** and **paper** sold to paper industry;
- hard and soft **plastics** re-used or incinerated;
- biodegradable organics transformed into **compost** and **biogas**.

The philosophy of this new type of municipal waste treatment plant, of which the first are being operated in Germany, The Netherlands and Belgium, is to minimise the non-reusable residual fractions which have to be landfilled or incinerated (Fig. 24.9). Since the biodegradable organics constitute ca. 60% of municipal solid wastes, the last item in the above list deserves special attention. The composting of these biodegradable organics is already widely applied in regions where this waste fraction (the vegetable, fruit, and garden waste or bio-waste) is selectively collected.

The organic fraction of municipal solid waste is composted either aerobically or anaerobically. While aerobic composting is a well-known technology and has traditionally been applied, recent developments in anaerobic composting are conferring several advantages to this new technology, making it increasingly attractive and increasing steadily its market share (Table 24.3).

Different environmental companies commercialise various designs of anaerobic digesters of solid waste, differing in terms of:

- solids concentration in the reactor (from 50 to 400 $g\,l^{-1}$);
- temperature (from mesophilic, at 35 °C, to thermophilic, at 55 °C);
- number on stages (one or two).

One such design, the DRANCO process (**Dry Anaerobic Composting**), uses thermophilic temperature (55 °C) at high solid concentration (200–400 $g\,l^{-1}$) in a one-stage fermentation. It is in fact a similar process to that taking place in landfills, with the difference that it is carried out in a closed reactor under well-controlled conditions and at a much greater reaction rate. The very high reaction rates attained make it possible to complete the digestion process in two weeks (Table 24.2) instead of 20 years as in landfills. Key to the process is the high temperature and

Fig. 24.9 Closed reactor used for the anaerobic biological conversion of biowastes into biogas (mixture of methane and carbon dioxide). Biogas is then converted into electrical power which is sold to the network. The picture illustrates a plant in Salzburg, Austria, treating 20000 tonnes biowaste annually with a single-phase thermophilic solid state fermentation process. The conveyor belt in the foreground transports the shredded and screened (40 mm) biowaste to a dosing unit. After thorough mixing with digested waste (to ensure inoculation) and heating up to 55 °C via steam injection, the hot mixture is pumped to the top of the reactor via the left pipe, while the pipe at the right-hand side transports the biogas from the top of the reactor to gas motors. Each tonne wet waste produces about 135 m^3 biogas (250 kWh) after a retention time of 16 days (9 m^3 biogas $m^{-3}\cdot$day).

intense mixing through recirculation allowing much higher reaction rates and the feeding of the solids directly in the reactor without addition of dilution water (Fig. 24.10). Wet processes, on the other hand, require dilution water in order to feed a slurry in the reactor. This has the drawbacks of higher water usage and much larger reactor volumes. Because mechanical agitation is not possible in the dry process, the output of the reactor is recycled several times, with addition of fresh feed material at each passage (Fig. 24.10). This recycle loop ensures adequate mixing and inoculation of the feed material.

The humus end-product has proven an excellent soil conditioner, superior to conventional aerobic composts in terms of plant germination and yield. The reason is that aerobic composts may be phytotoxic due to their high salt content, while anaerobic composts contain much

Table 24.3 Comparison of aerobic and anaerobic composting. While aerobic composting is slightly cheaper and has often been preferred in the past, recent developments in anaerobic composting technology make this new technology more attractive. Because it is carried out in closed reactors, anaerobic composting requires less space, produces less odours, and kills pathogens more efficiently than aerobic composting

	Aerobic composting	Anaerobic composting
Cost	60 €/tonne wet	75 €/tonne wet
Floor area	large	small
Energy balance	consumes energy	produces energy[a]
Odours	problem	no problem
Quality final compost		
Salt content	high (toxic)	low
Pathogens	present	absent

Notes:
[a] 600 kWh energy is produced in the form of biogas per tonne wet bio-waste; upon combustion in a gas motor with 33% electrical conversion, it produces 200 kWh electricity.

less salts due to the fact that about half of these are eliminated with the water in the filter press (Fig. 24.10). Moreover, anaerobic composts contain far fewer weed seeds and microbial pathogens compared to aerobic composts. The market value of composts is however low and special post-treatments should be sought for targeted applications. The latter can be achieved by adding beneficial micro-organisms such as N-fixing and plant growth-promoting bacteria, mycorrhizae or biocontrol micro-organisms. The restoration of polluted soils can also benefit from compost addition as this can serve either as a source of inoculum and nutrients for the degradation of xenobiotic compounds or as an organic matrix promoting the binding of xenobiotics.

24.5 Treatment of waste gases

Waste gases polluted with a wide variety of organics – most of them at the $\mu g\ m^{-3}$ level or below – are inherent to domestic and industrial activities. In the coming years, air pollution control in general, and odour abatement in particular, will become of increasing importance. Biotechnological cleanup of waste gases, of odours and of in-house air is an area undergoing full development. The focal point of this technology is the possibility of growing and maintaining organisms capable of removing a wide spectrum of pollutants even at the extremely low concentrations at which they occur in the gas phase.

24.5.1 Removal of volatile organic compounds (VOCs)

Conventional physico-chemical treatment of polluted industrial waste gases, such as combustion or adsorption on activated coal filters, tend

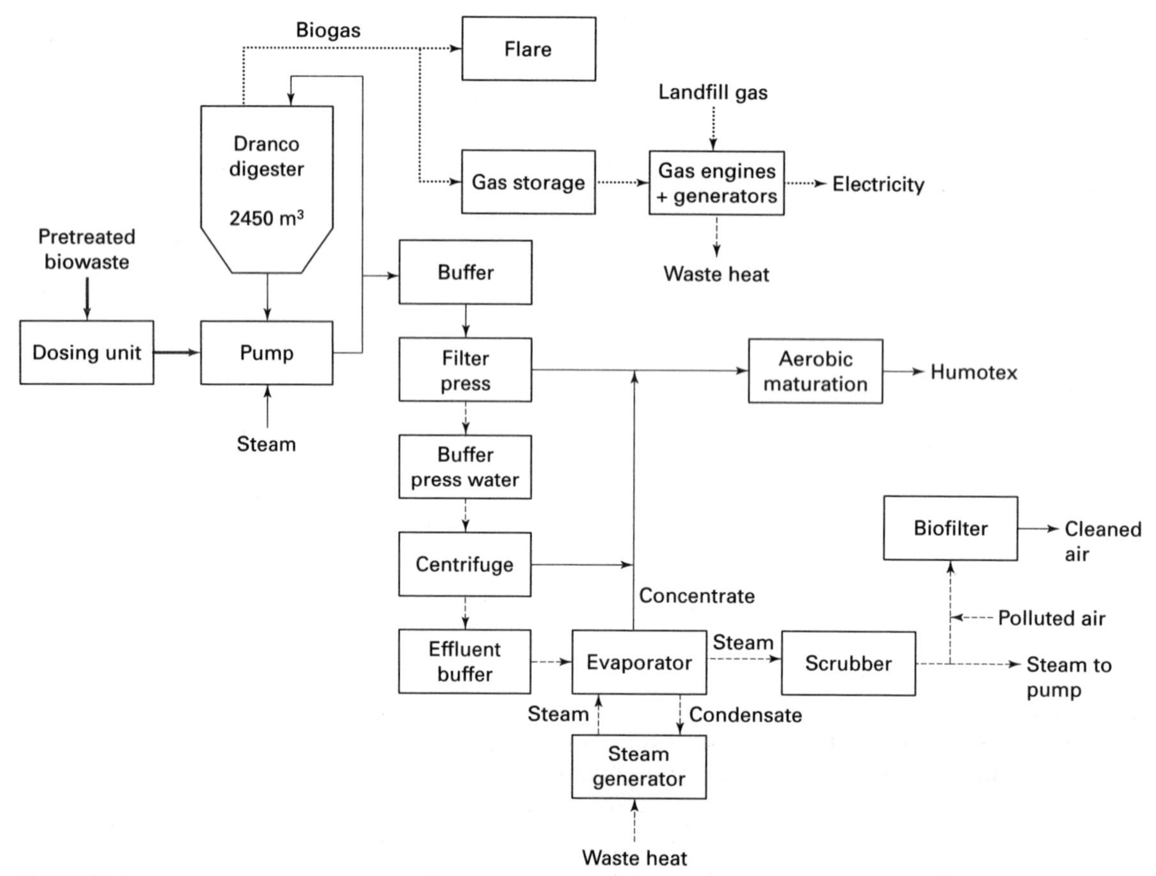

Fig. 24.10 Process flow diagram of the anaerobic composting plant treating 20 000 tonnes biowaste per year in Kaiserslautern, Austria. The waste is directly fed in 'solid state' (300 g solids l^{-1}) one-stage anaerobic digester, maintained at 55 °C, wherein each tonne wet waste produces ca. 150 m^3 biogas (60% methane). The biogas is converted to steam to warm up the digester and to electricity in a gas engine. Waste heat from the motors is re-used to evaporate the waste water generated during the mechanical dewatering of the digested paste. Dewatered paste (500 g solids l^{-1}) is subjected to a short (1–2 weeks) aerobic post-treatment yielding a humus-like material. The various points where odours are produced, e.g. the aerobic post-composting, are ventilated and the waste air is treated in a biofilter where volatile organic compounds are removed.

to waste a lot of energy and create secondary pollution. Pollutant concentrations in industrial emissions, for example, are of the order of 100 ml m^{-3}. To burn these gases in an incinerator, at least 50 litres methane need to be added per m^3 in order to ensure complete destruction. A bioreactor may, in most cases, achieve the same oxidation provided the VOCs are brought in close contact with degradative microbes, O_2, H_2O and nutrients. Biodegradation rates vary with the pollutant being degraded:

- quickly biodegraded: alcohols, ketones, aldehydes, organic acids, organo-N;
- slowly biodegraded: phenols, hydrocarbons, solvents (e.g. chloroethene);
- very slowly biodegraded: poly-halogenated and poly-aromatic hydrocarbons.

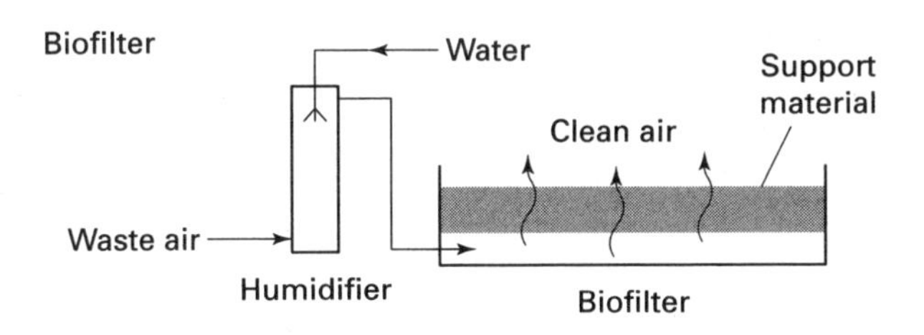

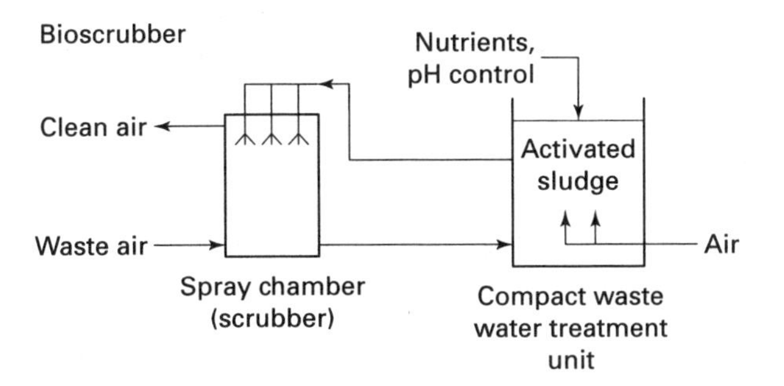

Fig. 24.11 Biofilters and bioscrubbers are used to remove volatile organic compounds (VOCs) from waste gases via biological means. Biofilters are simple, robust, and cheap, but require a large floor area. Bioscrubbers rely on classical waste water treatment processes (activated sludge or trickling filter) after having transferred the pollutants from a gaseous stream to an aqueous phase in a scrubber. Bioscrubbers are more amenable to process optimisation and require less floor area than biofilters. They are however more costly and are less efficient at removing poorly soluble VOCs, e.g. hydrocarbons.

Despite the broad spectrum of air pollutants amenable to biofilter treatment, the introduction of this new technology is slow, perhaps because its low cost does not ensure high profit margins and because the physico-chemical air pollution control industry is well entrenched.

Various types of reactor designs are used to treat air biologically (Fig. 24.11). In **biofilters**, contaminated air flows slowly through a wet porous medium – compost, peat, or wood chips – which support a degradative microbial population living in the thin water film coating the solid support material. The superficial gas flow varies from 1 to 15 cm s^{-1}. This yields a contact time, for a typical bed height of 1–3 m, of 10–100 s. For normally biodegradable compounds, removal efficiencies of 90% can be expected at volumetric loading rates of 0.1–0.25 kg organics m^{-3} reactor·day. The advantages of biofilters are:

- simple and cheap design (support material replaced every 2–4 years);
- high internal surface area makes biofilters ideally suited to remove poorly soluble pollutants, e.g. hydrocarbons;
- possibility to inoculate with bacteria especially adapted for the breakdown of xenobiotic compounds, e.g. chloromethane.

The most difficult problem is the control of the pH in the biofilter since H_2S is oxidised to H_2SO_4, NH_3 to HNO_3, and chloro-organics to HCl. For example, the removal efficiency of dimethyl sulphide in a compost biofilter seeded with the bacterium *Hyphomicrobium* dropped within two months of operation, from 1 to 0.1 g m^{-3}·day due to a pH drop to 4. Repeated dosing of 25 kg limestone powder ($CaCO_3$) per m^3 compost carrier eliminated the inhibition for a two-month period. The biggest disadvantages of biofilters are:

- large floor space necessary;
- not possible to control the process conditions, e.g. pH;
- support materials such as compost themselves generate odours.

The disadvantages of the biofilter can be avoided in a **bioscrubber** (Fig. 24.11). A conventional scrubber transfers a substance present in a gaseous stream to a liquid stream by spraying a liquid in a chamber though which the gas is passed. In a bio-scrubber, the sprayed liquid is a suspension of micro-organisms which cycles back and forth between the spray chamber and a waste water treatment unit where biodegradation takes place. The process parameters such as adequate nutrient supply and pH are much more easily controlled (in the circulating liquid) than in a biofilter, leading to fast reaction rates. While biofilters require a large footprint since their height preferably should not exceed 1 m in order to avoid clogging, bioscrubbers require much less space because the tank where biodegradation takes place can be several metres high.

Bioscrubbers appear best suited for large air flows because of their low back pressure and small size. They however can be employed only for the removal of gases which are sufficiently soluble because the mass transfer rate in a spray chamber is less than that attainable in a biofilter unit. In case the obtained contaminant concentration in the outlet gas is too high, a second bioscrubber inoculated with micro-organisms capable of degrading lower contaminant concentrations must be installed. This aspect requires further development.

At present, considerable research is devoted to the design of a system that can combine the adsorption of the gas onto a solid surface (e.g. activated carbon) and biodegradation of the sorbed compound. **Bio-trickle filters** are sheets of a plastic or other microbial support medium hung in the contaminated air stream. The sheets are bathed continuously by a re-circulating stream of water containing the nutrients required by the micro-organisms. Bio-trickle filters hold promise where space utilisation is paramount. Bio-oxidation rates per unit volume are high so that these filters can be as small as physico-chemical units. Being operated at higher loading rates, they are however more sensitive to peak loads and nutritional requirements need be monitored closely.

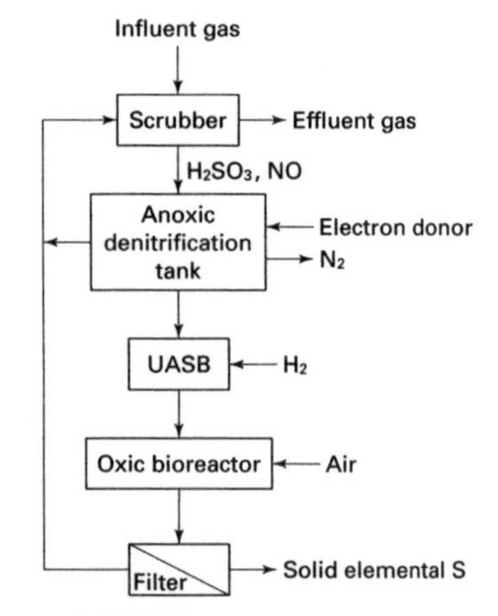

Fig. 24.12 A newly developed bioprocess for the simultaneous desulphurisation and NO removal from flue gases produced in thermic plants. The sequential steps are solubilisation in a scrubber, N removal in a bioreactor, sulphite reduction to sulphide in a UASB reactor, sulphide partial oxidation to elemental S° in a submerged oxic attached biofilm reactor and recovery of solid sulphur. The liquid phase is continuously recycled.

24.5.2 Biological removal of sulphur and nitrogen compounds from flue gases

Nitrogen oxides (NO_x) and sulphur dioxide (SO_2) are major air pollutants formed during the combustion of coal and oil and released in flue gases. There is considerable interest in the development of an efficient and low-cost biotechnology for the simultaneous removal of these air pollutants, since conventional physico-chemical technologies are either very expensive or inefficient. A new system is currently being proposed in which the flue gas is led through a scrubber in which >95% SO_2 and >80% NO_x dissolve in a solution of $NaHCO_3$ and Fe(II)-EDTA (the latter compound seems to raise the solubility of NO_x, the bottleneck of the process). The S- and N-laden solution is regenerated in three sequential biological steps (Fig. 24.12). The first step consists of an anoxic reactor wherein NO is converted to inert N_2 gas via biological denitrification:

$$2\,Fe^{II}(EDTA)\,(NO) + \text{electron donor} \rightarrow 2\,Fe^{II}(EDTA) + N_2 + CO_2 + H_2O$$

An electron donor, e.g. methanol or ethanol, needs to be added in order to sustain the reaction. In the two following steps, H_2SO_3 is sequentially reduced biologically to H_2S and finally partially re-oxidized to solid elemental sulphur:

$$H_2SO_3 + 3\,H_2 \rightarrow H_2S + 3\,H_2O$$

$$H_2S + \tfrac{1}{2}\,O_2 \rightarrow S^\circ + H_2O$$

The reduction of H_2SO_3 takes place in a UASB reactor (Fig. 24.5) seeded with sulphate-reducing bacteria. Flocculant polymers are added, together with the necessary nutrients and reducing equivalents (ethanol or H_2) to adjust the (BOD / H_2SO_3) molar ratio at a value of one. In the third bioreactor, aerobic bacteria oxidise sulphide back to solid S° (end-product). The further oxidation of S° to H_2SO_3 and H_2SO_4 is prevented by dosing limiting amounts of O_2. The overall process is fully automated with about 120 parameters being continuously analysed, most of them on-line. The water is continuously recycled.

This process of biodesulphurisation will undoubtedly also be applied in the future to treat other waste streams. There is a growing interest in depolluting waste waters through the activity of sulphate-reducing bacteria in sulphidogenic UASB reactors. Sulphate concentrations reach very high levels in effluents from the paper board industries (2 g l^{-1}), in molasses-based fermentation industries (2–9 g l^{-1}), and in edible oil refineries (up to 50 g l^{-1}). Very large amounts of sulphate are also present in acidic mine drainage where pyrite rock is being processed. When heavy metals are present, these can be very efficiently removed (>99%) via sulphide precipitation.

24.6 Soil remediation

One of the major problems facing the industrialised world today is the contamination of soils, groundwater and sediments. The total world hazardous waste remediation market is approximately US $16 billion per year. There are at least 350000 contaminated sites in Western Europe alone and it may cost as much as US $400 billion to clean just the riskiest of these sites over the next 20–25 years. The most common contaminants are chlorinated solvents, hydrocarbons, polychlorobiphenyls and metals. **Bioremediation**, i.e. the use of micro-organisms to degrade or detoxify pollutants, is becoming increasingly used mostly in cases of hydrocarbons pollutions. However, bioremediation is not yet universally understood or trusted by those who must approve of its use and its success is still an intensively debated issue. One reason is the lack of predictability of bioremediation, due to insufficient information on:

- **bioavailability**, i.e. how to obtain good contact between contaminant molecules and micro-organisms (see Box 24.2);
- **biostimulation**, i.e. how to supply the microbes with stimulating agents;
- **bioaugmentation**, i.e. how introduced microbes behave in the field.

Box 24.2 Bioavailability of pollutants

Under *in situ* soil conditions, contaminant removal is often exceedingly slow due to poor availability to micro-organisms which bring about their degradation. Compounds which tend to sorb onto solid particles, such as clay and humus, are less easily taken up by micro-organisms. This tendency is quantified by the Freundlich equation:

$$S_{eq} = (OM)\, K_{oc}\, C_{eq}^{\,1/n}$$

where S_{eq} is the concentration of the pollutant on the sorbent phase (mg g^{-1}), OM is the % organic matter in the soil, K_{oc} is the partition coefficient, C_{eq} is the concentration in the aqueous phase (mg L^{-1}), and n is a constant related to the sorbent. Compounds with $K_{oc} < 100$ are little sorbed and therefore readily available (e.g. benzene; $K_{oc} = 10^{1.9}$) whereas compounds with $K_{oc} > 1000$ are quite well sorbed (e.g. phenanthrene; $K_{oc} = 10^{4.4}$). The K_{oc} is easily estimated in the laboratory from a measurement of the octanol/water partition coefficient (K_{ow}), i.e. the ratio of compound concentration in n-octanol and in water. The K_{ow} value of hydrocarbons can be used to estimate their half-life in soils. Half-life is an indication of disappearance rate of substrates which are degraded according to first order kinetics (Fig. 24.13).

These concepts do not however explain the observation that certain pollutants in soil cannot be removed below a certain level. This phenomenon, common with pesticides, has been called **ageing**. It is thought that ageing results from the diffusion of the pollutant molecules through the interstitial micropores of aggregates and through the three-dimensional matrix of natural organic matter. As this process is not reversible, the Freundlich equation is no longer valid and different molecules of the same compound have different half-life values. Pesticides may persist in soil at non-removable residual levels as a result of this phenomenon.

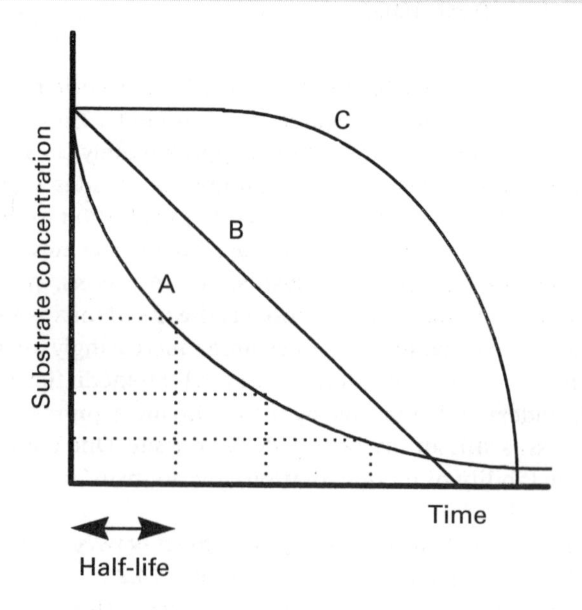

Fig. 24.13 The kinetics of biodegradation of pollutants in soil systems are usually first order, which means that the rate of substrate disappearance is proportional to the substrate concentration (curve A). With first order kinetics, the half-life defines the time during which half of the substrate is degraded. First order kinetics occur when the substrate concentration is low (smaller than the affinity constant K_s) as is often the case in soils. Zero order refers to a constant reaction rate which is independent of the substrate concentration and which typically occurs during co-metabolism (curve B). In cases of high pollutant concentrations, microbial growth can occur resulting in increasing rates with time and the curve C is typically observed.

Table 24.4 Different possible strategies to induce bio-remediation of soils and groundwater. While bio-stimulation relies on autochthonous micro-organisms, i.e. those already present in the contaminated site, bio-augmentation makes use of laboratory-grown bacteria, fungi, or pre-adapted consortia

Action	Mechanism	Example
Bio-stimulation, i.e. to stimulate the micro-organisms already present		
Add nutrients N, P	Optimise the chemical make-up for balanced growth	'Fertilise' oil slicks at sea (Exxon Valdez spill in Alaska)
Add co-substrates	Pollutant is degraded by an enzyme intended to process the co-substrate	Inject methane to degrade trichloroethylene in aquifers
Add electron acceptors	Oxidation of organics in groundwater typically limited by poor solubility of O_2	Bioventing of aquifers (air injection) or addition of nitrate
Add surfactants	Hydrocarbons and non-aqueous phase liquids (NAPLs) are not available to micro-organisms	Adding surfactants will disperse the hydrophobic compounds in the water phase
Bio-augmentation, i.e. to (re)introduce microbial cultures grown in the laboratory		
Add a pre-adapted strain	Certain sites may not contain adequate micro-organisms to degrade pollutants	Inoculation of soils and waste water treatment plants with chloroaromatic degraders
Add pre-adapted consortia	The presence of the right combination of micro-organisms is ensured	Seed sediments with PCB-dechlorinating enrichment cultures
Add genetically optimised strains	Existing degradation pathways may release dead-end or toxic intermediates	Construction of strains effecting complete simultaneous oxidation of chloro- and methyl-aromatics
Add genes packaged in a vector	Genes encoding for desirable functions are transferred into micro-organisms already present	Degradation of PCBs or pesticides

24.6.1 Biostimulation and bioaugmentation

The micro-organisms capable of biodegrading pollutants are usually already present in contaminated soils and groundwater. Thus, in the vast majority of cases, bioremediation of soils or groundwater will occur satisfactorily by stimulating the micro-organisms already in place with the required nutrients or other factors (**biostimulation**). Thus, oil spills at sea or hydrocarbon leakages in groundwater are remedied by 'fertilising' the sea or ground with nitrogenous and other nutritious compounds (Table 24.4). Surfactants may also be added in order to facilitate the mass transfer of poorly-soluble hydrocarbons into the water phase where the micro-organisms live. Another example of biostimulation is the injection of methane in aquifers polluted with certain chlorinated solvents or benzoic acid in aquifers polluted with certain polychlorobiphenyls (PCBs). The injected carbon sources methane and benzoic acid promote the growth of specific micro-organisms which possess enzymes that will degrade both the injected substrate and the pollutant already present. As the micro-organisms do not seem to draw any advantage from the breakdown of the pollutant, this reaction is called **co-metabolism.**

Inoculation with specific populations of micro-organisms (**bioaugmentation**) may be advantageous in certain polluted sites, for example when the pollutant is a complex molecule which can be broken down only by a particular combination of very specific micro-organisms (called a consortium). Such pollutants include polyaromatic hydrocarbons (PAHs), halogenated organic compounds, certain pesticides, explosives such as trinitrotoluene (TNT), polychlorobiphenyls (PCBs), etc. Proper conditions and appropriate microbial strains have been found that affect biodegradation of these compounds in laboratory setups. For example, the degradation of simple chlorinated aromatics in soils and waste water treatment plants can be accelerated by inoculating pure cultures of micro-organisms selected in the laboratory. More complex pollutants, e.g. PCB, may require the concerted action of several microbial strains. For this particular case, dechlorinating consortia developed from contaminated sediments have been mass cultured in granular form in methanogenic UASB reactors. These granules have been shown to accelerate the degradation of PCB *in situ* in soils and sediments.

For specific applications, bioaugmentation can be carried out with genetically engineered micro-organisms (GEMs). GEMs may be particularly useful to avoid the misrouting of degradation intermediates into unproductive dead-end pathways as may occur during the degradation of mixtures of chloro- and methylaromatics. GEMs may also help to prevent the formation of toxic intermediate products that may destabilise the community and inhibit biodegradative processes. One example is PCB degradation during which a first group of micro-organisms release chlorobenzoates, which are further attacked by a second group of micro-organisms. PCB degradation is normally blocked by a byproduct of chlorobenzoate degradation. GEMs have been constructed which do not produce the toxic intermediates, hence giving better survival of the PCB degraders. The greatest challenge is indeed to increase the chances of survival of inoculated strains, i.e. to make them ecologically competent. In this respect, chances of survival are usually greater when the inoculated strain was originally isolated from the site which is to be bioaugmented.

24.6.2 Soil remediation techniques

A great variety of biotechnologies are being used to treat polluted soil. In increasing degree of complexity and cost, the most commonly used techniques include:

- *in situ* bioremediation;
- landfarming;
- slurry-phase bioreactors.

***In situ* bioremediation** relies on biological clean-up without excavation. It is usually applied in situations where contamination is deep in the sub-surface or under buildings, roadways, etc. *In situ* biorestoration of pollutants is gaining interest since it avoids excavation costs and produces no toxic by-products as is the case with *ex situ* physico-chemical treatment. Water is cycled through the sub-surface using a series of recovery and recharge trenches or wells. Water may be oxygenated by sparging with air or via addition of H_2O_2. Microbial clean-up by enhancement of anaerobic degradative activity *in situ* has received less study. The obvious drawback of *in situ* bioremediation is that it is difficult to stimulate microbial activity throughout the contaminated soil volume because the injected water carrying the necessary nutrients and micro-organisms tends to flow through larger soil interstices, leaving substantial amounts of residual contaminant within more impermeable layers. It may take years for the contaminants to diffuse to the bio-active zones where biodegradation is occurring.

One specific type of *in situ* soil bioremediation, called **bioventing**, has emerged recently as one of the most cost-effective and efficient technologies available for the remediation of the vadose zone (unsaturated zone above the groundwater table) of petroleum-contaminated sites. Bioventing consists of stimulating aerobic biodegradation by circulating air through the sub-surface. High removal efficiencies (>97%) can be obtained for soluble paraffins ($<C_{16}$) and poly aromatic hydrocarbons (PAHs) after several years operation. Bioventing is however limited to homogeneous sub-surface formations since heterogeneities would cause the air to move through the most permeable areas causing treatment to occur only in limited areas.

Another success story of *in situ* soil bioremediation is **phytoremediation**. Here specific plants are cultivated which accumulate heavy metals in the above-ground plant tissue or stimulate organic breakdown in their rhizosphere (the zone immediately adjacent to the roots). While phytoremediation is elegant, 'clean' and cheap, its main drawbacks are that only the surface layer of soil (0–50 cm) can be treated and that the treatment takes several years and leaves substantial residual levels of contaminants in the soil. Phytoremediation is however undergoing full development at present.

Removal of oil slicks by so-called **landfarming** is an established method based on microbial degradation (Fig. 24.14). Given half-lives of the order of one year, it would take about 7 years of treatment to remove 6.4 g hydrocarbon per kg soil down to the clean-up goal of 50 mg kg^{-1}. This low-tech technology can be somewhat upgraded by mixing the soil with fresh organic residues (compost). Elevated temperatures and increased microbial diversity and activity increase reaction rates. Moreover specific co-substrates favour co-metabolism. Landfarming systems can be upgraded by including anaerobic pretreatment. For example, anaerobic tunnels are used to reduce compounds such as trinitrotoluene by adding nutrients and co-substrates for the indigenous bacteria. In a second aerobic stage, the reduced metabolites are either completely mineralised or polymerised and irreversibly immobilised in the soil matrix. This approach has also been used successfully to decontaminate soils polluted with chloroethene and BTX aromatics (mixtures of benzene, toluene and xylene).

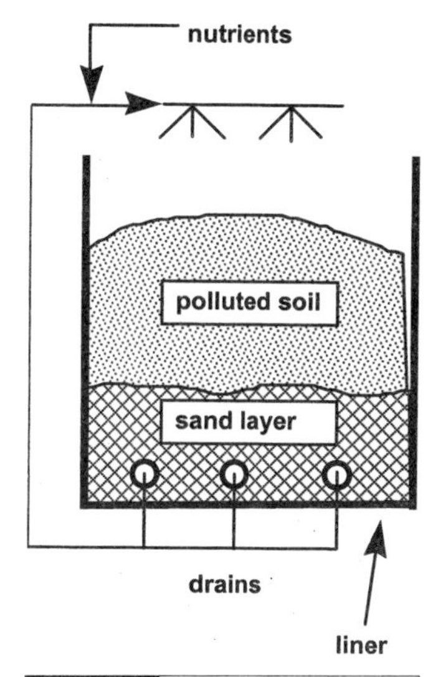

Fig. 24.14 Cross-sectional view of a solid-phase soil 'reactor', or landfarming system. The soil is excavated, mixed with nutrients and micro-organisms, and evenly spread out on a liner. With regular ploughing to favour mixing and aeration, mineralisation of petroleum hydrocarbons present at initial concentrations of tens of g kg^{-1} follows first order kinetics with a half-life of ca. 2 years. Landfarming is an established technique for the remediation of hydrocarbon-contaminated soil.

Slurry-phase bioreactors may achieve the same clean-up levels in considerably less time. In this case, excavated polluted soil is treated under controlled optimal conditions, ensuring effective contact between contaminant and micro-organisms. The latter are, in most cases, specific cultures of adapted micro-organisms. With overall degradation rates in the range 0.2–2 g oil per kg soil per day, solid residence times of 30 days, in place of several years, are sufficient to meet the cleanup levels. Treatment costs increase accordingly to ca. 130 euros per tonne of soil.

24.7 Treatment of groundwater

24.7.1 Active remediation

The predominant groundwater remediation strategy in the US and Europe has been the application of the so-called 'pump-and-treat' technology. This approach uses mainly physico-chemical techniques to remove the pollutants in the above-ground treatment units, via for example air stripping and activated carbon, while biological reactors are used in fewer than 10% of cases (Fig. 24.15). The limited use of biological treatment may be due to limited experience and demonstration data, limited acceptance of the technology, but also failures to achieve the cleanup levels required. To date, probably most experience with full scale *ex situ* and *in situ* applications of bioremediation has been acquired for the biodegradation of petroleum hydrocarbons, comprising straight and branched chain, saturated, unsaturated and cyclic aliphatics to mono-, di- and polyaromatic hydrocarbons. Recently, however, new types of bioreactor designs have been developed that eliminate polychlorinated solvents and aromatics as well. For example, UASB reactors seeded with granular methanogenic sludge have been shown to completely (>99%) dechlorinate tetrachloroethylene present at 4 mg l^{-1} in polluted groundwater. Acetate was used as carbon source and electron donor and process costs were competitive (US \$1.2 per m^3 treated). The UASB reactor technology is also being upgraded with granular sludge combining both anaerobic and aerobic bacteria.

The 'pump-and-treat' strategy fails however to achieve cleanup targets in most cases and moreover requires long cleanup times. Of the 77 pump-and-treat sites evaluated by a committee under the auspices of the US National Research Council (NRC) in 1992, only eight had reportedly reached the cleanup goals, which in all cases were the maximum contaminant levels for constituents regulated under the Safe Drinking Water Act. Of the eight successful sites, six were polluted with petroleum hydrocarbons which would also have been eliminated via natural attenuation.

The NRC Committee concluded that pump-and-treat methods were quite limited in their ability to remove contaminant mass from the subsurface because of sub-surface heterogeneities, presence of fractures, low-permeability layers, strongly adsorbed compounds, and slow mass transfer in the sub-surface. Even with the best extraction methods, very often only a small fraction of soil-bound contaminants can be mobil-

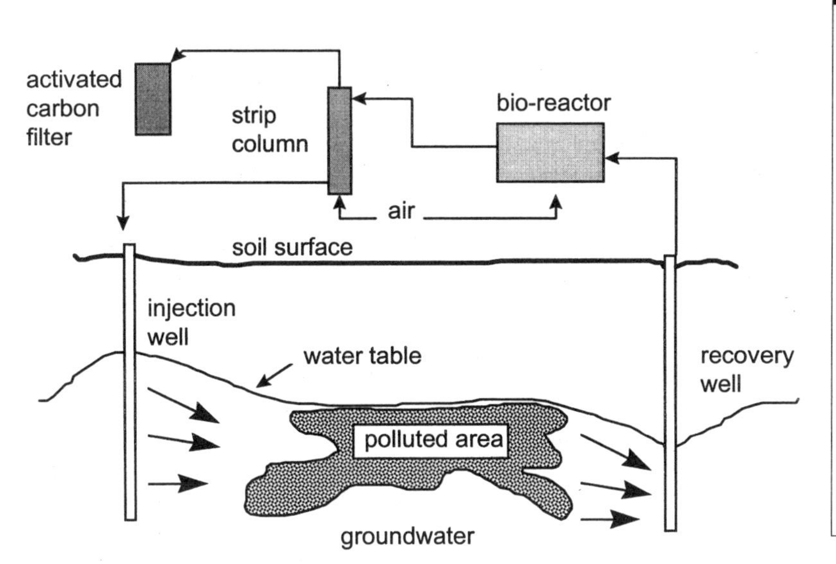

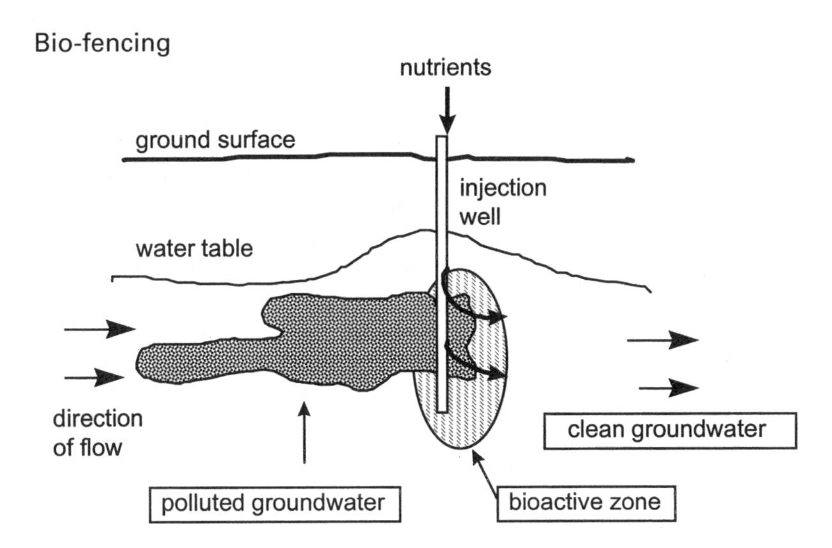

Fig. 24.15 The **'pump-and-treat'** remediation technology uses recovery and recharge wells that 'wash' the groundwater to the ground surface where it is treated by a combination of various physico-chemical and biological techniques. Treated water is re-injected several times to improve pollutant recovery. **Bio-fencing**, on the other hand, is only a containment technique. It consists of setting up a bioactive zone at the down-gradient edge of a contaminated groundwater area via nutrient injection. As impacted groundwater enters the bioactive zone, contaminants are biodegraded.

ised, leaving a large residual fraction in the soil. As a result of this failure, remediation policy and technical developments are shifting towards increased use of *in situ* containment practices, e.g. **biofencing** (Fig. 24.15), rather than full treatment scenarios. In cases where full treatment is necessary, less stringent cleanup goals are set, based on risk assessment taking into account type of land use.

Aside from the much-studied generic compounds discussed above in this chapter, there is a host of toxic compounds usually present at trace level and whose fate remains poorly studied. One example are the polychlorinated dioxins and furans which are formed as by-products of chemical synthesis processes. They are also produced by combustion of garbage, waste oils, soils polluted with oils, chemical wastes containing PCBs, and by various other high temperature processes. Because of the high toxicity of some dioxins and furans, these compounds are of major eco-toxicological concern. Ongoing research and development has attempted to minimise their formation in incinerators and emission via fly ashes. Yet, the biological breakdown of these compounds in the environment is of considerable importance. Indeed, they are often present in wastes which are extremely difficult to treat properly by incineration (e.g. polluted soils and river sediments). They are also present in fly ashes of incinerators which are deposited in landfills and, notwithstanding all precautions, can contaminate landfill leachates.

24.7.2 Natural attenuation and monitoring

Several factors have recently generated a lot of interest in new monitoring techniques. One such factor is the fact that remediation technologies are often insufficient to meet stringent cleanup targets. This limitation is making legislators reassess the target pollutant levels and making them consider the use of risk-based end-points in place of absolute end-point values. The new concept of risk-based end-points requires the development of new analytical tools which assess the bioavailable rather than the total pollutant concentration. These new tools typically rely on bioassays because the traditional analytical methods cannot distinguish pollutants that are available to biological systems from those that exist in inert, or complexed, unavailable forms. Subjecting a polluted soil to a period of intensive microbial activity can reduce the toxicity by a factor of 5 to 10. This ecotoxicological information can be easily deduced by running a simple bioassay with soil leachates. One type of bioassay is based on the inhibition of the natural bioluminescence of the marine organism *Photobacterium phosphoreum*, which is used, for example, in the Microtox, Lumistox and Biotox tests. These assays are, however, not specific since light inhibition will occur upon exposure to any toxicant. This limitation is circumvented in a new class of bacterial biosensors which are specific to certain types of toxicants. For example, biosensors able to detect bioavailable metals, were constructed by placing *lux* genes of *Vibrio fischeri* as reporter genes under the control of genes involved in the regulation of heavy metal resistance in the bacterium *Alcaligenes eutrophus*. The recombinant strains, upon mixing with metal-polluted soils or water, emit light in proportion to the concentration of specific bioavailable metals. Light emission is easily measured spectrophotometrically.

Another factor responsible for the recent interest in new monitoring tools is the high cost and slow pace of remediation technologies. A more pragmatic remediation approach, termed natural attenuation (or **intrinsic bioremediation**), is being advocated by the US Environmental Protection Agency. Intrinsic bioremediation relies on natural processes to remove, sequester or detoxify pollutants without human intervention. Intrinsic bioremediation has been observed most frequently with groundwater contaminated with hydrocarbons. If evidence suggests that a site is improving due to intrinsic bioremediation, and that the pollution does not pose a threat to human health, the environmental regulating agency may grant that site 'monitoring only' status. This strategy simply requires remote monitoring in order to follow subsurface contaminant concentrations *in situ*. Remote monitoring can be carried out with a ground-penetrating radar which monitors contaminant breakdown in the sub-surface based on the increase of solution conductance which accompanies the breakdown of hydrocarbons or chlorinated solvents. Another technique used in remote monitoring uses genetically engineered micro-organisms which produce light in response to the presence of specific contaminants. As these microorganisms are attached on a photocell connected to a radio chip, the light signals are converted into radio waves which are detected at a distance. These sensors can be scattered throughout polluted sites to monitor the progress of pollutant breakdown.

24.8 Further reading

Alexander, M. (1994). *Biodegradation and Bioremediation*. Academic Press, San Diego.

Baveye, Ph., Block, J.-C. and Goncharuk, V. V. (1999). *Bioavailability of Organic Xenobiotics in the Environment – Practical Consequences for the Environment*. Kluwer Academic Publishers, Dordrecht.

Haug, R. T. (1993). *The Practical Handbook of Compost Engineering*. Lewis Publishers, Boca Raton, Florida.

Hurst, C., Knudsen, G. R., McIncercy, M. J., Stetzenbach, L. D. and Walter, M. V. (1997). *Manual of Environmental Microbiology*. American Society for Microbiology, Washington, DC.

Grady, L. C. P., Daigger, G. T. and Lim, H. C. (1999). *Biological Waste Water Treatment, 2nd edition*. Marcel Dekker Inc., New York.

Sayler, G. S., Sanseverino, J. and Davis, K. L. (1997). *Biotechnology in the Sustainable Environment*. Plenum Press, New York.

Verstraete, W., de Beer, D., Pena, M., Lettinga, G. and Lens, P. (1996). Anaerobic bioprocessing of organic wastes. *World J. Microbiol. Biotechnol.* **12**, 221–238.

Index